The Law of the Sea

The Law of the Sea

*An Historical Analysis of
the 1982 Treaty and
Its Rejection by the United States*

JAMES B. MORELL

McFarland & Company, Inc., Publishers
Jefferson, North Carolina, and London

The present work is a reprint of the library bound edition of The Law of the Sea: An Historical Analysis of the 1982 Treaty and Its Rejection by the United States, *first published in 1992 by McFarland.*

LIBRARY OF CONGRESS CATALOGUING-IN-PUBLICATION DATA

Morell, James B., 1956–
    The law of the sea : an historical analysis of the 1982 treaty and its rejection by the United States / by James B. Morell.
        p.   cm.
    Includes bibliographical references and index.

    ISBN 978-0-7864-7710-4
    softcover : acid free paper ∞

    1. Maritime law. 2. Ocean bottom (Maritime law).
    3. Marine mineral resources—Law and legislation.
    4. United Nations Conference on the Law of the Sea.
    5. United States—Foreign relations—1981–1989.
    I. Title.
    JX4411.M67 2013
    341.4'5—dc20   91-52749

BRITISH LIBRARY CATALOGUING DATA ARE AVAILABLE

On the front cover: the Sea (iStockphoto/Thinkstock); United Nations emblem

Manufactured in the United States of America

McFarland & Company, Inc., Publishers
  Box 611, Jefferson, North Carolina 28640
    www.mcfarlandpub.com

For Hugo and Suzie

Acknowledgments

There are four University of Southern California professors without whose guidance this book would never have come into being: Robert Friedheim, Gerald Caiden, Edwin "Rip" Smith, and, especially, Arvid Pardo, without whose interest in the subject there would have been nothing to write about—or for. The help of Annelore Stern of the Von KleinSmid Library was invaluable to my research efforts.

The assistance of Lee Kimball of the Council on Ocean Law, Professor William Burke of the University of Washington School of Law, and Professor Louis Sohn of the University of Georgia School of Law has been greatly appreciated.

At the United Nations, Roy Lee, Jean-Pierre Levi, Mati Pal, and Moritaka Hayashi provided much helpful background information at various times over the ten-year period this treatise was written.

Thanks are also due to several federal administrators, whose cooperation and candor have also been appreciated: John Padan of the National Oceanic and Atmospheric Administration, Maureen Walker of the Office of Oceans Law and Policy, and Bob Paul and LeRon Bielak of the Minerals Management Service.

Special thanks to Judge Irving Shimer for his interest, to my family for their patience, and to Billie for her friendship.

Contents

Abbreviations

AGIP	Azienda Generale Italiana Petroli
ANC	African National Congress
BINC	Belgium, Italy, the Netherlands, and Canada
COCOM	Coordinating Committee for Multilateral Security Export Controls
OPUOS	Committee on Peaceful Uses of Outer Space
ECOSOC	(U.N.) Economic and Social Council
EEC	European Economic Community
FAO	Food and Agriculture Organization
GATT	General Agreement on Tariffs and Trade
IAEA	International Atomic Energy Agency
ICAO	International Civil Aviation Organization
ICNT	Informal Composite Negotiating Text
IFRB	International Frequency Registration Board
ILO	International Labor Organization
IMCO	Intergovernmental Maritime Consultative Organization
IMF	International Monetary Fund
IMO	International Maritime Organization
INTELSAT	International Telecommunication Satellite Organization
IOC	Intergovernmental Oceanographic Commission
ITU	International Telecommunication Union
IWC	International Whaling Commission
LOS	Law of the Sea
NATO	North Atlantic Treaty Organization
NIOC	National Iranian Oil Company
NOAA	National Oceanic and Atmospheric Administration
OAS	Organization of American States
OPEC	Organization of Petroleum Exporting Countries
PLO	Palestine Liberation Organization
RICNT	Revised Informal Composite Negotiating Text

RSA	Reciprocating States Arrangements
RSNT	Revised Single Negotiating Text
SALT	Strategic Arms Limitation Talks
SNT	Single Negotiating Text
SWAPO	South West African People's Organization
UNCITRAL	U.N. Commission on International Trade Law
UNCLOS	U.N. Conference on the Law of the Sea
UNCTAD	U.N. Conference on Trade and Development
UNDP	U.N. Development Program
UNEP	U.N. Environmental Programme
UPU	Universal Postal Union
WARC	World Administrative Radio Conference
WHO	World Health Organization
WMO	World Meteorological Organization

Introduction

Seventy-one percent of the surface of this planet is covered by seawater. The seabed beneath the oceans is divided into two distinct geographical regions: the continental margin and the deep ocean floor, the latter comprising approximately 85 percent of the entire seabed area. The continental margin consists of three geological subregions: the continental shelf, extending seaward from the coast on a gentle incline; the continental slope, dropping off steeply toward the deep ocean floor; and the continental rise, an accumulation of sediment sloping gradually outward from the base of the continental slope.[1] The deep ocean floor consists of several distinct geographical features, including mid-oceanic ridges, undersea mountains, deep trenches, and flat abyssal plains.

Several hundred thousand different types of living marine organisms are believed to inhabit the seabed and its superjacent waters, although their total number is difficult to estimate accurately.[2] Three general categories of marine mineral resources are located beyond the continental shelf: subsurface deposits associated with preexisting geological formations, surficial deposits, and deposits in solution. Although petroleum and other mineral resources have been recovered from the continental shelf in recent decades, deep-sea mineral exploitation has to date been regarded as technologically feasible only with respect to the millions of ferromanganese nodules which were discovered more than a century ago at the bottom of the world's largest ocean basins.[3] Clustered at depths of 400 to 3,400 fathoms (one-half to four miles), these nodules vary in composition but have been found to contain as much as 40 percent manganese, 4.6 percent nickel, 4.5 percent copper, and 0.9 percent cobalt, as well as trace amounts of other metals such as zinc and molybdenum.[4] It has been estimated that the nodules are growing at an average rate of one millimeter every 10 million years, although in some areas they are expanding at somewhat faster rates.[5]

On July 9, 1982, President Ronald Reagan announced that the United States would not sign the United Nations Convention on the Law of the Sea. The treaty, a comprehensive instrument intended to embody both codification and progressive development of international law with respect to use of the oceans, had been adopted 10 weeks earlier by more than 130 states following a 14-year period of negotiation. These negotiations had been described by Secretary of State Henry Kissinger in 1975 as "the world's last chance" to avoid mounting oceanic conflict through harmonization of competing practices and claims.[6]

   While expressing a general willingness to cooperate with other nations regarding use of the sea, President Reagan stated that the United States did not agree to certain terms and conditions of the treaty which were not viewed as "reasonable."[7] In light of its diverse marine uses and interests, the United States is unquestionably the world's leading oceanic state, and U.S. refusal to sign a document widely regarded as one of the most important agreements ever negotiated raises fundamental questions regarding not only the future legal regime applicable to the world's oceans, but also the leadership of the United States with respect to the promotion of international law and order. This treatise will examine the historical background to the international law of the sea and the negotiations which culminated in the decision of the United States to reject the treaty. The legitimacy of the U.S. position will be examined through an analysis of its objections to the Convention, of its ability to exploit the deep seabed under principles of customary international law, and of the costs of U.S. nonparticipation.

Chapter 1
Historical Development
of the Law of the Sea

Evolution of the Customary Legal Regime

Historically, the law of the sea has been established mainly by custom. There have traditionally been three basic types of territorial regime under customary international law: national sovereignty, *res nullius*, and *res communis*.[1] Sovereignty has been described as "legal shorthand for legal personality of a certain kind, that of statehood."[2] A *res nullius* regime is characterized by a lack of any sovereignty or ownership interest in an area or movable object which nevertheless remains available for appropriation by a claimant; *res communis*, in contrast, designates community ownership—an area or movable object which may not be appropriated nor used in a manner which impairs its use by others, except perhaps through general acquiescence.[3] "A true *res communis* cannot lawfully be reduced into sovereignty at all, whereas a *res nullius* can: but unless and until a *res nullius* is so reduced, it is open to common use and exploitation of the nationals and undertakings of all countries."[4]

Maritime trade made an important contribution to commerce among ancient civilizations, some of which sought to control portions of the Mediterranean Sea through the exercise of naval power.[5] There is little evidence that the exercise of such control constituted application of a recognized legal regime, however; whereas classical Roman jurisprudence specifically applied the *res communis* doctrine to the sea and the air.[6] The nature of the governing regime again became unclear with the assertion of exclusive maritime claims in the Middle Ages, although much commercial navigation was subject to maritime codes which had evolved among private traders since ancient times.[7] In 1493 Spain and Portugal divided most of the world's oceans between themselves, claiming exclusive navigation rights in a joint act of appropriation ratified by Pope Alexander VI and confirmed the following year in the Treaty of Tordesillas.[8] This *"mare clausum"* doctrine was soon challenged by emerging maritime states—particularly France, England, and the Netherlands—and the defeat of the

Spanish Armada off the English coast in 1588 effectively ended the naval dominance of these two Catholic states. In 1609 the Dutch jurist Hugo de Groot (Grotius) published a pamphlet, *Mare Liberum*, in which he attacked the closed sea concept and argued instead for the "freedom of the seas" as an adjunct of the *res communis* principle: "The sea can in no way become the private property of any one, because nature not only allows but enjoins its common use. . . . Nature does not give a right to anybody to appropriate such things as may inoffensively be used by everybody and are inexhaustible, and therefore, sufficient for all."[9] The Grotian doctrine prevailed, and was accepted by the international community for more than 300 years as the legal regime applicable to the high seas. *Mare Clausum*, published in 1635 by John Selden in defense of England's sovereignty over coastal fishing waters, received a more limited acceptance, as the closed sea regime was relegated to a limited band of territorial waters. In 1703 the Dutch jurist Cornelius van Bynkershoek fixed this territorial limit at the distance of a cannon shot from shore, and by the beginning of the 19th century most maritime states regarded territorial sovereignty as extending three miles seaward from the coast, although some states claimed wider territorial seas.[10]

Manganese nodules were first discovered on the bottom of the ocean by the HMS *Challenger* during its scientific voyage of 1872–76.[11] The question of what legal regime was applicable to the seabed was virtually ignored during the long evolution of the freedom of the seas doctrine, until enhancement of deep-sea survey techniques and increasing use of the sea for military and communications purposes in the second half of the 19th century began to focus attention on the issue.[12] Unlike the seabed areas of internal waters and territorial seas, which were clearly subject to the territorial sovereignty of coastal states, the legal status of the continental shelf area and the deep ocean floor gave rise to disagreement among international lawyers. Although it might simply have been presumed that the regime applicable to the superjacent waters also controlled the seabed below, there had been no history of either exploitation or affirmative abstention from exploitation, and it was generally believed that no seabed regime could have been established by operation of customary international law. Similarly, the SS *Lotus* case, which held that a state will be presumed to have authority to act in the absence of a prohibitory norm of international law, could not be considered controlling where, because of a lack of technological capacity to exploit the resources of the ocean floor, there had been no opportunity to establish such a norm.[13]

Although the conclusion of a positive multilateral agreement governing use of the oceans had been urged for decades,[14] no such treaty was negotiated until the inadequacy of the existing regime became inescapably clear with the significant advances in technology which occurred during the second half of the 20th century. Because of economic and technological limitations, exploitation of both living and nonliving marine resources had traditionally been centered in coastal areas, where most of these resources are concentrated. The British Colonial Act of 1811 set forth claims to living resources of the seabed beyond the limits of the territorial sea, but until the middle of the 19th century the main objects of this authority were pearls, sponges, oysters, sea cucumbers, and other renewable resources which traditionally had been harvested by prescriptive right under customary international law.[15] Other legislation extended customs jurisdiction beyond the limits of territorial waters, causing concern among maritime states anxious to preserve the commercial advantages of unrestrained navigation.[16] The freedom of the seas doctrine was codified in 1856 in the Treaty of Paris and its accompanying declaration, which sought to ban privateering and define the rights of belligerents and neutrals,[17] but a 1907 conference proved less successful in attempting to establish an international prize court to adjudicate such rights.[18]

The Conference on the Progressive Codification of International Law, held at The Hague under the auspices of the League of Nations in 1930 and attended by 42 mostly European states, discussed law of the sea issues but reached no decision with respect to the width of territorial waters or the extent of contiguous fisheries jurisdiction, although there was agreement on the need to regulate exploitation of "riches constituting the common patrimony."[19] The Codification Conference was sharply divided on the questions of coastal-state jurisdiction over offshore fisheries and the establishment of security zones in high-seas waters, although there was general agreement on a right of innocent passage through territorial waters and on a right of hot pursuit with respect to vessels departing the territorial sea.[20] Participants were able to agree upon a minimum width of three miles for the territorial sea, but general acceptance of a maximum limit of three miles was conditioned upon the adoption of a contiguous zone of as much as twelve miles, which proved impossible.[21] Several years earlier the League's Committee of Experts had recognized the "urgent necessity for international regulation of the exploitation of the biological wealth of the sea."[22] Indeed, a major impetus for the Conference was the increasing usage of the oceans and the resulting challenge to the assumption of inexhaustibility of ocean capacity and resources upon which the Grotian principle of freedom of the seas had been based.[23]

Following the collapse of the 1930 Conference, the rate of expansion of coastal-state jurisdiction increased markedly, further eroding the traditional freedom of the seas doctrine. In the 1939 Declaration of Panama, the United States and 20 other American states established a temporary 300- to 1200-mile-wide "security zone" around the Western hemisphere, asserting an inherent right to protect vessels within their coastal waters against belligerent attack.[24] By 1944 the width of the territorial sea had been extended to six miles by six states, to nine miles by one state, and to twelve miles by two states.[25] These unilateral extensions of sovereignty were prompted by the threatened depletion of coastal fisheries by foreign fishermen, and were seldom subject to serious challenge by maritime states; yet the primary method for extending jurisdiction was nevertheless changing — from simple increases in the width of the territorial sea to changes in the baselines from which the territorial sea is measured, as well as to the creation of functional contiguous zones which extended coastal-state jurisdiction over specific activities and resources.[26]

By far the most important of these functional extensions was the Truman Proclamation of September 28, 1945, whereby the United States unilaterally declared "the natural resources of the subsoil and sea-bed of the continental shelf beneath the high seas but contiguous to the coasts of the United States as appertaining to the United States, subject to its jurisdiction and control."[27] Like Britain's 1942 assertion of an exclusive right to control use of part of the shelf area in the Gulf of Paria, the Proclamation was intended to address "the long range world-wide need for new sources of petroleum and other minerals."[28] The Truman Proclamation was extremely significant, not only because it was the first general extension of seabed jurisdiction by a major maritime state, but also because it asserted a claim to the resources of the continental shelf rather than to the shelf itself or to its superjacent waters.[29] Like the right to impose fisheries conservation measures unilaterally in adjacent high-seas waters, which was asserted in a companion proclamation but never enforced,[30] the novelty of the continental shelf concept was attributable also to its reliance upon appurtenance to land — rather than upon the traditional concept of *res nullius* — as the basis for an assertion of exclusive jurisdiction with respect to use of sea resources. Furthermore, the U.S. claim was based only upon anticipated use, whereas the formation of rules of customary international law had historically required a long and continued practice of states.

Although the 1945 proclamations were at least partly a result of the emergence of influential new transnational interest groups within the United States — particularly scientists, ecologists, and the oil and mining industries[31] — nonetheless, they served as a legitimizing precedent for coastal states which had desired to expand their own offshore jurisdiction but had doubted the legality of such unilateral claims. Within five years, jurisdiction was extended farther offshore by more than 30 coastal states, several of which claimed expanded territorial sovereignty as far as 200 miles from shore.[32] Because such expansive claims restricted high-seas navigation and fishing rights traditionally exercised by foreign vessels in coastal waters beyond the territorial sea, they encountered vigorous protests from maritime states, which were willing to accept only limited coastal-state jurisdiction over the continental shelf.[33] Claims to the shelf and its resources surpassed territorial-sea extensions by a ratio of 10 to 1 as the preferred method of increasing coastal-state jurisdiction, but a majority of the new claims — particularly those of Latin American states — were based upon assertions of full sovereignty over the seabed itself rather than upon the carefully limited usufructuary rights embodied in the Truman Proclamation.[34] The opposition of the United States to such excessive claims proved unavailing, its protestations being undermined by the unilateral nature of its own proclamations.[35] The expansionary trend thus continued for more than a decade, giving rise to numerous disputes over the exercise of coastal-state jurisdiction. Following jurisdictional increases by 6 Latin American states during 1950, another 28 offshore claims were asserted unilaterally by coastal states from 1951 to 1958, 17 of which involved some form of sovereignty or functional jurisdiction over the continental shelf.[36] In August 1952 Chile, Ecuador, and Peru issued the Declaration of Santiago, proclaiming the inadequacy of their existing territorial seas and contiguous zones and asserting "sole sovereignty and jurisdiction" over the adjacent sea to a distance of 200 nautical miles from shore, including complete territorial sovereignty over the subjacent continental margin and ocean floor.[37] Although the U.S. version of the continental shelf regime was unanimously endorsed by 20 American states at Ciudad Trujillo in 1956, expansionist attitudes were becoming increasingly predominant within the Organization of American States (OAS) and in the mid-1950s the United States began looking to the U.N. system as a forum for resolution of law of the sea issues.[38]

Although in 1949 an international arbitrator refused to recognize the emerging "continental shelf doctrine" as an established rule of international law,[39] one year later Sir Hersch Lauterpacht argued in favor of its recognition,[40] and by the mid-1950s it was generally agreed that a new rule of international law had indeed become firmly established, although its precise content remained undefined.[41] After several regional organizations, including the OAS, proved unable to agree upon a uniform definition, the International Law Commission in 1949 selected the law of the sea as one of its three priority areas of discussion. Created two years earlier by the U.N. General Assembly pursuant to its power under article 13 of the U.N. Charter to promote the "progressive development of international law and its codification," the 15-member Commission met annually from 1949 to 1956 to formulate draft articles for submission to an anticipated law of the sea conference. Although agreement upon the width of the territorial sea eluded the Commission, it agreed that a width of more than 12 miles would be excessive.[42] Coastal states were to be granted usufructuary jurisdiction over resources of the continental shelf, the limits of which were defined by the Commission, after several draft revisions, as "the sea-bed and subsoil of the submarine areas adjacent to the coast but outside the area of the territorial sea, to a depth of 200 metres or, beyond that limit, to where the depth of the superjacent waters admits of the exploitation of the natural resources of the said areas."[43] This jurisdictional formula was at least partially the result of the efforts of interested oil companies, which were able to influence the Commission's work through their lawyers' membership in the International Law Association.[44] International jurisdiction over the resources of the continental shelf was rejected on the ground that international administration would pose overwhelming practical difficulties for the efficient exploitation of marine resources.[45] The Commission drafted 73 articles, with commentaries, which were transmitted to the General Assembly in 1956 along with a recommendation that a law of the sea conference be convened to act upon them. Article 67, containing the definition of the continental shelf, was qualified by articles 68 and 69, which respectively denied coastal state sovereignty over the shelf itself and separated the legal regime of the shelf from that of the superjacent waters. The grant of jurisdiction in article 68 was nevertheless sufficient to permit coastal states to exclude foreigners from the exploitation of offshore mineral resources, and a general lack of immediate economic interest in the seabed facilitated effective international ratification of the Truman Proclamation.[46]

UNCLOS I and II

The First United Nations Conference on the Law of the Sea (UNCLOS I) was held in 1958 in Geneva. The 73 articles recommended for adoption were referred to five separate committees for review, and the committee recommendations were in turn organized by the Conference into four complementary conventions, an optional protocol establishing compulsory dispute settlement procedures, and nine companion resolutions.[47] The treaties were largely drafted with a view to achieving political compromise rather than to establishing clear-cut legal rules, and numerous textual ambiguities virtually assured subsequent disputes over their interpretation.[48] The tasks of the Conference included not only codification of the customary law of the sea, but also the creation of new legal rules to fill the gaps in existing international law — i.e., *lex ferens* as well as *lex lata* — and the outcome of the Conference represented an important victory for coastal states at the expense of maritime interests, confirming more than 60 jurisdictional extensions. Although the dozen landlocked states attending the Conference were well aware that geography had placed them at somewhat of a disadvantage vis-à-vis coastal states with respect to the customary law of the sea, they took a common position only with respect to their rights of access to the sea.[49] The other major political division at the Conference was the East-West "Cold War" rivalry, which dominated discussion of security-related issues such as navigation through territorial and coastal waters. Forty-eight of the 86 participants were developing states, many of which had only recently gained their national independence and had had little experience in asserting and defending their interests in the context of multilateral negotiations, and there was little support among such states for rules of international law which had largely been developed to promote the interests of European colonial powers.[50]

The 1958 Continental Shelf Convention constituted formal international affirmation of the Truman Proclamation, as formulated in the draft articles of the International Law Commission. The definition of the continental shelf contained in draft article 67 was adopted virtually unchanged as article 1 of the Convention, the delegates overcoming any concerns which may have existed with regard to the inequity of basing jurisdiction over shelf resources upon nonuniform geographical and technological criteria (the slope of the adjacent shelf and the exploitability of shelf resources).[51] Instead of full territorial sovereignty over the continental shelf, coastal states were granted "sovereign rights for the purpose of exploring it and exploiting its natural resources," nonsedentary living resources being excluded from the regime so as to avoid the linkage of continental-shelf and superjacent-water regimes which had been vigorously sought by Latin American states troubled by foreign depletion of their offshore fishing stocks.[52] Coastal-state consent was required for the conduct of scientific research on the continental shelf, although the Convention provided that such consent "shall not normally" be withheld.[53] Other provisions of the Convention were designed to protect the exercise of high-seas freedoms against unjustifiable coastal-state restrictions.[54]

Article 2 of the High Seas Convention expressly identified four protected "freedoms of the seas" — navigation, fishing, overflight, and the laying of submarine cables and pipelines — as well as "other" freedoms "recognized by the general principles of international law." These freedoms were not absolute, however — indeed, much of the substantive language of the four conventions constituted regulation intended to limit the exercise of traditional high-seas freedoms. Notwithstanding their status as privileged activities, navigation and fishing had benefited from rapid technological advances which had increased their rates of growth so dramatically that the effects of pollution and overfishing increasingly called into question the Grotian assumption of resource inexhaustibility upon which the freedom of the seas regime was based. Furthermore, these freedoms were subject to coastal-state restrictions when exercised within internal waters, the territorial sea, or international straits. Yet high-seas areas beyond the limits of national jurisdiction generally continued to be governed by the traditional regime of unrestricted access.

There had been no agreement within the International Law Commission with respect to fishing rights or the width of the territorial sea, and alignments on these issues at the Conference tended to be based more upon national interest than upon custom or disinterested rationality, making compromise more difficult than it had been with regard to the continental shelf, and the Conference adjourned without reaching agreement on either of these important issues.[55] The Convention on the Territorial Sea and the Contiguous Zone did define the interests of coastal and maritime states in the territorial sea, particularly with respect to rights of innocent passage, which were to be enjoyed by warships within international straits but not within territorial waters.[56] The Convention also contained provisions governing the use of bays, islands, straits, and baselines, as well as a 12-mile contiguous zone within which coastal states were authorized to enforce their customs, fiscal, immigration and sanitary regulations.

In addition to the various widths claimed for territorial waters, contiguous fishing zones of as much as 100 and 200 miles had been asserted unilaterally by more than a dozen states.[57] The Convention on Fishing and Conservation of the Living Resources of the High Seas required states to cooperate in the establishment of conservation measures with respect to high-seas fisheries, but lacked a uniform allocation of rights and duties with respect to fisheries exploitation in waters beyond the territorial sea. Article 7 authorized coastal states to impose unilateral conservation measures for high-seas fisheries adjacent to their territorial waters, such measures to be "valid as to other states" — pending review by a special commission of five experts — when based on "appropriate scientific findings" of urgent need and when adopted on a nondiscriminatory basis vis-à-vis foreign fishermen.

Disputes continued over the proper extent of fisheries jurisdiction and territorial waters, and between 1958 and 1960 several more states claimed territorial seas and contiguous zones of 12 miles, despite the protests of major maritime states.[58] The U.N. General Assembly followed the recommendation of UNCLOS I in convening the Second United Nations Conference on the Law of the Sea (UNCLOS II), noting that resolution of the remaining jurisdictional issues would "contribute substantially to the lessening of international tensions and to the preservation of world order and peace."[59] During the six-week Conference, held at Geneva in the spring of 1960, the same negotiating alignments predominated, with developing states generally participating only as clients, allies, or dependents of the United States or the Soviet Union.[60] The United States and Canada proposed the establishment of a six-mile territorial sea with an additional six-mile contiguous fishing zone within which foreign fishing would be phased out over a 10-year period, but the proposal fell one vote short of the two-thirds majority vote necessary for adoption,[61] and the Conference adjourned without having resolved any significant outstanding issues. The participating states also failed to establish an effective procedure for the compulsory settlement of disputes relating to uses of the sea.

Between 1960 and 1967 several important trends were at work to undermine the adequacy of the legal framework erected by the four Geneva conventions. The treaties themselves attracted relatively meager support: they were slow in coming into force — the last of them did not enter into effect until 1966 — and none were ever ratified by a majority of the international community. Indeed, most of the newly independent developing states chose to disregard the agreements, which in their view had been negotiated by — and for the benefit of — the dominant colonial powers.[62] The emergence of Soviet maritime power caused a reevaluation in both the United States and the Soviet Union of national interests with regard to the law of the sea. President John Kennedy proposed a long-range oceanic strategy, and President Lyndon Johnson oversaw the enactment and implementation of a national oceans program which included Sealab, Sea Grant, and other new oceanographic projects. Although other developed states also experienced an increasing interest in ocean science and technology during these years, their interest was much less a function of perceived national security concerns.[63]

Two extremely important and interrelated developments affecting the postwar evolution of the law of the sea have been the accelerating increase in the rate of technological change and the corresponding expansion of virtually all ocean uses in response to rapidly increasing population growth. Worldwide trends of population increase, industrialization, and urbanization have led to greater demand for food, minerals, and energy, and the impact has been particularly severe in fast-growing coastal communities, which have in turn placed unprecedented burdens upon sea resources.[64] World merchant tonnage was growing at an annual rate of 8 percent — with tanker tonnage increasing at a significantly higher rate[65] — necessitating greater international regulatory control to promote navigational safety and to prevent marine pollution, as the grounding of the *Torrey Canyon* off the coast of England in 1967 dramatically illustrated.[66] Similarly, freedom of fishing was increasingly called into question, as unrestricted fishing — in effect the specific application of a *res nullius* regime to living, nonsedentary marine resources — encouraged individual fishermen to maximize their current catch without regard to the effect on future fisheries stocks. Technological changes produced huge factory ships and sophisticated harvesting methods, brought distant-water commercial fishing to faraway coastal areas, and made severe depletion of a fisheries stock within a period of months possible.[67] In contrast to the relative equilibrium which prevailed prior to UNCLOS I, the annual worldwide fish catch doubled between 1955 and 1967,[68] and coastal fisheries began to be severely depleted by a rate of harvest significantly in excess of the maximum level of sustainable yield.[69] Technological advances also fostered development of offshore oil deposits during the 1960s, and by the middle of the decade coastal states had come to value their rights of access to offshore oil deposits nearly as much as their offshore fishing rights.[70] "[E]xtensive utilization of existing ocean use rights by more and more nations, and to a greater degree, demonstrated that when the number of uses and users increases significantly, difficult problems of accommodation are created."[71] Damage was being done to the oceans by the overexploitation of living marine resources, by pollution from sewage and industrial wastes, and by applications of new nuclear technologies;[72] yet long-standing international rivalries and diplomatic practices prevented the law of the sea from developing apace.[73]

At the time of UNCLOS I, offshore hard mineral exploitation was relatively modest, occurring mainly within waters less than 100 feet deep, where sand, phosphates, and other surficial deposits had been mined for centuries. Exploitation of manganese nodules on the ocean floor remained both technologically and commercially impractical, although articles advocating deep-sea nodule exploitation were published in 1952 and 1959 by an entrepreneurial scientist named John Mero.[74] Encouraged by the results of several oceanographic studies, Mero presented an optimistic appraisal of the commercial potential of deep-sea mining operations in his 1965 book, *The Natural Resources of the Sea*. Although Mero's projections were seriously questioned at the time, several large mining companies nevertheless undertook preliminary surveys of the deep ocean floor in the early 1960s.[75] "Projections in the 1960s about feeding the world from the sea, extracting vast amounts of mineral wealth, and other solutions from the new field of oceanography were over-dramatized."[76] The potential economic advantages of seabed mining vis-à-vis land-based mining were perceived to include low labor costs, lack of drilling or excavation expenses, and relatively low transportation costs; the major disadvantages were identified as the high initial costs of the new mining technology and the questionable efficiency of the land-based separation processes required to produce marketable quantities of nickel, copper, cobalt, and manganese.[77]

The major political and legal obstacle to seabed mining was the uncertain nature of the applicable legal regime, an uncertainty which was only one aspect of the general instability of the various regimes known collectively as the law of the sea. In the years immediately following UNCLOS I and II, the unresolved issues of fisheries jurisdiction and the width of the territorial sea continued to undermine the work of the conferences, providing a strong impetus for further unilateral expansion of offshore claims. Such claims were encouraged by the legitimizing effect of the customary international law process and by the dynamics of geopolitical rivalries, as well as by expectations of new natural-resource reserves and of the consequent enhancement of national economic, political, and military power.[78] Although many jurisdictional claims were advanced by newly independent developing states, many were also asserted by older Latin American states, several of which — precluded from basing claims to coastal waters upon the continental shelf doctrine as they had done prior to UNCLOS I — asserted 200-mile territorial seas in an effort to obtain full control over coastal fisheries.[79] Most territorial-sea extensions were limited to 12 miles, however, and soon the number of states making 12-mile claims equaled those adhering to a distance of three miles.[80] Many of the latter group of states extended their fisheries jurisdiction beyond their territorial waters by enacting contiguous zones, and by the time the United States increased its own fisheries zone to 12 miles from shore in 1966 as many states claimed fisheries jurisdiction to a distance of 50 miles or more as adhered to the traditional three-mile limit.[81] The growing number and variety of offshore claims caused a further increase in the number of jurisdictional disputes between coastal and maritime states, the latter fearing a loss of navigation rights as a result of expanded territorial sovereignty and generally refusing to recognize such claims despite enhanced coastal-state military capabilities.[82] Official concern was also heightened among landlocked and other geographically disadvantaged (i.e., shelf-locked) states, which were unable to exploit offshore resources of their own and which therefore suffered particularly severe losses insofar as the unilateral claims detracted from their interests in the shrinking high-seas regions.[83]

The Exploitability Clause

In addition to the territorial sea and fishing issues, a third major area of dispute arose during the 1960s with respect to delimitation of the continental shelf. The definition of the continental shelf set forth in article 1 of the Continental Shelf Convention had been based on three criteria: adjacency, depth, and exploitability. Although the Truman Proclamation, upon which the Convention was modeled, had grounded its jurisdictional claim upon the geographical adjacency of the shelf, in the years of rapid technological development following UNCLOS I most states chose instead to base their claims upon the more expansive — indeed potentially unlimited — criterion of exploitability.[84] Reliance upon this latter element was justified by the language of article 1, which plainly permitted the exercise of exclusive coastal-state jurisdiction beyond the 200-meter isobath when exploitation of shelf resources proved possible.[85] Under general principles of textual interpretation, however, there was a strong presumption that the other two criteria were not intended to be meaningless, but rather to be "essential elements of a whole."[86] Although several international lawyers were of the opinion that technological advances had already caused an automatic division of the entire ocean floor among coastal states, such a view was generally regarded as unsound.[87] There was a growing apprehension, however, that states would nevertheless eventually seek to use the exploitability clause as a justification for enclosure of the entire deep sea-bed, which would lead in turn to new forms of international conflict.[88]

In formulating the definition of the continental shelf which was eventually adopted at UNCLOS I, the International Law Commission had decided not to rely solely upon the geographical criteria used by scientists, largely because the differing widths and depths of the shelf throughout the world would have resulted in widely disproportionate benefits among coastal states.[89] In 1950 the Commission did agree that the area of coastal-state jurisdiction over the shelf "should be limited," but that "where the depth of the waters permitted exploitation, it should not necessarily depend on the existence of a continental shelf."[90] At its next session the Commission unanimously adopted a 200-meter depth criterion, with a limit of 20 miles beyond the territorial sea for states lacking a physical shelf;[91] after reconsideration, however, this dual definition was replaced by a single compromise formula intended to anticipate future technological development: "the sea-bed and subsoil of the submarine areas contiguous to the coast, but outside the area of territorial waters, where the depth of the superjacent waters admits to the exploitation of the natural resources of the sea-bed and subsoil."[92] In 1953, after the reactions of numerous states had been received, the 200-meter depth criterion was reinstated because of its certainty, its practical sufficiency "at present," and its resemblance to the geographical definition of the shelf; it was explained that this criterion was to be subject to exceptions where considerations of equity so required.[93]

This geographically based definition survived until the 1956 session, at which time the Commission, prodded by the Latin American states and citing a recent formulation of the Inter-American Specialized Conference on Conservation of Natural Resources,[94] added the exploitability clause, creating the hybrid definition which was eventually ratified as article 1 of the Continental Shelf Convention. In its commentary, the Commission emphasized that "exploitation of a submarine area at a depth exceeding 200 metres is not contrary to the present rules, merely because the area is not a continental shelf in the geological sense,"[95] implying that such exploitation might be unlawful under some circumstances for other reasons — such as the adjacency requirement — and that the exploitability clause was intended to provide the necessary flexibility to avoid inequitable results in specific cases rather than to be a generally applicable alternative criterion. Furthermore, the commentary to article 27 noted that the Commission had not "made specific mention of the freedom to explore or exploit the subsoil of the high seas. It considered that apart from the case of the exploitation or exploration of the soil or subsoil of a continental shelf . . . such exploitation had not yet assumed sufficient practical importance to justify special regulation."[96] By distinguishing the continental shelf from the subsoil of the high seas and expressly leaving the latter area unregulated, the

Commission thus by implication confirmed the existence of a portion of the ocean floor which would remain beyond the limits of national jurisdiction, the exploitability clause notwithstanding. The compromise language formulated by the Commission withstood several amendment attempts at UNCLOS I.[97]

The seemingly definite limit embodied in the 200-meter depth criterion was thus rendered uncertain by the exploitability clause. Moreover, the concept of exploitability is itself ambiguous. It is unclear, for example, whether scientific (as well as economic) activities might satisfy the exploitability criterion — and whether actual exploitation is required, or if the capacity for exploitation is itself sufficient for the exercise of jurisdiction.[98] Indeed, if the exploitability test were construed to refer to the particular technology available to (or used by) each individual coastal state — rather than to the best available anywhere at any specific point in time — a state with technology adequate to exploit the deep ocean floor would be able to claim exclusive resource rights clear up to the outer territorial limits of an opposite state lacking similar technology. The meaning of the adjacency criterion is similarly imprecise, although it was evidently intended to assure some linkage of the legal definition of the continental shelf to its geographical counterpart, as well as to the territorial jurisdiction of the adjacent coastal state.[99] According to the Commission, the adjacency clause was also intended to forestall premature operation of the 200-meter standard where "submerged areas of a depth less than 200 metres, situated fairly close to the coast, are separated from the part of the continental shelf adjacent to the coast by a narrow channel deeper than 200 metres";[100] inclusion of the adjacency requirement thus implied that the exploitability clause was to be strictly construed — applicable only where operation of the 200-meter depth criterion would produce similarly inequitable results, particularly for narrow-shelf coastal states. Finally, the absence of any evidence in the legislative history of article 1 of a clear intent to authorize a complete division of the world's ocean seabed among coastal states is a further indication that no such division was contemplated; rather, the subsequent rapid technological advances which made the exploitability clause potentially supportive of deep-sea claims were simply not foreseen at the time of UNCLOS I and II, when petroleum was the only marine mineral resource of interest to policymakers.[101]

The development of economically and technologically feasible means of exploiting the manganese nodules of the deep ocean floor made some solution of the novel jurisdictional questions raised by article 1 increasingly imperative in order to prevent technologically advanced states from using its ambiguous language to support new and politically unacceptable claims to seabed resources well beyond the continental margin. Since, by the mid–1960s, the United States and several other states had extended their functional jurisdiction over the continental shelf beyond the 200-meter depth limit in pursuit of offshore oil deposits, repeal of the exploitability clause was generally viewed as impracticable, and the potential threat of repeal may in fact have accelerated the rate of offshore claims. The many states which failed to ratify the Continental Shelf Convention — together easily comprising a majority of the world's coastal states — did not consider themselves bound to observe even the exploitability limitation, their jurisdictional claims being instead governed by customary international law.[102] The nature of the legal regime applicable to the seabed beyond territorial waters under principles of customary international law remained in dispute, however: the doctrines of res nullius and res communis had each been forcefully advocated by eminent international jurists in the years prior to UNCLOS I,[103] and the debate assumed new importance as commercial exploitation of the deep ocean floor seemed increasingly possible during the 1960s. Although the question of the nature of the legal regime applicable to the seabed beyond the continental shelf was not formally addressed by the International Law Commission or the UNCLOS negotiations, the Commission did note that, in addition to the four primary high-seas freedoms, "there are other freedoms, such as freedom to explore or exploit the subsoil of the high seas"[104] — presumably one of the "other" recognized freedoms mentioned in article 2 of the High Seas Convention. This concept of the seabed and subsoil as subject to exploitation — but not appropriation — was advocated by proponents of unrestricted access on the ground that manganese nodules are analogous to other marine resources traditionally subject to appropriation under the doctrine of res nullius (e.g., fish, pearls, and sponges), the seabed itself being incapable of occupation under the res communis regime applicable as well to the superjacent waters.[105] Largely because of the exploitability clause, which had found some support in state practice, many legal scholars supported application of a res nullius regime to the deep seabed prior to 1967.[106]

The Movement Toward Internationalization

The increasing recognition of the potential commercial importance of manganese nodules resulted in preliminary consideration in early 1966 by the U.N. Economic and Social Council (ECOSOC) of a proposal by the Secretary-General for a five-year resource development survey program, part of which was to be directed toward the study of offshore mineral resources, which were widely believed to "hold the key to the world's metal supply."[107] With a view to speeding economic development and possibly providing "a source of international capital," the United States proposed that the Secretary-General undertake an additional "survey of the information currently available concerning the resources of the sea and the techniques for exploiting them,"[108] and on March 7 ECOSOC adopted Resolution

1112, formally requesting the Secretary-General to prepare such a survey and, in particular, "to identify any gaps in available knowledge which merit early attention by virtue of their importance to the development of ocean resources, and of the practicality of their early exploitation."[109] The U.S.-sponsored resolution recognized that "a compilation of available knowledge on known resources and techniques for exploiting them would be an indispensable tool for developing countries to improve their programmes to extract the riches of the sea."[110] In December 1966 the U.N. General Assembly adopted Resolution 2172, also sponsored by the United States, which reaffirmed Resolution 1112 and requested the Secretary-General to prepare a comprehensive survey of marine science activities and to develop plans for "ensuring the most effective arrangements for an expanded programme of international co-operation to assist in a better understanding of the marine environment through science and in the exploitation and development of marine resources."[111] In support of the resolution, which was passed by a vote of 100 votes to 0, with 11 abstentions, the United States expressed its "strongly held" hope that the international community would promote the "utilization of the sea for mankind."[112]

For much of the previous 20 years a small but vocal minority of prominent international legal scholars had opposed unilateral extensions of coastal state jurisdiction, favoring instead some measure of international control over the seabed and its resources. These internationalists generally viewed the seabed as a form of common property, which was not subject to appropriation through exploitation and which should be exploited for the benefit of developing states within the United Nations system.[113] Arguments advanced in favor of international control included the inadequacy of the existing legal regime for investment purposes, including the ambiguity of the exploitability clause and the apparent impracticality of exclusive appropriation by occupation;[114] the perception that conflict could best be avoided by the employment of international machinery to allocate and guarantee exclusive rights to the exploitation not only of manganese nodules but also of other marine resources such as fisheries;[115] the need for enforceable production controls to prevent the market disruptions expected to result from large-scale seabed mining operations; the macroeconomic efficiency made possible by a uniform, comprehensive system of international regulation; the promotion of international goodwill through accommodation of the interests of developing states;[116] and the importance of a potential source of new revenue to the U.N. system as a mechanism for the promotion of world order.[117] Opponents of internationalization, on the other hand, argued that international institutions were too weak to administer such a regime effectively,[118] and that international control would therefore lead to inefficiency and new forms of conflict.[119]

International administration of the seabed had been proposed in 1934 by the Institut de Droit International; it was again advocated at meetings of the International Law Association and the International Law Commission in 1951.[120] The idea resurfaced in 1955 with the backing of Georges Scelle, a member of the Commission who advocated creation of a supranational organization with jurisdiction over fisheries, seabed nodules, and other marine resources.[121] In the same year the World Association of Parliamentarians for World Government called for international administration of the oceans, citing potential revenues for the United Nations from the leasing of exploitation rights.[122] U.N. jurisdiction was also advocated in the 1957 report of the Commission to Study the Organization of Peace, which called for the exercise of "ownership and control" of the seabed by a U.N. agency also empowered to regulate pollution and to issue recommendations for fishing quotas.[123] It was recognized that seabed exploitation might proceed under a generally accepted alternative regime of flag-state or coastal-state jurisdiction.[124]

Despite occasional mention of internationalization at the initial UNCLOS negotiations,[125] and in academic works, the idea did not gain widespread support or provoke significant controversy until the effects of technological change upon the environment began to be more fully appreciated in the mid–1960s, particularly within the United States. "With the world population outpacing its food supply, with industrial requirements for energy and minerals growing faster than population, and with increasing concentrations of waste being unwittingly injected into the marine environment, it had become clear that neither the problems ahead nor the solutions proposed for them could terminate at the water's edge."[126] In May 1966 the Commission to Study the Organization of Peace elaborated upon its previous proposal and recommended establishment of "a special agency of the United Nations to ... control and administer international marine resources; hold ownership rights; and grant, lease or use these rights in accordance with the principle of economic efficiency.... It should distribute the returns from such exploitation in accordance with directives issued by the United Nations General Assembly."[127] At the first annual conference of the Law of the Sea Institute, held at the University of Rhode Island in the summer of 1966, support was voiced for the convening of a third U.N. Conference on the Law of the Sea to promote rational regulation of common marine resources.[128] On July 13, 1967, the World Peace Through Law Conference, attended by 2,500 lawyers from 100 states, unanimously adopted a resolution asserting that "the high seas are the common heritage of all mankind" and calling upon the U.N. General Assembly to proclaim that the seabed and its mineral resources "appertain to the United Nations and are subject to its jurisdiction and control."[129]

In 1965, at a White House Conference on International Cooperation, several participants urged the establishment of a U.N. agency with jurisdiction over fisheries as well as the deep seabed, and a draft proposal outlining such an agency was prepared within the State Department.[130] At the United Nations, the United States declared that "what [man] finds beneath the sea may be used for international benefit — without infringing on the sovereign rights of nations. . . . It is not too early . . . to start dreaming and thinking exciting thoughts about the role the U.N. can take."[131] The following year, in a widely quoted speech at the commissioning of the research ship *Oceanographer*, President Johnson announced U.S. policy with respect to the deep seabed, observing that

> great accomplishments in oceanography will require the cooperation of all the maritime nations of the world. . . . Under no circumstances, we believe, must we ever allow the prospects of rich harvest and mineral wealth to create a new form of colonial competition among the maritime nations. We must be careful to avoid a race to grasp and to hold the lands under the high seas. We must ensure that the deep seas and the ocean bottoms are, and remain, the legacy of all human beings.[132]

The presidential statement was an elaboration of the new national oceans policy set forth in the 1966 Marine Resources and Engineering Development Act, which called for the promotion of a "national program in marine science for the benefit of mankind."[133] The Act established the Commission on Marine Science, Engineering, and Resources — known as the Stratton Commission — to examine the use and preservation of the marine environment and "to formulate a comprehensive, long-term, national program for marine affairs designed to meet present and future national needs in the most effective possible manner."[134] A February 1967 report to the Senate Committee on Foreign Relations recommended that the United Nations be granted title to deep-sea nodules as a means of enhancing the financial independence of the organization.[135]

Following the decline of British naval power, the United States and the Soviet Union had become the major maritime competitors for global naval supremacy. In 1966, concerned over increasing restrictions on navigation resulting from the expansion of territorial waters, the United States and the Soviet Union agreed to consult with other maritime states regarding the convening of a new international conference on the law of the sea which would fix the maximum width of the territorial sea at 12 miles while guaranteeing rights of transit through newly enclosed straits.[136] A third element of the regime, proposed by the United States as an inducement for Latin American states to accept the package, provided for extended offshore fisheries jurisdiction.[137] Early in 1967 the Soviet Union, which historically had opposed efforts to strengthen international organizations, proposed that a working group of the Intergovernmental Oceanographic Commission (IOC) be empaneled to draft a convention to govern exploration and exploitation of deep-sea mineral resources, with a view to the convening of a new law of the sea conference to act upon it.[138]

On August 17, 1967, in a surprise move, the delegation of Malta requested the inclusion in the agenda of the 22nd session of the U.N. General Assembly of a supplementary item entitled "Declaration and treaty concerning the reservation exclusively for peaceful purposes of the sea-bed and of the ocean floor, underlying the seas beyond the limits of present national jurisdiction, and the use of their resources in the interests of mankind." The accompanying memorandum, the work of Arvid Pardo, Malta's U.N. Ambassador, outlined the potential danger posed by the lack of a legal regime to govern the ocean floor and proposed a far-reaching solution:

> . . .2. In view of the rapid progress in the development of new techniques by technologically advanced countries, it is feared that . . . the sea-bed and the ocean floor . . . will become progressively and competitively subject to national appropriation and use. This is likely to result in the militarization of the accessible ocean floor through the establishment of fixed military installations and in the exploitation and depletion of resources of immense potential benefit to the world, for the national advantage of technologically developed countries.
> 3. It is, therefore, considered that the time has come to declare the sea-bed and the ocean floor a common heritage of mankind and that immediate steps should be taken to draft a treaty embodying, *inter alia*, the following principles:
> (a) The sea-bed and the ocean floor . . . are not subject to national appropriation in any manner whatsoever; . . .
> (c) The use of the sea-bed and of the ocean floor . . . shall be undertaken with the aim of safeguarding the interests of mankind. The net financial benefits derived from the use and exploitation of the sea-bed and of the ocean floor shall be used primarily to promote the development of poor countries;
> (d) The sea-bed and the ocean floor . . . shall be reserved exclusively for peaceful purposes in perpetuity.[139]

It was further asserted that an international agency should be established

> (a) to assume jurisdiction, as a trustee for all countries, over the sea-bed and the ocean floor . . . ;
> (b) to regulate, supervise and control all activities thereon; and
> (c) to ensure that the activities undertaken conform to the principles and provisions of the proposed treaty.[140]

The initiative, largely a product of Pardo's own interest in the subject,[141] was based upon information already well known to experts in the field of ocean law, but was nevertheless new to most U.N. delegates, many of whom regarded consideration of the issue as premature.[142] The memorandum provoked opposition within the U.S. Congress, where an existing anti–U.N. prejudice was reinforced by the introduction of more than two dozen resolutions intended to safeguard perceived U.S. national security and economic interests.[143] Senator Claiborne Pell submitted a resolution calling for the convening of a multilateral conference to draft a comprehensive treaty placing deep-sea mining and other ocean uses under direct international control, but the idea failed to attract significant support,[144] and the Johnson administration withdrew from its initial support for Ambassador Pardo's proposal.[145]

At the United Nations, the item was considered in the First Committee of the General Assembly. In his note of October 31 introducing the new agenda item, the Secretary-General reported that the preliminary results of the study being conducted pursuant to ECOSOC Resolution 1112 indicated that the legal status of deep-sea minerals and the institutional machinery whereby their exploitation would accrue to the benefit of developing states constituted "major gaps" in the law which might delay seabed exploitation.[146] On November 1 Ambassador Pardo opened the First Committee debate with a three-and-one-half-hour lecture, citing history, law, science, and a variety of scholarly publications in defense of his proposal. After assuring the assembled delegates that the Maltese proposal had been "formulated entirely without the benefit of advice from other countries and . . . that we are not a sounding-board for any State," Pardo described the geophysical features of the deep ocean floor and the "vast" resources lying on and below its surface.[147] He then recounted the recent technological developments which had altered man's historical relation to the oceans and necessitated the negotiation of a new legal regime, dwelling in particular upon the dangers posed by the impending spread of the arms race and nuclear waste disposal to submarine areas and upon the inadequacy of the definition of the continental shelf contained in article 1 of the Continental Shelf Convention:

> The process has already started and will lead to a competitive scramble for sovereign rights over the land underlying the world's seas and oceans, surpassing in magnitude and in its implication last century's colonial scramble for territory in Asia and Africa. The consequences will be very grave: at the very least a dramatic escalation of the arms also by the intolerable injustice that would reserve the plurality of the world's resources for the exclusive benefit of less than a handful of nations. The strong would get stronger, the rich richer, and among the rich themselves there would arise an increasing and insuperable differentiation between two or three and the remainder. Between the very few dominant powers, suspicions and tensions would reach unprecedented levels. Traditional activities on the high seas would be curtailed and at the same time the world would face the growing danger of permanent damage to the maritime environment through radio-active and other pollution; this is a virtually inevitable consequence of the present situation.[148]

Arguing that the existing specialized agencies of the U.N. system lacked competence to respond effectively to these novel problems, and that the United Nations was itself institutionally ill-equipped to administer a seabed regime, Pardo outlined as a long-term goal the creation of an independent agency which would assume jurisdiction "not as a sovereign, but as a trustee for all countries," and would be "endowed with wide powers to regulate, supervise and control all activities on or under the oceans and the ocean floor."[149] Based upon optimistic revenue projections for deep-sea oil drilling and nodule mining operations, Ambassador Pardo estimated that an agency established by 1970 could be collecting gross income of more than $5 billion annually by 1975, but he emphasized that "the concurrence of all is essential" for the creation of a workable legal regime.[150]

Reaction to the speech was wide-ranging and often intense as delegates scrambled to reevaluate — or, in some cases, to formulate for the first time — their ocean policies.[151] By separating the problem of seabed exploitation from other ocean issues and by defining the issue in the context of the efforts of developing states to achieve a more equitable distribution of international economic power vis-à-vis the more technologically advanced states, the Maltese proposal provoked immediate controversy and substantially determined the subsequent handling of the issue in the United Nations.[152] In the 60 speeches delivered in the First Committee by representatives of member states and international organizations in the days following Pardo's speech, most developing states expressed general support for the Maltese initiative, although many were reluctant to proceed without giving further consideration to their own interests, and Latin American states opposed any actions which might jeopardize their extensive offshore claims.[153] Despite Pardo's evident disdain for their economic dominance and the manner of its exercise, most developed states did not reject the Maltese position outright, but instead adopted an attitude of caution — generally supporting further study of the matter.[154] With the sole support of its Eastern-bloc allies, the Soviet Union opposed the initiative outright, despite its previous IOC proposal addressing largely the same issues.[155] Several new alignments of special interests emerged during the debates, including mineral-consuming states (consumer states), mineral-producing states (producer states), multinational mining companies, and environmentalists.[156] It became clear that deep-sea nodule exploitation could not proceed under an unregulated freedom of the seas regime; yet the lack of a definite international regime raised many questions, one of the most important of which related to the unresolved dispute concerning delimitation of the area to which the regime would apply.[157] A U.S. proposal for the establishment of a permanent Committee on the Oceans was rejected in favor of a 35-state Ad Hoc Committee to Study the Peaceful Uses of the Sea-Bed and the Ocean Floor Beyond the Limits of National Jurisdiction

(Ad Hoc Committee), which was established by a unanimous vote of the General Assembly on December 18, 1967. The mandate of the Ad Hoc Committee recognized "the common interest of mankind in the sea-bed and the ocean floor" and directed it to prepare a study of the matter, including "[a]n indication regarding practical means of promoting international cooperation in the exploration, conservation and use of the sea-bed and the ocean floor, and the subsoil thereof, . . . and of their resources, having regard to the views expressed and the suggestions put forward by Member States."[158]

Chapter 2
Preparations for UNCLOS III

The Ad Hoc Committee

Prior to the initial meetings of the Ad Hoc Committee, several developments occurred which had a significant effect upon the future course of the negotiations. In early 1968 the United States and the Soviet Union informally agreed to seek general approval of a 12-mile territorial sea with rights of free passage through international straits enclosed by the extension; it was further agreed that in order to obtain the necessary support of other states, such a zone should be accompanied by an extended functional zone over coastal resources.[1] On February 19 the Secretary-General distributed the report mandated by ECOSOC Resolution 1112, which discussed political and economic issues relating to the development of marine resources beyond the continental shelf and identified two major problem areas: the need for further scientific research and the need for resolution of outstanding jurisdictional issues.[2] With regard to the latter, the report found that "a major deterrent to initiative in advancing marine mineral development is the absence of a proper jurisdictional framework to guarantee to mining ventures the economic security they are entitled to expect, as well as safeguard the interests of other legitimate activities."[3] The report also concluded that the mining of manganese nodules could have a "particularly harmful" effect upon those developing producer states which were economically dependent upon mineral exports, and that to control such effects any international machinery established to administer deep-sea mineral exploitation should be "very broadly based, with extensive powers and responsibilities."[4]

The Ad Hoc Committee was composed of 19 developing and 16 developed states, giving the latter group a disproportionately strong representation; landlocked and geographically disadvantaged states on the other hand were underrepresented.[5] It held three separate sessions during 1968, the first of which was held March 18–27 at U.N. headquarters in New York. Faced with a general lack of information regarding the subject matter of its somewhat open-ended mandate, the Committee divided the major portion of its work between two working groups, dealing respectively with legal

22

and with economic and technical aspects of the item. It reserved for the Plenary consideration of scientific issues and methods for promoting international cooperation in the exploration, conservation, and use of the deep seabed and subsoil.[6] At the insistence of the developed states, who feared being outvoted on important issues, the Committee decided to proceed by consensus, thereby ensuring that the negotiations would proceed slowly but also that political confrontations would be minimized in favor of compromises acceptable to all parties.[7] The Ad Hoc Committee had before it several proposals submitted by member states, the most notable of which was a draft treaty of the Netherlands calling for the establishment of an independent authority within the U.N. system to lease mining sites and supervise nodule exploitation.[8] Under President Johnson, the United States assumed a leadership role as the major power most receptive to calls for international cooperation on ocean matters, formally proposing an International Decade of Ocean Exploration and urging that "efforts should be made forthwith to avoid unfair competition to appropriate the sea-bed and the ocean floor, which [are] the legacy of all human beings."[9] The convening of a new law of the sea conference was formally proposed in order to address questions relating both to deep-sea mining and offshore jurisdiction, and the U.N. Secretariat was requested to prepare a study of existing and potential rules of international law applicable to exploration and exploitation of the seabed beyond the limits of national jurisdiction.[10]

On April 24, 1968, the Secretary-General distributed the report on marine science and technology which had been requested by General Assembly Resolution 2172. Citing the need for increased international cooperation in order to promote greater understanding and use of ocean resources, the report observed that "questions of international concern which arise in the development and exploitation of marine resources . . . are specialized and largely technical in nature," and "are therefore best considered within the framework of the appropriate specialized agencies concerned, wherein lies the necessary technical competence and where Governments can most easily contribute effectively to international coordination through their appropriate national bodies."[11] The Secretary-General accordingly recommended that the role of such specialized agencies be strengthened, but noted that "special problems of a predominantly political nature" were under discussion in the General Assembly and that he therefore had "no proposal to offer in those domains for the time being."[12] The Secretary-General's report went on to urge greater international cooperation in the development of technology and scientific knowledge with regard to deep-sea mineral exploitation, as well as in the "resolution of associated economic, administrative and legal issues."[13] After noting that the role of intergovernmental organizations in regard to such matters is normally a limited one, and that, with the exception of "scientific research and preliminary studies," very little was being done within the U.N. system regarding mineral resources beyond the continental shelf, the Secretary-General nevertheless proposed that the General Assembly

> take steps to expand further the existing activities in the continental shelf area to ensure that, as far as the whole ocean is concerned, the United Nations is given adequate responsibility for systematic collection and diffusion of information regarding economic marine mineral deposits, techniques appropriate for their development, as well as for resolving related juridical, general administrative and political issues.[14]

The Ad Hoc Committee held only three plenary meetings during its second session, most of its work being conducted by the Legal Working Group and the Economic and Technical Working Group, which met 14 and 11 times, respectively, between June 17 and July 9. Observing in his opening remarks that the seabed was "a geographical area to which no clear legal principles yet applied," the chairman of the Legal Working Group urged the delegates to seek not only "the adaptation of existing legal norms," but also "the establishment of new norms to govern new relationships."[15] Although the Legal Working Group was able to agree that seabed minerals should be used in the interest of mankind, there was substantial disagreement among the delegates concerning the nature of the norms and interests in question, and the Group adjourned without having completed its program of work.[16] The report which the Committee had requested from the Secretariat was distributed to the delegates at the June session, identifying three possible types of "not necessarily mutually exclusive" legal regimes: "First, alternative regimes involving some extension of state sovereignty to areas presently beyond the limits of national jurisdiction; second, freedom of exploration and exploitation and scientific investigation; and, third, some form of international control."[17] The study suggested that the Maltese proposal might be implemented by an international organization with the power to apply the principles embodied in the common heritage concept — specifically, to adopt resolutions, to promulgate binding regulations, to license seabed mining operations, and to administer a tax levied upon mining operations on a flat fee or royalty basis.[18] The report concluded that seabed mining operations would be likely to exacerbate pollution of the superjacent ocean waters, and that a future seabed regime should therefore include regulatory measures for the control of such pollution.[19]

    The Economic and Technical Working Group was able to make satisfactory progress at the second session, and its report indicated that a "stable" seabed regime "providing for orderly progress and security of title" would promote development of mineral deposits.[20] The report noted that the costs of ocean mining were greater than those of land-based mining operations, and suggested that the former mineral source would therefore remain economically impracticable "for some time."[21] Deep-sea recovery of nickel, copper, and cobalt — if not of manganese itself — was expected to prove economical eventually, but the high ratio of cobalt present in nodules relative to the entire world supply made it likely that seabed mining operations would have a substantial depressant effect upon cobalt prices, suggesting to Group members a "need for arrangements for the exploitation of mineral resources byond [sic] the continental shelf that will avoid adverse consequences for the world market in general and the economy of developing countries in particular."[22] The Group's report further pointed out that, although a wide range of ocean interests had been accommodated in the past, seabed mining would probably conflict with ocean uses traditionally regarded as high-seas freedoms, including navigation, fishing, scientific research, and the laying of submarine cables; it was urged that such interference "should not discourage the development of marine resources, but rather bring about efforts to reconcile the conflicting interests in the regulatory framework to be set up for the purpose of mineral development."[23] The need for greater international cooperation with regard to marine mineral development was recognized by all delegations, but there was no agreement upon how such cooperation was to be achieved; proposals ranged from improved coordination of scientific and technical activities to the creation of a new international institution to govern all aspects of deep-sea mineral exploitation.[24] Only a few states were believed to possess the capability to conduct seabed mining operations, and it was widely feared that their technological and financial advantages "would accentuate the economic imbalance existing between developed and developing countries, and ... would also be an incentive for the former to grab and hold the areas which are most promising."[25] The participants agreed upon the need for early delimitation of a boundary to separate the areas of national and international jurisdiction.[26]

Some progress was made at the third negotiating session, which was held at Rio de Janeiro in late August, but the Ad Hoc Committee was unable to reach agreement on a set of principles to govern deep seabed mining. The Committee discussed scientific, political, military, and administrative issues in its plenary meetings, and it was concluded that a standing committee should be created as a forum for continued negotiations.[27] A major dispute had emerged between states favoring a set of general principles which could claim unanimous support and states seeking adoption of a detailed set of principles which could be more readily operationalized. Two draft resolutions, generally corresponding to the respective positions of developing and developed states, were introduced and were ultimately reported to the General Assembly as alternative proposals. The common heritage of mankind principle proved to be one of the most contentious issues, with developed states generally opposing, and developing states supporting, a legal significance for the phrase; the question of the type of international machinery to implement the common heritage concept gave rise to similar disagreement.[28] There was widespread support, however, for the International Decade of Ocean Exploration and for an expanded coordinating role for the IOC.[29] Most states also agreed that the principle of peaceful use should be applicable to the seabed area beyond the territorial waters of coastal states, and nuclear weapons states appeared willing to compromise in order to reach agreement on this issue;[30] developing states were nevertheless unhappy with the unwillingness of developed states to negotiate a complete ban on all military uses of the oceans.[31] Despite its eagerness to negotiate a treaty limiting military use of the seabed—and in contrast to the United States, which reserved a final decision on the question of international control of the seabed[32]—the Soviet Union quickly emerged as the most vociferous opponent of the common heritage of mankind concept.[33] Overall, developing states were somewhat successful in their attempt to exert pressure on the developed states to agree to a new international regime,[34] and momentum began to build for the renegotiation of all related law of the sea issues.

Following adjournment of the Ad Hoc Committee, the First Committee of the General Assembly was faced with a large number of draft resolutions concerning various aspects of the issues discussed in the Ad Hoc Committee. Despite protracted negotiation and debate, strong disagreement remained with respect to the wording of a declaration of principles, and the problem was relegated to a 42-member Committee on the Peaceful Uses of the Seabed and Ocean Floor Beyond the Limits of National Jurisdiction (Seabed Committee) which was created as a permanent replacement for the Ad Hoc Committee. The Committee was given a mandate to study several subjects: the legal principles and norms which would promote international cooperation in the exploration and exploitation of the seabed; the means of promoting such exploitation for the benefit of mankind as a whole; proposals to prevent the marine pollution expected to be caused by such exploitation; and reservation of the seabed exclusively for peaceful purposes.[35] At the insistence of the Soviet Union and its Eastern European allies, the First Committee rejected a proposal to require the Seabed Committee also to study "the question of establishing in due time an appropriate international machinery for the exploration and exploitation of this area . . . and the use of these resources in the interest of mankind, especially those of developing countries."[36] The study was instead assigned to the Secretary-General by a separate resolution, and the General Assembly also endorsed the International Decade of Ocean Exploration and the preparation of a study of the protection of marine resources from pollution.[37] By the end of 1968 it was apparent that the gap between the negotiating positions of the developed and developing states was widening and that creation of a new regime would take much more time than had been expected.[38]

The Seabed Committee (1969–73)

The composition of the new Seabed Committee was much the same as that of its predecessor: 3 developing states had been dropped and 10 others added, bringing the total to 26; landlocked and geographically disadvantaged states remained underrepresented, as were producer states,[39] and the work of the Committee was divided between a Legal Subcommittee and an Economic and Technical Subcommittee. The addition of new delegations further slowed the negotiations by diverting resources toward their education on the past course of the negotiations and the nature of the outstanding issues.[40] At the 1969 sessions of the Seabed Committee — held in New York in February, March, and August — the generation of new information, which included reduced estimates of commercially exploitable nodule resources, ensured that even veteran delegates were involved in an ongoing educational process and tended to dissipate the sense of urgency which Ambassador Pardo had created.[41] Two issues assumed new and lasting prominence: the transfer of marine technology from developed to developing states and the economic hardship on producer states likely to result from seabed mining operations.[42] It was generally understood that the question of the delimitation of the international area and that of the nature of the new international regime were mutually dependent; yet the common heritage principle remained a subject of contention despite movement by the developed states toward acceptance of the developing-state position.[43] All developed states except the Soviet Union were now willing to support some form of international machinery, although most, including the United States, remained opposed to the establishment of an international agency with an independent mining capability.[44]

The report of the Secretary-General, issued June 18, outlined possible registration, licensing, operational, and dispute settlement functions for the proposed machinery, as well as organizational issues relating to its establishment, membership, administration, voting procedures, and financial arrangements.[45] It concluded that "the main possibilities with respect to various forms of international machinery would appear to be the following: (i) a secretariat centre or unit, which might be established within an existing organization; (ii) a United Nations subsidiary organ; and (iii) an intergovernmental body having an independent legal status."[46] The Secretary-General also discussed the relation between the existing law of the sea and the proposed international machinery. In addition to the overriding issue of delimitation of the area beyond the limits of national jurisdiction, the report also addressed the need to respect existing high-seas freedoms, the threat of marine pollution, jurisdiction over mining ships and installations, and the general nature of the legal regime applicable to deep-sea mining operations.[47]

Progress was made in the Legal Subcommittee on the drafting of a declaration of principles, but the delegates continued to work toward complete agreement on a comprehensive statement, and the item was tabled until 1970.[48] General agreement was nevertheless reached on the need for special consideration for developing states and for equal treatment of coastal, landlocked, and geographically disadvantaged states.[49] Although a variety of interest group alignments developed on the Seabed Committee, the split between developed and developing states remained most pronounced, the latter group negotiating aggressively on broader political questions where use of moral suasion proved particularly effective.[50] Confrontationalist tactics were most frequently employed by African states, while the delegation of Malta assumed a leading role in the effort to negotiate an accommodation of interests.[51]

The 1969 report of the Economic and Technical Subcommittee noted that although the development of mining technology had been proceeding slowly, "industry [was] becoming increasingly aware of the vast mineral deposits contained in the ocean floor which could in the future become technically recoverable and economically exploitable."[52] The Subcommittee agreed that mankind as a whole would benefit from seabed mining operations, both by the expanded world supply of nodule minerals and by the sharing of the revenues derived from their exploitation, and that, to the extent possible, all states should participate in the exploitation and equitable sharing of the mineral resources of the seabed.

> It was therefore considered important (i) to promote international co-operation providing for the training of nationals of developing countries with a view to enabling developing countries to participate directly in such undertaking and (ii) to provide for international arrangements which will benefit all mankind, taking into account the special needs and interests of developing countries.[53]

The Subcommittee also discussed the Secretary-General's analysis of possible forms of international machinery, and while it was generally agreed that the complexity of the question and the inadequate preparation of the delegates prevented full consideration of the issue, the report noted that "a mere registry system would lack the authority necessary to protect the interests of mankind as a whole—and of the developing countries in particular—since it would be confined to the purely passive role of recording activities which were initiated and notified by [sic] States."[54] Similarly, it was argued with regard to the licensing alternative that "the international machinery should not be set up for the sole purpose of licensing the development of sea-bed resources, but rather should regulate, co-ordinate, supervise and control all activities relating to the exploration and exploitation of the resources."[55] The operational agency alternative was also criticized on several grounds, although further study of the concept was urged.[56] In contrast to the Legal Subcommittee, the Economic and Technical Subcommittee conducted its proceedings in relative harmony and was able to make significant progress on several issues.[57]

The question of the seabed regime was again a subject of further contention in the U.N. General Assembly. After passing two relatively non-controversial resolutions relating to U.N. support for a long-term program of oceanographic research and the study of marine pollution,[58] the General Assembly adopted four additional resolutions on December 15, 1969. Two of these, adopted without opposition, instructed the Seabed Committee to continue its efforts to draft a formal declaration and requested the Secretary-General "to prepare a further study on various types of international machinery, particularly a study covering in depth the status, structure, functions and powers ... including the power to regulate, co-ordinate, supervise and control all activities relating to the exploration and exploitation of [deep-sea] resources, for the benefit of mankind as a whole."[59] A third resolution, calling upon the Secretary-General to communicate with all states regarding the desirability of convening a new law of the sea conference, was passed by a vote of 65 to 12, with 30 abstentions.[60] The fourth, and most controversial because of its rejection of the open-access regime advocated by many developed states, was the so-called Moratorium Resolution, which was adopted by a vote of 62 to 28, with 26 abstentions.[61] The Moratorium Resolution declared that exploitation of the resources of the seabed and ocean floor was to be "carried out under an international regime including appropriate international machinery," and that

> pending the establishment of the aforementioned international regime:
>   (a) States and persons, physical or juridical, are bound to refrain from all activities of exploitation of the resources of the area of the sea-bed and ocean floor, and the subsoil thereof, beyond the limits of national jurisdiction;
>   (b) No claim to any part of that area or its resources shall be recognized.[62]

Although General Assembly resolutions are legally nonbinding under the U.N. Charter,[63] adoption of the Moratorium Resolution may nonetheless have contributed further to the legal uncertainty inhibiting investment in seabed mining technology, particularly within the United States, which, while denying its binding effect, characterized the Resolution as "an important statement to be given weight in the determination of United States policy."[64] Controversy surrounding the Moratorium Resolution caused a division in the ranks of developing states, and contributed to a devaluation of General Assembly resolutions and of the U.N. system in general.[65]

Although emplacement of weaponry on the seabed ultimately proved to be economically and technologically impractical, in the late 1960s it was believed that the military potential of seabed use surpassed its value as a source of mineral wealth.[66] Following Pardo's call for reservation of the seabed exclusively for peaceful purposes, the United States and the Soviet Union nevertheless undertook negotiations for demilitarization of the seabed within the Conference of the Committee on Disarmament. In exchange for agreement by the United States that the treaty should apply to the entire seabed beyond 12 miles from the coast, the Soviet Union dropped its insistence upon complete demilitarization of the seabed.[67] A final draft of the seabed demilitarization treaty was completed on September 1, 1970, and referred to the General Assembly, where it was adopted and recommended to states by a vote of 102 to 2, with 2 abstentions, despite some concern that the treaty failed to provide for coastal-state participation in verification procedures or for negotiations to achieve further reductions in military uses of the seas.[68] The Treaty on the Prohibition of the Emplacement of Nuclear Weapons and Other Weapons of Mass Destruction on the Seabed and the Ocean Floor and in the Subsoil Thereof was signed by 62 states on February 11, 1971, in an effort to address the "peaceful uses" element of Pardo's initiative.[69]

The Seabed Committee met again in March and August 1970, with a mandate to complete work on a declaration in time for its consideration by the fall session of the General Assembly. The continued slow progress of these negotiations in the Legal Subcommittee increasingly encouraged states to turn their attention from seabed issues to broader law of the sea subjects, however, and there was a noticeable reduction in political conflict between developed and developing states as the focus of discussion shifted to questions of coastal-state jurisdiction.[70] In August, 14 of 20 Latin American and Caribbean states meeting in Lima, Peru, asserted the right of states to extend their offshore sovereignty over the seabed and its superjacent waters as far as necessary to secure access to both living and nonliving marine resources.[71] Although proposals for expanded offshore jurisdiction were made in the Seabed Committee, the nature and extent of such jurisdiction remained a subject of disagreement.

Despite an encouraging start, the Economic and Technical Subcommittee was unable to resolve the complex economic and technical questions relating to the exploration and exploitation of deep-sea minerals. The Subcommittee did recognize the need for widespread "dissemination of the results of scientific research and exploration," particularly with regard to "knowledge of the ecology of the [seabed] area and its vulnerability to pollutants."[72] The report of the Seabed Committee emphasized the lack of "effective international safeguards . . . to prevent pollution of the oceans, apart from cases of oil spills from ships," and recommended improved coordination among national and international pollution control programs and assumption by the proposed international machinery of major responsibility for the prevention and control of pollution resulting from deep-sea mining activities.[73] Increasing worldwide concern over environmental preservation was also reflected in the unilateral enactment by Canada of a 100-mile-wide pollution control zone off its Arctic coasts.[74]

At its August session, the Seabed Committee considered the latest report of the Secretary-General on international machinery, as well as working papers submitted by France, the United Kingdom, and the United States. The proposals of France and the United Kingdom were both based upon a licensing regime under which states would lease block tracts of the seabed from an international authority and would in turn allocate exploitation rights among their nationals.[75] The U.S. working paper, a comprehensive draft convention also based upon such a three-tiered system, envisioned the creation of an International Seabed Resources Authority with three principal organs — an Assembly, a Council and a Tribunal — as well as a secretariat with three special commissions.[76] As outlined in a presidential statement of May 23, 1970, and in a formal Draft Treaty submitted to the Seabed Committee in August, the U.S. proposal also included a 12-mile territorial sea; a "trusteeship zone" extending limited coastal-state jurisdiction over resources of the continental shelf beyond the 200-meter depth limit; payment of "substantial mineral royalties" to the Authority, "to be used for international community purposes, particularly economic assistance to developing countries"; and promulgation of regulations by the Authority to govern exploitation of the deep seabed.[77] In contrast to states supporting a moratorium on seabed mining activities pending negotiation of a generally accepted international regime, the President called for seabed mining to proceed under interim arrangements, "subject to" the future regime under which "agreed international machinery would authorize and regulate exploration and use of seabed resources beyond the continental margins."[78] The trusteeship zone, to be administered by coastal states on behalf of the international community as part of the common heritage of mankind, was proposed in conjunction with a relatively strong Authority as a bargaining tool to secure strategically important rights of transit — particularly for the U.S. Navy — through the 116 international straits which would come under the sovereignty of coastal states with the extension of territorial waters to a width of 12 miles.[79] Additional provisions of the draft convention were designed to secure other important U.S. interests, including security of investment, stability of expectations, and compulsory settlement of disputes.[80]

Ambassador Pardo, among others, saw much merit in the U.S.

proposal,[81] and although it failed to attract immediate support in the Committee — largely because of opposition from Latin American states, which would have been forced to renounce many of their jurisdictional claims, and from the Soviet Union, which, despite intense pressure from many developing states, still objected on ideological grounds to the establishment of a strong international agency[82] — the draft treaty provided a new impetus to the stalled negotiations.[83] By the end of its August session the Committee had agreed that benefits accruing to the international community from seabed mining operations "should be shared equitably by all States and in particular the less developed among them."[84] There was also agreement that an acceptable convention should contain "general provisions governing the entities admitted to apply for licenses, the conditions under which licenses would be used, the size of the areas to which licenses might be applied, their duration, the minerals to be covered by the licenses and the amounts of fees and payments," and that the features of the machinery should not include truly supranational powers but should instead allow for direct regulation of individual states, which would in turn be responsible for control of their nationals.[85]

Efforts to agree upon a set of principles carried into the 1970 session of the General Assembly, where a Declaration of Principles was finally produced which commanded nearly universal support by including only the most general and noncontroversial points.[86] Underlying disagreements with respect to the operational significance of the enumerated principles, together with a growing movement toward the reopening of related questions within the context of a new law of the sea conference, detracted from the Declaration's intended status as a solution to the problems which had been outlined by Pardo.[87] The Declaration, adopted by a vote of 108 to 0, with 14 abstentions, affirmed that

> 1. The sea-bed and ocean floor, and the subsoil thereof, beyond the limits of national jurisdiction (hereinafter referred to as the area), as well as the resources of the area, are the common heritage of mankind....

> 3. No State or person, natural or juridical, shall claim, exercise or acquire rights with respect to the area or its resources incompatible with the international regime to be established and the principles of this Declaration....

> 5. The area shall be open to use exclusively for peaceful purposes by all States, whether coastal or land-locked, without discrimination, in accordance with the international regime to be established....

> 7. The exploration of the area and the exploitation of its resources shall be carried out for the benefit of mankind as a whole....

9. On the basis of the principles of this Declaration, an international regime applying to the area and its resources and including appropriate international machinery to give effect to its provisions shall be established by an international treaty of a universal character, generally agreed upon. The regime shall, *inter alia*, provide for the orderly and safe development and rational management of the area and its resources and for expanding opportunities in the use thereof, and ensure the equitable sharing by States in the benefits derived therefrom, taking into particular consideration the interests and needs of the developing countries, whether landlocked or coastal.[88]

Although arguably of greater legal force than the Moratorium Resolution by virtue of its nearly unanimous approval, the Declaration of Principles was still viewed by many developed states as being of doubtful validity as a statement of existing customary international law.[89] The General Assembly also adopted three other resolutions, two of which called for the preparation of further studies from the Secretary-General;[90] the third called for the convening of a general conference on the law of the sea in 1973 to negotiate all law of the sea issues, including "an equitable international regime" for the deep seabed, and expanded the membership of the Seabed Committee, which was instructed to prepare draft articles for consideration by the conference.[91] To ensure widespread participation — and thereby avoid the kind of legal fracturization which had resulted from the four separate UNCLOS I treaties — it had been agreed that the conference should produce a single comprehensive convention — a "package deal" for a new law of the sea.[92] Because of the increased complexity of the negotiations expected to result from this comprehensive approach, it was expected that the negotiations might continue for an additional five years or more.[93]

The Seabed Committee reconvened in 1971 for two more regular sessions, held in the spring and summer in Geneva, and for a special three-day concluding session in New York in October. The size of the Committee had been more than doubled, to 86 states, in order to accommodate the large number of requests for participation by interested states, and the increased membership further slowed the proceedings, making consensus more difficult to achieve and requiring additional time to be spent on the education of new delegates, many of whom represented developing states.[94] The Committee's new mandate emphasized the need for it to address nonseabed issues, encouraging the formation of new political alignments based upon national interests and negotiating goals, the strongest of which, under the leadership of the territorially aggressive Latin American states, represented the interests of coastal states.[95] Landlocked and geographically disadvantaged states remained underrepresented and internally divided, but nevertheless comprised one of the most active interest groups.[96] The Committee formed three subcommittees, each with a specific mandate, and reserved the familiar issues of peaceful uses of the seabed and delimitation of national and international zones for consideration by the Plenary.[97] With regard to the latter question, a general suspicion of U.S. motives had caused developing coastal states to seek a further seaward extension of national sovereignty rather than a trusteeship zone or other form of limited offshore jurisdiction, and there was by this time widespread support for the establishment of a 12-mile territorial sea.[98]

Subcommittee I, instructed to prepare draft treaty articles embodying the international seabed regime and machinery, made little progress in the face of continuing disputes between developed and developing states, despite the introduction of several new proposals. One of the most significant proposals was a draft ocean space treaty submitted by Ambassador Pardo on behalf of the delegation of Malta. Based on a conception of the world's oceans as an interrelated ecological system, the Maltese delegation sought replacement of the traditional freedom of the seas doctrine with "balanced international institutions with wide competence and strong powers," arguing that it was "essential to establish a new, equitable international legal order of an institutional character for ocean space (as distinguished from a regime including machinery purely for the seabed beyond national jurisdiction) based on the concept of the common heritage of mankind."[99] Although in his original initiative Pardo had opposed extensions of coastal state jurisdiction, he was now convinced that no progress at the approaching conference would be possible without acceptance of such extensions,[100] and the 1971 Maltese proposal thus called for the creation of a functional zone of exclusive coastal-state jurisdiction over ocean resources to a fixed distance of 200 miles from shore — a so-called "exclusive economic zone."[101] The draft treaty was greeted with interest by some developing states, but with caution by the major Western states anxious to maintain customary high-seas freedoms in waters superjacent to the seabed, and with ridicule by the Soviet Union.[102]

A second draft proposal of major importance, introduced by 13 Latin American delegations, called for creation of a unitary system "in which mankind, in the capacity of owner, would participate directly in the administration and management of the area and the exploitation of its resources."[103] Specifically, the so-called Thirteen-Power Proposal provided for establishment of an

> International Seabed Authority . . . empowered: . . .
>
> (c) to undertake exploration of the area, and exploitation of its resources as well as all activities relating to production, processing and marketing; . . .
>
> (e) to take all necessary measures, including *inter alia*, control, reduction or suspension of production or fixing of prices of products obtained from

exploitation of the area, whenever it deems that such production may have adverse economic effects for developing countries, exporters of raw materials.[104]

The draft convention further proposed that the Assembly — "the supreme organ" of the Authority, consisting of all member states — would have the power to decide any questions arising under the proposed convention "or relating to the powers and functions of the Authority," as well as the power to approve regulations governing the formation of joint ventures and profit-sharing agreements.[105] The Council was to be composed of 35 members and was to have the power to "issue regulations pertaining to all activities undertaken in the area" and to supervise those activities, particularly with regard to "prevention of pollution and contamination of the marine environment from seabed activities."[106] The Thirteen-Power Proposal originated the concept of an International Seabed Enterprise to exploit the deep seabed on behalf of the entire international community, an idea which drew unexpectedly widespread support from developing states and relatively weak opposition from developed states.[107]

Despite its previous opposition to the vesting of broad powers in any international machinery, and despite — or, perhaps, because of — its own commitment to develop a capacity for exploration and exploitation of seabed resources,[108] the Soviet Union also submitted a draft convention to the Seabed Committee, calling for the creation of an International Seabed Resources Agency with licensing powers and with limited regulatory jurisdiction over the seabed and ocean floor — an organization substantially similar in its organizational structure to the machinery outlined in the Thirteen-Power Proposal, with the notable exception of the Enterprise, to which the Soviet Union offered no analogous operational organ.[109] Another point of divergence between the two texts related to freedom of scientific research, which was specifically protected in the Soviet draft, as were several other important Soviet interests.[110] Most developed states joined the Soviet Union in opposing the conduct of mining operations by the contemplated international machinery, and many were similarly disposed toward the exercise of production and price controls over minerals recovered from the seabed.[111] Faced with a lack of consensus on all issues, Subcommittee I postponed an attempt to prepare draft articles.

Subcommittee II, given a mandate to "prepare a comprehensive list of subjects and issues relating to the law of the sea ... and to prepare draft treaty articles thereon," also failed to make substantial progress, concluding that "detailed discussion and negotiation of specific subjects and issues at future sessions of the Sub-Committee will be necessary to ensure adequate preparation for the future conference on the law of the sea."[112] The results in Subcommittee III, which had been assigned the task of preparing draft articles relating to the marine environment and scientific research, were similarly disappointing, although there was general agreement that marine pollution presented a grave danger to the marine environment.

> It was generally agreed that adequate and effective measures should be taken within the context of the environment of the whole and that in adopting such measures due account should be taken of the interests of all States, particularly coastal States. Special attention should be given to the interests and needs of developing countries in participating in scientific research as well as in sharing of the results of such research and the benefits derived therefrom.[113]

Except for expanding the Seabed Committee to 91 members by the addition of the People's Republic of China and four other states, the General Assembly took little action with regard to law of the sea issues in its fall 1971 session.[114]

In 1972 the Seabed Committee met in New York during the month of March and in Geneva during July and August, general debate being followed by efforts within working groups to negotiate consolidated draft articles from the various proposals offered by delegations. The sessions were generally conducted with a greater degree of professionalism than had been present in prior years, as delegates became more familiar with the complex and detailed issues, but participating states were nevertheless unable to reach agreement on the major remaining questions.[115] A sense of stagnation pervaded the seabed negotiations following adoption of the Declaration of Principles, and attention turned to the question of coastal-state jurisdiction over the continental shelf and offshore waters, an issue of growing concern in light of the continuing technological advances in offshore oil recovery operations, the great disparity in the nature and extent of coastal-state claims, and the linkage of coastal jurisdiction issues to the question of deep-sea nodule mining.[116] Although negotiating alignments of states in the Plenary and Subcommittee I remained divided along developed- and developing-state lines, as they had been since 1967, Subcommittee II alignments were determined instead by geographical factors, and the continued slow pace of the negotiations tended to strengthen the bargaining power of coastal states at the expense of landlocked and geographically disadvantaged states, which favored a stronger international regime.[117] In June a group of 15 Caribbean states issued a declaration asserting sovereign rights over all natural resources within a "patrimonial sea" extending 200 miles offshore.[118]

It was becoming clear that most states accepted a basic assembly-council-secretariat-tribunal-commissions structure for the Seabed Authority, but also that there continued to be strong disagreement over its powers and functions.[119] Subcommittee I produced a draft text of seabed principles, but many of its articles were so general as to be virtually meaningless.[120] Subcommittee II drafted a comprehensive list of 107 unresolved issues and sub-issues for discussion at the Conference; issues which included marine pollution, scientific research, delimitation of the territorial sea and of coastal-state jurisdiction beyond, navigation rights in and over international straits, and the nature of coastal-state fishing rights, as well as the right to exploit seabed minerals located beyond the limits of national jurisdiction.[121] The main negotiating objectives of developing coastal states included seaward extension of their offshore resource jurisdiction, as well as the establishment of international machinery with exclusive authority to explore and exploit seabed minerals beyond the limits of that jurisdiction; maritime states, on the other hand, were primarily interested in preserving high-seas freedoms within as large an area of the sea as possible.[122] In August the United States submitted a draft convention on territorial seas and straits passage, and indicated its readiness to accept extended offshore resource jurisdiction if such jurisdiction was subject to compulsory dispute settlement procedures and conformed to international standards.[123] Subcommittee III made progress in developing a system of national and international standards for the control of marine pollution, and began preparations for consideration of issues relating to scientific research.

During 1972 the U.S. Congress held additional hearings on unilateral seabed mining legislation drafted by the American Mining Congress, but no further congressional action was taken because of strong tactical opposition by the State Department: "Not only could it be seen to be opposing unilateralism within its own country and supporting the international negotiating process, but also it could take advantage of the pressure created by a unilateral threat to serve its own interests in the negotiations."[124] The 1972 U.N. Conference on the Human Environment was followed with considerable interest in the Seabed Committee, as the Stockholm conferees endorsed a Statement of Objectives and Principles on the Marine Environment as guidance for UNCLOS III delegates.[125] In December the General Assembly, noting "further progress" in the preparatory negotiations, unanimously passed a resolution giving the Seabed Committee another year to complete its work and postponing the first substantive session of the Conference to 1974, with an organizational session planned for late 1973.[126]

At its 1973 sessions, held in New York and Geneva, the Committee continued its efforts to complete preparations for the Third United Nations Conference on the Law of the Sea (UNCLOS III). Subcommittee I continued its inconclusive debate on the relative merits of different forms of international machinery, with most nations favoring a strong but not completely autonomous Authority.[127] An emerging area of dispute in this regard concerned the relative powers of the proposed Assembly and Council, developed states hoping to advance their interests by securing a veto power in — if not outright control over — the latter organ.[128] A new approach to the negotiations, introduced by the developed states, focused discussion on the specific rights and duties of the Authority — with regard to agreed-upon functions and powers, such as collection of fees, allocation of mining areas, promulgation of operating standards, and duration of contracts — rather than on the possible forms of institutional arrangements.[129] Discussion of the controversial subject of production controls — limitations on nodule recovery designed to protect developing producer states — was studiously avoided, but there was widespread support for a tribunal with as yet undefined jurisdiction to resolve seabed mining disputes.[130] Although the U.S. delegation reported "largely satisfactory" results in Subcommittee I, observers were able to perceive little progress in the draft articles it produced, which were set forth in two or more versions.[131] Aside from the nearly complete polarization of states along developed-developing lines, the Subcommittee I negotiations were hindered by the complexity of its task, the use of the consensus procedure, the continuing development of new technologies, and the shift of attention to issues of offshore jurisdiction.[132] By the end of 1973, resolution of these latter issues had replaced the establishment of a nodule mining regime for the deep seabed as the most important task facing UNCLOS III, despite predictions by the U.S. delegation that commercial mining operations would begin within five years.[133]

The emerging agreement on a 12-mile territorial sea with rights of transit through international straits was increasingly viewed as contingent upon the creation of other offshore zones, including, in particular, an extended economic zone within which coastal states would exercise exclusive resource jurisdiction.[134] On July 2 the Organization of African Unity adopted a declaration "recognizing" coastal-state sovereignty over all living and nonliving resources within such an exclusive economic zone, which "shall not exceed 200 nautical miles, measured from the baseline establishing their territorial seas," yet recognizing also that "land-locked and other disadvantaged countries are entitled to share in the exploitation of living resources of neighbouring economic zones on equal basis [sic] as nationals of coastal States."[135] In a speech to Subcommittee II on August 8, Ambassador Pardo outlined a Maltese proposal providing for judicial settlement of disputes regarding jurisdictional claims and for payment of a portion of coastal-state benefits derived from the exclusive economic zone to the Authority for distribution as a part of the common heritage of mankind; noting again that legal concepts had failed to keep pace with the increasingly rapid changes in technology and ocean use, Pardo emphasized that the situation had completely changed since he had submitted his original initiative six years earlier:

> Widely supported proposals are submitted that would place under national control, either immediately or in the near future, almost 80 per cent of ocean space, virtually the totality of its living resources and even the greater part of the manganese nodules of the abyss. In short, from a regime of virtually total freedom over virtually the entire oceans we are now passing to a regime of virtually total sovereignty over virtually the entire oceans. Almost the only common trait between these apparently totally opposite regimes is the obstinate refusal of coastal States to assume any internationally enforceable obligations.[136]

Indeed, the 200-mile exclusive economic zone was generally understood to encompass virtually all exploitable ocean resources, including most offshore fisheries and most of the potentially vast petroleum deposits located beneath the continental margin,[137] and to include as well coastal-state jurisdiction over marine pollution and scientific research. Although a fixed limit of less than 200 miles, with extended continental shelf jurisdiction, was supported by approximately one-fourth of the delegations, the remainder generally favored a full 200-mile economic zone.[138]

Subcommittee II discussions of straits passage were further complicated by the emergence of a forceful bargaining group of straits states, several of which demanded the right to require prior notification and authorization for transit passage by ships bearing nuclear and other hazardous materials — a requirement which was expected to substantially impair U.S. and Soviet naval mobility, and possibly supertanker navigation as well.[139] Despite new proposals by the United States and others, Subcommittee III was unable to complete its evaluation of draft articles relating to the consent regime for marine scientific research and the control of land-based, vessel-source, and seabed-source pollution; as a result, the Subcommittee reported a list of 25 items and 90 subitems, mostly in alternative form.[140] At the 1973 session of the General Assembly, which was heavily lobbied by U.S. interest groups participating in the U.S. delegation, delegates set final dates for the first and second sessions of UNCLOS III, instructed the Conference "to adopt a convention dealing with all matters relating to the law of the sea," dissolved the Seabed Committee,[141] and concluded a Gentleman's Agreement to proceed by consensus at UNCLOS III in order to avoid divisive votes on controversial issues and, if possible, on the convention itself.[142] Between 1967 and 1973, 230 new offshore claims were asserted by 83 coastal states — many of them anticipating the validation of new jurisdictional zones at UNCLOS III — and by the mid–1970s a trend toward unilateral establishment of 200-mile exclusive economic zones had clearly emerged.

> The coastal, straits, and archipelagic states that led this territorialist trend did so for a variety of reasons: to extend control over living and nonliving

resources...; to stop any foreign research leading to commercial "exploitation"; to prevent pollution and accidents; to protect their security from real or imagined threats; and to respond to nationalist sentiments and internal political promptings.[143]

Economic and Technological Developments

A major reason for the slowness of the negotiations was the economic and technological uncertainty surrounding seabed mining operations. Assessment of nodule resources was hampered by a lack of hard scientific data: less than 4 percent of the ocean floor had been thoroughly surveyed, although intensive exploration had been conducted since the late 1960s.[144] The problem of identifying potential mining sites was further compounded by a wide geographical variation in the mineral composition of deep-sea nodules. High-grade, exploitable nodule deposits were identified in the central-eastern region of the Pacific Ocean known as the Clarion-Clipperton Fracture Zone, but deposits in most other regions of the sea were found to be of a relatively low-grade quality.[145] "The rapid escalation of private-sector interest in nodule mining in the early 1970s was due primarily to industry optimism about the future growth in metal demand and pessimism about the availability of low-cost land-based deposits, particularly for nickel."[146] During this period, it was discovered that Mero's early mineral resource projections had indeed been overly optimistic, particularly with regard to the renewability of manganese nodule deposits.[147]

Although the major existing consortia were composed of mining companies from the United States, Japan, West Germany, France, and the Soviet Union, companies from other states such as Canada and Australia were also seeking to develop a seabed mining capability, as were more than 30 additional companies from the United States.[148] The decision of these companies to proceed with production would depend primarily upon the existence of adequate financing, cost-effective technology, and legal protection for exclusive mining rights, as well as upon favorable market conditions.[149] Existing technology provided a mining capacity to depths of more than 6,000 feet, but the cost of such operations was becoming prohibitive.[150] High front-end capital costs made deep-sea mining a risky and highly speculative business venture under even the most favorable political and legal conditions, and banks were unwilling to provide the several hundred million dollars of capital initially required by each private mining operator until firm legal rules were instituted so that they could more accurately assess the risks involved; this was the case even though most of such costs were attributable to the processing plant and other land-based operations which must employ separation techniques devised for nodules of specific concentrations extracted from particular mining sites.[151] The risks involved in the financing of mining operations were compounded by three critical factors: the amount of economic rent captured by the governing authority, the profitability of nodule mining operations, and the degree of competition from land-based producers.[152] Additional financial uncertainties included the general state of the investment climate, the question of whether taxes would have to be paid to national as well as international authorities, and the possibility of liability for environmental damage caused by mining operations.[153] It was nevertheless expected that seabed mining would become commercially viable by 1985.[154]

Although it was argued by laissez-faire economists, among others, that economic efficiency would be best served by a system of unrestricted use of ocean resources, it was nevertheless generally acknowledged that "even when the sole criterion is efficiency, substantial controls will be needed" to prevent costly externalities.[155] It was recognized that seabed mining operations would lead to an expanded supply of — and lower prices for — copper, nickel, and cobalt, and that the export earnings of producer states would be adversely affected as a result; the benefit of lower prices to consumer states, on the other hand, would be relatively minor in light of the relatively low contribution of raw materials to the total cost of manufactured products.[156] Despite predictions that no developing state with an export dependency greater than 20 percent would be significantly affected by nodule mining, developing producer states remained apprehensive and continued to support production controls as a safeguard against severe economic injury, particularly in light of U.N. studies showing that net revenues collected by an International Sea-Bed Authority would be insufficient to compensate such states fully for their lost export earnings[157] — as well as a 1972 U.S. report indicating that recovery of only 1 percent of deep-sea nodules would satisfy world demand for nickel, copper, cobalt, and manganese for 50 years.[158] Compensation schemes, supported by some developed states as a form of insurance for vulnerable producer states, were generally opposed by developing states on the grounds that calculation of the amount of compensation awards would be extremely difficult and that the appropriate authorities would be unwilling or unable to disburse the funds; most developing states preferred a preventive approach, whereby seabed mining operations would be closely regulated and controlled, nodule production being tied to production rates for land-based mining operations — an approach opposed by states whose nationals were actively engaged in the development of a capability for seabed mining operations.[159]

Because the anticipated ratio of the recoverable minerals in manganese nodules does not comport to their supply ratio in world mineral markets, their processing and sale will have a differential impact upon these markets.[160] Although the long-term impact of seabed mining on nickel supplies was known to be potentially significant, it was expected that during the early years of mining operations rising demand would largely offset the effects of increased production,[161] and that, in any event, a drop in nickel prices might increase demand by permitting the substitution of nickel for copper in many uses.[162] Because of the large and increasing worldwide demand for copper, seabed mining was expected to have little effect upon copper prices.[163] However, the market for cobalt, an important element in the production of jet engines and turbines, would be substantially affected by seabed mining, which could double its relatively small world supply and cause its price to fall by two-thirds — at which point cobalt would become a feasible substitute for nickel.[164] Although manganese is essential for the production of iron, steel, oil, and gas, the form of manganese recoverable from seabed nodules has little commercial utility, and the impact of mining on the relatively large world manganese market is expected to be negligible, particularly since many private consortia may choose to avoid additional expense by not separating manganese during processing operations.[165] Overall, markets for nodule minerals increased steadily from the end of World War II until demand slowed toward the end of the 1970s.[166]

Technological development accelerated from 1967 to 1973 in other areas of ocean use, including weather forecasting and oceanographic research.[167] The rising world fish catch, which exceeded 70 million tons annually,[168] was giving rise to demands for new global conservation arrangements to replace the outdated and unobserved 1958 Convention, particularly in view of the expanding distant-water trawler fleets of Japan and the Soviet Union which were severely depleting fisheries stocks in large areas of the high seas.[169] Oil drilling was proceeding at increasingly greater depths off the coasts of more than three dozen states, and would soon far outstrip fishing as the most valuable use of exploitable marine resources.[170] Technological advances continued to widen the existing gap between the demand for and the supply of ocean resources generally, placing greater stress upon the marine environment even as the understanding of the many sources of marine pollution improved.[171] Seabed mining operations were expected to cause pollution of the bottom, the water column, and the surface of the high seas, and further damage to the marine environment was anticipated from at-sea nodule processing operations.[172] All told, new developments in marine science and technology had a significant political impact upon the preliminary UNCLOS III negotiations.[173] By 1972 the annual contribution of ocean products and uses to the U.S. economy had reached $32 billion — more than 2 percent of the U.S. gross national product — and was growing rapidly.[174]

U.S. Policy

At UNCLOS I and II the United States had been primarily concerned with the protection of its traditional interests in freedom of navigation and fishing.[175] At the time of Ambassador Pardo's original proposal, U.S. policy was aimed at achieving several major goals with respect to use of the world's oceans: production of wealth, maintenance of national security, acquisition of knowledge, promotion of public and private welfare, and advancement of world community interests.[176] As then Under Secretary of State Elliot Richardson has related, the United States was not overly concerned by the issues Pardo raised:

> Deep seabed mining was not a major concern that impelled the U.S. in 1969–70 to take the lead in promoting a new international conference on the Law of the Sea. At that time its paramount concern was preserving the navigational and overflight freedoms against the "creeping jurisdiction" of the coastal states. The U.S.'s next priority was the conservation of fishery resources, which had been threatened by the destructive practices of distant water factory ships. Next came protection of the marine environment and control over the hydrocarbon potential in the American outer continental shelf. Deep seabed mining was an interest to be sacrificed to achieve other interests.[177]

In 1969 the Stratton Commission issued its report on national marine policy, terming the pollution of U.S. coastal waters "a growing national disgrace," and calling for greater regulatory control of potentially harmful and conflicting ocean uses.[178] While acknowledging a general lack of sufficient knowledge "to permit hard decisions on alternative courses of action" for the development of ocean resources, the Commission emphasized the need for "a U.S. Navy capable of carrying out its national defense missions anywhere in the oceans, at any desired depth, at any time," called for "[n]ew international frameworks" to promote the efficient development of deep-sea mineral resources, and urged that the benefits of ocean use be shared with developing states in order to promote greater international stability.[179]

The Nixon administration followed the recommendations of the Stratton Commission in drafting its 1970 seabed proposals.[180] U.S. policy at the time

> was developed largely on premises of enhancing world order.... There was a firm belief that all nations would benefit.... The activist U.S. policy in international cooperation began to gain credibility. Then, late in 1970, the situation changed sharply, with decline in leadership and insufficient funds.... With U.S. motives suspect by the all-too-apparent anxiety of the U.S. Navy over restriction of navigation, the sprouts of comity over peaceful uses of the sea began to wither. The enlightened Nixon policy on the seabed regime became another trading stamp. And some oil and mining interests sought to drown the policy.[181]

Despite a lack of support within the international community, the U.S. government took the position that seabed mining beyond the limits of national jurisdiction was a high-seas freedom under customary international law,[182] and that individual nodules could therefore be appropriated under the doctrine of res nullius. Because of the legally nonbinding character of General Assembly resolutions, the United States had officially regarded the Moratorium Resolution and the Declaration of Principles as, at most, only evidence of international law.[183] During the preliminary U.N. negotiations on the seabed, the United States, while not actively opposing international control in the manner of the Soviet Union, had sought to limit the jurisdiction of the proposed international machinery to the licensing of seabed mining operations under nondiscretionary registration procedures, and had been wary of efforts to establish an Authority with the power to regulate other high-seas activities, such as navigation, fishing, and scientific research.[184] The inclusion of an effective veto power and dispute settlement procedure in the 1970 Draft Treaty reflected U.S. insistence that the Authority should be, to the extent possible, an apolitical institution.[185]

Mining companies within the United States have identified four rights which a seabed mining company would require for successful operation: the right to explore for a commercially attractive deposit; the right to exclude others from the use of such a deposit; the right to enjoy continued access to the deposit; and the right to use the site for ancillary purposes.[186] There was thus a perceived need for mutual accommodation between mining companies and the proposed Authority, but mining interests within the United States were strongly opposed to any treaty which would empower the Authority to conduct its own mining operations.[187] Urging the United States to proceed rapidly to exploit the seabed, and believing that UNCLOS III would be unable to reach agreement upon a satisfactory regime in the foreseeable future, mining companies sought enactment of less-restrictive domestic legislation to license interim mining operations unilaterally under a freedom of the high-seas regime. The American Mining Congress lobbied heavily on behalf of the legislation, arguing that it would not affect the rights of other states or the validity of a subsequent treaty, and that by failing to promote its seabed mining industry the United States would lose its technological lead to other potential mining states such as Japan and West Germany.[188] From 1971 until his death in 1978, Senator Lee Metcalf chaired a Senate subcommittee which held hearings on industry-backed bills intended to secure a more lucrative investment climate for U.S. seabed mining firms outside the UNCLOS III framework through a regime of mutual recognition by reciprocating states.[189] Although it was argued that the threat of unilateral action by the United States would speed up the UNCLOS III negotiations, the actual effect of the hearings may instead have been to polarize further the positions of many states and thus to contribute to the prolongation of the negotiations.[190] Moreover, the interim nature of the proposed legislation was doubted by many observers and foreign states, who expected that, once begun, unilateral mining operations would tend to prevent the adoption of a treaty, thereby undermining the creation of the generally accepted regime upon which the legitimacy of deep-sea nodule exploitation must ultimately depend.[191]

The State Department consistently and successfully opposed U.S. unilateral legislation, preferring instead to seek the adoption of a provisional regime at UNCLOS III so long as it appeared that a "timely and successful" conclusion to the negotiations was likely — a condition apparently violated by the failure of the Conference to complete its work by the summer of 1975.[192] The Ford administration continued to participate in UNCLOS III and to oppose unilateral legislation, believing that adverse international reaction — particularly from developing states which perceived the legislation as a deliberate attempt to undermine the common heritage principle — could jeopardize the entire Conference, including agreements reached upon other important issues during the negotiations.[193] There had also been concern that unilateral legislation might portend "a narrow nationalist approach," which could damage broader U.S. foreign policy objectives.[194] Indeed, some developing states had threatened retaliatory action should the United States adopt a unilateral approach to the law of the sea, including economic boycotts, lawsuits before the International Court of Justice, confiscation of nodule minerals, and even sabotage and expropriation of foreign U.S. assets.[195] In 1977 Ambassador Richardson, newly appointed as head of the U.S. delegation to UNCLOS III by President Jimmy Carter, nevertheless stated that the administration would support unilateral legislation under certain conditions.[196] Congressional support for such legislation had been building, and in the following year the Senate pledged to press for its enactment "in the next year or two," provided that UNCLOS III "was still unproductive of an acceptable treaty."[197]

In conjunction with viewpoints advanced by interested private parties, congressional sentiment had a significant impact upon the U.S. negotiating position, particularly after 1970.[198] Following the Maltese proposal of 1967, the United States advocated the establishment of a Committee on Oceans within the United Nations, "with a broad mandate to develop international law and promote interest and cooperation with respect to the ocean and ocean floor."[199] By 1968 a discernible gap had emerged between the positions of the United States and developing states regarding the appropriate attributes of an international seabed mining regime.[200] Detailed formulation of U.S. seabeds policy was nevertheless delayed for two years by a dispute between military and oil interests within the Nixon administration, with the former ultimately prevailing in the 1970 Draft Treaty.[201] The

decision to seek agreement on narrow zones of coastal-state jurisdiction –
coupled with relatively strong international regulation of the seabed area
beyond – thus resulted directly from a determination by the Department of
Defense that unilateral extensions of continental shelf jurisdiction were
leading to jurisdictional claims over the superjacent waters, that U.S.
security interests required the preservation of unrestricted navigation to the
maximum extent practicable consistent with generally accepted pollution-
control measures and the preservation of other common interests in ocean
use, and that seabed minerals were, by comparison, strategically unimpor-
tant.[202] Yet President Nixon's 1970 proposal was also prompted by a
broader concern that failure to achieve multilateral agreement on a new
legal regime would inevitably lead to international conflict: "At issue is
whether the oceans will be used rationally and equitably and for the benefit
of mankind or whether they will become an arena of unrestrained exploitation and conflicting jurisdictional claims in which even the most advan-
taged states will be losers."[203] Negative reaction by foreign states and within
the mineral-industry-dominated Interior Department eventually caused
withdrawal of the "trusteeship zone" proposal in favor of a 200-mile exclusive economic zone, however, as the focus of U.S. oceans policy shifted
to marine mineral resource development.[204]

At a time of economic recession, when U.S. national confidence and
prestige were also being damaged by the Watergate scandals and by withdrawal from Vietnam, the position of the United States with regard to deep-
sea mineral resources was affected dramatically by the 1973–74 oil embargo
instituted by the Organization of Petroleum Exporting Countries (OPEC).
Reflecting a newfound concern over inflation and economic dislocation,
U.S. negotiators at UNCLOS III suddenly found themselves in the position
of seeking to convince developing states that unrestricted unilateral seabed
mining would produce benefits for all states, and that such benefits could
in fact only be secured through the efficiency of private enterprise –
arguments "so transparently self-serving that the United States never has
gained the credibility required to convince others of their merits."[205] Fears
of a mineral embargo have been attributed to the importance of nodule
minerals to modern industrial society and to the vulnerability of the United
States – and of other Western developed states – to disruptions of foreign
mineral supplies upon which they have become economically dependent.[206]
In addition to the national apprehension caused by depletion of domestic
mineral sources and by the OPEC example, the "U.S. mineral problem" also
resulted from a perceived international reaction against the role of multinational oil companies, many of which were developing a deep-sea mining
capacity as members of private seabed mining consortia within the United
States.[207]

By 1980, in the wake of further oil price increases initiated by the

OPEC cartel, assured access to seabed nodules had emerged as the primary goal of the United States in the UNCLOS III negotiations.[208] Although there was some support within the United States for charitable treatment of developing states,[209] policy was largely determined by the symbolic value of seabed mining as a "last straw" issue at a time when U.S. interests were perceived as being under attack throughout the world:

> What is in dispute is not only deep seabed mining, but more importantly, alternative visions of a future economic and political international order.... The preferred U.S. vision of the future, not surprisingly, is one that retains U.S., or at least Western, hegemony over global economic transactions..., often cloaked in more technical terms like "economic efficiency" and "free-market assurances."[210]

Much of the U.S. opposition to a strong Authority was thus generated by speculative fears that such an institution might collude with land-based producer states to monopolize nodule mining operations.[211] The United States, nonetheless, has a unique multiple role in the world — as a coastal state, as a maritime state, as a major consumer of energy and raw materials, and as a world leader with important worldwide interests — and U.S. oceans policy has consequently been influenced by domestic pressures generated by industry lobbyists and bureaucratic in-fighting as well as by broader international needs and responsibilities.[212]

The Group of 77

Since the early 1960s, developing states have been seeking to wrest natural resources within their territorial sovereignty from the control of multinational corporations; in 1962 and 1966 the U.N. General Assembly adopted resolutions entitled Permanent Sovereignty Over Natural Resources, endorsing such efforts and setting forth recommendations and guidelines for the transfer of technology to developing states.[213] The Group of 77 was founded in 1964, in the context of the first U.N. Conference on Trade and Development (UNCTAD), as an instrument for the political mobilization and coordination of developing states — most of them former colonial territories — in international organizations.[214] Although it continued to promote the interests of developing states and experienced rapid growth, the Group of 77 did not begin to displace the East-West rivalry as the major focus of contention in international politics until the legal status of the seabed had become an issue of contention in the late 1960s.[215] The developing states soon refined their goals, calling for new international regimes to govern direct foreign investment and trade in raw materials and manufactures; developed states, not being represented by a similar organization, offered some concessions but consistently refused to consider making the kind of substantial changes sought by the Group of 77, and the so-called "North-South" confrontation became a fixture of UNCLOS III and other international negotiations.[216]

Developing states generally believe that under the existing international system they have been rendered permanently dependent upon the developed states as a result of unequal bargaining power and the de facto technological monopoly enjoyed by multinational corporations — "the political colonialism had merely been replaced by economic colonialism"[217] — and they have attempted to redress international economic and technological imbalances by seeking to negotiate new rules for trade and technology transfer at UNCTAD sessions.[218] By revealing the dependency of the developed states on imported minerals, the 1973 OPEC oil embargo fundamentally changed the perspective with which the Group of 77 viewed the negotiating process.[219] Although the embargo may in fact have been the product of factors unique to the petroleum industry, developing states nevertheless believed, as did the United States and other developed states, that the balance of world economic power was in fact shifting and that producer-state cartels could be utilized by developing states to remedy existing economic, political, and military imbalances.[220] The OPEC embargo, therefore, was widely supported within the Group of 77 — despite its harmful consequences for many oil-importing developing states — as an important step toward the achievement of a New International Economic Order to replace the inequitable international system inherited from Western Europe.[221] "The United States is undoubtedly the major target of Third World claims as well as the major necessary participant in any effort to implement the New International Economic Order."[222]

The Group of 77 was particularly active in the years immediately following the oil embargo — a period highlighted by the General Assembly's adoption on December 12, 1974, of the Charter of Economic Rights and Duties of States, article 29 of which related to the seabed:

> The sea-bed and ocean floor and the subsoil thereof, beyond the limits of national jurisdiction, as well as the resources of the area, are the common heritage of mankind. On the basis of the principles adopted by the General Assembly in resolution 2749 (XXV) of 17 December 1970, all States shall ensure that the exploration of the area and exploitation of its resources are carried out exclusively for peaceful purposes and that the benefits derived therefrom are shared equitably by all States, taking into account the particular interests and needs of developing countries; an international regime applying to the area and its resources and including appropriate international machinery to give effect to its provisions shall be established by an international treaty of a universal character, generally agreed upon.[223]

Despite the great variety of national and regional interests among its individual member states — which frequently made it difficult to achieve consensus on issues of major importance – the Group of 77 nevertheless exhibited strong cohesion during the early years of the law of the sea negotiations.[224] The opportunity to share in benefits from a newly discovered source of wealth created considerable interest in the seabed and in the concept of the common heritage of mankind, even though subsequent agreement on the adoption of an exclusive economic zone removed a large portion of the seabed from the jurisdiction of the proposed Authority.[225] With the active encouragement of the U.N. Secretary-General, developing states began to consider ways in which they might prospectively create an organization which would promote internationalism while addressing the worsening problem of global economic inequality.[226] The events of 1970 and 1971 – the adoption of the Declaration of Principles, the decision to convene UNCLOS III, the movement toward consensus on a 200-mile exclusive economic zone backed by Latin American coastal states, and the draft conventions submitted by the United States, Malta, and the 13 Latin American states – caused a policy disagreement within the Group of 77 similar in nature and scope to that which had divided U.S. policymakers two years earlier. Because members of the Group of 77 were unwilling to sacrifice their political unity by negotiating individually, this disagreement was responsible for much of the delay in the Seabed Committee negotiations from 1971 to 1973.[227]

While a substantial number of developing states viewed the law of the sea negotiations solely within the context of North-South relations, many others sought a strong international seabed regime as an end in itself.[228] The Group of 77 as a whole was naturally enticed by Pardo's vision of an international mechanism to redress the historical imbalances which remained following decolonization, and the common heritage of mankind principle was readily adopted as an important symbol of its struggle against economic domination.[229] Early negotiating positions on the seabed issue met with unexpected success, and by the mid–1970s the Group was presenting its demands forcefully, with states such as Algeria, Mauritania, and Tanzania having replaced Malta in a leadership role.[230] During the negotiations, the Group of 77 openly attacked the traditional freedom of the seas concept – a doctrine "long perceived . . . as no more than a code word for the protection of the naval requirements of the great powers"[231] – and similarly rejected U.S. contentions that unrestricted exploitation of the seabed by private mining companies would be of long-term benefit to all mankind.[232] The Group maintained that the prolonged drafting and unopposed approval of the Declaration of Principles rendered it – and the common heritage of mankind principle in particular – binding as an expression of international law, thereby transforming the legal character of deep-sea nodules from *res nullius* to *res communis*.[233] It "conceived of the common heritage of mankind in such a way that giving U.S. companies guaranteed

access to deep seabed mining would be a great concession on their part."[234]
Distrustful of self-serving pronouncements by developed states, the Group
of 77 sought to control the course of the UNCLOS III negotiations, favoring
a treaty of general principles which would give the Authority broad discretionary powers to regulate deep-sea nodule mining.[235]

Chapter 3

The Third United Nations Conference on the Law of the Sea

First Session: 1973 (New York)

The first session of UNCLOS III met in New York, December 3–15, 1973, to resolve procedural questions. After adopting an agenda and electing Hamilton Amerasinghe of Sri Lanka as President of the Conference, delegates of the 148 states in attendance were able to reach agreement on credentials, selection of officers, and use of the same three-committee structure which had been employed in the Seabed Committee. As in the Seabed Committee, the chairmanships of the respective committees were allocated to the African, Latin American, and Eastern European groups, each committee being assigned two geographically selected vice-chairmen, and 30 Conference vice presidencies were distributed on the basis of geographic criteria.[1] A 23-state Drafting Committee was also created, and membership on the General Committee was restricted to 48 states. Much of the session was consumed by a debate within the Western States and Others Group on the question of U.S. representation on the various committees, and it soon became apparent that additional meetings would be necessary to achieve complete agreement upon voting procedures, particularly with respect to operationalizing the Gentleman's Agreement, and the delegates decided to reconvene a week prior to the second session to resolve the issue.[2] All states were aware that the decision on this issue would be an important factor in determining the power of contending alignments, since the Group of 77 constituted a clear majority of the Conference participants and would thus benefit most directly from any procedure which allowed substantive issues to be determined by voting. The site of the second session was changed from Santiago to Caracas when the invitation to use the former site was withdrawn, following the overthrow of the Chilean government.[3]

Despite a greater number of delegations and issues than had been present at UNCLOS I and II, and despite the lack of a single set of recommended

51

draft articles such as had been prepared by the International Law Commission during the 1950s, few delegations openly voiced concern regarding a lack of adequate preparation.[4] A March preconference meeting of 19 landlocked and geographically disadvantaged states at Kampala reached agreement on a nine-point declaration, which included demands for proportional representation in the Authority, and for access both to the high seas and to resources accruing to coastal states upon the establishment of an exclusive economic zone.[5] The Group of 77 met from March 25 to April 5 in Nairobi, but internal divisions — particularly between the coastal and the landlocked and geographically disadvantaged states — prevented agreement upon a common negotiating position, and informal consultations held in late February and mid-June to discuss conference rules of procedure proved inconclusive.[6]

Second Session: 1974 (Caracas)

The first substantive session of UNCLOS III met June 20–August 29, 1974, and was attended by approximately 5,000 delegates representing 148 states, 10 U.N. agencies, 10 intergovernmental organizations, and 33 nongovernmental organizations.[7] Unable to match the technical and legal expertise available to delegations of developed states, many developing states were represented by their regular U.N. diplomatic personnel, who nonetheless proved to be skilled negotiators on behalf of the New International Economic Order.[8] In the week before the opening of the session, a compromise agreement was reached on rules of procedure which protected potential mining states and other minority interests by requiring that no votes be taken until all efforts to achieve consensus had been exhausted, in which case decisions were to be taken by a two-thirds-present-and-voting majority.[9] During the first two weeks of July, dozens of delegates addressed the Plenary, placing particular emphasis upon the need to clarify the seaward limits of national jurisdiction.[10] The Conference then divided into committees to consider agenda items individually.

Committee I, which at the opening of the Conference had progressed furthest in the drafting of alternative articles, held a week of general debate on the deep seabed regime, followed by an article-by-article review of the alternative proposals.[11] There was widespread support for an Authority consisting of four principal organs — an assembly, a council, an operational organ, and a secretariat — as well as several subsidiary commissions which would deal with economic, technical, and legal affairs.[12] There was also an emerging consensus that some form of dispute settlement procedures would be necessary, although the specific attributes of such procedures remained unresolved.[13] It was understood that the Authority would eventually be self-supporting, but there was fundamental disagreement on the question of the appropriate means for securing its financial independence, as well as uncertainty regarding its costs of operation.[14] Although the delegates agreed that the Assembly should be composed of all member states, issues related to decisionmaking procedures and to the composition of the Council were not formally discussed at Caracas, as the delegates focused instead upon the conditions under which mining would occur, as well as upon the possible economic impact of mining operations.[15] There was growing support among developing states for the creation of an Authority with broad discretionary powers and with an operational Enterprise substantially similar to the organ which had been described in the Thirteen-Power Proposal three years earlier, and an outline of "Basic Conditions" submitted to Committee I by the Group of 77 generated new discussion of alternative mining arrangements.[16] The developed states, on the other hand, concentrated their efforts upon an enumeration of specific conditions which could be met by a mining applicant in order to obtain automatic and exclusive rights of access, thereby avoiding the possibility of discrimination by the Authority against private mining operators.[17] As stated by the chairman of a working group, the major choice before Committee I seemed to be "between exploitation by the new international authority and the de facto monopoly of a few technologically developed countries under a licensing system."[18] Despite several days of debate on the economic implications of nodule mining, disagreement also continued over the appropriate method of dealing with the effects of nodule mining upon developing producer states.[19] Overall, Committee I produced few changes in the existing batch of alternative articles.[20]

Committee II discussed questions relating to offshore jurisdiction, a subject generally regarded as the most important before the Conference.[21] Negotiating groups and working papers were used to rework its 15 agenda items into a 13-part working paper containing 243 provisions.[22] There was near-universal agreement on a 12-mile territorial sea, conditioned upon general acceptance of an exclusive economic zone and rights of passage through newly enclosed straits.[23] Although the precise balance of rights and duties within the exclusive economic zone remained uncertain, there was nearly unanimous support for a fixed limit of at least 200 miles,[24] a limit which would remove as much as one-third of the world's oceans — including virtually all harvestable fisheries and all presently recoverable offshore oil and gas reserves — from the jurisdiction of the new international seabed regime.[25] There was also growing support for the inclusion in such a zone — or, more properly, in the continental shelf regime — of as much of the physical margin as extended beyond the 200-mile limit, subject, however, to payment of royalties to the Authority in recognition of the continuing interest of the international community in those areas.[26] Many states favored a proposal that coastal states be required to share the living resources of the zone with neighboring landlocked and geographically disadvantaged states "on an equal and non-discriminatory basis."[27] International management regimes for anadromous and highly migratory species such as salmon and tuna were strongly opposed by developing coastal states anxious to retain jurisdiction over all living resources within their exclusive economic zones.[28] Archipelagic states urged that they be permitted to delimit their offshore zones by the use of straight baselines drawn between their outermost islands; the special problem posed by islands was also recognized: "application of a 200-mile zone to every rock and reef in the ocean would result in enormous areas of seabed and superjacent water falling to island and island-owning states," depriving the Authority of jurisdiction over substantial portions of the deep ocean floor.[29] Although it held 46 formal and 23 informal meetings, Committee II only completed the preparatory phase of its work.

After adopting its program of work, Committee III divided into two working groups: one dealing with protection of the marine environment and the other discussing the conduct of scientific research. With respect to marine pollution, many delegations initially expressed a preference for a regime which would grant significant regulatory and enforcement powers to coastal states.[30] The United States argued strongly in favor of international regulation, fearing that a regime of diverse coastal-state standards would impede navigation, and developing coastal states eventually did agree to uniform international standards, largely out of concern for the free movement of international trade; generally, developing states were unenthusiastic about protection of the marine environment, believing that environmental damage is an inevitable by-product of economic development and that efforts by developed states to control pollution therefore threatened their developmental efforts.[31] Developing states also remained suspicious of marine scientific research, which they believed was generally conducted by developed states only for military or commercial purposes, leading inevitably to further economic domination and exploitation of developing states; developed states, on the other hand, argued that the full coastal-state jurisdiction over scientific research advocated by the Group of 77, requiring coastal-state consent prior to the conduct of all research, would tend to inhibit vital research and thereby hinder the development of new technology.[32] There was nevertheless general agreement that marine scientific research should be conducted for peaceful purposes, should comply with applicable environmental standards, and should not form the basis for any claim of exclusive right to portions of the sea.[33] In order to protect their technological advantage, the developed states resisted a proposal by 18 developing states that all technology employed in seabed mining activities be made available to any developing state upon request.[34]

The European Economic Community (EEC) and the Soviet Union

became increasingly active at Caracas, the latter for the first time accepting, in principle, the status of the seabed as the common heritage of mankind.[35] Within the Group of 77, the influence of Latin American states declined noticeably after they secured the adoption of an exclusive economic zone, and the resulting leadership vacuum was filled by the African and newly powerful Arab states; in contrast to the loose coalition of coastal states, which had difficulty maintaining cohesion during the second session, the landlocked and geographically disadvantaged states emerged as a more cohesive bargaining force, particularly in Committees II and III.[36] Most negotiating alignments lost power, however, as the discussions became focused on narrower issues and formal speeches were replaced by informal consultation.[37] After considerable discussion, it was decided that national liberation groups should be invited to participate in the Conference as observers, without the right to vote.[38] Two standing negotiating groups were formed: a disputes settlement group, which produced a working paper of alternative texts, and an informal working group chaired by Norwegian delegate Jens Evensen, which dealt with the more difficult issues raised in Committees II and III.[39]

Despite progress on some important issues, UNCLOS III continued to proceed at an unexpectedly slow pace, mainly because of the unwillingness of participants to make the compromises necessary to achieve agreement, but also because of institutional factors related to the large number of participating delegations, the number and complexity of issues, and the consensus procedure.[40] It was nevertheless believed that future negotiations would be facilitated by the experience gained at Caracas and by an improvement in the organization of work, as well as by the costs of prolonged negotiations in general, including the adverse effects of legal uncertainty upon economic activities related to the sea.[41] Despite the failure of the Caracas session to produce an agreed-upon set of draft articles, virtually all delegations were satisfied with the progress of the negotiations, and many informal consultations occurred during the subsequent intersessional period.[42] Most UNCLOS III participants exercised restraint with respect to offshore claims, only three states finding it necessary to extend their jurisdiction farther seaward prior to the third session,[43] which the U.N. General Assembly scheduled for the following spring in Geneva.

Third Session: 1975 (Geneva)

The third session of UNCLOS III, which met from March 17 to May 9, 1975, began in much the same manner as the Caracas session had ended, with delegates forsaking formal debate in favor of closed negotiations within informal groups. By the middle of the session, however, it had become evident that a willingness to make the difficult compromises necessary to achieve consensus was still lacking among the participants, and that a new method of procedure would have to be instituted if the session were to produce meaningful draft articles.[44] The Conference, therefore, adopted a procedure which had been employed successfully within the Evensen Group: the chairman of each committee prepared a compromise negotiating text to be used unofficially — representing the judgment of the chairmen based on their own assessment of the status of the negotiations — as the basis for further committee discussion.[45] It was initially expected that the chairmen would present their drafts before the end of the session, but it soon became apparent that release of the texts would provoke extreme controversy which "would shatter the facade of accomplishment that the session was trying to present"; distribution of the drafts was therefore delayed, causing delegates to forgo their negotiations in favor of lobbying efforts directed toward the newly powerful committee chairmen.[46] In mid–May the three committee texts were combined to form an Informal Single Negotiating Text (SNT).

The Committee I negotiations were conducted in an atmosphere of hope and accommodation during the first few weeks of the session, before degenerating into political posturing. A Working Group of 50 chaired by Christopher Pinto of Sri Lanka discussed two broad sets of issues: the conditions of nodule exploitation and the nature of international seabed mining machinery.[47] There was growing support for creation of joint-venture arrangements between state-sponsored mining operators and the Authority as the primary vehicle of shared access.[48] The SNT was noticeably responsive to U.S. interests, seemingly in deference to the significant portion of the U.S. delegation which was pressing for congressional enactment of unilateral seabed mining legislation.[49] It included provisions for compulsory dispute settlement by a tribunal with comprehensive jurisdiction over virtually all disputes arising under the convention, including those between the Authority and seabed mining operators, as well as an annex setting forth detailed rules and conditions for the conduct of mining operations.[50] One area in which the Group of 77 did prevail, however, concerned the Enterprise: the SNT contained an operational organ with virtually all the powers and functions which had been specified in the Thirteen-Power Proposal.[51] Although the United States formally agreed to the creation of such an operational organ, it did so conditionally, insisting that a parallel system of national access be implemented to give private companies an opportunity to mine the seabed simultaneously under flag-state licensing regimes.[52] After expressing initial interest, the Group of 77 rejected this compromise proposal on the ground that the Enterprise would be unable to attract mining companies as partners for joint-venture operations or to conduct independent mining activities without access to seabed mining technology.[53] An angry reaction by the U.S. delegation to the rejection polarized Committee I negotiations for the remainder of the session, but in September

Secretary of State Henry Kissinger indicated that, while U.S. willingness to accept such a parallel system was conditioned upon the adoption of balanced voting procedures and upon the exclusion of production controls from the treaty, the United States would be "prepared to explore ways of sharing deep seabed technology with other nations."[54]

There was a good deal of discussion regarding the relative powers of the Assembly and the Council. Under the SNT scheme, three interest groups would each be represented by six states in a 36-member Council, which was to share power with an Assembly designated "the supreme policy-making organ of the Authority."[55] Despite this language, most important powers and functions relating to the control of deep-sea mining activities were vested in the Council, which was to be responsible for approving contracts for mining operations and was to be assisted by a Technical Commission and an Economic Planning Commission.[56] The Group of 77 agreed to the representation of seabed mining states on the Council, which was to exercise full executive control over the conduct of nodule mining activities.[57] Both the Assembly and the Council were to proceed by a two-thirds-majority vote, but developed states continued to press for a higher voting requirement in the latter organ in order to assure a blocking power through the exercise of their combined votes.[58] Although there was general support for some form of production controls, the Group of 77 remained internally divided over the precise nature of the controls to be imposed on seabed mining operators.[59]

Committee II spent the first three weeks of the third session reviewing the alternative texts which it had formulated at Caracas, the remaining weeks being devoted to informal discussions in the various negotiating groups. Despite strong lobbying efforts by a handful of states in favor of a 200-mile territorial sea, there was general agreement upon a territorial sea of no more than 12 miles with specifically defined rights and duties relating to innocent passage, and the coastal-state group became less active, having largely achieved its objectives; in concluding its negotiations on straits passage, the Conference adopted a British proposal applying the transit passage regime to straits connecting high-seas areas, but reserving the regime of innocent passage for straits leading to territorial waters.[60] The new regime of transit passage was complemented in archipelagic waters by a similar regime of unrestricted passage within designated sea lanes, which would be available "solely for the purpose of continuous, expeditious and unobstructed transit" by aircraft and submarines as well as by surface vessels.[61] The SNT incorporated much of the Evensen text relating to rights and duties within the exclusive economic zone, coastal states being given comprehensive jurisdiction for fisheries management and conservation, subject to foreign-state access where fish stocks exceeded the harvesting capacity of the coastal state.[62] The SNT also provided for liability of both transiting vessels and coastal states where rights or duties within the zone were violated.[63] The continental shelf regime was defined as extending to the outer edge of the continental margin, subject to a system of revenue sharing where the natural prolongation extended more than 200 miles seaward.[64] The SNT did not provide for direct access by landlocked and geographically disadvantaged states to shelf resources, however;[65] nor did it specify a method of determining the outer edge of the margin — the "Hedberg formula," proposed by the United States, which would have extended jurisdiction to a point on the deep ocean floor 60 miles from the foot of the continental slope.[66] A 24-mile contiguous zone for enforcement of customs, immigration, and sanitary regulations attracted wide support and was included in the SNT. Several other important Committee II issues remained unresolved, including the regimes applicable to islands and to highly migratory species.[67]

Despite experimentation with various informal negotiating groups and the submission of 12 new proposals, limited progress was made in Committee III during 1975.[68] In an important victory for the United States, the SNT required flag states to establish pollution-control regulations no less effective than "generally accepted international rules"; coastal states were permitted to establish more stringent standards only within their territorial waters and only so long as they did not hinder innocent passage.[69] While jurisdiction to enforce such regulations with respect to vessel-source pollution within the exclusive economic zone generally remained with flag states rather than coastal states, the latter were permitted to enforce international discharge standards violated in offshore waters by vessels subsequently docking in their ports. The SNT left regulation of land-based sources of pollution to the domestic jurisdiction of coastal states, but it called for the adoption of international and regional standards to control such pollution.[70] Having decided to require coastal-state approval for publication of the results of resource-related research within the exclusive economic zone, Committee III discussed a Soviet proposal to require coastal-state consent for the conduct of research relating to marine resources but to permit non-resource-related research subject only to internationally agreed standards, with recourse to dispute settlement procedures to determine the character of contested research.[71] Although freedom of scientific research continued to exist beyond the limits of national jurisdiction, the Authority was given a nonexclusive competence to conduct general research on behalf of the international community. In contrast to the negotiating text produced by Committee I and despite objections by the United States and the Soviet Union, the Committee III SNT contained provisions requiring the mandatory transfer of technology and other proprietary information.[72]

More than 60 states participated in discussions in the disputes settlement negotiating group, which revised the Caracas texts into three annexes,

allowing states to choose from among four available procedures: conciliation, arbitration, or adjudication by either the International Court of
Justice or a new Tribunal.[73] The Group of 77 grew stronger during 1975 at
the expense of single-issue and regional blocs, but all sides showed new flexibility on important issues; despite partial success in gaining rights of access
to the exclusive economic zone, landlocked and geographically disadvantaged states suffered some loss of influence at Geneva as a result of their collective political and economic weakness.[74] Although, for many of the
developed states, the Evensen Group was the most effective negotiating
forum at UNCLOS III, the Group was criticized by developing states excluded from its limited membership and by other states impatient with the
slowness of its methods.[75] Although use of the informal negotiating texts
did eliminate some of the inflexibility previously displayed by developed
and developing states, and although it was widely perceived that agreement
had been reached on 85 percent of the treaty, it had also become clear that
difficult issues remained and that deep seabed mining was emerging as "the
principal stumbling block" to a comprehensive treaty agreement.[76] Because
the Conference was proceeding more slowly than had been expected under
its new negotiating procedure, a proposal by the United States that the
negotiations be concluded in 1976 was rejected by the General Assembly,
which determined that the negotiations should reconvene in the spring of
1976 and, if necessary, again in the summer.[77]

Fourth Session: 1976 (New York)

When UNCLOS III reconvened in New York from March 15 through
May 7, 1976, 149 states were in attendance. In an opening statement, U.N.
Secretary-General Kurt Waldheim commended the Conference's contribution toward a more equitable international economic system.[78] On April 8
Secretary of State Kissinger delivered a highly publicized speech on the law
of the sea which was welcomed "as an important contribution to achieving
an atmosphere of accommodation."[79] On the last day of the session the
committee chairmen and the President released a Revised Single Negotiating Text (RSNT) reflecting the results of the latest negotiations.[80] Intensive lobbying during the intersessional period had persuaded Chairman
Paul Engo to undertake private Committee I negotiations with moderate
developing states in an effort to promote a compromise on seabed mining
which would be more acceptable to developed states. These negotiations
continued during much of the fourth session, and most of the seabed provisions of the RSNT originated in these consultations, from which the more
vociferous developing states were excluded.[81] Disclosure of the secret
meetings provoked angry reactions from the uninvited delegations, and the
Chairman was heavily criticized for "selling out" to developed-state interests.[82]

The RSNT provided for a parallel mining system under which mining operations could be conducted not only by the Enterprise, but also by states parties, state enterprises, and natural or juridical persons sponsored by states parties. Annex I vested ownership rights to deep-sea mineral resources in mankind as a whole, and specified that title was to pass only upon recovery pursuant to a contract conforming to the convention and to the duly adopted regulations and decisions of the Authority; upon compliance with such contractual terms and conditions, as well as with specific duties with respect to transfer of technology, the applicant would automatically be granted exclusive rights and security of tenure with respect to the particular mining site.[83] Annex I outlined two revenue-sharing alternatives based upon royalties and profit sharing rather than upon fixed fees or front-end charges, and Annex II, which occasioned only limited discussion near the end of the session, provided for an Enterprise which would conduct mining operations under the same rules of access and revenue-sharing obligations applicable to state-sponsored mining operators.[84] Although the Enterprise was given the power to finance its operations by borrowing funds, no duty was imposed upon states parties to make such capital directly available.[85] A dispute emerged between the United States and other developed states as to whether quotas should be imposed on the number of mining operations which each state would be permitted to conduct or sponsor.[86]

In his April 8 speech, Secretary Kissinger announced that the United States would accept temporary aggregate production controls on nodule mining operations, and that the Authority should be given the power to participate in any international commodity agreements affecting nodule minerals.[87] His proposal, which was incorporated into the RSNT, was a compromise between the U.S. preference for a "free-market" regime and the demand by producer states that the Authority be given broad discretionary powers to set prices and production quotas: during a 20-year period beginning with the commencement of commercial mining operations, aggregate nickel production from nodule mining operations could not exceed the projected cumulative growth segment for the world nickel market.[88] The RSNT significantly curtailed the scope of the Authority's power by limiting its jurisdiction to deep-sea mining activities; in particular, the Authority's power to coordinate scientific research in the area beyond the limits of national jurisdiction (the Area) was limited and its jurisdiction over shipwrecks was eliminated.[89] The Authority retained power to control the pollution problems expected to arise from seabed mining operations.[90] The Assembly was redesignated "the supreme organ of the Authority," its policymaking powers were more narrowly restricted, and a provision was included permitting one fourth of the states members to delay a vote on any Assembly measure indefinitely pending review of its legality by the

Tribunal.[91] "While action in the assembly is made difficult, the council, which previously could only exercise the specific functions and powers entrusted to it, now acquires 'the power to prescribe the specific policies to be pursued by the Authority on any questions or matters within the competence of the Authority' and may totally bypass the assembly on important questions."[92] The composition and voting system of the Council were not discussed, and the RSNT specifically left these issues open. Although Part I of the RSNT was viewed as acceptable by a significant number of states, many developing states rejected the Committee I text, preferring instead to regard Part I of the SNT as a more suitable basis for further negotiations.[93]

In Committee II an article-by-article review of the SNT under a rule of silence produced more than 1,000 amendments,[94] but it soon became apparent that the basic provisions of the Committee II text enjoyed broad support.[95] Delegates nevertheless remained divided with regard to the legal status of the exclusive economic zone.[96] Coastal-state jurisdiction over shelf resources beyond 200 miles continued to gain support, and broad-margin states agreed to support a depth-of-sediment formula granting them sovereign rights seaward beyond 200 miles to a point at which the thickness of the sediment was equal to 1 percent of the distance from the base of the slope.[97] Landlocked and geographically disadvantaged states experienced a resurgence of political power as it became clear that developed coastal states would be the primary beneficiaries of expanded offshore jurisdiction, and their aggressive demands for access to resources within neighboring exclusive economic zones prompted a strong reaction by many coastal states.[98] Delimitation between opposite and adjacent states proved to be an extremely divisive issue; the RSNT favored the application of equitable principles, with consideration being given to use of a median or equidistance line where possible. Because of the rigorous article-by-article review, little time remained for use of informal negotiating groups, although consultations did occur on several difficult issues.[99]

Committee III did not review the first 15 articles of Part III of the SNT, to which there was no remaining disagreement, and much of the discussion was conducted in informal and ad hoc working groups.[100] Broader agreement on pollution questions became possible as ideological posturing gave way to more pragmatic considerations.[101] In order to prevent offshore vessel-source pollution without burdening navigation with nonuniform coastal-state regulations, it was decided that only generally accepted international regulations would be applied within the exclusive economic zone.[102] Although most major maritime states believed that the same international standards should be exclusively applicable within the territorial sea, the United States and many other coastal states advocated giving coastal states the right to adopt more stringent regulations in their territorial waters, subject only to limitations imposed by the right of innocent passage.[103] Enforcement of such regulations was to remain the responsibility of flag states, unless the violations were "gross or flagrant"—in which case coastal-state enforcement would be permitted, and flag states were given the right to preempt port-state and coastal-state enforcement proceedings unless the violation had caused major damage or the flag state proved unable or unwilling to take appropriate action.[104] In contrast to the SNT, the RSNT required coastal-state consent for the conduct of nonresource-related scientific research within the exclusive economic zone.[105]

> With respect to marine scientific research, a major change has been made which would require the consent of the coastal State for marine research activities in the economic zone or on the continental shelf, provided that consent shall not be withheld unless the project bears substantially upon the exploration and exploitation of resources, involves drilling or the use of explosives, unduly interferes with coastal State economic activities in accordance with its jurisdiction, or involves the construction, operation or use of artificial islands and structures subject to coastal State jurisdiction.[106]

Although Kissinger had proposed that the treaty include incentives for the transfer of seabed mining technology, extensive Committee III discussion of the question proved largely fruitless, with states differing on the nature and extent of any duty on the part of developed states to share such technology.[107]

In six days of plenary meetings, 72 delegations addressed the question of dispute settlement, producing RSNT provisions which required a ratifying state to choose from among four alternative procedures: arbitration, a Seabed Tribunal, the International Court of Justice, or specialized procedures available in particular types of cases.[108] The RSNT required states to attempt to resolve disputes by conciliation before resorting to any of these mechanisms, and disputes relating to a coastal state's exercise of sovereign rights in its exclusive economic zone were generally excluded from the system of binding dispute settlement.[109] Private entities were granted limited access to the dispute settlement procedures. Overall, the RSNT favored the interests of developed states, and it encountered strong objections from the Group of 77 during the last few days of the session.[110] Despite reservations by many developing states, which wanted more time to prepare for the next negotiating session, the United States persuaded the Conference to reconvene again in August.[111]

Fifth Session: 1976 (New York)

The delegates reassembled in New York on August 2 for a seven-week session, during which the remaining unresolved issues were discussed in informal meetings. In a belated effort to placate the Group of 77's hostility to the RSNT, Secretary of State Kissinger offered several new U.S. proposals, warning developing states that unless they were prepared to cooperate to resolve remaining issues equitably, "various governments may conclude agreement is not possible, resulting in unilateral action which can lead to conflict over the uses of ocean space."[112] Although Kissinger's initiative was well received,[113] time did not permit further revision of the negotiating text, and the results of the negotiations were set forth in reports by the committee chairmen and the President.

In Committee I there was general agreement on the need to ensure that the Authority would encourage nodule production to meet world mineral demand.[114] The Group of 77 initially rejected the system of parallel access contained in the RSNT, favoring instead broad discretion for the Authority to reject state-sponsored applicants and to withhold seabed sites from exploitation.[115] Such a system was unacceptable to most developed states, including the Soviet Union,[116] and in exchange for guaranteed access for private mining consortia, Kissinger indicated that the United States was prepared to finance the initial operations of the Enterprise in order to permit it to begin nodule exploitation at the same time as state-sponsored seabed mining operators, and would agree to periodic conferences to review the functioning of the seabed mining system.[117] The Chairman noted in his report that the committee was confronted with

> the central and most difficult problem of all . . . : should the new system of exploitation provide for a guaranteed permanent role in sea-bed mineral exploitation for States parties and private firms? Or should such a role for States parties and private firms be considered only at the option of and subject to conditions negotiated by the Authority? Or again, should their role be conceived of as essentially temporary, or be phased out over a defined period agreed to beforehand?
>
> With regard to these, there appears to be no indication that the proponents of any will accept the others. We thus find ourselves in an impasse.[118]

Because the delegates were unable to reach a compromise agreement on the system of exploitation, they did not discuss less fundamental issues, such as the powers and functions of the Authority's principal organs.

Committee II met only informally, and consultations occurred mainly within five negotiating groups which were established to discuss the more difficult remaining issues: transit passage through international straits, rights of landlocked and geographically disadvantaged states, boundary delimitation between opposite and adjacent states, revenue sharing and the definition of the outer edge of the continental margin beyond 200 miles

from shore, and the legal nature of the economic zone.[119] Although progress was made on revenue sharing from continental margin exploitation
beyond 200 miles, and although there was general acceptance of the articles
relating to transit passage,[120] the discussions generally proved unproductive in an atmosphere tainted by the acrimonious posturing in Committee
I.[121] Comments at the end of the session nevertheless indicated general
satisfaction with most of Part II of the RSNT,[122] and many states began to
extend their offshore resource jurisdiction following enactment by the
United States of a 200-mile exclusive fishing zone in April.[123]

Committee III met informally, focusing on several key unresolved
issues and producing several new articles. Two readings of the articles
relating to vessel-source pollution were completed in informal meetings
before the item was referred to a special negotiating group, and the few
amendments approved with respect to environmental standards were mostly
designed to resolve drafting ambiguities rather than to alter the substance
of the text.[124] The Chairman identified the most significant remaining Committee III question as the nature of coastal-state competence within the territorial sea, although disagreement also continued with regard to the
establishment of international standards and the nature of port-state
jurisdiction.[125] Thirteen meetings of a special negotiating group produced
no major changes in the text relating to scientific research, as the key issue
of coastal-state consent remained unresolved.[126] Despite the personal intervention of Secretary Kissinger, who indicated U.S. acceptance of the
principle of coastal-state control of resource-related research, "it was
becoming increasingly clear that the U.S. position was becoming more
difficult to defend. The proponents of the stronger regime had the advantage of strength of numbers, made even more serious by the reversal of the
Soviets. The task of those opposing the trend was to ensure that the conditions of refusal of consent were more carefully specified."[127] Four informal
meetings were devoted to a discussion of technology transfer, the delegates
focusing their attention on a proposal by several developing states that the
Authority act as a clearinghouse for the redistribution of marine
technology.[128]

An informal article-by-article examination of the dispute settlement
procedures produced some progress, although disagreement emerged
regarding the nature of the tribunal, with some states advocating creation
of a Law of the Sea Tribunal with comprehensive jurisdiction over all law
of the sea matters and others favoring a Seabed Tribunal with a more
limited competence.[129] Most states voiced support for provisional application of the Convention upon its being opened for signature.[130] Yet the session as a whole was generally regarded as a failure, and it was decided that
the first three weeks of the sixth session would be devoted exclusively to
a discussion of seabed mining questions and that a new negotiating text

would be prepared before the end of the 1977 session.[131] It had become apparent by the end of the year that seabed mining was the sole outstanding issue of major importance still confronting the delegates, and a two-week intersessional meeting in the spring of 1977 produced increasing support for the parallel mining system.[132] During 1976 20 coastal states extended their offshore jurisdiction to 200 miles, most of them emulating the unilateral adoption of the U.S. fisheries conservation zone.[133]

Sixth Session: 1977 (New York)

The Conference met again in New York from May 23 to July 15, 1977. During the first three weeks the delegates met two and three times a day in a Working Group of the Whole under the chairmanship of Jens Evensen, considering only Committee I issues and producing a draft text which drew strong protests from developed states.[134] Further discussions and subsequent revisions yielded a less controversial text, which enjoyed greater international acceptance by virtue of its compromise language.[135] After considerable discussion, it was decided that the committee chairmen, rather than the President of the Conference, would control the final content of the new Informal Composite Negotiating Text (ICNT).[136] When Chairman Engo used this power to alter the Evensen text at the behest of a group of developing states, the United States joined other developed states in protesting the new negotiating text.[137] Committees II and III were able to make significant progress in producing generally acceptable texts, however.[138]

Reflecting the general negotiating trend of the Conference, President Amerasinghe instructed Committee I to conduct discussions in informal working groups on the key remaining issues. Ambassador Elliot Richardson did not offer major new proposals on behalf of the newly installed Carter administration as had been expected, but rather sought to exact concessions from the Group of 77 in exchange for clarification of the 1976 U.S. initiatives.[139] The ICNT, as had the RSNT before it, declared the Area and its resources to be the common heritage of mankind and required that activities in the Area be carried out for the benefit of mankind as a whole. The Evensen text provided a more specific list of policies to be pursued by the Authority in order to promote international cooperation and global economic growth: efficient development and conservation of deep-sea mineral resources; equitable distribution of the resulting benefits; promotion of technology transfer; enhancement of world mineral supplies at "just, stable and remunerative" prices; prevention of monopolization; and provision of opportunities for all states parties to participate in seabed mining activities.[140] Both the Evensen text and the ICNT permitted the Authority to negotiate joint-venture agreements with respect to exploitation of reserved areas — as well as specific terms of access for state-sponsored applicants.[141] In contrast to the Evensen text, which provided that questions of technology transfer would be resolved in these negotiations, the ICNT conditioned private access to seabed nodules upon the transfer of mining technology and further provided that seabed mining operators would automatically lose their rights of access if a review conference to be convened 20 years after the commencement of commercial production proved unable to reach agreement within five years.[142] It was agreed that the Enterprise would be required to comply with the rules and regulations of the Authority, as well as the production controls set forth in the treaty, but despite the efforts of the special working group no consensus was reached on the issue of production controls.[143] Whereas the Evensen text limited mineral production to 100 percent of the increase in demand for nickel for the first seven years, and two thirds of the increase in subsequent years, the ICNT continued to limit production to three fifths of the growth in nickel demand after seven years.[144] Without specifying percentages, the ICNT called for revenue sharing in the form of a fixed annual fee, a production change, and a percentage of net revenues, to be paid to the Authority for administrative expenses and for distribution to designated developing states and national liberation movements.[145] In elaborating on Kissinger's 1976 offer, the United States proposed the use of loan guarantees to finance the initial operations of the Enterprise, a proposal which attracted widespread support among the delegates and was ultimately included in the Evensen text.[146]

The governing board of the Enterprise was to be composed of 15 members elected by the Assembly, some of whom would be selected in accordance with principles of equitable geographical representation. The powers and functions of other organs of the Authority remained in considerable dispute, with the United States emphasizing the need for economic efficiency and developing states seeking a more political institution.[147] The ICNT restored to the Assembly many of the powers which had been circumscribed in the RSNT, and reinstated a two-thirds-present-and-voting requirement for substantive questions.[148] The Assembly was to be empowered to elect the Secretary-General of the Authority, as well as the members of the Tribunal, and both the ICNT and the Evensen text also gave the Assembly a limited power to review the workings of the Authority every five years and to enact or recommend certain amendments.[149] The one-nation-one-vote principle was applied to the Council as well as to the Assembly, in recognition of the sovereign equality of states.[150] The Council was to be composed of 36 states elected by the Assembly, but only the complex chambered structure set forth in the Evensen text held out the possibility of an effective veto power for the United States and other developed states.[151] The Seabed Tribunal was replaced by a Sea-Bed Disputes Chamber operating as a subsidiary of a Law of the Sea Tribunal, and the compulsory jurisdiction of the Chamber was limited in order to secure the Group of 77's acceptance of its jurisdiction over disputes between the Authority and states parties or their nationals.[152]

Committee II negotiations during the last five weeks of the session were mostly conducted in small negotiating groups, which focused on three major unresolved issues: the continental margin beyond 200 miles, the nature of the exclusive economic zone, and the delimitation of maritime boundaries between opposite and adjacent states. A 15-state negotiating group was formed near the end of the session in an effort to resolve differences between Committee II and Committee III with respect to rights and duties within the exclusive economic zone; the discussions produced new compromise texts on scientific research, settlement of fisheries disputes, and the legal status of the zone — all of which were incorporated into the ICNT.[153] General support for some sharing of revenues from exploitation of the continental margin beyond 200 miles was reflected in ICNT provisions for payment to the Authority of 1 percent of revenues in the sixth year of production, with successive annual increases of 1 percent to a permanent rate of 5 percent in the tenth year.[154] Negotiations on the outer limit of the margin proved less fruitful, however, despite broad support for a compromise formula introduced by Ireland, and the ICNT omitted all references to such delimitation; continuing efforts to develop a formula for determining boundaries between opposite and adjacent states were no more successful.[155] Some progress was made in the development of a plan to regulate exploitation of highly migratory species such as tuna, and landlocked and geographically disadvantaged states unsuccessfully sought guaranteed access to fisheries within the exclusive economic zones of neighboring coastal states.[156]

Committee III met regularly beginning in the fourth week of the session, again conducting most of its work within informal working groups, where discussions focused upon vessel-source pollution. At the urging of the United States, the ICNT gave individual coastal states the authority to impose stricter pollution standards as a condition of access to national ports.[157] Coastal-state authority to restrict innocent passage in order to prevent pollution was also increased, and international liability was imposed upon states which caused environmental damage by failing to fulfill "their international obligations concerning the protection and preservation of the marine environment," but warships and other state-owned, noncommercial vessels and aircraft were granted immunity from all international pollution controls.[158] Developing coastal states continued to seek complete discretion to grant or withhold consent for scientific research within the exclusive economic zone, and only the United States and West Germany still insisted upon complete freedom of scientific research beyond the territorial sea; the ICNT rejected the distinction between resource-related and nonresource-related research, but provided that restrictions on publication could only be imposed prior to the actual conduct of research.[159] The role of the Authority as a clearinghouse for the transfer of marine technology remained unsettled.[160]

In plenary meetings it was determined that states should be permitted to select any of a revised group of dispute settlement procedures: traditional arbitration, a special arbitration panel, the International Court of Justice, and the Law of the Sea Tribunal.[161] The Tribunal was to have jurisdiction over most disputes arising under the convention, including those to which private entities were party.[162] Coastal states were exempted from the dispute settlement procedures where the dispute involved scientific research or fishing within the exclusive economic zone.[163] When the ICNT was finally released after the close of the session, the United States denounced it as "fundamentally unacceptable," citing the provisions on seabed mining in particular.[164] The United States met in November with six other developed states to consider alternatives to the UNCLOS III negotiations, and in January 1978 the Carter administration indicated that, for the first time, the executive branch might support unilateral seabed mining legislation.[165] Intersessional meetings were held in November and December and again in February in an effort to develop new procedures which would permit the formulation of satisfactory texts. During 1977, 36 additional states extended their offshore jurisdictional claims to 200 miles.[166]

Seventh Session: 1978 (Geneva and New York)

The seventh session of UNCLOS III met in Geneva from March 28 to May 18 and resumed in New York from August 21 to September 15. The Latin American states sought unsuccessfully in the first two weeks of the session to oust President Amerasinghe.[167] To prevent further abuse of the negotiating process, the rules of procedure were amended at the outset to allow revision of the ICNT only upon the recommendation of the President and committee chairmen and with the approval of the Plenary, a change which had "a very substantial formalizing effect" upon the undisputed portions of the ICNT.[168] The major issues upon which consensus was lacking were discussed in seven new negotiating groups dealing, respectively, with access to nodule minerals; financial aspects of deep-sea mining; the institutional structure of the Authority; rights of access by landlocked and geographically disadvantaged states; dispute settlement within the exclusive economic zone; definition of the outer limits of the continental margin beyond 200 miles; and delimitation of maritime boundaries between opposite and adjacent states.[169] By mid–May only the negotiating groups dealing with seabed access and the definition of the continental margin had failed to make significant progress, and it was decided, on a close vote and without much enthusiasm, to reconvene the session in

August.[170] The summer session proved generally fruitless, however, and no formal revisions were made to the ICNT, although several new articles were produced by the negotiating groups.

Meeting in both its main group and a smaller working group, Negotiating Group I undertook a review of the seabed mining provisions of the ICNT.[171] The principle of parallel access was generally accepted, and new draft texts clarified the independent status of state-sponsored mining operators.[172] Brazil, Tanzania, and the United States negotiated a new text on technology transfer requiring the transfer of seabed mining technology to the Enterprise on "fair and reasonable" commercial terms — subject to compulsory conciliation and arbitration — only after the granting of a mining contract and only upon formal request.[173] Over the objection of Western states, Brazil succeeded in generating substantial support for a provision (the so-called "Brazil clause") requiring the transfer of seabed mining technology, under similarly limited circumstances, to operators sponsored by developing states, and the United States took particular exception to an antimonopoly provision proposed by other seabed mining states which would limit the total number of mine sites to which each state could maintain access.[174] In exchange for agreement by the developed states to provisions on technology transfer, the exclusive competence of the Authority to control scientific research within the Area was eliminated.[175] Canada and the United States were able to agree upon a compromise formula permitting seabed mining to the extent of 60 percent of the growth segment for nickel consumption over a 25-year "interim" period beginning five years before the commencement of commercial nodule production.[176] Instead of automatic conversion to a unitary mining system under the Authority's control, there was growing support for imposition of a moratorium on all new seabed mining operations if the review conference was unable to reach agreement after five years.[177] During the resumed summer session in New York, strong disagreement emerged between developed and developing states regarding the amount of discretion which the Authority should be permitted to exercise in selecting from among state-sponsored applicants.[178]

Negotiating Group II examined the financing of the Authority and the Enterprise, as well as the nature and extent of revenue-sharing payments, the latter issue dominating discussion throughout the session.[179] The Group discussed the form, timing, and amount of such payments, and concluded that each contractor should pay an application fee, a fixed annual fee, and either a graduated production charge or a graduated production charge in combination with a graduated schedule of profit sharing.[180] No agreement could be reached upon the appropriate percentage amounts for the assessments, however, although the negotiations were now conducted in full consciousness of the difficulty and importance of balancing the need of mining

companies to secure an adequate ràte of return against the need of the
Authority to secure an adequate flow of working capital.[181] The Enterprise
was to obtain capital from voluntary contributions, loans, and loan
guarantees, as well as from the revenue-sharing funds received from contractors; it was decided that the direct financial assessments of individual
states would be based on their assessed share of U.N. contributions and
would be used only to cover the administrative expenses of the Authority
until income from other sources became sufficient to meet such expenses.[182]

Negotiating Group III, which discussed the composition and voting
procedures of the Council as well as its relationship to the Assembly, failed
to make significant progress and produced no new texts.[183] It was generally
agreed that representation on the Council would be allocated on the basis
of both economic interest and equitable geographical distribution,
although the exact identity and numerical strength of the representative
groups remained in dispute.[184]

> Such a voting system implies that if the states with major seabed mining
> and consumer interests cannot be assured of substantial positive decision-
> making power in the Authority, they must be assured that their basic in-
> terests will be protected in the treaty itself (absent improbable decisions
> by the Council). Needless to say, the land-based producers argue that they
> face similar problems.[185]

At the resumed session in New York, the Group consolidated the subsidiary
organs into an Economic Planning Commission and a Legal and Technical
Commission, which were to act in an advisory capacity to the Council.[186]

Negotiating Group IV produced new texts giving landlocked and geographically disadvantaged states rights of equitable participation in the exploitation of fisheries within neighboring exclusive economic zones.
Preference was to be accorded to developing landlocked and geographically disadvantaged states and to states which had become economically
dependent upon regional fisheries stocks. The new texts clarified the rights
of landlocked and geographically disadvantaged states — whose bargaining
power was being rapidly eroded by acceptance of the 200-mile exclusive
economic zone — and further changes were regarded as unlikely, although
acceptance of the compromise language was threatened by bargaining tactics in Negotiating Group VI.[187]

Consultations in Negotiating Group V were conducted primarily
within a 36-member working group and a smaller discussion group.[188]
Landlocked and geographically disadvantaged states and distant-water
fishing states favored compulsory and binding dispute settlement procedures for all disputes relating to foreign access to fisheries within the exclusive economic zone; most coastal states, on the other hand, were
adamantly opposed to any adjudication of what they regarded as their

sovereign rights to marine resources within the zone.[189] It was ultimately decided that such disputes should be submitted to compulsory conciliation rather than to binding arbitration or adjudication. Japan and the Soviet Union remained dissatisfied, but debate on the issue was cut short during the New York meetings.[190]

Discussions in Negotiating Group VI centered on the definition and legal status of the continental margin where it extended beyond 200 miles. Initially it appeared that a compromise formula offered by Ireland — essentially offering broad-margin states a choice between the Hedberg and depth-of-sediment formulas — would provide a generally acceptable solution; however, the Soviet Union proposed that a maximum limit on the outer extent of any such delimitation be established,[191] and a number of Arab states sought to limit coastal-state jurisdiction over the continental shelf to the area encompassed by the 200-mile exclusive economic zone. The Arab proposal was supported by the landlocked and geographically disadvantaged states, which conditioned their acceptance of the Irish formula upon greater concessions from the coastal states in Negotiating Group IV.[192]

The meetings of Negotiating Group VII were well attended, but the delegates remained deadlocked between the "equidistance" and "equitable principles" criteria for delimitation between opposite and adjacent states.[193] There was, nevertheless, general agreement

> that delimitation should be effected by agreement between the Parties; that "special or relevant circumstances" (such as the presence of islands) should be taken into account; that the result should be equitable (in the sense of fair) to both sides; that the Convention should contain a specific provision on interim measures to be applied pending agreement or settlement in delimitation cases (and there was some support for the suggestions that the interim measures should provide for mutual restraint as well as regulating economic activity); and that the Article in the ICNT on territorial sea delimitation was acceptable subject to minor amendments.[194]

Discussions in New York revealed a similarly deep division on the question of whether compulsory dispute settlement procedures should apply to these boundary delimitations.[195]

Committees II and III continued to meet informally to discuss unresolved minor issues not assigned to the negotiating groups. Committee II again conducted an article-by-article review, but only 3 of the 93 proposed changes achieved general acceptance.[196] An informal working group established to formulate new rules for the management of marine mammals produced new texts strengthening the powers of the state of origin over anadromous species.[197] Committee III continued to make progress in its discussions on pollution control and marine scientific research, with the former item receiving the bulk of the committee's attention as delegates focused on standard setting and enforcement powers within the territorial sea and the exclusive economic zone. In the wake of several major offshore oil spills, the United States and France were able to secure support for stronger coastal-state powers of intervention in cases of maritime casualties.[198] Despite U.S. efforts to introduce new texts on marine scientific research, most states opposed any changes to the qualified consent regime embodied in the ICNT.[199] The Committee III articles on technology transfer were considered broadly acceptable, pending the final outcome of negotiations on the subject in Committee I.[200]

President Amerasinghe conducted several plenary meetings in which preliminary discussions were held on the preamble and final clauses of the treaty, and the Drafting Committee met informally to begin the task of harmonizing and standardizing the language of the ICNT.[201] Although 90 percent of the ICNT reflected a genuine consensus, it was nevertheless recognized that failure to resolve the remaining 10 percent of the issues "would certainly prove disastrous to the Conference," and it was agreed that at the following session the ICNT should be revised with a view to its formalization into a draft convention.[202] Many states had already begun to reassess the need for continued negotiations, as the agreed-upon provisions of the text began to be adopted unilaterally in international state practice.[203] Although the United States had undertaken consultations with other Western seabed mining states in late 1977 regarding negotiation of an alternative "mini-treaty" regime, support for such discussions evaporated following the progress made at the seventh session; however, Ambassador Richardson announced that the Carter administration would support domestic seabed mining legislation which, while providing for nodule exploitation under an alternative regime if necessary, could be drafted both to encourage investment by private mining firms and to implement an eventual UNCLOS III regime.[204]

Eighth Session: 1979 (Geneva and New York)

UNCLOS III met twice again in 1979, from March 19 to April 27 in Geneva and from July 19 to August 24 in New York. "Intensive substantive work on major outstanding issues began on the first day and continued, almost without interruption, throughout both parts of the session."[205] Further discussions were conducted within the formal negotiating groups, except that Negotiating Groups IV and V were disbanded after satisfactory solutions to the problems of geographically-disadvantaged-state access to fisheries and settlement of fisheries disputes were achieved in New York. At the end of the Geneva meetings, the Conference officers produced a revised ICNT (RICNT), which consisted of 304 articles and 7 annexes and included new texts on seabed exploitation, technology transfer, boundary delimitation, protection of the marine environment, and foreign access to fisheries within the exclusive economic zone. Discussions on many of these issues continued in New York, but no further revisions were made to the main negotiating text even though the delegates had hoped to produce a formal draft text by the end of the session.[206]

A small Working Group of 21, established to facilitate negotiations on key seabed issues, coordinated the progress of Negotiating Groups I, II, and III.[207] It was decided that plans of work submitted by mining applicants would be considered in the order received, and that, in the absence of a blocking vote by the Council within 60 days, "the Authority shall approve such plans of work, provided that they conform to the uniform and non-discriminatory requirements established by the rules, regulations and procedures of the Authority, subject only to itemized exceptions" in cases where the production limitation or antimonopoly formulas were applicable or where a unique marine environment would be threatened by deep-sea mining activities.[208] The provisions on technology transfer were revised to ensure that transfer would be required only if the Enterprise could not obtain equivalent technology on the open market, and to protect proprietary information from unlawful disclosure by the staff of the Authority.[209] Several potential seabed mining states, including the United States, continued to seek deletion of the Brazil clause, however,[210] and it remained unclear whether the technology transfer obligations applied to processing as well as mining technology. A newly formed Group of Legal Experts negotiated a text which permitted a private contractor to submit any disputes with the Authority to commercial arbitration rather than to the Sea-Bed Disputes Chamber and which provided that members of the Chamber would be selected by the members of the Law of the Sea Tribunal rather than by the Assembly.[211]

Although producer states objected to a proposed floor on seabed production authorizations, negotiations between consumer states and producer states seemed to be nearing agreement on a compromise formula on production controls.[212] To assure the Enterprise of capital funding for mining and processing operations, states parties were to be required to contribute half of such funds directly in the form of long-term, interest-free loans, and to guarantee commercial loans to the Enterprise to finance the other half. The RICNT succeeded in clarifying the powers of the respective organs of the Authority, but there was no agreement upon either the composition of the Council or the number of votes to be required for important Council decisions.[213] Negotiating Group II experimented with various revenue-sharing formulas and the U.S. delegation reported "substantial improvements" in the continuing efforts to find an acceptable compromise text.[214]

There was continued disagreement in Negotiating Group VI over delimitation of the continental margin beyond 200 miles, and the RICNT formulation reflected a compromise between the Irish and Soviet proposals; the revenue-sharing provisions of the RICNT were changed to require a maximum royalty of 7 percent after 12 years, instead of 5 percent after 10 years.[215] Negotiating Group VII remained deadlocked on the issues of delimitation, provisional measures, and dispute settlement. At the end of the Geneva meetings, the negotiations on protection of the marine environment were completed, but Committee III continued to discuss scientific research and was able to clarify several provisions of the negotiating text, agreeing that consent would be necessary for research on the continental margin beyond 200 miles only where the area had been designated by the coastal state as an area of actual or potential exploration.[216] Committee II again met informally to discuss the conservation of marine mammals and other issues.[217] "The debates in the Second and Third Committees revealed a widespread desire to avoid tampering with the texts."[218]

The Drafting Committee reviewed the progress of each of its six language groups, and the Plenary conducted an in-depth discussion of the final clauses and produced a set of draft articles in conjunction with the Group of Legal Experts. The treaty was to be open to signature by all states, and there was general agreement on the need for provisional application of the convention, pending its entry into force, in order to ensure its widespread acceptance.[219] All states agreed that the nature of the convention would require that reservations to at least some articles be prohibited, and the choice before the delegates appeared to have been reduced to that "between a prohibition on all reservations and a prohibition on almost all reservations"; yet there was growing support for inclusion of an article permitting denunciation of the treaty.[220] Although there was agreement that the new convention generally should not supersede the specific provisions of bilateral and multilateral treaties between states parties, there was significant support for effective nullification of the 1958 conventions by the substitution of new provisions in the UNCLOS III treaty.[221] The U.S. delegation characterized the RICNT as "a substantial further step toward final agreement," but the Conference was still unable to conclude its work, and agreement on a final convention text was planned for August 1980.[222] By the end of 1979, 70 percent of coastal states had claimed offshore jurisdictional zones of 200 miles.[223]

Ninth Session: 1980 (New York and Geneva)

The ninth session of UNCLOS III was again split into spring and summer sessions, meeting in New York from March 3 to April 4, and in Geneva from July 28 to August 29, 1980. Although intensive discussions regarding the composition and voting procedures of the Council prevented the Conference from completing its work in August, another revision of the

RICNT was issued after the New York meetings, and further revision resulted in the issuance of a Draft Convention on September 22.[224] The Draft Convention included the work of the Drafting Committee, which held a special three-week intersessional meeting in June and largely completed its harmonization of the draft language for approximately 400 articles.[225]

At the resumed session in Geneva, Committee I employed four informal negotiating groups and the Working Group of 21, and was able to achieve "a generally advanced basis for accommodation on deep seabed mining."[226] Although little progress had been made in New York on the question of voting procedures in the Council, the summer session focused on the issue and achieved several breakthroughs which satisfied all major interest groups.[227] The resulting compromise agreement established a three-tiered voting scheme for substantive decisions: consensus ("the absence of any formal objection") for adoption of amendments, of measures for the protection of producer states, and of the rules, regulations, and procedures of the Authority; a three-fourths majority for most other operational decisions, for which a broad base of support was desirable; and a two-thirds majority for the remaining, less-controversial decisions.[228] In deference to the United States, the Draft Convention reduced the Council's discretion to approve or deny a plan of work, favoring instead a more objective review by the Legal and Technical Commission and requiring consensus agreement in the Council to overrule the Commission's recommendation that an application be approved.[229] It was agreed that technology transfer obligations would end 10 years after the commencement of mining operations by the Enterprise and that such obligations would not apply to processing technology, and the scope of the Brazil clause was narrowed to require the transfer of technology to developing states only where the technology was not also transferred to the Enterprise.[230]

A "split contracting system" was devised whereby the review of plans of work submitted by applicants was procedurally separated from the allocation of production authorizations, so that approval of plans of work would not be withheld on account of the production limitation.[231] In negotiations on the formula for determining the production limitation, the delegates agreed on a floor limitation equivalent to the lesser of 3 percent of the previous years' nickel consumption or 100 percent of the current annual increase, and a maximum annual tonnage allowance of 46,000 tons was imposed upon each mining site.[232] During the Geneva meetings, the Council was given power to compensate developing producer states for economic losses caused by nodule mining operations and to impose non-discriminatory production controls after expiration of the 25-year interim period. The RICNT was further changed to permit the Authority to participate only in commodity conferences attended by "all interested parties including both producers and consumers," and an antimonopoly formula was included to allow each sponsoring state and its nationals to exploit no more than 2 percent of the nonreserved Area.[233] It was decided that excess revenues accruing to the Authority would be distributed in accordance with the Authority's rules, regulations, and procedures.

Few changes were made to the revenue-sharing formulas during the session, but the provisions on the financing of the Enterprise were revised to ensure that funds would be made available on more flexible terms and would be repaid in accordance with sound financial practices, and the Statute of the Enterprise was revised to better reflect its independent commercial character.[234] With regard to the review conference, which was to be convened after 15 years of nodule production, the access moratorium was abandoned in favor of a system allowing amendments to limited portions of the seabed regime to enter into force upon ratification by two thirds of states parties; amendments not altering the basic structure of the parallel system could be adopted prior to the review conference by a consensus agreement of the Council and a two-thirds vote of the Assembly.[235] There was general agreement on the need for a system of preparatory investment protection to preserve rights of access for mining operators who had already undertaken substantial prospecting activities in particular areas of the seabed.[236] Although the Plenary was able to agree that a Preparatory Commission would meet from the end of the Conference until the entry into force of the convention and that active Commission membership should be restricted to states which signed the convention, delegates were unable to define the powers and duties of the Commission with precision.[237] It was recognized, however, that because of the consensus requirement for the adoption of regulations by the Council, any rules, regulations, and procedures adopted provisionally by the Preparatory Commission "would continue to apply for some time—perhaps a very long time."[238]

Committee II conducted a full review of proposed amendments, and completed most of its work by the end of the session. The language relating to suspension of innocent passage was clarified, and demands by developing states to require prior notification to coastal states of passage by warships through the territorial sea failed to achieve consensus support.[239] Although supported by more than 30 states, measures to protect "straddling" fisheries stocks, which overlap the high seas and the exclusive economic zone, were not included in the Draft Convention; proposals to establish a Common Heritage Fund financed by revenues from exploitation within the exclusive economic zone—as well as on the continental margin beyond—similarly failed to attract adequate support.[240] The United States succeeded in obtaining general acceptance of a provision permitting stronger conservation measures for whales and other marine mammals.[241] Negotiating Group VI largely completed its work, having reached agreement on a compromise formula for delimiting the outer edge of the continental margin which took account of oceanic ridges, and having provided for the creation of a Commission on the Limits of the Continental Shelf empowered to make authoritative recommendations regarding the extent of such limits.[242] Negotiating Group VII drafted a new text to govern delimitation between opposite and adjacent states, combining the "equitable principles" and "equidistance" criteria with a requirement that delimitation be effected "in conformity with international law."[243]

Although Committee III met in Geneva to consider drafting changes, its substantive work was completed during the New York meetings. In order to allow researchers to achieve compliance with applicable coastal-state regulations, the Draft Convention required suspension of foreign scientific research projects by coastal states prior to their complete termination.[244] Consent was to be implied if the research was sponsored by an international organization of which the coastal state was a member, although the coastal state would have four months within which to object to such a project; in a victory for the United States, it was further decided that the consent regime would apply to the continental margin beyond 200 miles from shore only where mineral resource development was involved.[245] With regard to technology transfer, there was virtually unanimous agreement that a state party, acting in good faith, should not be required "to supply information the disclosure of which is contrary to the essential interests of its security."[246]

With agreement that compulsory conciliation would be employed to resolve disputes regarding delimitation, scientific research, and fisheries exploitation within the exclusive economic zone, the Plenary completed its discussion of dispute settlement procedures. Agreement was also reached on a compromise preamble which exceeded the normal length but omitted much of the political language sought by the Group of 77.[247] The Plenary and the Group of Legal Experts further agreed that the treaty should remain open for signature for two years, and would enter into force one year after its ratification by 60 states.[248] It was also decided that no reservations would be permitted, except as specifically allowed in the convention, but that denunciation would be permitted at any time upon one year's notice.[249] Amendments to nonseabed provisions of the treaty would require ratification by two thirds of states parties and would be binding only upon ratifying states. The convention was to "prevail, as between States Parties, over the Geneva Conventions on the Law of the Sea of 1958."[250]

On June 28 the United States adopted the Deep Seabed Hard Mineral Resources Act (Seabed Act),[251] and West Germany enacted similar unilateral legislation on August 17.[252] The legislation, which would authorize commercial mining operations on a provisional and possibly permanent basis under a flag-state regime among reciprocating seabed mining

states, caused the Conference to delay discussion of preparatory investment protection.[253] The ninth session nevertheless ended with a renewed
sense of optimism and with an expectation that the Conference would complete its work in the spring of 1981; as Ambassador Richardson announced,
so far as the United States was concerned, "[a]ll the objectionable features
of the ICNT . . . have been replaced by fair and workable compromises."[254]
The tenth session, therefore, was scheduled to deal with final issues relating
to preparatory investment protection, the Preparatory Commission, and
participation in the convention by nonstate organizations.[255] The election
of Ronald Reagan on a platform critical of the emerging seabed mining
regime gave U.S. critics of the UNCLOS III process renewed vigor,
however,[256] and on March 2, 1981, the State Department unexpectedly announced that the United States would "seek to ensure that the negotiations
do not end at the present session of the conference, pending a policy
review."[257] While acknowledging that some treaty provisions would benefit
the United States, the Reagan administration regarded others as
ideologically offensive to domestic economic interests.[258]

> Although it was to be expected that a new Administration would want to
> review the complex issues in the conference, the timing of the announce-
> ment — just seven days before the conference was to begin and coinciding
> with an intensification of mining industry lobbying against the treaty —
> and the peremptory tone of the announcement, aroused bitter speculation
> among many conference participants that the United States was about to
> revise its objectives so radically as to pose a direct threat to the success
> of the conference.[259]

The top U.S. negotiators were dismissed on March 7, only two days before
the opening of the tenth session.[260]

Tenth Session: 1981 (New York and Geneva)

The delegates met in New York from March 9 to April 24 and again
in Geneva from August 3 to August 28, but were unable to complete the
work of the Conference without effective participation by the U.S. delegation, which sought to outline the concerns of the Reagan administration but
refused to engage in negotiations.[261] After electing Tommy Koh of
Singapore as the new President of the Conference during the first week,[262]
the delegates discussed some of the remaining substantive and drafting
issues for the remainder of the session. "Aside from the perceived need of
some delegates to explain the time and expense devoted to the session, the
main purpose was to demonstrate a determination to press ahead to completion of the Convention."[263] The Drafting Committee met from January
12 to February 27 and from June 29 to July 31, as well as during the sessional
meetings, and produced a revised Draft Convention which was approved

by the Plenary.[264] The decision of the United States not to take part in the session was widely criticized by UNCLOS III participants — including its Western allies — as well as by members of the U.S. delegation itself.[265] The decision raised questions as to whether the United States was violating its duty to negotiate in good faith; well-publicized negotiations with other potential seabed mining states, with the goal of establishing "reciprocating states arrangements" (RSA) as an alternative to the UNCLOS III regime, proceeded during the U.S. review and further damaged the credibility of the United States as a good-faith negotiator.[266] At an even more basic level, it was believed that no single state could be permitted to determine the outcome of such a multilateral conference.[267]

With the assistance of the Working Group of 21, Committee I discussed the Preparatory Commission. Australia garnered support for a provision invoking the General Agreement on Tariffs and Trade (GATT) to deter subsidization of seabed mining operations.[268] There was general agreement that the Commission would be established by resolution rather than by the treaty itself, and that Commission expenses would be funded by the United Nations; but the delegates were unable to resolve the issue of participation.[269] Developing states were unwilling to discuss preparatory investment protection without full U.S. participation, and opposed provisional application of the rules, regulations, and procedures adopted by the Commission.[270] Although developing states indicated a willingness to consider some modification of the seabed provisions in order to accommodate the practical — as opposed to ideological — concerns of the United States, the new U.S. ambassador, James Malone, emphasized that the objections of the Reagan administration were based upon "matters of principle" which "follow from widely shared American attitudes towards the creation of global regulatory institutions, the availability of resources for our industrial economy, the financial and economic burdens of an unbalanced budget, and the role the United States should play in decisions affecting its vital interests."[271] Specifically, at the request of other UNCLOS III delegates, Ambassador Malone indicated the administration's major areas of concern: lack of adequate representation and voting power within the decisionmaking structure of the Authority; need for greater overall encouragement of seabed mining activities; uncertainties regarding private access, particularly with respect to impediments posed by the convention's regulatory regime; amendment of the treaty without U.S. consent; and the budgetary impact of assessed financial contributions.[272]

Despite substantial opposition, Committee II approved a new formula whereby shelf delimitation between opposite and adjacent states "shall be effected by agreement on the basis of international law ... in order to achieve an equitable solution."[273] Efforts to require prior notification or authorization for passage of warships in the territorial sea and for the conduct of military exercises in the exclusive economic zone again failed, although such restrictions upon the traditional concept of freedom of navigation were supported by approximately half the delegations.[274] The revised Draft Convention included a provision allowing the EEC to participate in the treaty to the extent that it was granted the competence to do so by its member states, and it was also decided that the Authority and the Law of the Sea Tribunal would be located in Jamaica and Hamburg, West Germany, respectively.[275] A six-week intersessional meeting of the Drafting Committee was scheduled for early 1982, and it was agreed that a final draft of the Convention — along with any formal amendments — should be adopted in April 1982 at the final UNCLOS III session, followed by a September signing ceremony in Caracas.[276]

The U.S. review process, which began in March 1981 and included participation by Congress, defense interests, foreign states, and private industry, was not completed until late 1981, and in January 1982 the National Security Council recommended that the United States return to the spring session of UNCLOS III in an effort to negotiate more acceptable provisions.[277] Although there was serious disagreement among U.S. policymakers concerning the viability of a reciprocating states regime, President Reagan was apparently advised that it was indeed a realistic option for the United States.

> Accordingly, when a final decision was made and detailed instructions for the delegation were negotiated, there was an assumption in the Administration that if the United States adopted a tough uncompromising stance and as a result lost the opportunity to improve the treaty, it could afford to stay out of the treaty because there was a viable alternative — a separate mini-treaty with our allies.[278]

Because of its antagonistic negotiating posture, which alienated virtually all other delegations, the United States found itself diplomatically isolated at UNCLOS III during 1981, but the general desire to conclude a widely accepted treaty substantially increased U.S. bargaining power for the eleventh session.[279]

Eleventh Session: 1982 (New York)

The final session of UNCLOS III was convened in New York on March 8. Following the U.S. policy review, President Reagan announced on January 29 that "while most provisions of the draft convention are acceptable and consistent with U.S. interests, some major elements of the deep seabed mining regime are not acceptable," and that the United States would

> seek changes necessary to correct those unacceptable elements and to achieve the goal of a treaty that:

—Will not deter development of any deep seabed mineral resources to meet national and world demand;

—Will assure national access to these resources by current and future qualified entities to enhance U.S. security of supply, to avoid monopolization of the resources by the operating arm of the Authority, and to promote economic development of the resources;

—Will provide a decision-making role ... that fairly reflects and effectively protects the political and economic interests and financial contributions of participating States;

—Will not allow for amendments to come into force without approval of the participating States, including in our case the advice and consent of the Senate;

—Will not set other undesirable precedents for international organizations;

—Will be likely to receive the advice and consent of the Senate. In this regard, the Convention should not contain provisions for the mandatory transfer of private technology and participation by and funding for national liberation movements.[280]

Although these objectives did not on their face appear unreasonable to many delegates, a former U.S. negotiator has revealed that instructions received by the U.S. delegation—instructions which required the U.S. negotiators to insist upon a radical revision of the seabed mining regime, permitting nodule mining operators to "stake a claim" to nodule deposits on the deep seabed and thereby enjoy exclusive exploitation rights without any corresponding obligations except revenue sharing—allowed little negotiating leeway for their achievement.[281] While the inflexibility of the U.S. negotiating position may have been intended to enhance the prospect of a reciprocating states regime as an alternative to UNCLOS III, President Koh obtained a moratorium on discussions concerning establishment of such a regime during the session.[282] After holding intersessional consultations during the last week of February, the delegates proceeded through several stages of informal negotiations, considering three separate sets of amendments before adopting the entire United Nations Convention on the Law of the Sea by vote on April 30.[283]

An opening three weeks of informal meetings produced four reports containing proposed amendments, which were debated during three days of plenary meetings. The United States submitted a "Green Book" containing a comprehensive set of draft amendments, all of which were rejected,[284] whereupon 11 mid-sized developed states known as the Group of 11 interceded unsuccessfully with a set of compromise amendments designed to alleviate many of the U.S. concerns. "While the United States did not intend to reject these proposals out of hand, it stated its difficulties with these proposals in such strong terms as to lead Conference leaders to conclude that they had been rejected."[285] Two days of intensive consultations held by Koh produced a set of amendments which modified the seabed mining provisions in several respects in an attempt to secure the support of the United States and other potential mining states: the largest consumer state — i.e., the United States — was guaranteed a Council seat; the majority requirement for adoption of amendments at the review conference was raised from two-thirds to three-fourths; "development of the resources of the Area" was made the primary goal of the Authority's policy; the system of state-sponsored access was eased by elimination of a requirement that the Authority investigate each contractor's compliance with rules and regulations prior to approving plans of work; and a moratorium on exploitation of non-nodule resources was dropped in favor of a provision requiring the Authority to promulgate rules and regulations to govern the exploitation of such resources.[286] During the second week in April, delegations were invited to submit formal amendments to the Draft Convention "on any matters on which they were still not satisfied," and the merits of the 31 sets of amendments offered were debated inconclusively in six meetings of the Plenary.[287] On April 26 President Koh appealed to all states to withdraw their proposed amendments in favor of his own compromise texts, and most states, including the United States and the Group of 11, did so.[288] Although the United States adopted a somewhat more flexible negotiating position during the final weeks of the Conference, real progress eluded the delegates, and the session concluded with a general feeling of dissatisfaction at the inability to accommodate U.S. concerns.[289]

The negotiations produced a set of supplementary resolutions which were adopted separately from the treaty itself in order to allow them to be implemented before the treaty entered into force. With regard to preparatory investment protection, Resolution II, a product of proposals by potential mining states, the Group of 11, and the Group of 77,[290] provided that exclusive exploration rights, together with priority in the subsequent allocation of production authorizations, would accrue to pioneer seabed mining operators sponsored by Japan, France, India, and the Soviet Union, as well as to four established Western mining consortia and to any qualifying developing states — each of which would be required to transfer technology, train personnel, and explore its reserved area on a cost-reimbursable basis for the benefit of the Enterprise in return for guaranteed access.[291] The task of organizing the initial operations of the Authority, including implementation of the system of preparatory investment protection, was allocated to the Preparatory Commission, which was established by a second resolution.[292] It was decided that the decisionmaking procedures of the Conference would apply provisionally in the Commission, pending adoption of its own procedural rules, and that states signing only the Final Act would be allowed to "participate fully in the work, but not in the decision-making process" of the Commission.[293] National liberation movements and other organizations which had participated in UNCLOS III as observers and which signed the Final Act were also permitted to participate as observers in the Commission and in the Authority.[294]

On April 30 the United States formally expressed its continuing displeasure with the seabed mining provisions and called for a vote on the Convention, preventing its adoption by consensus in an effort to avoid becoming bound by its provisions under customary international law.[295] Despite intense U.S. diplomatic pressure, only Israel, Turkey, and Venezuela joined the United States in opposing the Convention, which was adopted by a vote of 130 to 4, with 17 abstentions; developing states affirming their support for the UNCLOS III agreement were joined by France, Japan, and most other Western states.[296] The Drafting Committee, which had met for six weeks in January and February, as well as during the session itself, met again during July and August to complete its work, and the Plenary was reconvened for three days in September to adopt the final drafting changes.[297] The eleventh session was reconvened in Montego Bay, Jamaica, on December 6, and four days later 117 states and 2 organizations signed the Convention. Although several Western European states joined the United States in declining to sign, the signatory states included five EEC members and most other Western states, the Soviet bloc, and most Asian, African, and Latin American states.[298]

> Never in the history of treaty-making ha[d] such a large and varied number of countries signed a convention on the day it was opened for signature. The number of signatories exceeded the most optimistic expectations and surprised those who had predicted that it would be difficult to obtain on the first day the fifty signatures that were required to convene the Preparatory Commission.[299]

UNCLOS III had been in session for a total of 585 days during a nine-year period; it had been the largest multilateral treaty-making conference in history, attended by some 3,000 delegates from 157 states who produced a Convention of more than 400 articles, including 9 annexes.[300]

The Preparatory Commission

The tasks of the Preparatory Commission included registration of pioneer investors, establishment of the Enterprise, preparation of provisional rules, regulations, and procedures for the Authority, and study of the impact of seabed mining upon developing producer states.[301] At its initial session in the spring of 1983, held in Kingston, Jamaica, and attended by delegations from more than 100 states, the Commission decided to hold two four-week sessions each year at Kingston and New York in an effort to ensure that the Authority and the Tribunal would begin timely operations.[302] After lengthy discussion, the Commission selected Joseph Warioba of Tanzania as its chairman and adopted a Consensus Statement of

Understanding, which required consensus for all decisions for which it is required in the Council under the Convention – including the adoption of provisional rules, regulations, and procedures governing seabed mining activities – and which established the basic structure of the Commission: four special commissions; the Plenary, as the principal organ of the Commission with overall responsibility for implementation of Resolution II; and a General Committee of 36 states to administer the system of preparatory investment protection.[303] At its resumed summer session, the Commission elected additional officers and a nine-member Credentials Committee, and agreed upon a division of responsibility among the special commissions: Special Commission 1 to identify methods of dealing with problems faced by developing producer states as a result of deep seabed mining; Special Commission 2 to oversee start-up operations of the Enterprise; Special Commission 3 to draft provisional rules, regulations, and procedures in the form of a Seabed Mining Code to govern deep-sea mining operations; and Special Commission 4 to make arrangements for establishment of the International Tribunal for the Law of the Sea. In the absence of the United States, which confirmed in July that it regarded the Convention as "hopelessly flawed" and that it would therefore not participate in the Commission,[304] there was a genuine desire among the participants to develop a set of provisional rules, regulations, and procedures which would accommodate the interests of nonsignatory states sufficiently to permit them to adhere to the Convention.[305] Most Western seabed mining states nevertheless made clear that their ultimate decisions with respect to ratification would depend upon the outcome of further negotiations within the Commission.[306]

Although paragraph 5 of Resolution II required sponsoring states to submit unresolved claim conflicts to binding arbitration during 1983, it was not entirely clear how resolution was to occur or whether this procedure would require – or even permit – the participation of potential pioneer investors who were delaying their applications to the Commission.[307] Nor was it clear how the interests of nonsignatory seabed mining states as potential sponsors of pioneer applicants could be preserved in anticipation of their possible signature prior to the December 1984 deadline – or, subsequently, in anticipation of their possible adherence to the Convention prior to its entry into force – since it had originally been expected that all pioneer investors would be registered simultaneously. "The problem is that Resolution II does not specify a deadline after which pending applications may be processed, regardless of whether or not all eight pioneers named have found a sponsor that qualifies by having signed the 1982 treaty."[308] Western mining states successfully sought a flexible interpretation of Resolution II in order to preserve their rights as pioneer investors, pending the outcome of overlap negotiations with all potential sponsoring states as well as the outcome of substantive negotiations within the Preparatory Commission.[309] Thus, when early efforts to resolve overlapping claims among pioneer investors proved unsuccessful, India and the Soviet Union were prevented from invoking the overlap resolution procedures contained in Resolution II.[310]

Work proceeded slowly during 1984, as the Preparatory Commission sought to focus its efforts on the drafting of rules and procedures to effect the registration of pioneer applicants under Resolution II.

> Delegates in the Commission and Commission leadership were growing increasingly concerned in 1984 that unless the private consortia joined the pioneer investor scheme established by Resolution II, it would be difficult to make the international mining regime work. The expertise and experience of the commercial mining groups were considered particularly important in launching the Enterprise.[311]

The Commission completed an initial reading of draft rules prepared by the U.N. Secretariat, and did adopt many of them provisionally, but it failed to complete work on procedural rules relating to the timing, review, and confidentiality of applications.[312] The Soviet Union and India had submitted applications in July 1983 and January 1984, respectively, and France and Japan formally applied for registration in August.[313] The Commission received no applications from the four private pioneer investors, but several groups of states sought to obtain eligibility for additional pioneer investors.[314] Faced with an August 3 agreement among Western seabed mining states which seemed to promote U.S. unilateral mining efforts by laying the groundwork for a reciprocating states regime,[315] the Commission agreed on August 31, 1984, to register simultaneously, during the 1985 session, all pioneer applicants certified by states which had signed the Convention by the December 10 deadline, and a timetable was established for pioneer investors to resolve overlapping claims prior to the convening of the Commission in the spring of 1985.[316] In order to minimize the burden of financing the initial operations of the Enterprise and to avoid possible application of the mandatory technology transfer provisions, and in keeping with the general consensus that the Authority should be "viable and cost-effective," the delegates agreed to examine seriously the possibility of joint-venture arrangements between the Enterprise and pioneer registrants, as well as the feasibility of establishing a "nucleus" mining operation.[317]

Although West Germany and the United Kingdom, along with 13 other states, ultimately decided not to sign the Convention,[318] all seabed mining states except the United States participated in the Commission discussions in an effort to ensure that private mining activities could be undertaken on a competitive basis, and by the end of 1984 the final roster of participants included a number of voting Commission members well situated to assert their interests as potential sponsors of the four private pioneer applicants: Belgium, Italy, the Netherlands, and Canada (the "BINC Group"), as well as Japan and the EEC.[319] Representatives of France, Japan, India, and the Soviet Union met several times prior to the spring 1985 session of the Preparatory Commission, but France and the Soviet Union were unable to resolve a substantial overlap of their mining claims in the Clarion-Clipperton region of the Pacific Ocean.[320] Although France joined the Soviet Union in seeking early registration upon resolution of the overlap, the BINC Group indicated it would oppose registration of any pioneer applicant until all overlaps with sites claimed by their own potential pioneer applicants had also been resolved.[321]

> The problem faced by those who would register, and by Commission leaders, [was] finding a solution to the Soviet/French/Japanese overlaps that provides the second site required for the Enterprise and avoids leaving either the Enterprise or the Soviets holding the bag with a mine site that conflicts with the areas licensed under U.S. (or U.K.) law. For despite protestations about the legality of [its] unilateral acts, no one [was] particularly interested in establishing a potential collision course over mine sites with the United States.[322]

During the summer of 1985, Soviet opposition to negotiations with nonsignatory states began to soften in the face of these more practical considerations.[323]

　　Meanwhile, slow progress continued to be made in the other work of the Commission, with agreement in Special Commission 1 that identification of affected producer states should await the commencement of commercial exploitation, at which time comprehensive studies would be undertaken.[324] The possibility of joint-venture operations was discussed further in Special Commission 2, where it was recognized that the pace of negotiations with respect to the start-up of the Enterprise would be largely determined by related developments beyond its immediate purview, including the registration of pioneer investors, the commencement of commercial mining operations, and the entry into force of the Convention, as well as the condition of mineral markets.[325] Special Commission 3 proceeded with preparation of a "mining code" for deep-sea nodules, focusing in particular on procedures for approval of plans of work, revenue sharing, financial incentives, and technology transfer.[326] Special Commission 4 began to negotiate instruments to govern the headquarters of the Tribunal and the privileges and immunities of judicial access.[327] The slow pace of the negotiations was attributed to the delay in registering the pioneer investors, which created an atmosphere of frustration and suspicion and led to renewed political posturing and an unwillingness to compromise, particularly in Special Commission 2 where it was feared that timely commencement of mining operations by the Enterprise would be jeopardized by the absence of the technical assistance which pioneer registrants were required to provide under Resolution II.[328] Work in the Preparatory Commission was facilitated by a Friends of the Convention Group, but many of the smaller developed states comprising this group became increasingly concerned about the size of their financial obligations under the Convention in the absence of participation by some of the wealthiest states, and a study of such obligations was requested from the U.N. Secretariat.[329]

In the weeks prior to the spring 1986 session, consultations among the four applicant states and the Chairman of the Preparatory Commission resulted in a procedural agreement known as the Arusha Understanding, whereby, upon registration, the first group of four applicants would relinquish portions of their overlapping claims to be reserved for the use of potential pioneer applicants – i.e., additional pioneer investors applying to the Commission for registration prior to the entry into force of the Convention.[330] In September the Understanding was unanimously confirmed by the Commission, with modifications establishing a procedure for registration of the first group, requiring pioneer registrants to assist the Commission and the Authority in the exploration of the first site worked by the Enterprise, authorizing a group of Eastern European states to sponsor a pioneer investor, and extending the deadline for developing states to qualify as pioneer investors until the entry into force of the Convention, although any such additional allocation of pioneer sites would be "without prejudice to . . . the interests of" the private consortia.[331] The new registration procedure called for the Soviet Union to resolve its remaining overlaps with the consortia and for the four pioneer applicants then to submit revised applications for review by a 15-member Group of Technical Experts, which was to make recommendations to the General Committee for final action during the spring 1987 session of the Commission.[332]

Special Commission 1 completed a preliminary review of procedures for identifying producer states likely to be affected adversely by nodule mining and began to review proposals regarding remedial measures.[333] Amid growing uncertainty about the economic viability of deep-sea mining, Special Commission 2 concentrated on defining an internal structure for the Enterprise which could be readily adapted to changing economic circumstances.[334] Special Commission 3 continued its "very detailed and technical" examination of the rules governing the Authority's application procedure and held important discussions on the financial terms of mining contracts, during which the Group of 77 indicated a willingness to seek ways to accommodate the concerns of Western states with respect to financial burdens created by revenue-sharing obligations.[335] Special Commission 4 reviewed draft rules for the Law of the Sea Tribunal and moved toward a requirement that the Federal Republic of Germany be one of the first 60 states to ratify in order to assure adequate time to find an alternative site for the Tribunal in the event of German nonparticipation.[336] The Plenary reviewed draft rules and procedures for the Economic Planning Commission and the Legal and Technical Commission, but postponed difficult decisions relating to their composition and decisionmaking procedures and to the creation of a Finance Committee as an additional subsidiary organ of the Council.[337]

The Preparatory Commission convened its fifth session in Kingston on March 30, 1987, and again in New York on July 27. With the resignation of Chairman Warioba, who had become the Prime Minister of Tanzania, the spring 1987 session became deadlocked over the choice of his successor — José Luis Jesus of Cape Verde was elected by acclamation at the resumed summer session — and made little progress on several other important issues.[338] Although the pioneer investors reported that progress had been made in their continuing consultations, timely resolution of the remaining site overlaps again proved impossible, and registration of the first group of applicants was therefore postponed until August,[339] at which time the Commission proceeded to register the Indian applicant in accordance with the terms of the 1986 Understanding.[340] Having finally resolved Soviet overlaps with all of the pioneer consortia in a "midnight agreement" on August 14, the pioneer investors sponsored by France, Japan, and the Soviet Union were registered by the General Committee on December 17, 1987, despite some misgivings, following a determination by the Group of Technical Experts that the pioneer and reserved sites were of equal commercial value.[341] Although the August agreement — which was without prejudice to the positions of the parties with respect to the Convention — was concluded between the Soviet Union and the BINC states, the Soviet Union also entered into bilateral companion notes with West Germany, the United Kingdom, and the United States, which in effect also made those states parties to the site-allocation agreement.[342] "The three non-signatories have thus been given an opportunity to accede to the Convention before its entry into force with full entitlement to apply on behalf of one or more of the consortia for their registration as pioneer investors with respect to the areas they claim."[343]

Commission negotiations continued during 1987, with delegates focusing on measures to ensure successful commencement of the parallel mining system. In Special Commission 1, the Group of 77 proposed the creation of a compensation fund, but EEC states expressed support instead for the use of existing preventive mechanisms, and Eastern European states advocated bilateral arrangements between seabed mining states and developing producer states. "There appeared to be a general understanding, however, that developing land-based producers seriously affected by seabed mining should be provided with some form of assistance."[344] Although all four pioneer registrants reiterated their intention to comply with their training obligations under Resolution II, issues relating to the timing and costs of the training programs created dissension in Special Commission 2, which established an ad hoc working group to assist in the formulation and implementation of the programs;[345] it was understood that it would be necessary to minimize the initial administrative expenses of the Enterprise, however, and most delegates agreed upon the desirability of establishing a small nucleus organization to commence preliminary mining activities on behalf of the Enterprise when the Convention entered into force.[346] Special Commission 3 reviewed a working paper on draft regulations for the contract application process, its chairman seeking to ensure that plans of work initially need only include sufficient data to enable the Authority to make an informed determination of the equal commercial value of the site to be reserved for the Enterprise.[347] Calls by Western mining states for changes in the revenue-sharing provisions of the Convention continued to be resisted by the Group of 77, which emphasized that modification of the terms of the Convention was beyond the Commission's mandate, but there was an increasing willingness to consider amelioration of such financial burdens in light of worsened economic conditions.[348] Special Commission 4 moved ahead with its consideration of a draft headquarters agreement prepared by the Secretariat, but was still unable to resolve the question of the future location of the Tribunal.[349] The Plenary provisionally adopted rules of procedure for the Economic Planning Commission and undertook a review of another set of such rules for the Council.[350] It was noted that the atmosphere at the Commission was improving, with states becoming more willing to enter into pragmatic discussions on important substantive issues, although consideration of the more difficult issues continued to be deferred: "The . . . session once again confirmed that financial matters and the decision-making procedures relating to them in the various organs of the Authority remain the hard-core issues."[351]

During its sixth session, which met in the spring and summer of 1988 following registration of the first group of pioneer applicants, the Preparatory Commission focused on the definition of their obligations under Resolution II—including personnel training and technology transfer, periodic reports and expenditures, exploration of a mining site for the benefit of the Enterprise, and payment of a $1 million annual fee—and delegated consideration of the subject to an informal consultative group where discussions proved inconclusive.[352] Disagreement remained in Special Commission 1 with respect to the issues of subsidization and creation of a compensation fund for producer states adversely affected by sea-bed mining; it was agreed that specific identification of such states should be deferred to the Authority.[353] A set of draft principles and policies prepared by the ad hoc working group was reviewed by Special

Commission 2 and approved for use in preparing guidelines and procedures to implement training programs.[354] Special Committee 3 conducted a detailed review of draft articles on technology transfer and of a set of amendments submitted by Western seabed mining states.[355] Having completed work on a draft headquarters agreement for the Law of the Sea Tribunal and on procedures for the prompt release of detained vessels and crews, Special Commission 4 took up consideration of a draft protocol on privileges and immunities.[356] In addition to discussing procedures for approval of plans of work, the Plenary provisionally approved most rules of procedure for the organs of the Authority, and was able to reach general agreement on the need for a Finance Committee and on its membership qualifications, although disagreement remained with respect to its competence and decisionmaking procedures.[357]

During the 1989 session of the Preparatory Commission, delegates attempted to pick up the pace of the negotiations in an effort to meet a target date of 1991 for completion of its work.[358] Special Commission 1 reported progress in developing criteria for the identification of adversely affected producer states and in devising a procedure for their compensation, and began reviewing a list of 66 provisional conclusions.[359] After receiving recommendations from its ad hoc working group, Special Commission 2 prepared a proposal for administration of the training program by a 15-member Training Panel — a proposal which was approved by the Plenary — and completed a review of articles relating to the organizational structure of the Enterprise.[360] Special Commission 3 completed its reading of the draft articles on technology transfer, and reported "good progress" in its initial review of draft rules governing the issuance of production authorizations.[361] Work on a draft protocol on privileges and immunities was completed in Special Commission 4, which began discussing jurisdictional agreements with other international organizations.[362] With assistance from Special Commission 2 and the Group of Technical Experts during the resumed summer session, Chairman Jesus secured agreement on a comprehensive three-stage plan for implementation of the obligations of pioneer registrants with respect to the Enterprise's primary exploration site in the Clarion-Clipperton Zone: a six-month preliminary phase involving data analysis and site examination; a two- to three-year exploratory phase, during which prime nodule deposits at the site would be identified and training of Enterprise personnel would begin; and a four- to seven-year period of more detailed exploration and data collection.[363] On the final day of the session, the chairman of the Group of 77 joined other negotiating groups in indicating a readiness to enter into discussions with any state, including nonsignatories such as the United States, in order to ensure universal support for the Convention.[364] There was general support within the Preparatory Commission for a cost-effective Authority of limited size in order to assure "that the structure of the Authority will be developed as the needs require and in accordance with the financial capacity of member states."[365]

When the Commission reconvened in March and August 1990, delegates focused their attention upon the training program, which was to be implemented beginning in 1991, and upon possible means of accommodating the remaining concerns of Western seabed mining states.[366] Special Commission 1 completed an initial review of its provisional conclusions and examined existing mechanisms of assistance for developing producer states.[367] The interim structure and operations of the Enterprise continued to be studied in Special Commission 2, which also discussed the commercial viability of seabed mining in the Clarion-Clipperton Fracture Zone.[368] Progress in Special Commission 3 was slowed by the resignation of its chairman, but work proceeded on draft regulations for the protection of the marine environment during mining operations — a matter of some importance to the delegates.[369] Having completed its consideration of a draft headquarters agreement, Special Commission 4 discussed the structure and cost of the Tribunal for the first time, agreeing that its operations should be guided by the principle of maximum economic efficiency.[370] The Plenary reviewed the draft headquarters agreement and draft protocol prepared by Special Commission 4, and reported substantial progress in its consideration of the functions and composition of the Finance Committee.[371] At a seminar held to discuss the environmental aspects of seabed mining, experts from developed seabed mining states urged that deep-sea mineral exploitation should be undertaken with caution, warning that the environmental consequences were unforeseeable.[372] With formal adoption of an agreement for implementation of the training program in August,[373] China formally applied for registration of a state-run pioneer mining operation in the eastern Pacific.[374] On March 6, 1991, following a December review by the Group of Technical Experts, the Chinese applicant was registered by the General Committee; a week later a consortium of Eastern European states submitted an application to the Commission on behalf of their own pioneer investor operation.[375]

Post-Conference Diplomacy

Following the adoption of the Convention in 1982, the United States worked actively to undermine support for it and to force a renegotiation of its seabed mining provisions, even withholding its U.N.-assessed share of the Preparatory Commission's administrative expenses on the grounds that the Commission negotiations "would be meaningless and arguably contrary to American interests."[376] The ratification process was further delayed by weakening demand for nodule minerals and by the slow pace of negotiations within the Preparatory Commission.[377] By the end of 1983 the Convention had been ratified by eight states and the U.N. Council for Namibia,[378] and in the next three years an additional 23 states deposited their instruments of ratification.[379] The number of ratifications declined after 1986, however: three in 1987, two in 1988, five in 1989, and three in 1990.[380] Moreover, the ratifying states were not generally representative of the international community as a whole: virtually all were developing states – approximately half of which were African – and none were major industrialized states or states with a demonstrated interest in deep-sea nodule mining. By 1987 there was growing doubt that the Convention would enter into force before 1990;[381] by 1989 there was a general expectation within the Preparatory Commission that the 60th ratification would not be deposited until 1992;[382] and by mid-1991 13 ratifications were still needed.[383]

After the 1982 announcement that it would neither sign nor attempt to renegotiate the Convention as a whole, the United States took steps to implement a reciprocating states regime under the 1980 Seabed Act, even though two of the four private consortia refused to oppose the Convention.[384] On September 3, having failed to generate significant support for such a regime, the United States concluded an interim agreement with France, West Germany, and the United Kingdom for the identification and resolution of overlapping claims.[385] On August 3, 1984, a Provisional Understanding Regarding Deep Seabed Matters was concluded by eight Western seabed mining states, providing for priority of right among claimants through a system of mutual registration and conflict resolution – an agreement which met with vociferous objections from the Group of 77 and Soviet-bloc states.[386] The conclusion of the Provisional Understanding was followed later that year by the unilateral issuance of exploration licenses – four U.S. and one British – to the four consortia, with two German licenses being granted in 1986.[387] As required by the Understanding, each of the signatory states enacted legislation authorizing national seabed mining operations, but most were careful to preserve their right to proceed with mining under the framework of the Convention and have worked within the Preparatory Commission to achieve an acceptable mining regime.[388] "These industrialized mining states have maintained an interest in an alternative multilateral agreement to govern deep sea-bed mining in the event that the Commission does not produce a workable regulatory system for mining. They would prefer a widely acceptable international mining regime."[389]

On August 30, 1985, after extensive debate and without a formal vote, the Commission adopted a declaration sponsored by the Group of 77 which recognized the UNCLOS III regime as "[t]he only regime for exploration and exploitation of the area and its resources" and rejected as "wholly illegal" any unilateral claim or alternative seabed mining regime.[390] In the U.N.

General Assembly that fall, the United States disputed the August 30 declaration, arguing that the Convention "will not create legal obligations for, nor abridge the legal rights of, those nations that have not expressly consented to be bound ... by ratification or accession,"[391] and in early 1986, in an apparent effort to preserve the claims of its consortia in the face of weak minerals markets, the United States formally notified the United Nations of the coordinates of its four licensed mine sites, admonishing member states to pay "reasonable regard" to states exercising their high-seas freedoms.[392] On April 11, 1986, in reaction to the issuance of unilateral seabed mining licenses by West Germany and the United Kingdom, the Commission adopted, by a vote of 59 to 7, with 10 abstentions, a resolution reaffirming the 1985 declaration and reiterating the Commission's formal objection to "any claim, agreement or action ... incompatible with the Convention and related resolutions."[393]

Beginning in the fall of 1984, the General Assembly adopted a series of annual resolutions which, in virtually identical language, endorsed the 1982 Convention and the efforts of the Preparatory Commission to implement Resolution II.[394] The resolutions called upon all states "to safeguard the unified character of the Convention and related resolutions" and "to desist from taking actions which undermine the Convention or defeat its object and purpose." Although their language was generally somewhat more conciliatory than that of their predecessors, the 1989 and 1990 resolutions, like the declarations adopted by the Preparatory Commission and the other resolutions adopted by the General Assembly since 1985, nonetheless declared that "the Convention provides the regime to be applied to the Area and its resources." Each of these resolutions was supported by at least 135 states—including Australia, Belgium, Brazil, Canada, China, France, Italy, Japan, the Netherlands, and the Soviet Union—and was opposed by only Turkey and the United States.[395] Responding to the overtures of the Group of 77 in the Commission earlier that year, the U.S. ambassador informed the 1989 session of the General Assembly that the United States was willing to engage in discussions with a view to achieving the goal of a universally acceptable treaty but that the "fundamental reform" demanded by the United States "exceeds the capabilities of the Preparatory Commission."[396]

Having expected more support among U.S. allies for its opposition to the Convention than was forthcoming, the Reagan administration had initially accepted ten applications for exploration licenses under the domestic regulatory regime established pursuant to the 1980 Seabed Act, but the number was subsequently reduced to five in amended applications which were submitted after overlapping claims had been resolved.[397] Despite the absence of an RSA regime, the United States ultimately certified and registered a single site for each of the four private consortia in 1984.

Estimates of start-up costs for a commercial mining operation in the eastern Pacific were rising as high as $2.5 billion, however,[398] and within two years the consortia had found it necessary to request amendment of their exploration licenses to reflect substantially scaled-back plans for deep-sea mining operations in light of new data, technological advances, and "prevailing market conditions."[399]

Although prices of nodule minerals had risen steadily with energy prices throughout the 1970s, causing consumer states to look to seabed mining as an important alternative source of supply, by 1980 the production capacity of land-based producers had been doubled in response to the higher prices, and mineral prices began to fall as world consumption slowed.[400] Faced with unanticipated difficulties in nodule recovery efforts, and with predictions that deep-sea mining operations would not become profitable until well into the 21st century, private mining firms were quick to reconsider their involvement in ocean mining efforts.[401] "Inconclusive results from mining system tests, recession and the associated downturn in the metal markets, combined with an uncertain legal situation with respect to tenure of seabed claims, resulted in almost total cessation of work by virtually all of the consortiums [sic] by the early 1980s."[402] Meanwhile, it was nevertheless increasingly anticipated that additional applications for pioneer investor status might be forthcoming in the Preparatory Commission prior to the entry into force of the Convention, particularly from Brazil and South Korea, which established serious national deep-sea mining programs during the 1980s.[403] Scandinavian states and several other mid-sized developed and developing states have also sought to acquire a seabed mining capability.[404] "European and Asian Governments have financed virtually the whole of the technical development that has taken place since 1981, their motive being that of long-term supply considerations."[405] Although by the end of the 1980s no prototype nodule mining operation had yet been established, Japan made known its intention to establish a pilot mining project at its eastern Pacific site in 1992, and India was expected to undertake a similar effort by 1995.[406]

While the UNCLOS III negotiations focused attention on deep-sea nodule resources, geological interest was also being generated by other marine minerals, including polymetallic sulfide accretions at hydrothermal vents along the submarine ridges surrounding mid-ocean spreading zones.[407] The sulfide deposits contain some of the same minerals as manganese nodules — frequently at higher concentrations[408] — and are forming continuously at a relatively rapid rate.[409] Cobalt-rich manganese crusts which have been discovered in waters off a number of Pacific islands, including the exclusive economic zone of the Hawaiian archipelago, constitute a potentially significant source of nickel, cobalt, and manganese, and are expected to provide a cost-competitive alternative to deep-sea nodule mining.[410] As interest in deep-sea nodules waned, these newly discovered offshore mineral deposits – within the exclusive resource jurisdiction of foreign coastal states as well as the United States – became the primary focus of the U.S. ocean mining industry.[411] By 1990, however, commercial interest in polymetallic sulfides had all but evaporated, and only the Japanese were actively pursuing manganese crust exploitation within the U.S. exclusive economic zone.[412]

By the time the Law of the Sea Convention was opened for signature, it was recognized that adoption of an exclusive economic zone would be necessary for the exercise of sovereign rights with respect to these non-nodule minerals, many of which were located beyond the continental shelf but within 200 miles of the U.S. coast.[413] Thus, in March 1983 President Reagan, noting that such deposits "could be an important future source of strategic minerals," proclaimed a 200-mile exclusive economic zone for the United States.[414] In so doing, Reagan announced that "the United States is prepared to accept and act in accordance with the balance of interests relating to *traditional* uses of the oceans – such as navigation and overflight" – set forth in the Convention.[415] The United States subsequently adopted the position that the nonseabed provisions of the Convention generally reflect customary international law, and sought to challenge non-conforming maritime claims militarily through an "exercise-of-rights program."[416] The Reagan administration sought to encourage the leasing of hard-mineral mining sites within the U.S. exclusive economic zone pursuant to regulations adopted under the Outer Continental Shelf Lands Act,[417] and the ocean mining industry has indicated that marine mineral exploitation is likely to proceed within U.S. jurisdiction – as well as, if not instead of, in the international seabed area beyond – when mineral markets recover sufficiently to make deep-sea mining economically attractive.[418] Although the election of George Bush as President in 1988 initially gave rise to some hope for a reappraisal of the U.S. position, the new administration maintained official U.S. opposition to the Convention.[419] By 1991, however, amid new activity by the second group of pioneer applicants within the Preparatory Commission, and under increasing diplomatic pressure from other Western states, the United States had entered into informal consultations at the United Nations with a view to drafting a protocol which would contain the adjustments regarded as necessary to render the Convention regime universally acceptable.[420]

Chapter 4

The U.S. Objections Examined

In his statement of January 29, 1982, President Reagan specified six negotiating goals for the U.S. delegation,[1] none of which were considered to have been achieved during the final UNCLOS III session. The United States based its rejection of the Convention on the seabed mining provisions set forth in Part XI and related annexes, which "do not even minimally meet U.S. objectives."[2] In order to evaluate the correctness of the decision, it is necessary to examine closely the validity of the objections to these provisions.

Technology Transfer

The Law of the Sea Convention imposes a qualified duty upon seabed mining operators to transfer their technology to the Enterprise or to developing states to which the Authority has granted the contractual right to exploit selected reserve areas. Under article 5 (3) of Annex III, each applicant for a plan of work must agree to make available, "on fair and reasonable commercial terms and conditions," the technology used in conducting seabed mining activities under the contract; an operator need only undertake to provide technology which he is entitled—under domestic law—to transfer, and the obligation may only be invoked if the transferee is "unable to obtain the same or equally efficient and useful technology on the open market and on fair and reasonable terms and conditions."[3] Technology which the operator is not legally entitled to transfer may not, however, be used in seabed mining unless a written assurance is obtained from the legal owner permitting its transfer "on fair and reasonable commercial terms and conditions, to the same extent as made available to the contractor," and an applicant must further agree to acquire the legal right to transfer such technology "if it is possible to do so without substantial cost" and if the technology "is not generally available on the open market."[4] Seabed exploitation technology is to be made available on the same terms to "a developing State or group of developing States" who have been granted exclusive mining rights with respect to the reserved site submitted to the Authority by the contractor, but only if the Enterprise has not already obtained technology from that contractor, and any technology so

96

transferred may not be retransferred to other states.[5] The technology transfer provisions of article 5 may be invoked only until 10 years following commencement of nodule production by the Enterprise,[6] and disputes regarding the fairness, reasonableness, and commercialism of offers may be submitted to binding commercial arbitration;[7] disputes regarding other aspects of technology transfer — including confidentiality and the definition of terms such as "the same or equally efficient and useful," "generally available," and "substantial cost" — may be submitted to the Sea-Bed Disputes Chamber for ultimate resolution, although such terms may be further elaborated in the rules, regulations, and procedures of the Authority.[8] Before the entry into force of the Convention, pioneer registrants must "provide training at all levels for personnel designated" by the Preparatory Commission, an obligation which the Preparatory Commission has determined may be satisfied by comprehensive training for as few as 12 trainees at the registrants' expense, and must "undertake" compliance with the Convention's technology transfer provisions.[9] Technology is defined as "the specialized equipment and technical know-how, including manuals, designs, operating instructions, training and technical advice and assistance, necessary to assemble, maintain and operate a viable system and the legal right to use these items for that purpose on a non-exclusive basis."[10] Where seabed mining operators enter into joint ventures with the Enterprise, technology transfer will occur only "in accordance with the terms of the joint venture agreement" to be negotiated.[11] Article 144 of the Convention imposes no specific obligations with respect to technology transfer, setting forth instead a general duty on the part of states to promote technology transfer and to provide "opportunities to personnel from the Enterprise and from developing States for training in marine science and technology and for their full participation in activities in the Area"[12]; other such general-obligation provisions of the Convention are similarly unenforceable with respect to particular technologies.[13]

Several articles were included in the Convention to protect the confidentiality of the proprietary data and technology to be made available to the Authority by contractors.[14] "Fair and reasonable commercial terms and conditions" may be construed to prohibit disclosure to third parties, and further restrictions may be expressly negotiated into technology transfer agreements.[15] The Convention provides that the archives of the Authority "shall be inviolable," that "proprietary data, industrial secrets or similar information and personnel records shall not be placed in archives which are open to public inspection," and that disclosure of proprietary data "to anyone external to the Authority" is forbidden.[16] Members of the Council's subsidiary commissions are forbidden to "disclose, even after the termination of their functions, any industrial secret, proprietary data which are transferred to the Authority..., or any other confidential information coming to their knowledge by reason of their duties for the Authority"; identical provisions are applicable to the Secretary-General of the Authority and his staff, and, as in the case of other provisions relating to the transfer of data and technology, the Authority would be liable "for the actual amount of damage" in the event of their violation.[17] The Convention also contains provisions prohibiting conflicts of interest,[18] and although the penalties which may be imposed upon employees of the Authority — disciplinary proceedings and dismissal — are less severe than those available to the U.S. government under federal law, the Reagan administration sought no changes in them.[19] Article 302 provides further protection against unwarranted technology transfer by ensuring that no state is required to make available "information the disclosure of which is contrary to the essential interests of its security."[20]

In its 1982 Green Book, the United States sought only three changes to the above provisions relating to technology transfer. First, it proposed a minor amendment which would have expressly extended the conflict-of-interest and confidentiality restrictions to include members of subcommissions of the Council.[21] Second, the United States sought to allow the review conference to consider amendments to article 144.[22] Finally, and most significantly, article 5 of Annex III (the annex entitled "Basic Conditions of Prospecting, Exploration and Exploitation") was entirely deleted, and an article relating to technology transfer was inserted in Annex IV (the annex entitled "Statute of the Enterprise"). The new article, rather than requiring mandatory technology transfer, would have recognized a "policy of the Authority to ensure that the Enterprise is able to become a viable commercial entity . . . at the earliest possible date," and provided for the convening of "a committee comprising all [seabed mining] States which shall have the responsibility . . . to facilitate the acquisition by the Enterprise of appropriate technology" by identifying and analyzing available technology, developing exploration plans for its site-specific application, preparing market analyses, and advising the Enterprise regarding the acquisition of technology on favorable terms and conditions.[23] The same package of assistance was to be made available to any developing state to which the Authority granted the right to exploit a reserved area.[24]

In a set of revised amendments submitted to the Conference on April 13, 1982, in association with six other industrialized states, the United States agreed to accept mandatory technology transfer, provided that article 5 of Annex III was modified to provide that contractors would only be required to transfer technology already available in a limited market and that such technology would be transferred "on terms and conditions no less favorable than the terms and conditions under which the contractor has made or is willing to make the technology available to third parties under similar circumstances."[25] Seabed mining operators were also to have been required to agree generally "to cooperate with the Authority in the acquisition by the Enterprise on fair and reasonable commercial terms and conditions of the technology necessary for the carrying out of its activities in the Area," and "to assist, if and when the Authority so requests, the Enterprise in obtaining on the free market efficient and useful technology"; the same obligations were to have been applied with respect to developing states to which the Authority granted a right to exploit selected reserved areas.[26] The amendments would also have substituted enforcement by states parties for the system of penalties set forth in the Convention.[27] Finally, the broad definition of seabed mining technology set forth in paragraph 8 of article 5 was deleted.

Although it is difficult to discover a complete and official statement of U.S. objections to the technology transfer provisions, various objections have been raised to the idea of technology transfer in principle. Thus, for example, it has been asserted that the technology transfer articles, despite their qualified character, "violate a basic principle that owners of technology have rights in its sale and use,"[28] and that mandatory transfer would "amount to a forced sale and could never be fair to the private contractors."[29] The ability of the United States to restrict the rights of U.S. citizens by treaty is not open to serious question, however.[30] Indeed, the power of compulsory appropriation of property is regarded throughout the world as an inherent power of the state, and the exercise of such power by an international organization to which it had been delegated would thus not expose international consortia to an unknown type of risk. "The strange English expression 'compulsory purchase' becomes 'expropriation' on the Continent of Europe, 'eminent domain' in North America, and 'resumption' in Australia, but the same basic problems have to be faced, it seems, in all the countries of the world."[31] In such states, the primary purpose of patent protection has been understood to be the public disclosure of inventions rather than the rewarding of inventors,[32] and abuse of monopoly has been considered grounds for patent revocation or compulsory licensing in Britain, Japan, and other advanced legal systems;[33] even under U.S. patent law, patentees enjoy exclusive property rights only to the extent such rights are expressly conferred by the relevant statutory and regulatory provisions.[34] Indeed, although most U.S. seabed mining technology has been patented,[35] the patent laws of the United States are generally unenforceable beyond U.S. jurisdiction in areas such as the high seas.[36]

With regard to the political and economic wisdom of technology transfer requirements in the context of deep-sea mining, it is significant that the 1970 Draft Treaty proposed by the United States would have authorized the Council to provide "technical assistance" to developing states upon request.[37] Former Secretary of State Henry Kissinger endorsed technology transfer to the Enterprise in 1975, acknowledging "that the time has come to put the technological and economic gains of mankind into the service of progress for all."[38] The following year, Kissinger reaffirmed that the United States was "prepared to make a major effort to enhance the skills and access of developing countries to advanced deep seabed mining technology in order to assist their capabilities in this field. For example, incentives should be established for private companies to participate in agreements to share technology and train personnel from developing countries."[39] Kissinger's proposal was offered as a negotiating "carrot" at a time when many developing states feared that the Enterprise would be deprived of the resources needed for it to conduct successful mining operations, and was viewed as an important concession in light of the traditionally conservative position of developed states on the issue of technology transfer;[40] the initiative was quickly adopted by EEC member states, and the need for technology transfer programs to assist developing states has been widely recognized, even by President Reagan and other outspoken free-market proponents.[41]

> The post–World War II experiences of Japan and, more recently, Hong Kong and Taiwan, have proven that if the necessary capital, technology and training become available, less-developed countries with lower labor costs and fewer governmental constraints can compete favorably with developed-country industries. At the same time, the [developing states] would be acquiring valuable marketing experience and making important international business contacts which would facilitate industrialization in other economic sectors.[42]

Indeed, technology transfer has been recognized as one of the most effective mechanisms for promoting economic development in developing states,[43] and could thus be expected to enable the Enterprise to function effectively. Sharing of resources has traditionally been favored under international law as a method of achieving an equitable distribution of wealth among states,[44] and since the end of World War II U.S. exports have been a principal source of advanced technologies throughout the world.[45] Domestic laws in both Europe and the United States are designed to prevent industrial monopolization, but such laws have done little to discourage the anti-competitive practices of multinational corporations within developing states.[46]

It has frequently been said that the Convention would require the transfer of technology largely controlled by the United States: "There is a deeply held view in our Congress that one of America's greatest assets is its capacity for innovation and invention and its ability to produce advanced technology. It is understandable, therefore, that a treaty would be unacceptable to many Americans if it required the United States, or more particularly, private companies to transfer that asset in a forced sale."[47] In

1973 33 firms in five states, including the United States, were actively developing seabed mining technology,[48] and by 1976 U.S. companies "reputedly" held a technological lead over other states.[49] Secretary Kissinger apparently took this technological lead into account, however, when he proposed technology transfer "so that the existing advantage of certain industrial states would be equalized over a period of time."[50] In 1982 the U.S. lead still was estimated to be 5 to 7 years over Japan and the United Kingdom, and 10 to 15 years over the Soviet Union.[51] By 1990, however, state-of-the-art seabed mining technology had become widely available — more than 100 firms throughout the world had actively undertaken the development of seabed mining technology — and a number of states had surpassed the United States.[52] Moreover, in the early 1980s the U.S. Bureau of Mines reported that there were "at least four manufacturers for every component of seabed-mining hardware," making it very unlikely that the Enterprise or developing states would need — or be able, in the absence of demonstrated need — to invoke the mandatory technology transfer provisions of the Convention with respect to mining equipment.[53] India, which began its deep-sea mining operations as the UNCLOS III negotiations were drawing to a close, has been able to obtain necessary seabed mining technology from several states, including Denmark, France, and the Federal Republic of Germany.[54]

Although it has been asserted that the acrimonious debate over these provisions during the Conference belies this argument,[55] it would seem that technology transfer is more properly viewed as a type of "safety net" for the Enterprise rather than as its primary means of acquiring seabed mining technology. Topography and other factors specific to each mine site will determine the particular mining technology necessary for each mining operation,[56] but once the Enterprise has commenced mining activities it is expected to be able to acquire needed technology on world markets.[57] The basic technology of seabed mining has already been developed, and research will continue into technological improvements, which can be expected to become more widely available as its developers seek to obtain the advantage of economies of scale.[58] Indeed, it is unclear what legitimate interest of the supplier (or its shareholders) would be served by its refusal to sell its technology to the operator for use by the Enterprise or developing states, particularly when it has an economic incentive to make such technology widely available to potential purchasers. Significantly, partnership agreements among members of the private multinational consortia provide for automatic transfer of technology from U.S. firms in the event of the consortia's participation in a multilateral seabed mining regime to which the United States is not a party.[59]

The Reagan administration argued that the technology transfer provisions were "commercially impractical," and that private industry would not proceed with mining operations under the terms of the Convention.[60] U.S. mining companies have complained that, in addition to losing their technological advantage, they would be unable to recover their full investment costs and unable to secure transfer authorization from the "thousands of consortium subcontractors" which own seabed mining technology.[61] Concern has also been expressed regarding the transfer of "know-how" to non-profit-seeking competitors[62] — a legitimate argument which, however, applies with equal force to subsidized competition in all industries. The role of private companies in international commodities markets has been decreasing, and they might be able to regain important footholds by exchanging their technology for assured access and clear title to the nodules. "It is difficult not to be sympathetic to those in society who take risks and achieve new technological breakthroughs which advance the interests of all mankind. Nevertheless some, even among the sympathetic, could argue that the rewards of corporate initiative could well, in this case, be sacrificed for the greater good."[63] Moreover, multinational corporations have, in recent years, become accustomed to the inclusion of technology transfer provisions in mining agreements with developing states, for example — in Latin America, "structural problems in various markets" have led host states to insist upon technology transfer as a condition of access.[64] It is thus unclear to what extent criticism by the seabed mining industry during the Conference may have been intended merely to enhance the U.S. bargaining position, particularly in light of acknowledgments by industry representatives that the Convention does not compare unfavorably with typical concession contracts in developing states.[65]

The United States had initially sought to limit the mandatory technology transfer provisions to a shorter period of time, but agreed to the ten-year requirement in exchange for the exclusion of marketing and processing technology from the transfer obligations.[66] Although the Reagan administration claimed that the third-party "written assurance" requirement would "cast grave doubt on the ability of miners to utilize vital technology owned by third parties,"[67] failure to honor the assurance could at most result in the supplier being barred from participation in future seabed mining operations but would not affect existing contractual rights of operators.[68] The obligation to make technology available to developing states — the Brazil clause — is likely to prove even less consequential, but is difficult to justify except as a form of involuntary foreign aid, and the United States sought unsuccessfully to eliminate the provision completely.[69] The "fair and reasonable commercial terms and conditions" standard should ensure that mining operators receive an adequate return on their investment, and would satisfy the constitutional requirement of just compensation.[70] The lack of any specific requirement for mandatory technology transfer in joint-venture agreements may be particularly significant, since joint ventures have become a preferred mechanism for effecting technology transfer, and since such agreements may supplant independent mining operations by the Enterprise.[71] Although the Convention's definition of technology may be somewhat expansive in comparison to traditional commercial formulations, its scope may not be inappropriate with respect to seabed mining technology: "The word 'technology,' normally used as convenient shorthand, conceals at least five important ingredients: (a) hardware, (b) operating procedures, (c) maintenance procedures, (d) operating and maintenance skills, and (e) management capacity. In some cases the hardware may be the least important ingredient while skills and management capacity may be the most important."[72] Because of the legal complexities associated with the mandatory technology transfer process, it is expected that the Authority would instead seek to facilitate the acquisition of needed technology through more informal, mutually agreeable arrangements.[73]

Treaty opponents have also cited "implications for our national security" in opposing the technology transfer provisions of the Convention, particularly in light of the determined efforts of the Reagan administration to stifle the flow of military technology to Warsaw Pact states during the 1980s.[74] Under a broad definition of national security such as was enforced by the Reagan administration, virtually all new technologies were regarded as having direct or indirect military application.[75] Even under these strict standards, however, only three types of exports were subject to control: nuclear technology, munitions, and "dual-use" items having both civilian and military applications.[76] In fact, even if seabed mining technology were properly classified as dual-use and even if Soviet access to such technology were still a genuine national security concern, article 302 would nonetheless exempt such technology from mandatory transfer.[77] Former Ambassador Elliot Richardson has thus described the claim that the Convention would require mandatory disclosure of "sensitive national-security-related technology" as one of the "most frequently repeated misstatements" concerning the Convention[78] — indeed, it may be inferred from the consensus support for article 302 at UNCLOS III that the professed concerns of the Reagan administration on this issue were not shared by U.S. allies on the Coordinating Committee for Multilateral Security Export Controls (COCOM) or even by the U.S. negotiating team.[79] Nor did national security appear to be a problem for the Soviet Union, which offered to convey its seabed mining technology to anyone on commercial terms.[80] "[T]he technology ... is relatively 'low tech' and more akin to assembling automobiles than making super computers."[81]

Revenue Sharing

In addition to revenue-sharing provisions relating to the continental margin beyond 200 miles, the Convention imposes, in article 13 of Annex

III, a three-part assessment upon seabed mining operators: an initial application fee, an annual fee, and an assessed portion of production
revenues. The Reagan administration criticized each of these assessments,
objecting, in particular, that such revenue-sharing provisions "would impose burdensome financial requirements," increasing the cost of nodule
mining and thereby placing seabed mining operators at a competitive
disadvantage vis-à-vis land-based producers.[82] As early as 1971, however,
it had been stated U.S. policy that the Convention "could and should make
a very important contribution to the economic and technological advancement of developing countries," and that the system of revenue sharing
should be "an integral part of an equitable settlement regarding rights and
responsibilities in the seabeds."[83] President Nixon confirmed this policy,
declaring that "the smaller and poorer nations of the world should be given
a fair share of the benefits from these resources, which are the common
heritage of mankind."[84] It has subsequently been argued that nodule mining
is a freedom of the high seas under customary international law, and that
consequently neither seabed mining states nor their nationals need share
any revenues; yet seabed mining states, including the United States, have
acknowledged an existing revenue-sharing obligation derived from the
principle of the common heritage of mankind, and the collection of
economic rent may help to ensure efficient exploitation of nodule resources
under a res communis regime.[85] The revenues are to be used to pay the expenses of the Authority — and, under the agreed-upon parallel system, of
the Enterprise, until it is able to contribute its own surplus earnings — as
well as to compensate developing producer states "which suffer serious
adverse effects on their export earnings or economies" as a result of deep-sea
mining operations.[86]

> There can be no doubt as to the propriety of collecting revenues for inter-
> national community purposes from the exploitation of the mineral
> resources of the international seabed area beyond the limits of national
> jurisdiction. At the same time, it is ... essential to keep these revenues
> within reasonable bounds so as to encourage, and not destroy, the incen-
> tive to develop these resources for the general benefit of the world com-
> munity.[87]

The decision to share revenues was thus both a political and economic one,
made in the belief that the macroeconomic interests of the international
community should be balanced against considerations of economic efficiency and other factors, including the risk of significant economic and
social dislocation within some developing states.[88]

    The Convention requires each seabed mining applicant to pay a
$500,000 administrative fee, refundable to the extent that actual processing
costs incurred by the Authority are lower[89] — a nominal amount to which

the Reagan administration voiced no objections.[90] Following approval of a plan of work, the contractor would be required to pay "an annual fixed fee" of $1 million, payment of which would be waived when other revenue-sharing payments exceed that amount.[91] "Obviously where production cannot be expected to commence more or less immediately, unrealistically high rental fees . . . could discourage potential investors"[92]; yet the amount is small in relation to the $10 billion in revenues each mining site is expected to produce during its lifetime, and imposition of minimum annual fees is necessary to discourage the "banking" of mining sites by speculators.[93] Federal regulations governing mineral leasing on the U.S. continental shelf also require payment of an annual rental fee, and the Green Book proposed no substantive changes in the annual fee, which the Preparatory Commission has waived for pioneer registrants.[94]

Article 13 provides that, "within a year of the date of commencement of commercial production," each nodule mining operator must choose between two alternative formulas for determining revenue assessments: a simple production charge or "a combination of a production charge and a share of net proceeds."[95] The production charge formula, intended for use by state-run mining operations such as those of the Soviet Union – and, apparently, France, Japan, and India – is fixed at a percentage of the market value of processed minerals produced from nodule mining operations: 5 percent during the first ten years of commercial production and 12 percent thereafter.[96] The production charge portion of the combination formula is similarly divided into two periods, but the payment percentage is lower: 2 percent during the first period, which ends "in the accounting year in which the contractor's development costs with interest on the unrecovered portion thereof are fully recovered by his cash surplus," and 4 percent in the second period thereafter except in years where the contractor's return on investment after payment of the 4 percent charge would fall below 15 percent[97] – a rate of return regarded as necessary by the U.S. seabed mining industry.[98] The same two-period division applies also to the profit-sharing revenues ("share of net proceeds"), which are calculated by application of a three-tiered progressive rate schedule to the contractor's net proceeds "attributable" to seabed mining operations – a calculation in which the capital investment ("development") costs of mining operations are amortized.[99] In the first period, the Authority's share is 35 percent if the operator's attributable net proceeds represent a return on investment of less than 10 percent, 42.5 percent if attributable net proceeds represent a return on investment of between 10 and 20 percent, and 50 percent if attributable net proceeds represent a 20 percent or greater return on investment. During the second period, the Authority's share in the same three return-on-investment brackets is 40 percent, 50 percent, and 70 percent, respectively.[100] However, because "attributable net proceeds" is based upon profits derived only from mining activities, which are expected to be calculated at 25 percent of total profits, the effective profit-sharing rates payable by most private contractors would actually be one fourth of these percentage amounts.[101] "Viewed in absolute terms, [the Authority's] income varies greatly with the profitability of the project, but in relative terms, the Authority always receives a 15 percent to 20 percent share of an integrated project's income."[102] Resort to binding commercial arbitration is specifically provided for "[i]n the event of a dispute between the Authority and a contractor over the interpretation or application of financial terms of a contract."[103]

In deriving the above revenue-sharing formulas from among the dozens of quantitative and qualitative variations proposed at UNCLOS III, Committee I delegates examined the historical development of mineral production agreements between multinational mining firms and developing states for guidance in developing a model for effective cooperation in the private development of publicly controlled resources.[104] Under the traditional concession contract, the mining operator paid consideration to the host state in exchange for title to resources recovered during the term of the contract. In recent decades the standard agreement has been modified to provide host states with greater benefits in exchange for their mineral resources:

> an increase in the fees paid by the concessionnaire upon signature of the contract, upon the discovery of commercially exploitable deposits, and in respect of production and export; an increase in the minimum amount to be spent on exploration; an increase in rentals and royalties; assurances of additional benefits, such as an undertaking by the contracting enterprise to finance industrial projects...; speedier return of areas explored, purchase of local goods and services, recruitment and training of national personnel, and so on.[105]

Profit-sharing agreements became common—host states usually take 50 percent of the mining operator's net revenues in exchange for a commensurate share of the risk—and host-state governments have now assumed more active roles in project management.[106] A 1957 agreement between Azienda Generale Italiana Petroli (AGIP) and the National Iranian Oil Company (NIOC), "a model for many subsequent contracts," provided NIOC, a company owned and operated by the host government of Iran, with 75 percent of net profits.[107] "The magnitude and nature of payments made under the convention compare favorably with those made in many developing countries by land-based mining operations, but, unlike many such terms, ... rates are not in practice renegotiable upwards."[108]

At UNCLOS III, developed states persuaded developing states to write the revenue-sharing provisions into the text of the Convention rather than allow the Authority to prescribe such terms in contractual negotiations or otherwise at its discretion.[109] Although the Conference did not achieve a revenue-sharing formula completely satisfactory to the United States and other developed states, these provisions do accommodate their expressed concerns.[110] "The developed countries seemed to expect a modest level of profits; accordingly, the high rates that apply for bonanza projects were not much of a concession. Many developing countries professed to expect high profits; agreeing to low rates of sharing for normal profit levels was no great concession."[111] Such "taxes" on net proceeds provide financial flexibility where some form of revenue sharing is desirable for high-risk ventures, but are difficult to calculate; simple production charges, on the other hand, are easier to administer but provide less flexibility in response to changing economic conditions.[112] Although production charges are regarded as necessary to prevent speculative practices by contractors, such charges could cause mining sites to become prematurely unprofitable if set at too high a level.[113] Because the profitability of seabed mining — at least initially — remains uncertain even under favorable economic conditions, the combination formula set forth in the Convention would impose relatively low production payments and relatively high percentage charges on net earnings.[114] The United States was successful in seeking to base the calculation of attributable net proceeds only upon mining operations rather than upon transportation or processing activities — a valuation basis which substantially reduced the amount of revenues payable to the Authority — although it had sought a minimum attribution ratio of 20 rather than 25 percent of a vertically integrated seabed mining operator's total profits.[115] The Convention includes a grace period on payments during the prospecting and exploration phases, and, by allowing interest deductions in the computation of operating costs, subsidizes the capital costs of seabed mining operators.[116] Although, after an initial grace period of up to ten years, the Enterprise is required to make payments to the Authority equivalent to those required of seabed mining operators, the United States favored complete equality of revenue-sharing obligations.[117]

Economists have suggested an optimal revenue-sharing percentage for the Authority of between 10 and 25 percent.[118] The 1970 U.S. Draft Treaty proposed that the Authority collect "substantial mineral royalties" of between 2 and 20 percent of production value,[119] and at UNCLOS III the United States supported production charges of 2 to 4 percent as part of the revenue-sharing formula.[120] A study by the Massachusetts Institute of Technology, begun in 1976 and completed in 1978, concluded that under the Convention seabed mining operators would be able to achieve a net return on investment of more than 18 percent, but the costs of seabed mining technology have risen substantially since then while mineral markets have softened, and it is now difficult to predict when seabed exploitation will become sufficiently profitable to justify further private investment and produce income for revenue sharing.[121] Nonetheless, it remains significant that the Outer Continental Shelf Lands Act of 1953, which authorizes tract leasing under a system of competitive bonus bidding and royalty payments, has long provided a framework for profitable U.S. offshore oil drilling operations.[122] Indeed, despite its public criticism of the revenue-sharing formula, the Reagan administration sought no substantive changes to article 13 in 1982.[123] U.S. seabed mining firms have, in fact, admitted that they would likely be able to operate profitably under the financial provisions of the Convention when demand for nodule minerals improves, particularly if they were given tax credits under domestic law.[124] "All that is necessary is to give revenue-sharing payments to the Authority the same treatment as taxes paid to a foreign government. Indeed, this has all along been advocated by the State Department."[125]

Production Controls

Article 150 of the Convention provides that mining activities in the Area are to:

> be carried out in such a manner as to foster healthy development of the world economy and balanced growth of international trade, ... with a view to ensuring: (a) the development of the resources of the Area; ... (f) the promotion of just and stable prices remunerative to producers and fair to consumers..., and of long-term equilibrium between supply and demand;... (h) the protection of developing countries from adverse effects on their economies or on their export earnings ... caused by activities in the Area.[126]

To implement subparagraph (h), the Authority may become a party to international commodity agreements restricting seabed mining production when "necessary to promote the growth, efficiency and stability of markets," but only if "all interested parties, including both producers and consumers, participate" in the negotiations.[127] Any such agreement would be "without prejudice to" the other policy objectives specified in article 150 — including development of deep-sea mineral resources and "increased availability of the minerals derived from the Area ... to ensure supplies to consumers" — and its application with respect to seabed mining operators would have to be uniform, nondiscriminatory, and "consistent with the terms of existing contracts."[128]

Before proceeding with nodule exploitation, each seabed mining applicant must prepare and submit an application for a production authorization, setting forth "the annual quantity of nickel expected to be recovered under the approved plan of work."[129] The Authority "shall issue a production authorization for the level of production applied for unless the sum of that level and the levels already authorized exceeds the nickel production ceiling, as calculated . . . in the year of issuance of the authorization, during any year of planned production falling within the interim period."[130] The amount of nickel recovery which the Authority may so authorize — essentially the full amount of projected growth in global nickel consumption during a five-year period ending one year prior to commencement of commercial production, plus 60 percent of projected growth in consumption thereafter — is to be derived by a complex logarithmic formula, but would, at a minimum, be set at the lesser of the amount yielded by a 3 percent annual growth in world nickel consumption or the full amount of growth in consumption.[131] During the "interim period," which could be expected to span the first 20 years of commercial production, each contractor may produce no more than the authorized quantity of nickel, although the annual production limit may be exceeded by as much as 8 percent in order to compensate for shortfalls during prior years.[132] Supplementary production authorizations may be made available, but "only after all pending applications by operators who have not yet received production authorizations have been acted upon and after due account has been taken of other likely applicants," and no seabed mining operator may be authorized to produce more than 46,500 metric tons of nickel annually.[133]

If the combined amount of production sought by competing applicants in any four-month application period would exceed the ceiling, authorizations are to be allocated by a three-fourths vote of the Council pursuant to criteria established in the rules, regulations, and procedures of the Authority, and unsuccessful applicants "shall have priority in subsequent periods until they receive a production authorization."[134] Except for two authorizations which are reserved for the Enterprise, pioneer registrants would enjoy priority with respect to all other applicants and would be able to apportion the available authorizations among themselves in the event the production ceiling became applicable.[135] The Enterprise is subject to the same production controls as private operators, but its applications — and those of developing states — for production authorizations for mining in reserved areas would be accorded priority vis-à-vis nonpioneer applicants "[w]henever fewer reserved areas than non-reserved areas are under exploitation."[136] Production levels established in the Authority's rules, regulations, and procedures for nodule minerals other than nickel "should not be higher than those which would have been produced had the operator produced the maximum level of nickel."[137] The Authority is to establish performance requirements, including limitations upon the duration of the exploration period,[138] and may limit production of non-nodule resources in the Area "under such conditions and applying such methods as may be appropriate."[139] The size of the mining areas is to be established in the rules, regulations, and procedures of the Authority, "taking into account the state of the art of technology then available for sea-bed mining and the relevant physical characteristics of the areas."[140]

President Reagan required that the treaty "not deter development of any deep seabed mineral resources to meet national and world demand,"[141] and objected to the production controls, which "would impose artificial limitations on seabed mineral production."[142] Economists had argued that such controls would harm consumers in both developed and developing states by preventing lower prices for nodule minerals, and, worse, "could discourage potential investors, thereby creating artificial scarcities."[143] The 1982 Green Book would have replaced the production ceiling with "programs for adjustment assistance" — essentially a compensation fund — implemented during a 15-year period beginning with approval of the first plan of work, and would have permitted the Authority to participate in international commodity agreements only for the purpose of imposing production controls upon the Enterprise;[144] the April 13 set of proposed amendments also would have rendered such agreements applicable only to the Enterprise, but would have retained the system of production controls with a higher trend line valuation method and with production authorizations issued on a first-come, first-served basis.[145]

Land-based sources of nickel consist of both sulfide and laterite deposits.[146] Although 60 percent of current production is taken from sulfide mines, more than two thirds of world reserves are contained in laterite deposits, most of which are located within developing states.[147] It was expected that the Convention's system of production controls would benefit Canada, France, and the Soviet Union — which together have provided three fourths of world nickel production — but developing-state production is expected to increase dramatically in coming years as sulfide deposits are depleted.[148] "Many new laterite projects are located in countries which need a regular inflow of hard currency.... In addition, they have high debt loads, and are dependent upon economies of large scale to partially overcome high energy cost disadvantages."[149] Because production from laterite deposits is more costly than sulfide mining or even deep-sea nodule mining, the lower nickel prices which are expected to result from nodule mining operations may have a particularly severe impact upon developing producer states.[150] A single 3-million-ton-per-year nodule mining operation yielding 42,000 tons of nickel each year would produce approximately 5 percent of world consumption.[151] It is thus expected that simultaneous operation of several seabed mining projects would exert significant downward pressure upon nickel prices — particularly since potential nodule reserves outnumber land-based reserves by a ratio of more than five to one[152] — and that revenues in several developing producer states would be substantially impaired as a result.

The world copper market is larger than that of nickel, and potential nodule reserves amount to only one half of presently existing land-based reserves.[153]

> [T]he two largest consumers of copper (United States and USSR) are also the major producers and are, to a large extent, self-sufficient in supply, whereas, the other two major industrialized groups (European Economic Community and Japan) are almost entirely dependent on imported copper. This demand, and also the shortfall in the other industrialized groups, is met by imports from Canada, Australia and South Africa but mainly from several developing countries who are major producers and exporters of copper. These are Chile and Peru in the American region, Zambia and Zaire in Africa and the Philippines and Papua New Guinea in the Pacific area.[154]

Developing states produced only 42 percent of world copper production in 1972, but have produced more than half of total copper exports, and could be expected to experience serious balance-of-payments difficulties as a consequence of successful seabed mining operations.[155] Zaire, with approximately 40 million tons of copper reserves, and Zambia, in which copper accounts for 93 percent of total exports,[156] would be significantly affected. Because of their industrial limitations, developing producer states have been unable to use their mineral resources to manufacture consumer goods, relying instead on raw material exports.[157] Although the 37,000 tons of copper expected to be produced by a 3-million-ton-per-year seabed mining operation would amount to less than 0.5 percent of total world production and the copper market is much larger than that for nickel, the impact of several seabed mining operations is not likely to be "clearly insignificant" as the United States has predicted.[158] The effect upon certain developing producer states would be particularly severe.

Because potential nodule reserves of cobalt are roughly 40 times as large as land-based reserves, even seabed mining on a small scale would be expected to lower its price to the point where it became substitutable for nickel in the production of steel alloys.[159] A single 3-million-ton-per-year seabed mining operation is expected to produce 3,500 tons of cobalt each year — more than 10 percent of the 30,000 tons consumed throughout the world in 1980.[160] It has been predicted that cobalt production could account for more than 30 percent of revenues from seabed mining operations, and that a 50 percent increase in world cobalt production might result in a two-thirds drop in market price, "as supply is relatively inelastic with respect to price."[161] It has further been claimed that, because cobalt is produced as a by-product of nickel and copper, revenue increases from production of the latter two metals would offset any losses accruing to cobalt production in developing producer states; yet analysis of the nickel and copper markets belies this argument, as do recent structural weaknesses in these markets.[162] Production of cobalt from seabed mining would affect Australia, Morocco, Cuba, New Caledonia, Zambia, and Zaire, and the economy of the last of these states could be severely damaged.[163] The percentage of world cobalt production provided by developing producer states has been increasing, as has its importance as an "export earner."[164] The impact of seabed mining would be even more severe if, as expected, exploitation of cobalt-rich manganese crusts were undertaken within national exclusive economic zones.[165]

A 3-million-ton-per-year seabed mining operation could produce 750,000 tons of manganese, approximately 8 percent of annual worldwide production.[166] Because the manganese contained in nodules is of relatively poor quality, however, the impact of seabed mining upon world manganese markets may be limited.[167] Manganese derived from seabed nodules may have long-term applications in steel production, but its more immediate uses in pure form are limited to about 3 percent of total manganese consumption, and cost-competitive production of manganese from nodule mining could thus be expected to more than displace land-based sources of pure manganese.[168] The major producers of manganese have included South Africa, the Soviet Union, Gabon, India, and Australia; Brazil, China, Ghana, Morocco, and other states would also face varying degrees of economic dislocation if there were substantial seabed production of the mineral.[169] The United States has argued that the improbability of manganese production makes any production controls unnecessary,[170] but potential nodule reserves of manganese exceed land-based reserves, and it evidently seemed prudent to most UNCLOS III delegations that such controls be maintained as a "safety net" in the event that technological progress were to make manganese recovery profitable for seabed mining operators.[171]

Although it is impossible to estimate accurately the impact of deep-sea mining operations upon developing producer states before commercial recovery of nodules actually begins, seabed mining is expected to become increasingly attractive as an alternative to land-based mineral sources which would otherwise remain profitable.[172] Although the first few seabed mining operations are expected to be relatively inefficient, "the future costs of extracting certain ocean minerals will fall significantly to levels below current costs for comparable mineral extraction on land,"[173] and land-based producers will probably be forced to make substantial economic adjustments during the second and third decades of seabed mining.[174] The United States has correctly asserted that the production controls would "cause prices to be higher than they would otherwise be,"[175] and that justification for such controls is not to be found in classical economic theory.[176] More questionable, however, is the U.S. contention that revenue losses in developing producer states would be more than offset by lower consumer prices for goods produced from nodule minerals.[177] Assuming that lower commodity prices were indeed passed on to consumers at the retail level, it is likely that the primary beneficiaries of such price reductions would be citizens of developed rather than developing states.[178] Studies of the subject have generally concluded that the aggregate benefit to developing states of unrestricted seabed mining is likely to be more than offset by the amount of economic damage to developing producer states.[179]

Although it has been pointed out that developing producer states "cannot realistically expect indefinitely to avoid all effects on their economies or export earnings" from seabed mineral exploitation,[180] unrestricted seabed mining could cause economic damage not only by lowering prices but also by diverting scarce capital away from land-based mining operations.[181] The impact would be most severe upon their balance of payments — "they typically depend more heavily on the minerals concerned for their export earnings and government revenues than do developed producing countries"[182] — and the aggregate cost to such states has been estimated at $200 million annually for each of the four nodule minerals.[183] The United States has claimed that only six developing states would benefit significantly from production controls,[184] but it was claimed by other states that more than thirty states would be protected by the controls.[185] The United States also predicted, incorrectly in retrospect, that the income of land-based producers would continue to increase "significantly" whether or not nodule minerals were introduced into world markets.[186] "In brief, it would appear that in the absence of special arrangements to protect the interests of developing countries, the availability of minerals from the sea-bed, while contributing to world economic development, may also result in a widening of the income gap between developed and developing countries."[187]

The system of production controls may also be justified — perhaps more convincingly — on economic grounds as a conservation measure analogous to the quotas and regulations established by the United States pursuant to international fisheries treaties and domestic offshore oil leasing programs; this economic need to regulate the rate of exploitation of exhaustible marine resources has been increasingly recognized since the end of the 1960s.[188] Indeed, such regulation may be necessary to avoid a free-market failure, in which unrestricted seabed mining activities would produce overcapitalization and market volatility.[189]

> One of the interesting things about the debate on the law of the sea is that those who oppose artificial restraints on market development of deep seabed resources have paid little attention to the fact that a primary motive for extending coastal state jurisdiction over the much more valuable resources of the economic zone and the continental shelf is precisely to impose artificial restraints.[190]

It has been argued that the high investment costs of deep-sea mining opera-tions would prevent overcapitalization,[191] yet the substantial impact of nodule mining upon mineral markets and the uncertain profitability of such operations raise the possibility that commercial production by the pioneer investors alone could result in overexploitation, further depressing mineral prices and forcing unsubsidized private consortia to write off their billion-dollar investments — indeed, the seabed mining industry may already have become overcapitalized.[192] The idea of mandatory conservation of marine mineral resources may nevertheless be expected to remain unpopular with "free enterprise" theorists, particularly within the Congress.[193]

At UNCLOS III, where the majority of delegations represented develop-ing states, the need for production controls was nonetheless regarded as "a question having to do as much with equity as with resource conserva-tion,"[194] and the original development of U.S. policy toward such controls was based upon equitable considerations. Thus, the United States stated as early as 1968 that "means must be available to prevent the economic dislocations which could result from the exploration and exploitation of the mineral resources of the sea,"[195] and this principle has been generally ac-cepted by the international community since that time.[196] In 1974 the United States reaffirmed its position, with the qualification that "consumers of goods made from raw materials found in the sea-bed should be protected from artificial price increases for such materials," and two years later Secretary of State Kissinger announced that the United States was "pre-pared to accept a temporary limitation, for a period fixed in the treaty, on production of the seabed minerals tied to the projected growth in the world nickel market."[197] The United States has thus effectively agreed that the potential economic impact of seabed mining should be regarded as a signifi-cant externality, the true cost of which would be best internalized by con-trolling the aggregate rate of nodule production. "Although the benefits of unrestricted production to the United States may be substantial, the political costs, due both to the disruption of world commodity prices and to the substantial decrease in the export revenues of land based producers, would also be significant."[198]

The impact of nodule mining upon world markets was "the over-whelming preoccupation of developing countries" from the beginning of the UNCLOS III negotiations,[199] and the Group of 77 initially had sought to give the Authority complete discretion with respect to the establishment of production controls.[200] Faced with a united developing-state bloc, the developed states were unable to achieve a common negotiating position on the issue; land-based producer states generally sided with the Group of 77, while states which import nodule minerals — including the United States and the EEC — were initially opposed to production controls.[201] Even developed consumer states acknowledged the potentially disruptive effect which deep-sea nodule mining might have upon producer-state markets, however, and Kissinger's 1976 compromise proposal was well received by UNCLOS III delegates.[202] In order to achieve a universally acceptable treaty, the United States and other consumer states thus sought to accommodate the interests of present and near-future producer states, whose support for the Convention would have been withheld in the absence of measures to ensure some degree of protection for their domestic mining industries.[203] Canada, itself a sponsor of seabed mining operations, strongly supported production controls and was an important contributor to the formula set forth in the Convention.[204]

> The main point made by Canada and other landbased producer states [was] that the formula . . . which allocated five-year cumulative growth to seabed mining at the outset, added by 60 per cent of further cumulative growth, would continue to be more than 100 per cent of actual growth in world nickel consumption for a period of ten years or so from the commencement of commercial production.[205]

It seems clear that these production controls, if they ever become applicable, could impose economic costs upon developing states – as well as upon the world economy as a whole – in the form of lost consumer surplus,[206] but it is not clear that these costs, which might be regarded as analogous to premiums on an insurance policy, ought to be avoided at the risk of severe economic dislocation to some developing states and to world mineral markets. The allocation of this risk is a decision for the international community as a whole, to be taken in the appropriate diplomatic forum – it is not for the United States to determine unilaterally that, as President Reagan has asserted, unrestricted seabed mining "should serve the interest of all countries."[207] Most other states have evidently concluded that their interests are best protected by the production controls, and have tended to regard U.S. arguments in support of "free-enterprise" economics as self-serving.[208]

From the end of World War II to the mid-1970s, annual growth in world nickel consumption exceeded 6 percent, but by the beginning of the 1980s the growth rate had fallen to 3.9 percent.[209] In 1980 the U.S. Bureau of Mines forecast a 3.4 percent average annual rate of growth for nickel consumption to the end of the century – a growth rate sufficient to permit the issuance initially of production authorizations to five or more 3-million-ton-per-year operators, with an additional two to three authorizations becoming available every four years thereafter[210] – and the State Department concluded "that the available production authorization quantities should be adequate to meet the needs of investors at the outset and for some time thereafter, although it is conceivable that companies might encounter delays in obtaining authorization in the middle and later years of the interim period if seabed mining turn [sic] out to be significantly more efficient than current projections indicate."[211] Application of the 3 percent minimum growth rate would permit the issuance of four or five production authorizations for 3-million-ton-per-year mining operations as soon as increased mineral consumption made nodule mining sufficiently profitable; although one would be claimed by the Enterprise, additional authorizations would become available biennially.[212] In 1987 the World Bank predicted annual growth of 1.2 percent in nickel demand during the 1990s — a rate too low to sustain profitable seabed mining operations, although the production ceiling would permit nodule production to the full extent of such growth — but even under such unfavorable market conditions all pioneer registrants desiring to do so would be able to obtain access initially through apportionment of the available authorization amount.[213]

> Notwithstanding the share of production taken up by the Enterprise, . . . there would still be sufficient tonnage under any reasonable set of circumstances to insure that private miners would get their authorizations when they need them. It is thus probable that market forces, not the production limitation formula, will determine how much nickel and, therefore, how much copper, cobalt, and manganese, will be produced by the first generation of seabed mining projects.[214]

Moreover, it now appears unlikely, for technical reasons, that operators would be able to sustain nodule production at 3 million tons per year, and by 1989 at least one had expressed an intention to recover only 1.5 million tons annually[215] — an amount which could double the number of authorizations available. Furthermore, as the United States has pointed out, seabed mining should lower the market price of nodule minerals,[216] and the elasticity of demand indicates that such price reductions would increase worldwide consumption, in turn raising the production ceiling and permitting a larger number of nodule mining operations in the future.

The Reagan administration argued that production authorizations would be allocated pursuant to "discretionary and discriminatory decisions by the Authority."[217] In fact, such allocations, when necessary, must be made "on the basis of objective and non-discriminatory standards," must be approved by three fourths of the Council, and must conform to the criteria specified in Annex III of the Convention, and disputes concerning conformity with these criteria, as with other aspects of the issuance of production authorizations, may be submitted to the Sea-Bed Disputes Chamber.[218] It has been reported that U.S. negotiators deliberately sought to obstruct progress at the final session of UNCLOS III by uncompromisingly seeking complete elimination of the production ceiling, "even though the ceilings actually negotiated appear to be so liberal as to have no practical effect."[219] To some extent, the market price of nickel would act — and already has — as "an automatic built-in stabilizer" for the seabed mining industry, even in the absence of formal production controls, and the principal effect of the production ceiling will thus be to protect both land-based producers and seabed mining operators from the possible effects of unwarranted overexploitation during the first 20 years of nodule production.[220] The system of production controls set forth in the Convention is somewhat inflexible;[221] yet seabed mining operators would be assured of annual increases in authorized production, and the 60 percent share of market growth represents a realistic political compromise.[222] It has been claimed that the performance requirements to be established by the Authority would cause "economic inefficiencies,"[223] but such requirements are commonly applied in laws governing mining operations, including domestic U.S. legislation, and Western states would be able to block the adoption of unfair and unworkable production standards in the Council.[224] Although the treaty text on commodity agreements is broader than that proposed by the United States in 1976, Secretary Kissinger did indicate at that time that the United States was "prepared to examine with flexibility the details of arrangements concerning the relationships between the Authority and any eventual commodity agreements,"[225] and the requirement that all interested consumer states participate in such agreements may be expected to prevent the formation of producer cartels.[226] Compensatory payments to developing producer states would be made only to developing states severely affected by seabed mining operations.[227] Although the Authority has discretionary power to establish production ceilings for non-nodule resources within the international Area, U.S. consent would be required for the enactment of such regulations by the Council,[228] and the Convention provides for the application of no production controls with respect to seabed mining operations within offshore zones of national resource jurisdiction.[229]

Supranationality

Supranationality — an elusive concept, the precise definition of which is "a matter of approach and interpretation" — implies a relinquishment of some measure of national sovereignty to an international organization.[230] One of the more meaningful criteria of genuine supranationality "is the direct binding effect of law emanating from the organization on natural and legal persons in the member states, i.e., a binding effect without implementation by national . . . organs"[231]; where the effect is binding upon member-state governments, but not upon their nationals directly, the power may be termed "quasi-supranational."[232] Opponents of supranationalism, seemingly imbued with a lingering ideological reverence for the familiar concept of national sovereignty, have been particularly critical of the Convention, asserting inaccurately, for example, that it confers "sweeping authority over almost all activities on the seas" upon "a large, complicated, highly organized, unelected, powerful government, with abundant funding that cannot be controlled or reduced or cut off."[233] It has been claimed that the Convention endows the Authority with supranational powers, including, according to the Reagan administration, "discretion to interfere unreasonably with the conduct of mining operations, and [to] impose potentially burdensome regulations on an infant industry."[234] According to the 1980 Seabed Act, U.S. evaluation of the Convention should include consideration of "the practical implications for the security of investments of any discretionary powers granted" to the Authority.[235]

Although the official records of UNCLOS III are rather sparse, and cannot provide a complete picture of the views of participating states, the records of the Conference and of the U.N. debates may be examined for evidence of intent with respect to supranationality.[236] In his 1967 speech before the First Committee of the General Assembly, Arvid Pardo called for creation of an "agency . . . endowed with wide powers to regulate, supervise and control all activities on or under the oceans and the ocean floor."[237] Pardo has stated that he was not seeking to create a supranational organization,[238] but in the ensuing debate several states endorsed the idea of an organization with strong regulatory powers,[239] and some delegations openly discussed the supranational implications of the Maltese initiative.[240] In 1968 the Netherlands formally proposed that the General Assembly be granted a quasi-supranational rule-making competence with respect to deep seabed resources, but the idea generated little support among other Western states,[241] and the Soviet Union expressed adamant opposition to "the establishment, with respect to the sea-bed, of an international regime of common ownership by the whole of mankind, ownership which would be managed by some sort of supranational body either within or outside the United Nations."[242] Developing states generally continued to seek stronger international regulation while downplaying the issue of supranationality,[243] and support for an organization competent to exercise regulatory control with respect to virtually all aspects of nodule mining operations became widespread during the Seabed Committee negotiations.[244] Despite strong rhetoric in favor of a comprehensive seabed mining regime, there were few specific proposals for an international enforcement competence;[245] Malta's 1971 draft ocean space treaty did provide for a relatively high degree of supranational control, but there was virtually no support for machinery with such broad jurisdiction.[246] By 1974, however, states had refined their positions on the issue of supranationality, generally with a view to granting broader powers to the new Authority,[247] and by 1975 even developed states supported an Authority with a general supervisory and rule-making competence.[248]

The Convention itself contains numerous compromises on the issue of supranationality. The Authority is granted those powers and functions "expressly conferred upon it," together with "such incidental powers, consistent with this Convention, as are implicit in and necessary for the performance of those powers and functions with respect to activities in the Area."[249] Because "activities in the Area" has been defined as "all activities of exploration for, and exploitation of, the [mineral] resources of the Area," any supranational or quasi-supranational powers of the Authority are applicable only with respect to a contractor's at-sea mining — as opposed to research, transportation, and processing — operations.[250] Moreover, the Authority, rather than being endowed with the independence characteristic of truly supranational institutions, is defined in the treaty as "the organization through which States Parties shall . . . organize and control activities in the Area."[251] The requirement that mining operators be sponsored by states parties imposes a further buffer against full and direct control by the Authority.[252] Contrary to criticisms voiced by opponents of the Convention, it is also clear that the Enterprise itself cannot properly be said to possess a supranational or even quasi-supranational competence.[253]

Although objections have been raised with respect to the exercise of "discretionary" powers by the Authority,[254] it is not entirely clear which of its powers may be properly so characterized.[255] There is no discretion in the granting of seabed mining contracts,[256] and the Authority is required to "avoid discrimination in the exercise of its powers and functions"[257]; moreover, the rules, regulations, and procedures of the Authority may only be applied prospectively, with no effect upon the rights and obligations of operators whose approved plans of work are already in effect.[258] At the insistence of Western states, the Authority's regulatory competence has been carefully defined by the terms of the Convention; it has been granted no direct power of taxation, and adoption of any additional production controls, while possible, would require the consent of the United States.[259] While successfully arguing that "investors must know whether and how the terms of an arrangement may be changed," private mining interests have acknowledged the need for regulatory flexibility in certain areas such as safety and environmental protection: "Wide power to regulate mining is functionally unnecessary if specific agreement is reached elsewhere on the questions of resource policy, technology transfer, revenue sharing and terms of admission."[260] The "unfriendly administrator" may thus be less of a potential problem under the Convention than it is under the Seabed Act, which permits the U.S. regulatory regime to be modified "at any time . . . as the Administrator determines to be necessary and appropriate."[261] Indeed, the 1970 Draft Treaty proposed by the United States would have provided the Authority with a substantial supranational regulatory competence with respect to such matters as qualifications of applicants and levels of revenue sharing.[262] Article 150, a subject of some criticism, merely sets forth a list of policy objectives without granting the Authority any additional power to implement them.[263]

To the extent that the Authority is empowered to promulgate rules, regulations, and procedures directly binding upon state-sponsored mining operators who become contractors, it may be said to possess a supranational legislative competence.[264] One of the most important aspects of the Authority's regulatory mandate is its ability to issue emergency orders to prevent serious environmental damage from nodule mining operations, a power reinforced by article 145: "Necessary measures shall be taken . . . with respect to activities in the Area to ensure effective protection for the marine environment from harmful effects which may arise from such activities."[265] The Authority may also take "necessary measures . . . to ensure effective protection of human life," as well as to regulate the stationing of seabed mining installations.[266] To permit regulatory flexibility, article 17 of Annex III provides a list of aspects of seabed mining operations which, "inter alia," are to be the object of the Authority's rules, regulations, and procedures, including inspection and supervision, avoidance of interference with other ocean activities, and enactment of mining standards and practices.[267] Such regulatory authority would not be unfamiliar to private U.S. seabed mining consortia,[268] and federal administrators have exercised a corresponding regulatory competence with respect to oil and gas drilling operations on the U.S. continental shelf.[269] The convention provides that the "rules, regulations and procedures drafted by the Preparatory Commission shall apply provisionally pending their formal adoption by the Authority" — an article which the United States insisted upon as an essential element of preparatory investment protection.[270] The Authority may also adopt rules, regulations, and procedures to promote conservation of seabed nodules,[271] as well as, despite U.S. objections, measures "adopted within three years from the date of a request to the Authority by any of its members" for the recovery of non-nodule deep-sea minerals.[272]

Although U.S. mining companies favored granting the Authority direct regulatory control only with respect to states parties — which in turn would have retained exclusive enforcement jurisdiction with respect to their national mining operations[273] — the Convention contains few instances of any such quasi-supranational legislative competence, except, of course, where the contracting seabed mining operator is itself a sovereign state (in which case it would be subject to the rules, regulations, and procedures of the Authority in the same manner as are private mining operators). In order to provide initial funding for the Authority's administrative budget, the Assembly, by a two-thirds majority vote, may assess states parties in proportion to their assessed U.N. contributions "until the Authority shall have sufficient income from other sources to meet its administrative expenses."[274]

Instances of quasi-supranational executive power consist mainly of sanctions which may be applied where a state fails to meet its responsibilities under Part XI of the treaty. A state party which is two years in arrears in the payment of its financial obligations to the Authority will lose its voting power unless "failure to pay is due to conditions beyond the control of the member."[275] Upon a finding by the Sea-Bed Disputes Chamber that a state party has "grossly and persistently violated" the seabed provisions of Part XI, and upon the recommendation of the Council, the state "may be suspended from the exercise of the rights and privileges of membership" by a two-thirds vote of the Assembly.[276] A sponsoring state may be held liable for damage resulting from its failure adequately to ensure that seabed mining activities carried out by it or its nationals comply with the Convention and the rules, regulations, and procedures of the Authority, unless such a state "has adopted laws and regulations and taken administrative measures which are, within the framework of its legal system, reasonably appropriate for securing compliance by persons under its jurisdiction."[277] Contractors and the Authority may also be held liable for compensatory damages resulting from their wrongful acts.[278]

The Authority possesses some elements of supranational executive power, particularly with respect to enforcement of Part XI and its rules, regulations, and procedures. Although sponsoring states retain primary responsibility for ensuring compliance by their nationals, the Authority exercises limited "control" over seabed mining activities and may take measures under Part XI "to ensure compliance with its provisions and the exercise of the functions of control and regulation assigned to it thereunder."[279] Such measures might include suspension or termination of mining rights where a contractor fails to comply with a binding decision rendered pursuant to the dispute settlement procedures, or where, "in spite of warnings by the Authority, the contractor has conducted his activities in such a way as to result in serious, persistent and willful violations of the fundamental terms of the contract, Part XI and the rules, regulations and procedures of the Authority."[280] In cases of less serious violations — or as an alternative to suspension or termination of the contract — "the Authority may impose . . . monetary penalties proportionate to the seriousness of the violation."[281] Sanctions may be recommended to the Assembly by the Council following a decision of the Sea-Bed Disputes Chamber,[282] or by a three-fourths majority vote in other "cases of non-compliance."[283] The U.S. government may impose similar sanctions under federal law, and the 1982 Green Book proposed no changes to the type of penalties available to the Authority.[284] In fact, the international character of the Authority's bureaucracy is expected to render it less difficult for private mining operators to deal with than are many foreign governments, particularly in developing states.[285]

The dispute settlement procedures contained in the Convention also contain supranational and quasi-supranational elements.[286] Although the American Bar Association has advocated access by individuals to international dispute settlement forums, terming purely quasi-supranational judicial competence "retrogressive," the Conference was generally unsympathetic to the idea of expanded private access, and nationals have been limited to appearances before the Sea-Bed Disputes Chamber.[287] The Chamber may not review the Authority's "exercise . . . of its discretionary powers in accordance with" Part XI, nor may it exercise a power of judicial review of the type wielded by the U.S. Supreme Court; but it may review "the *application* of any rules, regulations and procedures . . . in individual cases," may declare such enactments to be in excess of jurisdiction or a misuse of power, and may award damages to entities for the Authority's "failure . . . to *comply* with" the Convention, although the offensive enactments themselves cannot be declared invalid.[288] Overall, Western states were largely successful in obtaining broad jurisdiction for the Sea-Bed Disputes Chamber and other dispute settlement procedures.[289] Decisions of the dispute settlement bodies may be enforced within the territory of states parties, but lack binding precedential effect.[290]

The Convention also may indirectly bestow an effective supranational or quasi-supranational legislative competence upon existing organizations which, until now, have been unable to achieve universal observance of international standards of conduct. By requiring states and their nationals to observe such standards, and by providing coastal states with the power to enforce them, the Convention would force "flag-of-convenience" states to comply with rules and regulations no less effective than generally accepted international standards, even though they are not parties to the instrument establishing the standards.[291] These standards, which have been supported by the United States, would generally serve to strengthen the International Maritime Organization: "It is true that IMCO/IMO is nowhere expressly named, but many years' discussions and negotiations have brought about a consensus that IMCO/IMO is in mind in connection with safety of navigation and protection of the marine environment whenever a 'competent international organization' is referred to, at least when, significantly, 'organization' is used in the singular."[292] Other international organizations which could be rendered more effective by the Convention include the IOC, the International Civil Aviation Organization (ICAO), the World Health Organization (WHO), the International Atomic Energy Agency (IAEA), the Food and Agriculture Organization (FAO), and the United Nations Environmental Programme.[293] In a provision which significantly promotes the integration of the international legal system — and thus the peaceful resolution of potential conflict — the dispute settlement procedures of the Convention have been made applicable to "any dispute concerning the interpretation or application of an international agreement related to the purpose of this Convention."[294]

Although the Convention creates an Authority with relatively strong supranational and quasi-supranational powers, other existing international organizations already possess such powers in varying degrees.

> By a treaty an international agency may be established in which only a part of the contracting States are represented and this agency may be authorized by the treaty to adopt by majority vote norms binding upon all the contracting States. Such a treaty is not incompatible with the concept of international law or with the concept of the State as subject of international law; and such a treaty is a true exception to the rule that no State can be legally bound without or against its own will.[295]

Some organs of the EEC exercise powers of this type, as do the International Telecommunication Satellite Organization (INTELSAT), the International Telecommunication Union (ITU), and the World Bank.[296] This international regulatory competence also exists for ICAO, which may promulgate binding standards with respect to "any . . . matters concerned with the safety, regularity and efficiency of air navigation,"[297] and for the International Whaling Commission,[298] and has been strongly advocated for WHO.[299] Under the 1988 Antarctic Mineral Resource Convention negotiated by the Reagan administration, an Antarctic Mineral Resources Commission and affiliated Regulatory Committees would exercise a general supervisory responsibility with respect to Antarctic mineral exploitation similar to — and in some respects greater than — that exercised by the Authority's Council.[300] The Universal Postal Union has a quasi-supranational legislative competence, and the World Meteorological Organization (WMO) possesses similar powers.[301] The International Court of Justice exercises a quasi-supranational judicial power, and the U.N. Charter, to ensure compliance with a decision of the Court, provides for "recourse to the Security Council, which may, if it deems necessary, . . . decide upon measures to be taken to give effect to the judgment."[302] The quasi-supranational legislative and executive powers of the Security Council give it "almost unlimited discretion,"[303] and the Charter provides that to enforce its decisions the Security Council "may take such action by air, sea, or land forces as may be necessary to maintain or restore international peace and security."[304] Although the powers of the General Assembly are largely recommendatory, it does possess a quasi-supranational legislative competence with respect to assessments.[305] Even under traditional principles of customary international law, the better view acknowledges that private individuals may be made subject to rights and duties by the international community.[306]

Treaty opponents have claimed that the Convention creates "a new supergovernment,"[307] and "unprecedented steps toward supranationality" have been identified as "the prime force behind the U.S. rejection of the treaty."[308] Western developed states have indicated some concern that they would be unable to control such an organization and that their participation might somehow diminish their national sovereignty in relation to developing states.[309] The seabed mining industry has likewise manifested an unusually strong distrust of the Group of 77 in this regard, an attitude open to question in light of the lukewarm support among developing states for supranationalism in principle.[310] In fact, the Authority does not possess the broad regulatory powers common in the United States and elsewhere, and mining companies may be expected to overcome their philosophical distaste for the Authority if seabed mining appears profitable under its rules, regulations, and procedures.[311] Although the Reagan negotiating team sought elimination of what was vaguely described as "unreasonable interference with mining operations,"[312] the 1970 Nixon proposal had in fact endorsed "continuing and comprehensive international regulation" by an Authority with supranational powers,[313] and the U.S. Senate had in 1973 agreed on the need for "an effective International Seabed Authority to regulate orderly and just development of the mineral resources of the deep seabed as the common heritage of mankind, protecting the interests of both developing and developed countries."[314] Fear of supranationalism has traditionally been an element of Soviet rather than American foreign policy;[315] such fears, while certainly not confined to the context of seabed negotiations, are becoming increasingly dysfunctional in a world of growing interdependence.

Decisionmaking Procedures

In addition to the Enterprise, the Authority is comprised of three "principal organs" — a Council, an Assembly, and a Secretariat — an institutional structure largely typical among international organizations.[316] Agreement upon allocation of power generally tends to be difficult for larger institutions such as the Authority, and structural differences are frequently attributable to the specific purpose and tasks which the organization is designed to fulfill.[317] During the tenth session of the Conference, the United States acknowledged that an Authority with relatively strong powers might be acceptable if it safeguarded U.S. political and economic interests.[318] "The decisionmaking system should provide that, on issues of highest importance to a nation, that nation will have affirmative influence on the outcome. Conversely, nations with major economic interests . . . can prevent decisions adverse to their interests."[319] President Reagan indicated that these requirements had not been met.[320] In examining this ground for rejection of the treaty, it should be borne in mind that the United States would be better able to protect the interests of its private mining companies in the Authority than it is now able to do in many developing states.[321]

In order to remove, to as great an extent as possible, "political" considerations from the decisionmaking procedures of the Authority, developed states, including the United States, have sought to require that all important decisions regarding seabed mining operations be taken by a Council with few discretionary powers, and that its decisions be subject to review in accordance with the dispute settlement procedures set forth in the Convention.[322] The Convention requires a consensus of the Council for approval of the rules, regulations, and procedures governing exploration and exploitation of the Area and the "financial management and internal administration" of the Authority, for participation by the Authority in international commodity agreements, and for any distribution of revenues by the Authority to "peoples who have not attained full independence or other self-governing status."[323] A three-fourths-majority vote is required for the Council to establish its rules of procedure, to propose the annual budget of the Authority, to allocate production authorizations if necessary, to prevent seabed mining activities where there exists a threat of serious harm to the marine environment, to control mining operations by ensuring compliance with the Convention and the rules and regulations of the Authority, and to make recommendations to the Assembly concerning suspension of member states or "concerning policies on any question or matter within the competence of the Authority."[324] The Council may also, by a two-thirds-majority vote, issue directives to the Enterprise,[325] enter into agreements on behalf of the Authority with other international organizations,[326] and recommend measures to compensate producer states and to enforce decisions of the Sea-Bed Disputes Chamber.[327] In cases where it is unclear which of these voting majorities is applicable, the question is to "be treated as being within the subparagraph requiring the higher or highest majority or consensus as the case may be, unless otherwise decided by Council by the said majority or by consensus."[328]

The Council is to be composed of 36 member states "elected" by the Assembly: four states which either consumed or imported more than 2 percent of total world consumption or imports of nodule minerals "during the last five years for which statistics are available"; four of the eight states with the largest investment in seabed mining; four producer states which are "major net exporters" of nodule minerals, "including at least two developing States whose exports of such minerals have a substantial bearing upon their economies"; six developing states — to be chosen from among the most populous, the landlocked and geographically disadvantaged, the "major importers" of nodule minerals, the potential producer states, and the least developed of them — and "eighteen members elected according to the principle of ensuring an equitable geographical distribution of seats in the Council as a whole," including at least one state from each of five geographical regions — Africa, Asia, Latin America, Eastern Europe,

Western Europe, and others.[329] Council seats are guaranteed to two other states "from the Eastern (Socialist) European region" (in the seabed-investment and 2-percent-of-consumption-or-imports categories) and to "the largest consumer" state.[330] As stated by then Ambassador Richardson, "[t]he issue is not one of a 'majority' versus a 'minority.' It is a question, rather, of the identification and the balancing of the valid interests at stake—those of consumers, producers, investors, regional groups, and developing countries. 'Sovereign states' are only surrogates for such interests, and imperfect ones at best."[331]

Because each interest group is to be "represented by those members . . . nominated by the group,"[332] Western developed states would be able to control the six nonsocialist seats allocated to the seabed-investment and 2-percent-of-consumption-or-imports groups.[333] Indeed, even before the unraveling of the Warsaw Pact alliance in the late 1980s, Western states would have been able to control the nominating process for these two groups and thereby deny the Soviet Union and its erstwhile allies seats in favor of non–Soviet-bloc "Socialist" states of the Eastern European region.[334] This nominating procedure would also seem to belie contentions that the seat intended for the United States as "the largest consumer" might instead have been allocated to the Soviet Union.[335] It was thus inaccurate to claim, as did Assistant Secretary of State James Malone in 1981, that "the Soviet Union and its allies have three guaranteed seats, but the U.S. must compete with its allies for any representation."[336] Although two of the "major net exporters" must be developing states, it is likely that at least one of the remaining two seats in this group would be allocated to a developed state such as Canada or Australia[337]—indeed, the more economically powerful states have traditionally been accorded representation on such interest groups within the U.N. system.[338] Most of the remaining 24 Council seats would be allocated to developing states, giving them overall majority representation, with Western states likely to receive one additional seat.[339] The United States and its fellow Western states could thus expect to hold "at least seven, and probably eight or nine" seats,[340] although the exact composition of the groups represented would change every two years.[341]

Interest group representation was preferred to either of two other possible bases of composition for the Council—a committee of experts or a system of pure equitable geographical representation, which were rejected at UNCLOS III as premature and inadequate, respectively.[342] Although Secretary of State Kissinger proposed that Council seats should be allocated among all states parties on an equitable basis,[343] the 1976 U.S. negotiating position supported the principle of equitable geographical distribution only if the Council membership as a whole accounted for 90 percent of both world consumption and world production of the four primary nodule minerals.[344] Equitable geographical distribution has been generally accepted as a criterion for allocating seats on executive councils established in association with the United Nations, as well as nonpermanent seats on the U.N. Security Council and ECOSOC,[345] and some form of interest-group representation is also evident in many other such organs, including IAEA,[346] ICAO, IMO, the International Labor Organization (ILO),[347] the U.N. Trusteeship Council,[348] and the Consultative Parties to the 1959 Antarctic Treaty.[349] At the final session of the Conference, apparently believing the composition formula negotiated by Ambassador Richardson to be "clearly designed to benefit Third World and Socialist-bloc countries at the expense of the developed world,"[350] the United States and six other developed states sought to require that the four states selected to represent the seabed investment group be those accounting for the largest share of assessed financial contributions.[351] Any expectation that Western states could exert full control over the Council could only have been unrealistic, however, since such dominance would not have been acceptable to the rest of the international community.[352]

The United States and other developed states have looked to the voting procedures of the Council to protect their more important seabed mining interests. It is virtually impossible to design a voting system which perfectly reflects the various interests present in such an international organization, however,[353] and the voting procedure finally agreed upon was necessarily the product of political compromise. Although the Group of 77 had sought to maximize the one-nation-one-vote system, it was persuaded by Western states to accept a limited form of weighted decisionmaking procedure for the Council.[354] The Conference rejected pure weighted voting formulas based upon percentages of total world production and consumption of seabed minerals, or upon population;[355] the Group of 77 also rejected Western proposals for a complex system of concurrent voting majorities, including a 1982 plan to require approval by a majority — i.e., three of the four states — in each of the four interest groups "and a majority in each geographical region" in addition to the overall three-fourths-majority voting requirement.[356] Notwithstanding U.S. preferences, it was generally agreed that the kind of veto power exercised in the U.N. Security Council would be inappropriate in a Council intended to deal with economic rather than military and security issues,[357] and UNCLOS III delegates looked instead for guidance to the various procedures employed in other international organizations.[358]

It has been claimed that Western states "could not block many sorts of adverse decisions relating to access and budgetary matters, among many others," as the necessary "blocking fourth" could only be mustered with help from Eastern European states.[359] Assuming that support from traditional Soviet allies was indeed necessary, it is not clear that it would, in fact, be withheld, given improved U.S.-Soviet relations and the demonstrated interest of Soviet and other Eastern European states in assuring their own access to deep-sea nodule deposits.[360] Similarly, developing states have a strong interest in ensuring revenues for the Authority and success for the Enterprise by facilitating profitable seabed mining for state-sponsored operations under the system of parallel access. Moreover, the pursuit of national interests may be expected to undermine ideological solidarity within the Group of 77, particularly among the more developed Third World states; with a modicum of competent U.S. diplomacy, "this could lead to a cross section of support for issues rather than a strict north-south split as the United States fears."[361] Many developing states have historically tended to support Western-state interests, and the United States would be able to resort to a variety of political mechanisms to influence Council voting on critical issues.[362]

Consensus, defined in the treaty as "the absence of any formal objection," in effect provides each voting member with a veto power.[363] Although there has been some apprehension that the procedure could produce "institutional paralysis,"[364] failure to achieve a consensus for disapproval would result in the approval of any plan of work recommended to the Council by the Legal and Technical Commission,[365] and the number of other issues requiring consensus in the Council has been kept to a minimum in order to avoid unwarranted obstructionism.[366] The potential for abuse has been a matter of concern among developing as well as developed states, but the consensus procedure may nevertheless be preferable to majority rule or a formal veto system,[367] and the procedure has promoted slow but solid progress among divergent interests in negotiations at UNCLOS III and in the Preparatory Commission. "The consensus procedure is acceptable to the Group of 77 because it does not discriminate among nations.... Rather than giving a few powerful nations the strength to defeat a proposal by a negative vote, as in the traditional veto system, the consensus system gives every member of the Council the power to defeat a proposal."[368] By ensuring broad-based support for difficult decisions in a contentious political environment,[369] consensus has become an accepted and valued method of decisionmaking in a variety of international forums, including the General Assembly, the Security Council,[370] the Committee on Peaceful Uses of Outer Space (COPUOS),[371] and the Consultative Parties to the 1959 Antarctic Treaty.[372]

Levels of cooperative behavior vary among the different types of international organizations, but such organizations have generally succeeded in creating and sustaining international cooperation among states, and the levels of such cooperation have proven to be highest among functional organizations of universal membership such as the Authority.[373] "Although some differences exist between small and large states and between developed and developing nations, neither attribute discriminates among levels of cooperative behavior in a significant way. Even when we control for both of these factors . . . , the differences in cooperation among . . . types of states remain small and statistically insignificant."[374] In the U.N. General Assembly in recent years, the Group of 77 has outvoted the developed states on many issues,[375] frequently with the support of the Soviet bloc, but such symbolic votes do not necessarily portend a similar result in a Council with specific responsibilities and a more limited membership, and U.N. voting alignments may be expected to change significantly over time in any event.[376] Indeed, the U.S. State Department has concluded that the interests of the United States would not be better served by a system of weighted voting in the General Assembly.[377] With competent diplomacy, in light of these historical voting patterns, it is not at all unreasonable to expect that the United States and its fellow Western mining states would receive the voting support of one to three other members of the Council if necessary to prevent unwarranted three-fourths-majority votes — an outcome which led Richardson to conclude that this decisionmaking procedure "satisfies the imperatives of the United States and other industrialized countries."[378] "As the largest consumer, the United States would have a guaranteed seat . . . and veto power on decisions requiring consensus, those most pertinent to its interests. With these powers, and in light of the varied interests involved in deep sea mining, it is difficult to justify the United States position that its interests are not being proportionately represented."[379]

In reviewing plans of work submitted by seabed mining applicants, the Legal and Technical Commission "shall base its recommendations solely on the grounds stated in Annex III."[380] The 15 Commission members — nominated by states parties and elected by a three-fourths-majority vote of the Council, due account being taken of "the need for equitable geographical distribution and the representation of special interests"[381] — must possess "the highest standards of competence and integrity."[382] It has been correctly pointed out that the treaty contains "no guarantee" of Western representation on the Commission,[383] but developed states would be able to block election of any Commission members by marshaling 10 votes in the Council. It is likely that a representation "package" for the Commission would be negotiated prior to actual voting, and that Western states, with a relatively high proportion of technically competent personnel, would be well represented on the Commission, as they have been on the Group of Technical Experts established by the Preparatory Commission to process applications submitted by pioneer investors.[384] While the Legal and Technical Commission is to recommend to the Council the adoption and amendment of rules, regulations, and procedures for the Authority, the decisionmaking procedures of the Commission itself are to be set forth in the rules, regulations, and procedures initially adopted by the Preparatory Commission.[385] Although the decisionmaking procedures of all international organizations are ultimately governed by political values,[386] technical organs have succeeded in avoiding the type of outcome-oriented decisions characteristic of highly politicized institutions.[387] The prospects for impartial decisionmaking on the Legal and Technical Commission thus seem particularly good in light of its limited and strictly defined function with regard to review of plans of work, although "the possibility that the Commission will do a poor job or let political considerations play a role can not be entirely precluded."[388]

The Assembly is composed of all states parties to the Convention, each of which has one vote; although procedural decisions are to be taken by simple majority vote, substantive issues may be resolved only by a two-thirds majority of the members present and voting.[389] Developing states could theoretically muster such a two-thirds majority on each and every vote, but the voting system in the Assembly is in this respect no different from that in the assemblies of most other international organizations.[390] Moreover, one fifth of the Assembly members may delay an initial vote on each substantive issue for as many as five days,[391] and one fourth of the members may request an advisory opinion from the Sea-Bed Disputes Chamber "on the conformity with th[e] Convention of a proposal before the Assembly on any matter," thereby deferring consideration of the matter pending the court's decision.[392] Taken together, these procedures are expected to render it very difficult for developing states to take actions detrimental to the important interests of developed states.[393] The Assembly has no power to authorize or deny access to seabed mineral deposits. The Reagan administration made no attempt to alter the Assembly's composition or voting procedure in 1982.[394]

Whereas developing states sought to provide the Assembly with effective control over the other organs of the Authority, developed states insisted upon a clear separation of powers and functions.[395] The compromise language of the Convention has been criticized as providing the Assembly "enormous power."[396] However, while the most significant of the Assembly's enumerated powers might appear to be its authority to approve certain measures previously adopted by the Council,[397] rules, regulations, and procedures relating to seabed mining operations and to "the financial management and internal administration of the Authority" become applicable provisionally upon their adoption by the Council and cannot be disapproved by the Assembly alone.[398] The Convention does empower the Assembly, without the concurrence of the Council, to assess states parties, to authorize the distribution of revenues pursuant to the rules, regulations, and procedures of the Authority, to "consider problems of a general nature in connection with activities in the Area arising in particular for developing" and landlocked and geographically disadvantaged states, and, residually, "to discuss any question or matter within the competence of the Authority and to decide as to which organ shall deal with any such question or matter not specifically entrusted to a particular organ of the Authority, consistent with the distribution of powers and functions among the organs of the Authority."[399] Moreover, to the consternation of the U.S. Senate, the Assembly has been symbolically designated "the supreme organ of the Authority" — to which the Council and the Secretariat are to "be accountable as specifically provided" in the Convention — and has been granted "the power to establish general policies . . . on any question or matter within the competence of the Authority."[400] The Council, in contrast, has been designated "the executive organ of the Authority," and is empowered "to establish, in conformity with [the] Convention and the general policies established by the Assembly, the specific policies to be pursued by the Authority on any questions or matters within the competence of the Authority."[401] Because each principal organ must "avoid taking any action which may derogate from or impede the exercise of specific powers and functions conferred upon another organ," however, the Assembly has no competence with respect to the functions and powers specifically granted to the Council.[402] In fact, the powers of the Assembly, which are derived from the soverign equality of the members of the Authority,[403] are not significantly different from those which have been conferred upon the assemblies of other international organizations, including ICAO, WMO,[404] IMO,[405] INTELSAT,[406] ITU,[407] and the U.N. General Assembly.[408]

The Convention provides that the Assembly is to elect the Secretary-General, as well as the governing board and Director-General of the Enterprise, from among candidates submitted by the Council.[409] A three-fourths-majority vote of the Council is required for approval of all such candidates, giving developed states an opportunity to block their nomination, if necessary, and negating claims that the "majoritarian status of developing states in the Assembly ensures that the governing elements of the Enterprise will consist mainly of nationals from their group."[410] The 15 members of the governing board are to possess "the highest standards of competence," with "due regard" being paid to principles of rotation and equitable geographical distribution.[411] Similarly, the Secretary-General must appoint, pursuant to the rules, regulations and procedures of the Authority, a staff "of the highest standards of efficiency, competence and integrity," with "due regard . . . being paid to the importance of recruiting . . . on as wide a geographical basis as possible."[412] The strict responsibilities imposed upon the Secretariat by the Convention are designed to ensure the integrity of the Authority's administrative apparatus, and the Sea-Bed Disputes Chamber has jurisdiction to review the qualifications of any staff member.[413] Because the appointment, compensation, and discipline of staff members is to be governed by the rules, regulations, and procedures of the Authority, and because the Enterprise is "subject to the directives and control of" the Council — as well as to general policies adopted by the Assembly and to the Authority's rules, regulations, and procedures[414] — it may be expected that the new bureaucracy will be held accountable for its administrative efforts, despite fears of "inordinate control over important activities placed in the hands of officials who have been self-selected and who are accountable primarily to themselves and other international bodies and officials, but not to nations and peoples."[415]

Assured Access

Because private consortia will have expended hundreds of millions of dollars on site-specific seabed mining activities even before they apply for plans of work from the Authority, they have sought to ensure exclusive rights of access for themselves under the Convention.[416] In the 1980 Seabed Act, Congress has thus required that the Convention provide U.S. nationals with "assured and nondiscriminatory access, under reasonable terms and conditions," a requirement to be determined "by the totality of [its] provisions."[417] At the urging of the seabed mining industry, the U.S. delegation at UNCLOS III consistently sought well-defined procedures which would grant mining rights automatically under terms and conditions clearly set forth in the treaty.[418] "More than anything else, the companies feared that a decision might be made on political grounds long after they had made their investment, putting them at best in an impossible negotiating position, and at worst without the right to mine."[419] Although former U.S. Ambassador George Aldrich, just prior to his replacement in March 1982, described the Authority's contract approval procedure as "fair, clear, and well-nigh automatic,"[420] his successor claimed that the Convention provides "no reasonable protection for mine sites already in existence and no assurance of access to future sites."[421] In analyzing the approval process for applications submitted by seabed mining operators, one should keep in mind that the reciprocating states regime sought by the Reagan administration as an alternative to the Convention has failed to provide U.S. firms with exclusive mining rights.[422]

Annex III, which sets forth the basic conditions governing seabed mining activities, provides that each applicant for a plan of work must be sponsored by the state party or parties of which it is a national, as well as by those by which it is effectively controlled, and must tender the $500,000 application fee.[423] "On qualifications for applicants, it seems critical that the only qualifications required, other than being sponsored by a state party to the treaty, would be those related to technical, financial [and] managerial competence, and performance under any prior contracts."[424] Annex III, indeed, provides that qualification standards — to be set forth in the rules, regulations, and procedures of the Authority — must "relate to . . .

financial and technical capabilities ... and [to] performance under any previous contracts with the Authority."[425] Each applicant must also "undertake" to abide by the terms of the treaty and the decisions of the Authority, to accept control of mining activities by the Authority "as authorized" in the Convention, to provide "a written assurance that ... obligations under the contract will be fulfilled in good faith," and to comply with the technology transfer provisions.[426] As the U.S. State Department concluded in 1980, "[a]ny serious applicant for a seabed mining contract should have no difficulty with these undertakings and qualification standards."[427] The 1982 U.S. proposals would have substantially maintained these requirements – which are similar to the qualification standards imposed by the Seabed Act[428] – but would have specified the financial, technical, and prior-performance standards in the treaty itself rather than leaving their promulgation to the Preparatory Commission and the Authority.[429] Although former Ambassador Richardson has expressed satisfaction that the Convention "spells out the qualifications of applicants in clear, objective terms,"[430] treaty opponents have nevertheless objected, claiming that the provisions governing qualification standards lack predictability.[431]

Seabed exploration and exploitation may be conducted under the Convention "only in areas specified in plans of work ... approved by the Authority," each of which areas is to be "sufficiently large and of sufficient estimated commercial value to allow two mining operations."[432] The application must include coordinates dividing the area into two sites "of equal estimated commercial value," and must include data relating to "mapping, sampling, the abundance of nodules, and their metal content"; within 90 days the Authority must select one of the sites, which is to be banked as a reserved site upon approval of the applicant's plan of work.[433] The U.S. amendments of 1982 would have required the two submitted sites to be "of equivalent size and comparable value" – a standard more difficult for applicants to meet and one with which applicants already have an incentive to comply.[434] The United States also demanded that the Convention be revised to prevent the Authority from selecting its own reserved sites: such sites were to be designated by negotiation between the Authority and the applicant, or, failing agreement, at random by the Legal and Technical Commission.[435] The U.S. proposals were apparently motivated by a desire to minimize expenses and by fear that the Authority would be able to reserve the better site for itself,[436] or, worse, reject the application and use the data to initiate its own mining operations in the area,[437] concerns which would appear to be unwarranted if the applicant submits two areas of equivalent value and if the treaty's confidentiality provisions are observed. Certainly, the registration of the first group of pioneer applicants demonstrated that this step in the application procedure need pose no obstacle for seabed mining operators.

Seabed mining activities under the Convention must be conducted in accordance with a written plan of work conforming to the provisions of Annex III, which becomes a contract after review by the Legal and Technical Commission and approval by the Council.[438] The Commission will review plans of work in the order in which they are received, and its recommendations must be based "solely on the grounds stated in Annex III."[439]

> An applicant has only to be sponsored by a state party and to satisfy the financial and technical qualifications spelled out in the regulations. His plan of work must fulfill the specifications with respect to such matters as size of area, diligence requirements, and mining standards and practices, including those relevant to protection of the marine environment, that will also be set forth in the regulations. If these requirements are met, his plan of work *must* be approved; there is no discretionary basis for its rejection.[440]

Nevertheless, the United States in recent years has objected to the "substantial discretion" supposedly possessed by the Legal and Technical Commission with respect to the contract approval process,[441] and has insisted that the Commission be required to act "impartially and without delay."[442] Although such issues are expected to be addressed by the rules, regulations, and procedures of the Authority, some degree of interpretative discretion for the Legal and Technical Commission may be unavoidable and even desirable.[443] U.S. seabed mining regulations are themselves highly discretionary in a number of respects — for example, in requiring each applicant to demonstrate its financial and technical capability "to the Administrator's satisfaction," and in mandating a case-by-case review of proposed mining sites.[444] Representatives of U.S. mining interests have argued that the Authority, "an entirely new international agency, which may also become a competitor," could be more difficult to deal with than existing producer states;[445] because of the representation of Western seabed mining states in the Council and the Legal and Technical Commission, however, it is likely that the interests of private seabed mining operators under the Convention would be better protected than are the interests of their land-based counterparts under the national regulatory systems of developing producer states.[446] The Convention requires the Legal and Technical Commission — as well as other organs of the Authority — to promote "development of the resources of the Area,"[447] and there would seem to be every incentive for Commission members to seek to ensure that the system established by the Convention operates successfully. It is thus unlikely, absent actual corruption of Commission members, that the Commission would adopt a politically sophisticated antidevelopment policy in the face of the express command of a carefully negotiated and balanced treaty to which their home governments had become a party.[448] The Preparatory Commission may draft rules, regulations, and procedures which facilitate the approval process.[449]

After determining an application's compliance with the qualification standards of article 4 of Annex III, the Legal and Technical Commission will determine whether the proposed plan of work conforms to the other provisions of the Convention and to the rules, regulations, and procedures of the Authority.[450] After any deficiencies are remedied,[451] the Authority "shall approve" the application, unless part or all of the proposed site is unavailable or unless approval would exceed the quota allotted to the sponsoring state under the antimonopoly provisions.[452] These are objective criteria, compliance with which may readily be determined by an applicant prior to review of the application by the Commission, and the United States sought no changes in them during 1982. As the State Department pointed out in 1980, "[t]here is no discretion. . . . If the Authority fails to approve it, it acts unlawfully, and the applicant may sue the Authority in the Seabed Disputes Chamber of the Law of the Sea Tribunal."[453] As an organ of the Authority, the Commission is required to "avoid discrimination in . . . the granting of opportunities for activities in the Area," and must uniformly apply the Authority's rules, regulations, and procedures.[454] A consensus of the Council would be required for adoption, "on an equitable and non-discriminatory basis," of any procedures and criteria to govern selection among competing applicants for the same mining site.[455] The evident ability of U.S. seabed mining firms to gain access under the discretionary procedures embodied in the 1980 Seabed Act would seem to bode well for their efforts to obtain access under the more clearly defined terms of the Convention regime.[456]

In furtherance of the public policy of preventing unreasonable monopolization of mineral resources, which has "a long and unbroken history in English and Germanic mining customs, and in the common law, which can be traced back to the medieval period, if not earlier,"[457] article 150 of the Convention lists "prevention of monopolization of activities in the Area" as one of the policies applicable to seabed mining activities.[458] The antimonopoly provisions in Annex III may provide grounds for disapproval of a proposed plan of work if the sponsoring state is already sponsoring — or is itself already conducting — mining activities in 2 percent of the total area available for private exploitation, or if its existing mining activities, "together with either part of the area covered by the application. . . , exceed in size 30 per cent of a circular area of 400,000 square kilometres surrounding the centre of either part of the area covered by the proposed plan of work."[459] These provisions were originally included at the insistence of several developed states amid uncertainty over the number of available "first-generation" mine sites;[460] they were initially opposed by the United States, which had regarded the number of prime mining sites as adequate to support scores of seabed mining operations.[461] The State Department subsequently acknowledged that the provisions would not substantially interfere with access by U.S. firms, however, and the limitations on national access are therefore expected to be of only symbolic importance.[462] Moreover, the Authority may approve plans of work, notwithstanding the antimonopoly provisions, "if it determines that such approval would not permit a State Party or entities sponsored by it to monopolize the conduct of activities in the Area or to preclude other States Parties from activities in the Area."[463] The Reagan administration and its congressional allies nevertheless identified the antimonopoly provisions as one of the "chief" antidevelopment aspects of the Convention,[464] even though it sought no significant changes in them during the 1982 session, and even though U.S. seabed mining applications are themselves subjected to an antitrust review to determine the applicant's "market share . . . with respect to the mining or marketing of the metals proposed to be recovered."[465]

If approval of a plan of work is recommended by the Legal and Technical Commission, the Council is very unlikely to disapprove it – the assertions of the Reagan administration to the contrary notwithstanding[466] – since it is virtually certain that at least one U.S. ally would act to block the consensus necessary for disapproval.[467] If the Commission recommends disapproval, the Council may nevertheless approve the application "by a three-fourths majority of the members present and voting, provided that such majority includes a majority of members participating in that session."[468] The Commission is required to act in accordance with any guidelines and directives issued by the Council, but consensus would be necessary for their adoption.[469] The Convention also gives the Council power to make areas of the deep seabed unavailable for exploitation "in cases where substantial evidence indicates . . . risk of serious harm to the marine environment," a power also held by U.S. administrators under the Seabed Act;[470] because the Council could only exercise this power by a three-fourths-majority vote, and because the words "substantial" and "serious" constitute standards subject to judicial review,[471] frequent or unwarranted resort to this general disapproval procedure is unlikely. "The composition of the Council and the decision-making process entailed in the . . . Convention effectively grant access to the hard mineral resources of the deep seabed on a nondiscriminatory basis. The mandate of [the Seabed Act] in this respect is satisfied."[472]

Once a plan of work has been approved and thereby become a contract, a seabed mining operator must still obtain a production authorization before proceeding with commercial production.[473] As the Department of State noted, "[e]ach application for production authorization must be granted unless granting it, along with all other pending applications, would result in the quota being exceeded."[474] If a selection among contractors became necessary, the Council would award the available number of authorizations by three-fourths-majority vote, basing its decision on the objective standards set forth in the rules, regulations, and procedures of the Authority and in article 7 of Annex III: "The question would arise only if, at the end of any given 4-month period, approval of all the applications for a production authorization received during that period would result in production in excess of the interim production limitation or a production limitation arising under a commodity agreement to which the Authority has become party."[475] Although it is unquestionably true that any delays in mining caused by application of the production ceiling would constitute a temporary denial of access to seabed nodules, such delays remain unlikely, even under poor market conditions.[476] Apart from the priority enjoyed by the Enterprise and other operators mining reserved sites, the criteria to be employed in any allocation process tend to favor the more established and wealthy mining operators, and in any event the United States and its allies may be expected to be able to block any vote which would unfairly deny production authorizations to their nationals.[477] Because a production authorization, when issued, becomes part of an operator's contract, the adoption of any international commodity agreement could not affect existing contractors' rights of access.[478]

The 1980 Seabed Act requires the system of preparatory investment protection to recognize the

> rights of United States citizens who have undertaken exploration or commercial recovery ... before [the Convention] enters into force with respect to the United States to continue their operations under terms, conditions, and restrictions which do not impose significant new economic burdens ... with the effect of preventing the continuation of such operations on a viable economic basis.[479]

Resolution II authorizes each pioneer registrant to conduct exclusive exploration activities within a single mining area, which may initially encompass as much as 150,000 square kilometers, and requires each pioneer applicant to be certified by a state which has signed the Convention and to include an application fee of $250,000, as well as undertakings with respect to training, technology transfer, and performance requirements.[480] The antimonopoly provisions of the Convention are applicable, and each pioneer applicant must submit two prospective mine sites, one of which "is to be reserved in accordance with the Convention."[481] In order to maintain balance within the parallel system, pioneer registrants must also conduct—before the entry into force of the Convention, upon request by the Preparatory Commission—exploratory activities in the reserved site on a cost-reimbursable basis, and provide personnel training "at all levels" for designated individuals.[482] Despite Soviet objections, the Commission decided that registration of the private consortia would not require certification by nonsignatory states such as the United States.[483] Applying Resolution II flexibly in order to promote broad participation in the Convention, the Commission agreed in 1986 that "[t]he treatment to be accorded to potential applicants in respect of their applications shall be similar to the treatment given to the first group of applicants provided that potential applicants assume similar obligations . . . and submit their applications before the entry into force of the . . . Convention."[484] Once the Convention enters into force, a pioneer registrant will have at least six months within which to submit a plan of work, which the Authority must approve if it complies with "the relevant provisions of the Convention and the rules, regulations and procedures of the Authority," provided that the registrant's certifying state or states ratify the treaty.[485] Pioneer registrants whose plans of work are approved are to enjoy priority in the allocation of production authorizations with respect to all but two applicants of the Enterprise: unless the production ceiling is reached "[p]roduction authorizations shall be issued to each pioneer investor within 30 days of the date on which that pioneer investor notifies the Authority that it will commence commercial production within five years."[486] Although at least one U.S. negotiator has expressed satisfaction that Resolution II would effectively provide pioneer registrants with automatic access to seabed nodules,[487] it has elsewhere been argued by treaty opponents that the preparatory investment scheme is "defective in a number of respects"[488] — including, in particular, its alleged failure to "guarantee" approval of mining contracts once the Convention has entered into force[489] — even though at least half a dozen pioneer investor operations sponsored by other states have evidently found its protections satisfactory.

The 1980 Seabed Act requires that the Convention include "impartial and effective" dispute settlement procedures.[490] The International Tribunal for the Law of the Sea, to be elected initially at a meeting of all states parties to the Convention, will be composed of 21 members "enjoying the highest reputation for fairness and integrity and of recognized competence in the field of the law of the sea" and representing the principal legal systems and geographical regions of the world.[491] The Sea-Bed Disputes Chamber, consisting of 11 members selected by and from the full Tribunal,[492] will decide cases by majority vote.[493] "While the general dispute settlement system applies only to disputes between states, the seabed dispute settlement system applies in addition to disputes between states and the . . . Authority and to contract disputes involving states, the Authority, the Enterprise . . ., state enterprises, and natural and juridical persons."[494] Because private seabed mining operators lacked confidence in such "marvelously overblown" procedures,[495] commercial arbitration may be chosen as an alternative forum for settling more technical disputes, including those relating to technology transfer, revenue sharing, and other aspects of the interpretation or application of a contract.[496] The United States would have preferred that the Chamber possess jurisdiction to review the Authority's exercise of its discretionary powers and to invalidate rules, regulations, and procedures which conflict with the provisions of the Convention,[497] but its jurisdiction — extending to virtually all aspects of the contract-approval process, including technology transfer and revenue sharing, but excluding such inherently discretionary matters as the Authority's specific choice of a reserved mining site — is sufficient to provide seabed mining operators with broad protection against abusive practices by the Authority.[498]

U.S. mining firms have insisted that security of tenure under an acceptable treaty include prohibitions against any changes in the amount of nodule production, the location of mine sites, the duration of mining rights, or the revenue-sharing formulas.[499] Under the Convention, an approved plan of work will "confer on the operator . . . the exclusive right to explore for and exploit the specified categories of resources in the area covered,"[500] and will provide security of tenure by barring involuntary termination, suspension, or revision of contracts, except as a penalty for major violations.[501] Existing contractual rights of access would not be affected by subsequent adoption of rules, regulations, and procedures, by amendments to the seabed mining provisions, or by denunciation of the Convention.[502] The Council may, by a three-fourths-majority vote, issue 30-day emergency orders, "which may include orders for the suspension or adjustment of operations, to prevent serious harm to the marine environment"[503] — an important power which is nonetheless more limited than that enjoyed by federal bureaucrats under U.S. law, which permits any seabed exploration license or exploitation permit to be terminated, suspended, or modified "[i]n the Administrator's discretion" in cases of substantial noncompliance with U.S. law or with "any term, condition, or restriction in the license or permit."[504] Although U.S. mining firms would have preferred that the duration of mining rights under an approved plan of work be specified in the treaty itself,[505] the length of tenure secured by contracts is to be established in the rules, regulations, and procedures of the Authority.[506]

> Unlike the terms of many Third World mineral agreements, early seabed contracts with the . . . Authority are relatively fixed because the key provisions are written into the treaty itself. Such contractual stability will be attractive to private mining companies, many of whose frequent experiences with expropriation and forced renegotiation have led them to invest in developed countries or a few "safe" developing countries rather than be subject to "political risk" in many countries that have geographically superior deposits.[507]

Indeed, the long UNCLOS III negotiations yielded greater security of access than the United States had originally sought in the 1970 Draft Treaty, which would have permitted regulations and qualification standards to be both adopted and amended without U.S. consent.[508]

Exploration and exploitation of non-nodule resources, including revenue sharing and the applicability of any form of production controls, are to be governed by rules, regulations, and procedures of the Authority to be adopted within three years of a request by a state party, enactment of which will require consensus approval of the Council.[509] Because any failure to enact such rules, regulations, and procedures would constitute a justiciable omission by the Authority "in violation of" Part XI, the United States would be able to recover from the Authority any damages suffered by U.S. firms as a result of such delay.[510] The Authority, moreover, is required to give special consideration to plans of work with respect to multiple resources in the same area when mining methods are similar and when the resources "can be developed simultaneously without undue interference between operators developing different resources."[511] The Reagan administration nevertheless demanded that deep-sea mining be permitted to proceed with respect to non-nodule minerals—presumably under a freedom of the high-seas regime—"pending the development of rules and regulations,"[512] despite the similarity of the treaty's procedure to that faced by mining firms under U.S. domestic law,[513] and despite the ability of private seabed mining operators to conduct prospecting activities for such resources at any time under the Convention.[514]

Amendments to the seabed mining provisions of the Convention may be adopted by a consensus of the Council and a two-thirds-majority vote of the Assembly at any time,[515] or by a three-fourths-majority vote of the review conference to be convened 15 years after commencement of commercial production.[516] Certain basic aspects of the parallel system cannot be amended,[517] and any amendments adopted by these procedures "shall enter into force for all States Parties twelve months after ... ratification or accession by three fourths of the States Parties."[518] Although it had been feared that the review conference would trigger automatic conversion to a unitary system in which only the Enterprise would be assured of access to seabed nodules, the Convention does not impose a moratorium on state-sponsored mining operations; moreover, any amendments adopted "shall not affect rights acquired under existing contracts"—a greater measure of security of tenure than U.S. seabed mining operators enjoy under the Seabed Act.[519] As Secretary of State Kissinger recognized, the review conference would present an opportunity for seabed mining states to negotiate remedial changes before many of the objectionable provisions took full effect, particularly if private consortia had proven themselves able to operate most efficiently "for the benefit of mankind as a whole"[520]; yet

U.S. officials have indicated that "massive amendments" to Part XI would be expected from developing states.[521] The Reagan administration further objected that the review conference would "have the power to impose treaty amendments on the United States without its consent,"[522] even though such power is a legitimate procedural mechanism under international law.[523] Indeed, President Reagan himself described this procedure as "clearly incompatible with the U.S. approach to such treaties,"[524] when, in fact, the constitutive instruments of most international organizations of which the United States is a member provide for amendment by a similar procedure[525] — a quasi-supranational process necessitated by the practice of including in such instruments detailed regulatory provisions requiring frequent adjustment in light of changing circumstances. "The maintenance of a single set of rules applicable to all parties to a convention or treaty following its amendment, is of particular importance in regard to the constitutional rules of an international organization, as they govern the composition, functions and operation of the organs thereof, which obviously cannot operate at one point in time according to varying sets of rules."[526] A state party may avoid the application of unacceptable amendments by denunciation, effective 12 months after written notice — a period which may be made to coincide with the one-year period between three-fourths ratification and entry into force of such amendments.[527] Denunciation is the procedure traditionally available for disgruntled states to avoid application of treaty provisions which have become unacceptable;[528] yet the United States has in effect already denounced the Convention in anticipation of injury which may never materialize, a move which can only be characterized as premature.[529]

Competitive Balance

The Seabed Act expresses the intent of Congress that the Convention not significantly affect the economic viability of U.S. seabed mining operations, and instructs that the Convention's compliance with this directive "should be determined by the totality of the provisions of such agreement, including, but not limited to, ... any features that tend to discriminate against exploration and commercial recovery activities undertaken by United States citizens."[530] Although Congress contemplated that an internationally agreed regime might be somewhat less advantageous for U.S. seabed mining operators than the regulatory framework established by the Seabed Act, free enterprise proponents have expressed concern that discriminatory provisions within the Convention might confer unwarranted competitive advantages upon land-based producer states as well as upon the Enterprise and mining operations sponsored by developing states.[531] In addition to the production limitation formula, land-based producer states could benefit from "necessary and appropriate measures" which may be adopted by a consensus of the Council to provide protection against the adverse economic effects specified in article 150 (h), as well as from any "system of compensation or other measures of economic adjustment assistance" which may be implemented for the protection of developing producer states.[532] The Reagan administration complained that the Convention "creates a system of privileges which discriminate against the private side of the parallel system," providing the Enterprise with "substantial competitive advantages,"[533] and potentially resulting in monopolization of seabed mining activities by the Enterprise.[534] At least one Congressman has emphasized that failure to maintain a competitive balance within the parallel system — including "economic provisions compatible with free market precepts," — could in effect cause a denial of access to profit-oriented private mining firms.[535] Similar arguments have been made by the U.S. seabed mining industry, which has expressed particular concern that the Authority faces an inherent conflict of interest with respect to its regulation of the Enterprise,[536] and that the Convention "would mandate competitive advantages to those developing countries that seek to monopolize worldwide the supply of mineral commodities."[537] Several commentators have criticized the "competitive superiority" of the Enterprise,[538] as well as potential discretion by the Authority which "could ... render seabed mining less economic."[539]

While the Convention requires the Authority to avoid discrimination in favor of the Enterprise, it nonetheless calls for the promotion of "effective participation of developing States..., having due regard to ... in particular the special needs of the land-locked and geographically disadvantaged States among them," as specifically provided in the seabed mining provisions.[540] Contrary to some assertions, such provisions would not affect the ability of U.S. firms to obtain assured access or security of tenure under the Convention.[541] Although the Convention clearly would require U.S. mining firms to provide financial and technical assistance to the Enterprise at their own expense, the Authority — and the Enterprise itself — nonetheless has a strong incentive to promote the success of pioneer investors and other early state-sponsored mining operations in order to secure the benefit of such assistance, as well as revenue-sharing payments for the Authority.[542] Treaty opponents, ignoring the inherently discriminatory effect of any regulatory framework, have gone so far as to claim that "the favoritism shown the Enterprise ... amounts to discrimination against private individuals, violating the fundamental American abhorrence of governmental 'discrimination in any form',"[543] even though such favoritism was originally proposed by Secretary of State Kissinger "so that the existing advantage of certain industrial states would be equalized over a period of time."[544] Developing states, on the other hand, believe that there is a serious risk of discrimination against the Enterprise.[545] Outside the

UNCLOS III regime, U.S. firms must also compete, disadvantageously, against state-subsidized operators — including those sponsored by traditional allies such as France and Japan[546] — and it is significant that such states proceeded with development of their national seabed mining capabilities during the 1980s while private mining activities had all but ceased.

The economic riskiness of mining operations has traditionally been cited by private mining firms as justification for "supranormal returns" on their investments,[547] and the U.S. seabed mining industry has complained that the costs of the various forms of assistance to be rendered to the Enterprise would not only confer a significant competitive advantage upon the Enterprise, but would also consume a "substantial part" of total seabed mining investments.[548] Although there may be some justification for such concerns with respect to technology transfer and site reservation,[549] the net costs of these obligations to private mining firms would nevertheless be relatively small;[550] in fact, these costs "may be no more burdensome than certain non-business expenses associated with hard mineral extraction in developing countries, where contracts often require more of mining companies than merely profit-sharing or the payment of fees."[551] Contrary to the objections of the Reagan administration, the revenue-sharing provisions of the Convention would be largely inapplicable to U.S. mining firms until they have attained profitability, although the ten-year grace period enjoyed by the Enterprise with respect to such payments would provide it with somewhat of a short-term competitive advantage.[552] "In order to ensure that the Enterprise is able to carry out activities . . . in the same time-frame as States and other entities," pioneer registrants are in addition required to provide personnel training.[553]

In comparison to the Convention, the 1980 Seabed Act itself contains a number of requirements likely to impair the profitability of U.S. mining operations to a substantial degree, including compliance with the Clean Water Act and the National Environmental Policy Act.[554] Indeed, permittees may be required to employ "the best available technologies for the protection of safety, health, and the environment" in their nodule exploitation activities, and "[e]ach commercial recovery permit must contain [terms, conditions, and restrictions] established by the Administrator . . . which prescribe actions the permittee must take in the conduct of commercial recovery activities to assure protection of the environment."[555] The Seabed Act also requires each recovery plan to include detailed information regarding the site and its resources, the proposed commercial recovery schedule, environmental safeguards and monitoring mechanisms, waste disposal methods and technology, and "such other information as is necessary and appropriate."[556] Licenses and permits issued to U.S. seabed mining operators, and the terms, conditions, and restrictions included therein, may be modified for a variety of reasons which do not apply under the

Convention regime, and regulations issued pursuant to the Act may themselves be amended "at any time . . . as the Administrator determines to be necessary and appropriate in order to provide for the conservation of natural resources, protection of the environment, and the safety of life and property at sea."[557] All nodule recovery and at-sea processing vessels — as well as at least one ore carrier — must be registered under the U.S. flag "in order to insure that the advanced technology will remain available to the Nation and will not be freely exported to the site of cheapest construction."[558] Furthermore, U.S. seabed mining operators may process nodule minerals outside the United States only if they can convince federal regulators that domestic processing "is not economically viable" and if they are able to provide "satisfactory assurances" that the processed minerals may be returned to the United States upon request.[559] Together, these requirements are expected to increase the costs of seabed mining by several hundred million dollars for each U.S. mining firm.[560] "By including . . . vessel and processing requirements the advocates of unilateralism had been able to build a stronger domestic coalition."[561] Yet deep-sea mining under a freedom of the seas regime is considered so economically risky and legally uncertain that the U.S. seabed mining industry initially demanded compensatory government guarantees as protection against financial losses.[562]

The Convention contains several instances of procedural bias in favor of the Enterprise. Although its plans of work are subject to the same approval process as those submitted by private seabed mining operators,[563] the Enterprise need not be sponsored by a state party, need not submit a site of equal value for banking, and need not include any undertakings with respect to technology transfer.[564] To the extent it was able to sustain multiple mining operations, the Enterprise would be able to benefit from the priority it enjoys in the allocation of production authorizations for reserved sites — a preference opposed by the United States and unjustifiable on other than political grounds — but it may only claim two priority authorizations for its pioneer mining operations, and the Preparatory Commission has made plans for the Enterprise to commence its mining operations at a single "prime" site in the Clarion-Clipperton Zone.[565] The Enterprise has also been exempted from the Convention's antimonopoly provisions and from the penalties applicable to state-sponsored mining operations.[566] Otherwise, the Enterprise is subject to the same decisions, rules, regulations, and procedures of the Authority as private and state-operated mining firms.[567] The governing board of the Enterprise will make its decisions — including those relating to applications for production authorizations, the submission to the Council of formal plans of work, and the negotiation of technology transfer and joint-venture arrangements — by simple majority vote, and the Director-General is responsible for the day-to-day management of the personnel and operations of the Enterprise.[568] Although treaty critics have suggested that the Enterprise might unfairly benefit from institutional bias within the bureaucratic structure of the Authority,[569] such favoritism remains an unlikely possibility in light of its composition and institutional orientation.[570]

In order to ensure timely commencement of its mining activities, the Enterprise will rely initially on financial support from states parties to the Convention.[571] Concern has been expressed that such financial support would contribute to the "privileged competitive position" of the Enterprise,[572] and the "budgetary impact" of the treaty was identified by the Reagan administration as one of its objections to the Convention.[573] "Financing the Enterprise is disliked by almost all Western treasuries and economists; both land-based producers and potential seabed producers believe they are in effect being asked to finance a competitor."[574] The Enterprise would be able to borrow funds, but only after it had established its own creditworthiness through profitable mining operations.[575] States parties are therefore obligated to finance an initial mining operation by the Enterprise — through a combination of one-half long-term, interest-free loans and one-half loan guarantees, both to be assessed "in accordance with the scale of assessments for the United Nations regular budget"[576] — as well as to contribute a smaller amount of funds to meet the initial administrative expenses of the Authority.[577] Such financial arrangements have been employed successfully by the United States for the benefit of other international organizations as well as developing states.[578] "As an element of a mini-package for the system of exploration and exploitation, the negotiating states . . . agreed that the Enterprise should be provided with an amount of funds equivalent to that needed to carry out a four-metal integrated operation from mining to marketing."[579] Although these measures may be expected to put the Enterprise on a par with subsidized state-run operators, in a stronger financial position than private seabed mining firms, international financing of the Enterprise's first mining operation was first proposed by Secretary of State Kissinger in 1976 as part of the package deal offered by the United States for a parallel system of access,[580] and the loans and loan guarantees contained in the Convention are less burdensome than the mandatory fees sought by some developing states.[581] Even with such assistance, the Enterprise is expected to experience substantial and unavoidable delays in the commencement of mining operations.[582] The Enterprise may also obtain funding from voluntary contributions of states and the revenue-sharing receipts of the Authority, as well as from its own retained earnings.[583] As a member of the Council, the United States would have an effective veto power over the financial management and administration of the Authority, and developed states could expect to enjoy a blocking power in the Council with respect to the budget of the Authority, as well as the exercise of the Authority's borrowing power.[584]

U.S. seabed mining firms could themselves expect to enjoy competitive advantages under provisions of the Convention, as well as exclusive rights to relatively large mining sites and preferential treatment for pioneer registrants.[585] U.S. mining operators would benefit in particular from the deductibility of their financing costs,[586] from any bidding system employed in the allocation of production authorizations,[587] and also possibly from additional financial incentives adopted pursuant to the rules, regulations, and procedures of the Authority.[588] Unlike the Enterprise, contractors enjoy security of tenure and access to the Sea-Bed Disputes Chamber;[589] unlike contractors, the Enterprise is subject to the directives of the Council as well as the general policies of the Assembly, and is required to operate "on sound commercial principles."[590] Furthermore, the Enterprise may have difficulty selling nodule minerals in tightly controlled international markets.[591] A successful Enterprise could, in any event, represent an important new source of aid for developing states, possibly helping to supplant future U.S. foreign aid expenditures, and the United States could be expected to exercise a disproportionately strong influence within the Authority to the extent that it contributes the largest portion of initial financing for the Enterprise.[592] Developing states themselves are generally unlikely to devote scarce resources to costly and risky seabed mining operations when they are unable to meet the domestic needs of their own population; if they choose to do so, their operations could be expected to lag behind those of Western states.

The Enterprise may negotiate contractual agreements with state-sponsored operators "in the form of joint ventures or production sharing, as well as any other form of joint arrangement."[593] The claim of one Congressman notwithstanding, U.S. firms entering into joint-venture agreements would enjoy a status generally equivalent to that of independent private contractors — including complete security of tenure and access to the Sea-Bed Disputes Chamber — as would the Enterprise as their contractual partner, although the compulsory technology transfer provisions of the Convention would not apply.[594] Despite the evident willingness of half a dozen pioneer investors to proceed with independent deep-sea mining operations within the treaty framework, the Reagan administration claimed that the treaty's discrimination against private operators would provide U.S. mining firms with "little option but to enter joint ventures or other similar ventures."[595] The Authority may indeed adopt rules, regulations, and procedures which provide incentives for contractors to enter into joint arrangements with the Enterprise and developing states, but such incentives must be "uniform and non-discriminatory," must be uniformly applied, and must be approved by consensus in the Council;[596] moreover, the carefully crafted language of the Convention would protect private mining firms against the type of one-sided arrangements sometimes forced upon them by developing producer states.[597] "The terms of the joint venture would be established by negotiation between the Enterprise and the private firm; but, since there is always the opportunity to mine under the private system, the terms of the joint venture must improve on the provisions relating to the parallel system."[598] Joint-venture agreements could in fact be expected to provide significant additional benefits to private mining operators, including risk-sharing, economies of scale, and priority over state-sponsored operators in the allocation of production authorizations for mining operations conducted at reserved sites.[599] Joint-venture arrangements have been increasing in importance in recent years, although there has been a growing recognition that their success is dependent upon the profit-making ability of the private contracting party.[600] Such efforts "have had spectacular successes and embarrassing failures, depending on the skill and seriousness of the participants and the care of their advance planning."[601]

Since 1976, when the basic structure of the parallel system was agreed upon in principle, Western seabed mining states, land-based producer states, and developing states had sought to negotiate an equitable compromise package which would protect and promote each of their particular economic and political interests.[602] Developing producer states have believed from the beginning of the negotiations that unregulated access to seabed nodules would "allow the developed states to become autarchic and allow them to cease being dependent upon raw materials produced in certain developing states, thereby undercutting one of their scarce sources of foreign currency income. More generally, the developing fear that access to ocean raw materials by the developed will help widen the gap between developed and developing."[603] To dispel such fears, the United States during 1975 and 1976 agreed, on condition that the Convention protect "essential U.S. interests," to provide the Enterprise with technology and financing, to abide by revenue-sharing and production-limitation formulas, and to permit amendments to the parallel system at a future review conference.[609] Developing states have nonetheless remained genuinely concerned about the economic viability of the Enterprise, and it should not be readily presumed that they would deprive the Authority — and ultimately themselves — of the benefits of a fully operational parallel system by thwarting the efforts of private mining consortia.[605] "Acceptance of a parallel system is viewed as a great concession since it would place the Enterprise in the disadvantageous position of competing with private firms and state companies which have advanced technology and management experience."[606] It has been argued that their lack of a direct economic interest in deep seabed mining would permit developing states "to stand 'on principle' because they perceive that they have little to gain by development of the resource,"[607] but it has also been observed that the *relative* political and economic value of seabed exploitation may well be significantly greater to them than to the more developed states.[608] Unlike the Reagan administration, developing states have generally not viewed the Enterprise as a potential competitor with private mining firms, but rather as a vehicle for benefiting all states, including the United States and its allies.[609] Under such circumstances, the better course of action for the United States would seem to be to accept competition from the Enterprise, with its recognized handicaps, thereby avoiding any appearance of fearing to compete with a public international mining operation and any imputation of ulterior ideological motives.[610]

Precedent

The objections of the Reagan administration were, in large part, grounded in a concern that the Convention might "set other undesirable precedents for international organizations,"[611] and "would be justifiably perceived as a further sign of American weakness."[612] Congressional critics of the treaty too have claimed that it would set "a highly adverse precedent,"[613] and that its provision for "collective ownership" would constitute "a subterfuge for despotism."[614] The Authority would certainly be one of the more advanced international organizations, with full operational as well as regulatory powers[615] — competences regarded in some quarters as "dangerous" and as having a potentially "sweeping impact" upon future international negotiations.[616] "If the United States were to [participate in] this treaty, it would accept in principle standards of international law and economic justice that would undoubtedly be extended to other important areas. Once these standards are accepted in one context, it will be difficult to rationalize resistance to their application in another."[617] In fact, however, the United States has long acknowledged the need for an Authority with a limited managerial competence, and developing states have for many years demanded major changes in the United Nations and other international organizations without a good deal of success.[618] In many important respects, a model for the Enterprise was provided by INTELSAT,[619] an economically successful organization founded by the United States which free-market advocates within the Reagan administration also sought to undermine unilaterally.[620] The extent to which the Authority's powers fall short of the radical innovations originally sought by the Group of 77 provides further evidence that even major international initiatives are unlikely to produce major changes in the nationally oriented structure of international organizations.[621] Functionalist theory does predict that success by the Authority should spill over into other specialized international agencies, but many functionalists believe that the unique characteristics of the sea make such spillover less likely, and functionalism remains, in any event, an unproven method of promoting political integration.[622] Perception of conspiratorial trends and ominous precedents is, moreover, historically more characteristic of Soviet rather than U.S. foreign policy.[623] Concern has nevertheless been expressed that the common heritage of mankind principle, in particular, might be applied to the development of "other common resources,"[624] including outer space, the moon, the geosynchronous satellite orbit, and Antarctica.[625]

Although Part XI of the Convention applies only to the exploration and exploitation of mineral resources of the deep seabed, the nodule mining regime might eventually serve as a precedent for the exploitation of other marine resources.[626] It has been argued that the technology transfer provisions, in particular, "could undermine the negotiating position of the United States in similar conferences in the future"[627]; yet much of their substance is already commonplace in licensing agreements modeled on international codes of conduct,[628] and the precedent for this type of technology transfer was in fact established years earlier at INTELSAT.[629] Although the revenue-sharing provisions of the Convention have been described as a "tax" which would "represent a major step forward in the Third World's quest for an arbitrary redistribution of global wealth,"[630] these provisions may actually set a favorable precedent for private firms, in comparison with terms and conditions frequently obtained from individual developing states in concession contracts.[631] The system of production controls has also been cited as "a potentially dangerous precedence,"[632] but such controls have become a common resource management tool throughout much of the world, including the United States itself.[633] The limited supranational powers conferred upon the Authority have been a subject of similar criticism, even though other international organizations already exercise supranational competences, and even though such competences are expected to continue to expand in an increasingly interdependent world, regardless of the fate of the Law of the Sea Convention.[634] During the military and ideological escalation of the early 1980s, treaty opponents within the Reagan administration claimed to believe that the treaty would have "enormous adverse consequences ... as a precedent" by conceding "virtually unrestrained control over a major new international organization to the developing countries and Warsaw Pact states"[635]; such claims were asserted despite the similarity of the Council's representative structure to that of existing international institutions,[636] and despite consensus procedures which may actually enhance the relative voting power of Western states.[637] Concern was also expressed over the potential precedential impact of the review conference and the antimonopoly provisions, even though the former would operate by procedures substantially similar to those of other international conferences, and even though the latter were included in the Convention at the behest of developed rather than developing states.[638] Fear of competition from "international enterprises" has apparently been inspired by loyalty to "free market principles adhered to by the United States,"[639] but one must question whether such principles are protected by rejection of the treaty regime and whether it is wise to tie effective international cooperation to the fate of economic theories. Indeed, the U.S. position toward the 1982 Convention may have itself become a precedent for U.S. obstruction of other international efforts to respond contructively where use of common resources has given rise to a need for multilateral regulation, as in the case of global warming.[640]

National liberation movements which participated as nonvoting observers at UNCLOS III have been granted the same status in the Preparatory Commission and in the Authority, in accordance with the rules, regulations, and procedures of the Authority.[641] The Convention provides that activities in the Area are generally to take into consideration the interests of such "peoples who have not attained full independence or other self-governing status," who, in particular, may benefit from revenues distributed by the Authority in accordance with its rules, regulations, and procedures.[642] The Reagan administration complained of "commercial and political difficulty" created by these provisions,[643] specifically citing the possibility of "undesirable precedents."[644] "Some U.S. Congresspersons have voiced strong opposition to the concept that organizations such as the Palestine Liberation Organization (PLO) could receive money from an Authority which is funded largely by U.S. dollars."[645] In fact, however, any such distribution of revenues by the Authority would require consensus approval by the Council, where the United States and each of its allies would exercise an effective veto power.[646] Representative nongovernmental entities have historically been considered subjects of international law, and their influence has expanded significantly since the founding of the United Nations.[647]

> Whether an entity constitutes a subject of international law and thus becomes directly affected by rights and duties on the international plane will depend upon the actual practice of States. . . . It is clear that liberation movements as a class are not *ipso facto* endowed with international personality and the question of the status of particular organisations in the field will depend upon all the circumstances of the case, including for example the claims made by the organisation and the response of the international community.[648]

International organizations have in fact accorded increasing recognition to national liberation movements in recent years, as have multilateral conferences.[649] Because recognition of liberation movements may facilitate peaceful diplomatic solutions to potentially explosive conflicts,[650] U.S. foreign policy interests can actually be promoted through such recognition, and in recent years the United States has in fact adopted a more conciliatory approach toward some of the more prominent such organizations.[651]

Treaty critics have expressed particular concern over the precedential impact of the seabed mining provisions upon future regimes to govern activities in space,[652] where rapid advances in technology are expected to continue to improve capabilities for exploration and exploitation.[653] In late 1961, at the urging of the United States, the U.N. General Assembly adopted a resolution declaring that outer space and celestial bodies may be used freely by all states and may not be subjected to national appropriation, and that outer space should be used "only for the betterment of mankind and to the benefit of States irrespective of their economic or scientific development."[654] These principles were reaffirmed two years later in a formal declaration,[655] and were again reiterated in the Outer Space Treaty of 1967, which declared outer space to be "the province of all mankind" and provided for enforcement both by states parties and by a future international organization which would carry on activities in space.[656] Subsequent treaties to accommodate particular uses have been — and will continue to be — negotiated consistent with the general framework of the 1967 Treaty — a practice so different from the one-shot, comprehensive approach employed at UNCLOS III that the precedential impact of the latter upon future outer space regimes may well be negligible.[657] Indeed, many of the principles embodied in the Outer Space Treaty served as precedents for provisions subsequently included in the Law of the Sea Convention;[658] it has thus been observed that the 1967 Treaty "holds a legitimate paternity with respect to new disciplines, such as the law of the seabed and the ocean floor which, in a first generation, borrowed the concept of the common heritage of mankind from the law of outer space."[659] Finally, it should be pointed out, increasing pressures upon the customary freedom of the high-seas regime have necessitated the rather complex regulatory schemes contained in the 1982 Convention, and the absence of similar pressures in the relatively unexplored regions of space makes any extraterrestrial application of such schemes unlikely for decades to come.

The legal regime applicable to outer space resources was further refined in an agreement governing use of the moon and other celestial bodies,[660] and the precedential impact of the Law of the Sea Convention upon the 1979 Moon Treaty had been a source of some U.S. concern.[661] At the urging of the United States,[662] the 1979 Treaty declared the moon and its natural resources to be "the common heritage of mankind, which finds its expression in the provisions of this Agreement," and required that a regime be established to govern their exploitation "as such exploitation is about to become feasible."[663] As the State Department has properly acknowledged, "[t]he Law of the Sea experience with the common heritage concept, while relevant, would in no way be controlling regarding the negotiations of any such future agreement."[664] Indeed, Secretary of State Cyrus Vance himself pointed out that the Moon Treaty only prohibits appropriation of resources "in place," permitting title to be acquired by private industry over resources removed from the surface of the moon and other celestial bodies.[665] "In effect the developed States have agreed that the common heritage principle means that an international regime should control resource exploitation. In exchange for this concession, the developing countries agreed not to insist on a provision imposing a moratorium on exploitation pending the establishment of the international regime."[666] Technological developments continue to increase the potential uses of space resources, and the Moon Treaty language would permit the economical construction of space stations from asteroids and other celestial bodies.[667]

Advances in technology have also permitted rapid growth in use of the radio frequency spectrum and the geostationary orbit,[668] and international regulation of these resources has long been regarded as necessary for their effective use.[669] "As radio proliferated, self-interest drove states to avoid radio interference through international cooperation within the ITU. However, nations still had contradictory goals with respect to spectrum regulation; each sought primarily to satisfy radio users within its own jurisdiction."[670] The frequency spectrum and geostationary orbit, like deep-sea manganese nodule deposits, are finite, nonrenewable natural resources,[671] and the ITU's five-member International Frequency Registration Board (IFRB) has been given responsibility for allocating frequency assignments and orbital positions.[672] Because they generally lack their own domestic communications systems, developing states have been particularly eager to preserve opportunities for future access by their own nationals.[673] Although it has been argued that the seabed mining regime set forth in the Convention would have a "profound" effect on access to telecommunications resources,[674] ITU proceedings have historically been characterized by "a minimum of rhetorical or ideological confrontation and a strong emphasis on compromise and consensus."[675] In fact, although satellite communication is, like seabed mining, a billion-dollar business of great national economic importance for many states, and for the United States in particular, ITU's membership agreed at a 1982 plenipotentiary conference "that radio frequencies and the geostationary satellite orbit are limited natural resources and that they must be used efficiently and economically . . . taking into account the special needs of the developing countries and the geographical situation of particular countries."[676] Six years later, despite U.S. rejection of the Law of the Sea Convention, ITU members accommodated developing states by adopting a 20-year Allotment Plan which assures each state of national access to a position of its own in the geostationary orbit — an allocation of use accepted by the Reagan administration.[677]

Although scientific research has thus far been the primary use made of Antarctic resources, commercially significant deposits of petroleum and

hard-mineral resources may yet be found to exist,[678] and the waters surrounding Antarctica contain millions of tons of harvestable krill fish.[679]
The high costs and risks inevitably associated with any exploitation of Antarctic resources would necessitate a stable legal regime,[680] and U.S. opponents of the Law of the Sea Convention have criticized the Convention's
"profound" precedential implications for the "vital" Antarctic region.[681] Although there are indeed some similarities between the deep seabed and Antarctica,[682] the history of Antarctica favors a different legal regime to
govern the exploitation of its resources.[683] During the period of colonial expansion which preceded the twentieth century, Antarctica was visited by
several European states; seven states now claim sovereignty over more than
80 percent of the continent,[689] and more than half of the 40 states parties
to the 1959 Antarctic Treaty have qualified for Consultative Party status
by conducting "substantial scientific research activity" in the Antarctic
region.[685] The 1959 Treaty froze all territorial claims and prevented any
Antarctic activities from providing "a basis for asserting, supporting, or
denying a claim to territorial sovereignty."[686] The Treaty makes no
reference to the disposition of Antarctica's resources, however, and its provisions may not bind nonparty states in any event.[687] The regimes which
might ultimately be applied to Antarctica include division, internationalization, and consortium or condominium arrangements among the Consultative Parties.[688] Developing states have criticized the exclusivity of the
1959 Treaty,[689] and resolutions calling for regimes affecting the Antarctic
environment to be negotiated "with the full participation of all members of
the international community" have been adopted — without opposition, but
without the participation of many parties to the Treaty — in the U.N.
General Assembly.[690] Not only did such sentiments emerge within the
United Nations following U.S. rejection of the Law of the Sea Convention,
but the adoption in 1988 of the Antarctic Mineral Resource Convention —
and its subsequent rejection in the United States and elsewhere in favor of
an indefinite ban on Antarctic mining activities — would seem to have largely removed the possibility that implementation of the 1982 Convention
could adversely affect the future regime for Antarctic mineral resources.[691]
Even if the 1988 Convention were never to enter into force, it must be
regarded as significant that the United States has agreed to the application
in Antarctica of a number of provisions which it claimed to have found objectionable in the Law of the Sea Convention.[692]

   While similarities between Part XI of the 1982 Convention and other
potentially exploitable common-resource regimes have been noted, several
distinguishing features of the former serve to render its direct application
as a precedent for other negotiations problematic.[693] The most significant
precedent established by the Law of the Sea Convention may in fact be the
extensive allocation of offshore resources to coastal states,[694] a jurisdictional

expansion which has enormously increased U.S. national jurisdiction over marine resources.[695] International regulation of common-resource activities is certainly not a novel concept, and there are, in fact, existing precedents for most elements of the new seabed mining regime,[696] although such broad powers have been conferred upon international organizations only where necessary.[697] The most immediate precedential impact of the regime may be felt at the 15-year review conference, where much of Part XI of the Convention would be open for renegotiation.[698] Developing states remain dependent upon the United States and its Western allies, and, while a successful Authority and Enterprise would tend to promote the goals of the New International Economic Order, the Group of 77 would risk further frustration of these goals by pursuing them too aggressively in other forums.[699] There would seem to be little reason for future U.S. negotiators to be inhibited by the UNCLOS III example,[700] since fears of domination by developing states unwilling to permit exploitation of other common-resource areas are based upon a belief that these states are prepared to undermine their own national development in an effort to subvert the economic well-being of the developed states — a premise seemingly refuted by widespread international support for economic and military sanctions advocated by the United States in 1990 following the Iraqi invasion of oil-rich Kuwait,[701] as well as by the treatment of pioneer investors in the Preparatory Commission. A successful international deep seabed mining regime without U.S. participation would deal a severe blow to U.S. economic and foreign policy interests.[702] With U.S. participation, on the other hand, any failure of the new seabed mining regime could justifiably be regarded as a strong adverse precedent for similar efforts at international institution-building;[703] but the negative precedential effects of such failure would be limited in the absence of developed-state participation, since the United States and its allies might shoulder much of the blame for its failure, even if the system were inherently unworkable. The better approach would thus favor giving the treaty regime an opportunity to succeed or fail, on its own merits and with U.S. support, as outright rejection may itself constitute a most unfortunate precedent.

Chapter 5

The Legal Status of Deep Seabed Resources

The U.S. Position

The United States contends that deep seabed mining is a freedom of the high seas under customary international law,[1] that U.S. nationals enjoy a right of access to seabed minerals under an existing *res communis* regime,[2] and that this right may only be altered by U.S. acceptance of a different legal regime through processes of conventional or customary international law.[3] The 1980 Seabed Act affirms that "it is the legal opinion of the United States that exploration for and commercial recovery of hard mineral resources of the deep seabed are freedoms of the high seas," but denies any claim to "sovereign or exclusive rights or jurisdiction over" seabed minerals, instead treating nodules as analogous to high-seas fisheries — title to which has historically vested upon capture.[4] Although the 1958 High Seas Convention does not specify deep seabed mining as a high-seas freedom, the United States has argued that it may be deduced from article 2,[5] as well as other provisions.[6] Under article 2, every freedom of the high seas must be exercised "with reasonable regard to the interests of other states in their exercise of the freedom of the high seas,"[7] and the United States has attached particular importance to the concept of reasonableness in determining high-seas freedoms, claiming "that deep seabed mineral exploitation constitutes a reasonable use of the high seas."[8] Although structured as a transitional regime to a universally accepted international convention, the U.S. Seabed Act also provides for implementation of an alternative regime under a reciprocating states agreement,[9] and the United States has negotiated two conventions with other Western seabed mining states in an effort to implement an RSA regime — a 1982 Agreement Concerning Interim Arrangements (with France, Great Britain, and West Germany), and a 1984 Provisional Understanding (including also Belgium, Italy, Japan, and the Netherlands).[10] Four U.S.-based consortia received exploration licenses from the United States in 1984 and have been contemplating nodule mining outside the Law of the Sea Convention. It has been argued that the U.S.

155

position on deep-sea mining is supported by state practice prior to UNCLOS III, "by the greater weight of scholarly opinion and by dicta of the International Court of Justice"[11]; yet it is likely that any systematic exploitation of seabed nodules would be challenged before that Court,[12] and it is therefore necessary to examine closely the validity of the U.S. position.

While deep seabed mining is not expressly authorized by the 1958 High Seas Convention, it is certainly not specifically prohibited by that treaty,[13] and support for the U.S. position has been sought in the principle, originally set forth in the S.S. *Lotus* case, that restrictions on the exercise of maritime jurisdiction must be established "by the most conclusive evidence"[14] — a principle supported by commentators who perceive a "legal vacuum" with respect to deep ocean mining.[15] "Basically, the legal vacuum theory relates that if there is an occurrence that is not covered by existing international law, then the State affected by this transpiration is free to formulate rules to meet the problems thus created."[16] The lack of controlling customary law may be attributed to the previous technological incapacity to exploit the deep seabed,[17] however, and the legal vacuum theory has been subjected to serious criticism on the ground that existing principles of international law are sufficiently general and comprehensive to permit their application in virtually all circumstances.[18] Although it has been claimed that international "policy" has favored "use of the seas,"[19] the legality of seabed mining ultimately depends upon its acceptance by the international community, and it has also been pointed out that any "legal vacuum" might just as easily be filled by the common heritage of mankind principle.[20] It is, in any event, very doubtful whether a genuine legal vacuum exists with respect to the deep seabed, and the United States has itself denied the applicability of the theory to the high seas.[21]

Some have advocated a *res nullius* regime for the deep seabed, or its divisibility pursuant to the exploitability clause of the 1958 Continental Shelf Convention.[22] In 1974 Deepsea Ventures, one of the U.S.-based private consortia, asserted "exclusive rights to develop, evaluate and mine . . . all of the manganese nodules" within a 60,000-square-kilometer tract of seabed in the Pacific Ocean, a claim based largely upon traditional mining customs and practices among Western states.[23] "The historic reservation to the acceptance of the notion of claiming the seabed is based upon the physical incapability to occupy and exploit. Some authorities believe that exploitation without occupation may be sufficient to stake a legal claim to part of the seabed."[24] Unlike the case of sedentary fisheries, offshore rights to which are derived from historic usage and contiguity as well as effective occupation,[25] the validity of claims of constructive occupation to the seabed is open to serious question.[26] Indeed, the entire analogy to the regime of unoccupied land territories appears misplaced,[27] particularly when it is understood that application of a *res nullius* regime or the doctrine of constructive occupation to the deep seabed would be incompatible with existing high-seas freedoms and would provide an undesirable incentive for seabed mining operators to "bank" vast areas of the deep seabed for future exploitation.[28] The 1958 High Seas Convention, by defining the "high seas" as "all parts of the sea that are not included in the territorial sea or in the internal waters of a State," clearly implied that the deep seabed is a part of the high seas, of which "no State may validly purport to subject any part ... to its sovereignty," and the Pardo initiative succeeded in forestalling any attempt to divide the deep seabed under the exploitability clause of the Continental Shelf Convention.[29] In the 1968 Ad Hoc Committee, it "was generally agreed that there is an area of the sea-bed and ocean floor which is not subject to national jurisdiction"[30] — a view endorsed by the United States, confirmed in the Declaration of Principles and in the 1982 Convention, and now generally shared among most, if not all, states.[31] Notwithstanding the doctrine of constructive occupation and the "reasonable regard" language of the High Seas Convention, several other Western states joined the United States in rejecting the claim of Deepsea Ventures.[32] It nevertheless appears that the United States may now intend to claim that exploratory activities by U.S. consortia could be legally sufficient to establish exclusive mining rights with respect to their claimed areas of the deep seabed under the "reasonable regard" provision of the 1958 High Seas Convention — even though its counterpart in the 1982 Convention, which provides that high-seas freedoms must "be exercised ... with due regard for the interests of other States in their exercise of the freedom of the high seas," clearly distinguishes that freedom from "activities in the Area."[33]

Adequacy of the Alternative Regime

Despite frequent comparisons by various commentators, the analogy to high-seas fisheries offered in support of U.S. unilateral deep-sea mining is a poorly chosen one. Great emphasis has been placed upon the type of physical equipment used and the environmental impact of exploitation activities; less attention has been paid to the nature of the resources in question, it being observed in passing that both high-seas fisheries and deep-sea nodule deposits are "renewable, though not inexhaustible."[34] In fact, manganese nodules are produced by a process of concretion, requiring millions of years to form,[35] whereas most species of fish are biologically capable of replenishing themselves during life cycles of only a few years.[36] It has been mistakenly argued that the principle of freedom of fishing "has not undergone fundamental transformation to the present day, although the underlying Grotian theory of the inexhaustibility of fish in the high seas, offered in support of the principle, has lost its validity...."[37] In reality, the depletion of high-seas fisheries stocks has all but eliminated the traditional doctrine of freedom of fishing.[38]

The reasonableness standard relied upon by the United States as a crucial element of its freedom-of-the-seas position on seabed mining is of relatively recent origin.[39] The reasonableness of a particular high-seas use is supposedly determined by a balancing of affected interests — a procedure which is said to produce "the exact opposite of arbitrary decision,"[40] but which in fact simply masks the highly discretionary nature of the standard being applied.[41] At least one U.S. authority on the law of the sea "is unaware of any rule that 'states may make reasonable use of the high seas',"[42] and even the standard's strongest proponents have cautioned that "reasonableness" cannot be determined unilaterally: "the end of claimed unilateral competence can only be irresolvable conflict. Multilateral concurrence, whether by customary consensus or by explicit undertaking, is indispensable to accommodation."[43] Ultimately, the reasonable use standard proves to be a vague criterion, the application of which begs the question of whether a particular high-seas activity is permitted under general international law: if the use is permitted it is termed "reasonable," but if it is not permitted it is termed "unreasonable," as the U.S. Supreme Court's varying interpretations of the "unreasonable searches and seizures" language of the Fourth Amendment to the U.S. Constitution amply illustrate.[44] The term thus adds nothing to legal analysis and only tends to lead to dominance of the international legal system by states more adept at providing after-the-fact justification for their acts by resort to economic, statistical, and other generally quantitative factors which are "weighed" in the judicial balance. Such an approach is in reality founded upon empirical considerations which cannot constitute the basis for a true system of normative law.[45] Indeed, under such an ad hoc reasonableness approach, it is not even clear whether the value of an ocean use is to be measured in relative or absolute terms: arguably, the "strongest" state — economically, politically, culturally, militarily — should be favored to prevail in any dispute regarding ocean uses, since it would always be in a position to argue that its interests were quantifiably more significant. It has thus been noted that "some of the history of the use of the high seas appears to reflect the theme 'might makes right'."[46]

The legality of a use of the high seas will ultimately depend more upon its general acceptance by other states than upon any subjective or unilateral interpretation of its "reasonableness."[47] The United States, as a major seabed mining state, could have a significant influence upon the development of international practice,[48] and the amount of formal opposition thus becomes critical in determining the legality of unilateral nodule exploitation under customary international law.[49] "Where the question at issue is not title to a parcel of land territory but to a portion of what is alternatively claimed to be high seas, the attitude of all States, whether demonstrated in recognition or forms of acquiescence, is certainly relevant."[50] The doctrine of nonrecognition may operate to prevent recognition of "legally established facts . . . having their origin in a violation of law,"[51] thereby preventing a seabed mining operator from acquiring valid title to nodules reduced to possession by capture where the legal status of the nodules did not render them subject to the rule of capture in the first place. In contrast to the lack of protest with which the Truman Proclamation was received,[52] U.S. efforts to implement deep-sea mining as a high-seas freedom have been opposed as unlawful by developing and Eastern-bloc states, and by much of the rest of the international community as well.[53] Their protests have indicated a lack of general consent to unilateral seabed mining, and the state of customary law on the subject has been characterized as unsettled.[54]

The United States has acknowledged that exclusive access to seabed mining sites "can only be conferred by international agreement among at least most of the interested states"[55]; indeed, the very enactment of legislation authorizing a reciprocating states regime belies the adequacy of seabed mining under the freedom of the seas doctrine, since the essential purpose of the Seabed Act is to provide sufficient security of tenure for private mining firms to be able to raise the substantial amounts of capital needed—but not otherwise forthcoming—to undertake nodule mining operations.[56] The Act provides for the designation as a reciprocating state of any foreign state which "recognizes licenses and permits issued under [the Act] to the extent that such nation . . . (A) prohibits any person from engaging in exploration or commercial recovery which conflicts with . . . any such license or permit and (B) complies with the [July 1, 1981] date for issuance of licenses and the [January 1, 1988] effective date for permits."[57] Each of the national seabed mining laws enacted by other states asserts that deep-sea mining is a high-seas freedom and provides for establishment of an RSA on a similar basis.[58] To ensure its validity, and the ineffectiveness of the regime established in the 1982 Convention, such an alternative regime would have to include virtually all deep-sea mining states—Japan, Canada, France, Italy, Germany, Belgium, the Netherlands, the United States, the United Kingdom, India, and China—as well as Eastern European states and a number of other smaller developed and developing states.[59] The Interim Agreement merely established a conflict resolution procedure for the national mining operators of four states,[60] and the Provisional Understanding failed to provide the kind of recognition necessary to satisfy the legal requirements of the private consortia.[61] Indeed, most parties to the Provisional Understanding, as signatories to the Law of the Sea Convention, are under a duty provisionally to comply with article 137, which forbids unilateral appropriation of "any part of the Area or its resources"[62]; while claiming a right to proceed with seabed mining under an RSA regime, these states have kept open the possibility of their adherence to the Convention.[63] Developing states have protested the proposed RSA as a violation of the common heritage principle, and resolutions opposing seabed mining under a freedom of the seas regime have been repeatedly adopted by large majorities in the Preparatory Commission and in the U.N. General Assembly.[64] The legal framework established by the U.S. Seabed Act—like the seabed mining legislation enacted unilaterally by other developed states—was originally intended to be transitional, pending establishment of a permanent multilateral regime,[65] and there continues to be widespread recognition of the need for general agreement upon a single international regime for the deep seabed.[66] Although the United States has issued four exploration licenses, continuing legal uncertainty has significantly increased the risk of seabed mining for private companies,[67] and it remains possible that the consortia may ultimately be forced to abandon the U.S. regime in favor of foreign flag states which have decided to participate in the Convention.[68]

Legal Effect of the Declaration of Principles

Because the 1970 Declaration of Principles was adopted without opposition in the U.N. General Assembly, developing states have argued that this resolution is binding upon the United States and other Western states which voted in favor of it;[69] the United States, on the other hand, has taken the position that the Declaration constitutes a legally nonbinding political statement.[70] Under article 10 of the U.N. Charter, General Assembly resolutions are "recommendations," expressly not binding upon member states.[71] Although some legal scholars have argued that General Assembly declarations can be legally binding of their own force,[72] the better view would seem to be that certain declarations—particularly those adopted without opposition—may operate instrumentally to "crystallize" customary international law.[73] "The contents of such declaratory resolutions owe what legal force they may come to have, not to the decision of the Assembly taken pursuant to a competence which it does not as yet possess, but to the practice and attitude of member states embodied in these resolutions and which develop subsequently in respect thereof as elements in the formation of customary law."[74] Such resolutions may still be said to be of some legal consequence, notwithstanding their formally recommendatory character, a view taken of the Declaration of Principles by some states.[75] "If it be true that recognition or acquiescence may be important elements of a consolidated title, it seems hardly possible to dismiss as irrelevant a General Assembly Resolution."[76] Indeed, it has been widely recognized that the Declaration of Principles—together with the 1969 Moratorium Resolution and the preceding U.N. negotiations—had a significant impact upon the continental shelf doctrine by making clear that the deep seabed is not *res nullius.*[77] However, several Western states clarified their support for the Declaration by way of reservations indicating a lack of agreement with respect to certain of its provisions, particularly the principle of the common heritage of mankind—clarifications which could hinder their crystallization into customary norms.[78]

Although the common heritage of mankind principle has been broadly construed to include various elements of the Declaration of Principles, there was disagreement on its meaning at the time the Declaration was adopted.[79] Thus, while it seemed clear that the deep seabed was the common heritage of mankind, the precise legal significance of the phrase remained to be defined through subsequent negotiations—absent agreement on its meaning, the principle would lack effectiveness.[80] The common heritage principle had not previously appeared in a formal international instrument;[81] it has since been embodied in the 1979 Moon Treaty, although its meaning was specifically limited to the terms of that agreement.[82] The plain meaning of the phrase suggests a type of property held in trust, jointly and collectively, by the world community of nations,[83] but it has been elaborated within the UNCLOS III negotiations specifically to include international administration, nonappropriation by individual states or persons, and sharing of the benefits of exploitation with developing states unable to participate directly in seabed mining activities.[84] Although such a broad interpretation of the concept has been favored by developing states,[85] Western states have maintained that seabed mining remains a high-seas freedom under the common heritage principle.[86] Unlike developing states, which regard the principle as having the binding force of law,[87] developed states have generally argued that the common heritage of mankind is a political, philosophical, and moral, but not legal, concept.[88]

In addition to setting forth the common heritage of mankind principle, the Declaration also provides that deep seabed exploration and exploitation "shall be carried out for the benefit of mankind as a whole, irrespective of the geographical location of States, whether land-locked or coastal, and taking into particular consideration the interests and needs of the developing countries,"[89] a requirement which the United States regarded as strong evidence of the interests of the entire international community in the deep seabed, but which proponents of unilateral seabed mining argued could be met simply by the expanded supply of minerals obtained from the nodules.[90] Widespread support for the Declaration of Principles was achieved only after difficult negotiations, and it was generally recognized that a workable regime for the deep seabed could only be achieved through further negotiation and compromise based on the general provisions contained in the 1970 Declaration.[91] Indeed, the Declaration itself acknowledged that the "existing legal regime of the high seas does not provide substantive rules for regulating" seabed mining,[92] and further declared that all mining activities in the deep seabed "shall be governed by the international regime to be established."[93] On the basis of the principles set forth in the Declaration, which are to be given effect through "appropriate international machinery," the regime "shall be established by an international treaty of a universal character, generally agreed upon."[94] Although developing states have claimed that the Declaration prohibits unilateral seabed mining until there is agreement on an international regime,[95] its language is subject to differing interpretations, and Western mining states have continued to assert mining rights as a traditional high-seas freedom.[96]

Following the adoption of the Declaration of Principles, it was recognized that major disagreements remained to be resolved in subsequent negotiations, and the General Assembly undertook implementation of the Declaration by voting to convene UNCLOS III in order to negotiate the deep-sea mining machinery.[97] The Seabed Committee agreed in 1971 to negotiate the new treaty within the framework of the Declaration, and there was general agreement at the initial session of UNCLOS III that the seabed mining regime would include the principles of nonappropriation, equitable sharing of benefits, and acquisition of rights solely pursuant to the future international regime.[98] Developed and developing states continued to maintain their respective positions regarding the legality of unilateral seabed mining,[99] but the legal character of the common heritage principle was endorsed by the International Law Association and by several international conferences, as well as in a number of resolutions adopted in the United Nations and its affiliated agencies.[100] Even advocates of unrestricted seabed mining have acknowledged that the significance of the Declaration and the common heritage principle has increased with time.[101] Beyond their legal effect, the very existence and subsequent elaboration of these widely supported principles reflect an important political reality, and there certainly has existed an "expectation on the part of the vast majority of the international community" that seabed mining is not a high-seas freedom[102] — an expectation which was reinforced by the adoption of the 1982 Convention and which would be further crystallized into law by its entry into force for much of the international community.[103] Indeed, the UNCLOS III negotiations themselves had a substantial solidifying effect upon the Declaration of Principles: the Convention is entirely consistent with its provisions, and its preamble expressly states an intention "to develop the principles embodied in" the Declaration.[104]

The Convention as Binding International Law

Assuming that the Law of the Sea Convention is widely ratified, it is necessary to determine its effect upon nonparties such as the United States. Under traditional principles of international law, as codified in the 1969 Vienna Convention on the Law of Treaties, conventional rights and duties may become binding as such upon a nonparty only if they are intended to be binding and are expressly accepted as such by the affected state.[105] In light of the consistent U.S. opposition, it might thus be concluded that the United States cannot be bound by the provisions of Part XI, particularly in the absence of definitive evidence of an intention to bind nonparties.[106] Since the founding of the United Nations, however, it has been increasingly recognized that even nonconsenting nonparties may become legally bound by provisions of multilateral conventions which are ratified by the overwhelming majority of the international community and which are designed to promote international order through the formulation of generally applicable norms.[107] Indeed, the revised *Restatement* of U.S. foreign relations law confirms that multilateral conventions "may lead to the creation of customary international law when such agreements are intended for adherence by states generally and are in fact widely adopted."[108] The preamble to the 1982 Convention clearly states an intention to establish "a legal order for the seas and oceans," but it remains unclear at what point the Convention would have achieved sufficient ratifications for it to bind nonparties. Unlike the 1958 High Seas Convention, which only expressed an intention "to codify the rules of international law relating to the high seas,"[109] the preamble to the Law of the Sea Convention also identifies the 1982 Convention as a "codification and progressive development of the law of the sea."[110]

To satisfy the terms of the Declaration of Principles, the Convention must be "generally agreed upon," a somewhat ambiguous requirement which would appear to call for ratification by at least two thirds of the international community,[111] as well as by a majority of each of the world's major geographical and political groups.[112] Ratifications have been somewhat slow in forthcoming, given "the typical inertia of the ratification process" and other factors, even though the unusually large number of signatories would indicate that the Convention should ultimately attract the necessary general participation.[113] An informal survey conducted in the late 1980s indicated that more than 60 states intend to ratify — including Portugal, India, China, and Japan — but that participation might not be sufficiently widespread.[114] Although several important Latin American states are considered unlikely to ratify the Convention, ratification by many others is nevertheless expected.[115] Participation by most Asian states is also anticipated,[116] and the Organization of African Unity has urged its members to speed their ratification procedures.[117] Most former Soviet-bloc states appeared prepared to proceed with seabed mining under the Convention, but their support is not certain.[118] Western seabed mining states are expected to await the conclusion of negotiations in the Preparatory Commission,[119] but many seem likely to ratify the Convention eventually, both to obtain access to deep-sea nodules and to secure the benefit of its nonseabed provisions;[120] even Germany and the United

Kingdom may ultimately decide to accede.[121] The European Economic Community may play an important role in determining the amount of Western support for the Convention;[122] unless nonparticipation by larger developed states were to threaten them with excessive financial obligations, many mid-sized developed states are expected to ratify.[123] The United States is nevertheless in a position diplomatically to be able to work to prevent the Convention from becoming generally accepted, even if it should enter into force at some time in the future,[124] although U.S. opposition alone could not stop it from becoming legally binding upon nonobjecting states.[125]

The United States has taken the position that the seabed mining regime set forth in Part XI of the Convention, because of its structural dependence upon the creation of the Authority, is "clearly contractual in nature," and therefore binding only upon states parties.[126] The International Court of Justice has, however, recognized the objective existence of open-membership international organizations such as the United Nations, as well as their power to enforce norms upon nonmember states,[127] and the Convention would confer upon the Authority "international legal personality and such legal capacity as may be necessary for the exercise of its functions and the fulfillment of its purpose."[128] Moreover, the effectiveness of the Convention vis-à-vis nonparty seabed mining states does not necessarily depend upon the objective validity of its institutional provisions:

> Certainly, an obligation to create or join an international organization would not be likely to find its way into a rule of general international law. On the other hand, an obligation not to exploit the deep seabed resources outside of a general international agreement might form the basis of international law on the subject. Arguments that this second obligation creates a basis for international law arise from normative statements found in the Convention as well as U.N. resolutions, resolutions of other international organizations, the negotiating history of the Conference, and state practices up to this date (no commercial exploitation has yet been conducted).[129]

Indeed, Section 2 of Part XI, entitled "Principles Governing the Area," is logically distinct and severable from the institutional sections;[130] although ambiguous in some respects, taken together these principles do indicate that the freedom of the high seas doctrine cannot provide an adequate legal basis for unilateral deep seabed mining operations.[131]

Significantly, the Convention provides that "[a]ll rights in the [nodule] resources ... are vested in mankind as a whole, on whose behalf the Authority shall act," and article 137 requires that deep-sea mineral resources be exploited only "in accordance with" the provisions of Part XI.[132] Recognition of any claimed rights to seabed nodules mined outside the treaty framework is prohibited, for nonparties as well as parties.[133] Indeed, a clear distinction is maintained throughout the Convention between such generally binding obligations — applicable to "all States" or simply "States"[134] — and those obligations expressly binding only upon "States Parties."[135] "Activities in the Area," defined as "all activities of exploration for, and exploitation of, the resources of the Area," are to be carried out "for the benefit of mankind as a whole," in accordance with the parallel system set forth in article 153.[136] Furthermore, the non-seabed-mining provisions of the Convention clearly imply that seabed mining activities are not a freedom of the high seas.[137] It is thus not unreasonable to conclude that the U.S. "refusal to accept the deep seabed mining part of the Convention does not prevent even that part from becoming customary international law, in full or in part."[138]

U.S. efforts to reject Part XI while claiming rights under the remainder of the treaty have required it to reject characterization of the Convention as a "package deal," in which trade-offs throughout the long UNCLOS III negotiations were intended to produce an integrated instrument totally satisfactory to none but mutually advantageous to all.[139] Under the package deal, use of which has been traced to the conclusion of the Congress of Vienna in 1815, the key question is the interrelatedness and inseparability of the provisions of the Convention.[140] Although the general validity of the package deal concept finds support in article 44 of the Vienna Convention on the Law of Treaties, which prohibits partial denunciation of a treaty unless expressly permitted in the instrument itself, its validity as a general principle of international law remains unproven with respect to nonparties to multilateral conventions.[141] The Reagan administration argued that the package deal concept was simply a negotiating technique which did not survive the conclusion of the Conference;[142] yet the consensus procedure — which certainly ended with the Conference — has been construed as enduring evidence of *opinio juris*, if not of an intent to produce an integrated treaty.[143] Arguably, U.S. application of the nonseabed provisions of the Convention in practice may prevent it from denying the binding effect of at least the noninstitutional seabed mining provisions.[144]

In fact, the United States and the Soviet Union met with Latin American states in June 1970 and agreed that the latter should support key navigational rights for maritime states in exchange for greater developing-state control over seabed mining.[145] At the urging of the United States, the Seabed Committee subsequently recommended the convening of UNCLOS III for the purpose of negotiating such a "package deal,"[146] and this objective was pursued throughout the negotiations.[147] "The major industrial powers accepted, within the total balance of the Law of the Sea package, the general principle of sharing. In return, they were guaranteed access to the Area for commercial purposes and guaranteed rights of transit passage with definitive territorial sea jurisdiction limits."[148] Among the statements and declarations made during the 1982 signing ceremony in Montego Bay, a spokesman for the Group of 77 clearly and emphatically endorsed the continuing significance of the package deal embodied in the Convention, rejecting "selective" application of its provisions[149] — a view shared by the President of the Conference, as well as by most UNCLOS III delegations, and subsequently reiterated in General Assembly resolutions urging "all States to safeguard the unified character of the Convention."[150] Further evidence of a package deal — and of an intention to create generally applicable principles of customary international law — may be found in the Convention's express prohibition of reservations,[151] of amendments to the basic principles of Part XI,[152] or of any "derogation . . . incompatible with the effective execution of the object and purpose of" the Convention or with its basic principles.[153] Evidence of a "grand" package deal thus exists in the origin, purpose, and procedure of the UNCLOS III negotiations, in the expressed understanding of many of the delegations, including the United States, and in the terms of the Convention itself. The strongest argument for a package deal cannot yet be fully made: all of the provisions of the Convention which represent progressive development of international law will have emerged as a direct result of UNCLOS III, and general acceptance of the entire Convention would make it extremely difficult for individual states to claim exemption from particular provisions.

Customary Law Outside the Convention

Should the Convention fail to become binding in its entirety through general acceptance,[154] it would still be necessary to determine whether unilateral seabed mining operations were permissible under customary international law apart from application of the Convention *per se*. Custom is determined from uniformity of state practice,[155] which must be widespread but not necessarily universal.[156] International law has traditionally also required a showing of *opinio juris*, whereby states indicate that the practice is followed out of a sense of legal obligation "and not for reasons of comity or incapacity to act."[157] The practice must have been supported by *opinio juris* for a period of time,[158] and must include "specially affected states."[159] Formation of customary law thus "requires the consent, express or tacit, of the generality of States, as was taught by Grotius,"[160] and any state objecting persistently might exempt itself from a new rule's application.[161]

In 1969 the International Court of Justice acknowledged that a treaty provision may generate "a rule which, while only conventional or contractual in origin, [passes] into the general corpus of international law, and is . . . accepted as such by the *opinio juris*, so as to . . . become binding even for countries which have never, and do not, become parties to the

Convention."[162] Five years later, the Court confirmed that customary international law regarding offshore fisheries jurisdiction had indeed "evolved through the practice of States on the basis of the debates and near agreements at the [1960] Conference,"[163] and the Court has since emphasized the binding nature of any provision of the unratified 1982 Convention which "embodies or crystallizes a preexisting or emergent rule of customary law."[164] Any such provision should have been agreed to by consensus,[165] and "should, at all events potentially, be of a fundamentally norm-creating character such as could be regarded as forming the basis of a general rule of law."[166]

> Whatever theoretical objections there may be to treating the convention ... as a statement of generally applicable rules, the examination of past state practice has lost its primacy in the methodology of international law and has begun to atrophy. Correspondingly, *opinio juris* is no longer seen as a consciousness that matures slowly over time (and finally imparts obligatory force to a practice once motivated by habit, convenience, or moral sentiment) but instead as a conviction that instantaneously attaches to a rule believed to be socially necessary or desirable.[167]

Since there is a presumption in favor of international freedom of action on the high seas, the burden of proving the existence of a rule of customary law prohibiting unilateral deep-sea mining would have to be overcome,[168] but such a rule may continue to crystallize until such time as nodule exploitation is undertaken outside the framework of the Convention.[169]

In the North Sea Contintental Shelf case, the International Court of Justice stated that the validity of a rule of customary international law derived from a treaty depends upon its acceptance by "States whose interests [are] specially affected."[170] This requirement has been cited by proponents of the U.S. position, who contend that, in the case of deep-sea mining, such states must necessarily comprise Western states "having the technological capability to mine the deep seabed."[171] The qualifying special interests pursued by the United States through nodule mining have been specifically identified as "new and politically immune sources of essential raw materials,"[172] economic benefits from an improved trade balance,[173] and the protection and augmentation of an existing advantage in the development of mining technology.[174] The ability of the United States to promote these interests through unilateral deep seabed mining is doubtful, however: a growing number of other states have been able to obtain seabed mining technology on the open market;[175] the benefit to the U.S. balance of payments position is likely to be relatively insignificant;[176] whatever unique beneficial effect the United States might enjoy from seabed mining as the largest consumer state would be largely offset by the corresponding loss of export markets among "specially affected" producer states;[177] threats of loss of access to land-based sources of nodule minerals appear to have been exaggerated by domestic seabed mining interests;[178] demand for these minerals has become so weak that private firms may be unable to conduct mining operations profitably until well into the next century;[179] and newly discovered mineral deposits within the U.S. exclusive economic zone make it likely that when seabed mining does become economically attractive it will occur within U.S. jurisdictional waters as well as beneath the high seas.[180] To the extent that the development of customary international law was based upon the practice of only those states deemed economically capable of establishing a practice, the procedure would be inherently biased in favor of developed states and inconsistent with the principle of sovereign equality enunciated by Grotius, and would thus threaten to undermine the foundation of modern international law.[181] Instead, because of the sea's status as a common resource, all states are interested in its use, and the consent of the international community as a whole should be necessary for an allocation of rights to the use of limited marine resources, although the practice of a few states might suffice for the creation of special international law where the interests of third-party states were not affected.

In the North Sea Continental Shelf case, the International Court of Justice applied the "specially affected" test in the context of a determination as to whether landlocked states had an interest in the delimitation provisions of the 1958 Continental Shelf Convention, implying that the criterion was intended to be applied only with respect to unalterable characteristics of states, such as geography,[182] and in fact the Court has refused to take into account "economic considerations" in subsequent delimitation cases.[183] In cases dealing with historical rights of coastal states to offshore fisheries, the Court has recognized that "a coastal State's exceptional dependence on fisheries may relate not only to the livelihood of its people but to its economic development,"[184] but such factors are to be considered only where their "reality and importance . . . are clearly evidenced by a long usage."[185] Moreover, the Court does not require the agreement of all "specially affected" states;[186] indeed, the United States, despite its undisputed status as a leading maritime power, proved unable to prevent expansion of the territorial sea beyond its traditional three-mile limit during the 20th century.[187] Under the original "cannon-shot rule," the outer limits of the territorial sea were actually determined by the range of coastal artillery,[188] but the impracticality of such a technology-based rule soon became evident, and states eventually settled on the fixed three-mile limit instead.[189] The inadequacy of technologically based legal rules was also demonstrated by the international community's rejection of the exploitability clause during the early UNCLOS III negotiations.[190]

In general, state practice "can only be made by those states having the capacity to act,"[191] but it is generally accepted that, as codified in the 1958

High Seas Convention, all states have an equal right to use the high seas,[192] and there are few, if any, states which are truly incapable of undertaking deep seabed mining operations.[193] Like China and South Korea, India — a developing state which has obtained technology and a "soft loan" from West Germany — became a seabed mining state in only a few years,[194] and a good number of other states have taken steps toward the commencement of their own nodule mining operations, particularly since the deadline for developing and Eastern European states to apply for pioneer investor status was extended.[195] Landlocked states have traditionally enjoyed access to the sea,[196] and both the 1958 High Seas Convention and the 1982 Convention make clear that they too have an interest in the use of ocean resources.[197] It would seem inherently inappropriate for a state's legal rights to be determined by its economic status, since the absolute cost of a seabed mining project is the same for all states but the relative costs are higher for economically weaker states; economically stronger states would thus be able to obtain rights at lower relative costs for activities to which they may attach a lower relative economic value, causing market inefficiencies where limited resources are not allocated to their highest-valued uses. Law, ultimately, cannot be determined by economic power: hardship, inefficiency, and injustice are bound to result when a state is forced to allocate scarce resources to projects such as seabed mining in order to protect its own interests and rights of access to finite natural resources.[198]

Some commentators, often citing the common heritage of mankind principle, argue that unilateral deep-sea mining would violate customary international law;[199] others have supported the U.S. position.[200] The rapid acceptance of the continental shelf doctrine and other jurisdictional novelties in recent years nevertheless indicates the primacy of state practice in establishing customary norms under the modern international law of the sea.[201] The requirement that deep-sea mining proceed only under a generally accepted regulatory regime constitutes a generalizable, norm-creating principle capable of developing into customary law.[202] To determine the validity of such a rule under customary international law, "it is necessary to examine the status of the principle as it stood when the Convention was drawn up, as it resulted from the effect of the Convention, and in the light of State practice subsequent to the Convention."[203] Although seabed mining efforts under an RSA regime would have provided state practice in opposition to the rule of necessary multilateral regulation,[204] current practice — perhaps best described as forbearance by all seabed mining operators to some degree, although the more active have in fact been proceeding under the Convention regime — is generally consistent with that rule,[205] and the continued development of deep-sea mining operations within the UNCLOS III regime would further strengthen the norm.[206]

To establish multilateral regulation of deep-sea mining activities as a customary norm of international law, the current practice of forbearance must be shown to result from a recognized duty to abstain, pending agreement on procedures for mining activities to proceed under a generally accepted regime, and not merely from economic or other considerations.[207] The 1970 Declaration of Principles provided for the negotiation of such a regime, but there was disagreement as to whether states were under a duty to refrain from interim seabed mining activities.[208] The Convention itself, as an instrument negotiated by consensus at a law-making conference in which virtually all members of the international community participated, provides significant but inconclusive evidence of the necessary *opinio juris*,[209] and many states have indicated that they regard article 137 as customary international law.[210] Since 1982 most states have made clear their view that multilateral regulation of nodule mining is indeed legally required,[211] a position supported by declarations emanating from the Preparatory Commission as well as from the U.N. General Assembly.[212] Although Western seabed mining states have generally continued to maintain that deep-sea mining is a freedom of the high seas,[213] most have joined the vast majority of states rendering their forbearance obligatory by signing the Convention and working within the Preparatory Commission to achieve a generally acceptable regime.[214] It may therefore be concluded that the norm of necessary multilateral regulation has crystallized into customary international law, at least provisionally pending final decisions on ratification of the Convention.[215]

The United States as Persistent Objector

Even if unilateral deep-sea mining were ultimately determined to be prohibited by a newly emergent rule of customary international law, the United States may still not be bound if it has objected persistently to the rule from its inception.[216] Indeed, the revised *Restatement* acknowledges that article 137's ban on appropriation of seabed minerals may become binding customary law if the Convention becomes widely accepted, but asserts that the objections of the United States would prevent it from being bound by the new rule.[217] Arguably, U.S. insistence upon its right to mine the seabed as a high-seas freedom may have exempted it from the application of any norm of necessary multilateral regulation which may have crystallized since 1967.[218] As early as 1968, the United States asserted in the U.N. General Assembly that "exploration and exploitation activities should and will continue," pending agreement on international regulation.[219] The United States formally opposed the Moratorium Resolution in 1969, maintaining its right to proceed with deep-sea mining on an interim basis, pending negotiation of international machinery.[220] By 1979, however, the U.S. delegation was asserting that any UNCLOS III treaty would only bind states parties.[221] The 1980 Seabed Act formally declared "the high seas freedom to engage in exploration for, and commercial recovery of, hard mineral resources of the deep seabed in accordance with generally accepted principles of international law recognized by the United States."[222] The repeated protestations of the United States during the final session of UNCLOS III, its refusal to vote for or to sign the Convention, its unwillingness to participate in the Preparatory Commission, and its diplomatic efforts to establish an RSA regime have been intended, at least in part, to qualify the United States as a persistent objector, at least with respect to the application of the institutional structure embodied in Part XI.[223]

Nevertheless, it remains unclear to what extent the U.S. position may have represented a bargaining position, particularly since other positions taken by the United States during the UNCLOS III negotiations may be said to have indicated acquiescence in the development of customary norms prohibiting unilateral nodule mining either before or after general agreement upon an international regulatory regime.[224] As early as 1968, the United States, in furtherance of President Johnson's 1966 policy statement, acknowledged that deep-sea resources "should be exploited in a manner reflecting the interest of the world community in their development,"[225] and U.S. efforts to maintain persistent objector status with respect to deep-sea nodule mining are likely to be additionally hampered by a 1977 acknowledgment "that these minerals are in a sense the shared property of all people."[226] Although the U.S. delegation consistently maintained that the common heritage of mankind principle set forth in article 136 of the Convention did not imply common ownership of deep-sea nodules by the international community, until the final 1982 session it did not seek any changes in article 137, which prohibits appropriation of deep-sea mineral resources and vests title to these resources in "mankind as a whole"[227] — a phrase which does imply an affirmative obligation to share the proceeds of nodule mining operations with the international community, as the United States has admitted.[228] Former Ambassador Richardson has acknowledged that revenue sharing by U.S. mining firms operating under unilateral legislation is "not only a matter of making it look better," but is "consonant with recognition of the concept of the common heritage."[229] As early as 1968, the United States acknowledged the need to dedicate a portion of the value of recovered deep-sea mineral resources to "international community purposes,"[230] and the revenue-sharing provisions of the 1980 Seabed Act have been widely regarded as obligatory, if perhaps inadequate.[231] The Seabed Act was expressly intended to be an interim measure, with renunciation of unilateral mining rights to occur upon U.S. ratification of an UNCLOS convention,[232] but commercial mining activities were nevertheless prohibited until 1988.[233] Although the United States consistently endorsed the package deal concept throughout the UNCLOS III negotiations,[234] it remains true that the obligation not to defeat the object

or purpose of a treaty begins upon signature rather than during its negotiation.[235] Moreover, the International Court of Justice has held that a state
may repudiate an emerging customary norm "by refraining from ratifying"
a convention embodying that norm,[236] although the Court subsequently
seemed to imply that more would be required for a state to qualify as a persistent objector.[237]

Even if the United States may have exempted itself from the application of any new prohibitory norm which might emerge directly from the
UNCLOS III negotiations, it remains possible that the 1970 Declaration of
Principles nevertheless gave rise to a duty to implement in good faith a
generally agreed-upon treaty regime to govern deep-sea mining, a duty
which the United States has arguably accepted and may yet be found to
have violated.[238] "President Nixon, followed by President Gerald Ford and
President Jimmy Carter, entered into active negotiations at the United Nations preparatory meetings and at UNCLOS III to develop a treaty that
would create an international regime . . . consistent with the Declaration
of Principles. At no time did our negotiators express any reservations about
these Principles."[239] At the beginning of the Conference, the United States
was clearly committed to the negotiation of a multilateral regulatory
regime, and as late as 1980 the U.S. "expectation" of such a negotiated
regime was traceable to the Declaration of Principles.[240] The Declaration
expressly requires that a "generally agreed upon" regime be negotiated to
govern deep-sea mining and that states "ensure that activities in the area . . .
be carried out in conformity with the international regime to be established."[241] The United States acknowledged the legal force of such obligations, and it thus may be said to have contemplated and agreed that the
regime would become universally binding upon becoming generally accepted.[242] The U.S. vote in favor of the Declaration, as well as subsequent
U.S. implementation of the Declaration through the UNCLOS III negotiations, may well have prejudiced its ability to claim a persistent objector
status with respect to these provisions.[243] It may be regarded as significant
that in his statement of May 23, 1970, President Nixon endorsed the view
that seabed mining was to proceed on an "interim" basis pending negotiation of a regulatory framework and that such interim mining activities were
to be authorized "subject to the international regime to be agreed upon."[244]
This position became the basis of U.S. policy with respect to seabed mining, it being accepted that "the establishment of an international regime and
international machinery for the exploitation of seabed resources beyond
the limits of national jurisdiction required agreement on a clear, precise and
internationally-accepted definition of the areas involved."[245] By the end of
the 1970s, however, the United States had significantly altered its position
with respect to deep-sea mining, from the Nixon policy — whereby nodule
mining would ultimately be undertaken in conformity with whatever

regime was agreed upon—to a full-blown freedom-of-the-seas position whereby unilateral mining was to occur notwithstanding any multilateral regulatory regime.[246] Because it seems virtually certain that the "procedural" obligations of the Declaration of Principles crystallized into customary law without U.S. objection prior to this change of position, however, the United States may be legally bound to comply with any international regime which ultimately achieves general acceptance, even if the ambiguity of its position might otherwise have preserved its rights as a persistent objector.[247]

The validity of the persistent objector rule itself is subject to serious dispute among jurists.[248] Historically, the rule has proven useful to dissenting states only as a temporary, stopgap method of avoiding the application of prevalent international norms, and its primary function has been described as "a safety valve for the 'losers' in the global process."[249] "One can look in vain through writers' discussions of the persistent objector rule for references to State practice that clearly support the rule."[250] In fact, the United States has itself been notably unsuccessful in its efforts to claim persistent objector status with respect to coastal-state jurisdiction over highly migratory species and with respect to expansion of the territorial sea beyond three miles.[251] Finally, it remains possible that the legal doctrine underlying the freedom of the seas may not permit unilateral deep-sea mining, particularly in light of the erosion of the freedom of fishing during this century. Indeed, the entire Law of the Sea Convention may itself be taken as evidence of the fading relevance of the freedoms of the high seas, and U.S. acquiescence to the unprecedented regulation of these freedoms embodied in the 1982 Convention bodes poorly for the legality of unilateral nodule mining.

Grotius Reconsidered

Quoting Ferdinand Vasquez with approval, Grotius argued that the sea "is and has always been a res *communis*," noting that its legal status was determined by "a law derived from nature, the common mother of us all, whose bounty falls on all, and whose sway extends over those who rule nations."[252] "*Res communis*" may be best translated as "common property"[253]; Grotius defined the term as "ownership or possession . . . held . . . jointly according to a kind of partnership or mutual agreement."[254] *Res communis* thus implied an element of positive regulatory control, although Grotius traced the origin of property rights to a gradual allocation of rights to common resources, which in "the primitive law of nations" had been shared by the entire human race.[255] "The distribution of goods were inevitable: their scarcity prevented everything from remaining owned by all. This was an important point for Grotius with regard to his position on the 'open seas'."[256] Grotius regarded actual possession as a necessary element of property rights, seemingly supporting the position that good title to deep-sea minerals should vest upon capture; yet he also emphasized that "discovery *per se* gives no legal rights over things unless before the alleged discovery they were *res nullius*," and that, unlike individual fish, the seas as a whole were "forever exempt from such private ownership on account of their susceptibility to universal use."[257] Indeed, while Grotius noted that the expanse of the sea precludes its effective occupation,[258] his central argument rested upon the principle that "all that which has been so constituted by nature that although serving some one person it still suffices for the common use of all other persons, is today and ought in perpetuity to remain in the same condition as when it was first created by nature.... All things which can be used without loss to anyone else come under this category."[259] A part of the sea, including any "movable" subject to possession by capture, could thus become the property of one occupying it, but only insofar as such occupation would not affect the common use.[260] "And if it were possible to prohibit any of those things, say for example, fishing, for in a way it could be maintained that fish are exhaustible, still it would not be possible to prohibit navigation, for the sea is not exhausted by that use."[261] Where such uses did indeed prove to be limited, Grotius saw justification for regulation by positive law, "so that all men might use common property without prejudice to any one else."[262]

The extent to which contemporary classical jurists agreed on the application of this *res communis* concept to the sea has been largely overlooked. In his argument for freedom of the seas, Grotius was to some extent merely amplifying the opinions of his immediate juristic predecessors, including Vasquez,[263] Francis Alphonsus de Castro,[264] and Alberico Gentili.[265] William Welwood, a Scottish jurist who wrote a response to *Mare Liberum* in the early 17th century, accepted Grotius's argument that the high seas were open to free use by all, but contended that the depletion of fishery stocks off the coast of Britain justified a claim of sovereign authority to exclude foreigners from coastal waters.[266] Similarly, Selden, after asserting that marine resources "may through a promiscuous and common use of the Sea, be diminished in any Sea whatsoever," was able to conclude that any such sea is susceptible to national appropriation.[267] "Welwood, Selden, and many others, held, in opposition to Grotius and his school, that the fisheries along a coast might be exhausted or injured by promiscuous fishing, and that the inhabitants of the coast had a primary right to the *fructus* of the adjacent sea, as against the intrusion of foreigners."[268] Although his writings have been cited repeatedly by treaty opponents in support of a natural-law theory of economic rights,[269] John Locke referred to the sea as "that great and still remaining common of mankind" and emphasized that rights of use might be said to arise under natural law "where there is enough and as good left in common for others," conditions of scarcity giving rise to a requirement that such rights be allocated through a positive-law regulatory process.[270] Citing exhaustibility as the key determinant in the allocation of ocean resources, Samuel Pufendorf, supported by Christian Wolff,[271] observed that at any point in time particular uses of the sea might create conditions of scarcity, and that such conditions gave rise to a division of rights among users.[272] In the mid–18th century, Emmerich Vattel was able to cite state practice in support of inexhaustibility as the decisive factor precluding appropriation of the high seas, "since, every one being able to find ... what was sufficient to supply their wants, to undertake to render themselves sole masters of them, and exclude all others, would be to deprive them without reason, of the benefits of nature."[273] Later authorities also cited exhaustibility as the key determinant of ocean use,[274] and depletion of oceanic resources during the 20th century prompted calls by Gilbert Gidel and others for international regulation of the high seas.[275] Indeed, in recent years the exhaustion of high-seas uses has given rise to demands for greater protection of "inclusive claims,"[276] as well as to the Pardo initiative itself.[277] Even one of the foremost defenders of unilateral deep-sea mining has acknowledged that any U.S. usufruct could only provide "the right to enjoy the property of another and to take the fruits, but not to destroy it, or fundamentally alter its character."[278]

A reformulation of the Grotian *res communis* principle would thus emphasize that the oceans, as a collective resource of the world community, may be used freely for any purpose, provided such use does not impair the interests of other users; where such impairment does occur, use of the sea must be allocated through regulation, express or implied, by the international community. In essence, the *res communis* regime permits "exclusive use of limited duration and scope,"[279] so long as that use does not constitute an abuse of right.[280] The sea — including the surface of the water, the water column, the seabed, and all minerals and organisms found therein — is regarded as a juridical unity, subject to a single constitutive regime governing its use, a view supported by traditional jurisprudence as well as by the language of the 1958 and 1982 conventions;[281] indeed, both the claim of maritime sovereignty and the appropriation of limited marine resources may be properly regarded analytically as processes of unilateral exclusion, since other users are necessarily deprived of the benefit of common access. Both theoretical and empirical considerations point to this reformulated *res communis* regime as a norm of natural law,[282] providing an international framework within which subordinate regulatory mechanisms evolve where exhaustion of ocean resources is threatened. Such positive-law regulation may be customary or contractual in form,[283] but it must always result from consent and acceptance by the international community as a whole, since all states share an interest in the sea as a common resource.[284] Thus, as even Selden was careful to point out,[285] exclusive functional or territorial jurisdiction over sea resources results from a positive allocation by the international community rather than from isolated acts of claim and counterclaim among completely dissociated sovereign states.[286]

The Theoretical Case for a Res Communis Regime

Although the foregoing reformulation of the Grotian *res communis* concept has rarely been set forth in explicit detail, there has been widespread recognition in modern international law of shared interest in the sea as a potentially exhaustible common resource.[287] Arvid Pardo has pointed out that the freedom of the high-seas doctrine of nonregulation "is rooted in the assumption of abundance,"[288] and the need for international regulation to prevent overexploitation of common marine resources has been emphasized by other 20th-century jurists, including C. John Colombos,[289] Francisco Garcia-Amador,[290] and Sir Gerald Fitzmaurice.[291] Recognizing the seeming inexhaustibility of use of the sea for navigation, Myres McDougal pointed out that one ship may readily follow another over a single area of the sea's surface.[292] In 1926 the League of Nations Committee of Experts concluded that international regulation of high-seas fisheries would be necessary to avert their extinction, and there was considerable support within the International Law Commission for inclusion in the 1958 High Seas Convention of a provision proscribing "any acts which might adversely affect the use of the high seas by nationals of other States."[293] In agreement with other states,[294] the United States has itself acknowledged that under the *res communis* principle the entire international community shares an interest in deep-sea minerals.[295]

In order to withstand intellectual scrutiny, the *res communis* theory must provide an explanation for the doctrine of the freedom of the seas. The United States, officially acknowledging no limitation on the doctrine based on the exhaustibility of ocean resources and interpreting "freedom" as an absence of physical restraint, has, in effect, derived an equality of meaning between the *res communis* principle and the doctrine of freedom of the seas.[296] The freedom of the seas principle originally evolved as a legal doctrine for the protection of navigation rights against unwarranted encroachment by individual states, the essential inexhaustibility of this particular use of the sea justifying a minimal regulatory regime for high-seas navigation.[297] It has nevertheless been pointed out that, insofar as it may be said to embody the restrictions on exhaustible uses which inhere in a *res communis* regime, the freedom of the seas principle "has also a negative side," which operates as a restraint upon such uses.[298] "When it becomes clear that a use is exhaustible, the traditional concept of freedom of the seas ceases to control because it is theoretically and practically inapplicable to a limited resource."[299] A memorandum prepared for the International Law

Commission by the U.N. Secretariat in 1950 thus emphasized that the freedom of the seas doctrine provided no positive regime to govern utilization of high-seas resources,[300] and the Commission, with a nod to the criterion of exhaustibility, subsequently concluded that a high-seas freedom, "to be exercised in the interest of all entitled to enjoy it, must be regulated."[301]

The "freedom" of the seas may best be understood as an economic concept, signifying cost-free use under conditions of inexhaustibility, rather than as a right of unfettered access to limited ocean resources.[302] The externalities inherent in many ocean uses interfere with other users — both present and future — of marine resources, rendering international regulation necessary to prevent the inefficiency which would otherwise result from unrestricted access.[303] Maximization of the present economic value of a given ocean resource requires that exploitation proceed only up to the point where marginal revenue equals marginal extraction costs, both private and social;[304] under an open-access regime, however, the social and environmental costs of ocean use are fully discounted, and usage will continue to increase until marginal private costs have equaled marginal revenues, resulting in overexploitation to the extent of the unaccounted-for externalities.[305] Where such overexploitation will cause depletion of a common marine resource, it therefore becomes necessary to allocate rights and conditions of access both to preserve the resource base and to promote the maximization of net revenue by capturing economic rent which would otherwise be wasted.[306] The particular type of regulatory regime adopted by the international community with respect to a given use of the sea will tend to reflect the specific characteristics of that use; jurisdiction may be vested in the international community or in individual states, but "the central purpose of any institutional arrangement designed to enhance economic efficiency is the creation of rules, regulations and other constraints that correspond to the actual costs of using resources."[307] Although nonmarket resource allocation has been criticized as inefficient, such allocation may be justified where the value of the economic benefit obtained exceeds the costs of assigning and enforcing rights and conditions of access.[308]

It has been recognized that unrestricted access to ocean resources generally results in overexploitation, overcapitalization, reduced revenues, and low rates of return on investment, as well as depletion of limited resources.[309] Crowded sea lanes require restrictions upon shipping traffic as well as upon the discharge of marine pollutants; although high-seas fisheries are a self-renewing resource, they are nevertheless finite and subject to overexploitation and exhaustion where the rate of depletion exceeds their rate of renewal.[310] "This occurs because, under free and equal access, no individual fisherman can afford to restrain his own effort in the interest of future returns, since anything he leaves in the sea for tomorrow will be taken by others today."[311] It has been argued that fish and oil — unlike sea-bed nodules — are "common-pool" resources for which joint ownership and regulation may be necessary in order to avoid overexploitation, which would otherwise result given "the difficulty of identifying, keeping track of, and asserting property rights."[312] Clearly, it has been the need to accommodate conditions of exhaustibility which has led to the "enclosure" of common lands in England and in the western United States, as well as to other forms of regulation of common resources.[313]

Notwithstanding early U.S. representations that deep-sea nodule deposits were sufficiently abundant to permit unlimited mining operations,[314] it is now clear that manganese nodules are an exhaustible, essentially nonrenewable marine resource concentrated in only a few potentially viable mining sites on the deep ocean floor.[315] Because nodule mining sites cannot be replenished, their exploitation would permanently deplete the existing nodule stock, inevitably affecting the rights and interests of other potential seabed mining operators; whether or not nodules constitute a "common-pool" resource, the key determinant is exhaustibility, particularly in light of the varying composition, abundance, and exploitability of nodule deposits.[316] Under a regime of unrestricted access the limited supply of mining sites may be expected to lead to overcapitalization,[317] claim jumping,[318] and conflict among seabed mining states.[319] The need for international regulation under such circumstances is clear. Polymetallic sulfides, in contrast, require only decades — or less — to form, and may in fact be inexhaustible.[320]

The Historical Case for a Res Communis Regime

As Grotius pointed out, under "the primitive law of nations, which is sometimes called Natural Law," prehistoric communities recognized "no particular right" to exclusive use of common resources, including the sea.[321]

> Open access and free use regimes were universally regarded as both rational and just when it was generally assumed that the usable resources of the nonland areas were vast and abundant. Apart from some coastal navigation and fishing areas, where users might get in each other's way, the high seas seemed abundant enough for everyone's use and therefore not in need of any kind of regime to allocate their use.[322]

Although several ancient Mediterranean civilizations — including the Phoenicians and the Carthaginians — sought to control areas of the sea,[323] it is generally agreed that the exercise of such control by military force could not have constituted genuine legal dominion, since there was as yet no effective system of public international law.[324] Faced with threats from foreign powers as well as from pirates, a succession of Greek city-states achieved naval preeminence in the eastern Mediterranean Sea,[325] and Alexander the Great dominated much of the ancient world in the fourth century B.C.,[326] but it was not until several decades later that the Rhodian Sea Laws achieved general acceptance as the first internationally recognized code of maritime conduct.[327] The earliest known formal description of the legal status of the sea itself dates from the second century A.D., when Marcianus identified the seas of the Roman Empire as *res communis* under the law of nature,[328] a classification maintained in subsequent codifications of Roman law, which preserved the freedom of the seas in an era when marine resources were rarely depleted by use.[329] "According to this body of law, then, the sea is common to all, both as to ownership and as to use. It is owned by no one. It is incapable of appropriation, just as is the air. And its use is open freely to all men."[330] Freedom of the seas also existed – in practice if not in law – among ancient Asian maritime cultures,[331] although exclusive exploitation rights to exhaustible pearl fisheries off the coast of Bahrain and Ceylon were recognized as lawful.[332]

Although much of Roman law was lost during the Dark Ages which followed the fall of the western Empire,[333] classical law was revived by the glossators of the 12th century,[334] and the juridical supremacy of natural law was fully acknowledged in the late Middle Ages as the Catholic Church sought to reclaim the political unity of the Roman Empire.[335] With the expansion of maritime commerce, freedom of navigation was initially exercised throughout medieval Europe;[336] yet growth in fish consumption and evolution of feudal doctrine soon gave rise to restrictions upon freedom of fishing as exclusive rights were claimed to valuable offshore fisheries.[337] Although Henri de Bracton restated the freedom of the seas doctrine in the 13th century,[338] under the Stuarts England's long-standing practice of requiring a flag salute from foreign ships in offshore waters was transformed into a claim of sovereignty over "the British Seas."[339] In the late Middle Ages, sovereignty of the seas was also claimed by Italian city-states and Scandinavian kingdoms,[340] as well as by Portugal and Spain.[341] "It is thus clear that long before the beginning of the seventeenth century, the original simplicity of the Roman law regarding the appropriation of the sea had undergone a change. . . , and that the doctrine of sovereignty or dominion over a very considerable maritime zone was widely held by jurists."[342] Yet the defenders of maritime sovereignty were unable to formulate a generally recognized rule of law in support of such acts of appropriation; the process of acquisitive prescription would not have been inconsistent with a natural-law *res communis* regime where sea resources were threatened with exhaustion from competing users, but the threat of such exhaustion was generally lacking with respect to navigation, and the historical lack of general acquiescence to these claims cast serious doubt upon the validity of any prescriptive theory.[343] Medieval claims to maritime sovereignty were sanctioned by the Pope, and a papal competence to allocate rights to scarce marine resources was similarly consistent with a *res communis* regime;[344] yet there is little doubt that corrupt popes abused their authority by approving appropriation of the sea where conditions of scarcity did not in fact threaten.[345] Without offering any satisfactory explanation for a fundamental change in the legal status of the sea, some 17th-century jurists, reading Selden too broadly, sought to justify claims of maritime sovereignty based upon the alternative theory of occupation, whereby the sea was regarded as *res nullius* rather than *res communis*.[346]

The decline of feudalism and the triumph of the Protestant Reformation in 1648 gave rise to the modern system of international law, a system built upon a community of independent sovereign states and the teachings of Hugo Grotius.[347] "Being sovereign, being the supreme power, did not mean being free from rules, being above the law. The concept of a world community, no longer visibly represented by the Holy Roman Empire and the unifying factor of the Roman Catholic Church, was nevertheless maintained — as a legal community."[348] Under this decentralized system, rights to limited marine resources were allocated by international consent, which, before multilateral regulatory agreements became widespread in the 20th century, was manifested by customary state practice and given the force of law by *opinio juris*.[349] During the 18th century, freedom of the high seas became an accepted rule of international law, in recognition that it was in the common interest "to minimize exclusive rights and to admit all states in complete equality to exploit the available resources, at least up to the point of exhaustibility"[350]; with the expansion of colonialism and mercantilist trade, maritime states developed a greater appreciation for free navigation, subject to limited restrictions intended primarily to secure the safety of commercial shipping traffic where congestion developed.[351] The extensive and varied claims to maritime sovereignty, never having attained the full force of law, were transformed into a generally recognized legal principle: a territorial sea which geographically limited the exercise of such sovereignty, initially to the distance of a cannon shot from shore and then, by the end of the 18th century, to a width of three miles.[352]

> The various uses of the sea near its coast render it very susceptible of property. People there fish, and draw from thence shells, pearls, amber, etc. Now in all these respects its use is not inexhaustible; so that the nation to whom the coasts belong may appropriate to itself an advantage which it is considered as having taken possession of, and made a profit of it, in the same manner as it may possess the domain of the land it inhabits.[353]

The territorial sea was also intended to function as a security zone — a coastal-state use of the adjacent sea which limited the navigation rights of foreign vessels[354] and the general recognition of a concomitant right of innocent passage is evidence that the legal character of the "territorial" sea may be properly understood as a positive allocation of rights to exhaustible uses of the sea rather than as a simple seaward extension of territorial sovereignty by coastal states.[355]

Because the boundaries of the territorial sea did not correspond perfectly to those areas of the sea increasingly threatened with exhaustion from overuse, coastal states sought further regulation of ocean uses through an expansion of territorial waters and the creation of more limited zones of functional jurisdiction, a disorderly process which led to the convening of UNCLOS I in an effort to negotiate a uniform allocation of maritime rights and duties.[356] States parties to the 1958 Fisheries Convention have formally acknowledged "that the development of modern techniques for the exploitation of the living resources of the sea . . . has exposed some of these resources to the danger of being over-exploited," and that the traditional right of freedom of fishing must be subject to "necessary" measures for the conservation of living high-seas resources;[357] in 1974 the International Court of Justice confirmed that "preferential rights" to high-seas fisheries had accrued to coastal states under customary international law, as had the right to assert a 12-mile-wide exclusive fishing zone and the duty to conserve living marine resources.[358] When the exploitation of finite offshore oil deposits became economically feasible in the mid–20th century, the international community quickly allocated exclusive resource rights by developing the legal concept of the continental shelf, a positive-law doctrine which rested upon neither territorial sovereignty nor geological contiguity and which preserved unrestricted access to the same physical area for essentially inexhaustible uses such as navigation, scientific research, and the laying of submarine cables and pipelines.[359] The approach taken by the International Law Commission in drafting the High Seas Convention — permitting use on a nonexclusive basis, subject to regulation for the purpose of protecting the interests of the international community — was wholly consistent with the *res communis* principle;[360] the Convention thus prescribed a general duty of states to control vessel-source pollution and to "take measures to prevent pollution of the seas from the dumping of radioactive waste, taking into account any standards and regulations which may be formulated by the competent international organizations."[361] Although nuclear weapons testing, like navigation, excluded other users only temporarily, the magnitude and duration of the exclusion — as well as the danger posed by the release of radiation — nevertheless favored regulatory control in order to avoid interference with other ocean uses, and the 1963 Test Ban Treaty ended most oceanic testing of nuclear weaponry.[362] Problems relating to the accommodation of conflicting users became increasingly common with the development of new technologies, further undermining support for the traditional rule of unrestricted access.[363]

Because certain interests — fiscal, customs, sanitation, and immigration — are by their nature exclusive to the coastal state, use of offshore areas of the high seas to promote such interests was properly accommodated through the creation of functional contiguous zones which minimize interference with freedom of navigation.[364] By 1969 the need for international regulation of ocean uses was fully appreciated by the Stratton Commission, which concluded that the sea "cannot be understood, utilized, enjoyed or controlled by diffuse and uncoordinated efforts."[365]

The 1982 Convention remains consistent with an overriding natural-law *res communis* regime, establishing greater regulatory control over newly exhaustible uses of the sea while safeguarding the shared interests of the international community in such uses.[366] "Thus there has been a shift of emphasis in the direction of treating the ocean as an ecological whole, subject to multiple uses which must be regulated together."[367] Of the six high-seas freedoms listed in article 87 (1), all but two (navigation and overflight) are now "subject to" conditions and limitations set forth elsewhere in the Convention. The Convention imposes a general duty upon all states to conserve high-seas fisheries and sets "maximum sustainable yield" as the optimal exploitation level to avoid exhaustion, both within the exclusive economic zone and beyond, and states are required to cooperate to promote "the objective of optimum utilization" of highly migratory species;[368] although the allocation of exclusive managerial authority to a fixed distance of 200 miles is somewhat arbitrary, the international community has required coastal states to make surplus stocks available to foreign fishing states.[369] Scientific research and the construction of artificial islands and installations are specifically identified as high-seas freedoms, but these freedoms too have been eliminated within the exclusive economic zone and upon the continental shelf in order to avoid interference with coastal-state security interests and exploitation rights.[370] The failure to list production of energy from the sea as a high-seas freedom may be understandable in light of the interference with navigation which may result from the semi-permanent installations associated with such energy production and from the wide "safety zones" which states would need to establish around these installations.[371] The freedom to lay submarine cables and pipelines, which poses no exhaustibility problem and only minimal interference with other ocean uses, has been preserved on the continental shelf as well as on the deep ocean floor, subject to the limited coastal-state regulatory authority set forth in article 79.[372] Although pollution has not been expressly identified as a high-seas freedom, it may properly be regarded as such — a freedom which has necessarily become subject to international regulation as its harmful effects on the marine environment have become more fully appreciated.[373] The Convention requires states to establish and comply with international rules and standards for the prevention and control of vessel-source pollution, in effect delegating primary authority for the formulation of such regulations to the International Maritime Organization and conferring enforcement jurisdiction upon coastal as well as flag states.[374] Even coastal-state sovereignty over the 12-mile territorial sea is to be exercised "subject to" the limitations prescribed in the Convention, including transit passage and newly defined rights of innocent passage.[375] The U.S. Department of State Legal Adviser pointed out in 1972 that the resource rights of coastal states beyond their territorial waters would have to be "based on an express delegation of authority from the international community,"[376] and the positive-law, "*sui generis*" character of the exclusive economic zone is made clear by the partial incorporation of high-seas freedoms in article 58 of the Convention, as well as by the extension of the continental shelf regime to a distance of 200 miles from shore where no prolongation of the natural shelf exists.[377] By requiring that a percentage of the production value of exploitation be paid to the Authority by coastal states, the international community has retained an even greater interest in those areas of the continental margin which extend beyond 200 miles. International regulation of scarce deep-sea minerals was implicit in the concept of the common heritage of mankind,[378] and has been elaborated in the specific provisions of Part XI as well as in the 1969 and 1970 General Assembly resolutions;[379] indeed, the United States would have little need for assured access if seabed nodules were not an exhaustible ocean resource.

Res Communis as a Peremptory Norm

Since states are generally free to vary rules of international law through prescriptive or conventional processes,[380] it is necessary to determine whether the United States may avoid application of a *res communis* regime by undertaking deep-sea mining operations either unilaterally or within the framework of a reciprocating states agreement. A large majority of the international community expressly approved the legal concept of *jus cogens* set forth in article 53 of the Vienna Convention, which defines "a peremptory norm of general international law" as "a norm accepted and recognized by the international community of States as a whole as a norm from which no derogation is permitted and which can be modified only by a subsequent norm of general international law having the same character."[381] Just as national legislative and executive acts cannot exceed constitutional limits and private contracting parties cannot make agreements among themselves in contravention of public laws,[382] states ultimately cannot be completely free to create potentially inconsistent norms among themselves if a unified system of international law is to be maintained.[383] Notwithstanding the positive-law flavor of the *jus cogens* principle embodied in the Vienna Convention,[384] the element of consent would appear to be inherently inconsistent with the peremptory character of the *jus cogens* concept, which rather corresponds to the classical doctrine of natural law,[385] the primacy of which was recognized by Grotius,[386] Vattel,[387] and other classical jurists.[388]

> While in the theory of international law the term *jus cogens* has appeared rather recently (from the beginning of the 1930s), an idea of absolutely compulsory rules of law serving as criteria of the validity of international treaties has existed in the doctrine of international law for centuries. . . . The positivists of the nineteenth and twentieth centuries, except some most radical ones . . ., did not accept full freedom of the will of States making a treaty.[389]

Although most jurists agree upon the existence of peremptory norms,[390] there nevertheless has been little evidence in modern state practice of norms accepted and recognized as peremptory.[391] Indeed, since article 53 itself has not been "accepted and recognized by the international community of States as a whole" as *jus cogens*, preexisting natural-law peremptory norms must continue to be respected as such under the Convention's own terms, and the effort to limit the scope of *jus cogens* to "accepted and recognized" peremptory norms may therefore be properly understood as an impermissible derogation from the peremptory rule of *jus cogens* itself.[392] Since states are not permitted to agree among themselves to derogate from peremptory norms, it follows *a fortiori* that unilateral acts — prescriptive or otherwise — in derogation therefrom must also be invalid, as the revised *Restatement* of U.S. foreign relations law has recognized.[393]

Despite general agreement upon the existence of peremptory norms, the specific identity of such norms became a matter of considerable controversy at the Vienna Conference,[394] as well as among jurists.[395] *Jus cogens* status has been attributed to certain axiomatic rules relating to the structural foundation of the system of international law, such as *pacta sunt servanda*.[396] Norms prohibiting slavery, genocide, and other crimes against humanity have also been widely regarded as peremptory.[397] The International Law Commission cited the prohibition on the use of offensive force set forth in article 2 (4) of the U.N. Charter as an example of *jus cogens* — a view reaffirmed by a number of states at the Vienna Conference and by the United States itself.[398] More generally, the Commission recognized the invalidity of agreements which infringe upon the rights of nonconsenting third parties,[399] *jus cogens* being particularly applicable where the norms in question function to safeguard the common interests of the entire international community.[400] As an *in rem* regime comprising rights and duties vis-à-vis all states with respect to resources located beyond the limits of national sovereignty, the principle of *res communis* may be regarded as perhaps the clearest example of a peremptory norm.[401] Grotius himself emphasized that any attempt to derogate from this principle would be "dealing with the property of others" and therefore invalid, noting that "any one who uses a *res communis* does so evidently by virtue of common and not private right, and because of the imperfect character of possession he can therefore no more set up a legal title by prescription than can a usufructuary."[402]

Some commentators have identified the doctrine of the freedom of the high seas as a peremptory norm.[403] Other jurists, however, have correctly pointed out that states may lawfully agree to limit their own use of the sea.[404] Because the exercise of rights of unrestricted access under the governing *res communis* regime is fundamentally voluntary in nature and is conditioned upon inexhaustibility, such *inter se* derogation from the freedom of the seas doctrine is consistent with that regime's natural-law basis; derogation is in fact required — on a general, multilateral basis — where ocean use proves exhaustible.[405] Yet, as the International Law Commission recognized, an attempt to impair such rights of access with respect to *other* states under conditions of inexhaustibility would be invalid: "An agreement to appropriate or assert exclusive jurisdiction *erga omnes* on or over the high seas, would *per contra* be directly contrary to international law, the freedom of the seas as *res communis* being *jus cogens*."[406] The nonappropriation element of the *res communis* principle — mandatory in nature regardless of whether an ocean use is exhaustible or not, although some exclusive rights of use may be allocated under an international regulatory regime in the former case — is thus a true peremptory norm, prohibiting derogation even *inter se*; similarly, the norm of general international regulation, applicable under the *res communis* regime where conditions of exhaustibility arise, allows for no *erga omnes* derogation without impairment of the interests of the entire international community.[407] States may, however, agree among themselves to alter the application of certain positive-law regulatory regimes so long as the rights of other states remain unaffected, as when a foreign maritime state agrees with a coastal state to observe particular rules with respect to navigation in the latter's offshore jurisdictional zone.[408] As classical jurists recognized, violation of the *res communis* principle governing use of the sea will result in international conflict: "Can any thing be more just, than for the whole world to rise up against a nation that wishes to prevent others from sharing those common advantages, which, by the laws of nature and of nations, belong equally to all?"[409]

Although there have been relatively few instances in the modern era of attempts to assert exclusive rights with respect to inexhaustible uses of the sea — in part because such uses have themselves become increasingly uncommon, and in part, perhaps, because states have implicitly recognized the validity of the natural-law *res communis* regime applicable to the sea — ancient and medieval history provides a number of examples of military conflict engendered by exclusive maritime claims. "In most cases, . . . the appropriation of the sea was effected by force and legalised afterwards, if legalised at all, and the disputes on the subject between different nations not infrequently led to sanguinary wars."[410] There was clearly an ongoing contest among ancient Mediterranean civilizations over control of the sea, as successive claimants were challenged militarily by neighboring nations; Carthage's exclusive claim, whereby Romans were not even permitted to wash their hands in the Sicilian Sea, has been well documented as a cause of the Punic wars.[411] Indeed, resistance to such claims was so strong that Isocrates was prompted to observe that "this command of the sea which we affect, and term a sovereignty, though it be more properly named a misfortune, is sufficient to bring indiscriminate ruin on all who possess it."[412] In the late Middle Ages, Danish and Norwegian claims to northern waters led to wars with other Baltic states, as well as with Holland and England — in 1602 Elizabeth I formally notified Christian IV of Denmark that the seas were to be kept open for free and peaceful use unless the English consented to Denmark's navigational restrictions.[413] Gustavus Adolphus's alarm over German efforts to secure exclusive control of the Baltic Sea brought Sweden into the Thirty Years' War.[414] The medieval claims of Italian city-states were disputed by France, Spain, the Hanseatic League, and others, although the role of Venice as a buffer against piratical Saracens and Turks in the eastern Mediterranean made maritime states reluctant to undertake a direct military challenge to the tribute it imposed upon vessels navigating the Adriatic Sea.[415] England's 17th century claims to exclusive fishing rights and maritime sovereignty provoked several wars with the Dutch and ongoing conflict with the French;[416] its interference with high-seas navigation eventually caused France and Spain to enter the American War of Independence, precipitated the War of 1812,[417] and led Talleyrand to declare the intention of France "to fight on the sea, not for herself alone, but to liberate the ocean and to emancipate all peoples who are victims of England's cupidity."[418] In 1941, several months before the United States entered World War II, President Franklin Roosevelt publicly protested "the Nazi design to abolish the freedom of the seas and to acquire absolute control and domination of these seas for themselves," warning of U.S. retaliation against Germany for its continuing attacks on nonbelligerent vessels.[419]

The classic historical example of derogation from the res communis principle was, of course, the purported division of the world's oceans between Spain and Portugal at the end of the 15th century — an act of appropriation which provoked armed conflict with emerging European maritime states throughout the following century.[420] Indeed, Columbus's voyage violated exclusive maritime rights claimed by Portugal pursuant to previous papal grants, and the Treaty of Tordesillas (1494) was itself a product of the Portuguese threats of war which ensued when Pope

Alexander VI was persuaded to confirm Spanish claims to such rights in the Western hemisphere.[421] Less than a year after the Pope acted, France invaded Italy, marching through Rome and threatening Spanish interests in Naples before the Pope was able to organize a defensive league headed by Spain; the ensuing state of open hostility between the two neighboring Catholic states continued almost uninterrupted until 1659, despite the common threat presented by the Protestant Reformation.[422] Lacking the military resources available to France at the time, England avoided open involvement in the war, but Henry VII defiantly commissioned several expeditions to sail through the northern Atlantic waters claimed by Spain and handsomely rewarded mariners who provided new discoveries for England.[423] Although such voyages proved commercially unsuccessful and royal support for overseas expeditions waned after 1505, English fishermen continued to sail through Spanish waters to exploit rich Newfoundland fisheries, as did the French and the Portuguese,[424] and Henry VIII undertook a major expansion and modernization of England's navy.[425] Francis I of France maintained a more active challenge to Spanish control of the Western hemisphere, dispatching Giovanni da Verrazano and Jacques Cartier across the Atlantic and informing the Spanish ambassador that the seas were open to free use by all and that the papal bulls could have no binding *erga omnes* effect.[426] Bolstered by letters of reprisal issued by Francis, French corsairs attacked Spanish and Portuguese shipping throughout the Atlantic.[427] When the Treaty of Cateau-Cambrésis was concluded in 1559, France and Spain sidestepped the contentious issue of access to the New World by agreeing that the peace treaty would not apply west of the Azores or south of the Tropic of Cancer, and attacks "beyond the line" were continued by Huguenots with the support of French rulers and in alliance with Protestant seamen from England.[428] During the second half of the 16th century, western voyages by English merchants and privateers – including those of John Hawkins and Sir Francis Drake – were backed financially and politically by Queen Elizabeth,[429] whose unwavering support for the principle of the freedom of the seas was one of the main causes of the breakdown of the Anglo-Spanish alliance and of the war which ensued between these two states during the last 15 years of her reign.[430] "Drake did this, Hawkins that, Frobisher the other, with consequences which, somewhat astonishingly, added up to the foundation of the British Empire, accounted for as an unintended by-product of contempt for Spain and hatred of the Pope."[431] Although James I canceled the letters of marque Elizabeth had issued and made peace with Spain, England still refused to recognize the exclusive claims of Spain and Portugal; freedom of navigation was asserted throughout the 17th century by colonists and by merchants eager to trade with Spanish colonies and to establish commercial ties with southern Asia.[432] Near the end of the 16th century, Dutch privateers joined their English and French counterparts in raiding the Spanish Indies, and by the middle of the following century the Dutch East India Company and the Dutch West India Company were making substantial inroads into the Iberian trading monopolies.[433] "The later seventeenth century was a period of constant imperial clashes in America.... Both France and England made use of the international buccaneers of the West Indies whenever policy and opportunity coincided."[434] As late as the 18th century, war resulted when an English sailor's ear was cut off by Spaniards searching for contraband on the high seas.[435]

While other causes also contributed to the rise of Protestantism, it is clear that the global maritime division effected by papal bull on behalf of Spain and Portugal was widely regarded within the non–Iberian states where the Reformation subsequently flourished as a gross and unacceptable abuse of papal authority.[436] In challenging papal excesses, theologians of the early 16th century emphasized, in particular, the invalidity of papal acts which contravened divine – i.e., peremptory – natural law.[437] "Machiavelli seems to have regarded the deposition of the pope by international action as a practical possibility."[438] Certainly, it is no coincidence that Spain, which absorbed Portugal in 1580, assumed a leadership role in the Counter-Reformation on behalf of its papal benefactor, upon whose continued authority the legitimacy of its economically vital New World claims ultimately depended.[439] Grotius, writing in defense of the right of the newly independent and largely Protestant Dutch republic to sail to the East Indies through waters allocated to Portugal by the papal donation, argued adamantly "that the authority of the Pope has absolutely no force against the eternal law of nature and of nations, from whence came that liberty which is destined to endure for ever and ever."[440] Notwithstanding the religious overtones of such natural-law references, the identification of the res communis principle as a peremptory norm should not be construed as an implication of divine ordinance beyond the comprehension of the human intellect, but rather as the embodiment of a profound perception of the shortcomings of human nature and the consequences of unrestrained self-interest, achieved in a former era of enlightenment during which mankind was temporarily able to transcend the limitations of irrationality.

While most of the foregoing situations involved the assertion of exclusive claims with respect to inexhaustible ocean uses – largely because, except for fisheries, the sea and its resources were sufficiently vast to accommodate all users until the 20th century – the peremptory character of the res communis principle would also operate to prohibit exclusive use of exhaustible marine resources except pursuant to an allocation of rights by the international community.[441] Because the positive-law regime required by conditions of exhaustibility was traditionally established by state practice rather than by a process of multilateral negotiation, some degree of conflict was often present during the process of normative development, as was the case in the long, disorganized effort to achieve a generally accepted definition of the territorial sea; during the second half of the 20th century, the United Nations has played an important role in the peaceful and relatively rapid development of general norms to govern the exploitation of exhaustible marine resources, as has been the case with respect to the continental shelf doctrine and the exclusive economic zone. Although diplomatic pressure has frequently forced overfishing states to implement conservation measures, the voluntary nature of such measures under traditional law of the sea doctrine — which regarded such regulatory measures as fundamentally inconsistent with the freedom of the seas doctrine, when in fact both are complementary aspects of the governing *res communis* principle, their respective applicability being determined by the exhaustibility of the resource — has occasionally caused irreparable damage to some fisheries stocks and more frequently led to open conflict.[442] The renewability of fisheries stocks nevertheless made their depletion possible without exhaustion of the resource base and without complete elimination of the freedom of fishing; where scientific measurement techniques have permitted a determination that severe depletion actually threatened to produce species exhaustion, as in the case of whales, fishing states have generally been required to implement international conservation standards — an approach embodied in the 1982 Convention.[443] Manganese nodule deposits, on the other hand, will be permanently depleted by any amount of exploitation activity, and it is therefore significant that the Convention would establish the type of multilateral regulatory regime required by the *res communis* regime with respect to exhaustible marine resources.[444] Because the "midnight agreement" of August 1987, like the 1984 Provisional Understanding and the U.S. Seabed Act of 1980, is potentially consistent with a generally accepted seabed mining regime, condemnation by non-Western states of the preliminary mining activities of the private consortia as violative of *jus cogens* is premature, although a limited-participation RSA regime would clearly provide an inadequate legal framework for future unilateral seabed mining efforts in the face of sustained opposition by a majority of the international community.[445] Unilateral seabed mining operations pursuant to such an "alternative" RSA regime would constitute an invalid derogation from the peremptory requirement of a multilateral regulatory regime, even if the requirement of such a "generally agreed upon" regime had failed to crystallize into customary international law through the 1970 Declaration of Principles and the 1982 Convention. Because the existence of the *res communis* regime governing use of the sea preceded the founding of the United States in the 18th century, and because of the regime's peremptory nature, the United States would be unable to claim persistent objector status with respect to the requirement of multilateral regulation.[446]

Chapter 6
U.S. Interests in the Balance

Despite the negotiating history of UNCLOS III, the protestations of developing states, the doubts of some traditional allies, and in many instances a notable lack of supporting state practice, the United States has claimed that, except for Part XI, the provisions of the Law of the Sea Convention have attained the status of customary international law, and has sought to enforce favorable provisions through state practice.[1] If customary law were indeed fully embodied in the nonseabed provisions of the Convention, however, the nonseabed provisions would have little value as such. In fact, the preamble to the 1982 Convention makes clear that it constitutes a progressive development as well as a codification of general international law, and the United States may claim rights under the former sort of treaty provisions, as such, only where the intent to benefit nonparties is clear.[2] "To summarize, rights for third states under the stipulation *pour autrui* doctrine are circumscribed by the intent of states parties to the ... Convention, should it enter into force. On the basis of the expressed intent of the prospective parties to the Convention at Montego Bay, it may be difficult to apply the doctrine widely."[3] Absent such expressions of intent, use of terms such as "all States" or "States" — rather than "States Parties" — in some provisions of a treaty which entered into force without general acceptance might have been construed to permit application of those provisions to nonparty states as third-party beneficiaries of the agreement, rather than through the generation of customary international law.[4] This construction is consistent with use of the package deal to promote participation in the Convention, and at the December 1982 signing ceremony many states emphasized that the treaty was indeed a package deal and that they would oppose U.S. efforts to claim the selective benefit of its nonseabed provisions.[5]

The shortcomings of the U.S. position have been noted by the Chairman of the Preparatory Commission:

> While many legal concepts and legal institutions found in the 1982 Convention, such as the archipelagic State concept, the twelve-mile territorial

190

sea principle, and the two-hundred mile [exclusive economic zone] rule, amongst others, are considered by many of us as general international law today, the same cannot be said in respect of the detailed provisions in which these legal concepts and institutions are translated in the Convention.[6]

While some states have acted to bring their laws into compliance with the Convention, a greater number have sought to negotiate further expansions of coastal-state jurisdiction at specialized and regional law-making conferences.[7] Coastal states have also increasingly sought to expand their offshore jurisdictional zones indirectly, by lengthening the baselines from which such zones are measured.[8] Significantly, several Latin American states have actively opposed efforts to establish certain provisions safeguarding navigational rights in coastal waters as customary international law.[9] While the legal status of individual treaty provisions may vary, the uncertainty inherent in the customary law formation process — in terms of both the quantity and quality of state practice needed to generate a norm of general international law — renders it increasingly likely that the United States would find it necessary to deploy military forces or offer economic and political concessions bilaterally in order to secure important rights which would otherwise be available to it under the Convention.[10] Where the U.S. Government is either unwilling or unable to allocate resources necessary for such deployments or for the conclusion of such quid pro quo arrangements, the benefit of carefully negotiated treaty provisions may be lost for the United States. In this regard, it is significant that in 1973 the United States regarded continental-margin mineral resources, transit passage, offshore fisheries management, deep seabed mining, preservation of the marine environment, and freedom of marine scientific research as "equally important national interests."[11]

Marine Resources

In 1969 the Stratton Commission rejected "the idea that self-sufficiency in natural resources is a desirable goal for American policy," noting that "[e]fforts to favor certain domestic industries are not in the national interest if they raise production costs to levels which burden other segments of the domestic economy or provoke retaliatory action by other nations."[12] While national access to seabed minerals has been a matter of importance to many states,[13] in embracing a policy of autarky with respect to imported mineral resources the Reagan administration relied heavily upon an ideological conception of U.S. national security and a free-market economic theory deemed "vital to national well-being and the principles that form the foundation of American democracy."[14] The Reagan administration's underlying premise — that access to seabed nodules is of such vital interest to the United

States that an imperfect seabed mining regime would justify rejection of the entire Convention – does not appear to have undergone critical examination during the final UNCLOS III sessions, and prominent experts in the field of ocean law have subsequently concluded "that, for the foreseeable future, the United States has limited economic and security interests in deep seabed mining."[15] Indeed, the argument that reliance upon foreign mineral sources renders the United States dangerously vulnerable to supply disruptions was belied by a dramatic drop in the price of oil during the 1980s,[16] as well as by the willingness of the United States in 1990 to endure a doubling of oil prices in order to restore the government of Kuwait;[17] these events were particularly significant since the argument originated in the early 1970s with the lobbying efforts of the U.S. seabed mining industry on behalf of unilateral mining legislation following supply disruptions attributed to the OPEC oil embargo.[18] By the mid-1980s nonfuel mineral resources were regarded as adequate to ensure economic growth beyond the year 2030.[19] Under the Bush administration, only copper, the nodule mineral to which U.S. access has been least threatened, was still considered a "strategic" mineral; no special effort was being made to obtain national access to nickel or cobalt.[20]

    Although the United States did resolve overlapping claims with other pioneer investor states and issue exploration licenses to four private consortia, the Reagan administration failed in its efforts to implement an effective RSA regime among Western seabed mining states, and legal and economic uncertainties have all but halted the deep-sea nodule mining activities of private mining firms; despite the supposed strategic and economic importance of nodule minerals, banks have been unwilling to provide needed financing in the face of threatened foreign retaliation.[21] Recently discovered mineral deposits within the U.S. exclusive economic zone may provide a geologically more attractive and legally more secure source of marine minerals than deep-sea nodules,[22] rendering the Reagan administration's objections to Part XI of the Convention academic (assuming that national rights to such resources have become generally accepted as customary law apart from the Convention itself) – indeed, one might wonder why, if seabed minerals were of such great strategic importance to the United States, at the end of the 1980s it was Japanese rather than U.S. mining firms which were making serious efforts to proceed with seabed mining operations within the U.S. exclusive economic zone, as well as in the deep ocean beyond.[23] Thus, by 1990, active opposition by the U.S. seabed mining industry to the treaty regime had all but evaporated,[24] and the American Bar Association was able to conclude that "[a] realistic view of the current prospects for deep seabed mining and of the importance to the world community of the 1982 United Nations Convention . . . should allow countries to agree that Part XI should not present an obstacle to widespread international support for the Convention."[25] Moreover, it remains significant that a 1983 report by the General Accounting Office concluded "that reasonable access can be provided under provisions of the ... treaty, especially since the specific implementing rules and regulations remain to be negotiated" in the Preparatory Commission.[26] It would certainly be ironic — as well as short-sighted — if professed champions of free enterprise and national security, by opposing U.S. participation in the treaty, were to prevent U.S. firms from competing effectively with state-run mining operations for access to strategic minerals. In fact, by further delaying deep-sea mining by all states, U.S. efforts to undermine the UNCLOS III regime may indirectly serve to deprive U.S. consumers of access to nodule minerals which might otherwise be supplied to world markets by foreign seabed mining operators.[27]

Although it would seem that the basic concept of the 200-mile exclusive economic zone has become fairly well established in customary international law,[28] the binding effect of a number of detailed provisions of the Convention relating to the specific rights and duties of maritime and coastal states within the zone remain uncertain.[29] With respect to the continental margin beyond 200 miles from shore, there is little meaningful state practice; UNCLOS III President Tommy Koh indicated on the final day of the Conference that the 1982 Convention "expanded the concept of the continental shelf to include the continental slope and the continental rise," but only conditionally.[30] Coastal-state rights with respect to mineral resources of the outer continental margin may be expected to remain in doubt in the absence of general acceptance of the 1982 Convention, article 82 of which requires annual royalty payments to the Authority of 7 percent (phased in, after a five-year grace period, in 1 percent increments over a seven-year period) as a condition of exclusive resource jurisdiction.[31] The United States supported such payments in exchange for clear title to the mineral deposits located upon the outer margin, in order "to resolve disagreements over the use of this area,"[32] although it unsuccessfully opposed the exemption of developing states from revenue sharing with respect to mineral resources of which they are net importers.[33] The 1970 U.S. Draft Convention had provided for revenue sharing in the amount of 2 to 20 percent of the gross value of mineral production within the "trusteeship area" beyond the 200-meter isobath, one half to two thirds of which was to be paid to the Authority.[34] The royalty rates specified in article 82 would not represent an unusual burden for U.S. oil companies — revenue maximization has been a basic policy objective of federal offshore tract leasing under the Outer Continental Shelf Lands Act, and the U.S. government has historically collected royalties of more than 15 percent on offshore drilling operations, in addition to bonus-bid revenues.[35] The costs of petroleum production beyond 200 miles from shore are projected to be significantly higher than those of existing near-shore drilling operations, but technological advances and new discoveries may render oil recovery operations on the outer margin increasingly attractive in coming years.[36] In fact, although 140 million acres of the U.S. continental margin beyond 200 miles from shore was included in an offshore oil and gas leasing plan for 1987–92 proposed by the Reagan administration,[37] by 1990 the Department of the Interior still considered U.S. resource jurisdiction to end at the 200-mile limit and had no plans to lease mineral rights to the margin area beyond.[38] The revenue-sharing provisions of article 82 have been criticized by petroleum interests,[39] but in the absence of a generally accepted customary regime the United States is effectively precluded from the exercise of jurisdiction over resources of the outer continental margin beyond 200 miles from shore.[40]

Exclusive coastal-state fisheries management rights within the exclusive economic zone would appear to be rather well settled under international law. The Convention's requirement that coastal states "promote the objective of optimum utilization" is nevertheless expected to help U.S. distant-water fishermen gain access to foreign surplus stocks.[41] The Convention also contains important procedural protections for U.S. fishermen detained by foreign states: imprisonment may not be imposed as a penalty for violation of coastal-state fisheries laws within the exclusive economic zone; flag states are to be properly notified of actions taken; detainees and their ships are to be released "promptly . . . upon the posting of reasonable bond or other security"; and in the event they are not so released the flag state may seek immediate relief through the binding dispute settlement procedures established by the Convention.[42] Because the institutional mechanisms contained in such procedures cannot become generally available as part of customary international law, U.S. fishermen would continue to be subject to prolonged detention by territorially aggressive Latin American states.

Navigation

In 1970 the United States agreed in principle to accept greater international control over deep-sea mining operations in order to obtain the support of developing states for rights of unimpeded passage through the dozens of straits which would otherwise be enclosed by 12-mile territorial seas, as well as through coastal waters encompassed by the new exclusive economic zone.[43] Faced with a need to protect navigational rights and freedoms claimed for the United States under customary international law — approximately 90 percent of U.S. foreign trade is transported by sea[44] — the Reagan administration undertook the largest peacetime naval build-up in U.S. history, including the development of expensive new antimissile weapons systems to protect against coastal-state attack.[45] Even when rapprochement between the United States and the Soviet Union permitted significant reductions in U.S. military expenditures at the end of the 1980s, the Defense Department, which supports the navigational provisions of the Convention, continued to emphasize the need to maintain a strong naval capability to deal with "Third World brush fires."[46] "There [has been] a significant antiship missile threat to the U.S. Navy in international straits resulting from arms transfers to the Third World motivated by trends in negotiations on the Law of the Sea."[47] Significantly, multinational oil companies which have held interests in the U.S. seabed mining consortia believe that the benefits of the Convention to maritime commerce outweigh any benefits which deep-sea nodule mining might produce.[48]

In the absence of a transit passage regime for international straits connecting nonterritorial waters, more than 100 such straits would be governed by a regime of innocent passage, requiring coastal-state consent for overflight and submerged passage and permitting restrictive safety and environmental regulation.[49] Because U.S. negotiators made clear at the early sessions of UNCLOS III that U.S. navigation rights might indeed be restricted absent creation of a new transit regime, claims by the Reagan administration during the 1980s that the United States enjoys rights of transit passage under customary international law must be regarded as doubtful in the absence of supporting state practice.[50] "Retention of these passage and overflight rights is necessary to assure operational mobility for U.S. naval and air forces, to permit them to respond quickly and effectively to an unpredictable variety of strategic or political situations."[51] In 1973 Spain and Morocco denied overflight rights to U.S. planes seeking to fly vital aid to Israel through the Straits of Gibraltar, and have subsequently strengthened their straits defenses through the acquisition of missile-launching patrol boats.[52] At the 1982 signing ceremony, Iran asserted that nonparties might be denied transit passage through international straits, and in 1987 the United States was forced to incur considerable military expense in order to ensure passage of U.S.-flagged Kuwaiti oil tankers through the Straits of Hormuz, which are in places narrower than 24 miles, following Iranian deployment of Silkworm anti-ship missiles.[53] Apart from the budgetary expense of maintaining and deploying U.S. naval forces to challenge expansive claims of coastal-state jurisdiction, diplomatic costs and overall risks associated with use of such forces are likely to be substantially higher in the absence of a specific, binding treaty obligation, and the long-term effectiveness of challenge programs remains doubtful.[54]

Long after the conclusion of the UNCLOS III negotiations, a number of commentators have maintained that state practice continues to support application of the regime of innocent passage to international straits, and any customary transit passage regime which did evolve would lack the navigational protections carefully negotiated into the 1982 Convention by maritime states.[55] Similarly, the carefully crafted provisions governing navigation within the exclusive economic zone, which favor maritime interests by expressly preserving the exercise of high-seas freedoms of navigation and overflight and by providing procedural limitations upon the exercise of coastal-state enforcement jurisdiction, could not be fully incorporated into customary international law.[56] As was pointed out at the time new coastal-state enforcement powers were proposed in the UNCLOS III negotiations, "the very existence of a right of arrest necessitates the inclusion in the treaty of adequate provision for the prompt release under bond of vessels so arrested. Otherwise, the entire purpose of the international standards could be effectively vitiated by arbitrary detentions."[57] The United States also recognized at an early point that an effective system of dispute settlement would be necessary to prevent the exclusive economic zone from becoming "the functional equivalent of the territorial sea,"[58] and the applicability of the Convention's binding dispute settlement procedures to navigational disputes occurring in the zone, in international straits, and in archipelagic waters is particularly noteworthy.[59]

Use of force may occasionally prove counterproductive, and the Reagan administration recognized that it could not rely entirely on military power to safeguard U.S. navigational interests.[60] Thus, U.S. aid to Indonesia — which, with Malaysia, had declared in 1971 that the busy Straits of Malacca should be governed by a regime of innocent passage rather than transit passage, in order to permit application of restrictive environmental standards to commercial tanker traffic — was increased substantially in 1982, as that state assumed "strategic" importance in the wake of U.S. rejection of the Convention.[61] In requesting Congress to increase bilateral assistance, the State Department pointedly described Indonesia as "a nation located astride vital interocean sealanes."[62] Nevertheless, in 1988, despite ongoing U.S. efforts to obtain the compliance of archipelagic states with the navigational provisions of the 1982 Convention, Indonesia closed the straits of Lombak and Sunda in order to conduct military exercises, and Indonesian officials have continued to assert that these straits are not subject to any customary regime of transit passage.[63] In order to ensure its continued rights of passage through these straits, the United States has sought to maintain large military bases in Southeast Asia at considerable financial and political cost.[64] According to a former chairman of the Joint Chiefs of Staff, the cost of obtaining coastal-state acquiescence for such navigational "rights" through bilateral negotiations "may be expected to increase with the magnitude of the known U.S. interest involved."[65]

The 1958 Convention on the Territorial Sea and the Contiguous Zone did not define the navigational rights of foreign warships within the territorial sea. An increasing number of coastal states — according to one count, more than 60, before, during, and after the UNCLOS III negotiations — have asserted the right under international law to bar or restrict the passage of foreign warships through their offshore waters, including the exclusive economic zone.[66] Article 58 of the 1982 Convention makes clear that high-seas "freedoms ... of navigation and overflight" are enjoyed by all states within the exclusive economic zone, and among the coastal-state rights listed in Part V no mention is made of any power to impede warships; indeed, under articles 95 and 96, warships and other state-run vessels exercising their high-seas freedoms "have complete immunity from the jurisdiction of any State other than the flag State."[67] Within territorial waters, warships are to enjoy the right of innocent passage — i.e., passage "not prejudicial to the peace, good order or security of the coastal State," as specifically defined — so long as they comply with applicable coastal-state laws and regulations.[68] The Convention would further ensure that non-suspendable rights of transit through international straits and archipelagic sea lanes are available to "all ships and aircraft"[69]; in contrast to the territorial sea, where submarines are specifically required "to navigate on the surface and to show their flag," underwater navigation may not be barred within the exclusive economic zone or straits governed by the transit passage regime.[70] Because the Convention provides added maneuverability for transiting warships by permitting 25-mile deviations from "axis lines" designated by archipelagic states, a customary regime of archipelagic sea lanes passage may prove unsatisfactory to the United States.[71] Although a state might exempt itself from application of the dispute settlement procedures with respect to "military activities," coastal-state efforts to impair the peaceful offshore navigation or transit passage of warships would be specifically justiciable under the Convention.[72]

Scientific Research

The United States has traditionally sponsored scientific research in offshore waters which are now included within the exclusive economic zone, but developing states, suspecting that "research" vessels are frequently engaged in activities detrimental to their security and economic interests, have been increasingly unwilling to permit such research without prior coastal-state authorization.

> Throughout the 1960s, it was reasonable to assume that a great deal more military research was going on than ever broke surface, and that a good deal of "pure" scientific research had military connections.... Besides these suspicions, the developing countries felt that, as a result of these programmes, the gap was widening between them and the industrialized countries in research capability.[73]

Many developed states are similarly sensitive to the conduct of foreign research activities in their offshore waters.[74] Customary international law is rooted in the 1958 Continental Shelf Convention, which subjected such activities to coastal-state consent:

> the coastal state shall not normally withhold its consent if the request is submitted by a qualified institution with a view to purely scientific research into the physical or biological characteristics of the continental shelf, subject to the proviso that the coastal state shall have the right, if it so desires, to participate or to be represented in the research, and that in any event the results shall be published.[75]

Nevertheless, requests for coastal-state authorization under the 1958 regime were frequently denied, to the detriment of the scientific community.[76] The United States has by far the greatest number of marine scientists interested in conducting potentially valuable scientific research activities throughout the world's oceans, and U.S. negotiators worked hard at UNCLOS III, without much support even from traditional U.S. allies, to obtain substantive and procedural improvements in the regime in order to free researchers from having to engage in time-consuming negotiations in which coastal states have been able to exact costly concessions.[77]

Although the 1982 Convention maintains a coastal-state consent regime with respect to the conduct of scientific research within the exclusive economic zone, the circumstances under which consent may be withheld have been specifically limited to cases where the research involves artificial islands, shelf drilling, or use of environmentally harmful substances; where the research "is of direct significance for the exploration and exploitation of natural resources" within 200 miles from shore; or where the researcher submits inaccurate information.[78] Specific information regarding the nature of the proposed research is to be submitted to the coastal state at least six months in advance; if adequate information is provided and the researcher has complied with obligations relating to previous research projects, consent will be implied unless the coastal state expressly denies consent within four months of the date of application.[79] Suspension of research activities may occur only where the researcher has failed to comply with the authorization procedures specified in the Convention, and only until the noncompliance has been remedied.[80] Although decisions of the coastal state with respect to the granting or withholding of consent and to the suspension of research activities are excluded from binding dispute settlement procedures, such procedures do apply to the remainder of the regime governing scientific research within zones of coastal-state jurisdiction.[81] When issuing his 1983 proclamation, President Reagan apparently sought to preserve U.S. ability to reshape the consent regime through state practice, stating that the United States would recognize the jurisdiction of coastal states over scientific research activities within their exclusive economic zones "if that jurisdiction is exercised reasonably in a manner consistent with international law."[82]

Because of high anticipated costs and uncertain benefits, the United States has been unwilling to employ military force in support of freer scientific access to coastal-state waters.[83] In fact, prior to 1981 the U.S. delegation had sought to develop a cooperative international atmosphere at UNCLOS III in the expectation that greater trust and understanding would facilitate the operation of the consent regime upon which the rest of the international community insisted.[84] Although most coastal states have apparently conformed their practice to the provisions of the 1982 Convention,[85] it is not clear that they have done so out of a sense of legal obligation — i.e., *opinio juris* — rather than out of a sense of comity or convenience, and the binding effect of novel procedural provisions such as implied consent remains highly uncertain under customary international law. Significantly, following U.S. rejection of the Convention a number of coastal states adopted policies which have further obstructed U.S. researchers and driven up project costs by delaying decisions on consent or by withholding consent entirely without explanation, causing marine scientists to complain publicly that during the 1980s "there has been a net loss with regard to our ability to conduct oceanographic research" and that "we are gradually being excluded from areas where oceanographic research is extraordinarily important."[86] Although the Convention provides a streamlined implied consent procedure for research projects undertaken by an international organization of which the coastal state is a member, a number of developing states have opposed research activities by the IOC because of the involvement of states opposing the Convention.[87] Many broad-margin coastal states would like to extend the discretionary consent regime beyond the 200-mile exclusive economic zone to the entire continental margin, and would certainly seek to do so if the provisions of the Convention are not generally accepted as customary law.[88] Most marine scientists believe that their research would be facilitated under the provisions of the Convention.[89]

Environmental Protection

The inclusion of strong environmental protection measures in the Convention was an early and enduring goal of the United States at UNCLOS III. A series of accidental oil spills and tanker wrecks brought the problem of vessel-source pollution inescapably to the attention of national delegations,[90] yet four times as much oil was being intentionally discharged into the sea from ships in the absence of an effective prohibitory rule of international law.[91] Flag states, traditionally vested with exclusive jurisdiction over their registered vessels, found it economically advantageous to avoid the adoption and enforcement of costly pollution-control regulations.[92]

"Problems related to pollution of the marine environment cannot be effectively resolved solely by national action. Whatever the causes of the marine pollution, the consequences often extend beyond the jurisdiction of a single state."[93] Because most accidental discharges occur in the shallow waters near ports, it was envisioned that a regime of port-state jurisdiction over entering vessels would permit coastal states to exert a substantial degree of control over local and regional pollution activities.[94] The 1958 Convention had all but ignored the problem of marine pollution, and customary rules of international law relating to pollution of the marine environment were fragmentary or nonexistent.[95]

With a growing domestic environmental movement, the United States sought to negotiate into the Convention a number of carefully balanced provisions designed to promote environmental protection and economic efficiency while minimizing coastal-state interference with navigational freedoms.[96] A working paper submitted to the Seabed Committee by the United States in 1973 pointed out that, because of the nature of the vessel-source pollution problem, these objectives could only be achieved through implementation of a system of generally accepted international regulatory standards.[97]

> Uniform international standards for vessel operation can help to reduce the expenses inherent in varying regulations which might be applicable in this worldwide area. Minimum standards regarding environmental controls on resource development could promote uniformity in this area as well.... Additionally, such standards could preclude a competitive disadvantage for U.S. firms subject to more stringent requirements than other nations might apply to their firms in the absence of these standards.[98]

As President Nixon stated in 1972, "it is not possible for any nation, acting unilaterally, to ensure adequate protection of the marine environment. Unless there are firm minimum international standards, the search for relative economic advantage will preclude effective environmental protection."[99]

Under the Convention, minimum standards and practices which are established by competent international agencies with respect to pollution of the marine environment from a number of sources — dumping, vessels, and exploitation of the deep seabed and continental shelf — are made generally applicable, and disputes relating to compliance with such standards and practices are subject to the binding dispute settlement procedures set forth in the Convention;[100] because states are themselves expressly required to enforce such minimum standards and practices within their jurisdiction with respect to these sources of pollution — as well as with respect to pollution of the marine environment from land-based sources and from the atmosphere, when "applicable international rules and standards" have been established with respect to these sources — a state's failure to undertake enforcement action could render it liable to states suffering resultant environmental damage.[101] More importantly, the Convention would render flag states liable for pollution damage resulting from failure to ensure that their vessels comply with internationally accepted regulatory standards.[102] Coastal states are authorized to inspect vessels, to institute enforcement proceedings where there is clear evidence that the vessel has violated applicable international environmental standards and practices within the territorial sea or exclusive economic zone, and to enforce international standards and practices with respect to vessel-source pollution caused within their jurisdiction or on the high seas by vessels which have voluntarily entered their ports.[103] The Convention also extends coastal-state jurisdiction over dumping activities to the outer edges of the exclusive economic zone and continental margin.[104]

Together, these provisions could be expected to create a diffuse but widely effective international mechanism for the control of a significant amount of marine pollution. Because support for such measures outside the United States is relatively weak, particularly among developing states, they have been particularly jeopardized by U.S. rejection of the Convention; although the United States has sought to implement the new provisions for coastal-state jurisdiction under color of customary international law, by 1987 no other state had asserted such jurisdiction within its exclusive economic zone.[105] Significantly, the revised *Restatement* does not identify the Convention's novel, specifically defined jurisdictional arrangements as customary international law, but instead asserts a general, qualified obligation of states to comply with generally accepted international rules and standards relating to environmental protection.[106] "It is evident that some rules, such as those relating to port State jurisdiction..., were not customary law prior to UNCLOS III and did not achieve that status during the negotiations."[107]

World Order

An effective system for the settlement of disputes was identified by the United States in the early 1970s as one of its primary negotiating goals at UNCLOS III — in particular to prevent coastal and straits states from violating the navigation rights which have been so carefully defined in the Convention — and the U.S. delegation pursued that goal throughout the negotiations as an integral part of the treaty package: "Without a provision for compulsory settlement of disputes, the substantive provisions of the Convention would be subject to unilateral interpretation and the delicate balance of rights and duties ... would be quickly upset."[108] Historically, states have not been inclined to restrict their national sovereignty by voluntarily submitting to binding compulsory adjudication procedures, and the dispute settlement procedures embodied in the Convention have been described by a former U.S. negotiator as an "unprecedented step in global multilateral diplomacy."[109] Following exhaustion of alternative dispute settlement procedures, a state may bring "any dispute concerning the interpretation or application of" the Convention before the appropriate judicial or arbitral forum for a binding decision, excepting only certain disputes relating to coastal-state jurisdiction over marine scientific research and fisheries, disputes relating to maritime boundary delimitation, disputes relating to military activities, and disputes being heard by the U.N. Security Council.[110] These procedures would be unavailable to the United States if it were to remain outside the Convention,[111] and customary law is incapable of producing the kind of precise international rules of conduct widely recognized as necessary for stability and predictability in the law of the sea: "A major reason for seeking a rule of law in the first place is a concern that in its absence behavior might not conform to a desired norm, or any norm."[112] Successful implementation of the system of dispute settlement procedures embodied in the 1982 Convention would constitute an important positive precedent for further development of comprehensive mechanisms for peaceful enforcement of international legal obligations, and would in itself deter excessive claims of coastal-state jurisdiction and other departures from the carefully negotiated provisions of the Convention.[113]

In its determination to oppose "efforts to impose a collectivist ideology,"[114] the Reagan administration apparently attached little, if any, positive value to the treaty as an instrument for the promotion of international law and world order. In fact, as Secretary Kissinger pointed out, the United States has historically been a leader in the ongoing effort "to harness the conflict of nations by standards of order and justice."[115] The late Senator Robert Taft, while criticizing politicization at the United Nations, recognized that "in the long run the only way to establish peace is to write a law, agreed to by each of the nations, to govern the relations of such nations with each other and to obtain the covenant of all such nations that they will abide by that law and by the decisions made thereunder."[116] The Nixon administration's support for a comprehensive multilateral law of the sea agreement was an important part of its stated policy of promoting the advancement of international law, a recognition of the need to accommodate "the interests and needs of a wide variety of nations" in order to achieve "a clear system of law respected by the members of the international community."[117] As the world's wealthiest state, the United States has the most to gain from the international stability provided by such a system.

An effective international legal system is inconsistent with a political ideology based on autarky and cultivated positivist nationalism. U.S. opposition to the 1982 Convention has not only diminished the influence of the United States within regional and international organizations dealing with ocean issues, but has also tended to undermine the effectiveness of such organizations generally.[118] The disregard of the Reagan administration for the process of international law was further manifested in its withdrawal of the United States from the compulsory jurisdiction of the International Court of Justice following a decade-long propaganda campaign directed against the United Nations and its affiliated agencies,[119] and in its general disdain for multilateral law-making conferences. The Soviet Union, on the other hand, sought to improve its diplomatic position by reversing its traditionally uncooperative posture toward international organizations.[120] By the end of the 1980s, however, warming U.S.-Soviet relations had produced a new atmosphere of cooperation within the international community and a new U.S. appreciation for the United Nations as an instrument of world order.[121] In calling for development of a New World Order based upon the rule of international law, President Bush emphasized his "vision of a new partnership of nations . . . based on consultation, cooperation, and collective action, especially through international and regional organizations."[122] There may yet appear an understanding that an expectation that the United States should prevail in every international dispute is as unrealistic as it is arrogant and that U.S. criticism of renegade developing states is undercut when the United States is itself openly hostile to the institution of international law.

In calling for negotiation of a comprehensive convention for the benefit of the entire international community in 1970, President Nixon recognized that in the absence of multilateral regulation of ocean uses, "unilateral action and international conflict are inevitable."[123] In 1972 he reiterated that the situation was "urgent. . . . Competition among nations for control of the ocean's resources, and the growing divergence of national claims, could constitute serious threats to world peace."[124] During the Ford administration, Secretary of State Kissinger continued to emphasize that "the extension of legal order is a boon to humanity and a necessity," describing the UNCLOS III negotiations as the world's last chance to avoid "unrestrained military and commercial rivalry and mounting political turmoil," and pointing out that "[i]f every state proclaims its own rules of law and seeks to impose them on others, the very basis of international law will be shaken, ultimately to our own detriment."[125] Its extensive public and private interests in ocean use render the United States particularly vulnerable to the instability and destruction associated with such conflict,[126] and creation of equitable mechanisms for assistance to developing states was regarded as essential for conflict avoidance.[127] Multilateral development of the international Area is expected to discourage coastal states from asserting new offshore jurisdictional claims,[128] but because U.S. objections to Part XI of the Convention have been translated into opposition to the treaty as a whole, lack of widespread support for the Convention may lead to a premature reopening of the entire UNCLOS process — an outcome in which developing states could not be expected to be as accommodative of U.S. nonseabed interests.[129] To the extent the United States is able to influence the number of ratifications obtained by the Convention, it may still have a choice between jurisdictional instability with no unilateral deep-sea mining and jurisdictional uniformity with potentially profitable mining operations under a multilateral regulatory regime.[130]

Conclusions

The Third United Nations Conference on the Law of the Sea was convened after decades of unilateral claims had produced a fractured law of the sea which hampered peaceful and efficient use of the sea and its resources, and after rapid technological advances rendered new forms of international regulation necessary in order to prevent such use from exhausting a common resource base and thereby depriving others of their equal rights of access. Following contentious and protracted negotiations, the Conference produced a Convention which was widely supported within the international community but which was vehemently opposed by a new U.S. administration anxious to recreate a distinctive sense of national political purpose and economic independence at the expense of multilateral diplomacy.

Compromises with respect to Part XI of the Convention produced a number of restrictive provisions – particularly those relating to technology transfer and preferences enjoyed by the Enterprise – which are more justifiable on political than economic grounds. Yet such compromises are an unavoidable aspect of any legislative effort – the U.S. Seabed Act of 1980 itself contains economically burdensome requirements inconsistent with the theoretical ideal of unrestricted access – and it is generally agreed that the conditions of access applicable to state-sponsored seabed mining under the Convention are no more burdensome than those faced profitably by mining firms within foreign states. The degree to which the Reagan administration's objections to Part XI were overstated would seem to imply either that the terms of the Convention had not been closely examined or that Congress and the public were being intentionally misled as to its content. The failure of the Reagan administration to establish a viable U.S. seabed mining industry based upon the freedom of the high seas doctrine is attributable to the unwillingness of bankers to finance nodule mining in the absence of exclusive mining rights as well as to depressed mineral markets; meanwhile, seabed mining operators sponsored by other states are proceeding within the framework established by the Convention, in implicit recognition that exploitation of exhaustible resources governed by a peremptory *res communis* regime can only occur in accordance with such a generally accepted positive-law regime.

To the extent that seabed minerals are of genuine strategic importance to the United States, plentiful quantities are expected to be available within the U.S. exclusive economic zone as an alternative source of supply when market prices improve, whether or not deep-sea nodule mining is ultimately undertaken by private consortia. The importance of nodule mining too thus appears to have been overstated — indeed, it would appear that U.S. rejection of the Convention was based more on a desire to prevent the international community from establishing an effective regulatory regime than on a genuine interest in securing national access to nodule deposits. Support for unilateralism in this context contravenes the governing *res communis* principle and may be expected to produce international conflict, as President Nixon perceived and, the massive U.S. naval build-up of the 1980s attests, as the Reagan administration also apparently understood. U.S. opposition to the Convention thus not only jeopardizes substantial U.S. interests in the treaty without any significant offsetting benefit, but also constitutes an implicit rejection of the promotion of world order through international law as a foreign policy goal. At the very least, such a policy would seem to be founded on the unproven and undemocratic assumption, traditionally predominant within the Soviet Union rather than the United States, that nation-states are unable to overcome self-interest through enlightened cooperation — a premise belied by the united efforts of the international community in 1990 to preserve the flow of oil for the benefit of the world economy. Viewed less charitably, rejection of the Convention by the United States reflects a belief that unilateralism is a viable policy alternative when backed by military force, the *res communis* principle and budget deficits notwithstanding — an idea which is readily susceptible to ideological manipulation and which could be expected to lead to a further acceleration of competition for military power and, ultimately, to open warfare on an expanding scale.

Notes

Introduction

1. See Barry Buzan, *Seabed Politics* (New York: Praeger, 1976), p. xiv. The edge of the continental shelf ranges worldwide from 20 to 550 meters in depth, averaging 133 meters, and from 0 to 1,500 kilometers in width, averaging 78 kilometers. Martin Ira Glassner, "Developing Land-Locked States and the Resources of the Seabed," *San Diego Law Review* 12 (1975): 642. See also Alexandra M. Post, *Deepsea Mining and the Law of the Sea* (Boston: Martinus Nijhoff, 1983), p. 5.

2. United Nations, *Marine Science and Technology: Survey and Proposals: Report of the Secretary-General* (U.N. Doc. E/4487, April 24, 1968), pp. 11, 13. Approximately 80 percent of the world's living organisms are located in the oceans.

3. See Buzan, *Seabed Politics*, pp. xv–xvi. Their origin has been debated for many years and remains uncertain. See U.S. Department of Commerce, *U.S. Ocean Policy in the 1970s: Status and Issues* (Washington: U.S. Government Printing Office, 1978), part 6, p. 24. Measuring up to four inches in diameter, the potato-shaped nodules are strewn across as much as 25 percent of the deep ocean floor. Kent M. Keith, "Laws Affecting the Development of Ocean Resources in Hawaii," *University of Hawaii Law Review* 4 (1982): 276.

4. Rolf Akesson, "The Law of the Sea Conference," *Journal of World Trade Law* 8 (1974): 284; United Nations, *Resources of the Sea: Report of the Secretary-General* (U.N. Doc. E/4449, February 21, 1968), Summary, p. 7; and Post, *Deepsea Mining*, pp. 11, 23–24. The nodules contain traces of more than twenty metallic elements.

5. Dennis W. Arrow, "The Proposed Regime for the Unilateral Exploitation of Deep Seabed Mineral Resources by the United States," *Harvard International Law Journal* 21 (1980): 340. See Chap. 5 n. 315.

6. Secretary of State Henry Kissinger, address before American Bar Association, Montreal, August 11, 1975, reprinted in *U.S. Department of State Bulletin* 75 (1975): 355.

7. Statement of President Ronald W. Reagan, July 9, 1982, reprinted in *U.S. Department of State Bulletin* 82 (August 1982): 71.

Chapter 1

1. Ian Brownlie, *Principles of Public International Law*, 3rd ed. (Oxford: Clarendon, 1979), p. 109.

2. Ibid., p. 110.

3. See ibid., pp. 180–81.

4. Gerald Fitzmaurice, "The General Principles of International Law," *Recueil des Cours* 92 (1957): 143.

5. See Edgar Gold, *Maritime Transport* (Lexington, Mass.: D.C. Heath, 1981), p. 2 et seq.

6. See chap. 5 nn. 323–30 and accompanying text; and also R.P. Anand, *Origin and Development of the Law of the Sea* (Boston: Martinus Nijhoff, 1983), p. 11.

7. See chap. 5 nn. 333, 336–39, and accompanying text. "In olden times, the voice which

articulated the general maritime law was that of international commercial traders speaking through the Rhodian Laws, the Rules of Oleron and Visby, the Hanseatic League, and the staple courts of the English 'Cinque Ports.' The uniformity which had developed in this way devolved into many voices with the rise of nationalism and international rivalry." Gordon W. Paulsen, "An Historical Overview of the Development of Uniformity in International Maritime Law," *Tulane Law Review* 57 (1983): 1086.

8. See Chap. 5 n. 420 et seq. and accompanying text. Papal bulls during the preceding four decades had granted Portugal similar exclusive rights to the waters of the south Atlantic and Indian oceans. See Chap. 5 n. 341.

9. Hugo Grotius, *The Freedom of the Seas*, trans. Ralph van Deman Magoffin, orig. pub. 1633 (New York: Oxford University Press, 1916), p. 30. "In terms of the fishing effort and techniques which could have been fielded in Grotius' day, when the principle was supposedly established, it was perhaps reasonably accurate to say that the living resources of the sea were inexhaustible and therefore represented 'free goods'." Jon L. Jacobson, "Bridging the Gap to International Fisheries Agreement: A Guide for Unilateral Action," *San Diego Law Review* 9 (1972): 458.

10. See Chap. 5 nn. 188–89, 352 and accompanying text; and, generally, Thomas Wemyss Fulton, *The Sovereignty of the Sea* (Edinburgh: William Blackwood and Sons, 1911). As used throughout this treatise, "miles" refers to nautical miles, unless otherwise indicated.

11. See Clyde Sanger, *Ordering the Oceans: The Making of the Law of the Sea* (London: Zed Books, 1986), pp. 123–24, 159; and Jack N. Barkenbus, *Deep Seabed Resources: Politics and Technology* (New York: Free Press, 1979), p. xii.

12. See Barry Buzan, *Seabed Politics* (New York: Praeger, 1976), p. 2.

13. Dennis W. Arrow, "The Proposed Regime for Unilateral Exploitation of Deep Seabed Mineral Resources by the United States," *Harvard International Law Journal* 21 (1980): 353. See *The Case of the S.S. Lotus, P.C.I.J. Reports 1927*, ser. A, no. 10, p. 19. "This decision caused considerable controversy for a number of years and led to very involved study and discussion . . . which finally resulted in a 1952 convention that basically reversed the court's decision." Gold, pp. 206–7.

14. See Barkenbus, p. 33; and also C. John Colombos, *The International Law of the Sea*, 4th rev. ed. (London: Longmans, Green, 1959), p. 366.

15. Buzan, *Seabed Politics*, p. 2. See Chap. 5 n. 332 and accompanying text. The influence of such legislation upon the law of the sea during the 18th and 19th centuries was particularly significant in light of the role played by the British Admiralty Courts in the development of state practice.

16. Beginning in 1736, Britain adopted "Hovering Acts" permitting the seizure as far as 12 miles from shore of foreign vessels suspected of smuggling. See Fulton, p. 593. The Acts were repealed in the late 19th century out of concern that freedom of navigation would be threatened by corresponding expansions of offshore jurisdiction by other states. Anand, *Origin and Development*, pp. 141–42. Enactment of such zones nevertheless continued. In 1935 the United States extended its own customs jurisdiction more than 60 miles from shore. See Ann L. Hollick, *U.S. Foreign Policy and the Law of the Sea* (Princeton, N.J.: Princeton University Press, 1981), p. 20.

17. See Gold, p. 148.

18. "The court never sat, because no agreement could be reached as to what was the law of the sea, and before any settlement was found the [First World War] broke out." Louise Fargo Brown, *The Freedom of the Seas* (New York: E.P. Dutton, 1919), p. 6.

19. Final Act, quoted in Theodore G. Kronmiller, *The Lawfulness of Deep Seabed Mining* (New York: Oceana Publications, 1980), vol. 1, p. 365. See Colombos, pp. 21–22, 367. Cf. League of Nations Committee of Experts for the Progressive Codification of International Law, "Questionnaire No. 7 – Exploitation of the Products of the Sea: Annex: Report on the Exploitation of the Products of the Sea," *American Journal of International Law* 20 (Special Supplement, 1926): 236.

20. See Sanger, p. 14.

21. See Anand, *Origin and Development*, pp. 140–43.

22. League of Nations Committee of Experts, "Report on the Exploitation of the Products of the Sea," p. 232. The Assembly of the League also recognized the desirability of negotiating

an international agreement to govern exploitation of the resources of the sea. See U.S. 95th Congress, 2nd Session, Senate, Committee on Commerce, Science, and Transportation, *Report: The Third United Nations Law of the Sea Conference* (Washington: U.S. Government Printing Office, 1978), p. 5.

23. See Buzan, *Seabed Politics*, p. 6. The League's Committee of Experts had determined that regulation of the exploitation of marine resources "could not fail to be in the general interest; since, if the present confusion persists for a few years longer, the extinction of the principal species will be the inevitable consequence of their unrestricted exploitation." League of Nations Committee of Experts, "Report on the Exploitation of the Products of the Sea," p. 239. See also Chandler P. Anderson, "Editorial Comment: Exploitation of the Products of the Sea," *American Journal of International Law* 20 (1926): 752.

24. See Myres S. McDougal and Norbert A. Schlei, "The Hydrogen Bomb Tests in Perspective: Lawful Measures for Security," *Yale Law Journal* 64 (1955): 669–70. The zone had been proposed by the United States. Hollick, *U.S. Foreign Policy*, p. 20.

25. These states were, respectively, Uruguay (1930), Colombia (1930), Iran (1934), Cuba (1934), Greece (1936), and Italy (1942); Mexico (1944); and Guatemala (1940) and Venezuela (1941). Buzan, *Seabed Politics*, p. 6.

26. Ibid., pp. 4, 6–7. See n. 16 above. Unless otherwise indicated, references to the offshore width of coastal-state jurisdictional zones in the 20th century do not include the distance from the shoreline of the baselines from which such zones are measured, a distance which in some areas may extend several miles. See D.P. O'Connell, *The International Law of the Sea* (Oxford: Clarendon, 1982), vol. 1, pp. 171–230.

27. Presidential Proclamation No. 2667, September 28, 1945. Although the Proclamation did not specify the outer limits of the continental shelf regime, an accompanying White House press release indicated that U.S. jurisdiction would be exercised to a depth of approximately 200 meters, a depth reached in some places only at a distance of 250 miles from shore. Sanger, pp. 14–15. See also S. H. Amin, "The Regime of the Sea-Bed and Ocean Floor: A Legal Analysis," *Juridical Review* (1983), pp. 69–70. The size of the shelf area encompassed by the claim was large — approximately 700,000 square miles.

28. Presidential Proclamation No. 2667. See Buzan, *Seabed Politics*, p. 7. "The war was depleting U.S. oil reserves at an alarming rate, and the U.S. Department of the Interior consequently suggested to President Roosevelt in 1943 that an interdepartmental study should examine how the United States might best obtain access to the petroleum reservoirs known or expected to exist beneath the continental shelf off the U.S. coast. At the same time, there was growing concern for the conservation of another important resource, fisheries in the U.S. contiguous seas; and an interagency study was quickly mounted." Gold, pp. 251–52. The United Kingdom had divided the seabed of the Gulf of Paria with Venezuela and adopted an annexation order purporting to establish exclusive rights of use vis-à-vis the international community. "[T]he 1942 Treaty and Order in Council, taken together, represented a major and significant stepping-stone from which the exclusive right of coastal states to adjacent continental shelves evolved." Amin, p. 68.

29. Buzan, *Seabed Politics*, pp. 7–8. "The character as high seas of the water above the continental shelf and the right to their free and unimpeded navigation are in no way thus affected." Presidential Proclamation No. 2667.

30. Presidential Proclamation No. 2668, September 28, 1945. Such measures were to be authorized "in those areas of the high seas contiguous to the coasts of the United States wherein fishing activities have been or in the future may be developed and maintained on a substantial scale." Ibid. See also Gold, p. 252. Adverse international reaction and the possibility of reciprocating territorial-sea extensions by Latin American states — which would have threatened U.S. navigation rights — nevertheless persuaded the United States not to impose such conservation measures unilaterally, and the issue was ultimately addressed instead in the International Law Commission and the UNCLOS negotiations.

31. See Robert O. Keohane and Joseph S. Nye, *Power and Interdependence: World Politics in Transition* (Boston: Little, Brown, 1977), p. 110. "When viewed in the perspective of international legal concepts and the world's technological capabilities at that time, the Truman Proclamation might have been considered unnecessary. Ocean technology was then almost exclusively possessed by the United States, and no other nation had the technical

capability to exploit the resources of the U.S. continental shelf." U.S. Congressional Research Service (George A. Doumani), *Exploiting the Resources of the Seabed* (Washington: U.S. Government Printing Office, 1971), p. 68.

32. See Buzan, *Seabed Politics*, pp. 7–8; and also Lewis M. Alexander, "The Ocean Enclosure Movement: Inventory and Prospect," *San Diego Law Review* 20 (1983): 565.

33. Although all major maritime states are also coastal states, their role as principal users of the sea traditionally allowed them to exert predominant control over the formation of the customary law of the sea. See John D. Negroponte, Assistant Secretary for Oceans and International Evironmental and Scientific Affairs, "Who Will Protect Freedom of the Seas?" *U.S. Department of State Bulletin* 86 (October 1986): 42.

34. Buzan, *Seabed Politics*, pp. 7–9. See H. Lauterpacht, "Sovereignty Over Submarine Areas," *British Year Book of International Law* 27 (1950): 382–83. Some of these Latin American states lacked any real geographical shelf but regarded the U.S. proclamations as an opportunity to assert sovereign rights with respect to other ocean resources such as fisheries. Hollick, *U.S. Foreign Policy*, pp. 117–18.

35. Hollick, *U.S. Foreign Policy*, p. 61. In fact, the United States had actually invited other states to claim jurisdiction over continental shelf resources. Ibid., pp. 45–47. The 1939 security zone was also cited as a precedent by expansionary coastal states. Edward Wenk, Jr., *The Politics of the Ocean* (Seattle: Washington University Press, 1972), p. 254.

36. See Buzan, *Seabed Politics*, pp. 12–13; and also Hollick, *U.S. Foreign Policy*, pp. 83–84.

37. See United Nations, *Laws and Regulations on the Regime of the Territorial Sea* (U.N. Doc. ST/LEG/SER.B/6, November 1957), pp. 723–24; and also Gold, p. 254. Costa Rica acceded to the Declaration in 1955.

38. Hollick, *U.S. Foreign Policy*, pp. 94, 121–22, 377. See also Kronmiller, vol. 1, pp. 117–19.

39. See Buzan, *Seabed Politics*, p. 10.

40. Arvid Pardo, "The Future of the Sea," in *The Future of the Law of the Sea*, ed. J. Bouchez and L. Kaijen (The Hague: Martinus Nijhoff, 1973), p. 2. See Lauterpacht, "Sovereignty Over Submarine Areas," pp. 376–433.

41. Arvid Pardo, "Address," in *Proceedings of the 62nd Annual Meeting*, American Society of International Law (1968), p. 218.

42. Buzan, *Seabed Politics*, p. 25.

43. Draft article 67, submitted to the General Assembly in 1956, quoted in United Nations, *Legal Aspects of the Question of the Reservation Exclusively for Peaceful Purposes of the Sea-Bed and the Ocean Floor, and the Sub-Soil Thereof, Underlying the High Seas Beyond the Limits of Present National Jurisdiction, and the Use of Their Resources in the Interests of Mankind: Study Prepared by the Secretariat* (U.N. Doc. A/AC.135/19, June 21, 1968), p. 8. The average distance of the 200-meter isobath from shore is approximately 50 miles, but the actual width of the continental shelf varies greatly throughout the world. Statement of John R. Stevenson, Legal Adviser, Department of State, at White House news conference, May 23, 1970, reprinted in *International Legal Materials* 9 (1970): 810. See Introduction n. 1.

44. Keohane and Nye, p. 110.

45. *Yearbook of the International Law Commission, 1951*, vol. 2, p. 142.

46. Buzan, *Seabed Politics*, pp. 24–25.

47. Convention on the Territorial Sea and the Contiguous Zone, reprinted in *United Nations Treaty Series*, vol. 516, no. 7477, pp. 206–24; Convention on the High Seas, reprinted in *United Nations Treaty Series*, vol. 450, no. 6465, pp. 82–102; Convention on Fishing and Conservation of the Living Resources of the High Seas, reprinted in *United Nations Treaty Series*, vol. 559, no. 8164, pp. 286–300; and Convention on the Continental Shelf, reprinted in *United Nations Treaty Series*, vol. 499, no. 7302, pp. 312–20. See U.S. Senate, *The Third U.N. Law of the Sea Conference*, pp. 5–8. The resolutions addressed such issues as nuclear weapons tests, fisheries conservation, historic waters, and the convening of a second round of UNCLOS negotiations.

48. Lewis M. Alexander, "National Jurisdiction and the Use of the Sea," *Natural Resources Journal* 8 (1968): 376. See generally, Guenter Weissberg, "International Law Meets

the Short-Term National Interest: The Maltese Proposal on the Sea-Bed and Ocean Floor—Its Fate in Two Cities," *International and Comparative Law Quarterly* 18 (1969): 62–69.

49. See Buzan, *Seabed Politics*, p. 50.

50. U.S. Senate, *The Third United Nations Law of the Sea Conference*, p. 87. See Gold, 236–38; and also Robert L. Friedheim, "Understanding the Debate on Ocean Resources," Occasional Paper No. 1 (Kingston, R.I.: Law of the Sea Institute, February 1969), p. 6.

51. *Legal Aspects of the Question* (U.N. Doc. A/AC.135/19), p. 8. The Conference's Fourth Committee indicated the importance attached to the 200-meter depth criterion by approving article 67, with a minor amendment clarifying its application to islands, by a vote of 51 to 9 with 10 abstentions, after refusing, by a vote of 13 to 42 with 13 abstentions, to define the shelf solely in terms of exploitability. Buzan, *Seabed Politics*, p. 39. In the plenary session, the "exploitability clause" was retained by a vote of 48 to 20, with 2 abstentions, and the full article was adopted by a vote of 51 to 5, with 10 abstentions. UNCLOS I, *Official Records*, vol. II, 8th plenary meeting, para. 44.

52. Convention on the Continental Shelf, arts. 2, 3, in *United Nations Treaty Series*, vol. 499, no. 7302, pp. 312, 314. The Convention extended jurisdiction to sedentary living resources, but not to bottom fish.

53. Convention on the Continental Shelf, art. 5 (8), in *United Nations Treaty Series*, vol. 499, no. 7302, p. 316. Coastal states were given the right to participate in such research.

54. See U.S. Senate, *The Third United Nations Law of the Sea Conference*, p. 7. The Convention also provided for conservation of the living resources of the shelf.

55. See Buzan, *Seabed Politics*, pp. 32, 44. The conferees did agree that the width of the territorial sea should not exceed 12 miles. Hollick, *U.S. Foreign Policy*, p. 144.

56. Convention on the Territorial Sea and the Contiguous Zone, arts. 14–23, in *United Nations Treaty Series*, vol. 516, no. 7477, pp. 214–20.

57. Buzan, *Seabed Politics*, p. 32. Traditionally, such contiguous zones had been limited to no more than 12 miles from the baseline from which the territorial sea was measured. See L. Alexander, "National Jurisdiction," p. 375.

58. See Hollick, *U.S. Foreign Policy*, p. 153; and Sanger, p. 17.

59. U.N. General Assembly, 13th Session, Resolution 1307, December 10, 1958.

60. Buzan, *Seabed Politics*, pp. 49–50. UNCLOS II was attended by essentially the same states which had participated in UNCLOS I, Afghanistan and Nepal having been replaced by Cameroon, Guinea, Ethiopia, and Sudan, for a total of 88 participants.

61. See U.S. Senate, *The Third United Nations Law of the Sea Conference*, p. 8. The proposal was blocked by a coalition of Arab states, which found the enclosure of the Straits of Tiran by 12-mile-wide territorial seas strategically advantageous in their struggle with Israel and which were supported by the Soviet bloc. Sanger, pp. 15–16.

62. Buzan, *Seabed Politics*, p. 53. See also Wenk, *The Politics of the Ocean*, p. 253.

63. Buzan, *Seabed Politics*, p. 59. Development of the U.S. oceans program was at least partially a response to the loss of U.S. nuclear equipment at sea during the 1960s, losses which included the nuclear-powered submarines *Thresher* and *Scorpion* in 1963 and 1968, respectively, and the loss of nuclear bombs off the coast of Spain in 1966. See ibid., pp. 53, 56–57.

64. See Pardo, "The Future of the Sea," p. 3.

65. Ibid., p. 4. The number of ships in the world merchant fleet doubled, and total tonnage quadrupled, from 1950 to 1970. Hollick, *U.S. Foreign Policy*, p. 6. Between 1951 and 1956 the percentage of 17,000-ton tankers in operation more than doubled, to one-half the world tanker fleet. Gold, p. 328.

66. Sanger, pp. 100–2. "The *Torrey Canyon* incident ... ended the *laissez-faire* attitude of governments toward oil pollution and stimulated attempts to curb marine pollution." John Warren Kindt, "Special Claims Impacting Upon Marine Pollution Issues at the Third U.N. Conference on the Law of the Sea," *California Western International Law Journal* 10 (1980): 435. Regulatory standards were negotiated within the Inter-governmental Maritime Consultative Organization (IMCO), which became the International Maritime Organization (IMO) on May 22, 1982.

67. See Seyom Brown et al., *Regimes for the Ocean, Outer Space, and Weather* (Washington: Brookings Institution, 1977), pp. 50–51; and Jacobson, p. 476.

68. United Nations, *Marine Science and Technology: Survey and Proposals: Report of the Secretary-General* (U.N. Doc. E/4487, April 24, 1968), p. 18. The world fish catch began to stabilize during the 1970s.

69. S. Brown et al., p. 51. Cf. U.S. Commission on Marine Science, Engineering, and Resources, *Our Nation and the Sea: A Plan for National Action* (Washington: U.S. Government Printing Office, 1969), p. 89. "Indeed, a gradual accumulation of evidence pointed irresistibly to the conclusion that, in certain areas, specific stocks of fish were being overfished to such a degree that, unless stringent measures were taken to conserve these resources, these stocks would dwindle to the detriment of all concerned." D.W. Bowett, *The Law of the Sea* (Dobbs Ferry, N.Y.: Oceana Publications, 1967), p. 21.

70. Buzan, *Seabed Politics*, p. 270.

71. W. Frank Newton, "Seabed Resources: The Problems of Adolescence," *San Diego Law Review* 8 (1971): 556.

72. Such applications included nuclear-powered ships, oceanic testing of atomic weapons, and radioactive wastes. See Charles Fincham, "Upheaval in the Law of the Sea," *Nuclear Active* (July 1972), p. 25.

73. "It became apparent that since the law is by its practice conservative, and technology in its capacity to induce change is radical, the two tend to advance at vastly different rates." Wenk, *The Politics of the Ocean*, p. 253.

74. Buzan, *Seabed Politics*, p. 7. See John L. Mero, "Manganese," *North Dakota Engineer* 27 (1952): 28.

75. See Barkenbus, pp. 7–9. For a discussion of the technology and methods initially considered for manganese nodule mining, see *Marine Science and Technology* (U.N. Doc. E/4487), p. 30.

76. U.S. Department of Commerce, *U.S. Ocean Policy in the 1970s*, part 2, p. 1. The National Academy of Sciences estimated the total quantity of manganese nodules in the "most promising" area of the Pacific Ocean at 15 billion tons. Ibid., part 2, p. 3. As late as 1970, Mero was still claiming that there were "about 1.5 trillion tons" of nodules on the floor of the Pacific. John L. Mero, "A Legal Regime for Deep Sea Mining," *San Diego Law Review* 7 (1970): 499.

77. Rolf Akesson, "The Law of the Sea Conference," *Journal of World Trade Law* 8 (1974): 287.

78. See Buzan, *Seabed Politics*, p. 276.

79. Ibid., pp. 53–55. From 1966 to 1970, 200-mile territorial seas were formally declared by Argentina (1966), Ecuador (1966), Panama (1967), Uruguay (1969), and Brazil (1970). Hollick, *U.S. Foreign Policy*, p. 162.

80. See James K. Sebenius, *Negotiating the Law of the Sea* (Cambridge, Mass.: Harvard University Press, 1984), p. 74.

81. Letter of May 18, 1966, from Douglas MacArthur II, Assistant Secretary of State for Congressional Relations, to Senator Warren G. Magnuson, reprinted in Charles I. Bevans, "Contemporary Practice of the United States: U.S. 12-Mile Fishing Zone," *American Journal of International Law* 60 (1966): 832.

82. John Lawrence Hargrove, "New Concepts in the Law of the Sea," *Ocean Development and International Law Journal* 1 (1973): 9. See Sebenius, p. 74.

83. See Hargrove, pp. 9–10. "Shelf-locked states are those whose continental shelves do not open out onto the high seas; examples are countries bordering certain large area seas such as the Mediterranean and Caribbean." U.S. Senate, *The Third United Nations Law of the Sea Conference*, p. 17.

84. See Pardo, "Address," p. 220.

85. See text accompanying n. 43 above.

86. Pardo, "Address," p. 220. The ambiguity might thus be attributable to poor drafting. See generally *Legal Aspects of the Question* (U.N. Doc. A/AC.135/19), pp. 19–20.

87. L.F.E. Goldie, "The Exploitability Test—Interpretation and Potentialities," *Natural Resources Journal* 8 (1968): 452. See also Kronmiller, vol. 1, p. 123.

88. Buzan, *Seabed Politics*, p. 58. Goldie, "The Exploitability Test," p. 439.

89. *Yearbook of the International Law Commission, 1950*, vol. 2, p. 384. See Lauterpacht, "Sovereignty Over Submarine Areas," pp. 383, 385–86; and also *Legal Aspects of the Question* (U.N. Doc. A/AC.135/19), pp. 7–8. In light of the loss of access suffered by

landlocked and geographically disadvantaged states as a result of the extensions of coastal-state jurisdiction, this rationale may seem unpersuasive. It must be remembered, however, that the negotiation of the articles was a political process, and that coastal states outnumbered landlocked states by a ratio of 4 to 1. Buzan, *Seabed Politics*, p. 21. See also Northcutt Ely, "A Case for the Administration of Mineral Resources Underlying the High Seas by National Interests," *Natural Resources Lawyer* 1 (1968): 81. "Thus the exploitability test was adopted at the third session as a way of satisfying States that had narrow or small continental shelves." Anand, *Origin and Development*, p. 162. See also A.E. Gotlieb, "The Impact of Technology on the Development of Contemporary International Law," *Recueil des Cours* 170 (1981): 169.

90. United Nations, *Report of the International Law Commission to the General Assembly* (U.N. Doc. A/1316, July 1950), para. 198, reprinted in *Yearbook of the International Law Commission, 1950,* vol. 2, p. 384.

91. *Yearbook of the International Law Commission, 1951,* vol. 1, 1117th meeting, para. 65. Chile and Peru are among the narrow-shelf states.

92. *Yearbook of the International Law Commission, 1951,* vol. 2, p. 141.

93. *Yearbook of the International Law Commission, 1953,* vol. 2, pp. 213–14.

94. *Yearbook of the International Law Commission, 1956,* vol. 2, p. 296 (commentary, para. 4).

95. Ibid., vol. 2, p. 297 (commentary, para. 7). The Commission determined that "some departure from the geographical meaning of the term 'continental shelf' was justified." Ibid. (commentary, para. 9).

96. Ibid., vol. 2, p. 278. See Buzan, *Seabed Politics*, p. 24; and also *Yearbook of the International Law Commission, 1955,* vol. 1, pp. 282–83.

97. See Buzan, *Seabed Politics*, p. 40. The Fourth Committee rejected proposals to define the shelf solely in terms of the exploitability of the geographical shelf, to delete the exploitability clause and employ a simple 200-meter depth criterion, and to abandon the Commission's formulation entirely and rely instead upon a depth criterion of 550 meters. UNCLOS I, *Official Records,* vol. VI, Fourth Committee, 19th meeting, paras. 10, 12, 14 (U.N. Docs. A/CONF.13/C.4/L.4, and A/CONF.13/C.4/L.29/Rev.1). See also Goldie, "The Exploitability Test," pp. 447–48.

98. See Jacques Patey, "La Conférence des Nations Unies sur le Droit de la Mer," *Revue Générale de Droit International Public* 62 (1958): 460.

99. *Legal Aspects of the Question* (U.N. Doc. A/AC.135/19), p. 18. See Paul Lawrence Saffo, "The Common Heritage of Mankind: Has the General Assembly Created a Law to Govern Seabed Mining?" *Tulane Law Review* 53 (1979): 499; and also Goldie, "The Exploitability Test," pp. 447–48.

100. *Yearbook of the International Law Commission, 1956,* vol. 2, p. 297 (commentary, para. 8).

101. Barkenbus, p. 31. Saffo, p. 499. See *Legal Aspects of the Question* (U.N. Doc. A/AC.135/19), p. 18; and also Goldie, "The Exploitability Test," p. 451. It was anticipated that exploitation capacity would increase gradually, allowing a progressive seaward extension of coastal-state jurisdiction toward the deep ocean floor. Jon Van Dyke and Christopher Yuen, "'Common Heritage' v. 'Freedom of the High Seas': Which Governs the Seabed?" *San Diego Law Review* 19 (1982): 506, 512. See also Buzan, *Seabed Politics*, p. 43. "[E]xperts maintained that it was not possible to work the resources of the sea at a depth of more than 200 metres, so under those conditions the question did not arise." Statement of J.P.A. François in *Yearbook of the International Law Commission, 1951,* vol. 1, p. 270 (para. 55). Even in the mid–1960s, few offshore oil drilling operations had been undertaken at depths greater than 100 meters: "industry interest in areas as deep as 200 meters (656 feet) was remote." Hollick, *U.S. Foreign Policy*, p. 182.

Most commentators, nevertheless, agreed that, theoretically, the language of article 1 of the Continental Shelf Convention would have permitted each coastal state to claim exclusive exploitation rights "out to the median line" of the ocean floor. Gotlieb, p. 168. Such a division, which would have produced grossly disproportionate and inequitable results, "would have been even more unpalatable to the majority of the world community in the late sixties than it was in the fifties." Saffo, p. 500. See Wenk, *The Politics of the Ocean*, pp. 256–57.

102. U.S. Senate, *The Third U.N. Law of the Sea Conference*, p. 8.

103. Pardo, "Address," p. 217.
104. *Yearbook of the International Law Commission, 1955,* vol. 1, p. 282; vol. 2, pp. 21–22. See n. 96 above and accompanying text; and also *Yearbook of the International Law Commission, 1955,* vol. 1, pp. 222, 263–64.
105. See, e.g., Arrow, "The Proposed Regime," pp. 352–65; and also Chap. 5 n. 1 et seq. and accompanying text. This doctrine apparently corresponds to the traditional position of the Soviet Union regarding the nature of the seabed regime. Christopher C. Joyner, "Towards a Legal Regime for the International Seabed: The Soviet Union's Evolving Perspective," *Virginia Journal of International Law* 15 (1975): 878.
106. See William L. Griffin, "The Emerging Law of Ocean Space," *International Lawyer* 1 (1966): 585. For an exhaustive discussion, see Kronmiller, vol. 1, pp. 106–206. Properly construed, the adjacency requirement in article 1 of the Continental Shelf Convention should prevent the exploitability clause from authorizing appropriation of any part of the deep seabed not geographically linked to the coastal-state land mass, whereas a genuine *res nullius* regime would permit such appropriation by effective occupation. "If 'occupation' thus conceived were the true basis of the legal claim to the adjacent submarine areas then there would be nothing – save the extra-legal remedies of intervention or self-preservation on the part of the coastal state – to prevent distant and strategically and economically powerful states from 'occupying' the adjacent submarine areas of other states. . . . Wide and disturbing possibilities of friction would thus be opened." Lauterpacht, "Sovereignty Over Submarine Areas," p. 420.
107. Statement of U.N. Under Secretary for Economic and Social Affairs, February 28, 1966, in *Official Records,* U.N. ECOSOC, 40th Session, 1408th meeting, para. 8. In furtherance of the general purposes of providing developing states with the benefit of new technological advances and of formulating a policy for "the integrated development of natural resources," the surveys were intended to promote four specific objectives: "first, to obtain new information and work out new ideas and approaches on developing the non-agricultural resources of the developing countries; secondly, to gather data giving a perspective of the long-term potential world supplies and needs in the resources selected; thirdly, to assist Governments to build the legal and organizational framework required for the optimum development of natural resources; and fourthly, to prepare specific projects for investment from multilateral and bilateral sources of financial assistance." Ibid., 1408th meeting, paras. 4, 11. For a discussion of ECOSOC's role in promoting an examination of issues relating to marine resource use, see Wenk, *The Politics of the Ocean,* pp. 426–27.
108. Statement of the representative of the United States in *Official Records,* U.N. ECOSOC, 40th Session, 1408th meeting, para. 29.
109. U.N. ECOSOC, 40th Session, Resolution 1112, March 7, 1966, para. (c). The resolution was cosponsored by Ecuador, Pakistan, and the United States, and supported by the Soviet Union.
110. Ibid., preambular para. 4.
111. U.N. General Assembly, 21st Session, Resolution 2172, December 6, 1966, paras. 1–3.
112. Statement of the representative of the United States in *Provisional Verbatim Records,* U.N. General Assembly, 21st Session (U.N. Doc. A/PV.1485 December 6, 1966), paras. 56, 63. It has been asserted, without a great deal of corroborative evidence, that U.S. delegates were acting without official authorization during these U.N. proceedings. Zdenek J. Slouka, "United Nations and the Deep Ocean: From Data to Norms," *Syracuse Journal of International Law* 1 (1972): 67 n. 6.
113. See Buzan, *Seabed Politics,* p. 81. "Inherent in the laws governing mineral development, both ashore and on the continental shelf, is the firm conviction that somewhere in the scheme of things there must exist an omnipotent if not omniscient sovereign, who owns the minerals and can grant an exclusive right to develop them, good against the world; that this sovereign is going to protect the miner's investment against claim jumpers; and, of course, that this law-making power is going to collect taxes or other exactions in exchange for the exercise of the privileges that it grants." Ely, "A Case for the Administration of Mineral Resources," p. 81.
114. Brownlie, *Principles of Public International Law,* p. 234.
115. See Francis T. Christy, Jr., "Comments on International Control of the Sea's

Resources," in *The Law of the Sea: Offshore Boundaries and Zones*, ed. Lewis M. Alexander (Columbus: Ohio State University Press, 1967), p. 302.

116. Francis T. Christy, Jr., "Alternative Regimes for Marine Resources Underlying the High Seas," *Natural Resources Lawyer* 1 (1968): 65–66, 76.

117. Barkenbus, p. 33.

118. Ely, "A Case for the Administration of Mineral Resources," pp. 83–84.

119. Daniel S. Cheever, "The Role of International Organization in Ocean Development," *International Organization* 22 (1968): 648. See also Goldie, "The Exploitability Test," p. 442.

120. See, e.g., *Yearbook of the International Law Commission, 1951*, vol. 1, pp. 304, 407; and also Nasila S. Rembe, *Africa and the International Law of the Sea* (Germantown, Md.: Sijthoff & Noordhoff, 1980), p. 36.

121. *Yearbook of the International Law Commission, 1955*, vol. 1, p. 10.

122. Friedheim, p. 47.

123. Ibid.

124. "One alternative is to do nothing—to 'wait and see' what emerges from actual development before attempting to establish a regime. A second alternative—that might be called the 'coastal state' or 'national lake' approach—is to divide up the oceans along lines that are equidistant from the shores of the coastal states. The third system—referred to as the 'flag nation' approach—would permit exploitation under the jurisdiction, and with the protection, of the nation whose flag is flown by the discoverer and exploiter. And the fourth approach is to permit exploitation under the jurisdiction, and with the protection, of an international authority." Christy, "Alternative Regimes," p. 67. See generally Friedheim, p. 1 et seq.

125. Barkenbus, p. 33. See, e.g., statement of the representative of Monaco in *Official Records*, UNCLOS I, vol. VI, p. 18.

126. Edward Wenk, Jr., "Toward Enhanced Management of Maritime Technology," in *Pacem in Maribus*, ed. Elisabeth Mann Borgese (New York: Dodd, Mead, 1972), p. 147.

127. Commission to Study the Organization of Peace, *New Dimensions for the United Nations: The Problems for the Next Decade*, Seventeenth Report (Dobbs Ferry, N.Y.: Oceana Publications, 1966), p. 39, quoted in William T. Burke, "A Negative View of a Proposal for United Nations Ownership of Ocean Mineral Resources," *Natural Resources Lawyer* 1 (1968): 43.

128. Senator Claiborne Pell, "Preface," in *The Law of the Sea: Offshore Boundaries and Zones*, ed. Alexander, p. xi. See "Discussion," in ibid., p. 24.

129. Third World Conference on World Peace Through Law, Resolution 15, Geneva, 1967, *Proceedings* (Geneva: World Peace Through Law Center, 1969). See also Wenk, *The Politics of the Ocean*, p. 259. The phrase "common heritage of mankind" had been employed less than a month earlier in the U.N. Committee on Peaceful Uses of Outer Space by the representative of Argentina. Aldo Armando Cocca, "The Advances in International Law Through the Law of Outer Space," *Journal of Space Law* 9 (1981): 15.

130. Friedheim, pp. 47–48. Hollick, *U.S. Foreign Policy*, p. 195. See also Evan Luard, *The Control of the Sea-Bed* (New York: Taplinger, 1977), p. 83.

131. Rembe, p. 37. The speech, which "laid the groundwork for UN self-education on the political seabed issue," led directly to the adoption of Resolution 2172. Wenk, *The Politics of the Ocean*, p. 262.

132. White House press release, July 13, 1966, quoted in Griffin, p. 587. The speech "received widespread press coverage and favorable comment." Wenk, *The Politics of the Ocean*, p. 258.

133. 33 U.S.C. §1101 (a). "The United States sought to implement its 1966 Marine Sciences Act in the context of multinational cooperation toward practical as well as altruistic objectives of a community of nations at peace." Wenk, *The Politics of the Ocean*, p. 253.

134. U.S. Commission on Marine Science, Engineering, and Resources, *Our Nation and the Sea*, p. vi.

135. Frank Church, "The U.N. at Twenty-One: Report to the Committee on Foreign Relations, U.S. Senate, 90th Congress, 1st Session," February 1967, p. 25, cited in Wenk, *The Politics of the Ocean*, p. 259.

136. See Leigh S. Ratiner, "The Law of the Sea: A Crossroads for American Foreign

Policy," *Foreign Affairs* 60 (1982): 1007; "[T]he Russians approached the Americans . . . at a time when the Pentagon was growing impatient with the State Department's practice of sending out ineffectual notes of protest at each expanded claim." Sanger, p. 21.

137. Sanger, p. 21. See also Hollick, *U.S. Foreign Policy*, pp. 174–75.

138. Burke, "A Negative View," pp. 44–45. See Luard, p. 84.

139. United Nations, *Request for the Inclusion of a Supplementary Item in the Agenda of the Twenty-Second Session: Note Verbale Dated 17 August 1967 from the Permanent Mission of Malta to the United Nations Addressed to the Secretary-General* (U.N. Doc. A/6695, August 18, 1967), pp. 2–3.

140. Ibid., p. 3.

141. Buzan, *Seabed Politics*, p. 67. See also Wenk, *The Politics of the Ocean*, pp. 261–62. Pardo's goals included prevention of international conflict, promotion and dissemination of scientific research, efficient exploitation of seabed resources, and establishment of "norms for the management and control of ocean space for human benefit." Arvid Pardo, "Development of Ocean Space – An International Dilemma," *Louisiana Law Review* 31 (1970): 59–60.

142. Cheever, "The Role of International Organization," p. 629. See also Buzan, *Seabed Politics*, p. 69.

143. See U.S. Congressional Research Service, *Exploiting the Resources of the Seabed*, pp. 61–64. "The objections to any U.N. action stemmed primarily from fears that the United States might be giving away some valuable assets and rights the extent of which were not yet known." Ibid., p. 63. See also Weissberg, pp. 42–48.

144. Shigeru Oda, *The Law of the Sea in Our Time – I: New Developments, 1966–1975* (Leyden: A.W. Sijthoff, 1977), p. 11. Senator Pell's efforts did, however, play an important role in promoting the idea of internationalization within the U.S. government. See Wenk, *The Politics of the Ocean*, p. 276.

145. Weissberg, p. 47. See Hollick, *U.S. Foreign Policy*, pp. 198–200. The State Department "agreed that while its idealism was constructive, his proposition went too far, too fast, with insufficient hard data." Wenk, *The Politics of the Ocean*, p. 266. The initial positive U.S. reaction had been prompted by President Johnson's 1966 policy statement, as well as by the relatively low strategic importance assigned to seabed minerals in the years prior to the OPEC oil embargo. See n. 132 above and accompanying text, and Chap. 2 nn. 201–2 and accompanying text.

146. United Nations, *Note by the Secretary-General* (U.N. Doc. A/C.1/952, October 31, 1967), p. 4.

147. Statement of Arvid Pardo, representative of Malta, in *Provisional Verbatim Records*, U.N. General Assembly, 22nd Session, First Committee, 1515th meeting (U.N. Doc. A/C.1/PV.1515, November 1, 1967), pp. 1–5. In addition to manganese nodules, Ambassador Pardo discussed the potential for exploitation of phosphorite nodules, calcareous and siliceous oozes, pelagic clays, and deep-sea deposits of petroleum, gas, and sulfur.

148. Ibid., 1515th meeting (U.N. Doc. A/C.1/PV.1515), pp. 5–13.

149. Ibid., 1515th meeting (U.N. Doc. A/C.1/PV.1515), p. 14, and 1516th meeting (U.N. Doc. A/C.1/PV.1516, November 1, 1967), p. 1.

150. Ibid., 1516th meeting (U.N. Doc. A/C.1/PV.1516), p. 2. The estimate included revenues from petroleum exploitation of the continental shelf beyond a 12-mile territorial sea. Conversation with Arvid Pardo, Los Angeles, August 3, 1987. Pardo nonetheless tantalized his contemporaries by predicting that the mineral content of seabed nodules was sufficient to satisfy world demand for nickel, copper, and manganese for thousands of years. See Finn Laursen, *Superpower at Sea* (New York: Praeger, 1983), p. 3. The economic efficiency of mineral mining and the amount of projected revenues potentially available therefrom were, in fact, subjects of considerable disagreement. Buzan, *Seabed Politics*, pp. 86–87.

151. Slouka, p. 76. "There was a general feeling that the U.N. had here become involved in a new subject, of profound importance but great complexity and fascination, which would command the attention of delegates and officials for years to come." Luard, p. 87.

152. See Buzan, *Seabed Politics*, pp. 67–68.

153. See Luard, p. 88; and also Pardo, "Development of Ocean Space," p. 60.

154. Pardo, "Development of Ocean Space," p. 60.

155. See Weissberg, p. 55.

156. Buzan, *Seabed Politics*, p. 82.
157. See ibid., pp. 84–86.
158. U.N. General Assembly, 22nd Session, Resolution 2340, adopted by a vote of 99 to 0.

Chapter 2

1. John R. Stevenson, "Lawmaking for the Seas," *American Bar Association Journal* 61 (1975): 187.
2. United Nations, *Resources for the Sea: Report of the Secretary-General* (U.N. Doc. E/4449/Add.1, February 19, 1968), pp. 87–94.
3. Ibid., p. 89.
4. Ibid., p. 91.
5. Barry Buzan, *Seabed Politics* (New York: Praeger, 1976), p. 70.
6. United Nations, *Report of the Ad Hoc Committee to Study the Peaceful Uses of the Sea-Bed and the Ocean Floor Beyond the Limits of National Jurisdiction* (U.N. Doc. A/7230, 1968), p. 1.
7. Buzan, *Seabed Politics*, pp. 70–71.
8. United Nations, *United Nations Ad Hoc Committee to Study the Peaceful Uses of the Sea-Bed and the Ocean Floor Beyond the Limits of National Jurisdiction, Replies from Governments* (U.N. Doc. A/AC.135/1, March 11, 1968), pp. 26–30. In addition, the Committee considered numerous proposals drafted by private organizations with the aim of influencing the negotiations. See Robert L. Friedheim, "Understanding the Debate on Ocean Resources," Occasional Paper No. 1 (Kingston, R.I.: Law of the Sea Institute, February 1969), pp. 50–51.
9. Statement of the representative of the United States in *Summary Records*, U.N. General Assembly, 23rd Session, *Ad Hoc* Committee, 3rd meeting (U.N. Doc. A/AC.135/SR.3, August 28, 1968), p. 3. In his State of the Union Message in early 1968, the President had proposed that the United States "launch, with other nations, an exploration of the ocean depth to tap its wealth and its energy and its abundance." Guenter Weissberg, "International Law Meets the Short-Term National Interest: The Maltese Proposal on the Sea-Bed and Ocean Floor — Its Fate in Two Cities," *International and Comparative Law Quarterly* 18 (1969): 95.
10. United Nations, *Legal Aspects of the Question of the Reservation Exclusively for Peaceful Purposes of the Sea-Bed and the Ocean Floor, and the Subsoil Thereof, Underlying the High Seas Beyond the Limits of Present National Jurisdiction, and the Use of Their Resources in the Interests of Mankind: Study Prepared by the Secretariat* (U.N. Doc. A/AC.135/19, June 21, 1968), p. 4.
11. United Nations, *Marine Science and Technology: Survey and Proposals: Report of the Secretary-General* (U.N. Doc. E/4487, April 24, 1968), p. 75.
12. Ibid., p. 76.
13. Ibid., p. 81.
14. Ibid., p. 82. The Secretary-General also proposed that the mandate of the IOC be broadened to permit it to formulate and coordinate an expanded program to assist in a better understanding of the marine environment through science; that there be further improvement, within the existing international framework, of international collaboration with regard to fisheries development and conservation; and that states adopt more stringent multilateral agreements regarding the prevention and control of marine pollution. Ibid., pp. 77, 81, 84.
15. Statement of the Chairman of the Legal Working Group in *Summary Records*, U.N. General Assembly, *Ad Hoc* Committee, Legal Working Group, 1st meeting (U.N. Doc. A/AC.135/WG.1/SR.1, June 18, 1968), pp. 3–4.
16. *Report of the Ad Hoc Committee* (U.N. Doc. A/7230), p. 9. See Weissberg, p. 101.
17. *Legal Aspects of the Question* (U.N. Doc. A/AC.135/19/Add.2), p. 4.
18. Ibid., pp. 18–20.
19. Ibid., p. 29.
20. *Report of the Ad Hoc Committee* (U.N. Doc. A/7230), Annex I: Report of the Economic and Technical Working Group, p. 29.

21. Ibid., Annex I, p. 28. It was widely expected that actual mining operations might not begin for 10 to 20 years. See *Summary Records*, U.N. General Assembly, *Ad Hoc* Committee, Economic and Technical Working Group, 8th meeting (U.N. Doc. A/AC.135/WG.2/SR.8, June 28, 1968), pp. 3–4.

22. *Report of the Ad Hoc Committee* (U.N. Doc. A/7230), Annex I, pp. 31–32.

23. Ibid., Annex I, pp. 32–33.

24. Ibid., Annex I, pp. 33–34.

25. Ibid., Annex I, p. 34.

26. Ibid., Annex I, p. 35.

27. Ibid., pp. 3, 16.

28. See Buzan, *Seabed Politics*, pp. 72–73.

29. *Report of the Ad Hoc Committee* (U.N. Doc. A/7230), p. 7. See U.N. ECOSOC, 45th Session, Resolution 1380, August 2, 1968.

30. Statement of the representative of Ceylon in *Provisional Verbatim Records*, U.N. General Assembly, 23rd Session, First Committee, 1588th meeting (U.N. Doc. A/C.1/PV.1588, October 28, 1968), p. 61. See *Report of the Ad Hoc Committee* (U.N. Doc. A/7230), p. 10. During the *Ad Hoc* Committee negotiations, the United States and the Soviet Union exchanged draft proposals for demilitarization of the seabed, the former favoring a ban only on "weapons of mass destruction." Christopher C. Joyner, "Towards a Legal Regime for the International Seabed: The Soviet Union's Evolving Perspective," *Virginia Journal of International Law* 15 (1975): 883.

31. Buzan, *Seabed Politics*, p. 73.

32. Friedheim, p. 35.

33. Buzan, *Seabed Politics*, p. 77.

34. See Zdenek J. Slouka, "United Nations and the Deep Ocean: From Data to Norms," *Syracuse Journal of International Law* 1 (1972): 71.

35. U.N. General Assembly, 23rd Session, Resolution 2467A, December 21, 1968. See U.S. 95th Congress, 2nd Session, Senate, Committee on Commerce, Science, and Transportation, *Report: The Third United Nations Law of the Sea Conference* (Washington: U.S. Government Printing Office, 1978), p. 10.

36. See United Nations, *Report of the Committee on the Peaceful Uses of the Sea-Bed and the Ocean Floor Beyond the Limits of National Jurisdiction*, U.N. General Assembly, 24th Session (U.N. Doc. A/7622, 1969), Annex II, p. 83.

37. U.N. General Assembly, 23rd Session, Resolutions 2467B and 2467C, December 21, 1968.

38. Buzan, *Seabed Politics*, p. 79.

39. Ibid., p. 75. The developing states represented on the Seabed Committee now included eight African, five Asian, four Arab, and seven Latin American states, as well as Malta and Yugoslavia.

40. Jack N. Barkenbus, *Deep Seabed Resources: Politics and Technology* (New York: Free Press, 1979), p. 34. The Committee's 1969 report to the General Assembly stated that "it would be desirable in future that it be allowed more time to carry out its programme of work." *Report of the Committee* (U.N. Doc. A/7622), p. 7.

41. Buzan, *Seabed Politics*, p. 95.

42. Ibid., p. 94.

43. Ibid., pp. 85, 92.

44. Ibid., pp. 93–94.

45. See *Report of the Committee* (U.N. Doc. A/7622), pp. 102–17, 120–45.

46. Ibid., p. 119.

47. Ibid., pp. 146–53.

48. Ibid., pp. 11–12. See Buzan, *Seabed Politics*, p. 91.

49. *Report of the Committee* (U.N. Doc. A/7622), pp. 13, 21.

50. John Ludvik Løvald, "In search of an Ocean Regime: The Negotiations in the General Assembly's Seabed Committee 1968–70," *International Organization* 29 (1975): 708–9. Buzan, *Seabed Politics*, pp. 127–28. Negotiating groups representing the maritime states and the landlocked and geographically disadvantaged states also became particularly active. See Ann L. Hollick, *U.S. Foreign Policy and the Law of the Sea* (Princeton, N.J.: Princeton University

Press, 1981), pp. 250–51. Ironically, many of the arguments advanced by the developing states originated in a debate which was then being carried on in the United States among scholars and policymakers. Buzan, *Seabed Politics*, p. 77.

51. Løvald, pp. 699, 706.

52. *Report of the Committee* (U.N. Doc. A/7622), p. 50.

53. Ibid., p. 65.

54. Ibid., pp. 70–71.

55. Ibid., p. 71.

56. Ibid., pp. 70, 72. It was argued that an organization with exclusive mining rights would require exorbitant capital outlays; would have a monopoly power over seabed resources; would infringe upon protected patent rights and trade secrets; would be too risky for an international institution to undertake; and would actually harm developing states by preventing them from conducting their own national mining operations.

57. Arvid Pardo, "Development of Ocean Space — An International Dilemma," *Louisiana Law Review* 31 (1970): 66–67.

58. U.N. General Assembly, 24th Session, Resolutions 2560 and 2566, December 13, 1969.

59. U.N. General Assembly, 24th Session, Resolutions 2574B and 2574C, December 15, 1969. The resolutions were adopted by votes of 109 to 0, with 1 abstention, and 100 to 0, with 11 abstentions, respectively. See Slouka, p. 65.

60. U.N. General Assembly, 24th Session, Resolution 2574A, December 15, 1969.

61. U.N. General Assembly, 24th Session, Resolution 2574D, December 15, 1969. "The bloc of sixty-two read like a roster of the less developed and non-aligned countries of Africa, Asia, and Latin America, and Finland, Sweden and Yugoslavia. The twenty-eight against represented the developed North, including the United States and the Soviet Union with their respective allies and South Africa. Only Ghana and Malta disturbed the homogeneity of this group. The abstaining bloc and the absentees reflected various special interests or, perhaps, disinterest." Slouka, p. 72.

62. U.N. General Assembly, Resolution 2574D.

63. See Chap. 5 nn. 71–73 and accompanying text.

64. Letter from John R. Stevenson, Legal Adviser, Department of State, to Senator Lee Metcalf, January 16, 1970, reprinted in *International Legal Materials* 9 (1970): 832.

65. Buzan, *Seabed Politics*, pp. 99–100.

66. William E. Butler, *The Soviet Union and the Law of the Sea* (Baltimore: Johns Hopkins Press, 1971), p. 153. See also Hollick, *U.S. Foreign Policy*, p. 186. Military uses envisioned for the seabed included installation of guided missiles and submarine detection systems. Butler, p. 153. Arvid Pardo was particularly concerned that emplacement of anti-ballistic missiles on the ocean floor would be used to justify exclusive claims to large seabed areas. Clyde Sanger, *Ordering the Oceans: The Making of the Law of the Sea* (London: Zed Books, 1986), p. 19.

67. Pardo, "Development of Ocean Space," p. 67.

68. Ibid., p. 69. See also Joyner, "The Soviet Union's Evolving Perspective," pp. 883–84.

69. Sevinc Carlson, "Soviet Policy on the Sea-Bed and the Ocean Floor," *Syracuse Journal of International Law and Commerce* 1 (1972): 108. For a more complete history of the treaty, see Bennett Ramberg, *The Seabed Arms Control Negotiations: A Study of Multilateral Arms Control Conference Diplomacy*, University of Denver Monograph Series in World Affairs, vol. 15, bk. 2 (1978). "The treaty was a direct outcome of the original question on the reservation of the seabed for peaceful purposes raised in the General Assembly in 1967, and indirectly it helped to precipitate . . . arms limitations talks (SALT) between the USSR and the United States in which the whole problem of arms control [was] reviewed." Gerald J. Mangone, "The Effects of a Successful Ocean Regime on the United Nations System," in *The Fate of the Oceans*, ed. John J. Logue (Villanova, Penn.: Villanova University Press, 1972), p. 164. By the end of 1982, 70 states had become parties to the treaty.

70. Slouka, p. 72. See Buzan, *Seabed Politics*, p. 107. "Considering the novelty of the seabed regime, the gaps in technical knowledge of the sea, problems in interweaving technology with statecraft in the tapestry of legal, scientific, economic, political, and social questions involved, the national prestige, security, and territorial claims at stake, a long

gestation period was to be expected." Edward Wenk, Jr., *The Politics of the Ocean* (Seattle: Washington University Press, 1972), p. 287.

71. See Aaron L. Danzig, "A Funny Thing Happened to the Common Heritage on the Way to the Sea," *San Diego Law Review* 12 (1975): 659 and n. 14. The Lima Declaration also endorsed application of the common heritage of mankind principle to the deep seabed.

72. United Nations, *Report of the Committee on the Peaceful Uses of the Sea-Bed and the Ocean Floor Beyond the Limits of National Jurisdiction* (U.N. Doc. A/8021, 1970), pp. 5, 9.

73. Ibid., pp. 10–11. The report recommended that the declaration of principles include a statement "of the need to prevent and control pollution," and concluded that "action on a world-wide scale" was necessary. Ibid.

74. See John Warren Kindt, "Special Claims Impacting Upon Marine Pollution Issues at the Third U.N. Conference on the Law of the Sea," *California Western International Law Journal* 10 (1980): 437–38. Although the Canadian legislation was widely regarded as contrary to existing principles of customary international law, the Soviet Union claimed similar jurisdiction in 1971 with respect to its own Arctic waters. Ibid., p. 438.

75. United Nations, *Proposals Concerning the Establishment of a Regime for the Exploration and the Exploitation of the Seabed* (U.N. Doc. A/AC.138/27, August 5, 1970). United Nations, *International Regime: Working Paper Submitted by the United Kingdom* (U.N. Doc. A/AC.138/26, August 5, 1970). See Barkenbus, pp. 111–12.

76. United Nations, *Draft United Nations Convention on the International Seabed Area* (U.N. Doc. A/AC.138/25, August 3, 1970). Six members of the Council were to be the most industrially advanced states. The 18 other states were to include at least 12 developing states and 2 landlocked or shelf-locked states. To prevent developing states from controlling the Authority, the convention would have required decisions of the Council to be approved by concurrent majorities of the 6 industrially advanced states and the other 18 states as well as by a majority of the Council as a whole. Mangone, p. 165. See Samuel C. Orr, "Soviet, Latin Opposition Blocks Agreement on Seabeds Treaty," *CPR National Journal* (1970), p. 1977.

77. President Richard M. Nixon, "United States Policy for the Seabed," *U.S. Department of State Bulletin* 62 (1970): 737–38. See also John R. Stevenson, "Legal Regulation of Mineral Exploitation in the Deep Seabed," *U.S. Department of State Bulletin* 65 (1971): 51–52.

78. Nixon, "United States Policy for the Seabed," p. 737.

79. Orr, p. 1974. See Hollick, *U.S. Foreign Policy*, pp. 235–36. The United States was supported in this strategy by other major maritime states which had opposed the creation of strong international machinery. Such a strategy "explains why the major champion within the United States government for such a regime was the Department of Defense." Seyom Brown et al., *Regimes for the Ocean, Outer Space, and Weather* (Washington: Brookings Institution, 1977), p. 74.

80. See John J. Logue, "Introduction: The Trillion Dollar Opportunity," in *The Fate of the Oceans*, ed. Logue, p. xxiv.

81. Orr, p. 1977. Pardo particularly agreed with the provisions in favor of broad international jurisdiction and a strong international agency with effective regulatory powers, although he preferred a limit based upon a fixed distance from the shore rather than the outmoded 200-meter depth criterion.

82. Ibid., pp. 1974, 1977. See Wenk, *The Politics of the Ocean*, p. 283. Many newly independent states also opposed the draft convention because of connotations of colonialism associated with the term "trusteeship." Martin Ira Glassner, "Developing Land-Locked States and the Resources of the Seabed," *San Diego Law Review* 11 (1974): 653. See also Logue, "The Trillion Dollar Opportunity," p. xx.

83. See Hollick, *U.S. Foreign Policy*, p. 234.

84. *Report of the Committee* (U.N. Doc. A/8021), p. 6.

85. Ibid., p. 15.

86. See Barkenbus, p. 35. The disputed issues omitted from the Declaration included the operative definitions of "the common heritage of mankind" and "peaceful purposes"; the structure and functions of the proposed machinery; the transfer of technology from developed to developing states; and the delimitation of national and international jurisdictional zones. Buzan, *Seabed Politics*, p. 109.

87. Buzan, *Seabed Politics*, pp. 110-11.

88. U.N. General Assembly, 25th Session, Resolution 2749, December 17, 1970. Abstaining states — comprised mostly of the Soviet Union and its allies — "felt that the concept of common heritage was neither realistic nor practical and that the activities of states in the area should be in conformity with international law." Stephen Gorove, "The Concept of 'Common Heritage of Mankind': A Political, Moral, or Legal Innovation," *San Diego Law Review* 9 (1972): 399.

89. Dennis W. Arrow, "The Proposed Regime for the Unilateral Exploitation of Deep Seabed Mineral Resources by the United States," *Harvard International Law Journal* 21 (1980): 375-77. The United States and most of the other developed states maintained that deep seabed mining was permitted under customary international law. David P. Stang, "Political Cobwebs Beneath the Sea," *International Lawyer* 7 (1972): 14. Still, "for the United States to support the 1970 Declaration was a noticeable advance on the position it had taken two years earlier in its draft list of principles." Sanger, p. 35. See Chap. 5 nn. 269-79 and accompanying text.

90. U.N. General Assembly, 25th Session, Resolutions 2750A and 2750B, December 17, 1970. The resolutions were adopted by votes of 104 to 0, with 16 abstentions, and 111 to 0, with 11 abstentions, respectively. See Slouka, p. 65.

91. U.N. General Assembly, 25th Session, Resolution 2750C, December 17, 1970. See United States, "Reports of the United States Delegation to the Third United Nations Conference on the Law of the Sea," Occasional Paper No. 33, ed. Myron H. Nordquist and Choonho Park (Honolulu: Law of the Sea Institute, 1983), p. 9. The resolution was adopted by a vote of 108 to 7, with 6 abstentions, opposing votes being cast by the Soviet Union and its allies. Slouka, pp. 65, 72.

U.S. efforts to maintain the seabed negotiations in a forum separate from other law of the sea issues were abandoned in the face of developing-state insistence upon linkage—a favorable negotiating opportunity for the Defense Department, which "envisioned an agreement for continued navigational freedoms mainly brought about by accommodating Third World interests in continental shelves, fishing rights, and the nodule regime" — and widespread recognition of the physical and conceptual links between the deep seabed and its superjacent waters. James K. Sebenius, *Negotiating the Law of the Sea* (Cambridge, Mass.: Harvard University Press, 1984), pp. 106-7. See Orr, p. 1977.

92. See Elliot L. Richardson, "Law of the Sea: A Reassessment of U.S. Interests," *Mediterranean Quarterly* 1 (Spring 1990): 8-9.

93. Orr, p. 1977. See also Stevenson, "Lawmaking for the Seas," p. 187.

94. See Buzan, *Seabed Politics*, pp. 112-13; and "Reports of the United States Delegation," ed. Nordquist and Park, p. 10.

95. Buzan, *Seabed Politics*, pp. 140, 142-43, 149. Other important new political alignments included technologically developed and developing states, maritime and nonmaritime states, distant-water fishing states, and states with coastal fisheries. See U.S. Senate, *The Third United Nations Law of the Sea Conference*, p. 18.

96. Buzan, *Seabed Politics*, pp. 112, 202-3.

97. United Nations, *Report of the Committee on the Peaceful Uses of the Sea-Bed and the Ocean Floor Beyond the Limits of National Jurisdiction* (U.N. Doc. A/8421, 1971), p. 5. Each subcommittee followed a similar pattern of work: a general debate on the major issues, followed by negotiations within working groups on draft treaty articles. U.S. Senate, *The Third United Nations Law of the Sea Conference*, p. 13.

98. Bernard H. Oxman, "The World Outlook for the International Law of the Sea." Keynote address to Marine Technology Society, February 19, 1971. By 1972 63 states already claimed territorial seas of 12 or more miles. Sanger, p. 84.

99. *Report of the Committee* (U.N. Doc. A/8421), pp. 20-21. See United Nations, *Draft Ocean Space Treaty: Working Paper Submitted by Malta* (U.N. Doc. A/AC.138/53, August 5, 1971).

100. William T. Burke, "Comments on Issues Relating to the Law of the Sea," *National Resources Lawyer* 4 (1971): 663. Cf. statement of Arvid Pardo, representative of Malta, in *Provisional Verbatim Records*, U.N. General Assembly, 22nd Session, First Committee, 1515th meeting (U.N. Doc. A/C.1/PV.1515, November 1, 1967), p. 9.

101. *Draft Ocean Space Treaty* (U.N. Doc. A/AC.138/53). The Maltese delegation "was

among the first to accept the 200-mile overall limit to coastal state jurisdiction in the oceans. [It] did so mainly because [its] analysis showed that a limit much beyond 200 miles would benefit only a minority of the international community, while a limit much less than 200 miles was unlikely to be internationally acceptable in view of already existing claims of coastal States." Arvid Pardo, "A Statement on the Future Law of the Sea in Light of Current Trends in Negotiations," *Ocean Development and International Law Journal* 1 (1974): 330–31.

102. Monty Hoyt, "Sea Panel Races Time to Form International Control," *Christian Science Monitor*, October 13, 1971, p. 6.

103. *Report of the Committee* (U.N. Doc. A/8421), p. 21. See Finn Laursen, *Superpower at Sea* (New York: Praeger, 1983), p. 98.

104. United Nations, *Working Paper on the Regime for the Sea Bed and Ocean Floor and Its Subsoil Beyond the Limits of National Jurisdiction, Submitted by Chile, Colombia, Ecuador, El Salvador, Guatemala, Guyana, Jamaica, Mexico, Panama, Peru, Trinidad and Tobago, Uruguay, and Venezuela* (U.N. Doc. A/AC.138/49, August 4, 1971), art. 14.

105. Ibid., arts. 21, 23, 24 (g). The Assembly was also to have the power to require contribution to capital and to allocate the shared benefits of exploitation. Ibid., art. 24 (d), (i).

106. Ibid., arts. 26, 32 (f) (j). Each member state was to be entitled to one vote, and substantive decisions were to be taken by a two-thirds rather than a simple majority vote. Ibid., art. 29.

107. Barkenbus, pp. 114–15. See also "Reports of the United States Delegation," ed. Nordquist and Park, p. 41. The proposal had been intended as a bargaining chip to be used as a basis for negotiations with the United States, with a view toward securing two concessions of importance to Latin American states: seaward extension of national jurisdiction and international control over seabed mining operations. Barkenbus, p. 115. It has been claimed that the Enterprise was named after the spaceship in the television series "Star Trek." See Peter Grier, "Staking a Claim to the Ocean's Bed," *Christian Science Monitor*, April 9, 1981, pp. 12–13.

108. See Carlson, p. 109.

109. United Nations, *Union of Soviet Socialist Republics: Provisional Draft Articles of a Treaty on the Use of the Sea-Bed for Peaceful Purposes* (U.N. Doc. A/AC.138/43, July 22, 1971), arts. 9, 26. See Joyner, "The Soviet Union's Evolving Perspective," pp. 888–89.

110. See Joyner, "The Soviet Union's Evolving Perspective," p. 891. Cf. *Union of Soviet Socialist Republics: Provisional Draft Articles* (U.N. Doc. A/AC.138/43), art. 27.

111. Logue, "The Trillion Dollar Opportunity," p. xxii.

112. *Report of the Committee* (U.N. Doc. A/8421), pp. 4, 37.

113. Ibid., p. 44.

114. See U.S. Senate, *The Third United Nations Law of the Sea Conference*, p. 13.

115. Stang, p. 2.

116. Buzan, *Seabed Politics*, pp. 177, 180. "Resounding idealism had deteriorated to contests of bold self-interest." Wenk, *The Politics of the Ocean*, p. 286.

117. Buzan, *Seabed Politics*, pp. 205–6.

118. Ibid., pp. 187–88. See also Hollick, *U.S. Foreign Policy*, p. 252.

119. Hollick, *U.S. Foreign Policy*, p. 170.

120. Stang, pp. 1–4.

121. Ibid., pp. 1, 9–10. The list was organized under 25 main headings. See U.S. Senate, *The Third United Nations Law of the Sea Conference*, p. 15.

122. Stang, p. 10. Major maritime states "are preponderantly global rather than coastal. The system of law they have developed seeks to restrict the assertion of coastal authority, both quantitatively and qualitatively, in order to maintain an essentially *laissez-faire* regime conducive to the operation of their great navies and merchant and fishing fleets. The other littoral nations of the world in general may be characterized as 'coastal states' since their maritime interests are primarily centered on their own coasts and since many of them have in varying degrees sought to extend the scope of coastal authority." L.H.J. Legault, "The Freedom of the Seas: A License to Pollute?" *University of Toronto Law Journal* 21 (1971): 211.

123. See Wenk, *The Politics of the Ocean*, pp. 278, 284–85; Glassner, pp. 653–54; and also Ann L. Hollick, "United States Oceans Politics," *San Diego Law Review* 10 (1973): 490.

124. Buzan, *Seabed Politics*, p. 157. See Stang, p. 14. "The bill served as a convenient

'worst possible model' for the developing countries by raising a concrete possibility of the kind of technological imperialism that had originally inspired Pardo to make his proposal." Buzan, *Seabed Politics*, p. 157.

125. Sanger, p. 38. While the Action Plan adopted at the Conference merely called for consultations among "interested states" when appreciable harm to the environment was foreseen, the Conference nonetheless helped to raise environmental consciousness throughout the world and promoted the development of international environmental law. Frederic L. Kirgis, Jr., "Editorial Comment: Standing to Challenge Human Endeavors That Could Change the Climate," *American Journal of International Law* 84 (1990): 527.

126. U.N. General Assembly, 27th Session, Resolution 3029A, December 18, 1972.

127. An Authority with powers of direct control over seabed mining operations was supported by some 70 states, including several mid-sized developed states. Laursen, pp. 98–99. At the end of 1973 the idea of an Authority with the power to monopolize nodule exploitation completely was supported by only 15 states, however. Barkenbus, p. 115.

128. Buzan, *Seabed Politics*, p. 173. See also "Reports of the United States Delegation," ed. Nordquist and Park, pp. 39–40.

129. Buzan, *Seabed Politics*, p. 172.

130. "Reports of the United States Delegation," ed. Nordquist and Park, p. 43.

131. "Careful examination of the alternative texts . . . reveals that in fact there was no meaningful advance on the level of agreement achieved in the declaration of principles. In terms of advancing toward consensus, Sub-Committee I had been paralyzed for the entire three-year period of 1971–73." Buzan, *Seabed Politics*, pp. 175–76. Cf. "Reports of the United States Delegation," ed. Nordquist and Park, pp. 10, 53.

132. Canada and Malta were two of the "very few" states which avoided confrontationalist stances. Buzan, *Seabed Politics*, p. 176.

133. Ibid., p. 206. See René-Jean Dupuy, *The Law of the Sea* (Dobbs Ferry, N.Y.: Oceana Publications, 1974), p. 27.

134. See "Reports of the United States Delegation," ed. Nordquist and Park, p. 43; and also Stang, p. 9 nn. 9, 10. "The Soviet Union and the United States, allied on this issue of free navigation, competed in 1973 to be the first to offer the bargaining terms." Sanger, p. 63.

135. United Nations, *Organization of African Unity Declaration on the Issues of the Law of the Sea — CM/Res. 289 (XIX)* (U.N. Doc. A/AC.138/89, July 2, 1973), pp. 5–6 (paras. 6, 7, 9), quoted in Danzig, pp. 658–59.

136. Pardo, "A Statement on the Future Law of the Sea," pp. 323–24, 332, 333.

137. U.S. Senate, *The Third United Nations Law of the Sea Conference*, p. 26. It was generally understood that coastal states would be under a duty to manage and conserve coastal fisheries, although special regimes to govern anadromous and highly migratory species were advocated by the United States and others. See "Reports of the United States Delegation," ed. Nordquist and Park, pp. 46–47.

138. Buzan, *Seabed Politics*, pp. 194–96.

139. Sanger, pp. 90–93. Some Southeast Asian states had begun protesting straits navigation by military vessels. Ibid., p. 86.

140. U.S. Senate, *The Third United Nations Law of the Sea Conference*, p. 13. See "Reports of the United States Delegation," ed. Nordquist and Park, pp. 38, 48–49, 54. Opposition to mandatory pollution-control standards had begun to emerge among developing states, which sought to ensure that such standards would not impair their opportunity for economic growth. Major differences also existed between maritime-state proposals for preemptive international pollution-control standards in offshore waters and coastal-state demands for discretionary legislative authority. "Reports of the United States Delegation," ed. Nordquist and Park, pp. 48–50.

141. U.N. General Assembly, 28th Session, Resolution 3067, December 16, 1973. See Michael J. Berlin, "Some Diverse Views at UN on the Sea Bed," *New York Post*, March 23, 1973, p. 3. The resolution was adopted by a vote of 117 to 0, with 10 abstentions.

142. See Buzan, *Seabed Politics*, p. 216. "Multilateral treaties drafted by an international Conference are, upon completion, adopted by a vote of the membership of the Conference on the treaty as a whole. According to the rules of the Third Law of the Sea Conference, a decision on the adoption of the text of the Convention . . . as a whole will require a two-thirds

majority of the representatives present and voting." U.S. Senate, *The Third United Nations Law of the Sea Conference*, p. 101.

143. Sebenius, p. 74. See John Temple Swing, "Address," in *Current Issues in the Law of the Sea*, ed. Christopher C. Joyner (Allentown, Penn.: Muhlenberg College, 1980), p. 111.

144. UNCLOS III, *Economic Implications of Sea-Bed Mineral Development in the International Area: Summary and Conclusions of the Report of the Secretary-General* (U.N. Doc. A/CONF.62/25, May 22, 1974), reprinted in *Official Records*, UNCLOS III, vol. III, p. 6.

145. See Barkenbus, pp. 19–20; and Chap. 5 n. 315 and accompanying text.

146. Charles J. Johnson and Allen L. Clark, "Potential of Pacific Ocean Nodule, Crust, and Sulfide Mineral Deposits," *Natural Resources Forum* 9 (1985): 180.

147. Barkenbus, pp. 8–11. See also Sanger, pp. 159–61; and Richardson, "A Reassessment of U.S. Interests," pp. 4–5. Mero had asserted that the nodules were actually growing at a faster rate than they could be mined. Barkenbus, p. 8.

148. Gregory DeSousa, "Ocean Management and World Order," *Columbia Journal of World Business* 9 (1974): 126. See Buzan, *Seabed Politics*, p. 151; and S. Brown et al., p. 82. It was expected that, except in the case of socialist states, mining operations would actually be conducted by multinational consortia rather than by the states themselves. Barkenbus, p. 77.

149. See *Economic Implications: Summary and Conclusions* (U.N. Doc. A/CONF.62/25), in *Official Records*, UNCLOS III, vol. III, p. 6. 'The time when it will be as economic to recover hard minerals from the sea as from the land depends on such factors as rate of consumption, exploration and recovery technology, transportation costs, the availability of substitutes, re-use techniques, discovery of new land deposits, and the reliability of import sources." U.S. Commission on Marine Science, Engineering, and Resources, *Our Nation and the Sea: A Plan for National Action* (Washington: U.S. Government Printing Office, 1969), p. 13.

150. Lewis M. Alexander, "National Jurisdiction and the Use of the Sea," *Natural Resources Journal* 8 (1968): 388.

151. Barkenbus, pp. 14, 20–22.

152. Ibid., p. 23.

153. Ibid., p. 24.

154. Roy S. Lee, "Machinery for Seabed Mining: Some General Issues Before the Geneva Session of the Third United Nations Conference on the Law of the Sea," in *Law of the Sea: Caracas and Beyond*, ed. Francis T. Christy et al. (Cambridge, Mass.: Ballinger, 1975), pp. 122–23. See also Buzan, *Seabed Politics*, p. 153.

155. S. Brown et al., p. 23.

156. See Barkenbus, pp. 24, 57.

157. Ibid., pp. 49, 52–53, 58. See, e.g., UNCLOS III, *Statement Made by Mr. G.D. Arsenis on Behalf of the Secretary-General of the United Nations Conference on Trade and Development* (U.N. Doc. A/CONF.62/32, July 15, 1974), reprinted in *Official Records*, UNCLOS III, vol. III, p. 62.

158. See U.S. 92nd Congress, 2nd Session, House of Representatives, Committee on Foreign Affairs, Subcommittee on International Organizations and Movements, *Law of the Sea and Peaceful Uses of the Seabed* (Washington: U.S. Government Printing Office, 1972), p. 63.

159. Barkenbus, pp. 49, 54.

160. 'The probable ratio of minerals in a minimally commercial manganese nodule would be the following: 24 percent manganese, nearly 1 percent nickel, 0.5 to 1 percent copper, 0.35 percent cobalt." Ibid., p. 47.

161. Buzan, *Seabed Politics*, p. 159. Nickel is the mineral of primary interest in manganese nodules because of historically increasing demand, high market prices, and an oligopolistic market. See ibid., pp. 48–49.

162. *Economic Implications: Summary and Conclusions* (U.N. Doc. A/CONF.62/25), in *Official Records*, UNCLOS III, vol. III, p. 7.

163. See Barkenbus, p. 49.

164. Ibid., pp. 47–48. It has been asserted elsewhere, however, that there is no effective substitute for either nickel or cobalt. Arrow, "The Proposed Regime," p. 341.

165. Barkenbus, p. 48. See Chap. 4 nn. 167–68 and accompanying text.

166. See William D. Siapno, "Manganese Nodules: Overcoming the Constraints," *Marine Mining* 5 (1986): 458–59.

167. See S. Brown et al., pp. 20, 91; and also Wenk, *The Politics of the Ocean*, p. 27.

168. Mangone, p. 168. By the end of the 1960s the rate of increase in the world fish catch had diminished, although there were expectations that national fisheries management programs could improve the total catch still further. Sanger, p. 141. See also U.S. Commission on Marine Science, Engineering, and Resources, *Our Nation and the Sea*, p. 88.

169. See Francis T. Christy, Jr., "Fisheries and the New Conventions on the Law of the Sea," *San Diego Law Review* 7 (1970): 463–64. By 1970 high-seas fisheries were being exploited by some 900 large freezer and factory trawlers, approximately 400 of which were Soviet and 125 of which were Japanese. Sanger, p. 139. See also Carlson, p. 104.

170. See Buzan, *Seabed Politics*, pp. 124–25. In 1970 oil was being produced in waters 110 meters deep, and exploration had occurred at depths of more than 200 meters. Statement of John R. Stevenson, Legal Adviser, Department of State, at White House news conference, May 23, 1970, reprinted in *International Legal Materials* 9 (1970): 818. See also Mangone, p. 167. By the early 1980s, the monetary value of U.S. offshore oil production exceeded the value of the annual U.S. fish catch by a ratio of 10 to 1. Michael A. Champ, William P. Dillon, and David G. Howell, "Non-Living EEZ Resources: Minerals, Oil and Gas," *Oceanus* 27 (Winter 1984–85): 32.

171. See S. Brown et al., p. 21.

172. Richard Frank, "Environmental Consequences of Deep-Sea Mining," in *Law of the Sea: Conference Outcomes and Problems of Implementation*, ed. John King Gamble, Jr., and Edward Miles (Cambridge, Mass.: Ballinger, 1977), p. 319. Particularly worrisome were the unknown effects of the discharge back into the ocean of bottom water, sediment, and nodule tailings which would occur during mining operations. It has been estimated that such a discharge might sink at the rate of only 20 meters per year and that hundreds of thousands of square miles of the Pacific Ocean might be muddied. Ibid., p. 320. "[T]he expansion of scientific knowledge about ocean ecologies is undermining the traditional assumption that the ocean is a vast, self-equilibrating system that will rebound to its normal state no matter what is done to it." S. Brown et al., p. 19.

173. See Slouka, p. 74. The failure of developing states to seek a general transfer of this technology from developed states was noted by Ambassador Pardo during the final session of the Seabed Committee. Pardo, "A Statement on the Future Law of the Sea," p. 329.

174. Giulio Pontecorvo, "Contribution of the Ocean Sector to the United States Economy: Estimated Values for 1987 – A Technical Note," *Marine Technology Society Journal* 23 (June 1989): 11. Partly because Pontecorvo's calculations were done in 1987 dollars, official estimates were substantially lower: $8 billion, out of an aggregate annual worldwide economic value of $60 billion in 1972. U.S. Department of Commerce, *U.S. Ocean Policy in the 1970s: Status and Issues* (Washington: U.S. Government Printing Office, 1978), part 2, p. 2. United Nations, *Uses of the Sea: Study Prepared by the Secretary-General* (U.N. Doc. E/5120, April 28, 1972), p. 5.

175. See Hollick, *U.S. Foreign Policy*, p. 137.

176. Alexander, "National Jurisdiction," p. 398. See also Stevenson, "Legal Regulation of Mineral Exploitation," p. 50. Obviously, these goals were not always consistent. See Hollick, "United States Oceans Politics," p. 467.

177. Richardson, "A Reassessment of U.S. Interests," p. 4.

178. U.S. Commission on Marine Science, Engineering, and Resources, *Our Nation and the Sea*, pp. 2–3, 19, 72, 86, 88–90.

179. Ibid., pp. 3, 14–15, 21, 83.

180. David W. Proudfoot, "Guarding the Treasures of the Deep: The Deep Seabed Hard Mineral Resources Act," *Harvard Journal on Legislation* 10 (1973): 596. From 1970 onward, U.S. policy with respect to the law of the sea was coordinated by an Interagency Task Force within the executive branch. See Hollick, *U.S. Foreign Policy*, pp. 217–18, 257–58.

181. Wenk, *The Politics of the Ocean*, p. 416.

182. Barkenbus, p. 39. See Chap. 5 nn. 1–4 and accompanying text. "130 to 140 mostly

less developed states consider such activity violative of international law." Arrow, "The Proposed Regime," p. 350.

183. U.S. Senate, *The Third United Nations Law of the Sea Conference*, pp. 11–12. See n. 64 above and accompanying text.

184. Barkenbus, pp. 111, 144–45.

185. The United States "indicated that a system was needed that ensured, to the maximum extent possible, uniform interpretation and immediate access to dispute settlement machinery in urgent situations while at the same time preserving the flexibility of States to agree to resolve disputes by a variety of means." "Reports of the United States Delegation," ed. Nordquist and Park, p. 53.

186. Barkenbus, p. 40.

187. Glassner, p. 654. "Mining groups need the Authority to gain international legitimacy for their endeavor. The Authority, on the other hand needs mining groups to provide the expertise and technology necessary to carry out its mandate effectively." Barkenbus, p. 146.

188. Buzan, *Seabed Politics*, pp. 154–55. See also Barkenbus, pp. 36, 96–97.

189. Barkenbus, p. 157. See Laursen, pp. 94, 104–5, 107–10, 113–17. For a nominal fee of $5,000, the 1971 Deep Seabed Mineral Resources Act would have provided private mining firms with guaranteed access to nodule sites under licenses recognized by "reciprocating states." See David D. Caron, "Municipal Legislation for Exploitation of the Deep Seabed," *Ocean Development and International Law Journal* 8 (1980): 266. The act "had essentially been drafted by the American Mining Congress, a fact which Senator Metcalf made perfectly clear." Caron, "Municipal Legislation," p. 265. The Nixon administration took no position on the legislation until 1973, when it announced its opposition on grounds that passage might jeopardize the UNCLOS III negotiations, which it believed could be concluded by 1975. Laursen, p. 105.

In 1969, in response to the movement toward internationalization of the seabed, Senator Metcalf's Special Subcommittee on the Outer Continental Shelf had held hearings which sought to placate the petroleum industry by endorsing the exploitability clause of the 1958 Continental Shelf Convention as "positive, reliable and adequate." Wenk, *The Politics of the Ocean*, pp. 276–77. See U.S. Congressional Research Service (George A. Doumani), *Exploiting the Resources of the Seabed* (Washington: U.S. Government Printing Office, 1971), pp. 65–68.

190. Barkenbus, p. 157. The length of the negotiations was largely a result of political, conceptual, and institutional problems rather than deliberate delaying tactics, and unilateral threats were therefore generally ineffective as measures to increase the pace of the proceedings. Buzan, *Seabed Politics*, p. 159.

191. Barkenbus, p. 93.

192. U.S. 93rd Congress, 1st and 2nd Sessions, House of Representatives, Committee on Merchant Marine and Fisheries, Subcommittee on Oceanography, *Deep Seabed Hard Minerals* (Washington: U.S. Government Printing Office, 1974), p. 9. See also Barkenbus, p. 88.

193. Buzan, *Seabed Politics*, p. 156.

194. See Victor Basiuk, "Marine Resources Development, Foreign Policy, and the Spectrum of Choice," *Orbis* 12 (1968): 48.

195. See Lilliana Torreh-Bayouth, "UNCLOS III: The Remaining Obstacles to Consensus on the Deepsea Mining Regime," *Texas International Law Journal* 16 (1981): 89; and also Richardson, "A Reassessment of U.S. Interests," p. 7. The threat of such actions might itself deter banks from extending credit to mining companies under a unilateral regime.

196. Richardson stated that such legislation must designate no specific sites for mining, must contain detailed regulations for exploration but not for exploitation, and must contain none of the government investment or loan guarantees sought by mining companies. Barkenbus, p. 95.

197. U.S. Senate, *The Third United Nations Law of the Sea Conference*, pp. 97, 100.

198. Ibid., p. 1. Senator Metcalf's subcommittee had concluded that "modifications" should be sought in the 1970 Draft Treaty "to conform to our interpretation of the President's intent and with our recommendations." Report of the Special Subcommittee on the Outer

Continental Shelf, 1970, quoted in U.S. Congressional Research Service, *Exploiting the Resources of the Seabed*, p. 68.

199. U.S. 90th Congress, 2nd Session, U.S. House of Representatives, *The 22nd Session of the United Nations General Assembly: Report by Hon. William S. Broomfield and Hon. L.H. Fountain* (Washington: U.S. Government Printing Office, 1968), p. 21.

200. See Buzan, *Seabed Politics*, p. 79.

201. Orr, p. 1974. See Hollick, *U.S. Foreign Policy*, pp. 197, 208–9, 227–29, 380; and also Sebenius, pp. 76–77. At the time, Professor Louis Henkin was able to state with certainty that "oil is important to the United States, but oil is not most important." Louis Henkin, "Seabed Pact Would Help U.S.," *Washington Post*, August 8, 1971, p. B2. Undaunted, the petroleum industry turned to Congress to promote expanded national jurisdiction over offshore mineral resources. See n. 189 above.

202. Henkin, "Seabed Pact Would Help U.S.," p. B2. Richardson, "A Reassessment of U.S. Interests," p. 5. See also Hollick, *U.S. Foreign Policy*, pp. 187–90, 213, 216. U.S. naval interests include force projection, as well as unimpeded overflight and submarine passage. Sebenius, pp. 74–75.

203. Nixon, "United States Policy for the Seabed," p. 737.

204. U.S. Senate, *The Third United Nations Law of the Sea Conference*, p. 27. See also Hollick, "United States Oceans Politics," pp. 468, 471–72, 475, 477–78, 497. Ironically, much of the international opposition to the proposal was generated by concern among coastal states that U.S. oil companies would be able to exploit petroleum deposits on their continental shelves beyond the 200-meter isobath. See Sanger, p. 169.

The strong reaction in opposition to the Nixon proposal was not counterbalanced by support from any notable domestic constituency. Hollick, *U.S. Foreign Policy*, p. 259. See Torreh-Bayouth, p. 111 n. 185. Yet as late as 1973, the U.S. Senate unanimously agreed that the 1970 draft treaty "offer[ed] a practical method of implementing" U.S. negotiating goals at UNCLOS III. U.S. Senate, Resolution 82, July 16, 1973, reprinted in U.S. 93rd Congress, 1st Session, Senate, Committee on Interior and Insular Affairs, Subcommittee on Minerals, Materials, and Fuels, *Status Report on the Law of the Sea Conference* (Washington: U.S. Government Printing Office, 1973), p. 219.

205. Barkenbus, p. 45.

206. See ibid., pp. 45, 61–62. In 1970 the United States imported 86 percent of its manganese consumption, 92 percent of its cobalt consumption, and 100 percent of its nickel consumption, at an aggregate cost in excess of $500 million. "Oil, gas, nickel, and copper are commodities so basic to the continuous functioning of our society as we know it, that it would be difficult to describe the state of affairs which would exist in our society and other similarly situated societies were these commodities to be in short supply or obtainable only at substantially higher prices." Statement of Leigh S. Ratiner, Director of Ocean Resources, Office of the Assistant Secretary for Mineral Resources, U.S. Department of the Interior, in *Deep Seabed Hard Minerals*, U.S. House of Representatives, p. 31.

207. Barkenbus, p. 62. "The most powerful private interest in the seabed debate has been the petroleum industry." Hollick, "United States Oceans Politics," p. 471.

208. Torreh-Bayouth, p. 111. See G. Kevin Jones, "Outer Continental Shelf Petroleum Resources and the Nation's Future Energy Needs," *Syracuse Journal of International Law and Commerce* 15 (1989): 321–22. By the end of the 1970s, nonseabed issues, having been largely resolved in the negotiations, "came to be taken for granted," allowing attention to be focused on the seemingly intractable seabed mining problem. Richardson, "A Reassessment of U.S. Interests," p. 4.

Oil prices fell substantially during the 1980s, despite predictions of future OPEC-induced oil shortages. See Chap. 6 n. 16 and accompanying text. "By itself, perhaps it is not so extraordinary that presumed experts misjudged events. The more serious part of the equation is that these forecasts and forecasters had an enormous influence on the shaping of foreign and economic policy in the United States, Europe and Japan." Hobart Rowen, "Oil Well That Ends Well," *Washington Post National Weekly Edition*, June 18–24, 1990, p. 5.

209. See U.S. Senate, *The Third United Nations Law of the Sea Conference*, p. 101.

210. Barkenbus, p. 152.

211. Ibid., p. 72. Ironically, the OPEC embargo itself had an opposite effect, increasing

conservation measures and stimulating development of new land-based mineral deposits. See Joseph Haggin, "Marine Mining to Improve Its Organization, Direction, Financing," *Chemical and Engineering News*, November 18, 1985, p. 64.

212. Barkenbus, p. 152. "The more industrialized and complex a society, the more likely it is to be subject to such multifarious pressures. This condition tends to discourage the advanced maritime countries from escalating particular conflicts on questions of ocean use to matters of high national interest." S. Brown et al., p. 27.

213. U.N. General Assembly, 17th Session, Resolution 1803, December 14, 1962. U.N. General Assembly, 21st Session, Resolution 2158, November 15, 1966. See Ian Brownlie, "Legal Status of Natural Resources in International Law (Some Aspects)," *Recueil des Cours* 162 (1975): 256–57, 262. "So far as it is possible to indicate the historical origin of the New International Economic Order, it is to be found in this resolution of 1966." Brownlie, "Legal Status of Natural Resources," p. 262.

214. Buzan, *Seabed Politics*, p. 54. See Edgar Gold, *Maritime Transport* (Lexington, Mass.: D.C. Heath, 1981), pp. 276, 346. See also C. Clyde Ferguson, "The New International Economic Order," *University of Illinois Law Forum* (1980): 696. UNCTAD, an organ of the U.N. General Assembly, had been convened specifically to seek ways to address the growing economic gap between developed states and the newly independent developing states, but by the end of the decade continuing economic disparities were giving rise to frustration and animosity within the latter group. See Sanger, p. 7; and Hans Peter Kunz-Hallstein, "Patent Protection, Transfer of Technology and Developing Countries – A Survey of the Present Situation," *IIC International Review of Industrial Property and Copyright Law* 6 (1975): 444. Developing states soon became more vociferous, calling for economic and legal reform of the international system to accommodate their emergent interests. See Mark E. Ellis, "The New International Economic Order and General Assembly Resolutions: The Debate Over the Legal Effects of General Assembly Resolutions Revisited," *California Western International Law Journal* 15 (1985): 654.

215. Buzan, *Seabed Politics*, p. 54. See Hollick, *U.S. Foreign Policy*, pp. 11, 170. By 1968 membership in the Group of 77 consisted of 83 developing states. Ten years later, its membership had grown to 106 states. The Group's leaders included Algeria, China, and Tanzania. Laursen, p. 100.

216. Barkenbus, p. 168. Specific demands of the Group of 77 included commodity agreements, price stabilization plans, increased market access for exports, monetary reform, and transfer of technology. Ibid., p. 164. Many of these demands were pressed at an April 1974 special session of the United Nations.

The United States, like other Western states, enacted preferential tariffs for developing states. John Quigley, "Law for a World Community," *Syracuse Journal of International Law and Commerce* 16 (1989): 9. See, e.g., Trade Act of 1974, 19 U.S.C. §§2461–66.

217. Ellis, p. 654. See Helen E. Weidner, "The United States and North-South Technology Transfer: Some Practical and Legal Obstacles," *Wisconsin International Law Journal* (1983): 205–6.

218. See Weidner, pp. 211–15, 225–26; and Kunz-Hallstein, pp. 444–49. In addition to the negotiation of international commodity arrangements, the demands of increasingly impatient and frustrated developing states have included mandatory technical assistance from multinational corporations and modification of international rules protecting patent rights. Weidner, pp. 214–15. The Group of 77 has thus supported UNCTAD's efforts to establish an international code of conduct providing for the transfer of technology by multinational corporations on fair and reasonable conditions. Kunz-Hallstein, p. 445. The United States, while acknowledging the monopoly problems faced by developing states, has opposed adoption of such codes of conduct and other measures which might impair the functioning of the profit-oriented U.S. economic system, preferring instead to seek to promote international development through financial incentives. Weidner, pp. 216–20, 226. See Homer O. Blair, "Technology Transfer as an Issue in North/South Negotiations," *Vanderbilt Journal of Transnational Law* 14 (1981): 320–21.

219. Barkenbus, pp. 162–63.

220. Ibid., pp. 72, 163.

221. Hollick, *U.S. Foreign Policy*, pp. 170–71, 350. See A.A. Fatouros, "The Interna-

tional Law of the New International Economic Order: Problems and Challenges for the United States," *Willamette Law Review* 17 (1980): 94; and also Buzan, *Seabed Politics*, p. 212. "The movement for a New International Economic Order (NIEO) is based on the simple proposition that a system that allows, and even strengthens, the inequalities rampant in the current world economy cannot be considered just or even tolerable. It follows that the entire system, its structures and its constitutive principles—not only some particular features and manifestations—must be radically changed." Fatouros, p. 93. See also Ferguson, pp. 696–98. "The main point is not which particular norms or principles are to apply; the issue is one of process and of the identity of the decision makers." Fatouros, p. 103.

222. Fatouros, p. 106.

223. U.N. General Assembly, 29th Session, Resolution 3821, December 12, 1974. The resolution, adopted by a vote of 120 to 6, with 10 abstentions, has been described as "replete with provisions aimed at the economic growth of developing states and at preventing their future exploitation." Arrow, "The Proposed Regime," p. 382. Other related General Assembly resolutions adopted at the behest of the Group of 77 include the Declaration and Programme of Action on the Establishment of a New International Economic Order, as well as the International Development Strategy for the Second United Nations Development Decade. United Nations, *A New United Nations Structure for Global Economic Cooperation: The Report of the Group of Experts on the Structure of the United Nations System* (U.N. Doc. E/AC.62/9, May 28, 1975). See also Barkenbus, p. 164.

224. See Buzan, , *Seabed Politics*, pp. 129–30. "The Third World is not ideologically or politically monolithic, but its members have much in common. . . . By the 1960's its members came to recognize a substantial identity of condition, principally under-development, and a substantial identity of interest." Louis Henkin, "The Changing Law of the Sea: Technology, Law and Politics," in *Marine Technology and Law: Development of Hydrocarbon Resources in Offshore Structures*, ed. Ocean Association of Japan (Tokyo: Ocean Association of Japan, 1978), p. 139.

225. Barkenbus, p. 162.

226. Ibid., pp. 59, 162. Cf. Kurt Waldheim, *Building the Future Order: The Search for Peace in an Interdependent World*, ed. Robert L. Schiffer (New York: Free Press, 1980), p. 160.

227. Buzan, *Seabed Politics*, pp. 199–200.

228. Roy S. Lee, "Deep Seabed Mining and Developing Countries," *Syracuse Journal of International Law and Commerce* 6 (1978): 216.

229. See Buzan, *Seabed Politics*, p. 290.

230. Barkenbus, pp. 165–66.

231. Arrow, "The Proposed Regime," p. 414.

232. Barkenbus, p. 59.

233. Arrow, "The Proposed Regime," pp. 369, 381. See also R. Lee, "Deep Seabed Mining and Developing Countries," p. 214. For the contrary view, see Arrow, "The Proposed Regime," pp. 382–84.

234. Laursen, p. 102.

235. Barkenbus, p. 42. Developed states, in contrast, pushed for a convention containing detailed rules and regulations. See R. Lee, "Machinery for Seabed Mining," p. 139.

Chapter 3

1. Marion David Tunstall, *The Third United Nations Conference on the Law of the Sea* (Charlottesville, Va.: Michie, 1980), p. 22. For a list of the original UNCLOS III officers see United States, "Reports of the United States Delegation to the Third United Nations Conference on the Law of the Sea," Occasional Paper No. 33, ed. Myron H. Nordquist and Choon-ho Park (Honolulu: Law of the Sea Institute, 1983), pp. 79–80. "The chairmen were not deliberately chosen from smaller countries, although it is clear that a candidate from a big power would probably not have been acceptable. Nor were they chosen from any particular country, so much as for their ability and contacts." Clyde Sanger, *Ordering the Oceans: The Making of the Law of the Sea* (London: Zed Books, 1986), p. 41.

2. See Barry Buzan, *Seabed Politics* (New York: Praeger, 1976), pp. 216–18; and Tunstall, pp. 22–23. "[W]hile provisions for voting were needed as part of the rules package, actual resort to their use on substantive issues would in all likelihood have indicated a breakdown in the negotiations." Barry Buzan, "Negotiating by Consensus: Developments in Technique at the United Nations Conference on the Law of the Sea," *American Journal of International Law* 75 (1981): 331–32.

3. Tunstall, p. 3.

4. See Anthony Astrachan, "Who Owns Seabed Riches?" *Washington Post*, December 2, 1973, p. A22. The UNCLOS III agenda included nearly 130 items, "many of which could have formed the subject matter of a separate conference." Edgar Gold, *Maritime Transport* (Lexington, Mass.: D.C. Heath, 1981), p. 316. For a breakdown of the allocation of agenda items among the committees, see UNCLOS III, *Statement of Activities of the Conference During Its First and Second Sessions* (U.N. Doc. A/CONF.62/L.8/Rev.1, October 17, 1974), para. 40, reprinted in *Official Records*, UNCLOS III, vol. III, pp. 97–98.

5. United Nations, *The Kampala Declaration* (U.N. Doc. A/CONF.62/23, May 2, 1974), reprinted in *Official Records*, UNCLOS III, vol. III, p. 24. See also Ann L. Hollick, *U.S. Foreign Policy and the Law of the Sea* (Princeton, N.J.: Princeton University Press, 1981), p. 288. The coalition of landlocked and geographically disadvantaged states at UNCLOS III consisted of 29 landlocked states, almost half of which were African, and 53 states claiming various degrees of geographic disadvantage in their use of the sea, including Belgium and West Germany. Sanger, p. 32.

6. See Hollick, *U.S. Foreign Policy*, pp. 280–83, 288–89.

7. U.S. 95th Congress, 2nd Session, Senate, Committee on Commerce, Science, and Transportation, *Report: The Third United Nations Law of the Sea Conference* (Washington: U.S. Government Printing Office, 1978), p. 20.

8. Hollick, *U.S. Foreign Policy*, pp. 379–80.

9. See "Reports of the United States Delegation," ed. Nordquist and Park, p. 11.

10. Hollick, *U.S. Foreign Policy*, p. 289.

11. See Hollick, *U.S. Foreign Policy*, p. 290.

12. Roy S. Lee, "Machinery for Seabed Mining: Some General Issues Before the Geneva Session of the Third United Nations Conference on the Law of the Sea," in *Law of the Sea: Caracas and Beyond*, ed. Francis T. Christy et al. (Cambridge, Mass.: Ballinger, 1975), p. 127.

13. Hollick, *U.S. Foreign Policy*, p. 299.

14. R. Lee, "Machinery for Seabed Mining," p. 132.

15. Jack N. Barkenbus, *Deep Seabed Resources: Politics and Technology* (New York: Free Press, 1979), p. 125.

16. Ibid., pp. 106, 115. Article 9, which was endorsed by more than two-thirds of the Group of 77 and applauded by the United States as a basis for further negotiation, provided that seabed mining activities would be subject to the direct control of the Authority but that the Authority might contract with private mining firms for the performance of certain mining activities. Buzan, *Seabed Politics*, p. 225. When pressed, however, developing states appeared unwilling to make further concessions, even with respect to specific aspects of the proposal. Hollick, *U.S. Foreign Policy*, p. 26. See also "Reports of the United States Delegation," ed. Nordquist and Park, pp. 59, 61.

17. Buzan, *Seabed Politics*, p. 227. See also Barkenbus, p. 106. The United States continued to press for guaranteed access to mining sites, mining company control of mining operations, and limitations upon the discretionary powers of the Authority. Hollick, *U.S. Foreign Policy*, pp. 235–36, 355–56.

18. United Nations, *First Committee of Conference on Sea Law Ends Work for Year* (U.N. Doc. SEA/145, August 28, 1974).

19. See "Reports of the United States Delegation," ed. Nordquist and Park, pp. 61–62; and also Buzan, *Seabed Politics*, p. 231. A study by the U.N. Secretariat discussed prevention and compensation as two possible methods of protecting producer states against the economic impact of seabed mining operations. UNCLOS III, *Extract from Report of the Trade and Development Board on Its Thirteenth Session, Held at the Palais des Nations, Geneva, From 21 August to 11 September 1973* (U.N. Doc. A/CONF.62/26, June 6, 1974), Annex, reprinted in *Official Records*, UNCLOS III, vol. III, p. 42. The United States argued that the impact of

seabed mining operations on producer states would be minimal and that a system of production controls would be impracticable. Buzan, *Seabed Politics*, p. 231.

20. Hollick, *U.S. Foreign Policy*, p. 290. Officially, it was reported that Committee I had "made useful progress." *Statement of Activities of the Conference* (U.N. Doc. A/CONF.62/L.8/Rev.1), Annex I, para. 24, in *Official Records*, UNCLOS III, vol. III, p. 103.

21. Hollick, *U.S. Foreign Policy*, p. 292.

22. Tunstall, p. 28. See UNCLOS III, *Second Committee: Main Trends* (U.N. Doc. A/CONF.62/C.2/WP.1, October 15, 1974).

23. Hollick, *U.S. Foreign Policy*, p. 293. The transit passage regime, which was largely agreed upon during a four-day debate, provided for the application of generally accepted international regulations in international straits in order to prevent arbitrary coastal-state restrictions on navigation. Sanger, pp. 93–94. There was agreement that a similar regime would have to be negotiated with respect to archipelagic waters. "Reports of the United States Delegation," ed. Nordquist and Park, pp. 66, 67.

Because neither overflight nor submerged passage is permitted under the doctrine of innocent passage, and because such rights are considered essential to U.S. strategic interests, the United States has sought the application of a more liberal transit regime to the 116 straits between 6 and 24 miles in width which would otherwise become territorial waters under the treaty. See Chap. 2 n. 79 and accompanying text; and John Norton Moore, "The Regime of Straits and the Third United Nations Conference on the Law of the Sea," *American Journal of International Law* 74 (1980): 79–81. According to the Department of Defense, the United States "could not accept merely innocent passage through the Straits of Gibraltar from a security standpoint." U.S. 92nd Congress, 2nd Session, Senate, Committee on Commerce, Subcommittee on Oceans and Atmosphere, *Law of the Sea* (Washington: U.S. Government Printing Office, 1972), p. 27. Nor could it agree to distinctions between military and commercial vessels with respect to rights of navigation through such areas. "Reports of the United States Delegation," ed. Nordquist and Park, p. 67.

24. Buzan, *Seabed Politics*, p. 221. See also Tunstall, p. 28. At Caracas, the United States, the Soviet Union, and the United Kingdom each announced its support for such a zone for the first time, provided their respective maritime interests in freedom of navigation were adequately safeguarded. Hollick, *U.S. Foreign Policy*, p. 294.

25. David A. Kay, "Operational Aspects of Managing the Oceans," *Columbia Journal of World Business* 10 (1975): 30. See Chap. 2 n. 137 and accompanying text. However, almost 40 percent of the zone would be divided among 10 coastal states, including the United States, the Soviet Union, Canada, Australia, and New Zealand. See Robert E. Clute, "The African Perspective of the Law of the Sea," in *Current Issues in the Law of the Sea*, ed. Christopher C. Joyner (Allentown, Penn.: Muhlenberg College, 1980), p. 15. The zone would also provide coastal states with jurisdiction over marine pollution and scientific research, as determined in Committee III.

26. See Buzan, *Seabed Politics*, pp. 220–21, 223.

27. Hollick, *U.S. Foreign Policy*, p. 295.

28. The United States sought international management of tuna and coastal-state jurisdiction over its salmon stocks; Japan sought international management of salmon, a longstanding target of its distant-water fishing fleets. Ibid., pp. 20–28.

29. Buzan, *Seabed Politics*, pp. 232–33. See Hollick, *U.S. Foreign Policy*, p. 296. Archipelagic states include Fiji, Mauritius, Indonesia, Malaysia, the Philippines, and the Bahamas.

30. Tunstall, p. 28. "Vessels introduce pollutants into the marine environment in three principal ways — through oil and other cargoes entering the water due to collisions or other maritime casualties, through loading, unloading and bunkering operations, and through the intentional operational discharge of oil." "Working Paper on Competence to Establish Standards for the Control of Vessel Source Pollution, Submitted by the United States to the UN Seabed Committee, April 2, 1973," reprinted in *Status Report on the Law of the Sea Conference*, U.S. 93rd Congress, 1st Session, Senate, Committee on Interior and Insular Affairs, Subcommittee on Minerals, Materials, and Fuels (Washington: U.S. Government Printing Office, 1973), p. 536. However, the largest source of damage to the marine environment is land-based pollution, which enters the oceans via polluted rivers, factory discharges, and sewage treatment plants, accounting for between 80 to 85 percent of total marine pollution,

5 to 7 percent of which is attributable to vessel-source pollution. Norman Wulf, "Protecting the Marine Environment," in *Current Issues in the Law of the Sea*, ed. Joyner, pp. 90–91.

31. U.S. Senate, *The Third United Nations Law of the Sea Conference*, pp. 31–32. Developing coastal states were distrustful of IMCO, which they believed to favor the traditional maritime states in the establishment of international vessel design standards. Ibid., p. 32.

32. Ibid., p. 32. See Tunstall, p. 28; and also Sanger, pp. 129–30.

33. "Reports of the United States Delegation," ed. Nordquist and Park, p. 74.

34. Buzan, *Seabed Politics*, p. 233. See also Hollick, *U.S. Foreign Policy*, pp. 298–99.

35. Christopher C. Joyner, "Towards a Legal Regime for the International Seabed: The Soviet Union's Evolving Perspective," *Virginia Journal of International Law* (1975): 879.

36. Hollick, *U.S. Foreign Policy*, pp. 235–36.

37. U.S. Senate, *The Third United Nations Law of the Sea Conference*, p. 36.

38. Tunstall, p. 25. For a list of the nine organizations which accepted the invitation, see *Statement of Activities of the Conference*, U.N. Doc. A/CONF.62/L.8/Rev.1), para. 38, in *Official Records*, UNCLOS III, vol. III, p. 96.

39. See Hollick, *U.S. Foreign Policy*, p. 299. Although the formation of these negotiating groups was viewed as important evidence of an emerging attitude of cooperation, the 26-state Evensen Group excluded landlocked and geographically disadvantaged states, and was mistrusted by a substantial portion of the Group of 77. Buzan, *Seabed Politics*, pp. 237–38. Despite their lack of representativeness, such informal groups proved to be convenient negotiating vehicles and were heavily employed, often on an *ad hoc* basis, throughout UNCLOS III. Tunstall, pp. 60–61.

40. Buzan, *Seabed Politics*, p. 280. See U.S. Senate, *The Third United Nations Law of the Sea Conference*, p. 36; and also "Reports of the United States Delegation," ed. Nordquist and Park, p. 57.

41. Buzan, *Seabed Politics*, pp. 238, 280.

42. Ibid., pp. 219, 242. "The tone of the general debate and the informal meetings was moderate and serious." "Reports of the United States Delegation," ed. Nordquist and Park, pp. 56–57.

43. The three states were Oman, Guinea-Bissau, and Norway. Buzan, *Seabed Politics*, p. 243. In the United States the Ford administration succeeded in preventing congressional enactment of a Fisheries Extension Bill, as well as a Hard Minerals Bill. Ibid., p. 258.

44. Ibid., p. 244. Under the existing consensus procedure it was "rational strategy to hold out as long as possible in the hope that the other side will concede first." Buzan, "Negotiating by Consensus," p. 333.

45. U.S. Senate, *The Third United Nations Law of the Sea Conference*, p. 37. In the final analysis, "the most equitable utilization of the process depended upon the ability of the Chairman to assess impartially the trends and upon his honesty in expressing the assessment." Tunstall, p. 34.

46. Buzan, *Seabed Politics*, pp. 246, 252.

47. Hollick, *U.S. Foreign Policy*, p. 301. Because participation in the Group was open to all interested states, its size soon more than doubled.

48. Statement of Leigh S. Ratiner in *Law of the Sea*, U.S. 94th Congress, 1st Session, Senate, Committee on Foreign Relations, Subcommittee on Oceans and International Environment (Washington: U.S. Government Printing Office, 1975), p. 17.

49. Buzan, *Seabed Politics*, p. 253.

50. Barkenbus, p. 106. Developing states nevertheless continued to prefer an Authority vested with broad discretionary powers. Statement of Leigh S. Ratiner in *Law of the Sea*, U.S. Senate, 1975, p. 20.

51. U.S. Senate, *The Third United Nations Law of the Sea Conference*, p. 48. Pinto's text on the machinery was only partly accepted by the Committee I chairman, Ambassador Paul Engo of Cameroon, whose SNT contained provisions somewhat more favorable to the developing states. Buzan, *Seabed Politics*, p. 253.

52. Buzan, *Seabed Politics*, p. 251. Under U.S. and Soviet proposals, each state-sponsored mining applicant would submit two potential sites of equal value, one of which would be "banked" by the Authority for exploitation in the manner provided in the treaty. Statement of Leigh S. Ratiner, in *Law of the Sea*, U.S. Senate, 1975, p. 17.

53. Barkenbus, p. 119. See "Reports of the United States Delegation," ed. Nordquist and Park, p. 92.

54. Secretary of State Henry Kissinger, address before American Bar Association, Montreal, August 11, 1975, reprinted in *U.S. Department of State Bulletin* 73 (1975): 357. See Hollick, *U.S. Foreign Policy*, pp. 303, 356.

55. UNCLOS III, *Informal Single Negotiating Text: Part I* (U.N. Doc. A/CONF.62/WP.8/I, May 7, 1975), arts. 26, 27, reprinted in *Official Records*, UNCLOS III, vol. IV, pp. 141–42. It was agreed that each state party to the convention should be entitled to a seat and a single vote in the Assembly. Most delegates favored a Council with between 36 and 48 seats, although executive organs of international institutions had traditionally been composed of a lesser number of states. See Barkenbus, pp. 125, 128.

56. Although the Council was required to comply with "general guidelines and policy directions" of the Assembly, the SNT provided that the Assembly could not overrule the Council on matters specifically within the latter organ's competence. See Buzan, *Seabed Politics*, pp. 253–54.

57. "Reports of the United States Delegation," ed. Nordquist and Park, p. 92.

58. Buzan, *Seabed Politics*, p. 254. Although the Pinto text had called for a three-fourths-present-and-voting majority, such a blocking power was opposed by developing states, which feared domination of the Authority by the United States. Barkenbus, pp. 131–33.

59. U.S. Senate, *The Third United Nations Law of the Sea Conference*, pp. 39–40. See Barkenbus, p. 140.

60. "Reports of the United States Delegation," ed. Nordquist and Park, pp. 92–96.

61. UNCLOS III, *Informal Single Negotiating Text: Part II* (U.N. Doc. A/CONF.62/WP.8/II, May 6, 1975), art. 38, reprinted in *Official Records*, UNCLOS III, vol. IV, p. 158. See also George P. Smith, III, *Restricting the Concept of Free Seas* (Huntington, N.Y.: Robert E. Krieger, 1980), p. 102.

62. U.S. Senate, *The Third United Nations Law of the Sea Conference*, p. 37. In preparing the SNT articles, the Committee II chairman largely ignored calls for preferential access to coastal fisheries stocks by landlocked and geographically disadvantaged states, which were instead granted access on "an equitable basis." See Hollick, *U.S. Foreign Policy*, p. 306; and also "Reports of the United States Delegation," ed. Nordquist and Park, pp. 103–7. "When a consensus was achieved in 1975, it came because the coastal states accepted the compromise that, if they were to have exclusive sovereign rights over the resources . . . , these should be linked to the principle of 'optimum utilization.'" Sanger, p. 143.

63. See *Informal Single Negotiating Text: Part II* (U.N. Doc. A/CONF.62/WP.8/II), art. 23, in *Official Records*, UNCLOS III, vol. IV, p. 156. Many states sought to remove the sovereign immunity enjoyed by government vessels.

64. U.S. Senate, *The Third United Nations Law of the Sea Conference*, p. 49. The United States proposed a royalty equal to 1 percent of production value, beginning five years after exploitation and increasing by 1 percent annually to a rate of 5 percent in the 10th year. "Reports of the United States Delegation," ed. Nordquist and Park, p. 99.

65. "To all appearances, the landlocked and geographically disadvantaged states had lost this battle at the 1958 Law of the Sea Conference, when they did not contest the Continental Shelf Convention." Buzan, *Seabed Politics*, p. 248.

66. See "Reports of the United States Delegation," ed. Nordquist and Park, pp. 99–100; and also Buzan, *Seabed Politics*, p. 247.

67. Statement of Ambassador Thomas A. Clingan in *Law of the Sea*, U.S. Senate, 1975, p. 33. See "Reports of the United States Delegation," ed. Nordquist and Park, pp. 107–8.

68. See Buzan, *Seabed Politics*, p. 249; and Tunstall, p. 32.

69. States which favored the vesting of strong regulatory powers in coastal states, including Canada, nevertheless prevailed upon the chairman of Committee II to provide for such powers in the Committee's SNT, creating a conflict between the Committee II and Committee III texts. See Hollick, *U.S. Foreign Policy*, p. 310.

70. UNCLOS III, *Informal Single Negotiating Text: Part III* (U.N. Doc. A/CONF.62/WP.8/III, May 6, 1975), art. 22, reprinted in *Official Records*, UNCLOS III, vol. IV, p. 174. See "Reports of the United States Delegation," ed. Nordquist and Park, pp. 108–9.

71. U.S. Senate, *The Third United Nations Law of the Sea Conference*, p. 38. See also Hollick, *U.S. Foreign Policy*, p. 311. The Committee also considered a proposal by the Group of 77 to require coastal-state consent for all marine scientific research and a proposal by Western states that all such research conforming to international standards be permitted. See "Reports of the United States Delegation," ed. Nordquist and Park, pp. 109–12.

72. See Buzan, *Seabed Politics*, pp. 259–60.

73. See Hollick, *U.S. Foreign Policy*, p. 312.

74. Buzan, *Seabed Politics*, pp. 258, 261, 263, 284.

75. Ibid., pp. 264–65.

76. Statement of John R. Stevenson in *Law of the Sea*, U.S. Senate, 1975, p. 4. See Tunstall, p. 35; and also "Reports of the United States Delegation," ed. Nordquist and Park, pp. 82–83.

77. Statement of John R. Stevenson in *Law of the Sea*, U.S. Senate, 1975, p. 2. "Reports of the United States Delegation," ed. Nordquist and Park, p. 84. Although this new procedure facilitated progress on important issues, it "was naturally slow and time-consuming, as well as error prone, and contributed significantly to the plodding pace of the conference," which had already been slowed by the consensus procedure. Buzan, "Negotiating by Consensus," pp. 333, 336.

78. Finn Laursen, *Superpower at Sea* (New York: Praeger, 1983), p. 103.

79. "Reports of the United States Delegation," ed. Nordquist and Park, pp. 118, 125. See Kissinger, address before American Bar Association, August 11, 1975, p. 540. Kissinger's intervention was part of a general effort to defuse the North-South confrontation. See Hollick, *U.S. Foreign Policy*, p. 352.

80. "Reports of the United States Delegation," ed. Nordquist and Park, p. 118. See UNCLOS III, *Revised Single Negotiating Text* (U.N. Doc. A/CONF.62/WP.8/Rev.1, May 6, 1976), reprinted in *Official Records*, UNCLOS III, vol. V, pp. 125–85. As had been the case with the SNT, the RSNT was issued as a procedural device, without prejudice to the negotiating position of any state. See U.S. Senate, *The Third United Nations Law of the Sea Conference*, p. 46. More than 3,700 criticisms of the SNT were voiced. G. Smith, p. 95.

81. Hollick, *U.S. Foreign Policy*, p. 315. In exchange for concessions to the developed states, Engo apparently secured their political support in his personal rivalry with Ambassador Pinto, and the latter's Working Group of 50 was disbanded. Tunstall, p. 37.

82. Hollick, *U.S. Foreign Policy*, p. 38.

83. See U.S. Senate, *The Third United Nations Law of the Sea Conference*, pp. 65–67.

84. See ibid., pp. 54, 72; and "U.S. Delegation Report," in *Law of the Sea*, U.S. 94th Congress, 2nd Session, Senate, Committee on Foreign Relations, Subcommittee on Oceans and International Environment (Washington: U.S. Government Printing Office, 1976), p. 33.

85. U.S. Senate, *The Third United Nations Law of the Sea Conference*, p. 54. The financing of the Enterprise remained a controversial subject, however. Statement of T. Vincent Learson in *Law of the Sea*, U.S. Senate, 1976, p. 4.

86. "The developing countries sidestepped this issue rather than take sides in a dispute among and between developed countries." "U.S. Delegation Report," p. 34.

87. See Secretary of State Henry Kissinger, "The Law of the Sea: A Test of International Cooperation," *U.S. Department of State Bulletin* 74 (1976): 540–41.

88. "U.S. Delegation Report," p. 33. It was proposed that the production limit should be calculated to reflect a minimum annual growth rate of 6 percent. Statement of T. Vincent Learson, in *Law of the Sea*, U.S. Senate, 1976, p. 4. Although the U.S. mining industry objected to Kissinger's efforts at compromise, it was evident "that they could in fact live with the RSNT provisions, and that they were essentially basing their objections upon the precedent of encroachment upon market forces." Barkenbus, p. 141.

89. See Arvid Pardo, "Banquet Address: The Emerging Law of the Sea and World Order," in *Law of the Sea: Conference Outcomes and Problems of Implementation*, ed. Edward Miles and John King Gamble, Jr. (Cambridge, Mass.: Ballinger, 1977), p. 406.

90. Statement of T. Vincent Learson in *Law of the Sea*, U.S. Senate, 1976, p. 10. The Authority also retained certain powers to require data transfer and personnel training by mining entities in the course of their mining operations, and to inspect ongoing mining operations. See U.S. Senate, *The Third United Nations Law of the Sea Conference*, p. 65; and Barkenbus, p. 109.

91. The SNT had allowed only a six-month delay upon written petition by one third of the Assembly membership. Whereas the SNT had authorized the Assembly to "issue directions," the RSNT only permitted it to adopt resolutions and make recommendations. See Pardo, "Banquet Address," pp. 407–8.

92. Ibid., p. 408.

93. Ibid., p. 408. See statement of T. Vincent Learson in *Law of the Sea*, U.S. Senate, 1976, p. 7.

94. Tunstall, p. 38. "Delegates who remained silent were taken as supporting an article while those that spoke out were recognized as seeking change." Hollick, *U.S. Foreign Policy*, p. 313.

95. Statement of T. Vincent Learson in *Law of the Sea*, U.S. Senate, 1976, pp. 4–5. The text relating to transit passage and the exclusive economic zone was retained virtually without alteration.

96. See "U.S. Delegation Report," p. 28.

97. Buzan, *Seabed Politics*, p. 301. Although the RSNT did not include such a definition, it did provide for the sharing of revenues from exploitation of the outer shelf area. Precise numerical percentages were omitted from the text, however.

98. See Tunstall, p. 38; and Hollick, *U.S. Foreign Policy*, pp. 286–87.

99. "U.S. Delegation Report," p. 34.

100. Tunstall, p. 39.

101. U.S. Senate, *The Third United Nations Law of the Sea Conference*, p. 32.

102. "U.S. Delegation Report," p. 37. More stringent regulations were permitted only in ecologically fragile Arctic areas and in "special areas" approved by IMCO.

103. Ibid., p. 38. Although the U.S. view prevailed in Committee III, the Committee II text permitted only the application of international standards.

104. Coastal- and port-state enforcement powers were also limited by substantive and procedural safeguards, "including release on bond of vessels, liability for unreasonable enforcement, and sovereign immunity." Ibid., p. 37.

105. Many developing coastal states argued, successfully, that the distinction would prove impracticable. Hollick, *U.S. Foreign Policy*, p. 320. They were joined by the Soviet Union, which abruptly abandoned its own proposal. See Tunstall, p. 40.

106. "U.S. Delegation Report," p. 32.

107. Ibid., p. 39. At the request of several developed states, the RSNT expanded the class of potential transferees to include developed states "which may need and request technical assistance." Australia, Department of Foreign Affairs, *Third United Nations Conference on the Law of the Sea, Seventh Session, Geneva: Report of the Australian Delegation* (Canberra: Australian Government Publishing Service, 1978), p. 88.

108. "U.S. Delegation Report," pp. 32, 39. See also Hollick, *U.S. Foreign Policy*, p. 322.

109. See U.S. Senate, *The Third United Nations Law of the Sea Conference*, pp. 61–62. States would also be able to "opt out of" the compulsory procedures on matters concerning maritime boundaries and military activities, as well as issues under review by the U.N. Security Council. See "U.S. Delegation Report," p. 30.

110. Buzan, *Seabed Politics*, p. 302.

111. See U.S. Senate, *The Third United Nations Law of the Sea Conference*, p. 43; and Tunstall, p. 42. "The United States delegation made abundantly clear the volatile condition of American politics . . . , in effect threatening a breakdown of the conference unless . . . issues were resolved in favor of American interests." Buzan, *Seabed Politics*, p. 302.

112. Statement of Secretary of State Henry Kissinger, September 17, 1976, quoted in "Reports of the United States Delegation," ed. Nordquist and Park, p. 142. See Secretary of State Henry Kissinger, "Secretary Kissinger Discusses U.S. Position on the Law of the Sea Conference," *U.S. Department of State Bulletin* 75 (1976): 397–99; and also Hollick, *U.S. Foreign Policy*, p. 356.

113. "Reports of the United States Delegation," ed. Nordquist and Park, p. 118. "[A] more constructive willingness to address the entire complex of issues involved in the system of exploitation appeared." Ibid., p. 146.

114. UNCLOS III, *Report by Mr. P.B. Engo, Chairman of the First Committee, on the*

Work of the First Committee (U.N. Doc. A/CONF.62/L.16, September 6, 1976), reprinted in *Official Records*, UNCLOS III, vol. VI, p. 134.

115. U.S. Senate, *The Third United Nations Law of the Sea Conference*, p. 45. See also Laursen, p. 103. Many developing states maintained that a system of parallel access would be meaningless if the Authority did not possess the financial and technological resources necessary for it to conduct its own mining operations. U.S. Senate, *The Third United Nations Law of the Sea Conference*, p. 45. Kissinger's 1976 proposals were intended to address these concerns by bolstering the parallel system, but while doing so they shifted attention away from efforts to address the dual access problem through creation of joint ventures between the Authority and private contractors. Sanger, p. 173.

116. U.S. Senate, *The Third United Nations Law of the Sea Conference*, p. 45. "Under this proposal, the facade of a parallel system was maintained; but the Enterprise was given clear preeminence and State and private access was not guaranteed." "Reports of the United States Delegation," ed. Nordquist and Park, p. 144.

117. Hollick, *U.S. Foreign Policy*, p. 356.

118. *Report by Mr. P.B. Engo* (U.N. Doc. A/CONF.62/L.16), in *Official Records*, UNCLOS III, vol. VI, p. 132.

119. See UNCLOS III, *Report by Mr. Andres Aguilar, Chairman of the Second Committee, on the Work of the Committee* (U.N. Doc. A/CONF.62/L.17, September 16, 1976), reprinted in *Official Records*, UNCLOS III, vol. VI, pp. 135–36; and also Tunstall, p. 44. The most important of these issues for the United States was the legal status of the exclusive economic zone, which U.S. negotiators sought to ensure would remain residually part of the high seas. "Reports of the United States Delegation," ed. Nordquist and Park, p. 148.

120. See U.S. Senate, *The Third United Nations Law of the Sea Conference*, p. 44; and Tunstall, p. 45.

121. *Report by Mr. Andres Aguilar* (U.N. Doc. A/CONF.62/L.17), in *Official Records*, UNCLOS III, vol. VI, p. 139.

122. Hollick, *U.S. Foreign Policy*, p. 318. Cf. "Reports of the United States Delegation," ed. Nordquist and Park, p. 119.

123. See John Logue, "The Receding Heritage: UN Law of the Sea Conference," *Transnational Perspectives*, November 2, 1976; and also James K. Sebenius, *Negotiating the Law of the Sea* (Cambridge, Mass.: Harvard University Press, 1984), pp. 77–78. The Fishery Conservation and Management Act of 1976 "was the major cause of approximately 65 unilateral extensions of jurisdiction — most of which were more comprehensive and onerous" than the claims set forth in the Act. John Warren Kindt, "Special Claims Impacting Upon Marine Pollution Issues at the Third U.N. Conference on the Law of the Sea," *California Western International Law Journal* 10 (1980): 423. "Tight primary politics led Gerald Ford to sign the legislation over the vigorous opposition of the State and Defense Departments." Sebenius, p. 78. The efficient, high-seas fishing operations of the Soviet Union and Japan had caused the depletion of many fisheries stocks beyond the U.S. 12-mile contiguous fishing zone.

124. UNCLOS III, *Report of Mr. O. Yankov, Chairman of the Third Committee, on the Work of the Committee* (U.N. Doc. A/CONF.62/L.18, September 16, 1976), paras. 9, 10, 12, in *Official Records*, UNCLOS III, vol. VI, p. 140.

125. Ibid., para. 24, in *Official Records*, UNCLOS III, vol. VI, p. 141.

126. Ibid., para. 32, in *Official Records*, UNCLOS III, vol. VI, p. 142. See also U.S. Senate, *The Third United Nations Law of the Sea Conference*, p. 44.

127. Tunstall, p. 46. See Sanger, p. 130; and "Reports of the United States Delegation," ed. Nordquist and Park, pp. 142, 153–54.

128. *Report of the Australian Delegation*, p. 89. See *Report of Mr. O. Yankov* (U.N. Doc. A/CONF.62/L.18), para. 39, in *Official Records*, UNCLOS III, vol. VI, p. 143. It was recognized that the separate discussions of technology transfer in Committees I and III required some coordination. Ibid., para. 45, in *Official Records*, UNCLOS III, vol. VI, p. 143.

129. "Reports of the United States Delegation," ed. Nordquist and Park, p. 157. See also Tunstall, pp. 46, 52.

130. "U.S. Delegation Report," p. 33. The United States was particularly insistent that the seabed and fisheries regimes should apply provisionally. Ibid., p. 40.

131. See Tunstall, pp. 246–47; and U.S. Senate, *The Third United Nations Law of the Sea Conference*, p. 45.

132. Hollick, *U.S. Foreign Policy*, pp. 287, 324. See also Tunstall, p. 48.

133. Kindt, p. 421. Such zones were adopted by Canada, India, Mexico, the Soviet Union, and the EEC; Japan followed suit in 1977. See ibid., p. 420. Only three new 200-mile claims had been asserted during 1975. "It would be erroneous to contend that these 200-mile extensions would have occurred anyway and that the [U.S. Fisheries Conservation and Management Act] merely anticipated a trend toward 200-mile claims." Kindt, p. 421. See n. 123 above.

134. Tunstall, p. 48. The developed states objected that although Kissinger's compromise proposals had been incorporated into the text, corresponding concessions by the Group of 77 were omitted. "Reports of the United States Delegation," ed. Nordquist and Park, p. 171.

135. Hollick, *U.S. Foreign Policy*, p. 323. The Group of 77 was unable to reach agreement on the acceptability of the text. See "Reports of the United States Delegation," ed. Nordquist and Park, pp. 170–71.

136. Hollick, *U.S. Foreign Policy*, p. 323. See UNCLOS III, *Informal Composite Negotiating Text* (U.N. Doc. A/CONF.62/WP.10, July 15, 1977), reprinted in *Official Records*, UNCLOS III, vol. VIII, pp. 1–63. The chairmen were given, "in effect, a veto which Amerasinghe resignedly and diplomatically accepted in deference to those more acquainted with the particular substantive issues." Tunstall, p. 53. See also Jens Evensen, "Banquet Address," in *Law of the Sea: Neglected Issues*, ed. John King Gamble, Jr. (Cambridge, Mass.: Ballinger, 1979), p. 529.

137. See Laursen, pp. 110–13. Engo's contributions to the ICNT were developed privately, without benefit of open discussion and debate. See G. Smith, p. 100.

Noting that the Evensen text had offered a "real prospect" for an acceptable resolution of seabed mining issues, Ambassador Richardson recommended "a most serious and searching review of both the substance and procedures of the Conference." Statement of Ambassador Elliot L. Richardson, July 20, 1977, reprinted in "Reports of the United States Delegation," ed. Nordquist and Park, pp. 164, 166. The review subsequently concluded "that agreement on the large number of issues dealt with by the Conference will introduce into the international system a greater measure of stability and predictability and a commitment to peaceful settlement of disputes"—an achievable result which justified continued efforts to resolve the remaining difficult issues. Statement of Ambassador Elliot L. Richardson, March 16, 1978, reprinted in *The Third United Nations Law of the Sea Conference*, U.S. Senate, p. 117.

138. Tunstall, p. 53. See "Reports of the United States Delegation," ed. Nordquist and Park, p. 163.

139. Hollick, *U.S. Foreign Policy*, pp. 325, 352, 361.

140. See U.S. Senate, *The Third United Nations Law of the Sea Conference*, p. 72. The ICNT was substantially similar.

141. See ibid., pp. 66, 69–70; and also Hollick, *U.S. Foreign Policy*, p. 326. The ICNT listed "certain objectives . . . to guide negotiations and adoption of rules and regulations: (1) insure optimum revenues for the Authority; (2) attract investment and technology into area exploration and exploitation; (3) insure equality of financial treatment and comparable financial obligations on the part of all states and other entities obtaining contracts; (4) provide incentives on a uniform and nondiscriminatory basis for contracts to undertake joint arrangements and to stimulate technology transfer; and (5) enable the enterprise to engage in seabed mining effectively from the time the convention enters into force." U.S. Senate, *The Third United Nations Law of the Sea Confernce*, p. 71.

142. See U.S. Senate, *The Third United Nations Law of the Sea Conference*, pp. 64, 67–68.

143. Tunstall, p. 48. See U.S. Senate, *The Third United Nations Law of the Sea Conference*, p. 66. However, the Enterprise was to be exempted from payment of fees and production charges. See Myron H. Nordquist, "The Law of the Sea Conference and Deep Seabed Mining Legislation," in *Current Issues in the Law of the Sea*, ed. Joyner, p. 75.

144. See U.S. Senate, *The Third United Nations Law of the Sea Conference*, p. 74; and Hollick, *U.S. Foreign Policy*, p. 326. The Group of 77 had sought a limitation of 50 percent of growth in nickel demand after 7 years, and the United States favored maintaining the 100

percent ceiling throughout the 20-year period of production controls. Hollick, *U.S. Foreign Policy*, p. 325. The 100 percent ceiling, which had been agreed upon by the United States and Canada – major producers of copper and nickel, respectively – would have allowed for no growth in land-based production.

145. U.S. Senate, *The Third United Nations Law of the Sea Conference*, p. 71. See also Barkenbus, p. 108. Distribution to nonparty states was also permitted by both texts, despite U.S. opposition. U.S. Senate, *The Third United Nations Law of the Sea Conference*, p. 78.

146. "Reports of the United States Delegation," ed. Nordquist and Park, p. 170. See also Tunstall, p. 49. U.S. support for such a scheme remained conditioned upon an acceptable system of parallel access. Barkenbus, p. 143.

147. Barkenbus, p. 173.

148. See U.S. Senate, *The Third United Nations Law of the Sea Conference*, pp. 75, 77; and also Barkenbus, p. 127. The Authority was once again given control over scientific research within the Area. Hollick, *U.S. Foreign Policy*, p. 326.

149. See U.S. Senate, *The Third United Nations Law of the Sea Conference*, pp. 75, 78.

150. "Practical implementation of this principle is one of the most difficult elements in the seabed negotiating texts for the industrial nations, and particularly, the United States, to accept." Barkenbus, pp. 166–67.

151. See U.S. Senate, *The Third United Nations Law of the Sea Conference*, pp. 64, 77. In contrast to the position of the developing states, which called for a straightforward system of geographic representation, Evensen followed European models in proposing that majority approval be required in four of five chambers, composed of, respectively, seabed mining states, consumer states, producer states, developing states, and geographically representative states. Hollick, *U.S. Foreign Policy*, p. 325.

152. Hollick, *U.S. Foreign Policy*, pp. 331–32. The Chamber would have no power to review the discretionary decisions of the Authority or the conformity of its rules and regulations to the convention; otherwise, the Chamber was to have exclusive and compulsory jurisdiction over all disputes relating to seabed mining activities.

153. See ibid., p. 324. The text on the legal status of the zone was left "deliberately ambiguous" in an effort to satisfy both coastal and maritime states. Ibid., pp. 326–27.

154. U.S. Senate, *The Third United Nations Law of the Sea Conference*, p. 81. The RSNT's blanket exemption for all developing coastal states was replaced by a narrower exception in favor of states which were net importers of a mineral resource produced from their own shelves. Hollick, *U.S. Foreign Policy*, p. 327. "There was complete agreement, however, that the revenues should be distributed to developing countries." "Reports of the United States Delegation," ed. Nordquist and Park, pp. 176–77.

155. Hollick, *U.S. Foreign Policy*, pp. 327–28. The Irish proposal was a hybrid of the Hedberg and depth-of-sediment formulas, both of which continued to enjoy substantial support, although landlocked and geographically disadvantaged states continued to oppose all such expansionary approaches. "Reports of the United States Delegation," ed. Nordquist and Park, p. 176.

156. See Hollick, *U.S. Foreign Policy*, p. 328; and Tunstall, p. 50. Both the RSNT and ICNT permitted access by landlocked and geographically disadvantaged states only to the extent of the available surplus catch, as determined by the coastal state.

157. See *Informal Composite Negotiating Text* (U.N. Doc. A/CONF.62/WP.10), arts. 25 (2), 221 (1), in *Official Records*, UNCLOS III, vol. VIII, pp. 9, 39. The ICNT also empowered a port state to seize unseaworthy vessels. "In the wake of a series of disastrous oil spills off U.S. shores, the U.S. delegation found itself under increasing domestic pressure to pursue environmental rather than maritime considerations." Hollick, *U.S. Foreign Policy*, pp. 329, 330.

158. *Informal Composite Negotiating Text* (U.N. Doc. A/CONF.62/WP.10), arts. 236–37, in *Official Records*, UNCLOS III, vol. VIII, p. 41. See also Tunstall, pp. 51–52.

159. See U.S. Senate, *The Third United Nations Law of the Sea Conference*, p. 93.

160. See *Report of the Australian Delegation*, p. 89.

161. If the parties did not designate a common forum, the dispute could be resolved only through arbitration. See Hollick, *U.S. Foreign Policy*, p. 331. "Many delegations did not favor

one or another of the four alternative procedures." "Reports of the United States Delegation," ed. Nordquist and Park, p. 157.

162. "U.S. Delegation Report," p. 30. See *Revised Single Negotiating Text* (U.N. Doc. A/CONF.62/WP.8/Rev.1/I), arts. 33–39, in *Official Records*, UNCLOS III, vol. V, pp. 135–36.

163. Use of the dispute settlement procedures would nevertheless be compulsory where the dispute related to navigation, overflight, or submarine cables or pipelines within the zone. See Hollick, *U.S. Foreign Policy*, p. 332.

164. See U.S. Senate, *The Third United Nations Law of the Sea Conference*, p. 64; and n. 137 above. A congressional report concluded, however, that there were "only a few clear differences that might render the ICNT provisions fundamentally unacceptable to the United States — technology transfer, voting rights, and perhaps scientific research." U.S. Senate, *The Third United Nations Law of the Sea Conference*, p. 80.

165. Barkenbus, pp. 94–95. See Chap. 2 n. 196 and accompanying text. In congressional testimony, Ambassador Richardson stated that the decision was based upon the need for such legislation in order to promote orderly development of deep-sea minerals, whether or not the United States ultimately became a party to the convention, and upon an expectation that its enactment would not hinder the UNCLOS III negotiations. Laursen, p. 115. See n. 204 below and accompanying text.

166. Kindt, p. 421.

167. "Reports of the United States Delegation," ed. Nordquist and Park, p. 186. See also Tunstall, p. 54. Although he had been removed from the Sri Lankan delegation by that state's new government, and was thus subject to removal from his Conference position, Amerasinghe was supported by African and Asian states, as well as by the United States, and his presidency was reaffirmed by a vote of 75 to 18, with 13 abstentions. Hollick, *U.S. Foreign Policy*, p. 333.

168. Buzan, "Negotiating by Consensus," p. 337. "Any modification or revisions to be made in the Informal Composite Negotiating Text should emerge from the negotiations themselves and should not be introduced on the initiative of any single person, whether it be the President or a Chairman of a Committee, unless presented to the Plenary and found, from the widespread and substantial support prevailing in Plenary, to offer a substantially improved prospect of a consensus." UNCLOS III, *Organization of Work: Decisions Taken by the Conference at Its 90th Meeting on the Report of the General Committee* (U.N. Doc. A/CONF. 62/62, April 13, 1978), quoted in ibid.

Discussion of procedural issues consumed the third week of the session. "During the ensuing four weeks, the Conference witnessed one of the most intensive periods of simultaneous negotiation in its history." "Reports of the United States Delegation," ed. Nordquist and Park, pp. 187, 189.

169. See Buzan, "Negotiating by Consensus," p. 337. The basic membership of the negotiating groups was limited in order to facilitate discussions, but the groups were open to participation by all interested delegations, and agreements reached were without prejudice to positions taken by any delegation. "Reports of the United States Delegation," ed. Nordquist and Park, p. 194.

170. *Report of the Australian Delegation*, pp. 11, 31. See Hollick, *U.S. Foreign Policy*, p. 335.

171. See Bernard H. Oxman, "The Third United Nations Conference on the Law of the Sea: The Seventh Session (1978)," *American Journal of International Law* 73 (1979): 8. Discussions in the working group were dominated by the United States and the Group of 77. Tunstall, p. 55.

172. For example, private consortia would no longer conduct their mining operations "on behalf of" the Authority. Oxman, "The Seventh Session," p. 9.

173. See "Reports of the United States Delegation," ed. Nordquist and Park, pp. 196, 264. The contractor would only be required to make available technology which it was legally capable of transferring, but failure to fulfill technology transfer obligations could result in loss of mining rights. See Hollick, *U.S. Foreign Policy*, p. 335. "To enable the Enterprise to decide whether or not to take advantage of its rights, the miner would now have to give a general description of his technology to the Enterprise at the outset, an obligation which appeared to

be reasonable if the necessary protection for proprietary information could be added in the future," William C. Brewer, Jr., "Transfer of Mining Technology to the International Enterprise," Oceans Policy Study 2:4 (Charlottesville, Va.: Michie, 1980), p. 15.

174. "Reports of the United States Delegation," ed. Nordquist and Park, pp. 190, 231.

175. Tunstall, p. 55. See also Oxman, "The Seventh Session," p. 8.

176. See Hollick, U.S. Foreign Policy, p. 335. The formula "represented a U.S. effort to get the assistance of the Peruvian and Chilean delegates in moving the Group of 77 toward negotiations as well as to mollify Canada's nickel concerns." Ibid., p. 366.

177. See Oxman, "The Seventh Session," p. 9; and also "Reports of the United States Delegation," ed. Nordquist and Park, p. 190. The moratorium would apply also to the Enterprise.

178. Oxman, "The Seventh Session," p. 10.

179. "One of the most intractable aspects of the matter derives from the difficulty of making assumptions about an activity that has not yet begun." Ibid., p. 11. See also Report of the Australian Delegation, p. 46.

180. Oxman, "The Seventh Session," p. 13. See Mati L. Pal, "Financial Arrangements," Syracuse Journal of International Law and Commerce 6 (1979): 295. The simple production-charge formula was included as an alternative for "[t]hose economic systems where there is no concept of profit." Pal, p. 296.

181. Remarks of Lawrence L. Herman, in "Symposium Panel Discussion, February 24, 1979," Syracuse Journal of International Law and Commerce 6 (1979): 300. See also Report of the Australian Delegation, p. 14.

182. Oxman, "The Seventh Session," p. 11. See Report of the Australian Delegation, pp. 46–47. The Enterprise and the Authority were to be financed separately. See Tunstall, p. 56.

183. Hollick, U.S. Foreign Policy, p. 336.

184. Barkenbus, p. 130.

185. Oxman, "The Seventh Session," p. 15. Some Western states were wary of participating in an interest group in which the Soviet Union might be represented. Ibid.

186. See "Reports of the United States Delegation," ed. Nordquist and Park, p. 228. A subsidiary organ on financial matters was also discussed. See Oxman, "The Seventh Session," pp. 15–16.

187. See Hollick, U.S. Foreign Policy, p. 337; Oxman, "The Seventh Session," p. 6; and also "Reports of the United States Delegation," ed. Nordquist and Park, p. 191.

188. See Tunstall, p. 57.

189. Hollick, U.S. Foreign Policy, p. 337.

190. Ibid., pp. 337–38.

191. See Tunstall, p. 57; and Bernard H. Oxman, "The Third United Nations Conference on the Law of the Sea: The Eighth Session," American Journal of International Law 74 (1980): 19.

192. Hollick, U.S. Foreign Policy, pp. 337–38. See Report of the Australian Delegation, p. 92. By threatening to withhold their support for rights of access by landlocked and geographically disadvantaged states, broad-margin states also sought to hold the agreement achieved in Negotiating Group IV hostage to agreement on an acceptable definition of the continental margin. See "Reports of the United States Delegation," ed. Nordquist and Park, p. 192. "It had been clear for years that the elimination of jurisdiction beyond 200 miles would also eliminate any possibility of consensus on a treaty that included major broad-margin states." Oxman, "The Seventh Session," pp. 19–20.

193. See Report of the Australian Delegation, p. 15. "With Newtonian certainty, every substantive proposal to amend the ICNT articles met with a counter-proposal to amend the ICNT articles." "Reports of the United States Delegation," ed. Nordquist and Park, p. 227.

194. Report of the Australian Delegation, p. 61.

195. Hollick, U.S. Foreign Policy, p. 339.

196. Report of the Australian Delegation, p. 15. See also Tunstall, p. 56. The small number of changes "underlined the fact that many articles were close to their final formulation." Hollick, U.S. Foreign Policy, p. 337.

197. See Hollick, U.S. Foreign Policy, p. 337.

198. Ibid., pp. 339–40. The sinking of the Amoco Cadiz off the coast of Brittany in the

spring of 1978 "heightened concern of the international community with strengthening the legal regime for the protection of the marine environment." *Report of the Australian Delegation,* p. 17. The new texts included changes in ship routings as well as in coastal-state regulatory powers. Tunstall, p. 57.

199. Tunstall, p. 57. See Oxman, "The Seventh Session," p. 29.

200. *Report of the Australian Delegation,* p. 64.

201. See ibid., p. 26; and Oxman, "The Seventh Session," pp. 5, 30. The final clauses were to consist of provisions governing the entry into force of the convention, including articles relating to ratification, reservations, revision, duration, and provisional application.

202. Evensen, "Banquet Address," p. 530. See "Reports of the United States Delegation," ed. Nordquist and Park, p. 229.

203. Oxman, "The Seventh Session," p. 48.

204. See Hollick, *U.S. Foreign Policy,* pp. 363–64; and also Barkenbus, p. 98. The chairman of the Group of 77 warned that enactment of such legislation would be contrary to international law and would have serious adverse effects upon the UNCLOS III process, as well as upon other multilateral negotiations. The United States, supported by other Western seabed mining states, responded "that all States have a right to explore and exploit deep seabed resources as freedoms of the high seas, reiterated its commitment to negotiation of a Law of the Sea Convention, noted the interim character of the legislation, and explained the need for enactment of such legislation lest an infant mining industry feel compelled to abandon plans and investments looking toward mining several years hence, hopefully under a treaty." "Reports of the United States Delegation," ed. Nordquist and Park, pp. 228–29.

205. Oxman, "The Eighth Session," p. 1. See also "Reports of the United States Delegation," ed. Nordquist and Park, p. 252.

206. Oxman, "The Seventh Session," p. 38. See also Hollick, *U.S. Foreign Policy,* p. 341.

207. See Hollick, *U.S. Foreign Policy,* p. 341; and Oxman, "The Eighth Session," p. 11.

208. *Report of the Working Group of 21,* Annex II, art. 6 (3), quoted in Oxman, "The Eighth Session," p. 12. The short, "objective" list of exceptions satisfied U.S. demands for a system of guaranteed access for private mining operators. Hollick, *U.S. Foreign Policy,* p. 342. See "Reports of the United States Delegation," ed. Nordquist and Park, p. 255.

209. The United States expressed general satisfaction with these changes. Hollick, *U.S. Foreign Policy,* p. 342.

210. Ibid.

211. Oxman, "The Eighth Session," pp. 15, 17. Although the Soviet Union vehemently opposed equal access by private parties to international dispute settlement procedures, the assurance of a fair procedure for the settlement of disputes relating to contracts with the Authority was regarded by Western developed states as an essential element of the parallel system. Barkenbus, p. 109.

212. Oxman, "The Eighth Session," p. 5. Developed states were concerned that seabed mining investment could be unduly inhibited if the ceiling became too low. See Bernard H. Oxman, "The Third United Nations Conference on the Law of the Sea: The Ninth Session (1980)," *American Journal of International Law* 75 (1981): 216. Specific numbers remained to be agreed upon for the production ceiling, and developed states remained unhappy with a provision giving the Enterprise a preference over other mining operations in the allocation of production quotas. Hollick, *U.S. Foreign Policy,* p. 342.

213. See Oxman, "The Eighth Session," p. 15. The "supreme" powers of the Assembly were expressly limited to the instances specifically provided in the convention. See "Reports of the United States Delegation," ed. Nordquist and Park, p. 305. The United States argued in favor of a veto power by as few as 5 of the 36 Council members, but the Group of 77 insisted upon at least 7 to 9. Hollick, *U.S. Foreign Policy,* p. 343.

214. "Reports of the United States Delegation," ed. Nordquist and Park, pp. 305, 318–19, 322–23.

215. "The fixed points comprising the line of the outer limits of the continental shelf on the sea-bed ... either shall not exceed 350 miles from the baseline from which the breadth of the territorial sea is measured or shall not exceed 100 nautical miles from the 2,500 metre isobath." UNCLOS III, *Informal Composite Negotiating Text/Revision 1* (U.N. Doc. A/CONF.62/WP.10/Rev.1, April 28, 1979), art. 76, para. 5. During the New York meetings

it was realized that this definition might result in an allocation to coastal states of mid-oceanic ridges comprising much of the deep seabed, and that further modification of the formula would therefore be necessary. Hollick, *U.S. Foreign Policy*, p. 345.

The revenue-sharing percentages apparently were increased in order to secure the support of landlocked and geographically disadvantaged states – many of which continued to favor the Arab proposal during the eighth session – for the RICNT formulation of the outer limit of the margin. Hollick, *U.S. Foreign Policy*, p. 344.

216. See Hollick, *U.S. Foreign Policy*, pp. 345–46; and Oxman, "The Eighth Session," pp. 27–28.

217. See Hollick, *U.S. Foreign Policy*, p. 345.

218. "Reports of the United States Delegation," ed. Nordquist and Park, p. 303.

219. See Oxman, "The Eighth Session," pp. 32, 40.

220. Ibid., pp. 35, 37. Under article 19 of the 1969 Vienna Convention, reservations not "incompatible with the object and purpose of the treaty" are permitted in the absence of specific treaty provisions to the contrary. Vienna Convention on the Law of the Treaties, reprinted in *International Legal Materials* 8 (1969): 686–87. The underlying object and purpose of the UNCLOS III treaty, however, "is not so much its substance as widespread agreement on a comprehensive regime for the oceans embracing all interests at stake. With few exceptions, if any, it is difficult to conceive of reservations compatible with that object and purpose." Oxman, "The Eighth Session," p. 35.

221. Oxman, "The Eighth Session," pp. 36, 39.

222. "Reports of the United States Delegation," ed. Nordquist and Park, p. 254. See also G. Smith, pp. 108–9. Outstanding issues included decisionmaking by the organs of the Authority, revenue sharing, production controls, technology transfer, maritime boundary delimitation, and financing for the Enterprise.

223. Nordquist, "The Law of the Sea Conference," p. 78. "Almost all of the states that have not acted to date are ones which either have little area to claim, or such severe boundary problems with their neighbors that they are proceeding very cautiously." Ibid.

224. UNCLOS III, *Informal Composite Negotiating Text/Revision 2* (U.N. Doc. A/CONF.62/WP.10/Rev.2, May 11, 1980). UNCLOS III, *Informal Composite Negotiating Text/Revision 3* (U.N. Doc. A/CONF.62/WP.10/Rev.3, September 22, 1980). See Jeffrey Lee Gertler and Paul Wayne, "Synopsis: Recent Developments in the Law of the Sea 1979–1980," *San Diego Law Review* 18 (1981): 534.

225. Oxman, "The Ninth Session," p. 254.

226. Ibid., p. 213. See also Gertler and Wayne, p. 551 n. 124.

227. UNCLOS III, *Written Statement by the Delegation of Canada, Dated August 26, 1980* (U.N. Doc. A/CONF.62/WS/14, October 3, 1980), reprinted in *Official Records*, UNCLOS III, vol. XIV, p. 153. See Gertler and Wayne, p. 543; and also Oxman, "The Ninth Session," p. 213. The delegates agreed that potential land-based producers would be represented. "Semi-industrialized" states continued to press unsuccessfully for seating of their interests. See UNCLOS III, *Written Statement by the Delegation of Honduras, Dated August 25, 1980* (U.N. Doc. A/CONF.62/WS/13, October 3, 1980), reprinted in *Official Records*, UNCLOS III, vol. XIV, p. 152.

228. See "Reports of the United States Delegation," ed. Nordquist and Park, pp. 412–13. Failure to reach a consensus would give rise to compulsory conciliation. Procedural questions were to be resolved by simple majority vote.

229. See Gertler and Wayne, p. 554. The Council could overturn a negative recommendation by a three-fourths vote.

230. See "Reports of the United States Delegation," ed. Nordquist and Park, p. 353. The United States succeeded in obtaining improvements in provisions relating to blacklisting, protection of proprietary information, and dispute settlement procedures, and continued to seek complete elimination of the Brazil clause. Gertler and Wayne, p. 536.

231. Gertler and Wayne, p. 553 n. 142. "By separating the two procedures, it avoids artificial pressure to compete prematurely for portions of the production ceiling, reduces the risk that investment in exploration will be lost because production needs to be deferred, and permits more timely computation of the ceiling." Oxman, "The Ninth Session," pp. 216–17.

232. See Gertler and Wayne, p. 542; and also Sanger, p. 181.

233. See Hollick, *U.S. Foreign Policy*, p. 347; and Oxman, "The Ninth Session," p. 216.

234. See "Reports of the United States Delegation," ed. Nordquist and Park, pp. 354, 414; and Gertler and Wayne, pp. 541, 556–57. Private seabed mining firms were particularly concerned that the assistance initially provided to the Enterprise might give it an unfair commercial advantage.

235. See Gertler and Wayne, pp. 538, 556, 565.

236. "Reports of the United States Delegation," ed. Nordquist and Park, p. 415. See Gertler and Wayne, pp. 543, 559.

237. Gertler and Wayne, pp. 549–50. It was agreed that the Preparatory Commission would be empowered to draft provisional rules, regulations, and procedures for the Authority and to oversee the conduct of pioneer mining activities. Oxman, "The Ninth Session," p. 244.

238. *Written Statement by the Delegation of Canada* (U.N. Doc. A/CONF.62/WS/14), reprinted in *Official Records*, UNCLOS III, vol. XIV, p. 154. Provisional application of the rules and regulations adopted by the Preparatory Commission was regarded by the U.S. delegation as "[a] critical element" in the settlement of Committee I issues. "Reports of the United States Delegation," ed. Nordquist and Park, p. 412.

239. Gertler and Wayne, pp. 545–46, 560. Prior notification "was labeled a 'Conference breaker' by the major maritime powers, and was subsequently withdrawn." Ibid., pp. 560–61. See also "Reports of the United States Delegation," ed. Nordquist and Park, p. 356.

240. See Gertler and Wayne, pp. 560–61.

241. See Oxman, "The Ninth Session," p. 233. The Draft Convention also gave coastal states jurisdiction over removal of archaeological finds within the contiguous zone. See Gertler and Wayne, p. 562. Proposals to confer such jurisdiction with respect to the entire continental margin were unacceptable to maritime states. Oxman, "The Ninth Session," p. 240.

242. See Gertler and Wayne, pp. 544–45; and Oxman, "The Ninth Session," pp. 229–30. Coastal states would "submit particulars of such limits to the Commission along with supporting scientific and technical data," and "may agree or disagree with the recommendations. If it agrees, the coastal state is given an extraordinary power nowhere reproduced with respect to any other maritime limit: the continental shelf limits 'established by a coastal State on the basis of these recommendations shall be final and binding.'" *Informal Composite Negotiating Text/Revision 3* (U.N. Doc. A/CONF.62/WP.10/Rev.3), art. 76 (8). If a coastal state disagreed with the Commission's recommendation, it would be required to submit a revised application to the Commission.

243. The compromise formulation "was greeted with cries of anguish from the most vocal advocates of the two opposing points of view. The text changed the words of the ICNT. What else it changed is a matter upon which even the most courageous may hesitate to speculate." Oxman, "The Ninth Session," p. 231.

244. See Gertler and Wayne, p. 548. Suspension or termination of a research project was to be subject to compulsory conciliation. Ibid., p. 548 n. 107.

245. See Oxman, "The Ninth Session," pp. 236–37; and Sanger, p. 18.

246. *Informal Composite Negotiating Text/Revision 3* (U.N. Doc. A/CONF.62/WP.10/Rev.3), art. 302.

247. Gertler and Wayne, pp. 548, 562. "To the extent the Preamble is a guide to the values to be applied in interpreting the text of the Convention, those values include both equity and efficiency in utilization of ocean resources." Oxman, "The Ninth Session," p. 256 n. 206.

248. See Oxman, "The Ninth Session," pp. 243, 246. These provisions were designed to bring the convention into force at an early date while promoting broad participation.

249. See Gertler and Wayne, p. 564; and "Reports of the United States Delegation," ed. Nordquist and Park, p. 564. States were permitted to submit declarations and statements of understanding with their ratifications.

250. *Informal Composite Negotiating Text/Revision 3* (U.N. Doc. A/CONF.62/WP.10/Rev.3), art. 311 (1).

251. Laursen, pp. 117–22. See 30 U.S.C. §1401 et seq. The Act was passed "in order to prevent a further decline or a complete disintegration of the United States deep sea-bed mining industry." UNCLOS III, *Letter Dated July 30, 1980 From the Representative of the United*

States of America to the President of the Conference (U.N. Doc. A/CONF.62/103, August 1, 1980), reprinted in *Official Records*, UNCLOS III, vol. XIV, p. 109.

252. See Federal Republic of Germany, Act on Interim Regulation of Deep Seabed Mining, reprinted in *International Legal Materials* 19 (1980): 1330–39; and also Gertler and Wayne, p. 567.

253. See Oxman, "The Ninth Session," p. 212. While declaring seabed mining to be a high-seas freedom, the Act provided that the regime would be superseded by "the adoption of an international agreement at the Third United Nations Conference on the Law of the Sea, and the entering into force of such agreement, or portions thereof, with respect to the United States." Seabed Act, 30 U.S.C. §1441 (3) (A). In contrast to the 1976 U.S. fisheries legislation, the Seabed Act did not merely implement a regime which had already been agreed to at UNCLOS III, but instead provided for a flag-state licensing system potentially inconsistent with the seabed mining articles of the RICNT.

254. Ambassador Elliot L. Richardson, address before American Mining Congress, San Francisco, September 24, 1980, quoted in Linda Turbyville, "Plumbing the Depths," *Sea Power* (January 1981), p. 28. See Gertler and Wayne, p. 535; and "Reports of the United States Delegation," ed. Nordquist and Park, p. 410. It was "all but certain" that the negotiations would be concluded in early 1981. Paul Lewis, "After 6 Years Law of the Sea Parley Nears Mining Pact," *New York Times*, August 30, 1980, p. A4.

255. See "Reports of the United States Delegation," ed. Nordquist and Park, pp. 410–11. Further consideration of maritime boundary issues was also anticipated. See Oxman, "The Ninth Session," pp. 211–12.

256. "Law of the Sea . . . negotiations have served to inhibit U.S. exploration of the seabed for its abundant mineral resources. Too much concern has been lavished on nations unable to carry out sea-bed mining, with insufficient attention paid to gaining early American access to it." 1980 Republican National Platform, quoted in James L. Malone, "The United States and the Law of the Sea," *Virginia Journal of International Law* 24 (1984): 787. Malone, who served as U.S. ambassador to the final UNCLOS III sessions, had himself participated "on one of the principal advisory committees on international affairs and national security that was involved with the Republican National Committee and with the President's efforts." Statement of James L. Malone in *Nomination of James L. Malone*, U.S. 97th Congress, 1st Session, Senate, Committee on Foreign Relations (Washington: U.S. Government Printing Office, 1981), p. 28. Kennecott Copper and Lockheed Aircraft Corporation, each of which had substantial interests in a U.S. seabed mining consortium, were both clients of the late William J. Casey, who managed the 1980 campaign and subsequently served as Director of the Central Intelligence Agency, and who also held a significant ownership interest in Standard Oil of Indiana, which was also a partner in the Lockheed consortium. See Ronald Brownstein and Nina Easton, *Reagan's Ruling Class* (New York: Pantheon Books, 1983), pp. 630–32. The partisan criticism seemed particularly ironic in light of the fact that every chief U.S. negotiator at UNCLOS III had been a Republican. In a 1981 speech explaining U.S. objections to the seabed mining regime, Ambassador Malone indicated that the Reagan administration "undertook its review of the Draft Convention because the people of the United States, through their electoral process, have expressed their preference for a variety of policies that affect the work of the Conference." Statement of Ambassador James L. Malone before the Plenary, August 5, 1981, reprinted in "Reports of the United States Delegation," ed. Nordquist and Park, p. 524. "The president-elect was bombarded by letters from various components of the industry and their allies, including members of Congress." Laursen, p. 134. One such letter, on behalf of 14 members of Congress led by Representative John Breaux, was prepared by Theodore Kronmiller, who would soon be selected by the Reagan administration to chair its review of the Draft Convention. Laursen, pp. 135, 140.

257. U.S. Mission to the United Nations, press release, March 2, 1981, quoted in David L. Larson, "The Reagan Administration and the Law of the Sea," *Ocean Development and International Law Journal* 11 (1982): 298–99. The decision to undertake the policy review, made at a March 2 senior interagency meeting without prior consultation with U.S. allies, apparently surprised many U.S. as well as foreign diplomats. Statement of James L. Malone in *Nomination of James L. Malone*, U.S. Senate, p. 19. See also Peter Grier, "Staking a Claim to the Ocean's Bed," *Christian Science Monitor*, April 9, 1981, pp. 12–13.

258. See Turbyville, p. 23; and Larson, "The Reagan Administration," p. 298. Although "conventional wisdom appears to be that the Administration initiated the review at the behest of United States mining companies," there also were "important philosophical and pragmatic objections to the proposed treaty unrelated to the welfare of industry." Doug Bandow, "UNCLOS III: A Flawed Treaty," *San Diego Law Review* 19 (1982): 476–78. See also James L. Malone, "U.S. Policy and the Law of the Sea," *U.S. Department of State Bulletin* 81 (July 1981): 49.

259. Senator Claiborne Pell, "Introduction," *San Diego Law Review* 18 (1981): 392.

260. See Larson, "The Reagan Administration," p. 301. "A change in the leadership of the American delegation was essential in order to ensure that other countries clearly understood our seriousness of purpose with respect to the review. That action was also necessary in order to send the signal to other delegations that the United States could not be induced to return immediately and, thus, prematurely, to the bargaining table by offers of minor technical changes in the draft convention." Malone, "U.S. Policy and the Law of the Sea," pp. 48–49.

261. See Bernice R. Kleid, "Synopsis: Recent Developments in the Law of the Sea 1980–1981," *San Diego Law Review* 19 (1982): 641; and Malone, "U.S. Policy and the Law of the Sea," p. 50. "In acceding to the general desire to hold a session in August, the United States made clear that it viewed such a session as an opportunity for informal consultations rather than definitive negotiation on texts, and that it did not expect to complete its review until after the August session." "Reports of the United States Delegation," ed. Nordquist and Park, p. 459.

262. Hamilton Amerasinghe had died in December 1980, after presiding over UNCLOS III and the Seabed Committee since 1970. Koh was chosen by consensus, despite the reservations of Soviet-bloc states. See Bernard H. Oxman, "The Third United Nations Conference on the Law of the Sea," *American Journal of International Law* 76 (1982): 2 n. 6.

263. Oxman, "The Third United Nations Conference," p. 13.

264. See Kleid, pp. 637, 648 n. 124; and "Reports of the United States Delegation," ed. Nordquist and Park, pp. 456–57, 521. Although the text remained open to further amendment, its status was upgraded to that of an official Conference document, and the words "Informal Text" were deleted from its title. See UNCLOS III, *Draft Convention on the Law of the Sea* (U.N. Doc. A/CONF.62/L.78, August 28, 1981), reprinted in *Official Records*, UNCLOS III, vol. XV, pp. 172–240.

265. See Grier, pp. 12–13; Kleid, p. 637; and also "Reports of the United States Delegation," ed. Nordquist and Park, p. 523. Larson, "The Reagan Administration," pp. 301–2. For a sampling of official reaction to the U.S. reappraisal among the UNCLOS III delegates, see Oxman, "The Third United Nations Conference," pp. 4–6. "The motivations for these arguments are generally the same as those that underlie the 'delicate balance' argument whenever it is made in defense of hard-won compromise: genuine belief that the compromise is the best attainable, fatigue and impatience resulting from the time and effort devoted to working out the compromise, and tactical resistance to demands for new concessions." Oxman, "The Third United Nations Conference," p. 6.

266. Leigh S. Ratiner, "The Law of the Sea: A Crossroads for American Foreign Policy," *Foreign Affairs* 60 (1982): 1011. See also Lewis I. Cohen, "International Cooperation on Seabed Mining," in *United States Law of the Sea Policy: Options for the Future* (Oceans Policy Study Series, no. 6), ed. Sharon K. Shutler (New York: Oceana Publications, 1985), p. 148.

The International Court of Justice has identified good faith as one "of the basic principles governing the creation and performance of legal obligations, whatever their source." Nuclear Tests (Australia v. France), Judgment of 20 December 1974, *I.C.J. Reports 1974*, p. 253.

267. "Reports of the United States Delegation," ed. Nordquist and Park, p. 458.

268. "The plenary session . . . witnessed the strange sight of the British delegate speaking on behalf of the whole European community against the inclusion of this anti-subsidy clause, and Australian Ambassador Keith Brennan robustly retorting that he could see no reason for this opposition since the obligations under this clause were the same as those accepted by all industrialized countries under GATT." Sanger, p. 182.

269. "Some countries said it should be open to all who signed the Final Act . . . while others felt it should include only those who demonstrated an intention to be bound by the . . . Convention." Kleid, p. 638.

270. "Some developed countries, with the United States reserving its position at this session, have regarded this approach as essential to assuring those ratifying the treaty that the Seabed Resource Authority would operate in a foreseeable manner." Malone, "U.S. Policy and the Law of the Sea," p. 50.

271. Statement of Ambassador James L. Malone, August 5, 1981, in "Reports of the United States Delegation," ed. Nordquist and Park, p. 526. "Virtually all of the other delegates at the Conference warned the United States that the fundamentals of the treaty were non-negotiable." Bandow, p. 491. Cf. "Reports of the United States Delegation," ed. Nordquist and Park, pp. 457–58.

272. "Reports of the United States Delegation," ed. Nordquist and Park, pp. 526–29. "The perspectives presented showed clearly that the same old proposals that the Conference received from the United States at the first working session in Caracas were being revived, as if nothing had happened since in the negotiating process." UNCLOS III, *Report of the Chairman of the First Committee, March 29, 1982* (U.N. Doc. A/CONF.62/L.91, March 29, 1982), para. 24, reprinted in *Official Records*, UNCLOS III, vol. XVI, p. 206.

273. *Draft Convention* (U.N. Doc. A/CONF.62/L.78), art. 74 (1), in *Official Records*, UNCLOS III, vol. XV, p. 187. See Oxman, "The Third United Nations Conference," pp. 14–15.

274. Malone, "U.S. Policy and the Law of the Sea," p. 50.

275. See Oxman, "The Third United Nations Conference," pp. 13–15.

276. See Larson, "The Reagan Administration," p. 308. The United States agreed to the deadline. Ratiner, "A Crossroads for American Foreign Policy," p. 1015. President Koh emphasized that UNCLOS III would be concluded the following spring, regardless of U.S. preferences. Larson, "The Reagan Administration," p. 318.

277. Larson, "The Reagan Administration," pp. 303, 316.

278. Ratiner, "A Crossroads for American Foreign Policy," pp. 1010–11. U.S. officials also believed that the United States would be able to exploit nodules within its own exclusive economic zone, and that they would derive additional bargaining leverage from the desire of developing states to procure financing and technology for the Enterprise. G. David Robertson and Gaylene Vasatura, "Recent Developments in the Law of the Sea 1981–1982," *San Diego Law Review* 20 (1983): 686–87.

279. See Ratiner, "A Crossroads for American Foreign Policy," p. 1012; and Larson, "The Reagan Administration," p. 305.

280. Statement of President Ronald W. Reagan, January 29, 1982, reprinted in *U.S. Department of State Bulletin* 82 (March 1982): 54. A General Accounting Office study released March 10 concluded that while some objections of the seabed mining industry were valid, others "either are not as serious as portrayed or are premature," and it recommended that the United States negotiate appropriate changes in the Draft Convention. U.S. General Accounting Office, *Impediments to U.S. Involvement in Deep Ocean Mining Can Be Overcome: Report to the Congress* (Washington: U.S. Government Printing Office, 1983), Summary, pp. 4–5.

281. Ratiner, "A Crossroads for American Foreign Policy," pp. 1009–10, 1012–14. "The day-to-day negotiating process was monitored both within the delegation and back in Washington so closely by individuals who had supported the option of withdrawal from the Conference, that any negotiating move made by the American delegation was interpreted as a giant step down the slippery slope of compromise of principle and disaster.... [C]ounter-moves were executed as personal attacks on members of the American delegation, and as attacks on the process of negotiation itself, and were frequently marked by distortion and falsehood." Ibid., p. 1013. See n. 285 below; and also Sanger, p. 52. But see James L. Malone, "Who Needs the Sea Treaty?" *Foreign Policy* 54 (Spring 1984): 53. Ambassador Malone admitted that some modification of the instructions became necessary after the first three weeks of the session failed to yield any concessions from the Group of 77. Malone, "The United States and the Law of the Sea," pp. 793–94.

282. See Robertson and Vasatura, p. 703; and Ratiner, "A Crossroads for American Foreign Policy," p. 1014.

283. United Nations, *Press Release* (U.N. Doc. SEA/494, April 30, 1982), pp. 3, 6.

284. See UNCLOS III, *The U.S. Proposals for Amendments to the Draft Convention on the Law of the Sea* (U.N. Doc. A/CONF.62/WG.21/Informal Paper 18, 1982). So named

because of the color of its cover, the Green Book proposed "a multiplicity of sweeping amendments touching all the sections of part XI and annexes II and III" – i.e., all portions of the treaty
relating to seabed exploitation. "Apart from varying degrees of solidarity expressed by some
industrialized countries, all the other interest groups represented, including many Western
countries, expressed the view that the 'Green Book' could not possibly provide a basis for
negotiations." *Report of the Chairman of the First Committee* (U.N. Doc. A/CONF.62/L.91),
paras. 32–33, in *Official Records*, UNCLOS III, vol. XVI, p. 206. See "Reports of the United
States Delegation," ed. Nordquist and Park, pp. 533–34.

    285. Ratiner, "A Crossroads for American Foreign Policy," p. 1016. See Robertson and
Vasatura, p. 691; and "Reports of the United States Delegation," ed. Nordquist and Park, p.
534. President Koh had hoped to use these texts as a basis for accommodation between the
United States and the Group of 77. "For the United States, these papers moved significantly
toward meeting the President's publicly announced objectives, although they fell far short of
the American delegation's negotiating instructions." Ratiner, "A Crossroads for American
Foreign Policy," pp. 1015–16. While Secretary of State Alexander Haig sought to allow the
U.S. delegation greater negotiating flexibility, his efforts were undermined by White House
Counselor Edwin Meese and by treaty opponents within the Washington bureaucracy who
maligned U.S. delegates and the negotiating process itself on behalf of their constituency, the
seabed mining industry. Laursen, pp. 144–45.

    286. Ratiner, "A Crossroads for American Foreign Policy," p. 1016. See "Reports of the
United States Delegation," ed. Nordquist and Park, pp. 536–37; and *Press Release* (U.N. Doc.
SEA/494), p. 3.

    287. See Robertson and Vasatura, p. 8. The amendments related to 58 articles in the main
text, 3 annexes, and the proposed draft resolutions, and included an 18-page set of proposals
submitted by the United States and 6 other potential seabed mining states on April 13, a
shorter set offered by the Group of 11, and a proposal by 30 coastal states to require prior
notification and authorization for passage of warships through territorial waters. See "Reports
of the United States Delegation," ed. Nordquist and Park, pp. 535–36.

    288. *Press Release* (U.N. Doc. SEA/494), p. 9. Three amendments relating to nonseabed
issues were voted down. Supporters of prior notification for passage of warships through territorial waters withdrew their amendment only after Koh agreed to read a formal statement
that the withdrawal was without prejudice to their position on the issue. See Sanger, p. 152.
"The statement was compatible with the provisions of the draft text, which were carefully
negotiated to preserve pre-existing law, precluding coastal State discrimination against warships in their exercise of the general right of innocent passage." "Reports of the United States
Delegation," ed. Nordquist and Park, p. 547.

    289. Ratiner, "A Crossroads for American Foreign Policy," pp. 1013, 1016. "The Co-
Chairmen of the Working Group of 21 explored every possible avenue.... Opposing sides
and all interest groups were brought together in private consultations, but in vain." *Report
of the Chairman of the First Committee* (U.N. Doc. A/CONF.62/L.91), p. 7 (para. 35). So
far as the head of the U.S. delegation was concerned, however, "no meaningful negotiations
took place" on seabed issues during the eleventh session. Ambassador James L. Malone, "Law
of the Sea and Oceans Policy," *U.S. Department of State Bulletin* 82 (October 1982): 48.

    290. Robertson and Vasatura, p. 696. See also "Reports of the United States Delegation,"
ed. Nordquist and Park, pp. 538–40. The Group of 77 largely acquiesced to the demands of
developed states regarding preparatory investment protection, insisting only that pioneer
mining activities be limited to exploration and that all states sponsoring multinational consortia be required to ratify in order to obtain exploitation rights from the Authority after the Convention's entry into force. Sanger, pp. 187–88.

    291. See UNCLOS III, Final Act, Annex I, Resolution II: "Governing Preparatory Invest-
ment in Pioneer Activities Relating to Polymetallic Nodules," paras. 1 (a), 6, 8, 9 (a), 12 (a),
reprinted in *International Legal Materials* 21 (1982): 1254–57; and also "Reports of the United
States Delegation," ed. Nordquist and Park, pp. 540–42. Each pioneer investor was to be
granted exclusive mining rights in up to 150,000 square kilometers of seabed area, half of
which would be relinquished to the Authority over an eight-year period. Resolution II, para.
1 (e), in *Interational Legal Materials* 21 (1982): 1255.

    Each of the four private consortia included U.S. corporations. For the identity of the various organizations comprising each of the named pioneer investors, see Sanger, pp. 162–64. India and the Soviet Union had notified the Conference that they expected to qualify as pioneer investors. Unlike the claims of the other three state-operated pioneer investors, which were located at the Clarion-Clipperton Fracture Zone in the eastern Pacific, the Indian site was located in the Indian Ocean. Developing states were given until January 1, 1985, to qualify as pioneer investors by expending $30 million on seabed mining activities. Resolution II, para. 1 (a), in *International Legal Materials* 21 (1982): 1254. On April 13 the United States and four other Western states offered an amendment which, among other things, would have effectively prevented most developing states from qualifying as pioneer investors by changing the qualifying date to January 1, 1983. UNCLOS III, *Belgium, Federal Republic of Germany, Italy, United Kingdom of Great Britain and Northern Ireland and United States of America: Amendments* (U.N. Doc. A/CONF.62/L.122, April 13, 1982), Resolution II, proposed para. 1 (a), reprinted in *Official Records*, UNCLOS III, vol. XVI, p. 231.

292. UNCLOS III, Final Act, Annex I, Resolution I: "Establishment of the Preparatory Commission for the International Sea-Bed Authority and for the International Tribunal for the Law of the Sea," reprinted in *International Legal Materials* 21 (1982): 1253–54.

293. UNCLOS III, *Report of the Co-Ordinators of the Working Group of 21, T.T.B. Koh (Singapore) and Paul Bamela Engo (United Republic of Cameroon) to the First Committee* (U.N. Doc. A/CONF.62/C.1/L.30, March 29, 1982), para. 5, reprinted in *Official Records*, UNCLOS III, vol. XVI, p. 271. See Robertson and Vasatura, p. 698. The Final Act, which set forth the results of the Conference, was itself a document without any independent legal significance. See *Press Release* (U.N. Doc. SEA/494), p. 3.

294. "Reports of the United States Delegation," ed. Nordquist and Park, pp. 544–45. See UNCLOS III, Final Act, Annex I, Resolution I, para. 2, and Resolution IV, reprinted in *International Legal Materials* 21 (1982): 1253, 1258.

295. See Robertson and Vasatura, p. 693 n. 102; *Press Release* (U.N. Doc. SEA/494), pp. 9–10; and Ratiner, "A Crossroads for American Foreign Policy," p. 1006. Koh refused to extend the negotiating deadline beyond April 30 when he discovered that the United States was lobbying other states to vote against adoption of the Convention. Sanger, p. 53.

296. Israel objected to the capacity of the Palestine Liberation Organization, which had participated in the Conference as an observer, to sign the final act; Turkey and Venezuela found certain delimitation provisions unacceptable. Laursen, p. 148. The United States was able to persuade several of its allies—including West Germany, the United Kingdom, and Thailand—to abstain rather than support the Convention. Ratiner, "A Crossroads for American Foreign Policy," p. 1012. Ted L. McDorman, "Thailand and the 1982 Law of the Sea Convention," *Marine Policy* 9 (1985): 294. The Reagan administration also did everything possible to prevent any future administration from altering the U.S. position toward the Convention. Sanger, p. 197.

297. The United States participated in the September session "at a technical level." Malone, "Law of the Sea and Oceans Policy," p. 50.

298. See *International Legal Materials* 21 (1982): 1261. Japan signed several weeks later. The United States, like most other states which declined to sign the Convention itself, did sign the Final Act, entitling it to participate in the Preparatory Commission as a nonvoting observer. All told, 140 states and 9 other organizations registered their participation in the Conference by becoming signatories to the Final Act. See *International Legal Materials* 21 (1982): 1245.

299. Bernardo Zuleta, "The Law of the Sea After Montego Bay," *San Diego Law Review* 20 (1983): 475. "Eight states were represented by their Prime Minister or Deputy Prime Minister; six States were represented by the Attorney General or Minister of Justice; fourteen States were represented by their Foreign Minister; and 22 States were represented by Ministers of other portfolios or Vice-Ministers." Ibid., p. 475 n. 2.

300. Elisabeth Mann Borgese, "The Law of the Sea," *Scientific American* 248 (March 1983): 42.

301. See E.D. Brown, "Seabed Mining: From UNCLOS to Prep Com," *Marine Policy* 8 (1984): 159. "Despite the primarily technical nature of its mandate, the Commission has been placed by circumstances at the centre of a political process that may ultimately determine whether the Convention can become the universally accepted treaty it was intended to be."

Philippe Kirsch and Douglas Fraser, "The Law of the Sea Preparatory Commission After Six Years: Review and Prospects," *Canadian Yearbook of International Law* 26 (1988): 120.

302. See E. Brown, "From UNCLOS to Prep Com," p. 161; and David L. Larson, "Deep Seabed Mining: A Definition of the Problem," *Ocean Development and International Law* 17 (1986): 287.

303. See U.N. Department of Public Information, "Commission Preparing Sea-Bed Authority Recesses First Session," *UN Chronicle* 20 (June 1983): 12. Cf. text accompanying n. 228 above. Consensus was also required for adoption of the Commission's final reports and for additional financial assessments. Other decisions would be taken by a two-thirds-majority vote, while procedural issues would be determined by a simple majority vote.

304. See American Law Institute, *Restatement 3rd, Restatement of the Foreign Relations Law of the United States,* §523, reporters' note 4. The final decision to oppose the Convention was made at a National Security Council meeting on June 29, at which President Reagan stated that he "kind of thought that when you go out on the high seas you can do what you want." Laursen, p. 148. "The resounding pro–Convention vote in UNCLOS III in April 1982, subsequently confirmed by the procession of signatures in December, was undoubtedly a massive censure of U.S. foreign policy: but to the ideologues in Washington, the hostile votes of 'irresponsible' small states and the clients of an 'evil Empire' were taken as proof of the rightness of the administration's course, not the opposite." Ken Booth, *Law, Force and Diplomacy at Sea* (London: George Allen & Urwin, 1985), p. 30.

305. Jens Evensen, "Keynote Address," in *The 1982 Convention,* ed. Koers and Oxman, p. xxxix. U.S. critics of the Convention, on the other hand, doubted that accommodative agreement could be achieved, and felt that it would be inappropriate to try. Statement of Representative John Breaux in ibid., p. 16. Cf. Malone, "The United States and the Law of the Sea," p. 799 n. 31.

306. States declaring such views publicly included Belgium, France, Italy, and the Netherlands. "Even these statements represented only the tip of the iceberg." Kirsch and Fraser, p. 122.

307. Cf. Günther Jaenicke, "Conflicts Between Mine Sites of Signatories and Non-Signatories of the Law of the Sea Convention," in *The Law of the Sea: What Lies Ahead?* ed. Thomas A. Clingan, Jr. (Honolulu: Law of the Sea Institute, 1988), pp. 507–8. See also Moritaka Hayashi, "Registration of the First Group of Pioneer Investors by the Preparatory Commission for the International Sea-Bed Authority and for the International Tribunal for the Law of the Sea," *Ocean Development and International Law* 20 (1989): 4.

308. *Oceans Policy News* (May 1984), p. 2. See J.M. Broadus and Porter Hoagland III, "Conflict Resolution in the Assignment of Area Entitlements for Seabed Mining," *San Diego Law Review* 21 (1984): 546.

309. See *Oceans Policy News* (May 1984), p. 2; E. Brown, "From UNCLOS to Prep Com," p. 162.

310. See Hayashi, "Registration of the First Group," pp. 4 7; Broadus and Hoagland, p 546 n. 29, and also Kirsch and Fraser, p. 128. Beginning in July 1982, Canada had attempted unsuccessfully to negotiate a "Memorandum of Understanding" among pioneer investor states. Broadus and Hoagland, p. 547. See U.N. Department of Public Information, "Commission Studies Sea-Bed Mining Rules," *UN Chronicle* 21 (April 1984): 46. "Although the Canadians did not 'disinvite' the U.S., they made it clear that there was little utility in our attending their meetings." Cohen, p. 154.

311. Lee Kimball, "Turning Points in the Future of Deep Seabed Mining," *Ocean Development and International Law* 17 (1986): 378. The Commission met March 19–April 13, August 13–September 5, 1984.

312. *Oceans Policy News* (September 1984), p. 2. See also U.N. Department of Public Information, "Commission Studies Sea-Bed Mining Rules," pp. 45, 47.

313. See Peter Bruckner, "Preparatory Investment Under the Convention and the PIP Resolution," in *The 1982 Convention,* ed. Koers and Oxman, pp. 193–94 n. 9.

314. See Kimball, "Turning Points," p. 378. West Germany sought its own pioneer site in exchange for its signature of the Convention, Eastern European states demanded one "as a trade-off" if West Germany received one, and developing states sought an extension of the

January 1, 1985, deadline for them to qualify as pioneer investors. *Oceans Policy News* (September 1984), pp. 2–3, 4.

315. See nn. 386–89 below and accompanying text, and Chap. 5 n. 13 and accompanying text. Most of these states denied any attempt to establish an alternative seabed mining regime, several asserting that the Understanding was merely intended to implement a 1982 agreement reached by Western seabed mining operators. *Oceans Policy News* (January/February 1985), p. 4. U.N. Department of Public Information, "Preparing for Sea-Bed Regime: An Agreement on Claims Procedures," *UN Chronicle* 21 (July 1984): 30. Consideration of a draft resolution sponsored by Soviet-bloc states condemning the Understanding was delayed until 1985. See n. 390 below and accompanying text.

316. See Mina Mashayekhi, "The Present Legal Status of Deep Sea-Bed Mining," *Journal of World Trade Law* 19 (1985): 240; Larson, "A Definition of the Problem," pp. 288–89; and Hayashi, "Registration of the First Group," p. 8. The agreement was understood to be an effort to induce recalcitrant Western seabed mining states to sign the Convention before the deadline. Kimball, "Turning Points," p. 378.

317. U.N. Department of Public Information, "Commission Studies Sea-Bed Mining Rules," pp. 45, 47, 49. It was expected that a professional staff of 30 would be able to commence the operation of such a nucleus Enterprise within three years after the entry into force of the Convention. U.N. Department of Public Information, "An Agreement on Claims Procedures," pp. 32–33.

318. See Sanger, pp. 199–200; and also Larson, "A Definition of the Problem," pp. 297–303. Both West Germany and the United Kingdom almost signed the Convention, despite U.S. diplomatic efforts, the latter state in fact reversing an earlier decision to sign after the United States made its opposition clear. Statement of Tommy T.B. Koh in *Law and Contemporary Problems* 46 (Spring 1983): 27. It had been hoped that West Germany would sign, until Chancellor Helmut Kohl received a personal letter from President Reagan urging rejection of the treaty. John Tagliabue, "Bonn Cabinet Said to Oppose Sea Treaty," *New York Times*, November 24, 1984, p. 3. "That the F.R.G. did not sign when there had been high expectations that it would . . . was a particular blow to the [Preparatory] Commission." Kimball, "Turning Points," p. 379. The nonsignatory states also included "Albania, which is not usually regarded as a close friend of the Americans; Turkey, which fears an encouragement of Greece's ambitions in Aegean waters; Ecuador and Peru, which claim full sovereignty in their 200-mile zones; and a collection of other states—Syria and Israel, for once united, among them—that dislike various specific parts of this wide-ranging convention." "Oddfellows," *Economist* 293 (December 22, 1984): 30.

319. See Kimball, "Turning Points," pp. 375, 379; Chap. 5 n. 122; and also Sanger, p. 5. Western seabed mining states nevertheless continued to assert a right to conduct nodule mining under an alternative regime, and their participation in the Commission negotiations was generally regarded by other states as imperative for a successful outcome. See Mati L. Pal and Lee Kimball, "Planning for the International Sea-Bed Authority," in *The 1982 Convention*, ed. Koers and Oxman, p. 146. Several EEC members referred to the "considerable difficulties and flaws" of the Convention's seabed mining provisions, but there was a general willingness to work within the Preparatory Commission. See Malcolm J. Forster, "Law of the Sea Convention: Signatories Express Problems," *Environmental Policy and Law* 15 (1985): 2. Signature had been urged by the European Parliament. See European Parliament, "Law of the Sea Resolution," adopted June 9, 1983, reprinted in *Environmental Policy and Law* 11 (1983): 81. See also Council of Europe, "UN Convention on the Law of the Sea" (resolution adopted May 11, 1984), reprinted in *Environmental Policy and Law* 13 (1984): 80.

320. See Larson, "A Definition of the Problem," pp. 290–91. "This was advance notice that de facto amendments to Resolution II would be required if registration were to take place." Kirsch and Fraser, p. 130. A relatively small overlap between the Japanese and Soviet sites in the Pacific was tentatively resolved, pending resolution of the large overlap between France and the Soviet Union. Kimball, "Turning Points," p. 382. "The overlaps were such as to make it particularly difficult for all three to submit two areas of equal commercial value as required by Resolution II—one to be given to the applicant and the other to be reserved for the Enterprise." United Nations, *Preparatory Commission for Sea-Bed Authority and Law of the Sea Tribunal Concludes Session at Kingston, 29 March–16 April* (U.N. Doc.

SEA/819, April 20, 1987), p. 5. See also Hayashi, "Registration of the First Group," p. 9.

The Clarion-Clippterton Zone, located in the eastern Pacific, comprises approximately 1.4 percent of the Pacific Ocean—6 million sq. km, of which 2.5 million sq. km is believed to be mineable. Wolfgang Hauser, *The Legal Regime for Deep Seabed Mining Under the Law of the Sea Convention*, trans. Frances Bunce Dielmann (Deventer, Netherlands: Kluwer, 1983), p. 12 n. 9.

321. Hayashi, "Registration of the First Group," p. 10. See also Kimball, "Turning Points," pp. 380, 383. Believing, correctly, that the Soviet claim also overlapped to some extent with those of the private consortia, the BINC states persuaded the chairman of the Commission to include them in negotiations to resolve overlap conflicts. *Oceans Policy News* (January 1986), p. 3.

322. Kimball, "Turning Points," p. 382. See also Larson, "A Definition of the Problem," p. 291. Efforts to resolve the Soviet/French/Japanese overlaps were constrained by the prior agreement which had resolved overlapping claims among Western seabed mining states, making changes in their agreed site boundaries difficult. Fernando Zegers Santa Cruz, "Deep Sea-Bed Mining Beyond National Jurisdiction in the 1982 UN Convention on the Law of the Sea: Description and Prospects," *German Yearbook of International Law* 31 (1988): 115. See n. 385 below and accompanying text. Despite general agreement that its claim to a mine site in the Indian Ocean was uncontested, India declined to proceed alone in the registration process. *Oceans Policy News* (September/October 1985), p. 3.

323. Hayashi, "Registration of the First Group," p. 15.

324. See *Oceans Policy News* (September/October 1985), pp. 4–5.

325. "Various speakers pointed out that the ... Convention requires the Enterprise to operate in accordance with sound commercial principles, precluding early mining operations unwarranted by commercial markets." *Oceans Policy News* (October/November 1984), p. 2.

326. J.A. Walkate, "Developments in Special Commission 3 of the Preparatory Commission for the International Sea-Bed Authority and for the International Tribunal for the Law of the Sea: Drafting the Future Deep Seabed Mining Code," *Netherlands International Law Review* 36 (1989): 154. It was agreed that the code would contain chapters relating to prospecting; applications for plans of work; processing of applications; production authorizations; revenue sharing and financial incentives; and technology transfer. Ibid., p. 153.

327. U.N. Office for Ocean Affairs and the Law of the Sea, "Report on the Fifth Session of the Preparatory Commission for the International Sea-Bed Authority and for the International Tribunal for the Law of the Sea, Kingston, 30 March–16 April 1987," *Law of the Sea Bulletin* 10 (November 1987): 118.

328. See "UNCLOS: Preparatory Commission Meets," *Environmental Policy and Law* 15 (1985): 80.

329. *Oceans Policy News* (May 1984), p. 3. See Hayashi, "Registration of the First Group," p. 24. The Group consisted of Australia, Austria, Canada, Denmark, Finland, Iceland, Ireland, Norway, Sweden, and New Zealand states which, as members of the Group of 11, had worked to promote compromise at the final session of UNCLOS III. "These states reemphasized the need for a cautious and cost-effective approach to the implementation of the Convention and supported the establishment of 'lean' institutions." Kimball, "Turning Points," p. 381.

330. Rather than submitting two individual mining sites from which the Commission would select one to be reserved for the Authority, each of the four pioneer applicants was to be permitted to "pre-select" 52,300 sq. km of its mining site, with the remainder being allocated by the Commission and a single "prime" seabed area of equal size being reserved for the Enterprise from the overlap area in the eastern Pacific Ocean. In effect, the four applicants agreed to accelerate the eight-year relinquishment period set forth in Resolution II. "Since the maximum size of the site allocated to an applicant would be 75,000 sq. km after relinquishment, this formula would help to reduce considerably the size of overlaps with the areas claimed by the consortia, and thereby minimize the concern of the potential applicants." Hayashi, "Registration of the First Group," p. 15. See nn. 291, 341 above and below; and also Kirsch and Fraser, p. 132 n. 37.

The Arusha Understanding, named for the Tanzanian city where Chairman Warioba conducted the negotiations, was generally regarded by Commission delegates as a potentially acceptable solution to the problem of overlapping claims, although some modification was anticipated in order to accommodate the interests of the four unregistered pioneer consortia and of other potential applicants. See *Oceans Policy News* (May 1986), p. 5. After the Understanding was concluded, the Preparatory Commission abandoned efforts to draft uniform rules of procedure to govern the registration of pioneer applicants — rules which would have established uniform rules of procedure with respect to matters such as data requirements, confidentiality, payment of application fees, and periodic expenditures. Hayashi, "Registration of the First Group," p. 20.

331. Understanding of the Preparatory Commission for the International Sea-Bed Authority and for the International Tribunal for the Law of the Sea for Proceeding With Deep Sea-Bed Mining Applications and Resolving Disputes of Overlapping Claims of Mine Sites, paras. 10–14, 20–22, reprinted in *International Legal Materials* 25 (1986): 1327–29. See Preparatory Commission, *Statement Made by the Acting Chairman of the Preparatory Commission* (U.N. Doc. LOS/PCN/L.41/Rev.1, September 5, 1986), reprinted in *Law of the Sea Bulletin*, Special Issue (April 1988), 138–45; and also United Nations, *Law of the Sea: Report of the Secretary-General* (U.N. Doc. A/41/742, October 28, 1986), p. 28. "These procedures and mechanisms shall not be construed as setting a precedent for the implementation of the regime for sea-bed mining under the Convention, nor do they purport to alter or amend that regime in any way." Understanding of the Preparatory Commission, para. 18, in *International Legal Materials* 25 (1986): 1329. A spokesman for the Group of 77 stated that the Group had made "painful" concessions which "would have been unthinkable" two years earlier, in order to permit agreement on a procedure for implementing the pioneer system. U.N. Department of Public Information, "Preparatory Commission Agrees on Procedures for Registering Pioneer Investors in Deep Sea-Bed Mining," *UN Chronicle* 23 (November 1986): 89.

The four private consortia were given until the entry into force of the Convention to register as pioneer investors on "similar" terms, provided they complied with similar obligations. Understanding of the Preparatory Commission para. 15, in *International Legal Materials* 25 (1986): 1329. See Kirsch and Fraser, p. 142. On September 3 the Soviet Union and the BINC states agreed to seek to resolve remaining overlaps between the Soviet and private claims. Kirsch and Fraser, p. 133.

332. See Understanding of the Preparatory Commission, paras. 1–7, in *International Legal Materials* 25 (1986): 1326–27; and also Lee Kimball, "Introductory Note," *International Legal Materials* 26 (1987): 1502. The Group assembled in 1987 was composed of experts from 15 states: Brazil, Cameroon, Canada, Colombia, the People's Republic of China, Cuba, France, German Democratic Republic, India, Indonesia, Japan, Kenya, Norway, the Soviet Union, and Zambia. See *Oceans Policy News* (May 1987), p. 4. "The Group of Technical Experts was of high professional calibre." Kirsch and Fraser, p. 136 n. 53.

333. *Report of the Secretary-General* (U.N. Doc. A/41/742), p. 30. See also U.N. Department of Public Information, "Preparatory Commission Agrees on Procedures," p. 91. It was agreed that the possibility of a compensation fund should be studied. *Oceans Policy News* (September 1986), p. 8. There was increasing doubt in the Special Commission, however, about its ability to deal with such issues in advance of the actual commencement of seabed mining activities. *Oceans Policy News* (May 1986), p. 2.

334. See *Report of the Secretary-General* (U.N. Doc. A/41/742), p. 30; and *Oceans Policy News* (May 1986), pp. 9–10.

335. *Report of the Secretary-General* (U.N. Doc. A/41/742), p. 30. See also *Oceans Policy News* (May 1986), p. 10. The special committee's new chairman, Jaap Walkate of the Netherlands, employed informal negotiating groups to achieve progress on difficult issues. *Oceans Policy News* (May 1986), p. 2. It was expected that most of the concerns of Western states could be accommodated through the drafting of financial incentives for joint ventures, although "the number of issues raised . . . that will require additional clarifying rules appears somewhat daunting." *Oceans Policy News* (September 1986), pp. 6, 9. Although the Convention seemed to require submission of a single plan of work, based upon which the Authority would select one site to bank for the Enterprise, it was pointed out that preparation of a costly plan of work for the reserved area would be wasteful since the plan would likely be out of date when the Enterprise began its mining activities. Walkate, pp. 155–56.

336. *Oceans Policy News* (September 1986), p. 10. Although the Tribunal was to be fully operational six months after the Convention's entry into force, West Germany indicated that it had already selected a site and argued that it could preserve its right to be the host government so long as it acceded to the Convention prior to its entry into force. *Oceans Policy News* (May 1986), pp. 7, 11.

337. Western states proposed that a majority of such subsidiary organs be nominated by the 8 states "most substantially engaged in" nodule mining, and, with respect to the Finance Committee, by the 15 largest contributor states. U.N. Department of Public Information, "Sea-Bed Commission Condemns Issuing of Licenses for Exploration of International Area," *UN Chronicle* 23 (August 1986): 108.

338. See *Preparatory Commission* (U.N. Doc. SEA/819), p. 1.

339. Ibid., pp. 1–3, 6–8. See Preparatory Commission, *Statement of Understanding on the Implementation of Resolution II Made by the Chairman of the Preparatory Commission at the 34th Plenary Meeting, Held on 10 April 1987* (U.N. Doc. LOS/PCN/L.43/Rev.1, April 10, 1987), reprinted in *Law of the Sea Bulletin*, Special Issue II (April 1988): 151–53. The Soviet Union held discussions with the BINC Group, West Germany, the United Kingdom, and the United States, as well as with the consortia themselves. Hayashi, "Registration of the First Group," p. 18. It was agreed that the Indian applicant might be registered separately during the resumed summer session but that the other three applicants would be registered simultaneously — a procedure which, the Acting Chairman emphasized, "will not affect the priority and equal treatment of all the applicants of the first group and are without prejudice to the interests of the potential applicants." *Preparatory Commission* (U.N. Doc. SEA/819), pp. 1–2. See also U.N. Office for Ocean Affairs and the Law of the Sea, "Registration of Pioneer Investors in the International Sea-Bed Area in Accordance With Resolution II of the Third United Nations Conference on the Law of the Sea," *Law of the Sea Bulletin* Special Issue II (April 1988): 7. The 1987 Understanding represented a further modification of Resolution II, although it remained consistent with the provisions of the Convention. See Hayashi, "Registration of the First Group," pp. 19–20.

340. See U.N. Office for Ocean Affairs and the Law of the Sea, "Registration of Pioneer Investors," pp. 40–63. "The experts' group unanimously concluded that the application had been submitted in accordance with the relevant instrument and that the two areas identified by India were of equal estimated commercial value." Kirsch and Fraser, p. 136.

341. See U.N. Office for Ocean Affairs and the Law of the Sea, "Registration of Pioneer Investors," pp. 9–39, 64–129; Hayashi, "Registration of the First Group," pp. 20–22; and U.N. Office for Ocean Affairs and the Law of the Sea, "Report on the Meeting of the Preparatory Commission for the International Sea-Bed Authority and for the International Tribunal for the Law of the Sea, New York, 27 July–21 August 1987," *Law of the Sea Bulletin* 10 (November 1987): 120. Meeting December 7–16 to consider the report of the Group of Technical Experts, the General Committee raised questions with respect to the sufficiency of the data provided by the applicants and the manner in which the Group had evaluated the pair of mining sites submitted by each applicant. "In the end, the discussion petered out, both as a result of the lack of technical expertise of the General Committee members and of the straightforwardness of the spokesman for the Group of Technical Experts, who readily admitted there could be no absolute certainties at this stage." Kirsch and Fraser, pp. 137–38. Each of the three pioneer registrants received a mining site of 75,000 sq. km. A total of more than 437,000 sq. km of adjacent seabed area was reserved for the Authority, and more than 92,000 sq. km was relinquished for use by the potential pioneer applicants. Although the three registrants disclosed general descriptions of their seabed mining technology and agreed to comply provisionally with general environmental obligations set forth in the Convention, they did not agree to comply with provisional rules and regulations adopted by the Commission, as had India. Kirsch and Fraser, pp. 4–5.

342. See Agreement on the Resolution of Practical Problems With Respect to Deep Seabed Mining Areas, and Exchange of Notes Between the United States and the Parties to the Agreement, reprinted in *International Legal Materials* 26 (1987): 1504–15. The U.S.-Soviet agreement was expressly without prejudice to the parties' positions with respect to the Convention. Ibid., pp. 1510–11. The Western states also exchanged notes among themselves which

prevented termination of the operative agreement without mutual consent. Kimball, "Introductory Note," p. 1502.

Although the package of agreements represented somewhat of an enhancement of the viability of the licenses issued by the United States under the 1980 Seabed Act, insofar as the Soviet Union agreed to respect for an indefinite period the sites claimed thereunder, the remainder of the international community—with the exception of France and Japan, which had joined in the 1984 Provisional Understanding—would be under no obligation to respect such claims after entry into force of the Convention if the United States still refused to participate in the Convention's seabed mining regime.

343. Hayashi, "Registration of the First Group," p. 23. It has been asserted that the agreement effectively conferred the exclusive rights of pioneer registrants upon the four private consortia. Santa Cruz, p. 116. Because the 1986 Understanding provided that relinquished areas would be reserved (until entry into force of the Convention) for "potential applicants" generally, however, it remains possible that the four sites licensed to U.S. consortia under the Seabed Act could be infringed in pioneer investor applications submitted by developing states. See Kirsch and Fraser, pp. 132, 143.

344. U.N. Office for Ocean Affairs and the Law of the Sea, "Report on the Fifth Session," p. 117. See *Preparatory Commission* (U.N. Doc. SEA/819), p. 8. "[T]here was agreement that whatever measures were established would be implemented under the auspices of the [Authority] even if provided on a bilateral basis or through existing mechanisms." *Oceans Policy News* (May 1987), p. 7.

At the resumed summer session, there was a good deal of discussion about whether any such assistance would be necessary if seabed mining operations were not subsidized. U.N. Office for Ocean Affairs and the Law of the Sea, "Report on the Meeting," pp. 121–22. See n. 268 above. "Interestingly, Western European states and Japan, which would normally be thought of as supportive of free enterprise principles but which are also mineral consumers, are concerned about the idea of anti-subsidization provisions as an element of the seabed regime." Kirsch and Fraser, p. 149.

345. *Oceans Policy News* (May 1987), pp. 7–8. *Oceans Policy News* (September 1987), p. 8. Arguing that the obligations under Resolution II were not dependent on improved economic conditions, developing states were of the view that training should be provided by pioneer registrants without cost to the Commission. Although some states called for reimbursement of training costs by the Authority, India favored equitable sharing of training costs, and the other three pioneer applicants reserved their position on the issue. "On the question of timing, Japan and the Soviet Union linked the commencement of training to the beginning of exploitation. If training were to start now, it was argued, it would become obsolete by the time sea-bed mining began." *Preparatory Commission* (U.N. Doc. SEA/819), p. 9.

346. U.N. Office for Ocean Affairs and the Law of the Sea, "Report on the Meeting," p. 122. U.N. Department of Public Information, "Sea Law Commission Registers India as First Pioneer Investor," *UN Chronicle* 24 (November 1987): 58. See also *Oceans Policy News*, Special Report (March 1988), 4; and Kirsch and Fraser, p. 147.

347. Walkate, p. 157. See also *Preparatory Commission* (U.N. Doc. SEA/819), p. 10. "[A]fter the (provisional) designation of the two areas, the applicant must submit additional information *only* on the area which has been allocated to him in order to get the final approval of his plan of work." Walkate, p. 157 (emphasis in original).

348. *Oceans Policy News* (May 1987), pp. 3, 9. See Walkate, pp. 158–59. Six Western states—Belgium, Italy, Japan, the Netherlands, West Germany, and the United Kingdom—were supported by the Soviet bloc, which sought to ensure a minimum rate of return on seabed mining investment. *Oceans Policy News* (May 1987), p. 3. Privately, developing states were more willing to concede that changed circumstances favored modification of some aspects of the regime agreed to in 1982. Kirsch and Fraser, p. 152.

One possible improvement, from the point of view of the seabed mining states, would be to draft the regulations so as to limit the amount of nodule production subject to the revenue-sharing provisions. See *Preparatory Commission* (U.N. Doc. SEA/819), p. 10. The definition of "commercial production" remained to be determined; a liberal definition of "contractor's development costs" would reduce the amount of "attributable net proceeds" from which the Authority's "share of net proceeds" is calculated. See Chap. 4 nn. 99–100 and accompanying text; and also *Oceans Policy News* (September 1987), p. 9.

349. See U.N. Office for Ocean Affairs and the Law of the Sea, "Report on the Meeting," p. 123. Some states continued to seek to impose a deadline for German ratification, while others supported West Germany's claimed right to withhold its ratification without jeopardizing the seating of the Tribunal in Hamburg. *Preparatory Commission* (U.N. Doc. SEA/819), p. 11. See also *Oceans Policy News* (May 1987), p. 5.

350. *Oceans Policy News* (May 1987), p. 5. "Particular consideration was given to election of the President and four Vice-presidents of the Council and the need for representation of the five regional groups, as well as to the necessity of adequate consultation among the regional groups prior to presentation of a candidate for President, with a view to electing the President by consensus." *Oceans Policy News* (September 1987), p. 6. Much of the discussion focused on the question of decisionmaking procedures within the Economic Planning Commission with respect to recommendations for compensation to developing producer states; developed states sought a consensus requirement, while the Group of 77 favored a voting requirement of only two thirds. *Preparatory Commission* (U.N. Doc. SEA/819), pp. 4–5.

During the resumed summer session, the Plenary also reviewed proposals for the establishment of a Finance Committee to act in an advisory capacity to the Council as an additional subsidiary organ. U.N. Office for Ocean Affairs and the Law of the Sea, "Report on the Meeting," p. 121. The responsibilities, composition, and decisionmaking procedures of the Committee remained in dispute, however. See *Oceans Policy News* (September 1987), p. 6.

351. U.N. Office for Ocean Affairs and the Law of the Sea, "Report on the Fifth Session," p. 117. See *Oceans Policy News* (May 1987), p. 3. Several other issues, including subsidiary organs and the status of observers, were also deferred. U.N. Office for Ocean Affairs and the Law of the Sea, "Report on the Meeting," p. 121.

352. See U.N. Office for Ocean Affairs and the Law of the Sea, "Report on the Sixth Session of the Preparatory Commission for the International Sea-Bed Authority and for the International Tribunal for the Law of the Sea: Kingston 14 March–8 April 1988; New York 15 August–2 September 1988," *Law of the Sea Bulletin* 12 (December 1988): 47; and also United Nations, *Law of the Sea: Report of the Secretary-General* (U.N. Doc. A/43/718, October 20, 1988), pp. 41–42 (paras. 150–51). The pioneer registrants, pointing to changed economic factors and the lack of any such obligations on the part of the unregistered private consortia, sought adjustments in the requirements of Resolution II, including waiver of the annual fee. *Oceans Policy News*, Special Report (September 1988), p. 2.

353. "After commercial sea-bed production occurs, the developing land-based producers that consider themselves affected by sea-bed production would raise the issue with the Authority, presenting corroborating facts and studies." *Report of the Secretary-General* (U.N. Doc. A/43/718), p. 43 (para. 158). Difficult issues relating to compensation and subsidization were referred to a newly created working group, as Australia and Canada pressed for regulations which would deter seabed mining operations which were not justified by market conditions. See *Oceans Policy News*, Special Report (March 1988), pp. 2–4; and U.N. Office for Ocean Affairs and the Law of the Sea, "Report on the Sixth Session," p. 49.

354. U.N. Office for Ocean Affairs and the Law of the Sea, "Report on the Sixth Session," p. 49. It was agreed that training programs actually employed by the pioneer registrants would be made available on an equal basis to train Enterprise personnel. *Oceans Policy News*, Special Report (March 1988), p. 4. During the resumed summer session, it was further agreed that an advisory panel should administer the training program by matching qualified candidates nominated by member states with traineeship positions made available by pioneer registrants. *Oceans Policy News*, Special Report (September 1988), p. 5. See also U.N. Department of Public Information, "Training Programme for Enterprise to Be in Place by 1991," *UN Chronicle* 27 (June 1990): 38.

355. U.N. Office for Ocean Affairs and the Law of the Sea, "Report on the Sixth Session," p. 50. See also United Nations, *Law of the Sea: Report of the Secretary-General* (U.N. Doc. A/44/650, November 1, 1989), p. 38.

356. U.N. Office for Ocean Affairs and the Law of the Sea, "Report on the Sixth Session," p. 51. See also *Oceans Policy News*, Special Report (March 1988), p. 6. The draft protocol addressed the rights and duties of all states, parties, representatives, and witnesses with

respect to the Tribunal. *Oceans Policy News*, Special Issue (August/September 1989), pp. 3, 6.

357. U.N. Office for Ocean Affairs and the Law of the Sea, "Report on the Sixth Session," pp. 47–48. See *Oceans Policy News*, Special Report (March 1988), pp. 2–3; and also *Report of the Secretary-General* (U.N. Doc. A/43/718), p. 42 (paras. 152, 154). "Although formal decisions on the approval of the budget are to be taken by the Assembly, the decision-building role in the budgeting process is going to be played first and foremost by the Finance Committee." José Luis Jesus, "Statement on the Issue of the Universality of the Convention," University of Kiel, July 1990, reprinted in *Oceans Policy News*, Special Report (July 1990), p. 4. It was generally agreed that "major contributors" should be represented on the Committee. "For the industrialized states, in particular, the financial implications of the Convention regime will be an important factor in their evaluation, of whether to ratify the Convention." Kirsch and Fraser, pp. 147–48.

358. See *Report of the Secretary-General* (U.N. Doc. A/44/650), p. 40.

359. *Oceans Policy News*, Special Report (August/September 1989), p. 4. See *Report of the Secretary-General* (U.N. Doc. A/44/650), pp. 36–37; and also U.N. Department of Public Information, "Sea Law Commission Focuses on Pioneer Investor Duties, Training Programme Approved," *UN Chronicle* 26 (December 1989): 35. The provisional conclusions incorporated recommendations by an *ad hoc* working group, which had proposed that a 5 percent reduction in export earnings trigger a review of the actual or potential impact of nodule mining upon the affected producer state. *Oceans Policy News*, Special Report (February/March 1989), p. 4. The proposed compensation procedure would require producer states to exhaust other remedial measures, including bilateral assistance from seabed mining states, which previously imported nodule minerals from the claimant state, before being considered for economic assistance. *Oceans Policy News*, Special Report (August/September 1989), p. 4.

360. *Report of the Secretary-General* (U.N. Doc. A/44/650), p. 37. See n. 354 above; and also *Oceans Policy News*, Special Report (August/September 1989), p. 3. Rather than drafting rules and procedures, Special Commission 2 prepared annotated comments to several articles of the Convention.

361. *Report of the Secretary-General* (U.N. Doc. A/44/650), p. 38. U.N. Department of Public Information, "Sea Law Commission Focuses on Pioneer Investor Duties," p. 35.

362. *Oceans Policy News*, Special Report (August/September 1989), p. 6. It was recognized that the Tribunal would need to enter into formal relationship agreements with the Authority, the United Nations, and the International Court of Justice. Consultations continued with regard to the seat of the Tribunal.

363. *Oceans Policy News*, Special Report (August/September 1989), pp. 3–4.

364. See *Law of the Sea Bulletin* 15 (May 1990): 54–62; and also Elliot L. Richardson, "Law of the Sea: A Reassessment of U.S. Interests,"*Mediterranean Quarterly* 1 (Spring 1990): 10–11. The offer was made "without any preconditions, other than the fact that those willing to talk must indicate a positive approach to serious and meaningful talks." Statement of Group of 77 Chairman Mumba S. Kapumpa before the Plenary, September 1, 1989, quoted in *Report of the Secretary-General* (U.N. Doc. A/44/650), p. 40.

365. Jesus, "Statement on the Issue," p. 5.

366. U.N. Department of Public Information, "Training Programme for Enterprise," pp. 38–39. U.N. Department of Public Information, "Preparatory Commission Adopts 'Understanding' on Obligations of Pioneer Investors," *UN Chronicle* 27 (December 1990): 51–52. See *Oceans Policy News* (September 1990), pp. 7–8. The Chairman of the Commission discussed alternative formats for softening the impact of some of the more objectionable provisions of the Convention: informal "statements of understanding"—"a solid and secure procedure used to introduce major change to Resolution II in order to circumvent difficulties raised by changing circumstances"—or "a protocol or other visible legal instrument that, instead of purporting to change the formal text of the Convention, would rather embody a universal interpretation" effectively binding upon states parties. José Luis Jesus, Comments before Law of the Sea Institute, Tokyo, July 1990, reprinted in *Oceans Policy News* (September 1990), p. 7. Use of the latter procedure would create legal complications for states which had already ratified, and further delay the Convention's entry into force. In fact, the Convention specifically provides for the preservation of "rights and obligations of States

Parties which arise from other agreements compatible with th[e] Convention and which do not affect the enjoyment by other States Parties of their rights or the performance of their obligations." U.N. Convention on the Law of the Sea, art. 311 (2). See also Chap. 4 nn. 515, 518 and accompanying text.

367. U.N. Department of Public Information, "Training Programme for Enterprise," p. 39.

368. Ibid. See U.N. Department of Public Information, "Preparatory Commission Adopts 'Understanding,'" p. 52.

369. U.N. Department of Public Information, "Training Programme for Enterprise," p. 39. "Questions discussed . . . included: the definition of 'serious harm'; contingency plans for emergencies; environmental impact statements; monitoring; inspection; the use of reference zones; emergency response; the rights of coastal states; and the rights of contractors and sponsoring states in connection with environmental effects of mining." *Oceans Policy News* (September 1990), p. 2.

370. U.N. Department of Public Information, "Training Programme for Enterprise," p. 39.

371. Ibid. See U.N. Department of Public Information, "Preparatory Commission Adopts 'Understanding,'" p. 51.

372. U.N. Department of Public Information, "Preparatory Commission Adopts 'Understanding,'" p. 52.

373. The final agreement provided for waiver of the $1 million annual fee, for periodic expenditures in accordance with a schedule to be agreed upon in 1991, and for the training of at least 12 persons in all aspects of seabed mining operations. *Oceans Policy News* (September 1990), pp. 1–2. It was also decided that the deadline for pioneer registrants to submit plans of work following the Convention's entry into force should be reviewed by the Group of Technical Experts in light of economic conditions, a decision which could lead to an extension of the deadline by which certifying states must become parties to the Convention. Ibid., p. 2. See Chap. 4 n. 485 and accompanying text.

374. See U.N. Department of Public Information, "Preparatory Commission Adopts 'Understanding,'" p. 51. The August 22 application certified that the 300,000 sq. km site in the Clarion-Clipperton Zone did not overlap with sites allocated to registered pioneer investors or reserved for potential pioneer applicants. *Oceans Policy News* (September 1990), p. 1. See also U.N. Office for Ocean Affairs and the Law of the Sea, "Seminar on the Current Status of Developments in Deep Sea-Bed Mining Technology" (New York, 18 and 19 August 1988)," *Law of the Sea Bulletin* 12 (December 1988): 62.

375. See U.N. Department of Public Information, "Preparatory Commission Welcomes China as Fifth Pioneer Investor," *UN Chronicle* 28 (June 1991): 33. The consortium, formed in 1987 and headquartered in Poland, included Bulgaria, Cuba, Czechoslovakia, Poland and the Soviet Union. Ibid. See also Hayashi, "Registration of the First Group," p. 4. Members of the consortium had indicated in 1988 that they intended to apply for registration. See *Oceans Policy News* (March 1988), p. 1.

376. Malone, "Who Needs the Sea Treaty?" p. 58. See also Malone, "The United States and the Law of the Sea," p. 799. It has even been argued that the unexpectedly large number of signatories to the Convention may have resulted from foreign displeasure at U.S. diplomatic tactics. "The United States sought to organize a united front of the Western developed countries to oppose the Convention; this has not materialized. . . . The so-called 'mini-treaty' approach has also not matured." Jonathan I. Charney, "The Law of the Deep Seabed Post UNCLOS III," *Oregon Law Review* 63 (1984): 51–52.

By 1989 there was increasing discussion of a protocol to modify the seabed mining provisions of the Convention. See n. 366 and text accompanying n. 396 above and below; and also Santa Cruz, p. 118. However, there was little reason to believe that a new round of negotiations would produce the kind of changes sought by the United States. See Edward L. Miles, "Preparing for UNCLOS IV?" *Ocean Development and International Law* 19 (1988): 423–26.

377. David L. Larson, "When Will the UN Convention on the Law of the Sea Come Into Effect?" *Ocean Development and International Law* 20 (1989): 175–76, 177–78.

378. Fiji (December 10, 1982); Zambia (March 7, 1983); Mexico (March 18, 1983); Jamaica (March 21, 1989); Council for Namibia (April 18, 1983); Ghana (June 7, 1983); the Bahamas (July 29, 1983); Belize (August 13, 1983); and Egypt (August 26, 1983).

379. Ivory Coast (March 26, 1984); the Philippines (May 8, 1984); Gambia (May 22, 1984); Cuba (August 15, 1984); Senegal (October 25, 1984); Sudan (January 23, 1985); St. Lucia (March 27, 1985); Togo (April 16, 1985); Tunisia (April 24, 1985); Bahrain (May 30, 1985); Iceland (June 21, 1985); Mali (July 16, 1985); Iraq (July 30, 1985); Guinea (September 6, 1985); Tanzania (September 30, 1985); Cameroon (November 19, 1985); Indonesia (February 3, 1986); Trinidad and Tobago (April 25, 1986); Kuwait (May 2, 1986); Yugoslavia (May 5, 1986); Nigeria (August 14, 1986); Guinea-Bissau (August 25, 1986); and Paraguay (September 26, 1986).

380. Democratic Yemen (July 21, 1987); Cape Verde (August 10, 1987); Sao Tome and Principe (November 3, 1987); Cyprus (December 12, 1988); Brazil (December 22, 1988); Antigua and Barbuda (February 2, 1989); Kenya (February 3, 1989); Zaire (February 17, 1989); Somalia (July 24, 1989); Oman (August 17, 1989); Botswana (May 2, 1990); Uganda (November 9, 1990); and Angola (December 6, 1990).

381. *Oceans Policy News* (May 1987), p. 5.

382. Walkate, p. 165. "This would mean that in the course of 1993 the first session of the Assembly of the Authority should be held and that [the Preparatory Commission] will cease to exist." Ibid.

383. Grenada ratified on April 25, 1991; the Federated States of Micronesia on April 29, 1991. An informal survey of states conducted in the late 1980s indicated that the Convention could be expected to enter into force in the early 1990s, but possibly without widespread support among all groups of states. Larson, "When Will the UN Convention on the Law of the Sea Come Into Effect?" p. 179.

384. See Donald Cameron Watt, "Background to the Law of the Sea," *Marine Policy* 11 (1987): 243.

385. See Mashayekhi, pp. 245–47. The Agreement Concerning Interim Arrangements Relating to Polymetallic Nodules of the Deep Sea-Bed (Interim Agreement), which defined procedures for exchanging technical information and for arbitrating disputed claims, included elements of a Private Industry Arbitration Agreement concluded earlier in the year among the four private consortia and the French national mining firm which identified site overlaps and permitted their resolution by voluntary dispute settlement procedures. Cohen, pp. 151–52. The "settlement greatly facilitated the final resolution at the universal level of the pending conflicts between some of these Western enterprises and the public enterprise of the Soviet Union in 1987." Hayashi, "Registration of the First Group," p. 11.

386. Larson, "A Definition of the Problem," p. 281. See Provisional Understanding Regarding Deep Seabed Matters, reprinted in *International Legal Materials* 23 (1984): 1354–65; and also Chap. 5 nn. 10, 64. States parties to the Understanding – Belgium, France, Italy, Japan, the Netherlands, West Germany, the United Kingdom, and the United States – agreed not to authorize seabed mining by their nationals in areas previously allocated pursuant to the industry's conflict resolution procedures. The Understanding prohibited any nodule exploitation prior to January 1, 1988.

387. See Kimball, "Turning Points," p. 378; and Malone, "The United States and the Law of the Sea," p. 807.

388. See n. 319 above and accompanying text, and Chap. 5 nn. 62–63, 65, 68, 213–14 and accompanying text; and also *Oceans Policy News* (May 1984), p. 5. Each of the acts stated that deep-sea mining is a freedom of the high seas, and provided for noninterference with sites claimed by "reciprocating" states. Many contained revenue-sharing provisions comparable to those of the U.S. Seabed Act. Mashayekhi, pp. 243–44.

France and Japan both submitted the formal applications of their state-run mining operations to the Preparatory Commission less than three weeks after the signing of the Provisional Understanding had resolved potential overlaps with the claims of the private consortia. Two of the U.S. consortia also attempted to register as pioneer investors by seeking Canadian sponsorship in 1983 but were discouraged by the United States, which "chided the companies for undermining a U.S. position that they had urged the government to take." Kimball, "Turning Points," pp. 374–75.

389. Pal and Kimball, p. 146.

390. Preparatory Commission, *Declaration Adopted by the Preparatory Commission on 30 August 1985* (U.N. Doc. LOS/PCN/72, September 2, 1985). See "UNCLOS: Prepara-

tory Commission Meets," *Environmental Policy and Law* 15 (1985): 80. The declaration was adopted without the support of Western seabed mining states, which reserved their positions on the question of the legal status of Part XI of the Convention pending its entry into force. Hayashi, "Registration of the First Group," p. 13. See Kirsch and Fraser, pp. 140–41. The declaration had been introduced the previous year; it was prompted by the Soviet delegation's receipt of a letter from one of the private consortia asserting "exclusive rights" to the nodule resources of the seabed area licensed to it under the U.S. Seabed Act, as well as by a general feeling of frustration over the delay in the implementation of Resolution II. "UNCLOS: Preparatory Commission Meets," p. 80. See also *Oceans Policy News* (September/October 1985), p. 1. In an effort to minimize divisiveness and to maintain an informal consensus, the resolution was adopted without a formal vote. Hayashi, "Registration of the First Group," pp. 13–14.

391. Statement of the representative of the United States before U.N. General Assembly, quoted in *Oceans Policy News* (January 1986), p. 2.

392. United States, *Note Dated 13 January 1986 From the United States Mission to the United Nations Addressed to the Secretary-General of the United Nations*, reprinted in *Law of the Sea Bulletin* 7 (1986): 74–86.

393. Preparatory Commission, *Declaration Adopted by the Preparatory Commission on 11 April 1986* (U.N. Doc. LOS/PCN/78, April 21, 1986). See *Oceans Policy News* (May 1986), pp. 3–4. Virtually all the opposing and abstaining votes were cast by Western developed states. See Kirsch and Fraser, p. 141.

394. U.N. General Assembly, 39th Session, Resolution 39/73, December 13, 1984. U.N. General Assembly, 40th Session, Resolution 40/63, December 10, 1985. U.N. General Assembly, 41st Session, Resolution 41/34, November 5, 1986. U.N. General Assembly, 42nd Session, Resolution 42/20, November 18, 1987. U.N. General Assembly, 43rd Session, Resolution 43/18, November 1, 1988. U.N. General Assembly, 44th Session, Resolution 44/26, November 20, 1989. U.N. General Assembly, 45th Session, Resolution 45/145, December 14, 1990. See Larson, "A Definition of the Problem," pp. 293–94.

395. Peru, Venezuela, West Germany, and the United Kingdom consistently abstained.

396. Statement of the representative of the United States before U.N. General Assembly, November 20, 1989, quoted in *Oceans Policy News*, Special Issue (August/September 1989), p. 2.

397. See Joseph P. Flanagan, "Manganese Nodules: A Huge Resource for the Future," *Sea Technology* 25 (August 1984): 19; and R.P. Anand, "UN Convention on the Law of the Sea and the United States," *Indian Journal of International Law* 24 (1984): 176. All 10 applications were submitted by the four U.S. consortia designated as pioneer investors by UNCLOS III. "All applications ultimately were determined to be in full compliance with NOAA regulations, thus establishing an initial priority of right for each consortium for the issuance of licenses for the respective areas applied for. However, each application area overlapped with the area of at least one other application filed either in the United States or in another country during the same period." National Oceanic and Atmospheric Administration, U.S. Department of Commerce, *Deep Seabed Mining: Report to Congress* (Washington: U.S. Government Printing Office, 1983), pp. 9–10. See n. 385 above and accompanying text.

398. Larson, "A Definition of the Problem," p. 278. Most estimates were somewhat lower: $1.3 billion to $1.8 billion. James M. Broadus, "Seabed Materials," *Science* 235 (February 20, 1987): 857. See also Hauser, pp. 23–25. Approximately three fourths of these costs are attributable to on-land processing.

399. *Oceans Policy News* (February 1986), p. 7. See also *Oceans Policy News* (June 1986), p. 5. Although the intention of filing for a commercial recovery permit prior to expiration of the 10-year period of the exploration license was formally maintained, further exploratory surveys were eliminated, some at-sea testing was postponed until the second five-year licensing phase, and proposed expenditure levels were reduced by two thirds. *Oceans Policy News* (November 1985), p. 7. *Oceans Policy News* (June 1986), p. 5.

400. Sanger, p. 167. "Demand growth for all major metals slowed down; mineral prices dropped to historic lows, as a result of structural and long-term cyclical factors, and remain volatile; land-based production costs have been reduced, thus putting seabed activities relatively higher on the cost curve." Kirsch and Fraser, p. 123. Nodules were suddenly

regarded as a "medium-to low-grade" mineral resource in relation to land-based sources of nickel and copper. G.P. Glasby, "The Three-Million-Tons-Per-Year Manganese Nodule 'Mine Site': An Optimistic Assumption?" *Marine Mining* 4 (1983): 73.

401. Kurt Shusterich, "Mining the Deep Seabed: A Complex and Innovative Industry," *Marine Policy* 6 (1982): 178. See also Sanger, p, 165; and Kimball, "Turning Points," pp. 376–77. Rather than proceeding with development of a nodule mining capability in anticipation of eventual improvement in mineral markets, the private consortia focused instead on legal and regulatory maneuvering in an effort to secure exclusive rights to preferred mining sites. Broadus and Hoagland, p. 542. See nn. 341–43, 391–92 above and accompanying text. "The consortia learned through their field work that (i) areal coverage by the nodules is patchier, (ii) the sea-floor terrain is less uniform and holds more obstacles, and (iii) weather conditions are more of a factor than initially expected." Broadus, p. 860 n. 41. See also Glasby, pp. 74, 76.

By 1986 the ocean mining industry was "in a deep depression." Joseph Haggin, "Guest Editorial: An External View of Marine Mining," *Marine Mining* 5 (1986): 352. See also Sanger, p. 5. Two years later, however, the director of one of the four consortia indicated that an improving nickel market could lead to renewed private interest in nodule mining activities by 1991. Statement of Jan Markussen before Special Commission 3, August 18, 1988, quoted in Walkate, p. 164. See also Miles, p. 427. By the end of the 1980s, copper and manganese markets were also improving, but prices were still too low to sustain deep-sea nodule mining operations. U.N. Department of Public Information, "Sea Law Commission Focuses on Pioneer Investor Duties," p. 35. In 1990 a leading spokesman for the seabed mining industry estimated that such market improvement would require at least 10 years, followed by another decade of preliminary mining activity before full nodule production was achieved. *Oceans Policy News* (May 1990), p. 6. Distrustful of bearish market projections from developed states which had once championed the seabed mining industry, and mindful of the registration of state-run pioneer investors by the Preparatory Commission, developing states tended to discount the long-term effects of falling market prices. Larson, "When Will the UN Convention on the Law of the Sea Come Into Effect?" p. 178. See also Miles, p. 427.

402. David W. Pasho, "Canada and Ocean Mining," *Marine Technology Society Journal* 19, no. 4 (1985): 26. See Broadus, p. 857; and Kimball, "Turning Points," p. 376. The French national mining operation was similarly affected. Daniel Spagni, Luke Giorghiou, and Michael Gibbons, "French Marine Technology Policy," *Marine Policy* 9 (1985): 289. See also Joseph Haggin, "Marine Mining to Improve Its Organization, Direction, Financing," *Chemical and Engineering News* (November 18, 1985), p. 63.

403. *Oceans Policy News* (September 1988), p. 2. Broadus and Hoagland, pp. 563–64, 570. Hayashi, "Registration of the First Group," p. 4. See Chap. 5 n. 195. Korea was expected to apply for registration of its mining site in the Clarion-Clipperton region by the mid–1990s. Telephone conversation with Moritaka Hayashi, U.N. Office for Ocean Affairs and the Law of the Sea, March 11, 1991.

The possibility that it may apply for registration "explains the insistence of Brazil on the provisions of Annex III, Article 5, which gave to developing states rights to purchase technology which are very similar to those enjoyed by the Enterprise." William C. Brewer, Jr., "Deep Seabed Mining: Can an Acceptable Regime Ever Be Found," *Ocean Development and International Law Journal* 11 (1982): 64 n. 53. See nn. 174, 230 above and accompanying text.

404. See Broadus and Hoagland, p. 570; Shusterich, p. 178; and John King Gamble, Jr., "Assessing the Reality of the Deep Seabed Regime," *San Diego Law Review* 22 (1985): 789. Other potential seabed mining states include Australia, Egypt, Indonesia, Kuwait, Spain, and the Pacific archipelagic states. Broadus and Hoagland, p. 570.

405. Jan Markussen, quoted in U.N. Office for Ocean Affairs and the Law of the Sea, "Seminar on the Current Status of Developments," p. 62. "The real question then is who will mine irrespective of commercial profitability?" Miles, p. 427. See Charles J. Johnson and Allen L. Clark, "Potential of Pacific Ocean Nodule, Crust, and Sulfide Mineral Deposits," *Natural Resources Forum* 9 (1985): 186.

406. Statement of Jan Markussen, quoted in Walkate, p. 164. See also U.N. Office for Ocean Affairs and the Law of the Sea, "Seminar on the Current Status of Developments," pp. 60, 62. The willingness of these states to proceed with nodule mining under unfavorable

economic conditions reflects the premium they place upon national access to seabed minerals. See Sebenius, p. 96.

A test project is necessary to resolve major engineering problems anticipated in regard to the long-term operation of nodule recovery systems. Pal and Kimball, p. 147. Such testing will require the further investment of several hundred million dollars — funding which commercial bankers have been unwilling to provide until the legal rights of mining operators are clarified. E.J. Langevad, "Exploitation of the Mineral Resources of the Oceans as Affected by the Provisions of the Convention on the Law of the Sea," *Natural Resources Forum* 7 (1983): 234. See Hauser, pp. 21–22.

407. Peter A. Rona, "Perpetual Seafloor Metal Factory," *Sea Frontiers* 30 (1984): 141. See Jack Stringer, "Mineral Treasure from the Galapagos Ridge," *NOAA* 12 (Winter 1982): 14. First discovered in the late 1970s, the deposits are "concentrated by seawater that circulates through fractures in the seafloor and assimilates heat and metals by contact with hot volcanic rocks at sea-floor-spreading centers." Rona, p. 141. The value of a single deposit off the coast of Ecuador has been estimated at $2 billion. Stringer, p. 14. During the 1980s, the U.S. Geological Survey identified significant new deposits at Gorda Ridge, less than 200 miles off the coast of the state of Washington.

408. Some deposits have been found to contain as much as 10 percent copper and 50 percent zinc. Conrad G. Welling, "Polymetallic Sulfides: An Industry Viewpoint," *Marine Technology Society Journal* 16, no. 3 (1982): 6. See also Stringer, p. 14. "The polymetallic sulfide deposits are highly concentrated compared to manganese nodules, approximately one thousand times the mass per unit area. . . . Nevertheless, the task remains immense." Welling, p. 5. The sulfides also contain "commercially important" amounts of cobalt and molybdenum. Alexander Malahoff, "A Comparison of the Massive Submarine Polymetallic Sulfides of the Galapagos Rift with Some Continental Deposits," *Marine Technology Society Journal* 16, no. 3 (1982): 44–45. It was nevertheless understood that polymetallic sulfide deposits would not be considered a true "ore body" until exploitation had been proven feasible through further exploration and technological advances. Welling, pp. 5–6.

409. Malahoff, p. 44. Some deposits are "growing" at the rate of one foot (30 cm) every two years. "A New Ocean Treasure?" *Environmental Policy and Law* 10 (1983): 57. "Observations by marine geologists . . . studying the geology of the active hydrothermal vents of the mid-ocean ridges now suggest that a time frame of decades or less is a . . . reasonable time necessary to form the polymetallic sulfide bodies." Malahoff, p. 44.

410. Johnson and Clark, pp. 182, 185. See Jaenicke, "Conflicts Between Mine Sites," pp. 514–15 n. 1. Crusts form to a thickness of several centimeters upon seabed rock formations, generally at shallower depths than manganese nodule deposits. Haggin, "Marine Mining to Improve," p. 65. The composition of the crusts has been measured at 0.47 percent nickel, 0.73 percent cobalt, and 23.06 percent manganese. Broadus, p. 857. In some areas the cobalt content of the crusts is as high as 2 percent. Haggin, "Marine Mining to Improve," p. 65. "Assuming realistic recovery rates, annual production of 1 million tons of wet crust from such a deposit would supply virtually all U.S. annual cobalt needs and a significant portion of demand for other metals." Joseph L. Ritchey, "Assessment of Cobalt-Rich Manganese Crust Resources on Horizon and S.P. Guyots, U.S. Exclusive Economic Zone," *Marine Mining* 6 (1987): 241.

411. Sanger, p. 195. Broadus, p. 858. The discovery of these mineral resources in 1981 and 1982 "had a profound impact on U.S. policy in the UNCLOS negotiations and contributed significantly to the Reagan Administration's distaste for the deep seabed mining provisions." Kimball, "Turning Points," p. 377.

412. Telephone conversation with LeRon Bielak, Chief of Staff, Program Analysis and Issues Management, Office of Strategic and International Minerals, Minerals Management Service, U.S. Department of the Interior, November 9, 1990. See n. 417 below. Commercial interest in known sulfide deposits was "not really big." Telephone conversation with Bob Paul, Regional Supervisor, Office of Resource Evaluation, Minerals Management Service, Department of the Interior, May 30, 1990. Sulfide deposits were found to be too small and were located in waters too deep to be economically attractive, and market demand for sulfide minerals was expected to be too weak until the 21st century. Telephone conversation with LeRon Bielak, November 9, 1990. See also Johnson and Clark, pp. 184–85.

In late 1990 Japanese mining interests had completed the presale application process for mining rights with respect to manganese crusts off the Hawaiian islands, but Japan remained uncertain whether to proceed with crust mining operations or with deep-sea nodule mining. The German seabed mining industry also expressed interest in the crust deposits. Telephone conversation with LeRon Bielak, November 9, 1990.

413. See Michael R. Molitor, "The U.S. Deep Seabed Mining Regulations: The Legal Basis for an Alternative Regime," *San Diego Law Review* 19 (1982): 611–12.

414. Presidential Proclamation No. 5030, March 10, 1983, reprinted in *U.S. Department of State Bulletin* 83 (June 1983): 71. See also statement of President Ronald W. Reagan, March 10, 1983, reprinted in ibid., pp. 70–71. Development of mineral resources within the zone was expected to proceed slowly, with initial licenses being issued for exploitation of placers and phosphorites. Haggin, "Marine Mining to Improve," p. 66.

415. Presidential Proclamation No. 5030, in *U.S. Department of State Bulletin* 83 (June 1983): 71 (emphasis added). "In this respect, the United States will recognize the rights of other states in the waters off their coasts, as reflected in the convention, so long as the rights and freedoms of the United States and others under international law are recognized by such coastal states." Ibid.

416. John D. Negroponte, Assistant Secretary of State for Oceans and International Environmental and Scientific Affairs, "Current Developments in the U.S. Oceans Policy," *U.S. Department of State Bulletin* 86 (September 1986): 85. See *Rest. 3rd, Restatement of the Foreign Relations Law of the United States*, Introductory Note to Part V. "The United States is now engaged in a deliberate, methodical process of promoting the universal application of rules of international law reflected in the non-seabed parts of the Convention." Negroponte, "Current Developments," p. 85. In 1988 President Reagan proclaimed a 12-mile U.S. territorial sea, within which foreign vessels are to enjoy rights of innocent passage, "[i]n accordance with international law, as reflected in the applicable provisions of the 1982 United Nations Convention on the Law of the Sea." Presidential Proclamation No. 5928, December 27, 1988, reprinted in *U.S. Department of State Bulletin* 89 (March 1989): 72.

417. See Chap. 4 n. 513; and also Broadus, p. 858. The U.S. Department of the Interior has formed a joint task force with the state of Hawaii to promote the mining of manganese crusts. However, the domestic ocean mining industry has refused to proceed under the bonus-bid leasing system established by the Department under the authority of the 1953 Act, seeking instead special legislation which would vest exclusive mining rights, with no "front-end" payments, in any mining operator conducting prospecting activities. *Oceans Policy News* (March 1984), p. 3. See *Ocean Science News* 29, no. 25 (July 7, 1987): 1–2. Such legislation was introduced in Congress in 1987. *Oceans Policy News* (May 1987), p. 12. See also Broadus, p. 858. The legislation would also address the issue of foreign access to seabed minerals within the U.S. exclusive economic zone. Telephone conversation with LeRon Bielak, November 9, 1990. Evidently, the U.S. seabed mining industry would like to preserve prime offshore mineral deposits for itself, as it has sought to do with respect to deep-sea nodules, even though it is unprepared to proceed with mining operations.

418. Telephone conversation with LeRon Bielak, November 9, 1990. See Welling, pp. 5–7. "[T]he estimated value of metals in the investigated [mid–Pacific] seamount areas significantly exceeds that in prime deep-water nodule areas." Peter Halbach and F.T. Manheim, "Potential of Cobalt and Other Metals in Ferromanganese Crusts in Seamounts of the Central Pacific Basin," *Marine Mining* 4 (1984): 333.

419. See Francisco Orrego Vicuña, "The Deep Seabed Mining Regime: Terms and Conditions for Its Renegotiation," *Ocean Development and International Law* 20 (1989): 532. The Bush administration "has been marking time in the apparent belief that the U.S. is already getting all it needs from the convention through its embodiment of customary international law in the areas it cares about." Richardson, "A Reassessment of U.S. Interests," p. 8.

420. Telephone conversation with Moritaka Hayashi, March 11, 1991. U.S. participation in these consultations was prompted by a realization that important U.S. oceanic interests might be better safeguarded under the non-seabed provisions of the Convention, rather than by any immediate interest in a renewed effort to promote national seabed mining operations. Ibid. See Chap. 6. There appeared to be general agreement that such a protocol would not

significantly alter the basic system of parallel access. Telephone conversation with Moritaka Hayashi, March 11, 1991.

Chapter 4

1. Statement of President Ronald W. Reagan, January 29, 1982, reprinted in *U.S. Department of State Bulletin* 82 (March 1982): 54. See text accompanying Chap. 3 n. 280.

2. Ambassador James L. Malone, "Law of the Sea and Oceans Policy," *U.S. Department of State Bulletin* 82 (October 1982): 49.

3. U.N. Convention on the Law of the Sea, Annex III, art. 5 (3) (a). "Aside from the controversial Brazil Clause, the heart of Article 5 . . . is the fairly straightforward requirement that miners and their suppliers offer to sell to the Enterprise any equipment or process which they may themselves intend to use and which cannot be bought on the open market. Most of the detail of the article consists of limitations on this basic requirement which are intended to make it tolerable to the developed countries, and provisions designed to satisfy the Group of 77's concern that the transfer requirements cannot easily be avoided." William C. Brewer, Jr., "Transfer of Mining Technology to the International Enterprise," Oceans Policy Study 2:4 (Charlottesville, Va.: Michie, 1980), p. 19.

4. U.N. Convention on the Law of the Sea, Annex III, art. 5 (3) (b), (c). "Very likely, on the advice of counsel, this assurance will become a standard provision on the back of purchase orders used by the operators." Brewer, "Transfer of Mining Technology," p. 20. The terms of the assurance would be negotiable between the mining operator and its supplier, although the Enterprise might be unable to enforce its claim as a third-party beneficiary in many national legal systems. Wolfgang Hauser, *The Legal Regime for Deep Seabed Mining Under the Law of the Sea Convention*, trans. Frances Bunce Dielmann (Deventer, Netherlands: Kluwer, 1983), pp. 107–9.

After acquiring the right to transfer the technology from its supplier, the mining operator would then negotiate its transfer to the Authority on "fair and reasonable commercial terms and conditions." Alternatively, the Enterprise may negotiate directly with the owner of the technology for its transfer. See Hauser, pp. 109, 111.

5. U.N. Convention on the Law of the Sea, Annex III, art. 5 (3) (e). The obligation to transfer exploration technology was apparently omitted from this provision inadvertently. *Oceans Policy News*, Special Report (March 1988), p. 5. Developing-state transferees, as contractors, could be brought before the Sea-Bed Disputes Chamber and held liable for damages in the event of unauthorized retransfer. See U.N. Convention on the Law of the Sea, art. 187 (c); Annex III, art. 22. Additional safeguards against disclosure and retransfer may be included in the terms and conditions of transfer and in the rules, regulations, and procedures of the Authority.

6. U.N. Convention on the Law of the Sea, Annex III, art. 5 (7). See Hauser, p. 113. The 10-year limitation applies only to the obligations specified in paragraph 3. The more general provisions of article 144 relating to personnel training by states parties would thus remain in effect. See n. 12 below and accompanying text.

Mandatory technology transfer may be invoked by the governing board of the Enterprise, which is subject to the directives of the Council. U.N. Convention on the Law of the Sea, arts. 162 (2) (i), 170 (2); Annex IV, art. 6 (f). See nn. 507, 683 below and accompanying text.

7. U.N. Convention on the Law of the Sea, art. 188 (2); Annex III, art. 5 (4). It was agreed that the meaning of "fair and reasonable commercial terms and conditions" would be determined by reference to comparable cases. See Hauser, p. 103. "To the extent that comparable cases are unexpectedly lacking, the compensation paid by the Enterprise is to be based upon the Contractor's development costs. Favorable to the investor is the fact that unsuccessful development efforts as well as basic research are included in this definition of development costs." Hauser, p. 104. For a discussion of "fair and reasonable commercial terms and conditions" from the U.S. point of view, see United States, "Reports of the United States Delegation to the Third United Nations Conference on the Law of the Sea," Occasional Paper No. 33, ed. Myron H. Nordquist and Choon-ho Park (Honolulu: Law of the Sea Institute, 1983), pp.

293–94. The adjective "commercial," which was added in order to ensure that the terms and conditions of transfer comport with those generally included in commercial licensing transactions, "may prove to be of great significance in the long run, since it affects not only the price but also the range of conditions and restrictions on use." Brewer, "Transfer of Mining Technology," p. 15. See also Hauser, pp. 101–5.

Commercial arbitration may be invoked by either party, and will be conducted in accordance with arbitration rules established by the U.N. Commission on International Trade Law (UNCITRAL), "or such other arbitration rules as may be prescribed in the rules, regulations and procedures of the Authority." U.N. Convention on the Law of the Sea, Annex III, art. 5 (4). If the arbitral panel determines that the offer is not within the range of "fair and reasonable commercial terms and conditions," the contractor will have 45 days in which to submit a satisfactory revised offer; thereafter, the Authority may seek to invoke penalties, as in the case of other contractual violations. U.N. Convention on the Law of the Sea, Annex III, art. 5 (4). See nn. 341–46 below and accompanying text. The 1976 UNCITRAL arbitration rules have served well in the billion-dollar Iran-U.S. claims proceeding. See Stewart A. Baker and Mark D. Davis, "Arbitral Proceedings Under the UNCITRAL Rules – The Experience of the Iran-United States Claims Tribunal," *George Washington University Journal of International Law and Economics* 23 (1989): 267–347. A vocal critic of the Convention has claimed that the Authority's adoption of its own arbitration rules would give rise to a "considerable" possibility of "an unfair, politicized procedure." Representative John Breaux, "The Diminishing Prospects for an Acceptable Law of the Sea Treaty," *Virginia Journal of International Law* 19 (1979): 292. U.S. consent would be necessary for the adoption of any such rules, however. See n. 463 below and accompanying text. Breaux, now a Senator, has been regarded by the U.S. seabed mining industry as "their man on Capitol Hill." Finn Laursen, *Superpower at Sea* (New York: Praeger, 1983), p. 138. See Chap. 3 n. 256.

8. See Jon Van Dyke and David L. Teichmann, "Transfer of Seabed Mining Technology: A Stumbling Block of U.S. Ratification of the Convention on the Law of the Sea?" in *The Law of the Sea and Ocean Industry: New Opportunities and Restraints*, ed. Douglas M. Johnston and Norman G. Letalik (Honolulu: Law of the Sea Institute, 1983), pp. 532–33. Cf. U.N. Convention on the Law of the Sea, art. 187 (c); Annex III, art. 17 (1) (b) (x), (xi), (xiii). See nn. 315, 597–603 below and accompanying text. "Here the [Preparatory Commission] will seriously affect the acceptability of the Convention in all industrialized states." Ann L. Hollick, "Opportunities to Correct Defects in the Seabed Mining Regime," in *United States Law of the Sea Policy: Options for the Future* (Oceans Policy Study Series, no. 6, November 1985), ed. Sharon K. Shutler (New York: Oceana Publications, 1985), p. 49. Draft rules being prepared in the Commission would require that the Enterprise, once it has demonstrated that needed technology is unavailable on the open market, seek voluntary assistance from all contractors before the Council could invoke mandatory technology transfer. Satya N. Nandan, "The 1982 UN Convention on the Law of the Sea: At a Crossroad," *Ocean Development and International Law* 20 (1989): 517.

9. UNCLOS III, Final Act, Annex I, Resolution II: "Governing Preparatory Investment in Pioneer Activities Relating to Polymetallic Nodules," para. 12 (a) (ii), (iii), reprinted in *International Legal Materials* 21 (1982): 1257. See Chap. 3 n. 373. Surprisingly, a 1982 U.S. proposal would have required certifying states to ensure that pioneer investors "perform" rather than merely "undertake . . . to perform" the treaty's technology transfer obligations. UNCLOS III, *Belgium, Federal Republic of Germany, Italy, United Kingdom of Great Britain and Northern Ireland and United States of America: Amendments* (U.N. Doc. A/CONF.62/L.122, April 13, 1982), Resolution II, proposed para. 12 (b), in *Official Records*, UNCLOS III, vol. XVI, p. 232.

10. U.N. Convention on the Law of the Sea, Annex III, art. 5 (8).

11. Ibid., Annex III, art. 5 (6).

12. Ibid., art. 144 (2). See Van Dyke and Teichmann, p. 528. Implementation of article 144 will be defined more specifically in the rules, regulations, and procedures of the Authority. U.N. Convention on the Law of the Sea, Annex III, art. 17 (1) (b) (xi). See Hauser, pp. 111–12.

13. See, e.g., U.N. Convention on the Law of the Sea, Annex III, arts. 5 (3) (d), (5); 15. Part XIV of the Convention, entitled "Development and Transfer of Marine Technology," is entirely comprised of such general obligations. "Generally speaking, watering down is what

happened to Part XIV." Boleslaw A. Boczek, "Transfer of Technology and UNCLOS III Draft Convention," in *The Law of the Sea and Ocean Industry*, ed. Johnston and Letalik, p. 502. Because of their generality, such provisions could be violated only by flagrantly un-cooperative conduct — i.e., by refusal to provide even token assistance.

14. Each contractor is required to make available to the Authority "all data which are both necessary and relevant to the effective exercise of the powers and functions of the principal organs of the Authority in respect of the area covered by the plan of work." U.N. Convention on the Law of the Sea, Annex III, art. 14 (1). See also n. 519 below. Requirements for disclosure of such data are not unusual. "As it is generally known, industrial and developing countries alike require that all data on resource surveys or prospecting be divulged to the relevant authorities." UNCLOS VII, *Economic Implications of Sea-Bed Mineral Development in the International Area: Summary and Conclusions of the Report of the Secretary-General* (U.N. Doc. A/CONF.62/25, May 22, 1974), reprinted in *Official Records*, UNCLOS III, vol. III, p. 27. See also Van Dyke and Teichmann, p. 531.

The regulatory regime established pursuant to the Seabed Act of 1980 includes similar disclosure obligations. See n. 523 below, and n. 670 below and accompanying text. In fact, public disclosure of "confidential" proprietary data may be required unless the seabed mining operator specifically requests — and is able to justify — nondisclosure. See Deep Seabed Mining Regulations for Exploration Licenses, 15 CFR §970.212 (a); and Deep Seabed Mining Regulations for Commercial Recovery Permits, 15 CFR §971.802.

15. Hauser, p. 113 n. 348.

16. U.N. Convention on the Law of the Sea, art. 181 (1), (2); Annex III, art. 14 (3). It is thus inaccurate to claim, as has one commentator, that developing and Eastern-bloc states would have "indirect access" to transferred technology at below-market prices "through their participation in and control of the Authority." Guy M. Hicks, "The Law of the Sea Treaty: A Review of the Issues," *Journal of Social, Political and Economic Studies* 6 (1981): 111.

17. U.N. Convention on the Law of the Sea, arts. 163 (8), 168 (2); Annex III, art. 22. See n. 339 below and accompanying text. The Sea-Bed Disputes Chamber would have jurisdiction to adjudicate disputes regarding the Authority's compliance with such provisions. U.N. Convention on the Law of the Sea, art. 187 (b) (i). Although the Convention does not explicitly impose liability upon states parties for unauthorized disclosure by — or to — their nationals, it is likely that they too would be financially accountable for such disclosure. See n. 5 above.

18. The Secretary-General, his staff, and members of the commissions are not permitted to have any "financial interest in any activity relating to exploration and exploitation in the Area." U.N. Convention on the Law of the Sea, arts. 168 (2), 163 (8).

19. Cf. ibid., art. 168 (3); and Crimes and Criminal Procedure, 18 U.S.C. §1905.

20. U.N. Convention on the Law of the Sea, art. 302. Although article 302 does not refer expressly to technology transfer, the transfer of physical equipment without relevant "information" concerning its operation would be of little value to the transferee. "This Article stems from an American proposal and its importance was deemed to be linked to the transfer of technology within the provisions on deep sea mining." R.W.G. de Muralt, "The Military Aspects of the UN Law of the Sea Convention," *Netherlands International Law Review* 32 (1985): 97. It is likely that notification of any national-security-related technology would be included in the application for approval of a plan of work. *Oceans Policy News*, Special Report (March 1988), p. 5.

21. UNCLOS III, *The U.S. Proposals for Amendments to the Draft Convention on the Law of the Sea* (U.N. Doc. A/CONF.62/WG.21/Informal Paper 18, 1982), proposed art. 163 (8). The change amounted to a drafting correction to reflect the proposed division of the Legal and Technical Commission into two subcommissions. See ibid., proposed art. 163 (1) (a).

22. Ibid., proposed art. 155. The Green Book did not, however, propose any changes to article 144 itself.

23. Ibid., Annex IV, proposed art. 11 (bis) (2). Members of the committee would also have been required to "take appropriate measures, consistent with national law, to prevent persons subject to their jurisdiction from engaging in a concerted refusal to supply technology to the Enterprise on commercial terms and conditions." Ibid., Annex IV, proposed art. 11 (bis) (3).

24. Ibid., Annex IV, proposed art. 11 (bis) (4).

25. UNCLOS III, *Belgium, France, Federal Republic of Germany, Italy, Japan, United Kingdom of Great Britain and Northern Ireland and United States of America: Amendments* (U.N. Doc. A/CONF.62/L.121, April 13, 1982), Annex III, proposed arts. 3 (4) (b) (iv), 5 (3) (b), reprinted in *Official Records*, UNCLOS III, vol. XVI, pp. 228, 229.

26. Ibid., Annex III, proposed art. 5 (3) (a), (d), (e), in *Official Records*, UNCLOS III, vol. XVI, p. 229.

27. Ibid., Annex III, proposed art. 5 (5), in *Official Records*, UNCLOS III, vol. XVI, p. 229.

28. Malone, "Law of the Sea and Oceans Policy," p. 49.

29. Doug Bandow, "UNCLOS III: A Flawed Treaty," *San Diego Law Review* 19 (1982): 483. The technology transfer provisions "have aroused the greatest opposition of all the convention's provisions. In large part these objections draw force from the principle that technology should be sold only by the decision of the owner, and then on freely negotiated terms." James K. Sebenius, *Negotiating the Law of the Sea* (Cambridge, Mass.: Harvard University Press, 1984), p. 99.

30. "Since the state under general international law has lawmaking power with respect to its nationals, it may, in a treaty concluded with another state, dispose of the rights, especially of the property rights of its nationals." Hans Kelsen, *Principles of International Law* (New York: Rinehart, 1952), pp. 145–46.

31. J.F. Garner, "Introduction: Compensation for the Compulsory Acquisition of Land—Some General Observations," in *Compensation for Compulsory Purchase*, ed. J.F. Garner (London: United Kingdom National Committee of Comparative Law, 1975), pp. 3–4. The modern principle requiring payment of "appropriate compensation" to the owner where such appropriation occurs has been endorsed by the United Nations General Assembly. U.N. General Assembly, Resolution 1803, December 14, 1962, para. 4. See also American Bar Association, "Natural Resources of the Sea: Report of Section of Natural Resources Law," in *Status Report on the Law of the Sea Conference*, U.S. 93rd Congress, 1st Session, Senate, Committee on Interior and Insular Affairs, Subcommittee on Minerals, Materials and Fuels (Washington: U.S. Government Printing Office, 1973), p. 668.

In the United States the requirement of "just compensation" was included in the Fifth Amendment to the Constitution as a limitation upon the well-recognized governmental power of eminent domain. See n. 70 below and accompanying text; and also Susan Crabtree, "Public Use in Eminent Domain: Are There Limits After *Oakland Raiders* and *Poletown?*" *California Western Law Review* 20 (1983): 84. Subject to this constitutional limitation, eminent domain, as an attribute of sovereignty, may be delegated in whole or in part by the state. See People ex rel. Burhans et al. v. City of New York, *North Eastern Reporter* 92 (1910): 19–20. The power extends to intangible as well as tangible assets. See also American Law Institute, *Restatement 3rd, Restatement of the Foreign Relations Law of the United States*, §712, comment (e).

32. Thomas E. Ciotti, "A Comparative Overview of the Patenting Process" (paper presented at State Bar of California Conference on The Internationalization of Technology and the Law, San Francisco, January 19, 1990). See Friedrich-Karl Beier and Joseph Straus, "The Patent System and Its Informational Function—Yesterday and Today," *IIC International Review of Industrial Property and Copyright Law* 8 (1977): 393. Until the period of patent protection expires, patents function as a form of government-supported monopoly. See Helen E. Weidner, "The United States and North-South Technology Transfer: Some Practical and Legal Obstacles," *Wisconsin International Law Journal* (1983): 221.

33. See Yoel Tsur, "Compulsory Licensing in Israel Patents Law," *IIC International Review of Industrial Property and Copyright Law* 16 (1985): 541; and Ciotti. Some developing states require compulsory licensing of unused inventions after three years. Hans Peter Kunz-Hallstein, "Patent Protection, Transfer of Technology and Developing Countries—A Survey of the Present Situation," *IIC International Review of Industrial Property and Copyright Law* 6 (1975): 437–38.

34. See Dable Grain Shovel Company v. Flint, *United States Reports* 137 (1890): 43; and also David Silverstein, "Proprietary Protection for Deepsea Mining Technology in Return for Technology Transfer: New Approach to the Seabeds Controversy," *Journal of the Patent Office Society* 60 (1978): 164. "It has been the general practice, when inventions have been

made which are desirable for government use, either for the government to purchase them from the inventors, and use them as secrets of the proper department, or, if a patent is granted, to pay the patentee a fair compensation for their use." James v. Campbell, *United States Reports* 104 (1881): 358.

U.S. law provides patent protection for a period of 17 years for most inventions. Patent Law, 35 U.S.C. §154. "Obviously, . . . the length of time already consumed in the development process in the Law of the Sea negotiations . . . has seriously cut into the protected time of early patents in this field." Statement of Marne Dubs in *Deep Seabed Mining*, U.S. 95th Congress, 1st Session, House of Representatives, Committee on Merchant Marine and Fisheries, Subcommittee on Oceanography (Washington: U.S. Government Printing Office, 1977), p. 70.

35. Ross D. Eckert, "Exploitation of Deep Ocean Minerals: Regulatory Mechanisms and United States Policy," *Journal of Law and Economics* 17 (1974): 147.

36. Silverstein, pp. 150–59. See generally Dieter Stauder, "Patent Protection in Extraterritorial Areas (Continental Shelf, High Seas, Air Space, and Outer Space)," *IIC International Review of Industrial Property and Copyright Law* 7 (1976): 470–79. Cf. Patent Law, 35 U.S.C. §§100 (c), 272. Moreover, U.S. corporations generally do not apply for patent protection within developing states. Homer O. Blair, "Technology Transfer as an Issue in North/South Negotiations," *Vanderbilt Journal of Transnational Law* 14 (1981): 321–23. When they do, some form of technology transfer is often required as a condition of such protection. Statement of Marne Dubs in *Deep Seabed Mining*, U.S. House of Representatives, p. 70. "Even assuming that developed-country coastal states were both willing and able to extend their patent laws to cover their exclusive economic zones, their continental shelf, and their nationals anywhere in the world, [other entities] . . . would still be able to use 'pirated' ocean mining technology on a large portion of the seabed." Silverstein, pp. 159–60.

37. United Nations, *Draft United Nations Convention on the International Seabed Area* (U.N. Doc. A/AC.138/25, August 3, 1970), art. 41 (a).

38. Secretary of State Henry Kissinger, address before University of Wisconsin Institute of World Affairs, Milwaukee, July 14, 1975, reprinted in *U.S. Department of State Bulletin* 73 (1975): 155. See Chap. 3 n. 54. For a discussion of Kissinger's role in the development of the technology transfer provisions, see William C. Brewer, Jr., "Deep Seabed Mining: Can an Acceptable Regime Ever Be Found?" *Ocean Development and International Law Journal* 11 (1982): 64–65 n. 55.

39. Secretary of State Henry Kissinger, "The Law of the Sea: A Test of International Cooperation," *U.S. Department of State Bulletin* 74 (1976): 540.

40. For a discussion of the traditional position of developed states on the issue of technology transfer, see remarks by Gabriel M. Wilner, in *Proceedings of the 70th Annual Meeting*, American Society of International Law (1976), p. 6. For a discussion of the concerns of developing states, see Brewer, "Deep Seabed Mining," p. 46. "Few diplomats from industrial states could have anticipated this importance which developing states attached to this obligation, their difficulty in believing that the technology could be purchased or developed on the open market, and their insistence on treating the obligation as a condition precedent to mining." Brewer, "Deep Seabed Mining," p. 47. See Chap. 2 nn. 217–18 and accompanying text; and also Boczek, p. 501.

41. See Silverstein, p. 139 n. 12; and Tullio Treves, "The EEC and the Law of the Sea: How Close to One Voice?" *Ocean Development and International Law Journal* 12 (1983): 179–80. In 1983 the U.S. Ambassador indicated that the President "stands ready to help" with training and technological assistance for developing-state seabed mining operators. Speech by Ambassador James L. Malone, Montego Bay, January 6, 1983, reprinted in *Environmental Policy and Law* 10 (1983): 97. See also Brewer, "Deep Seabed Mining," p. 47; and Richard P. Brown, Jr., "Changing the Rules: International Law and the Developing Countries: The ABA Workshops of 1977," *International Lawyer* 12 (1978): 274.

43. William Sprague Barnes, "Technology Transfer Rules: A Study in Comparative Law," *Boston College International and Comparative Law Review* 3 (1979): 10.

44. Carl Q. Christol, "The Common Heritage of Mankind Provision in the 1979 Agreement Governing the Activities of States on the Moon and Other Celestial Bodies," *International Lawyer* 14 (1980): 453. "When . . . a nation is in absolute want of provisions, it may

oblige its neighbors who have more than they want for themselves, to deliver them up at a just price, or even to take them by force, if they will not sell them." M.D. Vattel, *The Law of Nations,* trans. Joseph Chitty, orig. pub. 1758 (Northampton, Mass.: Thomas M. Pomroy, 1805), p. 243.

45. Gerhard Mally, "Technology Transfer Controls," *U.S. Department of State Bulletin* 82 (November 1982): 53. "The United States has much to offer other nations in providing more effective techniques for tapping the sea's resources and will need their help in implementing international programs to permit all nations to use the sea to their benefit." U.S. Commission on Marine Science, Engineering, and Resources, *Our Nation and the Sea: A Plan for National Action* (Washington: U.S. Government Printing Office, 1969), p. 86.

46. Weidner, pp. 218-19. See Barnes, pp. 8-9; and also Robert Goldschneider, "The Technology Transfer Process: A Vehicle for Continuity and Change," *Vanderbilt Journal of Transnational Law* 14 (1981): 265. "The multinational corporation supplying the technology can use its powerful bargaining position to demand that the purchaser pay premium prices. The developing country does not have the option, usually envisioned in free market economic theory, to reject one seller's bargain and turn to another seller." Weidner, p. 218.

47. James L. Malone, "U.S. Participation in the Law of the Sea Conference," *U.S. Department of State Bulletin* 82 (May 1982): 62. See Kent M. Keith, "Laws Affecting the Development of Ocean Resources in Hawaii," *University of Hawaii Law Review* 4 (1982): 288. Yet, at a 1969 plenary session of the U.N. General Assembly, the United States had asserted that there was "simply no possibility of one country or group of countries having exclusive use of seabed exploitation technology, any more than one country or group of countries has exclusive use of the technology of exploiting resources on land." Statement of Ambassador Christopher Phillips, December 15, 1969, in *U.S. Department of State Bulletin* 62 (1970): 92.

48. Eckert, "Exploitation of Deep Ocean Minerals," p. 150.

49. Ann L. Hollick, "The Third UN Conference on the Law of the Sea: Caracas Review," in *The Law of the Sea: U.S. Interests and Alternatives,* ed. Ryan C. Amacher and Richard James Sweeney (Washington: American Enterprise Institute for Public Policy Research, 1976), p. 128.

50. Secretary of State Henry Kissinger, "Secretary Kissinger Discusses U.S. Position on the Law of the Sea Conference," *U.S. Department of State Bulletin* 75 (1976): 398.

51. Roger A. Geddes, "The Future of United States Deep Seabed Mining: Still in the Hands of Congress," *San Diego Law Review* 19 (1982): 618 n. 38. See Chap. 5 n. 171 and accompanying text; Dennis W. Arrow, "The Proposed Regime for the Unilateral Exploitation of Deep Seabed Mineral Resources by the United States," *Harvard International Law Journal* 21 (1980): 340; and also Virginia A. Pruitt, "Unilateral Deep Seabed Mining and Environmental Standards: A Risky Venture," *Brooklyn Journal of International Law* 8 (1982): 357 n. 74.

52. As development of seabed mining technology accelerated in other states, U.S.-based consortia virtually abandoned their nodule mining efforts. See Chap. 3 nn. 401-6 and accompanying text. As one federal regulator admitted, the U.S. seabed mining industry was now "in an infant stage compared to Europe and Japan." Telephone conversation with LeRon Bielak, Chief of Staff, Program Analysis and Issues Management, Office of Strategic and International Minerals, Minerals Management Service, U.S. Department of the Interior, November 9, 1990. See also J.A. Walkate, "Developments in Special Commission 3 of the Preparatory Commission for the International Sea-Bed Authority and for the International Tribunal for the Law of the Sea: Drafting the Future Deep Seabed Mining Code," *Netherlands International Law Review* 36 (1989): 163. "Knowledge and expertise invested in such extensive projects cannot be veiled in secrecy for long." F.M. Auburn, "The Deep Seabed Hard Mineral Resources Bill," *San Diego Law Review* 9 (1972): 492.

53. Tom Alexander, "The Reaganites' Misadventure at Sea," *Fortune* 106 (August 23, 1982): 144. See Lance N. Antrim and James K. Sebenius, "Incentives for Ocean Mining under the Convention," in *Law of the Sea: U.S. Policy Dilemma,* ed. Bernard H. Oxman, with Charles L.O. Buderi and David D. Caron (San Francisco: ICS, 1983), p. 92; E.J. Langevad, "Exploitation of the Mineral Resources of the Oceans as Affected by the Provisions of the Convention on the Law of the Sea," *Natural Resources Forum* 7 (1983): 234; and also U.N. Office for Ocean Affairs and the Law of the Sea, "Seminar on the Current Status of Developments in Deep Sea-Bed Mining Technology (New York, 18 and 19 August 1988)," *Law of the Sea*

Bulletin 12 (December 1988): 63. Economies of scale favor the sale of private seabed mining technology to the Enterprise and developing states, since multinational consortia would thereby be able to lower their per-unit costs of production. Several private firms have, in fact, already offered to provide the Enterprise with seabed mining systems. Bernard H. Oxman, "The Third United Nations Conference on the Law of the Sea," *American Journal of International Law* 76 (1982): 11 n. 27. The problem for the Enterprise "will be less that of prying technology loose from reluctant owners than of choosing from among the many eagerly proffered systems." Sebenius, p. 99. Nonetheless, it remains possible that some transfer of training "technology" might initially be required on a limited basis, if only to ensure proper use of seabed mining equipment. See UNCLOS III, *Costs of the Authority and Contractual Means of Financing Its Activities* (U.N. Doc. A/CONF.62/C.1/L.19, May 18, 1977), para. 99, reprinted in *Official Records*, UNCLOS III, vol. VII, p. 73.

54. "Taking a Dive," *Economist* 278 (March 21, 1981): 85. See Chap. 5 nn. 193–95 and accompanying text; and also Najmul Hassan, "Staking Its Claim: India Seeks Mineral Riches on Sea Floor," *Los Angeles Times*, September 15, 1983, part I-B, p. 4. "A developing country with the means to buy it will experience no problems in procuring technology, now or in the future." Statement of Jan Markussen before Special Commission 3, August 18, 1988, quoted in Walkate, p. 163.

55. Frank S. Brokaw and W. Scott Burke, "Ideology and the Law of the Sea," in *Law of the Sea: U.S. Policy Dilemma*, ed. Oxman et al., p. 55.

56. Geddes, p. 615 n. 15. See also Eckert, "Exploitation of Deep Ocean Minerals," p. 149.

57. UNCLOS III, *Description of Some Types of Marine Technology and Possible Methods for Their Transfer: Report of the Secretary-General* (U.N. Doc. A/CONF.62/C.3/L.22, February 27, 1975), para. 9, reprinted in *Official Records*, UNCLOS III, vol. IV, p. 203. The likelihood that seabed mining operators would sell their technology to the Enterprise and developing states has been conceded even by opponents of the Convention. See, e.g., Breaux, "The Diminishing Prospects," pp. 263–64. Moreover, because under the parallel system the Enterprise will acquire one reserved site for each site mined by a state-sponsored operator, the Enterprise will, to some extent, be able to select for initial exploitation those sites for which seabed mining technology is most readily available. The Enterprise would also have an incentive to avoid time-consuming litigation over the transfer of scarcer technology from reluctant, and possibly unreliable, private mining operators.

58. Auburn, p. 492. Technology transfer has been shown to promote additional voluntary sales of products by licensors to licensees. Goldschneider, p. 259.

59. Geddes, p. 627.

60. David L. Larson, "The Reagan Administration and the Law of the Sea," *Ocean Development and International Law Journal* 11 (1982): 310. Brokaw and Burke, p. 53.

61. Brokaw and Burke, p. 53; and Larson, "The Reagan Administration," p. 317. The industry has also objected to the Convention's definition of technology as "much more inclusive than anything used in current commercial practice," and has expressed concern about the security of proprietary information and technology. Larson, "The Reagan Administration," p. 311.

62. William J. Roche, Vice President, Secretary and General Counsel, Texas Instruments Incorporated, quoted in R. Brown, p. 274. It should be noted that the Enterprise is required to operate on sound commercial principles. See n. 709 below and accompanying text.

63. Leigh S. Ratiner and Rebecca L. Wright, "United States Ocean Mineral Resource Interests and the United Nations Conference on the Law of the Sea," *Natural Resources Lawyer* 6 (1973): 29.

64. Robert J. Radway, "Comparative Evolution of Technology Transfer Policies in Latin America: The Practical Realities," *Denver Journal of International Law and Policy* 9 (1980): 199. See Barnes, pp. 1–2, 9–10. "Investment is still continuing actively in the developing world despite mandatory technology transfer requirements, because the investments make economic sense even with these requirements." Van Dyke and Teichmann, p. 539. In a typical contract, concluded in 1969 between the Aluminum Company of America (ALCOA) and Indonesia, the former agreed "to make available to each operating company the secret processes and technical information relating to its activities." *Costs of the Authority* (U.N. Doc. A/CONF.62/C.1/L.19), in *Official Records*, UNCLOS III, vol. VII, p. 66 n. 24. China requires

licensors to agree to disclose future technological developments unconditionally and without charge. Bernard Detoit, "Joint Ventures and Intellectual Property," *IIC International Review of Industrial Property and Copyright Law* 20 (1989): 456.

65. T. Alexander, pp. 143–44. See also Weidner, pp. 207–8. Unlike the technology transfer provisions increasingly included in concession agreements, the Convention contains safeguards against retransfers to third parties. Ronald D. Katz, "A Method for Evaluating the Deep Seabed Mining Provisions of the Law of the Sea Treaty," *Yale Journal of World Public Order* 7 (1980): 126. The Convention also lacks other burdensome requirements commonly contained in such agreements, such as "infrastructure" clauses, "buy local" requirements, and extensive training of indigenous personnel – in Indonesia, for example, multinational mining operators are required to employ a work force comprised of 75 percent Indonesian nationals. Katz, pp. 126–27.

66. Jeffrey Lee Gertler and Paul Wayne, "Synopsis: Recent Developments in the Law of the Sea 1979–80," *San Diego Law Review* 18 (1981): 554. See Bernard H. Oxman, "The Third United Nations Conference on the Law of the Sea: The Ninth Session (1980)," *American Journal of International Law* 75 (1981): 223. Annex III of the Convention only requires the transfer of technology which the operator "uses in carrying out activities *in the Area* under the contract." U.N. Convention on the Law of the Sea, Annex III, art. 5 (3) (a) (emphasis added). If the Authority is unable to acquire "appropriate" processing technology, it may convene a group of seabed mining states to "consult together and . . . take effective measures to ensure that such technology is made available to the Enterprise on fair and reasonable commercial terms and conditions." U.N. Convention on the Law of the Sea, Annex III, art. 5 (5). See Hauser, pp. 113–14.

67. Statement of Theodore Kronmiller, Deputy Assistant Secretary of State for Oceans and Fisheries Affairs, September 15, 1982, in *Law of the Sea Negotiations*, U.S. 97th Congress, 2nd Session, Senate, Committee on Foreign Relations, Subcommittee on Arms Control, Oceans, International Operations, and Environment (Washington: U.S. Government Printing Office, 1983), p. 3.

68. See Antrim and Sebenius, p. 92. It is not even clear that sanctions could be imposed upon third-party owners refusing to transfer technology. See Hauser, pp. 107–8. Nor would suspension or termination of existing rights be likely where the contractor failed to obtain rights to the technology from third-party owners, in the absence of an outright refusal to seek such rights. Hauser, p. 110 n. 338. Where the contractor "exercises effective control over" the owner of the technology in question, failure to acquire the technology "shall be considered relevant to the contractor's qualification for any subsequent application for approval of a plan of work." U.N. Convention on the Law of the Sea, Annex III, art. 5 (3) (c).

69. Oxman, "The Ninth Session," p. 224. See Brewer, "Transfer of Mining Technology," pp. 15, 18. "For the individual mining company, . . . it probably makes no basic difference whether the required technology transfer is to the Authority or to a developing country." Hauser, p. 113. Most developing states are more likely to be interested in acquiring the technology needed to exploit and manage resources within their coastal waters than in undertaking large-scale deep-sea mining operations. See Hauser, p. 112; Brewer, "Deep Seabed Mining," p. 45; and also Elliot L. Richardson, "Law in the Making: A Universal Regime for Deep Seabed Mining?" *New York State Bar Journal* 53 (1981): 445.

70. See Arrow, "The Proposed Regime," p. 402. Cf. U.S. Constitution, Fifth Amendment. See *Rest. 3rd, Restatement of the Foreign Relations Law of the United States*, §712, comment (c), (d). "[U]nder *no* circumstance will technology have to be transferred free of charge." Walkate, p. 161 (emphasis in orig.). Because the technology transfer obligations are not truly condemnatory, but are rather in the nature of contractual conditions of access between the Authority and seabed mining operators, it is unlikely that payment of such compensation would even be constitutionally required under similar circumstances under U.S. law. At least one economist nevertheless has continued to claim that the Enterprise would be able to acquire seabed mining technology "on extremely soft terms." Federico Foders, "International Organizations and Ocean Use: The Case of Deep-Sea Mining," *Ocean Development and International Law* 20 (1989): 520.

71. Radway, p. 209. See also Wayne R. Smith, "Law of the Sea Treaty: Report on the Enterprise," *New York Law School Journal of International and Comparative Law* 3 (1981): 67.

72. *Description of Some Type of Marine Technology* (U.N. Doc. A/CONF.62/C.3/ L.22), para. 3, in *Official Records*, UNCLOS III, vol. IV, p. 202. See nn. 10, 61, 66 above and accompanying text; and also Van Dyke and Teichmann, pp. 535–36.

73. See Hauser, p. 115. "Those who think the United States did not sign the . . . Convention because of technology transfer have overblown the role of this one issue." Statement of Brian Hoyle, Director, Office of Oceans Law and Policy, U.S. Department of State, in *Consensus and Confrontation: The United States and the Law of the Sea Convention*, ed. Jon Van Dyke (Honolulu: Law of the Sea Institute, 1985), p. 263. "The issue . . . was a major concern only because President Reagan and Mrs. Thatcher chose to make it so in order to build up a plausible list of objections to a treaty they had decided not to sign." Clyde Sanger, *Ordering the Oceans: The Making of the Law of the Sea* (London: Zed Books, 1986), p. 178.

74. See statement of Theodore Kronmiller, September 15, 1982, in *Law of the Sea Negotiations*, U.S. Senate, p. 3; Gary C. Hufbauer and George N. Carlson, "United States Policy Toward the Transfer of Proprietary Technology: Licenses, Taxes, and Finance," *Vanderbilt Journal of Transnational Law* 14 (1981): 341; and also Larson, "The Reagan Administration," p. 301. One Senator apparently believed that the Convention would permit the Soviet Union to invoke the technology transfer provisions with respect to "certain technological things that have a national defense interest that are related to submarines and other types of activity under water." Statement of Senator Larry Pressler before U.S. Senate, March 8, 1982, reprinted in *Congressional Digest* 62 (January 1983): 17.

75. Statement of William Schneider, Jr., Under Secretary for Security Assistance, Science, and Technology, before Subcommittee on International Finance and Monetary Policy, Senate Committee on Banking, Housing, and Urban Affairs, March 2, 1983, reprinted in *U.S. Department of State Bulletin* 83 (June 1983): 72.

76. Mally, p. 52. In 1982 less than 200 such items were subject to export controls by the United States. Under the Export Administration Act of 1979, it was U.S. policy to restrict the export of technology which would significantly enhance the military capability of unfriendly states or which would significantly undermine U.S. foreign policy objectives — such as enhancement of regional stability, prevention of terrorism, and protection of human rights — rather than in support of general economic policies such as promotion of commercial advantage. Ibid., pp. 52–53. See also Hufbauer and Carlson, pp. 345–47.

77. See n. 20 above and accompanying text. By the end of the 1980s changing relations with the Soviet Union had caused the United States to reassess the need for export controls on dual-use technologies. See n. 79 below. "Significant changes . . . have occurred in the strategic environment." Ambassador Allan Wendt, Senior Representative for Strategic Technology Policy, "A Restructured COCOM," Current Policy No. 1290 (Washington: U.S. Department of State, 1990), p. 1. See also President George H.W. Bush, "The UN: World Parliament of Peace," Current Policy No. 1303 (Washington: U.S. Department of State, 1990), p. 1.

78. Statement of former Ambassador Elliot L. Richardson before House Foreign Affairs Committee, May 14, 1981, quoted in Oxman, "The Third United Nations Conference," p. 11 n. 27.

79. "There was very broad support from all regions and groups for the principle involved, and no dissent." Oxman, "The Ninth Session," p. 239. The United States, which first proposed the article, "had initially sought the inclusion of 'the words 'it considers' before the words 'contrary to the essential interests of its security.'" Ibid., pp. 238–39. Again, however, the Reagan administration sought no changes to article 302 in its Green Book or, apparently, elsewhere.

For a discussion of COCOM, see Mally, pp. 52–55. The collapse of the Warsaw Pact alliance and the general thaw in East-West relations caused the Bush administration in 1990 to reevaluate U.S. export control policies and to initiate "a complete overhaul" of COCOM's export control list, including "a new core list of goods and technologies . . . that is far shorter and less restrictive than the list that has served as the basis for our export control regime in the past" — changes which have been implemented by COCOM. Wendt, pp. 1–2.

80. U.N. Office for Ocean Affairs and the Law of the Sea, "Seminar on the Current Status of Developments," p. 59.

81. Giulio Pontecovo, "Musing About Seabed Mining, or Why What We Don't Know

Can Hurt Us," *Ocean Development and International Law* 21 (1990): 118. U.S. seabed mining technology could not have been patented, if its disclosure were "detrimental to the national security," but would instead have been "kept secret" pursuant to U.S. law "for such period as the national interest requires." Patent Law, 35 U.S.C. §181.

82. Malone, "Law of the Sea and Oceans Policy," p. 49. See James L. Malone, Assistant Secretary for Oceans and International Environmental and Scientific Affairs, "U.S. Policy and the Law of the Sea," *U.S. Department of State Bulletin* 81 (July 1981): 49; G. David Robertson and Gaylene Vasatura, "Recent Developments in the Law of the Sea 1981–82," *San Diego Law Review* 20 (1983): 683; and also statement of Theodore Kronmiller, September 15, 1982, in *Law of the Sea Negotiations*, U.S. Senate, p. 3.

83. John R. Stevenson, Legal Adviser, Department of State, "The Search for Equity in the Seabeds," *U.S. Department of State Bulletin* 64 (April 19, 1971): 530, 532. See also statement of the representative of the United States in *Provisional Verbatim Records*, U.N. General Assembly, 23rd Session, First Committee, 1590th meeting (U.N. Doc. A/AC.1/PV.1590, October 29, 1968). In 1975 Secretary of State Kissinger reaffirmed this policy. Secretary of State Henry Kissinger, address before American Bar Association, Montreal, August 11, 1975, reprinted in *U.S. Department of State Bulletin* 73 (1975): 357.

84. United States, President Richard M. Nixon, "U.S. Foreign Policy for the 1970s: The Emerging Structure of Peace: A Report to the Congress," reprinted in *U.S. Department of State Bulletin* 66 (1972): 409.

85. Paul Lawrence Saffo, "The Common Heritage of Mankind: Has the General Assembly Created a Law to Govern Seabed Mining?" *Tulane Law Review* 53 (1979): 518. See Chap. 5 nn. 229–31, 303–6, 365, 378 and accompanying text.

86. U.N. Convention on the Law of the Sea, arts. 151 (10), 173 (2).

87. American Bar Association, "Natural Resources of the Sea," p. 669.

88. Jonathan I. Charney, "The Equitable Sharing of Revenues from Seabed Mining," in *Policy Issues in Ocean Law*, American Society of International Law (St. Paul, Minn.: West Publishing, 1975), pp. 55, 89–90. "[I]t would only be by chance that the taxation and regulation designed to ensure efficiency will generate just exactly the amount of revenue for sharing that is 'fair.'" Richard James Sweeney, Robert D. Tollison, and Thomas D. Willett, "Market Failure, the Common-Pool Problem, and Ocean Resource Exploitation," *Journal of Law and Economics* 17 (1974): 180.

89. U.N. Convention on the Law of the Sea, Annex III, art. 13 (2). Pioneer investors are required to make an initial payment of only half the normal application fee, the remaining $250,000 being due upon the submission of a plan of work after the Convention has entered into force. Resolution II, para. 7 (a), in *International Legal Materials* 21 (1982): 1256.

By a two-thirds vote, the Council may raise the amount of the application fee "in order to ensure that it covers the administrative cost incurred," but the applicant could not be required to pay more than the actual administrative cost of processing the application. U.N. Convention on the Law of the Sea, arts. 161 (8) (b), 162 (2) (p); Annex III, art. 13 (2).

90. *The U.S. Proposals* (U.N. Doc. A/CONF.62/WG.21/Informal Paper 18), Annex III, proposed art. 13 (2). The Reagan administration agreed to an application fee of $500,000 for pioneer applicants. *Amendments* (U.N. Doc. A/CONF.62/L.122, April 13, 1982). Resolution II, proposed para. 7 (a), in *Official Records*, UNCLOS III, vol. XVI, p. 232. See also Lee Kimball and A.R.H. Schneider, "A Viable Convention?" *Environmental Policy and Law* 9 (1982): 70. Federal regulations promulgated pursuant to the Seabed Act of 1980 require "a fee payment of $100,000" with both license and permit applications, subject to similar adjustment provisions. Deep Seabed Mining Regulations, 15 CFR §§970.208 (b), 971.208 (b). By the terms of the act, this is "a reasonable administrative fee." Deep Seabed Hard Mineral Resources Act, 30 U.S.C. §1414. Such fees "are likely to have a minimal effect on production due to the large sums of money involved." Charney, "The Equitable Sharing of Revenues," p. 80.

91. U.N. Convention on the Law of the Sea, Annex III, art. 13 (3). Conversely, if the revenue-sharing charges should total less than $1 million, the contractor would only pay the annual fixed fee. Pioneer investors must pay the same $1 million annual fee. Resolution II, para. 7 (b), in *International Legal Materials* 21 (1982): 1256. See also Chap. 3 n. 472.

92. *Costs of the Authority* (U.N. Doc. A/CONF.62/C.1/L.19), para. 83, in *Official Records*, UNCLOS III, vol. VII, p. 71.

93. See Charney, "The Equitable Sharing of Revenues," p. 81; and Jon Van Dyke and Christopher Yuen, "'Common Heritage,' v. 'Freedom of the High Seas': Which Governs the Seabed?" *San Diego Law Review* 19 (1982): 546. "The fixed annual fee is . . . essentially equivalent to a security deposit or a bond for performance to insure that the mining company rapidly develops its operation. Such security deposits or performance bonds are common in mining contracts in developing countries." Katz, p. 123. The American Mining Congress has endorsed annual performance requirements of up to $1 million as a further method of discouraging speculative site banking. Laursen, p. 93.

94. *The U.S. Proposals* (U.N. Doc. A/CONF.62/WG.21/Informal Paper 18), Annex III, proposed art. 13 (3). Cf. 30 CFR §218.151. See Chap. 3 n. 373.

95. U.N. Convention on the Law of the Sea, Annex III, art. 13 (4). The alternative revenue-sharing formulas were designed to produce approximately equal revenue-sharing payments to the Authority. Fernando Zegers Santa Cruz, "Deep Sea-Bed Mining Beyond National Jurisdiction in the 1982 UN Convention on the Law of the Sea: Description and Prospects," *German Yearbook of International Law* 31 (1988): 110. By their terms, the two formulas only apply to nodule production and not to production of non-nodule deep-sea minerals. See n. 645 below and accompanying text. Contractors entering into joint-venture arrangements with the Enterprise are subject to the revenue-sharing provisions of article 13 in proportion to their partnership share. See n. 714 below.

96. U.N. Convention on the Law of the Sea, Annex III, art. 13 (5). See Antrim and Sebenius, p. 88.

97. U.N. Convention on the Law of the Sea, Annex III, art. 13 (6) (a), (d) (i). The combination formula was expected to be selected by the private consortia of which U.S. firms are members. The production charge is again "fixed at a percentage of the market value . . . of the processed metals produced." Ibid., Annex III, art. 13 (6) (a).

Annual "return on investment" is defined as "the ratio of attributable net proceeds in that year to the development costs of the mining sector." Ibid., Annex III, art. 13 (6) (m). "Development costs" are defined as "all expenditures incurred prior to the commencement of commercial production which are directly related to the development of the productive capacity of the area covered by the contract and the activities related thereto," as well as similar expenditures "incurred subsequent to the commencement of commercial production and necessary to carry out the plan of work, except those chargeable to operating costs." Ibid., Annex III, art. 13 (6) (h) (i), (ii), (n) (iv). "Cash surplus" is defined as a contractor's "gross proceeds less his operating costs and less his payments to the Authority." Ibid., Annex III, art. 13 (6) (d) (i).

98. Arrow, "The Proposed Regime," p. 401. See also Antrim and Sebenius, p. 84. By the early 1980s the projected rate of return had fallen below 10 percent — "too low to trigger investment." T. Alexander, p. 143. See n. 140 below; and also Hauser, p. 24. In 1972 the average return on investment earned by U.S. land-based mining operators was only 10.4 percent, however. The return on nonferrous mineral recovery was even lower — 7.2 percent — and the profitability of such operations also diminished substantially during the 1980s. *Economic Implications of Sea-Bed Mineral Development* (U.N. Doc. A/CONF.62/25), reprinted in *Official Records,* UNCLOS III, vol. III, p. 28. Hauser, pp. 24–25. The U.S. government has continued to collect large royalties from Outer Continental Shelf lease sales although a 1968 study found that the pretax rate of return on offshore oil and gas investment was only 7.5 percent. Charney, "The Equitable Sharing of Revenues," p. 72 n. 64. See n. 122 below and accompanying text.

The provision maintaining the production charge at 2 percent when return on investment falls short of 15 percent was included in recognition that the 4 percent charge "can be a heavy burden for the Contractor, even in the second period, if in a particular year the Contractor's project is doing badly." UNCLOS III, *Report on Negotiations Held by the Chairman and Coordinators of the Working Group of 21* (U.N. Doc. A/CONF.62/C.1/L.26, August 21, 1979), reprinted in *Official Records,* UNCLOS III, vol. XII, p. 79.

99. See Sebenius, p. 98. "Attributable net proceeds" is defined in the Convention as "that portion of the contractor's net proceeds which is attributable to the mining of the resources of the area covered by the contract," an amount which is to be imputed at a ratio of not less than 25 percent of such proceeds. U.N. Convention on the Law of the Sea, Annex III, art. 13 (6) (c) (i), (e). "Contractor's net proceeds" is defined as "the contractor's gross proceeds less his

operating costs and less the recovery of his development costs." U.N. Convention on the Law of the Sea, Annex III, art. 13 (6) (f), (n) (ii). Preproduction development costs are to be amortized in 10 equal annual installments, and interest is to accrue to the contractor's unrecovered development costs at the rate of 10 percent per annum. U.N. Convention on the Law of the Sea, Annex III, art. 13 (6) (d) (i), (j). All financial calculations are to be made in constant terms, adjusted for inflation, "implying that only real profits are taxed – not paper profits, as is common in inflationary times under domestic law." Sebenius, p. 98. Cf. U.N. Convention on the Law of the Sea, Annex III, art. 13 (13). See also Hauser, pp. 89–90.

In the Preparatory Commission, Special Commission 3 has prepared draft regulations which provide for payments semiannually, the amount being determined by the Authority based upon calculations submitted by the contractor, with resort to binding commercial arbitration (unless both parties agree otherwise) in the event of a dispute. Walkate, p. 160.

100. U.N. Convention on the Law of the Sea, Annex III, art. 13 (6) (c). See Sebenius pp. 39–40.

101. Antrim and Sebenius, p. 88. See n. 99 above and accompanying text; and also Santa Cruz, p. 111. In fact, the value of newly mined nodules is expected to be only 6 to 10 percent of their processed value. UNCLOS III, *Note by the Chairman of the First Committee* (U.N. Doc. A/CONF.62/C.1/L.2, July 26, 1974), para. 23, reprinted in *Official Records*, UNCLOS III, vol. III, p. 153. See also Hauser, p. 82.

102. Hauser, p. 90.

103. U.N. Convention on the Law of the Sea, Annex III, art. 13 (15). For a discussion of commercial arbitration procedures, see n. 7 above. To ensure the availability of these procedures, a contractor need only incorporate the revenue-sharing formula into its plan of work.

104. *Costs of the Authority* (U.N. Doc. A/CONF.62/C.1/L.19), para. 17, in *Official Records*, UNCLOS III, vol. VII, p. 60. See Robert L. Friedheim, "Understanding the Debate on Ocean Resources," Occasional Paper No. 1 (Kingston, R.I.: The Law of the Sea Institute, February 1969), p. 31.

105. *Costs of the Authority* (U.N. Doc. A/CONF.62/C.1/L.19), para. 32, in *Official Records*, UNCLOS III, vol. VII, p. 62.

106. Ibid., paras. 33, 36, in *Official Records*, UNCLOS III, vol. VII, pp. 62–63. See Sebenius, pp. 20–21. "The introduction into the petroleum industry of the concept of profit-sharing dates from 1950, when Saudi Arabia concluded with ARAMCO a contract incorporating what was considered the novel idea of apportioning the profits from exploitation equally between the Government and the concessionnaire." *Costs of the Authority* (U.N. Doc. A/CONF.62/C.1/L.19), in *Official Records*, UNCLOS III, vol. VII, p. 63 n. 13. Such profit-sharing arrangements are preferred by fiscal and business experts. *Economic Implications of Sea-Bed Mineral Development* (U.N. Doc. A/CONF.62/25), in *Official Records*, UNCLOS III, vol. III, p. 32.

107. *Costs of the Authority* (U.N. Doc. A/CONF.62/C.1/L.19), n. 21, in *Official Records*, UNCLOS III, vol. VII, p. 64. In exchange, AGIP received other benefits, including assumption by NIOC of the risk of nondiscovery of exploitable minerals. Ibid., n. 22, in *Official Records*, UNCLOS III, vol. VII, p. 64. Some agreements have allocated up to 95 percent of net revenues to the host state, after allowing for amortization of capital investment costs. Ibid., n. 31, in *Official Records*, UNCLOS III, vol. VII, p. 67.

108. Antrim and Sebenius, p. 88. See Katz, pp. 122–24.

109. Although developing states had argued that it would be impossible to set appropriate rates before seabed mining actually begins, developed states feared that the Authority might abuse any such discretionary power by imposing excessive charges. Jack N. Barkenbus, *Deep Seabed Resources: Politics and Technology* (New York: Free Press, 1979), p. 107.

110. See John Temple Swing, "Address," in *Current Issues in the Law of the Sea*, ed. Christopher C. Joyner (Allentown, Penn.: Muhlenberg College, 1980), p. 115. "As a general matter, the United States has pointed out that the Authority will be more likely to promote efficiency if the majority of the production fees were based on a share of profitability rather than a royalty based solely on production rates." Arrow, "The Proposed Regime," p. 401 n. 407. See International Law Association, American Branch, *Proceedings and Committee Reports, 1977–78* (New York: n.p., 1979), pp. 114–15. Developed states successfully advocated a progressive assessment schedule and a variable definition of attributable net

proceeds, but had sought maximum rates of 25 and 50 percent in the first and second periods, respectively. *Report on Negotiations* (U.N. Doc. A/CONF.62/C.1/L.26), in *Official Records*, UNCLOS III, vol. XII, pp. 78–79.

111. Sebenius, p. 57.

112. *Economic Implications of Sea-Bed Mineral Development* (U.N. Doc. A/CONF.62/25), in *Official Records*, UNCLOS III, vol. III, pp. 32, 33. See also Charney, "The Equitable Sharing of Revenues," p. 82. "If the revenue is based on profit sharing, the risks and costs inherent in ocean mining may result in a small amount of revenue. Since it is not a front end cost, it will not discourage entry into the field until the amount of profit sharing makes the return too low for the capital invested." Charney, "The Equitable Sharing of Revenues," p. 79. "This aspect must be weighed against the benefit to the Authority from a substantial amount of money at a very early stage in its operations." *Costs of the Authority* (U.N. Doc. A/CONF.62/C.1/L.19), para. 85, in *Official Records*, UNCLOS III, vol. VII, p. 71.

113. Charney, "The Equitable Sharing of Revenues," pp. 82–84. See David B. Johnson and Dennis E. Logue, "U.S. Economic Interests in Law of the Sea Issues," in *The Law of the Sea*, ed. Amacher and Sweeney, p. 55.

114. Mati L. Pal, "Financial Arrangements," *Syracuse Journal of International Law and Commerce* 6 (1979): 296. See James M. Broadus, "Seabed Materials," *Science* 235 (February 20, 1987): 857. "Mining projects are generally regarded as high risk ventures and as a result, acceptable rates of return for mining ventures tend to be higher than that for manufacturing, for example." *Costs of the Authority* (U.N. Doc. A/CONF.62/C.1/L.19), para. 72, in *Official Records*, UNCLOS III, vol. VII, p. 70. For a discussion of risks affecting rates of return on seabed mining investment, see Pal, pp. 297–98; and also Brewer, "Deep Seabed Mining," pp. 42–43.

115. Sebenius, pp. 25–26. Hauser, pp. 81–82. See also Gertler and Wayne, p. 541.

116. *Costs of the Authority* (U.N. Doc. A/CONF.62/C.1/L.19), paras. 75, 90, in *Official Records*, UNCLOS III, vol. VII, pp. 70, 72. Grace periods "are used to induce companies to exploit a resource by allowing them a lowering of initial costs by waiving either royalties or other special financial obligations or all or portions of the taxes that would eventually be collected." Ibid., para. 74, in *Official Records*, UNCLOS III, vol. VII, p. 70.

117. Gertler and Wayne, p. 538. See U.N. Convention on the Law of the Sea, Annex IV, art. 10 (1), (3). The grace period would be less than 10 years if the Assembly determines that the Enterprise is self-supporting. Concern that the Enterprise was being given an economic advantage was outweighed by the perception that the Enterprise would initially require a means of building its reserves. Oxman, "The Ninth Session," p. 225.

118. Robert D. Tollison and Thomas D. Willett, "Institutional Mechanisms for Dealing with International Externalities: A Public Choice Perspective," in *The Law of the Sea*, ed. Amacher and Sweeney, p. 95.

119. *Draft United Nations Convention* (U.N. Doc. A/AC.138/25), Appendix A, §10.2. The exact level of such royalty payments would have been established by a concurrent majority vote of the Council and a simple majority vote of the Assembly unless disapproved by more than one third of states parties, a procedure which would not have required the concurrence of the United States. Ibid., arts. 34 (5), 38, 67 (c)–(e), 68 (1) (a). See nn. 435, 475 below. "The American Mining Congress viewed the proposed fees as excessive and forming a 'large front load.'" Auburn, p. 504.

Within the trusteeship zone, the coastal state was to be permitted to retain one third to one half of the amount of royalties collected, the precise amount again to be determined without the necessity of U.S. approval. *Draft United Nations Convention* (U.N. Doc. A/AC.138/25), arts. 28 (d), 67 (c)–(e), 68 (1) (a); Appendix A, §10.3.

120. Van Dyke and Yuen, p. 533. The trust fund established for the benefit of developing states by the 1980 Seabed Act was to receive only 0.75 percent of the market value of processed nodule minerals, and no such royalties were required by the Act after June 28, 1990. Tax on Removal of Hard Mineral Resources from Deep Seabed, 26 U.S.C. §§4495 (a), 4497 (a), 4498 (a). Since no nodule exploitation by U.S. seabed mining acts had begun by 1990, the Act's revenue-sharing obligations were of no practical effect, although they are arguably of some legal significance. See Chap. 5 n. 261 and accompanying text. Seabed mining acts adopted by other developed states impose similar "taxes." See Chap. 3 n. 388.

121. Antrim and Sebenius, pp. 84–85. See Chap. 3 nn. 398–402 and accompanying text, and n. 98 above and accompanying text; Hauser, pp. 23–24, 181–84; and also Sebenius, pp. 45–49. A revision of the MIT model in 1981 yielded significantly lower profits, based in part on increased cost estimates for nodule mining operations. Sebenius, p. 95. For an extensive discussion of the role of computer models in deriving the revenue-sharing formulas of article 13, see Sebenius, pp. 27–39, 50–55. A 1979 study by Arthur D. Little predicted a rate of return of between 3 and 12 percent, based on capital costs of $600 to $900 million. See Charles J. Johnson and Allen L. Clark, "Potential of Pacific Ocean Nodule, Crust, and Sulfide Mineral Deposits," *Natural Resources Forum* 9 (1985): 181. Capital costs have risen substantially since then.

If the Chinese, Japanese, Soviet, French, Indian, and Eastern European state-run seabed mining operations prove successful, one might nonetheless expect that U.S. mining firms should be able to make an adequate profit under the Convention when demand for nodule minerals strengthens. Outside the Convention, "full recovery of the initial investments in deep seabed mining . . . could easily take ten years," if U.S. firms could convince reluctant bankers to finance such operations. Elliot L. Richardson, "The United States Posture Toward the Law of the Sea Convention: Awkward But Not Irreparable," *San Diego Law Review* 20 (1983): 510.

122. See Chap. 6 n. 40 and accompanying text. The U.S. oil industry has not sought reductions in these revenue-sharing rates. Charney, "The Equitable Sharing of Revenues," p. 85. Nor has the oil industry been discouraged by a system of land leasing wherein royalty rates are negotiated with the government. See statement of Representative James Oberstar in *Deep Seabed Mining*, U.S. House of Representatives, p. 340.

Use of bonus bidding to allocate exclusive rights to deep-sea mining sites would tend to discourage riskier nodule mining operations, as would other such front-end charges. See nn. 112, 114 above; and statement of Vincent McKelvey in *Syracuse Journal of International Law and Commerce* 6 (1978–79): 303–4.

123. Sebenius, p. 40. Cf. *The U.S. Proposals* (U.N. Doc. A/CONF.62/WG.21/Informal Paper 18), Annex III, proposed art. 13.

124. Richard A. Frank, "Jumping Ship," *Foreign Policy*, 43 (1981): 133. See also Charney, "The Equitable Sharing of Revenues," p. 78. "Capital outlay, operating costs, and metal prices — not payments for the right to mine such as fees, royalties, and profit shares — are the dominant factors governing the return on any mining investment. The latter, nevertheless, must not be excessively burdensome, and in this respect the . . . Convention's financial provisions are not worse than most other tax systems." Richardson, "Law in the Making," p. 442. Indeed, barring double taxation, the effective revenue-sharing rates payable to the Authority could, on the whole, be lower than those which a seabed mining operator would owe to the U.S. government under the Seabed Act or an RSA regime. See Leigh S. Ratiner, "Reciprocating State Arrangements: A Transition or an Alternative?" in *The 1982 Convention on the Law of the Sea*, ed. Albert W. Koers and Bernard H. Oxman (Honolulu: Law of the Sea Institute, 1984), p. 198.

125. Statement of Elliot L. Richardson, May 14, 1981, quoted in Oxman, "The Third United Nations Conference," p. 12. See Hauser, pp. 91–96; and also Elliot L. Richardson, "The Politics of the Law of the Sea," *Ocean Development and International Law Journal* 11 (1982): 22. The 1970 U.S. Draft Treaty proposed that revenue sharing be "in the nature of total payments ordinarily made to governments under similar conditions." *Draft United Nations Convention* (U.N. Doc. A/AC.138/25), Appendix A, §10.2. By 1980 two-thirds of more than $25 billion in foreign tax credits were being claimed by U.S. oil companies. Hufbauer and Carlson, p. 352. It has nevertheless been suggested, by a strong supporter of the petroleum industry, that "tax credits" should be granted instead by the Authority to offset national taxes imposed upon seabed mining operators. Statement of Senator Russell Long in *Congressional Record* 125 (December 14, 1979): 36033. With consensus approval in the Council, the Authority may, in fact, provide incentives for contractors to undertake seabed mining activities, as well as to optimize revenue-sharing proceeds for the Authority. U.N. Convention on the Law of the Sea, arts. 161 (8) (d), 162 (2) (o) (ii); Annex III, arts. 11 (2), (3), 13 (1) (a), (b), (d), (14). See n. 716 below and accompanying text; and also Walkate, p. 160.

126. The article contains a list of policy objectives rather than an authorization for actions by the Authority beyond those specifically provided for in the Convention. See Hauser, p. 38. Most of these policies had been included in the 1970 Declaration of Principles.

127. U.N. Convention on the Law of the Sea, art. 151 (1) (a), (b). Because the United States, as potentially both the largest consuming state and largest producing state, would necessarily be an interested party, it would be able to block the Authority's participation in any agreement which was not in its national interest by simply declining to participate. Given this broad participation requirement, and in light of the controversy stirred by this issue at UNCLOS III, it is very doubtful whether such an agreement could ever be achieved. One opponent of the Convention has nevertheless claimed that the United States would somehow "be deprived of the right to represent [its] production interest." Breaux, "The Diminishing Prospects," p. 269. Another appeared to maintain that an agreement of this nature "on its face would constitute a criminal violation of the antitrust acts of the United States." Statement of Senator Richard B. Stone in *Congressional Record* 125 (December 14, 1979): 36076. In this regard, it should be borne in mind that participation by the Authority in such agreements was proposed by the U.S. Secretary of State in 1976. See Chap. 3 n. 87 and accompanying text.

Agreements negotiated pursuant to this provision might provide for buffer stocks, export restriction quotas, marketing and distribution measures, pricing linkages, and financing arrangements. See UNCLOS III, *Statement Made by Mr. G.D. Arsenis on Behalf of the Secretary-General of the United Nations Conference on Trade and Development* (U.N. Doc. A/CONF.62/32, July 15, 1974), reprinted in *Official Records*, UNCLOS III, vol. III, p. 62; and also Hauser, p. 116. Commodity agreements have been sought by developing states in order to stabilize volatile markets in which fluctuations of 300 percent or more were frequent. See Dennis T. Avery, "International Commodity Agreements," Special Report No. 83 (Washington: U.S. Department of State, 1981), p. 8. By the early 1980s international commodity agreements had been negotiated with respect to coffee, cocoa, rubber, and tin, but not for any of the four primary nodule minerals. David Hegwood, "Deep Seabed Mining: Alternative Schemes for Protecting Developing Countries from Adverse Impacts," *Georgia Journal of International and Comparative Law* 12 (1982): 186. Francisco Orrego Vicuña, "The Deep Seabed Mining Regime: Terms and Conditions for Its Renegotiation," *Ocean Development and International Law* 20 (1989): 538. See Avery, pp. 2, 7–8; and also John R. Harry, "Deep Seabed Mining in the Law of the Sea Negotiation (I): The Contours of a Compromise," Oceans Policy Study 1:2 (Charlottesville, Va.: Michie, 1978), pp. 8–9. By the end of the 1980s, however, the tin, coffee, and cocoa agreements had collapsed with serious financial repercussions, calling into question the effectiveness of such arrangements. Ian A. Malloy, "Conduct Unbecoming: The Collapse of the International Tin Agreement," *American University Journal of International Law and Policy* 5 (1990): 835–36, 888. Seabed production would have to be included in any such comprehensive agreement eventually concluded between producer and consumer states. *Statement Made by Mr. G.D. Arsenis* (U.N. Doc. A/CONF.62/32), in *Official Records*, UNCLOS III, vol. III, p. 62.

The Council is empowered to "take, upon the recommendation of the Economic Planning Commission, necessary and appropriate measures" to implement subparagraph (h) of article 150. U.N. Convention on the Law of the Sea, art. 162 (2) (m). The United States, with a guaranteed Council seat, would have an effective veto power over such measures, since a consensus of Council members would be required for their adoption. See n. 463 below and accompanying text. The Authority may compensate developing states directly for economic damage suffered as a result of seabed mining operations. See n. 640 below and accompanying text.

128. U.N. Convention on the Law of the Sea, arts. 150 (a), (e), 151 (1) (a), (c).

129. Ibid., art. 151 (2) (a), (b).

130. Ibid., art. 151 (2) (d). The Legal and Technical Commission will calculate the production ceiling and issue production authorizations at four-month intervals. Ibid., art. 165 (2) (n); Annex III, art. 7 (1).

131. Ibid., art. 151 (4) (b) (ii). The production ceiling for each year of the interim period is calculated as the sum of (a) the increase in the "trend line values for nickel consumption" during the 5-year period ending the year before the first year of commercial production, and (b) 60 percent of the subsequent increase in the same trend line values. Ibid., art. 151 (4) (a). The trend line values are to be "derived from a linear regression of the logarithms of actual nickel consumption for the most recent 15-year period for which such data are available, time being the independent variable." Ibid., art. 151 (4) (b) (i). See Hauser, p. 118.

132. U.N. Convention on the Law of the Sea, art. 151 (6) (a). "Any increase over 8

per cent and up to 20 per cent in any year, or any excess in the first and subsequent years following two consecutive years in which excesses occur shall be negotiated with the Authority, which may require the operator to obtain a supplementary production authorization to cover additional production." Ibid. The interim period would begin 5 years prior to commercial production, and would last for 25 years unless it were superseded by an international commodity agreement or by amendments adopted at the review conference. Ibid., art. 151 (3).

133. Ibid., art. 151 (6) (b). This amount is expected to be sufficient for the conduct of large, 3-million-ton-per-year nodule mining operations. See n. 151 below and accompanying text.

134. U.N. Convention on the Law of the Sea, arts. 161 (8) (c), 162 (2) (q); Annex III, art. 7 (2), (4). "[S]uch a case is unlikely." Sebenius, p. 101. Priority would be accorded to those applicants which offer better assurances of performance and earlier revenue sharing with the Authority, and which "have already invested the most resources and effort in prospecting or exploration." U.N. Convention on the Law of the Sea, Annex III, art. 7 (3). The Council would also consider the need to promote diversity of national sponsorship and "to enhance opportunities for all States Parties ... to participate in activities in the Area and to prevent monopolization of those activities." U.N. Convention on the Law of the Sea, Annex III, art. 7 (5). It has been claimed that "[t]his is another way of saying that not all of the four American-created consortia can expect to be in production at any one time." Statement of Northcutt Ely in *Congressional Digest* 62 (January 1983): 31. In fact, all four would be guaranteed production authorizations; the only issue is which of them would have priority in the event a selection among pioneer registrants becomes necessary. See nn. 135, 590 below and accompanying text.

135. Resolution II, para. 9, in *International Legal Materials* 21 (1982): 1256-57. See nn. 574-77 below and accompanying text. If the pioneer registrants agreed upon apportionment, they would share priority in the allocation of subsequent authorizations: "the Authority shall award each of them a production authorization for such lesser quantity as they have agreed ... and their full production will be allowed as soon as the production ceiling admits of additional capacity sufficient for the applicants involved." Resolution II, para. 9 (f), in *International Legal Materials* 21 (1982): 1257. If the pioneer investors did not agree to apportion the available authorization amount within three months and were unable to agree upon priority among themselves, allocation of the available amount would be determined by binding commercial arbitration in accordance with UNCITRAL arbitration rules and the criteria set forth in Annex III of the Convention. Resolution II, para. 9 (c)-(e), (g), in *International Legal Materials* 21 (1982): 1256-57. See nn. 7, 134 above.

The first 38,000 metric tons of nickel available under the production ceiling are, in any event, reserved for the Enterprise. U.N. Convention on the Law of the Sea, art. 151 (5).

136. U.N. Convention on the Law of the Sea, Annex III, art. 7 (6). Unlike Resolution II, the Convention itself does not specify the precedence of priorities of the Enterprise via-à-vis those of previously unsuccessful state-sponsored applicants; the allocation of conflicting priority authorizations would be determined by the rules, regulations, and procedures of the Authority, by the Council, or by the Sea-Bed Disputes Chamber. See Hauser, pp. 72, 74-75.

137. U.N. Convention on the Law of the Sea, arts. 151 (7), 161 (8) (d), 162 (2) (o) (ii). Although the ceiling set forth in article 151 is to be calculated and applied in reference to nickel, the production limitations were primarily intended to protect developing-state producers of other nodule minerals. See Tullio Treves, "Seabed Mining and the United Nations Law of the Sea Convention," *Italian Yearbook of International Law* 5 (1980-81): 38-39; and Langevad, p. 231. But see n. 255 below and accompanying text.

138. Such requirements are to consist of "periodic expenditures ... which are reasonably related to the size of the area covered by the plan of work and the expenditures which would be expected of a bona fide operator who intended to bring the area into commercial production. U.N. Convention on the Law of the Sea, Annex III, art. 17 (2) (c). Again, consensus is required. Ibid., arts. 161 (8) (d), 162 (2) (o) (ii). "The absence of an effective performance requirement would practically invite a race to claim areas of potential value on the ocean floor. Such a race would be inefficient in the extreme." David B. Brooks, "Deep Sea Manganese Nodules: From Scientific Phenomenon to World Resource," *Natural Resources Journal* 8 (1968): 416.

139. U.N. Convention on the Law of the Sea, art. 151 (9). This provision may have been rendered superfluous by language added to article 162 at the final UNCLOS III session. See n. 332 below and accompanying text.

140. U.N. Convention on the Law of the Sea, Annex III, art. 17 (2) (a). The rules, regulations, and procedures must be applied to all operators "uniformly," and a consensus of the Council would be required for their establishment. Ibid., arts. 161 (8) (d), 162 (2) (o) (ii); Annex III, art. 17 (1). Each pioneer investor is to be granted exclusive exploration rights with respect to an area of up to 150,000 square kilometers, half of which is to be relinquished incrementally to the Authority over a five-year period. Resolution II, para. 11 (e), in *International Legal Materials* 21 (1982): 1257. This requirement of Resolution II has been applied flexibly by the Preparatory Commission. See Chap. 3 n. 330 and accompanying text.

141. Statement of President Reagan, January 29, 1982, in *U.S. Department of State Bulletin* 82 (March 1982): 54. The U.S. mining industry had complained that the treaty being negotiated was "anti-production." Marne A. Dubs, "International Industrial Interests and the Deep Seabed," in *Current Issues in the Law of the Sea*, ed. Joyner, p. 144. "Instead of encouraging seabed mining development, the goals of the Authority are anti-development, including 'orderly and safe development,' 'rational management,' 'just and stable prices remunerative to producers,' and 'the protection of developing countries from . . . adverse effects.'" Bandow, p. 483. It is unclear whether the United States would now support an Authority which promoted disorderly and unsafe development, irrational management, unjust and unstable prices, etc.

142. White House Fact Sheet, in *U.S. Department of State Bulletin* 82 (March 1982): 55.

143. Malone, "U.S. Policy and the Law of the Sea," p. 49; see Eckert, "Exploitation of Deep Ocean Minerals," p. 171. Thus it has been claimed that the system of production controls "translates to international governmental control over world commerce, founded on the conviction that government officials can manage a more rational system of production and distribution than can a free market." Robert A. Goldwin, "Locke and the Law of the Sea," *Commentary* 71 (June 1981): 49.

144. *The U.S. Proposals* (U.N. Doc. A/CONF.62/WG.21/Informal Paper 18), proposed art. 151 (1), (2) (a). The developing state would have been required to demonstrate "that it has already taken steps to alleviate its dependence on production of affected minerals." Ibid., proposed art. 151 (2) (d). Production controls upon nodule minerals other than nickel and upon non-nodule minerals were also to have been eliminated, as was the provision authorizing the Authority to compensate developing producer states. Ibid., proposed art. 151.

145. *Amendments* (U.N. Doc. A/CONF.62/L.121), proposed art. 151 (1); Annex III, proposed art. 7 (1), in *Official Records*, UNCLOS III, vol. XVI, pp. 226, 230. The 60 percent factor would have been increased by 2 percent annually over a 10-year period, beginning 5 years after commencement of commercial production. Ibid., proposed art. 151 (2) (b) (ii), in *Official Records*, UNCLOS III, vol. XVI, p. 226. Provisions authorizing compensation for developing producer states adversely affected by seabed mining would again have been eliminated, however. Ibid., proposed arts. 151 (4), 162 (2) (m), in *Official Records*, UNCLOS III, vol. XVI, pp. 226–28.

146. See UNCLOS III, *Possible Impact on the Convention, With Special Reference to Article 151, on Developing Countries Which Are Producers and Exporters of Minerals to Be Extracted From the Area* (U.N. Doc. A/CONF.62/L.84, March 2, 1982), Annex III, reprinted in *Official Records*, UNCLOS III, vol. XVI, p. 187.

147. "The principal known commercial sulphide ore-bodies are in Canada, the USSR, Australia, Zimbabwe, Botswana and Finland, though sulphide mineralization has been reported in other areas. The laterite deposits are more wide in occurrence and are being worked in New Caledonia, Cuba, the Dominican Republic, Colombia, Guatemala, the Philippines, Indonesia and Australia." Ibid. See table in United Nations, *The Nickel Industry and the Developing Countries* (U.N. Doc. ST/ESA/100, 1980); Hegwood, p. 181; and also Robert D. Cairns, "A Reconsideration of Ontario Nickel Policy," *Canadian Public Policy* 7 (1981): 528. "The major exporters of nickel are Canada, Norway, the Dominican Republic and New Caledonia." Pruitt, p. 350 n. 26.

148. See statement of the representative of the U.N. Secretariat, July 30, 1974, in *Official Records*, UNCLOS III, vol. II, p. 50 (para. 6); and UNCLOS III, *United States of America:*

Working Paper on the Economic Effects of Deep Sea-Bed Exploitation (U.N. Doc. A/ CONF.62/C.1/L.5, August 8, 1974), reprinted in *Official Records*, UNCLOS III, vol. XVI, p. 165. Most new land-based nickel production is expected to occur within developing states. Van Dyke and Yuen, p. 532.

149. Cairns, p. 527. Sulfide producers are better able to adapt their operations to changing market conditions. Ibid., p. 529.

150. See Santa Cruz, p. 111; *Possible Impact* (U.N. Doc. A/CONF.62/L.84), Annex III, in *Official Records*, UNCLOS III, vol. XVI, p. 187; and also J.C. Agarwal et al., "Comparative Economics of Recovery of Metals from Ocean Nodules," *Marine Mining* 2 (1979): 123, 130. Nickel produced from nodules is not expected to be price-competitive with terrestrial sulfide deposits. Saffo, p. 517 n. 114. See Agarwal et al., p. 120. "Cuba, Indonesia, and New Caledonia would be substantially affected, while the Philippines, Guatemala and the Dominican Republic would be affected to some degree." Third UNCLOS, *Chile: Working Paper on the Economic Implications for the Developing Countries of the Exploitation of the Sea-Bed Beyond the Limits of National Jurisdiction* (U.N. Doc. A/CONF.62/C.1/L.11, August 26, 1974), reprinted in *Official Records*, UNCLOS III, vol. III, p. 178.

151. Van Dyke and Yuen, p. 531 n. 163. An earlier estimate had been significantly lower. Eckert, "Exploitation of Deep Ocean Revenues," p. 151. See also Van Dyke and Yuen, p. 510 n. 70.

152. Pruitt, p. 345 n. 1. See Hauser, pp. 13–15. The actual effect of increased production upon the price of nickel will depend upon the elasticity of demand for the metal. *United States of America: Working Paper* (U.N. Doc. A/CONF.62/C.1/L.5), in *Official Records*, UNCLOS III, vol. III, p. 165.

153. Pruitt, p. 345. See also Hauser, p. 14.

154. *Possible Impact* (U.N. Doc. A/CONF.62/L.84), Annex III, reprinted in *Official Records*, UNCLOS III, vol. XVI, p. 186. See also Hauser, p. 16 n. 28. "In 1978, copper accounted for over 90% of Zambia's exports, 52% of Chile's, and 42% of Peru's. Due to the level of dependence on copper for export earnings, any impact on the world market could have a significant effect upon these economies." Hegwood, pp. 181–82. See also Robert E. Clute, "The African Perspective of the Law of the Sea," in *Current Issues in the Law of the Sea*, ed. Joyner, p. 17. "Haiti, Bolivia, Nicaragua, Mexico, Morocco, Cuba, the Republic of Korea and India would be affected to some degree." *Chile: Working Paper* (U.N. Doc. A/CONF.62/C.1/L.11), in *Official Records*, UNCLOS III, vol. III, p. 178. While not a copper exporter, the United States has itself been the largest producer state, providing 18 percent of world production in 1978. Avery, p. 2.

155. See *United States of America: Working Paper* (U.N. Doc. A/CONF.62/C.1/L.5), in *Official Records*, UNCLOS III, vol. III, p. 165; and Johnson and Logue, p. 49. Copper was the second largest source of foreign exchange for the independent states of Sub-Saharan Africa in 1973. Clute, p. 17.

156. *Possible Impact* (U.N. Doc. A/CONF.62/L.84), Annex IV, in *Official Records*, UNCLOS III, vol. XVI, pp. 189–90. Copper accounted for 36 percent of Zaire's exports in 1979. Kathryn E. Yost, "The International Sea-Bed Authority Decision-Making Process: Does It Give a Proportionate Voice to the Participant's Interests in Deep Sea Mining," *San Diego Law Review* 20 (1983): 668 n. 83.

157. See *Possible Impact* (U.N. Doc. A/CONF.62/L.84), Annex III, in *Official Records*, UNCLOS III, vol. XVI, p. 186.

158. Cf. *United States of America: Working Paper* (U.N. Doc. A/CONF.62/C.1/L.5), in *Official Records*, UNCLOS III, vol. III, p. 166. See Van Dyke and Yuen, p. 531 n. 163.

159. Eckert, "Exploitation of Deep Ocean Minerals," p. 151 n. 34. See Pruitt, p. 345 n. 1. A higher estimate of land-based reserves, made by the U.S. Geological Survey, yields a ratio of approximately 6 to 1. "These figures must be viewed as very approximate estimates because many cobalt-bearing copper and nickel ore-bodies are not accurately sampled for cobalt, nor is the estimated cobalt content recorded as reserves. Likewise, the cobalt content of many large cobalt-bearing tailings dams (waste dumps) and smelter slag dumps at operating copper and nickel mines is unrecorded." *Possible Impact* (U.N. Doc. A/CONF.62/L.84), Annex III, in *Official Records*, UNCLOS III, vol. XVI, p. 188.

160. *Possible Impact* (U.N. Doc. A/CONF.62/L.84), Annex III, in *Official Records*,

UNCLOS III, vol. XVI, p. 187. Elsewhere, this percentage has been estimated as high as 16.2 percent. Van Dyke and Yuen, p. 531 n. 163. See also Johnson and Clark, p. 183.

161. Hegwood, p. 182. See *Possible Impact* (U.N. Doc. A/CONF.62/L.84), Annex III, in *Official Records*, UNCLOS III, vol. XVI, p. 188 n. 16. "It is thought by some that the entire marketing system might be radically shaken by the introduction of cobalt from deep sea mining into the market." Van Dyke and Yuen, p. 63. See also Hauser, p. 14. The price of cobalt fluctuated dramatically during the 1970s, rising from less than $5 per pound to $50 per pound before falling back to less than $20 per pound.

162. See *Possible Impact* (U.N. Doc. A/CONF.62/L.84), Annex III, in *Official Records*, UNCLOS III, vol. XVI, pp. 187–88. Cf. *United States of America: Working Paper* (U.N. Doc. A/CONF.62/C.1/L.5), in *Official Records*, UNCLOS III, vol. III, p. 166.

163. *Chile: Working Paper:* (U.N. Doc. A/CONF.62/C.1/L.11), in *Official Records*, UNCLOS III, vol. III, p. 178. See also Hauser, p. 14. Cobalt production has accounted for approximately 41 and 11 percent of the mineral industries of Zaire and Zambia, respectively. Copper and cobalt together normally account for more than two thirds of the total value of Zaire's exports, and in 1979 accounted for more than 80 percent. *Possible Impact* (U.N. Doc. A/CONF.62/L.84), Annexes III, IV, in *Official Records*, UNCLOS III, vol. XVI, p. 188 nn. 16, 189.

164. *Possible Impact* (U.N. Doc. A/CONF.62/L.84), Annexes III, IV, in *Official Records*, UNCLOS III, vol. XVI, pp. 187–90. In 1980 Morocco, Zambia, and Zaire contributed almost 75 percent of world cobalt production. Clute, p. 17. In the late 1960s these states had provided for two thirds of world production. Johnson and Logue, p. 49.

165. See Chap. 3 nn. 410, 417–18 and accompanying text, and n. 229 below.

166. *Possible Impact* (U.N. Doc. A/CONF.62/L.84), Annex III, in *Official Records*, UNCLOS III, vol. XVI, p. 189. See Arrow, "The Proposed Regime," p. 343; and Van Dyke and Yuen, p. 531 n. 163.

167. See Eckert, "Exploitation of Deep Ocean Minerals," p. 151. Only one U.S. mining firm is interested in producing manganese from nodules, "and this manganese would be in the form of a highly pure metal which would serve only a very small segment of the manganese market." *United States of America: Working Paper* (U.N. Doc. A/CONF.62/C.1/L.5), in *Official Records*, UNCLOS III, vol. III, p. 166.

168. *Possible Impact* (U.N. Doc. A/CONF.62/L.84), Annex III, in *Official Records*, UNCLOS III, vol. XVI, pp. 188–89. See Agarwal et al., p. 121.

169. *Chile: Working Paper* (U.N. Doc. A/CONF.62/C.1/L.11), in *Official Records*, UNCLOS III, vol. III, p. 178. "Guyana, the Ivory Coast and the Philippines would be affected to some degree." Ibid. The manganese export earnings of Morocco, India, and Ghana could be completely eliminated by deep-sea mining operations. D. Brooks, p. 411. Gabon, the third largest producer, is the largest exporter of manganese, which accounted for 9 percent of its exports in the early 1980s, although petroleum is a more important national resource. *Possible Impact* (U.N. Doc. A/CONF.62/L.84), Annex III, in *Official Records*, UNCLOS III, vol. XVI, p. 189.

170. *United States of America: Working Paper* (U.N. Doc. A/CONF.62/C.1/L.5), in *Official Records*, UNCLOS III, vol. III, p. 168.

171. See Pruitt, p. 345 n. 1. The likelihood that such controls would never be implemented should make them less, not more, objectionable to the United States. Moreover, because manganese production is expected to begin, if at all, only after commercial production of other metals has already commenced, the effective "interim period" for manganese would be less than 20 years.

172. See Keith, p. 296. "The question of the probable impact of sea-bed mining on world markets centres on the degree of competitiveness between marine and land-based sources of metal supply. To be rigorous, a study of this question would require a comparison of the relative supply costs of these two sources" — a comparison which will not become possible until hard data has been generated by actual seabed mining operations. *Note by the Chairman of the First Committee* (U.N. Doc. A/CONF.62/C.1/L.2), part II, para. 7, in *Official Records*, UNCLOS III, vol. III, p. 152.

173. Kenneth W. Clarkson, "International Law, U.S. Seabeds Policy and Ocean Resource Development," *Journal of Law and Economics* 17 (1974): 117.

174. Statement by the representative of the U.N. Secretariat, July 30, 1974, in *Official Records*, UNCLOS III, vol. II, p. 48 (para. 44).

175. *United States of America: Working Paper* (U.N. Doc. A/CONF.62/C.1/L.5), in *Official Records*, UNCLOS III, vol. III, p. 168.

176. See Jonathan I. Charney, "The Law of the Deep Seabed Post UNCLOS III," *Oregon Law Review* 63 (1984): 28–29. To satisfy traditional economic criteria, the seabed regime "would have to refrain from creating anti-efficiency measures such as restricting entry by new mines, limiting output, or supporting metals prices." Eckert, "Exploitation of Deep Ocean Minerals," p. 165.

177. Cf. *United States of America: Working Paper* (U.N. Doc. A/CONF.62/C.1/L.5), in *Official Records*, UNCLOS III, vol. III, pp. 166, 168. It has been pointed out that such cost savings may not be passed on to consumers. Statement of Francis T. Christy, Jr., in *The Law of the Sea*, ed. Amacher and Sweeney, p. 117.

178. See *Note by the Chairman of the First Committee* (U.N. Doc. A/CONF.62/C.1/L.2), part IIIE, para. 1, in *Official Records*, UNCLOS III, vol. III, p. 156. See also statement of the representative of the U.N. Secretariat, July 30, 1974, in *Official Records*, UNCLOS III, vol. II, p. 50 (para. 7).

179. Van Dyke and Yuen, pp. 531–32. "The magnitude of these effects is highly speculative, depending upon a host of factors." Ibid., p. 531.

180. Arvid Pardo, "The Convention on the Law of the Sea: A Preliminary Appraisal," *San Diego Law Review* 20 (1983): 500–1 n. 47.

181. Van Dyke and Yuen, p. 531. See also Hegwood, pp. 186–87. "Because fewer alternative investment and employment opportunities exist in developing than in developed countries, particularly heavy economic and social costs will be incurred in any re-allocation of resources that may be necessitated by the competition from sea-bed production." *Note by the Chairman of the First Committee* (U.N. Doc. A/CONF.62/C.1/L.2), part IIIF, para. 1, in *Official Records*, UNCLOS III, vol. III, p. 157.

182. *Note by the Chairman of the First Committee* (U.N. Doc. A/CONF.62/C.1/L.2), part IIIE, para. 1, in *Official Records*, UNCLOS III, vol. III, p. 156. "The income effect can be visualized as the relative change in the economy of a country (incomes and employment) attributable to a change in the price of metals. For some developing country producers of minerals this effect can be quite serious." UNCLOS III, *Economic Implications of Sea-Bed Mining in the International Area: Report of the Secretary-General* (U.N. Doc. A/CONF.62/37, February 18, 1975), para. 8 (b), in *Official Records*, UNCLOS III, vol. IV, p. 124.

183. Dennis W. Arrow, "The Customary Norm Process and the Deep Seabed," *Ocean Development and International Law Journal* 9 (1981): 8. An earlier estimate had placed the cost to the export earnings of developing producer states at approximately $360 million annually. *Note by the Chairman of the First Committee* (U.N. Doc. A/CONF.62/C.1/L.2), Part IIIA, para. 4, in *Official Records*, UNCLOS III, vol. III, p. 154.

184. *United States of America: Working Paper* (U.N. Doc. A/CONF.62/C.1/L.5), in *Official Records*, UNCLOS III, vol. III, pp. 166–68. The United States identified the six states as Chile, China, Peru, the Philippines, Zaire, and Zambia, and asserted that six developed states would also be primary beneficiaries of production controls: Australia, Canada, Japan, South Africa, the Soviet Union, and the United States. Ibid.

185. See, e.g., *Chile: Working Paper* (U.N. Doc. A/CONF.62/C.1/L.11), in *Official Records*, UNCLOS III, vol. III, p. 178.

186. *United States of America: Working Paper* (U.N. Doc. A/CONF.62/C.1/L.5), in *Official Records*, UNCLOS III, vol. III, p. 166. In the early 1980s, a "two-year drop in commodity prices . . . left nations from Chile to Zambia staggering under debt, a situation that add[ed] to the most serious international banking problem in decades." John R. Emshwiller, "Material Change: Commodity Price Rises May Augur a Recovery But Not for Everyone," *Wall Street Journal*, February 11, 1983, p. 1.

187. *Statement Made by Mr. G.D. Arsenis* (U.N. Doc. A/CONF.62/32), in *Official Records*, UNCLOS III, vol. III, p. 62.

188. Clarkson, "International Law, U.S. Seabeds Policy and Ocean Resource Development," p. 118. The Outer Continental Shelf Lands Act requires offshore oil and gas production to be "at rates consistent with any rule or order issued by the President." Outer Continental

Shelf Lands Act, 43 U.S.C. §1334 (g) (1). "If access to a limited common property resource is permitted to increase without restraint — in terms of the numbers of users, the extent and efficiency with which each participant uses the resource or, usually, both — the capacity of that resource will eventually be exceeded, resulting in its desruction." E.W. Seabrook Hull, "The International Law of the Sea: A Case for a Customary Approach," Occasional Paper No. 30 (Kingston, R.I.: Law of the Sea Institute, 1976), p. 9. Performance requirements, which "are common to almost all lease or claim systems," perform a regulatory function by preventing the hoarding of valuable mining sites by claimants unprepared to exploit them. D. Brooks, p. 416. See n. 168 above. Concession contracts generally do not contain provisions similar to article 151, since developing states could impose production controls unilaterally if they chose to do so. See Katz, pp. 128–29; and n. 259 below. Such controls "are consistent with . . . economic theories in practice throughout the world." Charney, "The Law of the Deep Seabed Post UNCLOS III," pp. 28–29. See also n. 633 below and accompanying text.

189. L.F.E. Goldie, "A General International Law Doctrine for Seabed Regimes," *International Lawyer* 7 (1973): 800. See Chap. 5 nn. 342–44, 346–48, 355, 404 and accompanying text. In the case of fisheries, overexploitation has resulted in "an industry with excess capacity relative to what is required to catch the maximum sustainable yield. This situation restricts fishermen and vessel owners to low, unstable incomes and may result in total production less than that obtainable with less investment and effort." U.S. Commission on Marine Science, Engineering, and Resources, *Our Nation and the Sea*, p. 92.

190. Bernard H. Oxman, "The Two Conferences," in *Law of the Sea: U.S. Policy Dilemma*, ed. Oxman et al., p. 138. See nn. 694–95 below and accompanying text; and also Arvid Pardo, "An Opportunity Lost," in *Law of the Sea: U.S. Policy Dilemma*, ed. Oxman et al., p. 167 n. 14. The lack of adequate restrictions on access to U.S. coastal waters has resulted in overcapitalization and reduced profits for North American fishermen. Sanger, p. 153. "Even the state of Texas has traditionally and actively regulated the amount of hydrocarbons produced on private lands." Charney, "The Law of the Deep Seabed Post UNCLOS III," p. 29 n. 47.

191. Saffo, pp. 517–18. See also statement of Lawrence Herman, February 24, 1979, in *Syracuse Journal of International Law and Commerce* 6 (1978–79): 294.

192. See Chap. 3 nn. 399–405 and accompanying text; and also Hauser, p. 315. As a vocal critic of Part XI has acknowledged, a regime of unrestricted access would produce an efficient allocation of seabed resources only where such "serious market imperfections . . . can be assumed away." Foders, p. 521. In fact, once a substantial amount of capital has been committed, it may be economically impossible to discontinue mining operations. See Brewer, "Deep Seabed Mining," pp. 28–29.

193. See, e.g., Breaux, "The Diminishing Prospects," pp. 267–68; and also statement of Representative Jim Santini in *Congressional Record* 126 (June 9, 1980): 13692. Such opposition to production-control measures is particularly difficult to comprehend in light of congressional recognition of the need for "limited entry schemes" to address the problem of overcapitalization in the fishing industry. See *Oceans Policy News* (November 1985), p. 3.

194. *Economic Implications of Sea-Bed Mineral Development* (U.N. Doc. A/CONF.62/25), in *Official Records*, UNCLOS III, vol. III, p. 39.

195. Statement of the representative of the United States in *Summary Records*, U.N. General Assembly, 23rd Session, Ad Hoc Committee, Economic and Technical Working Group, 8th meeting (U.N. Doc. A/AC.135/WG.2/SR.8, June 28, 1968), p. 5.

196. See *Chile: Working Paper* (U.N. Doc. A/CONF.62/C.1/L.11), in *Official Records*, UNCLOS III, vol. III, p. 178. In 1970 the General Assembly unanimously reaffirmed "that the development of the area and its resources shall be undertaken in such a manner as to . . . minimize any adverse economic effects caused by the fluctuation of prices of raw materials resulting from [seabed mining] activities." U.N. General Assembly, 25th Session, Resolution 2750A, December 17, 1970.

197. Statement of the representative of the United States, July 17, 1974, in *Official Records*, UNCLOS III, vol. II, p. 43 (para. 76). Kissinger, "A Test of International Cooperation," p. 540. See Chap. 3 nn. 87–88 and accompanying text.

198. Arrow, "The Proposed Regime," p. 402.

199. Statement of the representative of Ceylon in *Provisional Verbatim Records*, U.N.

General Assembly, 23rd Session, First Committee, 1588th meeting (U.N. Doc. A/C.1/ PV.1588, October 28, 1968), para. 58. "Most developing countries tend to sympathize with and be protective of raw material producers, a tendency that has been encouraged artfully by Canada, the leading nickel producer. Thus, it has long been clear that there could not be a generally accepted Law of the Sea Convention that does not contain an interim production ceiling." Ambassador George H. Aldrich, "Law of the Sea," *U.S. Department of State Bulletin* 81 (February 1981): 58.

200. See Barry Buzan, *Seabed Politics* (New York: Praeger, 1976), p. 229. "To have these provisions withdrawn from the treaty, we agreed to this interim production control." Statement of Ambassador T. Vincent Learson in *National Ocean Policy*, U.S. 94th Congress, 2nd Session, House of Representatives, Committee on Merchant Marine and Fisheries, Subcommittee on Oceanography (Washington: U.S. Government Printing Office, 1976), p. 204.

201. See Treves, "The EEC and the Law of the Sea," p. 180; American Bar Association, "Natural Resources of the Sea," p. 664; and also Barkenbus, p. 140. It is significant that the three largest nickel-producing states – Canada, France, and the Soviet Union – signed the Convention, despite the restrictions placed upon their own nodule mining operations.

202. See Barkenbus, p. 141; and Hull, p. 7. U.S. mining firms, naturally, opposed such potential restrictions upon their profit-making ability. Barkenbus, p. 141.

203. Aldrich, "Law of the Sea," p. 58. See Richardson, "Law in the Making," p. 444; and also Ambassador George H. Aldrich, "A System of Exploitation," *Syracuse Journal of International Law and Commerce* 6 (1978–79): 249.

204. The production ceiling was negotiated by Canada, the world's largest nickel exporter, and the United States, the world's largest nickel importer. See Chap. 3 nn. 144, 176 and accompanying text. "Offending Canada, one of our strongest allies, is perhaps a greater threat than the possibility of Canada's closing our supply of nickel." Pruitt, p. 350 n. 27.

The International Nickel Company of Canada is a major partner of one of the most advanced seabed mining consortia. "It follows, of course, that the Canadian Government can support only solutions which are equitable from the point of view of the . . . Enterprise, other potential sea-bed mines, and land-based producers including potential land-based producers." UNCLOS III, *Written Statement by the Delegation of Canada, Dated 2 April 1980* (U.N. Doc. A/CONF.62/WS/4, April 10, 1980), in *Official Records*, UNCLOS III, vol. XIII, pp. 101–2.

205. S.P. Jagota, "Developments in the UN Conference on the Law of the Sea: A Third World Review," *Third World Quarterly* 3 (1981): 305.

206. See ibid. "Consumer surplus" has been described as "a nonrecoverable loss in the economy due to the inefficient allocation of resources." Johnson and Logue, p. 52. The consumer surplus lost could amount to several billion dollars annually, with the United States forgoing approximately 10 percent of this amount. Statement of Dennis E. Logue in *The Law of the Sea*, ed. Amacher and Sweeney, p. 119. The cost would include lost revenue-sharing and other benefits for the Authority. Santa Cruz, p. 111.

207. Statement of President Ronald W. Reagan, July 9, 1982, reprinted in *U.S. Department of State Bulletin* 82 (August 1982): 71.

208. "After all, the free market benefits the player who starts out with a large pile of chips." Statement of Joseph Nye in *The Law of the Sea*, ed. Amacher and Sweeney, p. 109.

209. Richardson, "Law in the Making," p. 444. See Hauser, p. 119; *Possible Impact* (U.N. Doc. A/CONF.62/L.84), Annex III, in *Official Records*, UNCLOS III, vol. XVI, p. 187; U.S. House of Representatives, *National Ocean Policy*, p. 205; and also Agarwal et al., p. 122. A growth rate of 4 percent would permit the initial issuance of production authorizations to 6 or more full-sized, 3-million-ton-per-year seabed mining operations, with at least two authorizations becoming available every 18 months during the interim period. Hauser, pp. 119–20.

210. Richardson, "Law in the Making," p. 444. See also Aldrich, "Law of the Sea," p. 59. A production ceiling based upon that level of consumption was expected to permit about 15 3-million-ton-per-year seabed mining operations by 2002. Van Dyke and Yuen, p. 511 n. 70. The number of contractors able to obtain authorizations would depend upon the size of their seabed mining operations, however. See n. 215 below and accompanying text.

211. U.S. Department of State, "Access to Seabed Minerals Under the Draft Convention on the Law of the Sea," *Environmental Policy and Law* 7 (1981): 40. See also Katz, p. 129.

Several years earlier, a U.S. ambassador to UNCLOS III had stated that the production controls "will in no way at all adversely affect the United States." Statement of Ambassador T. Vincent Learson in *National Ocean Policy*, U.S. House of Representatives, p. 204.

The seabed mining industry has nevertheless continued to criticize the controls, arguing that they might become effective if economic conditions became extremely poor or if too many seabed mining operators sought production authorizations. Brewer, "Deep Seabed Mining," p. 49. See American Mining Congress, "Statement of Policy on the Law of the Sea Treaty Negotiations and United States Minerals Policy," in *Deep Seabed Mining*, U.S. House of Representatives, p. 108. It is under precisely such conditions that restrictions to prevent overexploitation would be most necessary, however. See n. 192 above and accompanying text.

212. Hauser, p. 120. See n. 160 above and accompanying text; and also Sebenius, p. 98. A growth rate as low as 2.2 percent would still permit nodule mining to supply 18 percent of world consumption initially and 36 percent by the end of the interim period. Richardson, "Law in the Making," p. 444.

213. *Oceans Policy News*, Special Report (September 1988), p. 3. See nn. 160, 165 above and accompanying text; and Sebenius, pp. 98–99. Concern has been expressed that state-subsidized mining operations would proceed with seabed mining despite adverse market conditions, and thereby threaten to exhaust the available production authorizations before the U.S. consortia could expect to begin commercial exploitation. Richard G. Darman, "Legislative Policy for Deepsea Mining in the Context of International Law of the Sea Negotiations," in *Congressional Record* 125 (December 14, 1979): 36050. Such concerns have given rise to calls in the Preparatory Commission by developed producer states for further limitations upon the issuance of production authorizations under unfavorable market conditions. *Oceans Policy News*, Special Report (March 1988), p. 3. See Chap. 3 n. 353, and text accompanying n. 192 above. Even in the absence of any production controls necessitating an apportionment of authorizations among pioneer investors, however, economic and technological factors had still been expected to limit the number of seabed mining operations to no more than 10 by the year 2000. Van Dyke and Yuen, p. 510 n. 70. Indeed, even if the actual rate of growth in nickel consumption remained below 3 percent, it is doubtful that the Convention would impose restrictions upon nodule production significantly beyond those imposed by weak market conditions. "The inherent requirements for profitability in deep seabed mining will, in any case, dictate that existing land-based nickel production not be replaced but that marine production account for no more than the growth segment." Hauser, p. 121. Thus, in 1976, Secretary Kissinger readily agreed to limit seabed production to the growth in consumption of nodule minerals—the limit which would be applied under article 151 if projected growth were below 3 percent. See text accompanying nn. 160, 197 above.

214. Aldrich, "Law of the Sea," p. 59. One treaty opponent has asserted, without elaboration, that the mere existence of an inapplicable production limitation in the Convention would nonetheless "cause market distortions and affect investment patterns." Representative John Breaux, "The Case Against the Convention," in *The 1982 Convention*, ed. Koers and Oxman, p. 13.

215. *Oceans Policy News*, Special Report (August/September 1989), p. 6. See Orrego Vicuña, "The Deep Seabed Mining Regime," p. 76.

216. *United States of America: Working Paper* (U.N. Doc. A/CONF.62/C.1/L.5), in *Official Records*, UNCLOS III, vol. III, p. 168.

217. Malone, "Law of the Sea and Oceans Policy," p. 49.

218. U.N. Convention on the Law of the Sea, Annex III, arts. 7 (2), 187 (b). See n. 164 above and accompanying text; and Hauser, p. 74.

219. T. Alexander, p. 138.

220. See nn. 235–39, 265 above and accompanying text. "If expansion is too rapid, nickel prices will fall, thus reducing the economic incentive for further expansions." *Economic Implications of Sea-Bed Mining* (U.N. Doc. A/CONF.62/37), para. 18, in *Official Records*, UNCLOS III, vol. IV, p. 125.

221. See *United States of America: Working Paper* (U.N. Doc. A/CONF.62/C.1/L.5), in *Official Records*, UNCLOS III, vol. III, p. 168.

222. "If all future demand for the minerals that can be produced from nodules were to

be supplied by this new source, this would amount to a policy decision to halt further growth for traditional producers." *Economic Implications of Sea-Bed Mineral Development* (U.N. Doc. A/CONF.62/25), in *Official Records*, UNCLOS III, vol. III, p. 25.

223. Eckert, "Exploitation of Deep Ocean Minerals," p. 168. Performance requirements may, in fact, promote efficient nodule recovery. See nn. 170, 188 above.

224. See n. 168 above; Deep Seabed Mining Regulations, 15 CFR §§971.418, 971.503; Seabed Act, 30 U.S.C. §1418 (b), (c); and also *Economic Implications of Sea-Bed Mineral Development* (U.N. Doc. A/CONF.62/25), in *Official Records*, UNCLOS III, vol. III, pp. 38–39.

225. Kissinger, "A Test of International Cooperation," p. 541. Kissinger proposed "that the . . . Authority have the right to participate in any international agreements on seabed-produced commodities in accordance with the amount of production for which it is directly responsible." Ibid., pp. 540–51. This is similar to the position taken by the Reagan administration. See n. 176 above and accompanying text.

226. Oxman, "The Ninth Session," p. 217. See n. 127 above and accompanying text; and also Geddes, p. 621 n. 58. Fear of an OPEC-like cartel—and of other supply disruptions resulting from political instability within producer states—was one of the principal reasons nodule mining originally was considered to be a potentially important source of strategic minerals. Pruitt, p. 349. See Chap. 2 nn. 205–11 and accompanying text.

227. Yost, pp. 671–72. See n. 175 above and accompanying text. The United States apparently favored a compensation system which could be applied selectively to benefit developing—and not developed—producer states. *United States of America: Working Paper* (U.N. Doc. A/CONF.62/C.1/L.5), in *Official Records*, UNCLOS III, vol. III, p. 169.

228. U.N. Convention on the Law of the Sea, art. 161 (2) (o) (ii). See nn. 323, 379 below and accompanying text.

229. See Arvid Pardo, "Before and After," *Law and Contemporary Problems* 46 (Spring 1983): 103 n. 24. The amount of cobalt recoverable from a single manganese crust mining operation is expected to amount to approximately 20 percent of world consumption. Johnson and Clark, p. 183. Nodule deposits also exist within the exclusive economic zone. See text accompanying Chap. 2. n. 136. Ironically, the production controls established by the Convention with respect to deep-sea nodule mining could indirectly benefit seabed mining operations within the U.S. exclusive economic zone to the extent that the latter are not subject to similar limitations. But see n. 190 above and accompanying text.

230. Peter Hay, *Federalism and Supranational Organizations* (Urbana: University of Illinois Press, 1966), p. 53. See John R. Stevenson, "Lawmaking for the Seas," *American Bar Association Journal* 61 (1975): 190. "The term 'supranational' connotes something approaching a federal system, with central organs exercising certain governmental functions and endowed with real power in defined subject areas to execute their policies and to decide what member states shall do. At the same time, each member state retains its independence to a far greater extent than any member state in a true federal system." Frederic L. Kirgis, *International Organizations in Their Legal Setting* (St. Paul: West, 1977), p. 603.

231. Hay, p. 31. "In contradistinction to national law, which confers upon tribunals the competence of applying the law and upon special organs the exclusive power to use force in executing the sanctions, there are under general international law no special organs for the application of the law and especially no central agencies for the execution of the sanctions. These functions are left to the states, which are subjects of international law. But under particular international law, the creation as well as the application of the law may be—and actually is—centralized, and this process of centralization is steadily increasing." Kelsen, *Principles of International Law*, pp. 402–3.

232. The supranationality criterion may be applied to legislative, executive, and judicial powers of international organizations. See Edward Yemin, *Legislative Powers in the United Nations and Specialized Agencies* (Leyden: A.W. Sijthoff, 1969), pp. 4, 6. A thorough analysis of supranationality with respect to the Convention thus should employ a six-part classification scheme: supranational legislative power, supranational executive power, supranational judicial power, quasi-supranational legislative power, quasi-supranational executive power, and quasi-supranational judicial power.

233. Goldwin, "Locke and the Law of the Sea," p. 49. See also Brokaw and Burke, p. 43.

234. White House fact sheet, in *U.S. Department of State Bulletin* 82 (March 1982): 55. See Brokaw and Burke, p. 47. "Regulation generally limits experimentation and productivity and raises costs — in other words, makes it more difficult for the market to function." Bandow, p. 482. See also Johnson and Logue, pp. 47–48.

235. Seabed Act, 30 U.S.C. §1441 (2).

236. See Robert L. Friedheim and J.B. Kadane, with John King Gamble, Jr., "Quantitative Content Analysis of the United Nations Seabed Debate," *International Organization* 24 (1970): 481–82. These records have been described as "the single best source" of representative data on the issue. Ibid., p. 482. "Often it is argued that some states have a special set of propaganda responses prepared for use in the UN and quite different statements of policies for the 'real world.' While there is some truth in these accusations, a member state's general point of view comes through in the constant restatement of the case." Ibid., p. 481.

Supranationalism had been a major point of conflict at UNCLOS I and UNCLOS II. See John Ludvik Løvald, "In Search of an Ocean Regime: The Negotiations in the General Assembly's Seabed Committee 1968–1970," *International Organization* 29 (1975): 689 n. 11.

237. Statement of the representative of Malta, November 1, 1967, in *Official Records*, U.N. General Assembly, 22nd Session, First Committee, 1516th meeting, pp. 1–2.

238. Conversation with Arvid Pardo, Los Angeles, April 14, 1982. Pardo intended the Authority to constitute one of three major international organizations, along with the United Nations and an organization regulating activities in outer space. Ibid.

239. See, e.g., statement of the representative of Finland, December 8, 1967, in *Official Records*, U.N. General Assembly, 22nd Session, First Committee, 1544th meeting, p. 5 (para. 46). Liberia called for establishment of "an international consortium as a machinery for the purpose" of "collective action in the control, operation and disposal of [seabed] resources." Statement of the representative of Liberia, November 14, 1967, in ibid., 1528th meeting, p. 7 (para. 58).

240. Ceylon, for example, cited the Maltese proposal "as a harbinger of the grandest of all visions of human society — the vision of the supra-national or world state." Statement of the representative of Ceylon, November 13, 1967, in ibid., 1526th meeting, p. 10 (para. 118).

241. See United Nations, *United Nations Ad Hoc Committee to Study the Peaceful Uses of the Sea-Bed and the Ocean Floor Beyond the Limits of National Jurisdiction, Replies from Governments* (U.N. Doc. A/AC.135/1, March 11, 1968), p. 25. The United Kingdom, with U.S. concurrence, expressed opposition to "creation of extensive administrative machinery." Statement of the representative of the United Kingdom in *Summary Records*, U.N. General Assembly, Ad Hoc Committee, Economic and Technical Working Group, 8th meeting (U.N. Doc. A/AC.135/WG.2/SR.8), p. 5. Statement of the representative of the United States in *Summary Records*, U.N. General Assembly, Ad Hoc Committee, Economic and Technical Working Group, 8th meeting (U.N. Doc. A/AC.135/WG.2/SR.8), p. 5.

242. Statement of the representative of the Union of Soviet Socialist Republics in *Provisional Verbatim Records*, U.N. General Assembly, 23rd Session, First Committee, 1592nd meeting (U.N. Doc. A/C.1/PV.1592, October 31, 1968), pp. 16–17. See also statement of the representative of the Union of Soviet Socialist Republics in *Summary Records*, U.N. General Asembly, Ad Hoc Committee, 12th meeting (U.N. Doc. A/AC.135/SR.12, July 9, 1968), pp. 7–8.

The Soviet Union's position was generally supported by its Eastern-bloc allies. Barkenbus, p. 110. See, e.g., statement of the representative of Poland in *Provisional Verbatim Records*, U.N. General Assembly, First Committee, 1588th meeting (U.N. Doc. A/C.1/PV.1588), p. 6 (para. 62). But see statement of the representative of Czechoslovakia in *Summary Records*, U.N. General Assembly, Ad Hoc Committee, Economic and Technical Working Group (U.N. Doc. A/AC.135/WG.2/SR.13), p. 4.

After 1970 the Soviet Union abandoned its ideological opposition to "supranational" organizations, recognizing the inevitability of international regulatory control of deep seabed mining operations. Barkenbus, pp. 110–11. See n. 247 below.

243. See, e.g., statement of the representative of Brazil in *Summary Records*, U.N. General Assembly, Ad Hoc Committee, 13th meeting (U.N. Doc. A/AC.135/SR.13, August 19, 1968), p. 3. Apparently in an effort to mollify the Soviet Union, the United Arab Republic claimed that "the element of supranationality ha[d] never occurred" to it or to the other

members of the Ad Hoc Committee. Statement of the representative of the United Arab Republic in *Provisional Verbatim Records*, U.N. General Assembly, 23rd Session, First Committee, 1593rd meeting (U.N. Doc. A/C.1/PV.1593, October 31, 1968), p. 63. Only Trinidad and Tobago stated that it saw "no alternative other than ... to create supra-national arrangements" for seabed mineral development. Statement of the representative of Trinidad and Tobago in *Provisional Verbatim Records*, U.N. General Assembly, 23rd Session, First Committee, 1601st meeting (U.N. Doc. A/C.1/PV.1601, November 6, 1968), p. 15.

244. Robert L. Friedheim and William J. Durch, "The International Seabed Resources Agency Negotiations and the New International Economic Order" (paper presented at 1976 Annual Meeting, American Political Science Association, Chicago, September 2–5, 1976), cited in Barkenbus, pp. 116–17. Enthusiasm for a comprehensive regulatory agency was dissipated by growing nationalism among coastal states, however. Løvald, p. 689.

245. Stephen K. Glickman, "Enforcement Mechanisms of the Law of the Sea Treaty," *Suffolk Transnational Law Journal* 1 (1977): 12–13.

246. Edward Wenk, Jr., *The Politics of the Ocean* (Seattle: Washington University Press, 1972), p. 428. See Glickman, p. 13; and Chap. 2 nn. 99–103 and accompanying text. "Since 1970 there has been very little support for international control over activities in the water column superjacent to the area." Barkenbus, p. 145.

247. Aaron L. Danzig, "A Funny Thing Happened to the Common Heritage on the Way to the Sea," *San Diego Law Review* 12 (1975): 663. See, e.g., statement of the representative of Jamaica, July 15, 1974, in *Official Records*, UNCLOS III, vol. II, p. 17 (para. 57). Because many UNCLOS III delegates had not participated in the work of the Seabed Committee and were unfamiliar with the more technical issues, and because many states took advantage of the opening of the Conference to set forth a bargaining position, these 1974 statements had a relatively high ideological content. Lee Kimball, "Implications of the Arrangements Made for Deep Sea Mining for Other Joint Exploitations," *Columbia Journal of World Business* 15 (1980): 56.

Even the Soviet Union endorsed a quasi-supranational executive power for the Authority. Statement of the representative of the Union of Soviet Socialist Republics, July 17, 1974, in *Official Records*, UNCLOS III, vol. II, p. 38 (para. 28). Its socialist allies also supported an allocation of broad regulatory competence. See, e.g., statement of the representative of Poland, July 16, 1974, in *Official Records*, UNCLOS III, vol. II, p. 18 (para. 8). The Ukrainian Soviet Socialist Republic apparently opposed any grant of supranational or quasi-supranational legislative competence, however, asserting that all rules should be set forth in the treaty itself. Statement of the representative of the Ukrainian Soviet Socialist Republic, July 17, 1974, in *Official Records*, UNCLOS III, vol. II, pp. 40–41 (para. 50). This was in line with the negotiating goal of Western states with respect to the issue of access, although the United States had acknowledged the need for the Authority to possess some quasi-supranational legislative powers.

248. See, e.g., statement of the representative of the Federal Republic of Germany, April 28, 1975, in *Official Records*, UNCLOS III, vol. IV, p. 67 (para. 13); and also statement of the representative of Australia, April 28, 1975, in ibid., p. 69 (para. 34). China, which opposed vesting the Authority with "supra-national powers," nonetheless argued that it "should have broad powers, including the right to direct exploration and exploitation of sea-bed resources, and should regulate all activities in the international area, such as scientific research, production, processing and marketing." Statement of the representative of the People's Republic of China, April 28, 1975, in ibid., p. 68 (para. 25).

249. U.N. Convention on the Law of the Sea, art. 157 (2). These incidental powers are undefined, but arguably are analogous to the "necessary and proper" clause of the U.S. Constitution. The Green Book would have expressly denied such powers to the Authority – "except those specified in rules, regulations and procedures" – apparently in an effort to limit its enforcement power. *The U.S. Proposals* (U.N. Doc. A/CONF.62/WG.21/Informal Paper 18), proposed art. 157 (2). The April 13 set of jointly sponsored amendments sought no changes to this article, however. Any excessive use of such incidental powers would be open to challenge as an abuse of power. U.N. Convention on the Law of the Sea, art. 187 (b) (ii).

250. U.N. Convention on the Law of the Sea, art. 1 (3). The United States had sought this limitation upon the Authority's regulatory competence. "Reports of the United States

Delegation," ed. Nordquist and Park, p. 92. The U.S. seabed mining industry nevertheless cited "problems" with the nonexclusive capacity of the Authority to conduct its own scientific research in the Area, apparently on the theory that this too might somehow hamper private seabed mining operations. See "Summary of Detailed Ocean Mining Industry Comments," in *Congressional Record* 126 (June 9, 1980): 13683. Cf. U.N. Convention on the Law of the Sea, art. 263 (1).

251. U.N. Convention on the Law of the Sea, art. 157 (1). See n. 279 below and accompanying text.

252. See H. Gary Knight, "The Draft United Nations Conventions on the International Seabed Area: Background, Description, and Some Preliminary Thoughts," *San Diego Law Review* 8 (1971): 523.

253. See statement of the representative of the United States, August 30, 1976, in *Official Records*, UNCLOS III, vol. VI, p. 64 (para. 6). Cf. Malone, "U.S. Policy and the Law of the Sea," p. 49; and Goldwin, "Locke and the Law of the Sea," p. 49. Under the treaty, the Enterprise must operate under the rules, regulations, and procedures of the Authority, as must state-sponsored mining operators. U.N. Convention on the Law of the Sea, art. 170 (2); Annex III, art. 12 (1); Annex IV, art. 1 (2).

254. "Commercial interests want to reduce or define more clearly the regulatory discretion of the Authority." Harry, p. 7. See also T. Patterson Willsey, "The Deep Seabed Hard Mineral Resources Act and the Third United Nations Conference on the Law of the Sea: Can the Conference Meet the Mandate Embodied in the Act?" *San Diego Law Review* 18 (1981): 511. The United States thus sought "an exhaustive and clear listing" of the Authority's powers and functions. Aldrich, "A System of Exploitation," p. 255.

255. See Larianne P. Gaertner, "The Disputes Settlement Provisions of the Convention on the Law of the Sea: Critique and Alternatives to the International Tribunal for the Law of the Sea," *San Diego Law Review* 19 (1982): 591.

256. See n. 416 below et seq. and accompanying text. "If an applicant is a qualified applicant and if his proposed plan of work meets the requirements set forth in the rules, regulations and procedures, that plan of work *must* be approved and signed as a contract, with only three exceptions. There is no discretion." U.S. Department of State, "Access to Seabed Minerals," p. 39 (emphasis in original).

257. U.N. Convention on the Law of the Sea, art. 152 (1). The Authority may nevertheless give "special consideration" to developing states where specifically permitted by the Convention. Ibid., art. 152 (2).

258. Although most provisions of the Convention which relate to the application of the rules, regulations, and procedures of the Authority make no temporal distinction with respect to the status of plans of work, article 17 of Annex III clearly implies that the rules, regulations, and procedures in effect at the time of an approval will continue to apply for contractors and the Enterprise. Hauser, p. 125. See nn. 264, 506 below. Cf. U.N. Convention on the Law of the Sea, art. 153 (1), (4); Annex III, arts. 3 (4) (c), 21 (1). See also Richardson, "Law in the Making," p. 443.

U.S. senators, nevertheless, objected to "articles that require applicants . . . to agree to accept as enforceable all rules and regulations adopted by the Authority—without regard to the timing of such adoptions." Statement of Representative Jim Santini in *Congressional Record* 126 (June 9, 1980): 13691.

259. U.N. Convention on the Law of the Sea, arts. 161 (1) (a), (8) (d), 162 (2) (m). See Brokaw and Burke, p. 52; and also Richardson, "Law in the Making," p. 443. Discretionary production controls employed by developing states have been accepted by international mining companies. Statement of the representative of Zambia, April 25, 1975, in *Official Records*, UNCLOS III, vol. IV, p. 57 (para. 28). The United States, however, regarded such controls as "completely inimical," and agreed to the fixed quota system of article 151 as a compromise in order to avoid them. Statement of Ambassador T. Vincent Learson in *National Ocean Policy*, U.S. House of Representatives, p. 204.

260. Harry, pp. 7, 16. An Interior Secretary stated that "the resource manager should . . . be given the discretion to issue licenses on flexible terms and conditions so as to permit the adjustment of ocean mining activities under a license to protect the environment, the safety of operators and to provide for the possibility of correcting any mistakes which Government

regulations may make." Statement of Interior Secretary Thomas Kleppe in *National Ocean Policy*, U.S. House of Representatives, pp. 56–57.

261. Seabed Act, 30 U.S.C. §1468 (c). See also ibid., §§ 1411 (b) (2), 1412 (b) (1), 1416 (a) (2) (A); and Deep Seabed Mining Regulations, 15 CFR §§971.102, 971.601–971.602.

262. *Draft United Nations Convention* (U.N. Doc. A/AC.138/25), arts. 67 (c)–(e), 68, 71. See n. 508 below and accompanying text.

263. Oxman, "The Ninth Session," p. 215. See nn. 126, 128 above and accompanying text. In fact, these policies may only be applied "as specifically for in" Part XI of the Convention. U.N. Convention on the Law of the Sea, art. 150. Aside from several references to sub-paragraph (h), which calls for the protection of developing producer states from the effects of seabed mining operations, the only specific reference in Part XI to the policies of article 150 is in paragraph 1 of article 155, which instructs the review conference to consider how well these objectives have been achieved. An earlier draft of the treaty specifically empowering the Authority to implement the policies of article 150 was deleted. See Harry, pp. 9–10.

264. See Yemin, p. 16. Because the rules, regulations, and procedures of the Authority only become binding upon mining operators by virtue of the submission of plans of work for approval, it might be argued that their observance technically is more in the nature of a contractual condition of access. See U.N. Convention on the Law of the Sea, art. 153 (3); Annex III, art 3 (4) (b). Cf. Glickman, p. 21. In this regard, it is significant that the rules, regulations, and procedures of the Authority in effect at the time a plan of work is approved cannot be altered with respect to a nonconsenting contractor by any subsequent modification of the Authority's regulatory regime. See nn. 501–2 below and accompanying text. The U.S. regulatory regime employs a similarly contractual legal approach, but would permit modification of terms of access by federal regulators without the operator's consent. See nn. 261, 504, 557 above and below and accompanying text. The Authority's regulatory competence, moreover, is not an exclusive one: states may continue to regulate their nationals, provided the standards are no less stringent than, and not inconsistent with, those of the Authority. U.N. Convention on the Law of the Sea, Annex III, art. 21 (3).

265. U.N. Convention on the Law of the Sea, Annex III, arts. 145, 162 (2) (w). See n. 503 below and accompanying text. Seabed mining may produce potentially toxic elements, including arsenic, cadmium, and chromium. Statement of Senator Edmund Muskie in *Congressional Record* 125 (December 14, 1979): 36076. See Eckert, "Exploitation of Deep Ocean Minerals," pp. 149–50 n. 30; and also U.N. Convention on the Law of the Sea, Annex III, art. 17 (2) (f). Cf. Seabed Act, 30 U.S.C. §1402 (b) (1). Article 145 was unchanged in the 1982 Green Book.

Measures for the protection of the deep-sea environment are desirable from an economic as well as ecological point of view, since environment damage is more difficult to repair than to prevent. Arrow, "The Proposed Regime," p. 392. See statement of Ambassador T. Vincent Learson, May 20, 1976, in *Law of the Sea*, U.S. 94th Congress, 2nd Session, Senate, Committee on Foreign Relations, Subcommittee on Oceans and International Environment (Washington: U.S. Government Printing Office, 1976), p. 10. "It is generally agreed . . . that it is appropriate to allow deep seabed mining to commence before all environmental research is completed. . . . This flexibility is essential." Statement of Senator Warren Magnuson in *Congressional Record* 125 (December 14, 1979): 36034. The lack of retroactive effect for the rules, regulations, and procedures of the Authority could prove problematic in this regard. Richardson, "Law in the Making," p. 443. Yet newly adopted environmental regulations could be included in the terms and conditions of renewed or modified plans of work, and the Council would in any event be able to issue emergency orders. U.N. Convention on the Law of the Sea, art. 162 (2) (w); Annex III, arts. 17 (2) (b) (iii), 19. See nn. 503, 506 below.

266. U.N. Convention on the Law of the Sea, arts. 146, 147 (2) (a). The Green Book sought no changes in these articles. Federal regulations issued pursuant to the Seabed Act contain a similar rule-making power — including, in addition, protection of "property at sea" as a further object of the Administrator's regulatory authority. Deep Seabed Mining Regulations, 15 CFR §§970.800, 971.422.

267. U.N. Convention on the Law of the Sea, Annex III, art. 17 (1) (b). See Gertler and Wayne, p. 552 n. 132. The words *"inter alia"* were added as part of the compromise by which

a consensus of the Council was required for the adoption of these rules, regulations, and procedures. Oxman, "The Ninth Session," p. 220 n. 42. The United States nevertheless sought their deletion during the final UNCLOS III session without offering to forgo the consensus procedure. See *Amendments* (U.N. Doc. A/CONF.62/L.121), in *Official Records*, UNCLOS III, vol. XVI, p. 230.

268. See nn. 260–61, 456 above and below and accompanying text; and also statement of Secretary of Commerce Juanita Kreps in *Deep Seabed Mining*, U.S. House of Representatives, p. 474. The U.S. seabed mining industry has expressed satisfaction with the federal regulatory regime. See Chap. 5 n. 14. But see n. 269 below.

269. OCS Lands Act, 43 U.S.C. §§1333 (d) (1), (e), 1334 (a). See letter from Charles N. Brower, Acting Legal Adviser, Department of State, to Senator Henry M. Jackson, July 26, 1973 (Attachment 1: U.S. Coast Guard Response to the Senate Committee on Interior and Insular Affairs to Selected Questions and Policy Issues Related to Oversight Hearings on the Administration of OCSLA, April 1972; Attachment 5: McKelvey, "Summary of Procedures"), reprinted in *Status Report*, U.S. Senate, pp, 436, 471. Offshore petroleum and natural gas production, like deep seabed mining, is "one of the most complex and capital intensive marine industries." *Description of Some Types of Marine Technology* (U.N. Doc. A/CONF.62/C.3/L.22), para. 21, in *Official Records*, UNCLOS III, vol. IV, p. 204. Predictably, the seabed mining industry has nevertheless criticized the regulatory discretion available to federal administrators with respect to hard mineral exploitation within the U.S. exclusive economic zone. See Chap. 3 n. 417; and also William Perry Pendley, "It Ain't Broke — Don't Fix It: Mining in America's Exclusive Economic Zone Requires No New Legislation," *Marine Technology Society Journal* 23 (March 1989): 49.

270. U.N. Convention on the Law of the Sea, art. 308 (4). See Gertler and Wayne, p. 550 n. 117; and Walkate, p. 165. The procedure, which is likely to result in an effective reduction in the supranational legislative powers of the Authority, "constituted an integral part of the package." Statement of the representative of the United States, August 26, 1980, in *Official Records*, UNCLOS III, vol. XIV, p. 39 (para. 120). Pursuant to the 1983 Consensus Statement of Understanding, such provisionally applicable rules, regulations, and procedures are to be adopted by consensus in the Commission. See text accompanying Chap. 3 n. 303. "This means that if consensus is not reached, the rules and regulations drafted by the Prepcom would remain in force indefinitely." Uwe Jenisch, "Bridging the Gap for Seabed Mining: Preparatory Instruments for the New Law of the Sea Convention," *San Diego Law Review* 18 (1981): 411.

In their set of amendments offered on April 13, 1982, developed states had proposed that the Commission adopt the Authority's rules, regulations, and procedures by an actual consensus of prospective Council members — who apparently were to be selected before the treaty entered into force — or, after 18 months, by a simple majority of signatory states "whose nationals or state entities are within the scope of Paragraph 1 of Resolution II." *Amendments* (U.N. Doc. A/CONF.62/L.121), Resolution I, proposed para. 4 (bis), in *Official Records*, UNCLOS III, vol. XVI, p. 231. Significantly, "any developing State which signs the Convention" was potentially within the scope of paragraph 1. Resolution II, para. 1 (a) (iii), in *International Legal Materials* 21 (1982): 1254.

271. U.N. Convention on the Law of the Sea, Annex III, art. 17 (1) (b) (xii). "The objective is to avoid wasteful mining methods which tend to be adopted when a company holds the mineral right over a very large property. In these cases, the company attempts to maximize short-run profits by mining only the highest grade ores leaving behind ore which would be normally considered of commercial recovery grade." *Economic Implications of Sea-Bed Mineral Development* (U.N. Doc. A/CONF.62/25), in *Official Records*, UNCLOS III, vol. III, p. 39. The 1980 Seabed Act provides the United States with a similar regulatory competence with respect to U.S. mining companies. Seabed Act, 30 U.S.C. §1420.

272. U.N. Convention on the Law of the Sea, art. 162 (2) (o) (ii). Although the United States agreed that adoption of the rules and regulations should be within the competence of the Council and that the consensus procedure should apply, it sought to give the Preparatory Commission primary responsibility for dealing with the issue, apparently out of concern that the Council might later block the development of non-nodule mineral resources. See *Amendments* (U.N. Doc. A/CONF.62/L.121), Annex III, proposed art. 17 (1) (b) (xv); Resolution I, proposed paras. 4 bis (b), 5 (g), in *Official Records*, UNCLOS III, vol. XVI, pp. 230–31. Highly

concentrated mineral deposits are known to exist upon the slopes of mid-oceanic ridges. Silverstein, pp. 163–64. See Chap. 3 nn. 407–12 and accompanying text. U.S. industry was also concerned about access to oil and natural gas deposits below the deep ocean floor. Marlene Dubow, "The Third United Nations Conference on the Law of the Sea: Questions of Equity for American Business," *Northwestern Journal of International Law and Business* 4 (1982): 201.

273. Cf. *The U.S. Proposals* (U.N. Doc. A/CONF.62/WG.21/Informal Paper 18), proposed art. 153 (4) (b). "We would a whole lot rather deal with our own Government, both as to the payment of taxes, and the terms of the contract, and let the United States Government deal with the Authority." Statement of Northcutt Ely (representing Deepsea Ventures) in *Deep Seabed Mining*, U.S. House of Representatives, p. 343. It is unclear why the seabed mining industry would prefer to deal only with the U.S. government, apart from an expectation of favorable treatment and political influence.

274. U.N. Convention on the Law of the Sea, arts. 160 (2) (e), 159 (8). See also n. 89 above. This power of assessment is much more limited than that granted the General Assembly in article 17 (2) of the U.N. Charter. Although shortfalls in the initial funding of the Enterprise may be met by further assessments, approval by a consensus of the Assembly would first be necessary and such assessments, therefore, cannot be made binding upon any state without its consent. U.N. Convention on the Law of the Sea, Annex IV, art. 11 (3) (c).

275. U.N. Convention on the Law of the Sea, art. 184. This provision is virtually identical to article 19 of the U.N. Charter.

276. Ibid., arts. 159 (8), 160 (m), 185. Because no mention is made in article 160 of their effect, it cannot be said with certainty that the Assembly would be bound by such recommendations. See n. 282 below. The Council must act by a three-fourths-majority vote. U.N. Convention on the Law of the Sea, arts. 161 (8) (c), 162 (2) (t). The Green Book had proposed no changes to these provisions.

277. U.N. Convention on the Law of the Sea, art. 139 (2); Annex III, art. 4 (4). Article 139 (2) arguably would encompass failure to cooperate in promoting marine scientific research or technology transfer. Cf. ibid., arts. 143 (3), 144 (2). These duties to cooperate are so weak, however, that a state party would be able to avoid liability by making only limited efforts to fulfill these obligations. See n. 13 above and accompanying text. A more interesting question arises with respect to liability for violation, "in relation to the Area," of "the principles embodied in the Charter of the United Nations and other rules of international law." U.N. Convention on the Law of the Sea, art. 138. The Green Book requested no changes in these liability provisions.

278. U.N. Convention on the Law of the Sea, Annex III, art. 22. See n. 498 below and accompanying text.

279. See ibid., arts. 139 (1), 153 (4), (5), and also Resolution II, para. 5 (b), in *International Legal Materials* 21 (1982): 1255. A three-fourths-majority vote of the Council would be required for the adoption of such enforcement measures. U.N. Convention on the Law of the Sea, arts. 161 (8) (c), 162 (2) (1). The Council will "establish appropriate mechanisms for directing and supervising a staff of inspectors who shall inspect activities in the Area" to ensure compliance by mining operators, a function for which the Legal and Technical Commission has been given primary responsibility. U.N. Convention on the Law of the Sea, arts. 161 (8) (c), 162 (2) (z), 165 (2) (c), (h), (m), (3). The rules, regulations, and procedures by which these inspectors operate must be approved by consensus in the Council. U.N. Convention on the Law of the Sea, art. 161 (8) (d); Annex III, art. 17 (1) (b) (viii). The 1970 U.S. Draft Convention endorsed international supervision of mining operations in the Area, and the United States elsewhere indicated its support for such supervision. See *Draft United Nations Convention* (U.N. Doc. A/AC.138/25), arts. 38, 40; and statement of the representative of the United States, August 26, 1979, in *Official Records*, UNCLOS III, vol. XII, p. 35 (para. 111). Yet the 1982 Green Book would have eliminated the inspection function of the Commission, as well as part of its supervisory function. *The U.S. Proposals* (U.N. Doc. A/CONF.62/WG.21/Informal Paper 18), proposed arts. 153 (5), 165 (3). The April 13 set of amendments subsequently retained these functions essentially intact, however, while requiring a concurrent majority of interest groups in the Council to approve their implementation. See *Amendments* (U.N. Doc. A/CONF.62/L.121), proposed arts. 153 (5), 161 (1) (c bis), 162 (2) (k), in *Official Records*,

UNCLOS III, vol. XVI, pp. 227–28. It is unclear why the Reagan administration apparently believed that the Authority's inspectors would be more of a burden upon U.S. seabed mining operators than would federal inspectors, the latter having been authorized by the administration's own seabed mining regulations "to board and accompany" any vessel engaged in deep-sea mining "at such times and to such extent as the Administrator deems reasonable and necessary," to obtain "access to and use of the vessel's navigation equipment and personnel when the observer deems such access necessary . . .," and to demand food and accommodations "equivalent to those provided to officers of the vessel." Deep Seabed Mining Regulations, 15 CFR §970.1105 (a), (d), as adopted September 15, 1981. See also Seabed Act, 30 U.S.C. §1464 (b).

280. U.N. Convention on the Law of the Sea, Annex III, art. 18 (1) (emphasis added).

281. Ibid., Annex III, art. 18 (2). Monetary penalties provide an important intermediate sanction without which the Authority's enforcement system would be inflexible and possibly dysfunctional. Glickman, p. 19.

282. U.N. Convention on the Law of the Sea, art. 162 (2) (v). Although a two-thirds-majority vote would suffice for such a recommendation, a three-fourths-majority vote would be required for the Council to submit the issue to the Sea-Bed Disputes Chamber initially. Ibid., arts. 161 (8) (b), (c), 162 (2) (u). "This procedure gives the Contractor a high degree of protection." Hauser, p. 79. The Council is itself to receive nonbinding recommendations on the matter from its Legal and Technical Commission. U.N. Convention on the Law of the Sea, art. 165 (2) (i), (j). See n. 398 below.

It is unclear whether the Assembly would be bound to accept the specific "recommendations" of the Council, as the Convention contains no express mention of the Assembly's power to impose such penalties. See Hauser, pp. 78, 138. Cf. U.N. Convention on the Law of the Sea, art. 160. It has been claimed that the recommendations would "admit of a non-binding interpretation." Glickman, p. 21. The Green Book would have required the Council, rather than the Assembly, to determine the appropriate penalty. *The U.S. Proposals* (U.N. Doc. A/CONF.62/WG.21/Informal Paper 18), proposed art. 162 (2) (u). This change was not regarded as sufficiently important to warrant inclusion in the April 13 set of amendments, however. Cf. *Amendments* (U.N. Doc. A/CONF.62/L.121), in *Official Records*, UNCLOS III, vol. XVI, p. 228.

283. U.N. Convention on the Law of the Sea, arts. 161 (8) (c), 162 (2) (a). In such cases the contractor must first be given "a reasonable opportunity to exhaust the judicial remedies available to him" before any penalties are assessed, unless an emergency order has been issued by the Council. Ibid., Annex III, art. 18 (3). See nn. 288, 503 below and accompanying text.

284. See Deep Seabed Mining Regulations, 15 CFR §§971.417, 971.1003. These penalties become applicable whenever a seabed mining operator "substantially fails to comply with" any provision of the Seabed Act, any regulation or order issued pursuant to the act, or any term, condition, or restriction in the license or the permit. Ibid., 15 CFR §971.417 (a) (1). In addition, seabed mining ships and "any hard mineral resource which is recovered, processed, or retained . . . shall be subject to forfeiture to the United States." Seabed Act, 30 U.S.C. §1466 (a).

285. Katz, p. 119.

286. See U.N. Convention on the Law of the Sea, art. 291. "Judicial settlement in the traditional sense is limited to States as parties. . . . The appearance of parties other than States is a new phenomenon first established by the Peace Treaties after the First World War." Hermann Mosler, "Supra-National Judicial Decisions and National Courts," *Hastings International and Comparative Law Review* 4 (1980): 429. More recently, the Court of the European Communities and several Latin American tribunals have been granted power to issue decisions binding upon non-state entities. Mosler, p. 429.

287. U.N. Convention on the Law of the Sea, art. 187 (c), (d), (e). See American Bar Association, "Natural Resources of the Sea," p. 670.

288. U.N. Convention on the Law of the Sea, art. 189 (emphasis added). See nn. 497–98 below and accompanying text; and Charney, "The Law of the Deep Seabed Post UNCLOS III," p. 39. It thus seems that the Chamber may not invalidate the supranational or quasi-supranational legislative acts of the Authority, but that it may enjoin supranational or quasi-supranational executive acts, may award damages for *ultra vires* executive or legislative acts,

and review any act of the Authority alleged to be in excess of jurisdiction or a misuse of power. See Hauser, pp. 54–55. These powers of review, while not absolute, are sufficient to prevent the Authority from becoming "a law unto itself," as at least one treaty critic has feared.Cf. Luke W. Finlay, "The Proposed New Convention on the Law of the Sea – A Candid Appraisal," *Syracuse Journal of International Law and Commerce* 7 (1979–80): 142.

289. Most delegations at UNCLOS III favored only limited judicial review of decisions taken collectively by sovereign states through the institutional machinery of the Authority. See Hanser, p. 54. The Chamber also lacks jurisdiction to review disputes between organs of the Authority. See ibid., p. 139.

290. U.N. Convention on the Law of the Sea, art. 296 (2); Annex III, art. 21 (2).

291. Norman Wulf, "Protecting the Marine Environment," in *Current Issues in the Law of the Sea*, ed. Joyner, pp. 97–97. See U.N. Convention on the Law of the Sea, arts. 21 (2), (4), 39 (2) (a), (b), 41 (3), (4), 42 (1) (b), 53 (9), 60 (5), 208 (3), 211 (2), (5), (6) (a), (c), 213, 214, 216, 217 (1), 218 (1), 220 (2), (3), 220.

292. Wilhelm H. Lampe, "The 'New' International Maritime Organization and Its Place in Development of International Maritime Law," *Journal of Maritime Law and Commerce* 14 (1983): 329. "U.S. proposals on ship-generated oil pollution were characterized by the Canadian delegation as giving IMCO the status of 'a universal law maker.'" J.D. Kingham and D.M. McCrae, "Competent International Organizations and the Law of the Sea," *Marine Policy* (April 1979): 117.

293. Kingham and McCrae, pp. 116, 119, 123–24. See D.P. O'Connell, *The International Law of the Sea* (Oxford: Clarendon, 1982), vol. 2, pp. 997–1012.

294. U.N. Convention on the Law of the Sea, art. 288 (2). This language would extend the application of the dispute settlement procedures to bilateral and regional treaties governing use of the sea to which the disputants were party.

295. Hans Kelsen, *Peace Through Law* (Chapel Hill: University of North Carolina Press, 1944), p. 39. In 1953 the International Law Commission proposed the creation of a supranational organization to implement high-seas fisheries conservation measures – an "international authority . . . empowered to declare . . . regulations to be binding upon the States in question and upon their nationals." International Law Commission, *Report of the International Law Commission to the General Assembly* (U.N. Doc. A/2456, September 1953), reprinted in *Yearbook of the International Law Commission, 1953*, vol. 2, p. 218 (para. 98).

296. Eugene Brooks, "International Organization for Hydrospace," in *The Law of the Sea: International Rules and Organization for the Sea*, ed. Lewis M. Alexander (Kingston: University of Rhode Island Press, 1967), p. 389. See Harold K. Jacobsen, *Networks of Interdependence* (New York: Alfred A. Knopf, 1979), p. 91. INTELSAT has been described as "perhaps the strongest operational international organization yet created." Barkenbus, p. 124. Plenipotentiary conferences of the ITU may modify the organization's constitutive convention. Stephen E. Doyle, "Regulating the Geostationary Orbit: ITU's WARC-ORB – '85–'88," *Journal of Space Law* 15 (1987): 3. The World Bank and similar international financial institutions possess a quasi-supranational judicial power. See Yemin, p. 25.

297. Convention on International Civil Aviation, art. 37, reprinted in *United Nations Treaty Series*, vol. 15, no. II-102, pp. 320, 322. "This . . . clause gives the Organization an open-ended authority to adopt regulations on all matters falling within the general field of air navigation which it considers appropriate for international regulation." Yemin, p. 120. ICAO's rules and regulations governing civil air traffic over the high seas possess "the legal force necessary to be considered legislation, although express sanctions for their violation are not specified in the Convention." Yemin, pp. 150–51. In the 1970s the United States had sought "[a] rule-making procedure similar to ICAO's" for the Authority. "Reports of the United States Delegation," ed. Nordquist and Park, p. 93.

298. C. John Colombos, *The International Law of the Sea*, 4th revised ed. (London: Longmans, Green, 1959), p. 362. The Commission has lacked an effective enforcement mechanism, however. See Chap. 5 n. 488.

299. Eighty percent of WHO delegations have been found to favor giving the organization supranational or quasi-supranational legislative powers. Hanna Newcombe, "National Patterns in International Organizations," *Peace Research Reviews* 6 (November 1975): 312.

300. See Convention on the Regulation of Antarctic Mineral Resource Activities, arts.

12, 21, 31, 43, 52, reprinted in *International Legal Materials* 27 (1988): 874–75, 878–79, 883, 889, 893. "The Commission determines what areas may be identified for exploration and development and which areas will be protected from mineral activities. It has the capacity to adopt measures to insure the protection of the Antarctic environment and promote safe and effective exploration and development.... In addition, the Commission sets all fees and levies, adopts the budget and provides for the disposition of revenues." Mark P. Jacobsen, "Recent Developments: International Agreements," *Harvard International Law Journal* 30 (1989): 240–41.

301. Yemin, pp. 92, 107, 113, 168.

302. U.N. Charter, art. 94. See Statute of the International Court of Justice, art. 34 (1). The Security Council may require disputants to submit to the jurisdiction of the Court or of some other dispute resolution procedure. U.N. Charter, art. 33 (2).

303. Robert W. Tucker, "The Principle of Effectiveness in International Law," in *Law and Politics in the World Community*, ed. George A. Lipsky (Los Angeles: University of California Press, 1953), p. 335 n. 56. "The Members of the United Nations agree to accept and carry out the decisions of the Security Council." U.N. Charter, art. 25.

304. U.N. Charter, art. 42. "Under Chapter VII of the Charter, the Security Council is empowered, for the purpose of maintaining or restoring international peace and security, to enact general rules of conduct obligatory for all members." Yemin, p. 23. The Security Council relied on this legislative power in adopting binding sanctions against Iraq in 1990 following its invasion of Kuwait. See Chap. 6 n. 19.

305. U.N. Charter, art. 17. See Yemin, pp. 23–24. Because this authority to assess U.N. member states is theoretically unlimited, it may properly be described as a legal power of taxation.

306. See Kelsen, *Principles of International Law*, pp. 114, 115, 125, 128, 129.

307. Statement of Northcutt Ely before Subcommittee on Arms Control, Oceans, International Operations and Environment, Committee on Foreign Relations, U.S. Senate, September 30, 1982, reprinted in *Congressional Digest* 62 (January 1983): 29.

308. David D. Caron, "Reconciling Domestic Principles and International Cooperation," in *Law of the Sea: U.S. Policy Dilemma*, ed. Oxman et al., pp. 5–6.

309. Hicks, p. 113. Under the doctrine of sovereign equality, national sovereignty vis-à-vis other states could not be diminished—or enhanced—by such participation. See Chap. 5 n. 207.

310. See Robert L. Carneiro, "Political Expansion as an Expression of the Principle of Competitive Exclusion," in *Origins of the State: The Anthropology of Political Evolution*, ed. Ronald Cohen and Elman R. Service (Philadelphia: Institute for the Study of Human Issues, 1978), p. 217; and Harry, p. 7. Group of 77 support for a strong Authority is motivated by pragmatic rather than ideological considerations: "fear that if decisions and actions are left to the discretion of traditional mining entities, little express attention will be devoted to the concerns of developing nations." Barkenbus, p. 112.

311. Barkenbus, p. 137. "It would not have discretionary power to control entry, assign markets, fix prices, or otherwise intervene in opposition to market forces." Oxman, "The Ninth Session," p. 214. The Interstate Commerce Commission has exercised such regulatory functions in the United States since 1887. See Alexandra M. Post, "United Nations Involvement in Ocean Mining," in *Aspekte der Seerechtsentwicklung*, ed. Wolfgang Graf Vitzthum (Munich: Hochochule der Bundeswehr, 1980), p. 187.

312. Statement of James Malone before Plenary, August 13, 1981, quoted in Bandow, p. 491.

313. See statement of the representative of the United States before Seabed Committee, August 3, 1970, reprinted in *U.S. Department of State Bulletin* 63 (August 24, 1970): 211. The 1970 Draft Treaty would have empowered the Council, with the approval of two thirds of states parties, to regulate "virtually every aspect of resources exploitation operations" in the Area. Knight, "The Draft United Nations Conventions," p. 502.

314. U.S. Senate Resolution 82, adopted July 9, 1973, reprinted in *Status Report*, U.S. Senate, p. 219.

315. Newcombe, p. 23. See n. 242 above and accompanying text. Although the Soviet Union acquiesced in the creation of the Authority to govern seabed exploitation in the early

1970s, it continued to fear that the common heritage of mankind principle would lead to a loss of national sovereignty and the creation of a global supranational organization. Sylvia Maureen Williams, "International Law Before and After the Moon Agreement," *International Relations* 7 (1981): 1173.

316. See Bernard H. Oxman, "Summary of the Law of the Sea Convention," in *Law of the Sea: U.S. Policy Dilemma*, ed. Oxman et al., pp. 157–58; and also H. Jacobsen, p. 93.

317. Yost, p. 674; see Seyom Brown et al., *Regimes for the Ocean, Outer Space, and Weather* (Washington: Brookings Institution, 1977), p. 112.

318. See Bandow, p. 491. In January 1982 President Reagan stated that the decisionmaking procedure must be one which "fairly reflects and effectively protects the . . . financial contributions of participating states." Reagan, "U.S. Policy and the Law of the Sea," p. 54. In the mid-1970s the United States had simply required a "fair and thoughtful" decisionmaking procedure, with "realistic" voting arrangements in the Council. Statement of the representative of the United States, July 17, 1974, in *Official Records*, UNCLOS III, vol. II, p. 43 (para. 76). See also Barkenbus, p. 124.

319. Malone, "U.S. Participation," p. 62. Although the somewhat elaborate system of representation and voting procedures is designed to fulfill the second of these requirements, the first is somewhat ambiguous: any state represented by a delegation competent in diplomacy may be expected to exert some "affirmative influence" upon the negotiations; to expect each state to be able to determine the fate of legislation with respect to interests it subjectively defines as vital is unrealistic—if not Utopian—and incompatible with the principle of sovereign equality. See Chap. 5 n. 207 and accompanying text. A 1975 list of 12 "critical elements" for "an acceptable version" of the international machinery made no mention of either of these requirements. See "Reports of the United States Delegation," ed. Nordquist and Park, pp. 92–93.

320. Statement of President Reagan, July 9, 1982, in *U.S. Department of State Bulletin* 82 (August 1982): 71.

321. See Katz, p. 119. "This difference flows from the "common heritage" concept, which vests rights in the resources of the deep seabed in all mankind, whereas the rights to the resources in developing countries are, of course, vested in those countries." Ibid., p. 116.

322. See Julio Fuandez, "The Sea-Bed Negotiations: Third World Choices," *Third World Quarterly* 2 (1980): 494. The seabed mining industry understands that such considerations could not be completely eliminated from the decisionmaking process. Ratiner, "Reciprocating State Arrangements," p. 197. Indeed, political premises inhere in the values upon which even "technical" decisions are ultimately based. See n. 362 below and accompanying text.

323. U.N. Convention on the Law of the Sea, arts. 161 (8) (d), 162 (2) (m), (o), 150 (h), 151 (1) (a). Consensus is also required for the Council to take any actions "otherwise" within its competence which are not specifically assigned a voting majority requirement. Ibid., art. 161 (8) (f). Consensus is defined as "the absence of any formal objection," and a two-week conciliation procedure would be invoked when such an objection was asserted. Ibid., art. 161 (8) (e). See Hauser, p. 43.

324. U.N. Convention on the Law of the Sea, arts. 161 (8) (c), 162 (2) (a), (e), (l), (q), (r), (s), (t), (w), (x). See nn. 164, 276, 279, 282, 470, 503 above and below and accompanying text. The Assembly may only accept or reject—not modify—the recommendations forwarded to it by the Council. See Hauser, pp. 47, 138.

325. U.N. Convention on the Law of the Sea, arts. 161 (8) (b), 162 (2) (i). The United States would have deleted this provision and substituted a power to approve the budget of the Enterprise by a concurrent three-fourths-majority vote. *Amendments* (U.N. Doc. A/CONF.62/L.121), proposed arts. 161 (1) (c bis), 162 (2) (i), in *Official Records*, UNCLOS III, vol. XVI, pp. 227–28.

326. U.N. Convention on the Law of the Sea, arts. 161 (8) (b), 162 (2) (f). Such agreements must be within the Authority's competence and must be approved by the Assembly. Ibid., art. 162 (2) (f).

327. Ibid., arts. 161 (8) (b), 162 (2) (n), (v). Concurrent three-fourths-majority approval would have been required for these recommendations under the U.S. amendments of April 13. *Amendments* (U.N. Doc. A/CONF.62/L.121), proposed art. 161 (1) (c bis), in *Official Records*, UNCLOS III, vol. XVI, p. 227.

328. U.N. Convention on the Law of the Sea, art. 161 (8) (g).

329. Ibid., art. 161 (1) (a)-(e). The precise definition of "equitable geographical distribution," like that of the other categories of interest groups, is to be determined in the rules, regulations, and procedures of the Authority. See *Oceans Policy News* (July 1990), p. 2.

Similarly, the ITU's 36-member Administrative Council is elected every 8 to 10 years at its plenipotentiary conferences, which are comprised of all member states and which are vested with "supreme authority." Donna C. Gregg, "Capitalizing on National Self-Interest: The Management of International Telecommunication Policy by the International Telecommunication Union," *Law and Contemporary Problems* 45 (1982): 43.

330. U.N. Convention on the Law of the Sea, art. 161 (1) (a), (b). "The United States is, for the foreseeable future, the largest mineral consuming nation." Charney, "The Law of the Deep Seabed Post UNCLOS III," p. 31.

331. Elliot L. Richardson, "The Law of the Sea Conference: Is a Comprehensive Treaty Still Possible?" in *Report: The Third United Nations Law of the Sea Conference,* U.S. 95th Congress, 2nd Session, Senate (Washington: U.S. Government Printing Office, 1978), p. 113.

332. U.N. Convention on the Law of the Sea, art. 161 (2) (c). The Assembly "shall ensure" that the nominees of each interest group are given seats on the Council. Ibid., art. 161 (2). It is thus not accurate to state that Council members are "chosen" or "selected" by the Assembly. The size of the majority required to nominate members has not been specified, however. See Hauser, p. 42 n. 32.

333. Willsey, p. 516. See also Sanger, pp. 183-84. The United States would likely be joined in these two groups by five of its traditional allies: France, Japan, Germany, the United Kingdom, and either Italy, Belgium, or the Netherlands.

334. "Group," as used in article 161, apparently refers to the five categories of interest groups listed in paragraph 1. See text accompanying n. 401 above. This reading is implied from paragraph 3, which provides that the initial term for one half of each such "group" is to be two years. The "Eastern (Socialist) European region" must be construed as a "geographical region" rather than a "group," under article 161, and states of that region are therefore not entitled to select their own nominees under paragraph 2 (c). In contrast to the Convention, the September 1986 Understanding adopted by the Preparatory Commission specifically listed eight Soviet-bloc states as the "socialist States of Eastern Europe." Understanding of the Preparatory Commission for the International Sea-Bed Authority and for the International Tribunal for the Law of the Sea for Proceeding with Deep Sea-Bed Mining Applications and Resolving Disputes of Overlapping Claims of Mine Sites, para. 21 n. a, reprinted in *International Legal Materials* 25 (1986): 1330. Although a seat on the U.N. Security Council was reserved for the Eastern European region, that seat more often than not was given to states outside the Soviet bloc — including Greece, which was controlled by a democratically elected "Socialist" government throughout most of the 1980s. See Newcombe, p. 268.

335. See n. 330 above. Cf. statement of Theodore Kronmiller, September 15, 1982, in *Law of the Sea Negotiations,* U.S. Senate, pp. 17-18. No state has argued that the provision would not guarantee the United States a seat on the Council. Nandan, p. 517.

336. Statement of James Malone before House Committee on Foreign Affairs, April 29, 1981, quoted in Larson, "The Reagan Administration," p. 304. It has even been claimed that "the Soviet Union would have three votes *plus* all their satellites would have a vote." Statement of Senator Larry Pressler before U.S. Senate, March 8, 1982, reprinted in *Congressional Digest* 62 (January 1983): 17 (emphasis added). The U.S. proposals of 1982 explicitly retained these "guaranteed" seats. *Amendments* (U.N. Doc. A/CONF.62/L.121), proposed art. 161 (1) (a), (b), in *Official Records,* UNCLOS III, vol. XVI, p. 227.

337. Arrow, "The Proposed Regime," p. 397 n. 386. If Canada were instead given a seat by either the seabed-investment or 2-percent-of-consumption-or-imports group, other "Western" states — such as Australia — might be nominated by the major net exporters. The developing states selected to represent the interests of this group might include Brazil, Zaire, Gabon, Zambia, Chile, or Peru. The composition of this interest group is most difficult to predict, however, as the term "major net exporters" was left undefined.

338. Newcombe, p. 275. See also William B. Jones, "The International Sea-Bed Authority Without U.S. Participation," *Ocean Development and International Law Journal* 12 (1983):

156. "Clearly their cooperation is thought to be essential for achieving the purposes of the organization." H. Jacobsen, p. 95.

339. See Oxman, "Summary of the Law of the Sea Convention," p. 158; and Barkenbus, p. 128. The Assembly, apparently the nominating body for the 18-member "equitable geographical distribution" group, might give the "Western Europe and others" seat to one of the Scandinavian or Commonwealth states which frequently supported the developing states' position on the common heritage of mankind, or to one of the smaller developed states which have sought representation on the Council. Gertler and Wayne, p. 558 n. 174. Developing states most likely to hold Council seats include India, Mexico, China, Indonesia, Egypt, and Nigeria. W. Jones, "The International Sea-Bed Authority Without U.S. Participation," p. 156. Some smaller developing states may also be expected to obtain Council seats, particularly in light of a requirement that landlocked and geographically disadvantaged states be "represented to a degree which is reasonably proportionate to their representation in the Assembly." U.N. Convention on the Law of the Sea, art. 161 (2) (a).

340. Oxman, "Summary of the Law of the Sea Convention," p. 158. See also statement of Elliot L. Richardson, May 14, 1971, quoted in Oxman, "The Third United Nations Conference on the Law of the Sea," p. 11 n. 27; and Sanger, p. 184.

341. U.N. Convention on the Law of the Sea, art. 161 (3). Each member is to be elected to a four-year term, membership terms being staggered in order to provide new elections for half of each interest group every two years. "Members of the Council shall be eligible for re-election; but due regard should be paid to the desirability of rotation of membership." Ibid., art. 161 (4).

342. See Carol Barrett and Hanna Newcombe, "Weighted Voting in International Organizations," *Peace Research Reviews* 2 (April 1968): 68, 73; and Barkenbus, p. 128.

343. Statement of Ambassador T. Vincent Learson in *National Ocean Policy*, U.S. House of Representatives, p. 201.

344. Statement of Leigh S. Ratiner in *National Ocean Policy*, U.S. House of Representatives, p. 208. The actual composition of the Council under the Convention may be expected to approach this high degree of producer- and consumer-state representation.

An earlier U.S. Draft Treaty had called for equitable geographical distribution of Council seats, qualified only by assured representation for the six states parties with largest aggregate GNP—a condition also likely to be satisfied under the Convention. See Knight, "The Draft United Nations Conventions," p. 497.

345. See Louis B. Sohn, "Possible Future Regimes of the Sea-Bed Resources: International Regulatory Agency," in *Symposium on the International Regime of the Sea-Bed*, ed. Jerzy Sztucki (Rome: Accademic Nazionale dei Lincei, 1970), p. 397; and Newcombe, p. 236. Cf. U.N. Charter, art. 23, as amended.

346. Statute of the International Atomic Energy Agency, art. VI, reprinted in *United Nations Treaty Series*, vol. 276, no. 3988, pp. 12, 14.

347. Sohn, "Possible Future Regimes," p. 397. See also Colombos, pp. 186, 381. Employers' and workers' delegates each hold 25 percent of the voting power in the ILO's plenary conference. See Barrett and Newcombe, pp. 40–41.

348. U.N. Charter, art. 86 (1).

349. States parties to the 1959 Treaty may attain the status of Consultative Party by undertaking a substantial amount of scientific research on the Antarctic continent. See Edward E. Honnold, "Thaw in International Law? Rights in Antarctica Under the Law of Common Spaces," *Yale Law Journal* 87 (1978): 833.

350. Hicks, p. 114.

351. *Amendments* (U.N. Doc. A/CONF.62/L.121), proposed art. 161 (1) (a), in *Official Records*, UNCLOS III, vol. XVI, p. 227. The proposal also required two of the four seats in the 2-percent-of-consumption-or-net-imports group to be given to states eligible for representation in the seabed investment group. Ibid., proposed art. 161 (1) (b), in *Official Records*, UNCLOS III, vol. XVI, p. 227.

352. See Barrett and Newcombe, p. 94.

353. Tollison and Willett, pp. 99–100.

354. See Brewer, "Deep Seabed Mining," p. 50. The Council's system of interest-group

representation operates effectively as a system of weighted voting. See Barrett and Newcombe, p. 68; and also S. Brown et al., p. 119.

From the beginning of the UNCLOS III negotiations, developed states insisted upon a blocking vote. Harry, p. 6. Apparently in exchange for developing-state agreement to the system of interest-group representation and to a potential blocking power for developed states, the United States and other Western states accepted the Group of 77 plan for "a voting system keyed to the importance of specific issues to be decided by the Council." Richardson, "The Politics of the Law of the Sea," pp. 16–17. See text accompanying nn. 323–27 above.

355. See statement of Brian Hoyle in *Consensus and Confrontation*, ed. Van Dyke, p. 264; and also statement of Leigh S. Ratiner in *National Ocean Policy*, U.S. House of Representatives, p. 208. A system of pure weighted voting based upon the single factor of population is arguably more consistent with democratic theories of representation. See Barrett and Newcombe, p. 5. A congressional critic of the Convention has complained that the Council's voting procedure is "very undemocratic." Breaux, "The Diminishing Prospects," p. 281. However, the United States, with less than 5 percent of the world's population, would not be particularly well represented under a voting system weighted on the basis of population, and it has not encouraged such systems in other international organizations. Barrett and Newcombe, p. 28.

356. *Amendments* (U.N. Doc. A/CONF.62/L.121), proposed art. 161 (1) (c bis), in *Official Records*, UNCLOS III, vol. XVI, p. 227. See Chap. 2 n. 76; Richardson, "The Politics of the Law of the Sea," p. 77; and also Laursen, p. 92.

357. Yost, pp. 676–77. Decisions of the Security Council on substantive issues are taken by an affirmative vote of 9 of the 15 members, "including the concurring votes of the permanent members." U.N. Charter, art. 27 (3), as amended. Voting procedure was the most difficult issue for the drafters of the U.N. Charter, as it was for UNCLOS III delegates. See Felix Morley, *The Charter of the United Nations: An Analysis* (New York: American Enterprise Association, 1946), p. 19.

A formal U.S. veto power would inevitably have given rise to a demand by the Soviet Union and other states for their own veto rights and required a degree of cooperation which has been historically lacking; indeed, the inability of the Security Council to fulfill its envisioned role for the preservation of international peace and security has generally discredited the idea of a veto power. See Brewer, "Deep Seabed Mining," pp. 51, 67 n. 69. Moreover, the status of states in the international community may be expected to change with the passage of time, as recent events in the Soviet Union would seem to demonstrate. "We learn from the course of history . . . that Great Powers may fall back into the role of a small state." P.H. Kooijmans, *The Doctrine of the Legal Equality of States* (Leyden: A.W. Sijthoff, 1964), p. 243. Significantly, the United States agreed in 1971 to relinquish its veto power on INTELSAT's Board of Governors. Stephen E. Doyle, "Permanent Arrangements for the Global Commercial Communication Satellite System of INTELSAT," *International Lawyer* 6 (1972): 287.

358. Brewer, "Deep Seabed Mining," p. 50. The International Atomic Energy Agency takes some decisions by a two-thirds-majority vote and others by a simple majority. Statute of the International Atomic Energy Agency, art. VI (E), in *United Nations Treaty Series*, vol. 276, no. 3988, p. 16. ICAO requires only a two-thirds-majority vote of its council to promulgate regulatory standards unless a majority of member states register disapproval within three months, a procedure endorsed by the United States. Yemin, p. 128. See n. 297 above. Cf. *Draft United Nations Convention* (U.N. Doc. A/AC.138/25), art. 67 (e). In INTELSAT, which uses a capital-weighted voting procedure in its Board of Governors, the United States has forgone its original veto power and U.S. voting power declined from 61 percent in 1964 to 21 percent in 1980 as other states have increased their participation in the organization's activities. Donald Cameron Watt, "The Law of the Sea Conference and the Deep Sea Mining Issue: The Need for an Agreement," *International Affairs* (London) 58 (1981–82): 91. See also Doyle, "Permanent Arrangements," pp. 279–80, 287.

359. Statement of Theodore Kronmiller, September 15, 1982, in *Law of the Sea Negotiations*, U.S. Senate, p. 5. See W. Smith, p. 59. "This voting mechanism would effectively insulate most council decisions from the influence of the handful of Western industrialized nations that would be represented on the council, ensuring that the less developed nations would dominate the council as surely as they dominate the assembly." John Breaux, "What Course

to a Law of the Sea?" *Sea Power* 26, no. 5 (Special Edition, April 15, 1983): 35. See also statement of Senator James McClure in *Congressional Record* 125 (December 14, 1979): 36041.

360. See Jagota, "A Third World Review," p. 294; Yost, p. 667; and also Brewer, "Deep Seabed Mining," p. 40. The likelihood of ideological competition in the Authority has decreased with improving U.S.–Soviet relations. See n. 77 above. "Much has changed. . . . The Soviet Union has taken many dramatic and important steps to participate fully in the community of nations." President George H.W. Bush, "The UN: World Parliament of Peace," Current Policy No. 1303 (Washington: U.S. Department of State, 1990), p. 1.

361. Yost, pp. 668–69. See Kimball, "Implications of the Arrangements Made," p. 56. "It is a common fallacy to consider all developing countries as major world suppliers of minerals with interests identical to a Third World subcategory of land-based producers." Post, "United Nations Involvement," p. 194. See also Jagota, p. 305.

362. See Newcombe, pp. 102–3, 131–32, 295–96, 297, 298. Such political considerations may be expected to play a significant part in the Council's decisionmaking procedure. Watt, "The Need for an Agreement," p. 87.

363. U.N. Convention on the Law of the Sea, art. 161 (8) (e).

364. See James L. Malone, "Who Needs the Sea Treaty?" *Foreign Policy* 54 (1984): 55. But see n. 401 below and accompanying text. This fear may be traced to the failures of the U.N. Security Council, as well as to the arduous negotiating process at UNCLOS III. See n. 357 above; and also Hauser, p. 44.

365. U.N. Convention on the Law of the Sea, art. 162 (2) (j) (i). See Willsey, p. 524. The sponsoring state would be excluded from participation in the decision. U.N. Convention on the Law of the Sea, art. 162 (2) (j) (i). "The assumption is that a contract that satisfies the technicalities in the . . . judgment of the experts should proceed unless there is considerable opposition; an opponent would have to meet the high standard of assembling a consensus against it." Milner S. Ball, "Law of the Sea: Expression of Solidarity," *San Diego Law Review* 19 (1982): 468.

The 1982 Green Book would have eliminated the role of the Council in the approval process, requiring instead a consensus of the Technical Subcommission to overcome a presumption that each applicant is in fact qualified under Annex III. *The U.S. Proposals* (U.N. Doc. A/CONF.62/WG.21/Informal Paper 18), proposed arts. 153 (3), 165 (3) (g); Annex III, proposed art. 6 (2) (bis). The April 13 set of proposed amendments maintained the Council's power of final approval. *Amendments* (U.N. Doc. A/CONF.62/L.121), proposed art. 153, in *Official Records*, UNCLOS III, vol. XVI, p. 227.

366. Gertler and Wayne, p. 552. "Reports of the United States Delegation," ed. Nordquist and Park, p. 412. "The issues subject to the consensus requirement are of the type that will arise rarely, usually upon the occurrence of a predetermined event." Willsey, p. 522.

367. See Ball, pp. 463–64; and Kimball, "Implications of the Arrangements Made," pp. 54–55. "The consensus system grants the industrialized nations exactly the kind of protection they have negotiated for," and "may be the only procedure acceptable to all concerned." Willsey, pp. 520–22.

368. Willsey, p. 520. See also Ball, p. 465.

369. Barry Buzan, "Negotiating by Consensus: Developments in Technique at the United Nations Conference on the Law of the Sea," *American Journal of International Law* 75 (1981): 327. "If anyone is really attempting to build something that lasts, certainly something as important as a lawmaking instrument, it is now widely recognized that one should proceed by consensus." Alan Beesley, "The Negotiating Strategy of UNCLOS III: Developing and Developed Countries as Partners — A Pattern for Future Multilateral International Conferences?" *Law and Contemporary Problems* 46 (Spring 1983): 186.

370. Consensus has been used informally as a substitute for majority voting in the General Assembly and as a preliminary negotiating tool in the Security Council. Newcombe, p. 162.

371. Carl Q. Christol, "Telecommunications, Outer Space, and the New International Information Order (NIIO)," *Syracuse Journal of International Law and Commerce* 8 (1981): 348. See also Mark Orlove, "Spaced Out: The Third World Looks for a Way in to Outer Space," *Connecticut Journal of International Law* 4 (1989): 606–7.

372. See Honnold, p. 832. The consensus procedure "contributes to a business-like

atmosphere" at consultative meetings and "guarantees balanced recommendations and decisions reflecting the opinions and meeting the interests of all." Y. Deporov, "Antarctica: A Zone of Peace and Cooperation," *International Affairs* (Moscow) (1983, no. 11), pp. 34–35.

Under the 1988 Antarctic Mineral Resource Convention, the Reagan administration agreed that a consensus of the Mineral Resources Commission would be required for decisions relating to nondiscrimination, budgetary questions, and the opening of areas for exploration and exploitation; otherwise, decisions of the Commission were to be made by a three-fourths-majority vote. Convention on the Regulation of Antarctic Mineral Resource Activities, art. 22, in *International Legal Materials* 27 (1988): 879. Decisions by the various Regulatory Committees relating to access to Antarctic minerals would require, in addition to the normal two-thirds majority, concurrent approval by both the four-member claimant-state group and the six-member nonclaimant-state group. Convention on the Regulation of Antarctic Mineral Resource Activities, art. 32, in *International Legal Materials* 27 (1988): 883. See M. Jacobsen, pp. 241–42; and also Francesco Francioni, "Legal Aspects of Mineral Exploitation in Antarctica," *Cornell International Law Journal* 19 (1986): 186–87.

373. James M. McCormick, "Intergovernmental Organizations and Cooperation Among Nations," *International Studies Quarterly* 24 (1980): 85, 91. Thus the expansion of IMCO's membership to include newly independent developing states did not alter the "de facto domination of the organization by ocean transportation interests." S. Brown et al., pp. 48–49. See n. 636 below.

374. McCormick, p. 86. In fact, it was found that developing states are more cooperative in organizations such as the Authority than in regional or "high politics" organizations. Ibid., p. 89.

375. See Yost, pp. 673–75. Developing-state cohesion was highest on issues relating to colonialism, apartheid, and the Middle East. Richard J. Powers, "United Nations Voting Alignments: A New Equilibrium," *Western Political Quarterly* 33 (1980): 173. Unfortunately, votes upon certain of these sensitive issues in recent years have spawned U.S. reactions which might more appropriately have been reserved for clear-cut violations of international law. See Powers, p. 172.

Many U.N. voting alignments are in fact formed on an ad hoc basis. Newcombe, p. 8. "[E]ven in the worst of circumstances we do hold our own and in the long run keep our opponents from running the show." W. Jones, "The International Sea-Bed Authority Without U.S. Participation," p. 153.

376. Powers, pp. 167, 182. The Soviet Union was nevertheless historically unsuccessful in the General Assembly in relation to the United States and its allies. Newcombe, p. 67. See also Barrett and Newcombe, p. 78. Moreover, the Soviet Union was not always willing to prejudice its interests as a developed state in order to curry favor with the Group of 77. See Newcombe, p. 22.

377. Barrett and Newcombe, pp. 44, 49. See also *Oceans Policy News* (April 1987), p. 6.

378. Richardson, "The Politics of the Law of the Sea," p. 17. "Taken with the voting rules, . . . the composition of the Council places those states in a significantly advantageous position." E.D. Brown, "Seabed Mining: From UNCLOS to Prep Com," *Marine Policy* 8 (1984): 154. The United States—like other states—would face the possibility that its own efforts to seek affirmative Council action could be blocked by a similar coalition of opposing states. See G. Winthrop Haight, "Comments on Judge Oda's Approach to the Common Heritage of Mankind," *New York Law School Journal of International and Comparative Law* 3 (1981): 17.

379. Yost, p. 677. See nn. 323, 330 above.

380. U.N. Convention on the Law of the Sea, art. 165 (2) (b). See nn. 438–49 below and accompanying text.

381. U.N. Convention on the Law of the Sea, arts. 161 (8) (c), 163 (2), (4). Commission members serve five-year terms and are eligible for reelection. Ibid., art. 163 (6). The Council may, by a three-fourths-majority vote, expand the size of the Commission. Ibid., arts. 161 (8) (c), 163 (2).

It has been argued that the "special interests" represented would be "those of the Third World developing nations who have cleverly negotiated a political organization which they will control." Statement of Senator James A. McClure in *Congressional Record* 125 (December 14, 1979): 36039. The better view, however, is that "special interests" refers to the interest groups represented on the Council, and that the nationalities of Commission members would roughly correspond to the composition of the Council. The amendments offered by the United States on April 13, 1982, would have given each of the five geographical interest groups represented in the Council the power to appoint three Commission members. *Amendments* (U.N. Doc. A/CONF.62/L.121), proposed art. 163 (2), in *Official Records*, UNCLOS III, vol. XVI, p. 228.

382. U.N. Convention on the Law of the Sea, art. 163 (3). It is unlikely that three-fourths of the Council would decide to elect unqualified Commission members who would fail to apply the Convention and its rules and procedures faithfully. See Richardson, "Law in the Making," p. 442. Indeed, any failures of the Council to elect qualified Commission members could constitute "acts or omissions" within the jurisdiction of the Sea-Bed Disputes Chamber. U.N. Convention on the Law of the Sea, art. 187 (c) (i). See n. 498 below. The Green Book had sought to eliminate the treaty language calling for "the highest standards of competence and integrity." *The U.S. Amendments* (U.N. Doc. A/CONF.62/WG.21/Informal Paper 18), proposed art. 163 (3).

Commission members are also required to have "appropriate qualifications such as those relevant to exploration, exploitation and processing of mineral resources, oceanology, protection of the marine environment, or economic or legal matters relating to ocean mining and related fields of expertise." U.N. Convention on the Law of the Sea, art. 165 (1).

383. Dubow, p. 185. See also Larson, "The Reagan Administration," p. 300. Developed states had sought a consensus voting requirement for Commission elections. Jagota, p. 314.

384. See Chap. 3 n. 332. In fact, developing states have expressed concern that the Commission might be dominated by its more sophisticated Western members. Roy S. Lee, "Deep Seabed Mining and Developing Countries," *Syracuse Journal of International Law and Commerce* 6 (1978): 219. Western states have historically been allocated more than their share of important offices by the U.N. General Assembly. Newcombe, p. 265.

385. U.N. Convention on the Law of the Sea, arts. 161 (8) (d), 162 (2) (o) (ii), 163 (11), 165 (2) (f), (g). See n. 329 above. The Commission must act "in accordance with such guidelines and directives as the Council may adopt," probably by consensus. U.N. Convention on the Law of the Sea, arts. 161 (8) (f), 163 (9). The Council is not bound simply to accept or reject the Commission's recommendations. See n. 398 below.

The United States would prefer that decisions of the Commission be taken by a simple majority vote. Hauser, p. 68 n. 173.

386. "The economist's decision regarding development plans, the transportation engineer's concerning air routes, and the physicist's on nuclear weapons control are all ultimately political in nature despite the technical considerations." *The United Nations System and Its Functions*, ed. Robert W. Gregg and Michael Barkun (Princeton, N.J.: D. Van Nostrand, 1968), p. 403. See also Ball, p. 469.

387. Friedheim, p. 26. See also McCormick, p. 86. The ITU has been particularly successful in making decisions on a technical and scientific basis. Marvin S. Soroos, "The Commons in the Sky: The Radio Spectrum and Geosynchronous Orbit as Issues in Global Policy," *International Organization* 36 (1982): 672.

388. Hauser, p. 70. See also David M. Leive, "Essential Features of INTELSAT: Applications for the Future," *Journal of Space Law* 9 (1981): 47.

389. U.N. Convention on the Law of the Sea, art. 159 (1), (6), (7), (8). A quorum consists of a majority of states parties. Ibid., art. 159 (5).

The U.S. Draft Treaty of 1970 would have allowed most substantive decisions of the Assembly to be taken by a simple majority vote. *Draft United Nations Convention* (U.N. Doc. A/AC.138/25), art. 34 (5).

390. See Geddes, p. 617. The Health Assembly of the World Health Organization requires a two-thirds-majority vote for enactment of certain types of measures. Constitution of the World Health Organization, art. 60 (a), reprinted in *United Nations Treaty Series*, vol. 14, no. I-221, pp. 199–200. The one-nation-one-vote General Conference of the IAEA also requires a two-thirds vote for only a limited number of substantive issues. Statute of the International Atomic Energy Agency, art. V (A), (C), in *United Nations Treaty Series*, vol. 276, no. 3988, p. 10. INTELSAT's Assembly, its primary organ, takes its substantive decisions by a two-thirds-majority vote. See Doyle, "Permanent Arrangements," pp. 278, 286. The

ITU operates by the same rule: "This structure gives the developing countries, which represent two thirds of the ITU, a great deal of control in the proceedings." Harvey J. Levin, "Spectrum Negotiations and the Geostationary Satellites," *New York Law School Journal of International and Comparative Law* 4 (1982): 77–78 n. 3. In the U.N. General Assembly itself, some substantive issues are decided by a simple—rather than two-thirds—voting majority. U.N. Charter, art. 18 (2), (3).

391. U.N. Convention on the Law of the Sea, art. 159 (9). It may be expected that Western developed states would be capable of mustering this one-fifth requirement on many issues without help from the Soviet or Group of 77 blocs, and that the delay would provide the United States and its allies with the opportunity to dissuade the Assembly from hasty and ill-advised actions.

392. Ibid., art. 159 (10). "Such opinions shall be given as a matter of urgency." Ibid., art. 191. By submitting dubious legislation in this manner prior to its enactment, Western states could avoid the treaty's ban on after-the-fact judicial review and thereby also avoid the inefficiency of either complying with or challenging rules of questionable validity. See n. 288 above and accompanying text. The one-fourth requirement makes it unlikely that Western states would be able to initiate this review without support from other states. Nevertheless, the procedure could prove to be a potent tool of obstruction to a skillful parliamentarian whenever the Assembly's actions are even colorably open to legal challenge.

393. Statement of Leigh S. Ratiner in *National Ocean Policy*, U.S. House of Representatives, p. 194.

394. Cf. *The U.S. Proposals* (U.N. Doc. A/CONF.62/WP.121/Informal Paper 18), proposed art. 159; and *Amendments* (U.N. Doc. A/CONF.62/L.121), in *Official Records*, UNCLOS III, vol. XVI, p. 227.

395. Jagota, p. 310. See *The Third United Nations Law of the Sea Conference*, U.S. Senate, p. 24; and Willsey, p. 514. The Conference was faced with "the task of reconciling, within a single institutional framework, the requirements of sovereign equality . . . with the need to fulfill effectively an important and delicate economic function." Fuandez, p. 492.

396. Dubs, "Industrial Interests and the Deep Seabed," pp. 145–46.

397. These measures include the rules, regulations, and procedures of the Authority, its annual budget, and any system of compensation for developing producer states. U.N. Convention on the Law of the Sea, art. 160 (2) (f), (h), (l). Except for the budget, which requires a three-fourths-majority vote, these measures must first be enacted by consensus in the Council, and the United States would therefore be assured of a prior blocking power with respect to such Assembly actions. See nn. 378–79 above and accompanying text.

This type of limited bicameral legislative competence was initially proposed for the Authority in the U.S. Draft Treaty of 1970. See Knight, "The Draft United Nations Conventions," p. 497. It was maintained in the 1982 Green Book and in the proposals of April 13, although the Assembly's role in approving compensation for developing producer states was eliminated. *The U.S. Proposals* (U.N. Doc. A/CONF.62/WG.71/Informal Paper 18), proposed arts. 160 (2) (f), (g), (l), 162 (2) (m). *Amendments* (U.N. Doc. A/CONF.62/L.121), proposed art. 162 (2) (m), in *Official Records*, UNCLOS III, vol. XVI, p. 228.

398. U.N. Convention on the Law of the Sea, arts. 160 (2) (f) (ii), 162 (2) (o) (ii). "The Assembly can only send the regulations it dislikes back to the Council with a statement of its position and hope for a corresponding amendment." Hauser, p. 47. Because of the consensus procedure, any member of the Council would be able to prevent modification, leaving the regulation in effect indefinitely on a provisional basis. Ibid.

In contrast, the Council may only "recommend to the Assembly" rules, regulations, and procedures relating to the distribution of revenues received by the Authority; "if the Assembly does not approve the recommendations of the Council, the Assembly shall return them to the Council for reconsideration." U.N. Convention on the Law of the Sea, arts. 160 (2) (f) (i), 162 (2) (o) (i). Thus, only with respect to certain measures is it correct to state, as has one treaty opponent, that the Convention "in effect subjects these decisions of the Council to the veto power of the Assembly." Statement of Senator Richard Stone in *Congressional Record* 125 (December 14, 1979): 36076. See also n. 397 above.

The rules, regulations, and procedures adopted by the Preparatory Commission will similarly remain in effect on a provisional basis pending action by the Council. See n. 270

above; and also Richardson, "Law in the Making," p. 442. One vocal opponent of the Convention, apparently failing to understand the provisional application provision—or the requirement that the Commission act by consensus—has argued that the United States could not participate in the Preparatory Commission because the rules, regulations, and procedures adopted by the Commission might not be accepted by the Council. Statement of Representative John Breaux in *The 1982 Convention*, ed. Koers and Oxman, p. 16.

The Green Book would have required that the Legal and Technical Commission approve the rules, regulations, and procedures of the Authority prior to their adoption by the Council. *The U.S. Proposals* (U.N. Doc. A/CONF.62/WG.21/Informal Paper 18), proposed arts. 162 (2) (n) (ii), 165 (5) (a), (b), (6) (b), (7). The Convention itself provides that the Council need only make its decisions "taking into account the recommendations of the Legal and Technical Commission or other subordinate organ concerned." U.N. Convention on the Law of the Sea, art. 162 (2) (o) (ii).

399. U.N. Convention on the Law of the Sea, art. 160 (2) (e), (g), (k), (n). See n. 274 above and accompanying text. Issues "not specifically entrusted to a particular organ" include the conduct of scientific research by the Authority, the acquisition of technology and its transfer to developing states, the encouragement of prospecting activities in the Area, the transfer of funds from the Authority to the Enterprise, the approval of exceptions to the antimonopoly provisions, the selection of reserved areas, the withdrawal of preferences for plans of work, and the exercise of the Authority's incidental powers. Ibid., arts. 143 (2), 144 (1), 157 (2), 173 (2) (b); Annex III, arts. 2 (1) (a), 6 (4), (8), (10). Many of these issues, however, would seem to fall within the purview of the Council, which has been specifically empowered to "supervise and co-ordinate the implementation of" Part XI "on all questions and matters within the competence of the Authority." Ibid., art. 162 (2) (a). Any attempt by the Assembly to exercise its residual power under paragraph 2 (n) of article 160 in an unwarranted manner would be open to challenge in the Sea-Bed Disputes Chamber. See n. 392 above and accompanying text.

Exercise of this residual power does not in itself threaten U.S. interests: "if the developed states can identify and specifically protect their essential interests, then they can concede full constitutional control to the Group of 77, as well as a broad balance of residual powers." Harry, p. 14. The Green Book had sought to limit the Assembly's residual competence to one of discussion only. *The U.S. Proposals* (U.N. Doc. A/CONF.62/WG.21/Informal Paper 18), proposed art. 160 (2) (n). The April 13 set of amendments, however, proposed no changes to subparagraph (n). Cf. *Amendments* (U.N. Doc. A/CONF.62/L.121), in *Official Records*, UNCLOS III, vol. XVI, p. 227.

400. U.N. Convention on the Law of the Sea, art. 160 (1). "This designation has no independent authorizing function." Hauser, p. 45. Any "general policies" established by the Assembly would be subject to review by the Sea-Bed Disputes Chamber. See n. 392 above and accompanying text. The Senate nevertheless found article 160 (1) to be "objectionable and fundamentally unacceptable." Statement of Representative Jim Santini in *Congressional Record* 126 (June 9, 1980): 13692. It was "apparent" to one Senator that "the decisions of the council would be subject to a veto by the assembly through its general policymaking functions." Statement of Senator James McClure in *Congressional Record* 125 (December 14, 1979): 36038. Ambassador Malone also identified the first paragraphs of articles 160 and 162 as one of the major areas of concern to the Reagan administration. See Bernice R. Kleid, "Synopsis: Recent Developments in the Law of the Sea 1980–1981," *San Diego Law Review* 19 (1982): 636.

The SNT originally had described the Assembly as "the supreme policy-making organ of the Authority," but subsequent drafts reduced its power in relation to the Council. See Chap. 3 nn. 55–56, 91, 213 and accompanying text; Willsey, pp. 514–15; and Fuandez, p. 493. Although the Green Book sought to delete paragraph 1 of article 160, the April 13 set of proposed amendments left it substantially unchanged. *The U.S. Proposals* (U.N. Doc. A/CONF.62/WG.21/Informal Paper 18), proposed art. 160 (1). *Amendments* (U.N. Doc. A/CONF.62/L.121), proposed art. 160 (1), in *Official Records*, UNCLOS III, vol. XVI, p. 227.

401. U.N. Convention on the Law of the Sea, art. 162 (1). The Council could enact policies under this paragraph in cases where it is unable to reach a consensus with respect to more direct measures, since the "specific policies" of article 162 (1) may be established by a

three-fourths-majority vote. Ibid., art. 161 (8) (c). This would prevent single states from completely frustrating any action by the Council, while effectively preserving the blocking power of Western states.

It has been argued that the treaty establishes the Council as "the dominant policy-making organ.... The Council will be able to implement the general policies established by the Assembly in such a way as to temper or negate their impact." Willsey, p. 515.

402. U.N. Convention on the Law of the Sea, art. 158 (4). See Hauser, p. 46. Construed liberally, this provision alone would prevent the Assembly from establishing general policies with respect to any of the powers specifically reserved to the Council in article 162 (2). Cf. nn. 323–27 above and accompanying text. A strict interpretation would permit the Assembly to establish such policies so long as they did not "derogate from ... the *exercise* of" those powers, thus permitting the Assembly to act where the Council failed to do so. Paragraph 1 of article 160, moreover, provides that the general policies of the Assembly must be established "in conformity with the relevant provisions of [the] Convention" — such as article 162 (1) and article 170 (2), which enjoins the Enterprise to "act in accordance with ... the general policies established by the Assembly." It may thus be inferred that the absence of any reference to the Assembly's "general policies" in paragraph 2 of article 162 was intended to remove the powers listed therein from the scope of the Assembly's policymaking competence — a view reinforced by article 160 (1), which provides that the Council is accountable to the Assembly only "as specifically provided." Furthermore, the powers listed in paragraph 2 of article 162 are granted "[i]n addition" to the general policymaking power conferred upon the Council by paragraph 1, clearly indicating an intention that the enumerated powers are to enjoy a status separate from — and presumably equal to — the Council's policymaking power. The policies mentioned in the first paragraphs of articles 160 and 162 are thus applicable only to matters falling within the residual competence of the Authority and, possibly, to other matters where the Council — or the Assembly, if the policies are being established by the Council pursuant to paragraph 1 of article 162 — has failed to exercise its specific powers; it is therefore incorrect to claim, as did the American Branch of the International Law Association, that the Convention vests the Assembly with "very broad discretionary powers." Cf. International Law Association, *Proceedings and Committee Reports, 1977–78*, p. 117.

403. See U.N. Convention on the Law of the Sea, art. 157 (3). "[T]he doctrine of sovereign equality usually requires that, in legal theory at least, ultimate authority should be vested in the organ where all member states or constituent units are represented; this applies particularly to budgeting matters." H. Jacobsen, p. 94. See Chap. 5 n. 207; and also Kooijmans, *The Doctrine of the Legal Equality of States*, pp. 102, 242–43.

404. See Yemin, pp. 124, 170.

405. See UNCLOS III, *The Activities of the Inter-Governmental Maritime Consultative Organization in Relation to Shipping and Related Maritime Matters* (U.N. Doc. A/CONF.62/27, June 10, 1974), para. 9, reprinted in *Official Records*, UNCLOS III, vol. III, p. 44.

406. See Doyle, "Permanent Arrangements," pp. 277, 286; and also Richard R. Colino, "A Chronicle of Policy and Procedure: The Formulation of the Reagan Administration Policy of International Satellite Telecommunications," *Journal of Space Law* 13 (1985): 107.

407. See nn. 296, 329 above.

408. See U.N. Charter, arts. 10, 11 (1), (2) (4), 12, 13 (2), 14, 16, 17, 60, 63 (1), 66. "[T]he normal constitutional procedure, except when the preservation of peace is at issue, is for the Security Council to make recommendations and for the General Assembly to decide within the framework of these recommendations." Morley, p. 11. See also Hauser, pp. 138–39. Even with respect to the maintenance of international peace and security, the powers of the Security Council and the General Assembly are concurrent and supplementary, as the Uniting for Peace Resolution has demonstrated, rather than mutually exclusive. See Kelsen, *Principles of International Law*, pp. 181–82; and also Morley, p. 14.

409. U.N. Convention on the Law of the Sea, arts. 160 (2) (b), (c), 162 (2) (b), (c). The Director-General is to be nominated by the governing board and must be approved by both the Council and the Assembly. Ibid., Annex IV, art. 7 (1). See Hauser, p. 139; and Roy S. Lee, "The Enterprise: Operational Aspects and Implications," *Columbia Journal of World Business* 15 (1980): 63. The Secretary-General of the United Nations is nominated by the Security Council and appointed by the General Assembly. U.N. Charter, art. 97.

410. U.N. Convention on the Law of the Sea, art. 161 (8) (c). W. Smith, p. 60. In practice, it is likely that the membership of the governing board will be negotiated and agreed upon prior to the actual voting. R. Lee, "The Enterprise," p. 63.

411. U.N. Convention on the Law of the Sea, Annex IV, art. 5 (1), (2). In 1982 the United States had demanded that, until the Enterprise had repaid all its debts, the governing board "include members nominated by States Parties that account for at least one-half of the total amount of such obligations outstanding." The U.S. Proposals (U.N. Doc. A/CONF.62/WG.21/Informal Paper 18), Annex IV, proposed art. 5 (1). Amendments (U.N. Doc. A/CONF.62/L.121), Annex IV, proposed art. 5 (1), in Official Records, UNCLOS III, vol. XVI, p. 231.

412. U.N. Convention on the Law of the Sea, art. 167 (2), (3). This language is virtually identical to that of the U.N. Charter governing the selection of the staff of the U.N. Secretariat. U.N. Charter, art. 101 (1), (3). The system employed at the United Nations has resulted in a "vast number of American officials" in the Secretariat. Thomas George Weiss, International Bureaucracy (Lexington, Mass.: Lexington Books, 1975), p. 65. See also Newcombe, p. 272. The constitutive instruments of most other international organizations also specify both competence and equitable geographical distribution as twin criteria for staff recruitment. H. Jacobsen, p. 97. In appointing a separate staff for the Enterprise, the Director-General "shall, subject to the paramount importance of securing the highest standards of efficiency and of technical competence, pay due regard to the importance of recruiting personnel on an equitable geographical basis." U.N. Convention on the Law of the Sea, Annex IV, art. 7 (4). By becoming a member of the Authority, the United States could expect its nationals to fill "two top posts, probably of its choice"; by refusing to participate, the United States assures itself of none. W. Jones, "The International Sea-Bed Authority Without U.S. Participation," pp. 155–56.

413. U.N. Convention on the Law of the Sea, art. 187 (b) (i). See n. 498 below and accompanying text. "In the performance of their duties, the Secretary-General and the staff shall not seek or receive instructions from any Government or from any other source external to the Authority. They shall refrain from any action which might reflect on their position as international officials of the Authority responsible only to the Authority." U.N. Convention on the Law of the Sea, art. 168 (1). The Secretariat must "have no financial interest in any activity relating to exploration and exploitation in the Area," and similar restrictions apply to the Director-General and his staff, and to the governing board of the Enterprise. U.N. Convention on the Law of the Sea, art. 168 (2); Annex IV, arts. 5 (4), 7 (3). The same duties have been imposed upon the U.N. Secretariat by article 100 of the Charter, "and generally . . . have been honored" in most international organizations. H. Jacobsen, p. 97.

414. U.N. Convention on the Law of the Sea, arts. 167 (3), 168 (1), (4), 170 (2). See also ibid., Annex IV, art. 6 (1); and n. 325 above and accompanying text. Because paragraph 2 (i) of article 162 is a specific grant of power to the Council, article 158 would prevent the Assembly from establishing general policies which conflict with directives previously issued by the Council. See n. 402 above. Interestingly, the 1982 U.S. proposals would have removed the Council's power to issue directives, permitting it to control the Enterprise only by enacting policies pursuant to paragraph 1 of article 162, which specifically subordinates such specific policies to the general policies of the Assembly; because the same set of proposals would have required consensus for the promulgation of such specific policies, this change would also have had the effect of yielding effective control of the Enterprise to the Assembly. Amendments (U.N. Doc. A/CONF.62/L.121), proposed arts. 161 (7) (d), 162 (2) (i), in Official Records, UNCLOS III, vol. XVI, pp. 227–28. The Green Book had simply sought to eliminate the powers of both the Council and the Assembly to control the operations of the Enterprise. The U.S. Proposals (U.N. Doc. A/CONF.62/Informal Paper 18), proposed art. 170 (2).

415. Goldwin, "Locke and the Law of the Sea," p. 49. Cf. also Ratiner, "Reciprocating State Arrangements," p. 198. "Governments throughout the world are far from willing to accept the idea that the secretariat of an international governmental organization would be impartial." H. Jacobsen, pp. 98–99.

In any event, the United States could expect its nationals within the Secretariat and the Enterprise to protect its interests, even if no improprieties were committed. Weiss, pp. 54, 59. The United States and the Soviet Union have been particularly adept at ensuring that their nationals remain loyal to their homeland after joining the international civil service. See Weiss, pp. 64–66. Nevertheless, U.S. distrust of the Authority's Secretariat was at least partially responsible for the relatively wide range of tasks assigned to the Legal and Technical Commission, a subsidiary organ more directly accountable to the Council. Glickman, p. 20.

416. Frank, "Jumping Ship," p. 134. According to the American Mining Congress, security of tenure and a secure investment climate are the most important requirements of private seabed mining operators. Auburn, p. 310. "Lacking such a right, you will have no recourse against a competitor whose mining ship follows close to yours in the very same track that you had intended to cover in your next sweep." Elliot L. Richardson, "Law of the Sea: A Reassessment of U.S. Interests," *Mediterranean Quarterly* 1 (Spring 1990): 6–7. Banks have refused to finance private seabed mining in the absence of assured access to nodule deposits. See Michael R. Molitor, "The U.S. Deep Seabed Mining Regulations: The Legal Basis for an Alternative Regime," *San Diego Law Review* 19 (1982): 601. See also Brewer, "Deep Seabed Mining," p. 43.

417. Seabed Act, 30 U.S.C. §1441 (1) (A), (2). The "totality of . . . provisions" language "recognizes that a treaty negotiated by 150 countries with widely divergent interests cannot be expected to provide terms and conditions as favorable as U.S. legislation drawn specifically to promote the development of our seabed mining industry." Statement of Representative Jonathan Bingham in *Congressional Record* 126 (June 9, 1980): 13685.

418. U.S. 93rd Congress, 1st Session, House of Representatives, Committee on Merchant Marine and Fisheries, Subcommittee on Oceanography, *Ocean Affairs in the 93rd Congress* (Washington: U.S. Government Printing Office, 1975), p. 46. See Chap. 3 n. 17 and accompanying text; Yost, p. 660; and also Kissinger, "A Test of International Cooperation," p. 539. The seabed mining industry demanded "firm rights to explore and mine a given area of the seabed at an economic rate of production for the life of the mine." "Summary of Detailed Ocean Mining Industry Comments," in *Congressional Record* 126 (June 9, 1980): 13683.

419. Brewer, "Deep Seabed Mining," p. 43. See Statement of Senator Larry Pressler, March 8, 1982, in *Congressional Digest* 62 (January 1983): 17; and also Willsey, p. 524. To the detriment of their profits, mining companies have frequently been forced to renegotiate their contracts with developing producer states. See Mike Faber and Roland Brown, "Changing the Rules of the Game: Political Risk, Instability and Fairplay in Mineral Concession Contracts," *Third World Quarterly* 2 (1980): 11.

420. Aldrich, "Law of the Sea," p. 58.

421. Malone, "Who Needs the Sea Treaty?" p. 62.

422. "The minitreaty cannot by definition grant exclusive rights." Frank, "Jumping Ship," p. 134. It should also be noted that the Convention grants coastal states exclusive exploitation rights with respect to mineral resources within their exclusive economic zones, and that the Reagan administration did not object to the complete discretionary authority of such states to exclude U.S. corporations from the more strategically valuable and readily accessible mineral deposits within these zones.

423. U.N. Convention on the Law of the Sea, art. 153 (2); Annex III, arts. 4 (3), 13 (2). "The criteria and procedures for implementation of the sponsorship requirements shall be set forth in the rules, regulations and procedures of the Authority." Ibid., Annex III, art. 4 (3). The Green Book would have required sponsorship by only a single state party, which would have been required to have "effective control and jurisdiction" over only one consortium partner. *The U.S. Proposals* (U.N. Doc. A/CONF.62/WG.21/Informal Paper 18), proposed art. 153 (2) (b); Annex III, proposed art. 4 (2).

424. Aldrich, "A System of Exploitation," p. 248.

425. U.N. Convention on the Law of the Sea, Annex III, arts. 4 (1), (2), 17 (1) (b) (xiv). See Hauser, pp. 62–63. In the late 1970s the U.S. delegation expressed satisfaction with these standards. See Chap. 3 n. 208. The criteria set forth in the U.S. Seabed Act relate to similar areas of concern. See Seabed Act, 30 U.S.C. §1413 (c).

426. U.N. Convention on the Law of the Sea, Annex III, art. 4 (6). See nn. 3–8, 10, 14, 279 above and accompanying text. Each applicant must submit "a general description of the equipment and methods to be used . . . , and other relevant non-proprietary information about the characteristics of such technology and information as to where such technology is available." U.N. Convention on the Law of the Sea, Annex III, art. 5 (1). It is also similar to the information which applicants for exploration licenses must submit under regulations

adopted by the United States pursuant to the 1980 Seabed Act. See Deep Seabed Mining Regulations, 15 CFR §§970.202 (b) (1), 970.203 (b) (5). The description is similar to that required by developing states, and is expected to be brief and to consist mainly of technical, publicly available information. Van Dyke and Teichmann, p. 531. Brewer, "Transfer of Mining Technology," pp. 19–20. See also Hauser, p. 98. The applicant must also undertake to inform the Authority of any "substantial technological change or innovation" subsequently introduced. U.N. Convention on the Law of the Sea, Annex III, art. 5 (2).

427. U.S. Department of State, "Access to Seabed Minerals," p. 39. See also Willsey, p. 523.

428. *The U.S. Proposals* (U.N. Doc. A/CONF.62/WG.21/Informal Paper 18), Annex III, proposed art. 4 (bis). *Amendments* (U.N. Doc. A/CONF.62/L.121), Annex III, proposed art. 3 (4) (b), in *Official Records*, UNCLOS III, vol. XVI, p. 228. See Seabed Act, 30 U.S.C. §§1412 (c) (1) (E), (F), 1413 (c), 1414; and also statement of Representative Paul McClosky, Jr., in *Syracuse Journal of International Law and Commerce* 6 (1978–79): 257. The regulatory framework established by the Act permits the imposition of additional qualification standards, however. See n. 456 below.

429. *The U.S. Proposals* (U.N. Doc. A/CONF.62/WG.21/Informal Paper 18), Annex III, proposed arts. 4 (4), 17 (1) (b) (xiv). *Amendments* (U.N. Doc. A/CONF.62/L.121), Annex III, proposed art. 4 (4), in *Official Records*, UNCLOS III, vol. XVI, p. 229. The U.S. amendments would have required an applicant only to demonstrate access to "funds necessary to comply with the minimum annual expenditures for exploration established in the rules, regulations and procedures of the Authority," and to assure performance of its contractual obligations by posting a bond "in the amount of 50 percent of the minimum annual expenditures for the first three years of exploration." *The U.S. Proposals* (U.N. Doc. A/CONF.62/WG.21/Informal Paper 18), Annex III, proposed arts. 4 (4) (a), (b). *Amendments* (U.N. Doc. A/CONF.62/L.121), Annex III, proposed art. 4 (1) (a), (b), in *Official Records*, UNCLOS III, vol. XVI, p. 228. In addition, the applicant could not have had a previous contract terminated. *Amendments* (U.N. Doc. A/CONF.62/L.121), Annex III, proposed art. 4 (4) (c), in *Official Records*, UNCLOS III, vol. XVI, p. 11. See text accompanying n. 280 above. These changes would have had the effect of increasing the discretion of the members of the Legal and Technical Commission, despite the professed concern of treaty critics with respect to potential political bias in the Commission. See Larson, "The Reagan Administration," p. 300.

430. Statement of Elliot L. Richardson, May 14, 1981, quoted in Oxman, "The Third United Nations Conference on the Law of the Sea," p. 11 n. 27.

431. Breaux, "The Diminishing Prospects," p. 282.

432. U.N. Convention on the Law of the Sea, Annex III, arts. 3 (3), 8. See n. 170 above and accompanying text. The size of mining sites must "satisfy the requirements of article 8 . . . as well as stated production requirements consistent with article 151." Ibid., Annex III, art. 17 (1) (b) (i).

The Seabed Act provides for mining sites "of sufficient size to allow for intensive exploration," but limits the size of the site to an area within which mining activities can be conducted "in an efficient, economical, and orderly manner with due regard for conservation and protection of the environment, taking into consideration the resource data, other relevant physical and environmental characteristics, and the state of the technology of the applicant." Seabed Act, 30 U.S.C. §1413 (a) (2) (B), (E) (i), (ii). "Approval by the Administrator of a proposed exploration [area] will be based on a case-by-case review of each application"—a discretionary process which would presumably be objectionable to the United States if included in the rules, regulations, and procedures of the Authority. See Deep Seabed Mining Regulations, 15 CFR §970.601 (b).

Although pioneer registrants have been allocated sites of 75,000 sq. km, the United States indicated that it was seeking mining areas of 30,000 sq. km or less. U.S. Senate, *The Third United Nations Law of the Sea Conference*, p. 23. An operation producing 1 million dry tons of nodules annually might require a mining area of as few as 10,000 sq. km. Statement of John E. Flipse, President, Deepsea Ventures, in *Ocean Affairs in the 93rd Congress*, U.S. House of Representatives, p. 16.

433. U.N. Convention on the Law of the Sea, Annex III, art. 8. The selection must be made within 45 days unless "the Authority requests an independent expert to assess whether

all data required by this article has *[sic]* been submitted." Ibid. Reserved areas may be made available to developing-state-sponsored operators "if the Enterprise decides . . . that it does not intend to carry out activities in that area." Ibid., Annex III, art. 9 (4).

It is unclear which organ of the Authority is to select the reserved area, although the distribution of powers and functions would favor the exercise of such an executive function by the Council or its Legal and Technical Commission. See n. 399 above. Because this function is not specifically conferred upon the Council, however, the Assembly would be able to establish general policies to govern the selection. If, for example, it wished to promote the mining of nickel rather than copper or cobalt, it could direct the Council to select the site offering the greatest relative recovery potential for that particular mineral. See nn. 400–402 above and accompanying text.

434. *The U.S. Proposals* (U.N. Doc. A/CONF.62/WG.21/Informal Paper 18), Annex III, proposed art. 8. *Amendments* (U.N. Doc. A/CONF.62/L.121), Annex III, proposed art. 8, in *Official Records*, UNCLOS III, vol. XVI, p. 230. Because the Convention empowers the Authority to select the reserved area, the applicant can best prevent loss of a better site by submitting a pair of nearly equivalent sites. The Convention assists applicants to do so by providing that the total area submitted "need not be a single continuous area," either before or after division. U.N. Convention on the Law of the Sea, Annex III, art. 8.

435. *The U.S. Proposals* (U.N. Doc. A/CONF.62/WG.21/Informal Paper 18), Annex III, proposed art. 8. Alternatively, the United States proposed that the selection be made blindly, prior to submission of data by the applicant. *Amendments* (U.N. Doc. A/CONF.62/L.121), Annex III, proposed art. 8, in *Official Records*, UNCLOS III, vol. XVI, p. 230.

436. U.S. Steel, a partner in one of the consortia, claimed that prospecting a second site would cost it "another $20 million and 15 years, and you know which one we would get back — it is as simple as that." Statement of Phillip Hawkins, U.S. Steel, in *Deep Seabed Mining*, U.S. House of Representatives, p. 308. See also "Deep Seabed Mining: U.S. Steel's Investment," in ibid., p. 288. The cost of prospecting an additional mine site has been estimated elsewhere by a treaty critic at only $1.6 million, however, although more recent estimates are slightly higher. Cf. Breaux, "The Diminishing Prospects," p. 284 n. 124. See Sebenius, p. 97; and also Laursen, p. 100. Like the pioneer investors registered by the Preparatory Commission, the Kennecott consortium has managed to prospect a pair of mining sites — apparently for the purpose of complying with the banking provisions of the Convention, should it eventually enter into force — without undue economic hardship. Conversation with Joseph Flanagan, Ocean Minerals and Energy Division, National Oceanographic and Atmospheric Administration, U.S. Department of Commerce, November 26, 1984. Indeed, the regulations promulgated pursuant to the 1980 Seabed Act allow applications to include a second site, "[a]t the applicant's option, for the purpose of satisfying a possible obligation under a future Law of the Sea Treaty." Deep Seabed Mining Regulations, 15 CFR §970.601 (d).

437. An applicant, it has been claimed, might be required to "give away free a mine site . . . and then be told by the International Seabed Authority that it has no right to continue mining the seabed." Statement of Representative Edward Darwinsky in *Congressional Record* 126 (June 9, 1980): 13681. Of course, in reality, the Authority would gain no such site until and unless it approved the application. Indeed, "with the political pressure to get the Enterprise functioning at an early stage, the directors of the Enterprise will at least in the initial stages be doing their utmost to encourage activities on the other side of the parallel system." John S. Bailey, "The Future of the Exploitation of the Resources of the Deep Seabed and Subsoil," *Law and Contemporary Problems* 46 (Spring 1983): 74.

438. U.N. Convention on the Law of the Sea, art. 153 (3); Annex III, art. 3 (5).

439. Ibid., art. 165 (2) (b); Annex III, art. 6 (3). Plans of work are to be reviewed six months after the Convention enters into force and every four months thereafter, and the Commission must "report fully" to the Council concerning the grounds upon which its recommendation is based. Ibid., art. 165 (2) (b); Annex III, art. 6 (1).

440. Aldrich, "Law of the Sea," p. 58 (emphasis in original). See also U.S. Department of State, "Access to Seabed Minerals," p. 40. By 1985 Special Commission 3 of the Preparatory Commission was refining a procedure whereby the Legal and Technical Commission would designate the reserved area and negotiate needed contractual provisions with the applicant before recommending action to the Council. Walkate, p. 156.

441. Malone, "U.S. Policy and the Law of the Sea," p. 49. Elsewhere, Malone indicated that "the discretion whether to grant a contract should be tied exclusively to the question of whether an applicant has satisfied *objective* qualification standards." Malone, "U.S. Participation," p. 61 (emphasis added). The U.S. delegation had regarded this long-standing negotiating objective as satisfied by 1980. See Chap. 3 nn. 208, 229 and accompanying text.

442. Statement of Ambassador James L. Malone before the Plenary, August 5, 1981, quoted in Larson, "The Reagan Administration," p. 306. The Seabed Act requires U.S. bureaucrats only to "endeavor" to complete preliminary certification within 100 days after submission of the application; the subsequent drafting of terms, conditions, and restrictions may take 180 days, "or such longer period as the Administrator may establish for good cause shown in writing." Seabed Act, 30 U.S.C. §1413 (g). Deep Seabed Mining Regulations, 15 CFR §970.500 (b) (1). Yet the April 13 U.S. proposals called for the Commission to make its recommendation to the Council "within 120 days of taking up a plan of work for consideration." *Amendments* (U.N. Doc. A/CONF.62/L.121), Annex III, proposed art. 6 (4), in *Official Records*, UNCLOS III, vol. XVI, p. 230.

It has been claimed that under the Convention "it is not even clear that the plan of work will be acted upon at all." Kleid, p. 642. The Commission is in fact dutybound to "review formal written plans of work ... and submit appropriate recommendations to the Council," which must act within 60 days. U.N. Convention on the Law of the Sea, arts. 162 (2) (j), 165 (2) (b).

443. See Friedheim, p. 17. The Authority "is, like any regulator, a potential impediment to mining." Frank, "Jumping Ship," p. 132. See also Haight, p. 19. It is inevitable, however, "that some organ of the Authority would have to attest to conformity with the applicable standards of Annex III. Doubtless this would also have been true of the simple licensing system originally advocated by the industrial countries." Aldrich, "Law of the Sea," p. 58. See also Richardson, "The Politics of the Law of the Sea," p. 19.

444. Deep Seabed Mining Regulations, 15 CFR §§970.401 (b), 970.402 (b), 971.501 (c). See nn. 261, 456, 470, 504 above and below and accompanying text. Similarly, the Interior Secretary has had discretion to refuse offshore oil and gas leases to any bidder, "regardless of the amount offered." Outer Continental Shelf Minerals and Rights-of-Way Management, 30 CFR §256.47 (b).

445. Statement of John E. Flipse in *Deep Seabed Mining*, U.S. House of Representatives, p. 314. See also Haight, p. 19.

446. See Katz, pp. 117, 119; and also Frank, "Jumping Ship," p. 132. "If, as is likely, the system of governance of the Authority is not corrupt, that fact would give it a tremendous advantage over the bureaucracies in some developing countries." Katz, p. 119. Even in developing producer states, moreover, obstacles encountered by U.S. mining firms have "very little to do with anything which could normally be described as politics." Faber and Brown, p. 115.

447. U.N. Convention on the Law of the Sea, art. 150 (a). See text accompanying n. 128 above. But see n. 263 above and accompanying text.

448. See Jonathan L. Charney, "United States Interests in a Convention on the Law of the Sea: The Case for Continued Efforts," *Vanderbilt Journal of Transnational Law* 11 (1978): 68. See also nn. 373, 386–88 above and accompanying text. Even if a technical expert were able to determine with some confidence that a restricted supply of nodule minerals was indeed in the best interests of his home government, he would have also to consider that retaliation by Western states and their nationals within the Authority could impede mining operations by the Enterprise or by developing states and thereby also could substantially impair the economic advancement — as well as the prestige — of developing states in general.

449. For example, in late 1980 the chief U.S. negotiator indicated that U.S. fears of the potentially discretionary powers of the Legal and Technical Commission would be largely allayed if it were given the power to recommend approval of plans of work by a simple majority vote. See Aldrich, "Law of the Sea," p. 58. See also U.S. Department of State, "Access to Seabed Minerals," p. 39. The qualifications of applicants and the content of contracts have also been identified as important issues requiring clarification. Hauser, p. 80. See also Hollick, "Opportunities to Correct Defects," pp. 46–47. The elaboration of detailed rules and procedures not only reduces the discretionary power of the Authority, but also increases the

jurisdiction of the Sea-Bed Disputes Chamber. See nn. 497–98 below and accompanying text.

450. U.N. Convention on the Law of the Sea, Annex III, art. 6 (2), (3). See nn. 423–31 above and accompanying text. A similar determination of conformity to the provisions of the Seabed Act and the regulations issued pursuant thereto is to be made with respect to applications submitted by U.S. seabed mining operators. Seabed Act, 30 U.S.C. §1416 (a) (1).

The Green Book would have retained this determining function for the Legal and Technical Commission, but would have presumed compliance with article 4 and with the rules, regulations, and procedures of the Authority—a presumption which could only have been overcome by a consensus of the Commission. *The U.S. Proposals* (U.N. Doc. A/CONF.62/WG.21/Informal Paper 18), Annex III, proposed art. 6 (2) (b), (2) (bis). See also *Amendments* (U.N. Doc. A/CONF.62/L.121), Annex III, proposed art. 6 (1), (3), (4) (a), in *Official Records*, UNCLOS III, vol. XVI, pp. 229–30.

451. The Convention provides for a 45-day grace period during which an applicant may remedy any defect in the application procedure or in the undertakings required by paragraph 6 of article 4, Annex III, without losing its priority of right. U.N. Convention on the Law of the Sea, Annex III, art. 6 (2) (a).

452. Ibid., Annex III, art. 6 (3). But see n. 463 below and accompanying text. A mining site may be unavailable either because the area has been included in a prior plan of work or because the Council has withdrawn the area for environmental reasons. U.N. Convention on the Law of the Sea, Annex III, art. 6 (3) (a), (b). See n. 470 above and accompanying text.

The Convention's approach work is significantly different from that negotiated by the Reagan administration in the 1988 Antarctic Mineral Resources Convention, which requires affirmative votes by both the Antarctic Mineral Resources Commission (by consensus) and the appropriate Regulatory Committee (by concurrent majority) before a mining operator may be granted access to a particular site. Convention on the Regulation of Antarctic Mineral Resource Activities, arts. 3, 21 (1) (d), 22 (1), 31 (1) (b), (c), 32 (1), 43, 48, 54 (5), in *International Legal Materials* 27 (1988): 870, 878, 879, 883, 889, 891–92, 894. See nn. 300, 372 above; and also M. Jacobsen, pp. 240–44. "Because each Regulatory Committee includes a stated balance of four claimant states and six non-claimant states, only two claimant states are needed to prevent the approval of exploration and development permits." M. Jacobsen, p. 242 n. 49. See Convention on the Regulation of Antarctic Mineral Resource Activities, art. 29 (2), in *International Legal Materials* 27 (1988): 882. Mineral exporting states who are parties to the 1959 Antarctic Treaty and to the 1988 Convention would be in a position to deny access. See Bernard H. Oxman, "Antarctica and the New Law of the Sea," *Cornell International Law Journal* 19 (1986): 239. The inconsistency of the Reagan administration's position toward rights of access to Antarctic mineral resources with its position toward rights of access to deepsea mineral resources has been noted by at least one knowledgeable U.S. negotiator. See Oxman, "Antarctica and the New Law of the Sea," pp. 246–47.

453. U.S. Department of State, "Access to Seabed Minerals," p. 39. See also Hauser, p. 63.

454. U.N. Convention on the Law of the Sea, art. 152 (1); Annex III, art. 17 (1). "If the decision-making criteria of the Legal and Technical Commission are nondiscriminatory then access to the Area is assured." Willsey, p. 525.

The requirements set forth in the rules, regulations, and procedures must themselves be "uniform and nondiscriminatory." U.N. Convention on the Law of the Sea, art. 6 (3). This provision is stronger than the U.S. Seabed Act, which requires regulatory uniformity "except to the extent that differing physical and environmental conditions require the establishment of special terms, conditions, and restrictions for the conservation of natural resources, protection of the environment, or the safety of life and property at sea"—a potentially large loophole for bureaucrats. Seabed Act, 30 U.S.C. §1415 (b) (1).

455. U.N. Convention on the Law of the Sea, Annex III, art. 6 (5). The Council could only adopt such procedures and criteria at the end of the interim period defined in article 151. See n. 132 above and accompanying text. Since the United States would possess an effective veto power over the formulation of such measures in the Council, it is misleading to claim, as has one Congressman, that "equitable merit" necessarily "plays a role in the selection process." Cf. Breaux, "The Diminishing Prospects," p. 295. See n. 379 above and accompanying

text. The provision is in fact less potentially discriminatory than a provision of the Seabed Act which permits "the Administrator" to select among competing U.S. applicants by applying "principles of equity." Seabed Act, 30 U.S.C. §1411 (b) (3).

456. See nn. 432, 450, 454–55 above. Under U.S. law, exploration licenses and exploitation permits are issued subject to "terms, conditions and restrictions" imposed by federal regulators. Seabed Act, 30 U.S.C. §1412 (c) (1) (D). Deep Seabed Mining Regulations, 15 CFR §§970.500 (b) (1), 971.400 (b), 971.429. U.S. bureaucrats may deny a license or permit if they find that exploration or exploitation activities would "unreasonably interfere with the exercise of the freedoms of the high seas by other nations," would conflict with other uses of the mining area, would "create a situation which may reasonably be expected to lead to a breach of international peace and security involving armed conflict," would "reasonably be expected to result in a significant adverse effect on the quality of the environment," or would "pose an inordinate threat to the safety of life and property at sea." 15 CFR §§970.407 (a), 970.503 (a), (c), 970.505, 970.506, 970.507, 971.403 (a), (c), 971.405, 971.406, 971.407. The regulations as a whole "are lengthy, complex and quite burdensome; certainly not in the spirit of the [Reagan] Administration's enunciated policy of federal deregulation with the intent of promoting business." Molitor, p. 603. See, e.g., Deep Seabed Mining Regulations, 15 CFR §970.203.

457. Goldie, "A General International Law Doctrine," pp. 822–23.

458. U.N. Convention on the Law of the Sea, art. 150 (g). See n. 126 above. The task force which reviewed the Convention in 1981 apparently advised President Reagan that this provision had been included at the behest of the Soviet Union, and that it would provide "them" with a "near-certain capability of limiting United States access to deep seabed mineral resources." Report of the Foreign Policy Advisory Council and Strategic Minerals Task Force, quoted in Robertson and Vasatura, p. 633. Because the Legal and Technical Commission must base its decision solely on Annex III, however, its recommendation to the Council would be unaffected by the policy. See nn. 263, 439 above and accompanying text.

459. U.N. Convention on the Law of the Sea, Annex III, art. 6 (3) (c). Presumably the rules, regulations, and procedures will provide an objective formula for determining the centers of irregularly shaped sites. Because reserved areas are not included in the formulas, U.S. firms which exploit such areas under joint-venture arrangements with the Authority will not be affected by the antimonopoly provisions. See Jagota, p. 302. The application of these formulas may be determined objectively by seabed mining operators prior to their submission of plans of work, and it is therefore inaccurate to claim that the antimonopoly provisions "could inhibit siteholders from applying for licenses for newly-discovered sites for fear that valuable sites would be turned over to another nation." Cf. Geddes, p. 618.

460. See Chap. 3 n. 174 and accompanying text; and also Charney, "United States Interests," p. 62. A quota system had been proposed as early as 1974 by the EEC states, and was intended to prevent the United States from monopolizing the best seabed mining sites for its nationals. Harry, p. 5. Japan, too, supported limiting the number of U.S. mining sites. Ann L. Hollick, U.S. Foreign Policy and the Law of the Sea (Princeton, N.J.: Princeton University Press, 1981), p. 256. "The United States is virtually isolated in the world community in its insistence upon guaranteed access to the Area for all technically qualified enterprises on a nondiscriminatory basis." Barkenbus, p. 139.

Although the precise number of prime sites remains uncertain, estimates have been substantially reduced since the 1970s, from a high of 185 to a more realistic range of between 14 and 56 sites. Van Dyke and Yuen, p. 510 n. 70. Other recent estimates of the number of prime sites have been significantly lower. See Chap. 5 n. 353.

461. Barkenbus, p. 139. See "U.S. Delegation Report," in Law of the Sea, U.S. Senate, 1976, p. 34. Secretary of State Kissinger, believing there to be "many more productive seabed mining sites than conceivably can be mined for centuries to come," had rejected any "arbitrary or restrictive limitations on the number of mine sites which any nation might exploit," but he had also acknowledged that deep-sea nodules should "not be the exclusive preserve of only the most powerful and technologically advanced nations." Kissinger, "A Test of International Cooperation," pp. 539–40.

462. U.S. Department of State, "Access to Seabed Minerals," p. 39. The overall 2-percent limitation "would, as a practical matter, be almost impossible to violate." Ibid. See also Oxman, "The Third United Nations Conference on the Law of the Sea," pp. 12–13. There are

approximately 2.5 million square kilometers of minable seabed in the Clarion-Clipperton Fracture Zone — an area equal to 1.4 percent of the Pacific Ocean; the partial interests of U.S. firms would enable all four existing private consortia to operate full-sized sites. Hauser, p. 12 n. 9. See also Jagota, pp. 301-2. Cf. U.N. Convention on the Law of the Sea, Annex III, art. 6 (4). Indeed, by claiming geographically dispersed, noncontiguous sites, the potential pioneer applicants have in fact been able to avoid application of the anti-density formula by ensuring the inclusion of nonmining areas within the circles.

Treaty opponents have nevertheless claimed, without explanation, that the antimonopoly provisions would limit U.S. firms to "one or two minesites during the first 25 years of mining under the treaty." Statement of Senator James McClure in *Congressional Record* 125 (December 14, 1979): 36039. See also Breaux, "The Diminishing Prospects," p. 273. It has additionally been claimed that the provisions "would force" U.S. firms "to operate in conjunction with foreign corporations," causing "exportation of American jobs and technical expertise." Statement of Representative Jim Santini in *Congressional Record* 126 (June 9, 1980): 13692. See also statement of Senator James McClure in *Congressional Record* 125 (December 14, 1979): 36039. Of course, U.S. mining firms have already voluntarily allied themselves with foreign firms in the four private consortia, despite the apparent cost to domestic employment.

463. U.N. Convention on the Law of the Sea, Annex III, art. 6 (4). Such determinations could be made by the Council, acting by a three-fourths-majority vote, or by the Legal and Technical Commission. Ibid., art. 162 (2) (j). See nn. 466-68 below and accompanying text.

464. Malone, "U.S. Participation," p. 61. "Any such system threatens U.S. access, is inherently anticompetitive, and is certain to lead to misallocation of resources because inefficient producers are guaranteed entry at the expense of efficient producers." Breaux, "The Diminishing Prospects," p. 272. This statement is perhaps most noteworthy for its implicit acknowledgment that, apart from the antimonopoly provisions, applicants are indeed "guaranteed entry" to seabed nodule deposits under the Convention regime.

465. Deep Seabed Mining Regulations, 15 CFR §971.207. See also Post, "United Nations Involvement," p. 188. Cf. *Amendments* (U.N. Doc. A/CONF.62/L.121), in *Official Records, UNCLOS III*, vol. XVI, pp. 229-30. The Green Book proposed only minor changes in the wording of the antimonopoly policy, as well as, strangely, the elimination of the Authority's ability to grant exceptions. *The U.S. Proposals* (U.N. Doc. A/CONF.62/WG.21/Informal Paper 18), proposed art. 150 (f); Annex III, proposed art. 6 (4).

466. See Robertson and Vasatura, p. 684. Remarkably, while inaccurately protesting that the Convention would permit "disapproval by Third World and socialist countries of nonsocialist developed country applications," one treaty critic has complained that Western developed states would be unable to block "approval of possibly unqualified Third World and socialist applications." Breaux, "The Diminishing Prospects," p. 279.

467. Willsey, p. 524 n. 138. See n. 365 above and accompanying text. Unless all Western developed states on the Council happened to be sponsoring a single consortium — a situation easily avoided — rejection could only occur with the consent of such states. See Jagota, p. 313. Although disapproval by the Council could conceivably occur in truly extraordinary cases, it is in "the interest of most countries, developed or lesser developed, to make available the resources of the deep seabed." John G. Laylin, "The Law to Govern Deepsea Mining Until Superseded by International Agreement," *San Diego Law Review* 10 (1973): 440. Good faith compliance by applicants with their obligation to assist the Enterprise — e.g., by technology transfer — is expected to reduce further any inclination on the part of developing states to impede Council approval. W. Smith, p. 70.

468. U.N. Convention on the Law of the Sea, art. 162 (2) (j) (ii). "Therefore, while we may be able to block denial of a contract, we would not be able to insure that the contract would be granted." Aldrich, "A System of Exploitation," p. 250. Because Western mining states may be able to block such a three-fourths-majority vote, developing states could not be expected to override routinely the Commission's disapproval of applications submitted by the Enterprise or by developing states themselves, although political support for such applications might be expected to be more widespread in the Council. "The assessment of the licensing procedure thus depends conclusively on the provisions concerning the composition and operation of the Commission." Hauser, p. 68. See Chap. 3 n. 337 and accompanying text, and nn. 380-88 above and accompanying text.

469. U.N. Convention on the Law of the Sea, arts. 161 (8) (f), 163 (9).

470. Ibid., art. 162 (2) (x); Annex III, art. 6 (3) (b). See nn. 260, 265 above and accompanying text; and also Arrow, "The Proposed Regime," p. 393. Under the Seabed Act, an application is to be disapproved by federal regulators if exploitation of the proposed site "would result in a significant adverse impact on the quality of the environment which cannot be avoided by the imposition of reasonable restrictions." Seabed Act, 30 U.S.C. §1413 (2) (D) (ii). See also Deep Seabed Mining Regulations, 15 CFR §§970.506, 971.301, 971.600–971.606.

471. U.N. Convention on the Law of the Sea, arts. 161 (8) (c), 187 (b) (i). See nn. 497–98 below and accompanying text. Any applicant wrongfully denied access for environmental reasons could obtain injunctive relief or recover full monetary damages from the Authority. U.N. Convention on the Law of the Sea, art. 189. It is misleading to claim that the Council "may close seabed areas of its choosing" to seabed mining. Myron H. Nordquist, "The Law of the Sea Conference and Deep Seabed Mining Legislation," in *Current Issues in the Law of the Sea*, ed. Joyner, p. 76.

472. Willsey, pp. 531–32.

473. See U.N. Convention on the Law of the Sea, art. 151 (2) (a); and nn. 157–63 above and accompanying text.

474. U.S. Department of State, "Access to Seabed Minerals," p. 40.

475. Oxman, "The Ninth Session," p. 224. See nn. 164, 218 above and accompanying text. It is incorrect to state, as has one prominent treaty opponent, that the selection process would be applied "as the production ceiling is approached." Darman, p. 36050. Because of the relatively short four-month application window and the relatively small number of seabed mining operators competing for authorizations, the allocation system is in effect comparable to the type of first-come, first-served procedure preferred by U.S. firms. See Aldrich, "A System of Exploitation," p. 252. Moreover, it is unlikely that all pioneer registrants would apply for production authorizations simultaneously, or that each would initially produce at maximum capacity. UNCLOS III, *Report of the Co-Ordinators of the Working Group of 21, T.T.B. Koh (Singapore) and Paul Bamela Engo (United Republic of Cameroon) to the First Committee* (U.N. Doc. A/CONF.62/C.1/L.30 March 29, 1982), para. 25, in *Official Records*, UNCLOS III, vol. XVI, p. 272. Those mining operators which are denied authorizations pursuant to the Council's selection process may reapply immediately, and "shall have priority in subsequent periods until they receive a production authorization." U.N. Convention on the Law of the Sea, art 151 (2) (f); Annex III, art. 7 (4). See nn. 164–66 above and accompanying text.

476. See nn. 212–14 above and accompanying text. Cf. statement of the representative of the United States, July 17, 1974, in *Official Records*, UNCLOS III, vol. II, p. 43 (para. 76). Earlier drafts of the treaty had tied such selection more directly to the issue of assured access by applying the production ceiling during the contract-approval process. See Aldrich, "Law of the Sea," p. 58.

477. Brewer, "Deep Seabed Mining," p. 66 n. 64. See nn. 359–62, 378 above and accompanying text. One Congressman has expressed concern that the selection process would also be based upon the general policies set forth in article 150. Breaux, "The Diminishing Prospects," p. 267. None of these policies are to be applied by the Authority, moreover, except "as specifically provided" in Part XI. See n. 263 above.

478. Hauser, p. 117. See text accompanying nn. 128, 500–2 above and below; and U.N. Convention on the Law of the Sea, art. 151 (2) (e).

479. Seabed Act, 30 U.S.C. §1441 (1) (B). It is the intent of Congress that the treaty's compliance with this directive "should be determined by the totality of [its] provisions." Ibid., §1441 (2).

480. Resolution II, paras. 1 (a), (c), 2, 4 (e), 7 (a), (c), 12 (a) (iii), in *International Legal Materials* 21 (1982): 1254–57. See nn. 168, 170 above and accompanying text. "One can generally state that the preparatory investment protection in Resolution II adequately secures the exclusivity of permit sites prior to the approval to be granted to the Authority." Hauser, p. 50. After relinquishment of a reserved area for the Authority, each pioneer registrant retains a site as large as 75,000 sq. km, a size expected to be sufficient for a seabed mining operation of any size. See n. 432 above.

The Reagan administration complained that Resolution II "would allow the U.S.S.R., developing countries, and others to achieve pioneer investor status, despite the fact that their seabed mining activities to date have been extremely limited or nonexistent, and would thereby stimulate conflicting claims to minesites of bona fide pioneers." Statement of Theodore Kronmiller, September 15, 1982, in *Law of the Sea Negotiations*, U.S. Senate, p. 4. Of course, such conflicting claims would be no less likely if U.S. seabed mining firms were to proceed under a nondiscriminatory reciprocating states regime. See Van Dyke and Yuen, p. 546 n. 245.

481. Resolution II, para. 3 (a), (b), 15, in *International Legal Materials* 21 (1982): 1255, 1257. See n. 433 above and accompanying text. Data submitted with the two sites is protected "in accordance with the relevant provisions of the Convention and its Annexes concerning the confidentiality of data." Resolution II, para. 3 (a), in *International Legal Materials* 21 (1982): 1255. see n. 14 above et seq. and accompanying text.

482. Resolution II, para. 12 (a) (i), (ii), in *International Legal Materials* 21 (1982): 1257. See n. 168 and text accompanying n. 9 above. For a discussion of the training obligations agreed upon within the Preparatory Commission, see Chap. 3 n. 373. The Reagan administration criticized these provisions as "major problems," citing the lack of compensation for such training and the "artificially low" interest rate of 10 percent applicable to exploration costs until reimbursement. Statement of Theodore Kronmiller, September 15, 1982, in *Law of the Sea Negotiations*, U.S. Senate, p. 4. In fact, however, with the exception of the fixed rate of interest, these provisions were proposed by the United States during the 1982 session. *Amendments* (U.N. Doc. A/CONF.62/L.122), proposed para. 12 (a), in *Official Records*, UNCLOS III, vol. XVI, p. 232.

483. See UNCLOS III, *Letter Dated 29 April 1982 from the Chairman of the USSR Delegation Addressed to the President of the Conference* (U.N. Doc. A/CONF.62/L.144, April 29, 1982), reprinted in *Official Records*, UNCLOS III, vol. XVI, p. 250; and also statement of Tommy T.B. Koh in *Law and Contemporary Problems* 46 (Spring 1983): 24. Cf. Resolution II, paras. 1 (c), 8 (b), (c), in *International Legal Materials* 21 (1982): 1255, 1256. "Thus, for instance, U.S. companies may acquire rights by piggy-backing on, say, Canadian or Japanese signature." Kimball and Schneider, p. 70.

484. Understanding of the Preparatory Commission, para. 15, in *International Legal Materials* 25 (1986): 1329. See Chap. 3 nn. 330–31 and accompanying text; and also Santa Cruz, p. 117.

485. Resolution II, para. 8 (a), (c), in *International Legal Materials* 21 (1982): 1256. See "Reports of the United States Delegation," ed. Nordquist and Park, p. 542. The six-month period could be extended by the Preparatory Commission. See Chap. 3 n. 373. Absent ratification, multinational pioneer registrants would have to transfer sponsorship of their seabed mining operations to a certifying state or states which had ratified the Convention. Cf. Resolution II, paras. 1 (c), 8 (b), (c), in *International Legal Materials* 21 (1982): 1255, 1256.

Although Resolution II only provides for sponsorship of applicants for plans of work by "certifying" states—i.e., by the signatory states which sponsored the applicants as pioneer investors in the Preparatory Commission—in light of the Commission's demonstrated willingness to modify the terms of Resolution II it may be expected that nonsignatory states such as the United States would be able to secure plans of work for their pioneer investors by acceding to the Convention prior to its entry into force. See Chap. 3 nn. 309, 330–31, 343 and accompanying text. Cf. Resolution II, paras. 1 (c), 8 (b), (c), in *International Legal Materials* 21 (1982): 1255, 1256.

486. Resolution II, para. 9 (b), in *International Legal Materials* 21 (1982): 1256. See n. 165 above and accompanying text. In order to ensure its effectiveness under the Convention, this priority should be restated in the rules, regulations, and procedures governing issuance of production authorizations. See Jenisch, p. 412; and also U.N. Convention on the Law of the Sea, Annex III, art. 7 (2).

If competing registrants were unable to reach agreement within three months, the production authorizations would be awarded by an arbitral tribunal. Resolution II, para. 9 (d), (g), in *International Legal Materials* 21 (1982): 1256–57. See n. 165 above and accompanying text.

487. Leigh S. Ratiner, "The Law of the Sea: A Crossroads for American Foreign Policy," *Foreign Affairs* 60 (1982): 1014. See also Sebenius, pp. 100–1.

488. Lewis I. Cohen, "International Cooperation on Seabed Mining," in *United States*

Law of the Sea Policy, ed. Shutler, pp. 153–54. "For example, upon award of his so-called 'priority' the U.S. ocean miner could not know or foresee his future costs under whatever contract he might sometime acquire. . . . He could not know or foresee the contents of the rules and regulations governing non-financial matters under which he would be required to operate if he were awarded a contract. He could not know or foresee what technology he would be required to transfer to the Enterprise or to developing nations or what price, if any, he would be paid for such transfer." Statement of the American Mining Congress in *Congressional Record* 126 (June 9, 1980): 13680.

Although preparatory investment protection had originally been an important negotiating objective of the United States, the final version of Resolution II was not to its liking, as the amendments of April 13 indicated. See nn. 616, 617 above; and also Hollick, "Opportunities to Correct Defects," p. 52.

489. See, e.g., statement of Theodore Kronmiller, September 15, 1982, in *Law of the Sea Negotiations,* U.S. Senate, p. 4. It is, of course, not realistic to expect mining operators to be able to obtain guaranteed mining rights unless their applications comply with the Convention and with the rules, regulations, and procedures of the Authority. See text accompanying n. 438. Even under U.S. law an applicant obtains no rights — only a "priority" — until the application has been thoroughly reviewed and an exploration license issued, yet U.S. "pre-enactment explorers" have managed to surmount the uncertainties inherent in this procedure. Seabed Act, 30 U.S.C. §1411 (b) (1) (B), (3). See n. 456 above and accompanying text; and also Deep Seabed Mining Regulations, 15 CFR §970.301.

490. Seabed Act, 30 U.S.C. §1441 (2). "Without assurances that a fair and equitable process exists for the settlement of seabed mining disputes, U.S. industry cannot attempt to operate under this treaty." "Summary of Detailed Ocean Mining Industry Comments," p. 13683.

491. U.N. Convention on the Law of the Sea, Annex VI, arts. 2, 4 (4). Subsequent elections are to be held "by procedure agreed to by the States Parties." Ibid., Annex VI, art. 4 (4). Each state may seat no more than one of its nationals on the Tribunal, and each geographical region is guaranteed "no fewer than three members." Ibid., Annex VI, art. 3. The provisions of the Convention relating to the election and composition of the Tribunal, which are similar to provisions governing the election and composition of the International Court of Justice, "are not in all respects optimal but are acceptable." Hauser, pp. 52–53. See Statute of the International Court of Justice, arts. 2–13. "The Western European and Others Group . . . and the Eastern European . . . group are assured of no more than three members each and neither group may realistically expect to have more than four members. This means that thirteen to fifteen members will be from the developing countries." Finlay, p. 142.

The members will serve staggered nine-year terms and may be reelected, and each state party may nominate two qualified individuals for Tribunal membership. U.N. Convention on the Law of the Sea, Annex VI, arts. 4 (1), 5 (1). Tribunal members are subject to conflict-of-interest regulation. See U.N. Convention on the Law of the Sea, Annex VI, arts. 7 (1), 8 (1).

492. U.N. Convention on the Law of the Sea, Annex VI, art. 35 (1). Again, the membership is to ensure "representation of the principal legal systems of the world and equitable geographical distribution." Ibid., Annex VI, art. 35 (2). Thus each of the five major geographical regions could expect to hold two seats, with the 11th seat probably being allocated to a developing state.

493. Ibid., Annex VI, art. 29 (1). The Chamber may formulate its own rules of procedure, but it must base its decisions upon the terms of any contract at issue, the rules, regulations, and procedures of the Authority, and the Convention "and other rules of international law not incompatible" with it. Ibid., art. 293 (1); Annex VI, arts. 16, 38. The Chamber may also apply principles of equity, "if the parties so agree." Ibid., art. 293 (2). Each party to a dispute may select a Tribunal member to join the Chamber during the case. Ibid., Annex VI, art. 17 (3), (4).

494. Louis B. Sohn, "Peaceful Settlement of Disputes in Ocean Conflicts: Does UNCLOS III Point the Way?" *Law and Contemporary Problems* 46 (Spring 1983): 199. Disputes between states parties may be submitted at the request of any party to a three-member ad hoc chamber of the Sea-Bed Disputes Chamber, or, at the request of all parties, to a special three-member chamber of the Tribunal. U.N. Convention on the Law of the Sea, art. 188 (1). "While the

complex dispute settlement mechanism of the ... Treaty does not assure satisfactory decisions, the international context and choice of forum are factors not present in many developing countries and help to assure fairness." Katz, p. 122.

495. Aldrich, "A System of Exploitation," p. 255. See Finlay, p. 142; and also Fuandez, p. 497.

496. See nn. 7, 103 above and accompanying text; and U.N. Convention on the Law of the Sea, art. 188 (2) (a). Each arbitral tribunal will consist of five members — one of whom will be selected by each of the parties and three of whom will "be appointed by agreement of the parties," and all of whom are to be persons "experienced in maritime affairs and enjoying the highest reputation for fairness, competence and integrity." U.N. Convention on the Law of the Sea, Annex VII, arts. 2, 3 (a). Because some states had objected to any interpretation of the Convention by an arbitral panel in cases relating to a mining contract, legal issues requiring such interpretation must be referred to the Sea-Bed Disputes Chamber for an interlocutory ruling. Sohn, "Peaceful Settlement of Disputes," pp. 199-200. See U.N. Convention on the Law of the Sea, art. 188 (2) (a), (b); and also Hauser, pp. 56-57. An applicant for a contract may appeal a denial of his application only to the Chamber. U.N. Convention on the Law of the Sea, art. 187 (d).

U.S. mining firms objected to provisions permitting the President of the Tribunal, "who may be expected to be a developing-country national," to select the neutral members of the arbitral tribunal if the parties are unable to agree upon such selection within 60 days and if one of the parties — presumably the Authority — should so request. Finlay, p. 143. See U.N. Convention on the Law of the Sea, Annex VI, art. 35 (4); Annex VII, art. 3 (d), (e). It has nevertheless been observed that the dispute settlement procedures set forth in the Convention are in many respects superior to those available to U.S. mining firms in developing states. Katz, pp. 120-21. It may, perhaps, also prove significant that the treaty's dispute settlement procedures are available to private firms which enter into joint-venture contracts with the Authority, but not to the Enterprise in the event of a dispute with its parent organization. U.N. Convention on the Law of the Sea, arts. 153 (2) (b), (3), 187.

497. *The U.S. Proposals* (U.N. Doc. A/CONF.62/WG.21/Informal Paper 18), proposed art. 189. See nn. 254, 322 above and accompanying text. The discretionary acts of U.S. administrators under the 1980 Seabed Act are specifically exempt from judicial review. 30 U.S.C. §1427 (a) (2). The "discretionary function exemption" in U.S. domestic law has been narrowly construed. See, e.g., Jayman-Ruby, Inc. v. Federal Trade Commission, 496 F. Supp. 838, 844 (1980).

It has been argued that these exceptions to the Chamber's jurisdiction will leave "little effective judicial remedy for abuses by the Council or the Assembly." Breaux, "The Diminishing Prospects," p. 280. The U.S. proposals of April 13, 1982, sought no changes in the substance of article 189, however. See *Amendments* (U.N. Doc. A/CONF.62/L.121), proposed art. 189, in *Official Records*, UNCLOS III, vol. XVI, p. 228. Moreover, the Chamber has been expressly granted jurisdiction to adjudicate misuses of power by organs of the Authority. See n. 288 above and accompanying text. The insistence of developing states upon the exceptions would seem to indicate that they are less certain than Western states that the members of the Chamber will manifest any great sympathy for the goals of the New International Economic Order.

498. See nn. 7, 103, 218, 256, 288, 392 above and accompanying text; and also Hauser, p. 55. In addition to disputes between states parties regarding the interpretation or application of the seabed mining provisions of the Convention, the jurisdiction of the Chamber includes acts or omissions in violation of the Convention or the rules, regulations, and procedures of the Authority; misuse of power by the Authority; approval of plans of work; interpretation or application of contracts; and "acts or omissions of a party to the contract relating to activities in the Area and directed to the other party or directly affecting its legitimate interests." U.N. Convention on the Law of the Sea, art. 187 (a)-(d).

499. Laylin, p. 440. "They need assurance that if they live up to their obligations, they can be sure that they will retain their contracts and that those obligations will remain the same throughout the years remaining." Aldrich, "A System of Exploitation," p. 254. Security of tenure has been "common to all of the mining laws of the world that have succeeded, over the centuries, in attracting capital and have resulted in the efficient production of minerals to

meet world needs." Statement of John E. Flipse in *Deep Seabed Mining*, U.S. House of Representatives, p. 313.

500. U.N. Convention on the Law of the Sea, Annex III, art. 3 (4) (c). See also ibid., Annex III, arts. 3 (4) (c), 10, 16. Resolution II provides each pioneer registrant an "exclusive right to carry out pioneer activities in the pioneer area allocated to it." Resolution II, para. 6, in *International Legal Materials* 21 (1982): 1256. The American Mining Congress has nevertheless criticized Resolution II, arguing that "[m]ere knowledge of the location of a patch of seabed to which the U.S. ocean miner might possibly be awarded a contract if he is awarded one at all creates no rights to mine that patch of seabed." Statement of American Mining Congress in *Congressional Record* 126 (June 9, 1980): 13681. See nn. 486–89 above and accompanying text. Of course, such criticism applies with greater force to a freedom of the seas regime. Indeed, even under the U.S. Seabed Act, access is exclusive only "as against any other United States citizen or any citizen, national or governmental agency of . . . any reciprocating state." Seabed Act, 30 U.S.C. §1412 (b) (2). See also Deep Seabed Mining Regulations, 15 CFR §970.102 (b).

501. U.N. Convention on the Law of the Sea, art. 153 (6); Annex III, arts. 18, 19. See nn. 280, 282–83 above and accompanying text. "When circumstances have arisen or are likely to arise which, in the opinion of either party, would render the contract inequitable or make it impracticable or impossible to achieve the objectives set out in the contract or in Part XI, the parties shall enter into negotiations or revise it accordingly"; any modification would require the consent of both the mining operator and the Authority. U.N. Convention on the Law of the Sea, Annex III, art. 19. Joint ventures are governed by the terms of contracts, and private mining firms entering into such arrangements with the Authority will also enjoy security of tenure, contrary to the assertion of one Congressman. U.N. Convention on the Law of the Sea, Annex III, art. 11 (1). Cf. Breaux, "The Diminishing Prospects," p. 295. Plans of work submitted by the Enterprise do not become "contracts" upon approval, however, and its existing mining rights do not enjoy the same security. Cf. U.N. Convention on the Law of the Sea, art. 153 (3).

U.S. seabed mining operators have expressed concern that they might lose their rights of access because of failure to comply with the treaty's technology transfer provisions. See Aldrich, "A System of Exploitaiton," p. 253. Unless instances of such noncompliance were determined to be "serious, persistent and wilful," however—a determination which would be subject to judicial review in the Sea-Bed Disputes Chamber—mining operators will face no more than monetary penalties. U.N. Convention on the Law of the Sea, art. 187 (c) (ii); Annex III, art. 18 (1) (a), (2), (3).

502. See nn. 434, 634 above, and nn. 315, 625 and accompanying text above and below. Again, this protection does not extend to the Enterprise.

503. U.N. Convention on the Law of the Sea, arts. 161 (8) (c), 162 (2) (w). After 30 days the order would have to be confirmed by consensus. Ibid., art. 161 (8) (c). The issuance of such orders is subject to review by the Sea-Bed Disputes Chamber. Ibid., art. 187 (b). See n. 288 above and accompanying text. The Green Book would have permitted the Council to issue emergency orders only "upon the recommendation of the Technical Commission." *The U.S. Proposals* (U.N. Doc. A/CONF.62/WG.21/Informal Paper 18), proposed art. 162 (2) (v).

504. Seabed Act, 30 U.S.C. §1416 (a) (2) (A). See nn. 261, 265, 284, 470 above and accompanying text. Cf. Deep Seabed Mining Regulations, 15 CFR §§970.2503 (a), 971.417 (h) (2), 971.1003 (d) (4) (ii). Federal law also permits the President or "the Administrator" to suspend or modify any license or permit in order to ensure compliance with U.S. treaty obligations, to prevent "any situation which may reasonably be expected to lead to a breach of international peace and security involving armed conflict," or "to preserve the safety of life and property at sea." Seabed Act, 30 U.S.C. §1411 (b) (2). Deep Seabed Mining Regulations, 15 CFR §§971.414, 971.417 (a), (h) (2), 971.1003 (d) (4) (i). See also Seabed Act, 30 U.S.C. §1415 (c) (1).

Leases issued for offshore oil drilling on the U.S. continental shelf may be canceled, without compensation, for similar reasons. See OCS Lands Act, 30 U.S.C. §§1334 (a)–(d), 1351 (h) (2), 1356 (a) (2). Moreover, if they are unable to rectify conditions which bring about suspension, oil and gas lessees might "not be relieved of the obligation to pay rental, minimum royalty or royalty for or during the period of suspension." 30 CFR §218.154 (b). "Industry representatives have indicated for a number of years . . . that they were well satisfied with the

Outer Continental Shelf Lands Act and its administration and did not see it as a deterrent to further offshore exploration and exploitation activities." Knight, "The Draft United Nations Conventions," p. 548.

In 1988 the Reagan administration endorsed an Antarctic mineral resources regime providing for suspension, modification, or cancellation of a Management Scheme (the Antarctic equivalent of a plan of work) without prior judicial review if a Regulatory Committee determined that the environmental dangers of further mining activities were "beyond those judged acceptable." Convention on the Regulation of Antarctic Mineral Resource Activities, art. 51 (1), in *International Legal Materials* 27 (1988): 892.

505. Statement of Northcutt Ely, September 30, 1982, in *Congressional Digest* 62 (January 1983): 29. Seabed mining states had sought a contractual guarantee of at least 20 years of commercial exploitation. See Aldrich, "A System of Exploitation," p. 254. Cf. Seabed Act, 30 U.S.C. §1417 (a), (b).

506. U.N. Convention on the Law of the Sea, Annex III, art. 17 (1) (b) (ii). A representative of one of the U.S. consortia has incorrectly stated that a contract approved by the Authority "cannot last longer than 25 years from the date of the first commercial production by anyone, anywhere in the world." Statement of Northcutt Ely, September 30, 1982, in *Congressional Digest* 62 (January 1983): 31. "The total duration of exploitation, however, should ... be short enough to give the Authority an opportunity to amend the terms and conditions of the plan of work at the time it considers renewal in accordance with rules, regulations and procedures which it has adopted subsequent to approving the plan of work." U.N. Convention on the Law of the Sea, Annex III, art. 17 (2) (b) (iii). Theoretically, a short duration would allow the Authority to subject the contractor to burdensome and restrictive regulations not in effect when the plan of work was initially approved as a contract, but since their promulgation—as well as adoption of the duration itself—would require consensus approval in the Council, this provision is perhaps most significant for its implicit rejection of any retroactive effect for rules, regulations, and procedures adopted by the Authority subsequent to contract approval. Hauser, pp. 125–27. See n. 258 above and accompanying text.

507. Antrim and Sebenius, pp. 86–87. See Charney, "The Law of the Deep Seabed Post UNCLOS III," pp. 38–39; and also Sebenius, pp. 20–22, 101–2. Mining firms have found that much of the insecurity of their tenure "is either inherent in the nature of mining business, or a consequence of the way in which mining investment has to be fitted into a rigid contractual framework from a very early stage in the life of a project. It has very little to do with anything which could normally be described as politics.... Their complaint is that in the event of a disagreement about what changes should be made host governments can, and frequently do, resolve the matter by unilateral action or the threat of it." Faber and Brown, p. 115. "Even in developed countries where investment conditions are usually regarded as stable, taxes along with environmental and safety regulations are subject to unpredictable changes." Sebenius, p. 101.

508. *Draft United Nations Convention* (U.N. Doc. A/AC.138/25), art. 68 (1) (c). Following approval by a concurrent majority in the Council and by a simple majority vote in the Assembly, these "Rules and Recommended Practices" were to become binding unless disapproved by more than one third of states parties. Ibid., arts. 34 (5), 38, 67 (c)–(e), 71.

509. U.N. Convention on the Law of the Sea, arts. 137 (2), 151 (9), 161 (8) (d), 162 (2) (o) (ii); Annex III, art. 17 (1) (b) (iv). See nn. 169, 228, 272 above and accompanying text. "Priority shall be given to the adoption of rules, regulations and procedures for the exploration for and the exploitation of polymetallic nodules." U.N. Convention on the Law of the Sea, art. 162 (2) (o) (ii).

510. U.N. Convention on the Law of the Sea, arts. 187 (b) (i), 189. See n. 498 above. To the extent that the United States contributed to a stalemate within the Council, its recovery might be limited, however. U.N. Convention on the Law of the Sea, Annex III, art. 22.

511. U.N. Convention on the Law of the Sea, Annex III, art. 17 (2) (d).

512. Malone, "U.S. Participation," p. 62. Although the Green Book had proposed no changes to article 137, the April 13 set of amendments would have deleted the requirement that seabed resources be exploited only in accordance with Part XI. *Amendments* (U.N. Doc. A/CONF.62/L.121), proposed art. 137 (2), in *Official Records*, UNCLOS III, vol. XVI, p. 226.

513. The Outer Continental Shelf Lands Act provides, for example, that the Secretary

of the Interior may issue leases for exploitation "of any mineral other than oil, gas, and sulphur . . . upon such royalty, rental, and other terms and conditions as the Secretary may prescribe at the time of offering the area for lease." OCS Lands Act, 43 U.S.C. §1337 (k).

514. See U.N. Convention on the Law of the Sea, Annex III, art. 2 (1) (b), (c). "The Authority shall encourage prospecting in the Area," and prospecting "shall be without time-limit." Ibid., Annex III, arts. 2 (1) (a), 17 (2) (b) (i). Article 2 of Annex III "makes clear that no action by the Authority is necessary before a prospector may engage in prospecting. The Authority simply receives from the prospector the 'satisfactory written undertaking' prescribed by that article, the substance of which should be non-objectionable and a notification of the broad area or areas in which prospecting is to take place." U.S. Department of State, "Access to Seabed Minerals," p. 39.

515. U.N. Convention on the Law of the Sea, arts. 159 (8), 161 (8) (d), 314 (1). Every five years the Assembly is to conduct "a general and systematic review" of seabed mining activities, and may recommend to the Council "measures . . . which will lead to the improvement of the operation of the regime." Ibid., art. 154. The Council could only accept or reject, and not modify, such recommendations. See Hauser, pp. 47, 138.

516. U.N. Convention on the Law of the Sea, art. 155 (1), (3), (4). The review conference was proposed in 1976 by Secretary of State Kissinger; it was linked to the expiration of the interim period of production controls and accepted as an essential element of the parallel system. Jagota, p. 306. See Chap. 3 n. 153 and accompanying text. Most such conferences require only a two-thirds-majority vote. Yemin, p. 36.

517. These aspects include the legal status of the Area and its resources, the existence of "an Authority to organize, conduct and control activities in the Area," the general anti-monopoly policy, and "the basic principle relating to the common heritage of mankind," as well as the general provisions relating to "economic aspects of activities in the Area, marine scientific research, transfer of technology, protection of the marine environment, protection of human life, rights of coastal States, . . . and accommodation between activities in the Area and other activities in the marine environment." U.N. Convention on the Law of the Sea, arts. 155 (2), 311 (6). Cf. n. 22 above and accompanying text. Although article 155 (2) prohibits amendment of "principles . . . with regard to . . . the rights of states . . . and their participation in activities in the Area," it is unclear to what extent the review conference could limit the access of state-sponsored mining operators. See Hauser, p. 129.

Any amendments adopted by the Council which conflict with these fundamental elements of Part XI may be challenged before the Sea-Bed Disputes Chamber prior to their adoption by the Assembly. U.N. Convention on the Law of the Sea, arts. 159 (10), 314 (2). The Green Book would have permitted the review conference to adopt amendments to any aspect of Part XI. The U.S. Proposals (U.N. Doc. A/CONF.62/WG.21/Informal Paper 18), proposed art. 155 (2).

518. U.N. Convention on the Law of the Sea, arts. 155 (3), (4), 316 (5). See also ibid., arts. 306, 315, 316 (6). Although close to three fourths of the states attending the review conference could be developing states, the United States and other Western mining states would have ample time to exercise bilateral leverage — e.g., by threatening foreign aid cut-offs — in order to prevent the ratification of undesirable amendments adopted at the conference.

519. U.N. Convention on the Law of the Sea, art. 155 (5). See Antrim and Sebenius, p. 94. Because the Enterprise will conduct its mining activities under "plans of work" rather than "contracts," amendments may affect its existing mining rights. See nn. 501–2 above.

Although article 155 (5) refers only to amendments adopted at the review conference, the security of tenure provisions of the treaty would seem to govern amendments adopted by the Council and the Assembly as well, in which case it would be incorrect to claim, as has the U.S. seabed mining industry, that amendments adopted by these organs of the Authority "may clearly apply to operations earlier approved and may thus clearly change the circumstances . . . under which such operations have been conducted." "Summary of Detailed Ocean Mining Industry Comments," in Congressional Record 126 (June 9, 1980): 13683. See n. 501 above and accompanying text. In any event, as a member of the Council, the United States would be able to prevent the consensus necessary for adoption of such amendments, and any disputes concerning the applicability of amendments to existing contracts may be submitted to the Sea-Bed Disputes Chamber. See nn. 379, 514 above and accompanying text.

520. Ratiner, "A Crossroads for American Foreign Policy," p. 1015. See n. 516 above; remarks by Secretary of State Kissinger, September 1, 1976, in *U.S. Department of State Bulletin* 75 (1976): 398; and also Leive, p. 47. If the United States truly believes that private enterprise is inherently superior to and more efficient than public operations, it should not fear competition from the Authority's Enterprise. Any superiority of private exploitation should be evident by the time of the review conference, making further "socialist" mining endeavors less attractive for all states.

521. Statement of Theodore Kronmiller, September 15, 1982, in *Law of the Sea Negotiations*, U.S. Senate, p. 5. "It has always been the aim of the Group of Seventy-Seven to establish a unitary system of mining under the Authority and through the Enterprise once the necessary financing and technology has been obtained. While mining operations . . . could continue under existing contracts, future operations would be conducted by the Enterprise or pursuant to joint venture, service or other contracts with the Enterprise." Haight, p. 19. Conversion to such a unitary system could prevent private seabed mining firms from spreading the recovery of their capital investment costs over several generations of mining operations. Breaux, "The Diminishing Prospects," p. 285.

Western support for developing-state mining operations in reserved areas — possibly in the form of less-burdensome technology transfer arrangements — might nevertheless require developing states to tolerate, if not to encourage, future private mining operators. In fact, developing states are afraid that the Convention will fail to protect their interests, and they generally view the review conference as a mechanism for the prevention of "the worst scenario," in which the Enterprise is never given an opportunity to compete with state-sponsored mining firms. Jagota, p. 319.

522. White House fact sheet, in *U.S. Department of State Bulletin* 82 (March 1982): 55. It has even been argued that the U.S. Senate is constitutionally incapable of delegating its power of consent to such an amendment process. See Robertson and Vasatura, p. 685; and also Malone, "Who Needs the Sea Treaty?" p. 54. Prevention of the entry into force of such binding amendments was described by Ambassador Malone both as "a matter of constitutional principle for the United States" and as one of the "political goals" of President Reagan. Malone, "Who Needs the Sea Treaty?" p. 50. If the U.S. Senate were to endorse such an amendment process in the original treaty instrument, however, there would seem to be no legal reason why these amendments should not have binding legal effect for the United States; and if U.S. courts — it is ultimately for them, not the Senate, to resolve such issues — should determine that such amendments are not constitutionally valid, the Senate would have incurred no liability by authorizing them.

523. Yemin, pp. 51, 53, 208. See *Rest. 3rd, Restatement of the Foreign Relations Law of the United States*, §334, reporters' note 1; and n. 296 and text accompanying n. 295 above. Such procedures constitute a limited delegation of sovereignty by individual states to the organized international community. Yemin, p. 57.

524. Statement of President Reagan, July 9, 1982, in *U.S. Department of State Bulletin* 82 (August 1982): 71. See n. 522 above. However, the Green Book would have permitted binding amendments to enter into force for all states parties upon ratification by only two thirds of them, "provided that all . . . members of the Council ratify or accede." *The U.S. Proposals* (U.N. Doc. A/CONF.62/WG.21/Informal Paper 18), proposed art. 155 (4). See also *Draft United Nations Convention* (U.N. Doc. A/AC.138/25), art. 76.

525. See Yemin, p. 29; and Morley, pp. 26–27. "In only a minority of organizations do the constitutional amendment clauses make the validity of amendments dependent upon the contractual consent of members." Yemin, pp. 57–58. Constitutive conventions which confer quasi-supranational powers of amendment upon an international body include those of IMO, ILO, WHO, UNESCO, IAEA, FAO, IMF, and the World Bank. Yemin, pp. 51, 53, 208. Amendments adopted by plenipotentiary conferences of the ITU and the Universal Postal Union (UPU) enter into force provisionally "on a specified date for all members, irrespective of the fact that most of them may not have ratified by then." Yemin, p. 209. Amendments adopted by INTELSAT's Assembly enter into effect 90 days following their approval by 85 percent of its membership or by states which have invested two thirds of its total capital. Doyle, "Permanent Arrangements," pp. 283, 285, 287–88. See also Williams, pp. 1190–91.

Amendments generally enter into force for these organizations upon ratification by only two thirds of states parties. See, e.g., Constitution of the World Health Organization, art. 73, in *United Nations Treaty Series,* vol. 14, no. I-221, p. 202; and Statute of the International Atomic Energy Agency, art. XVIII (C) (ii), in *United Nations Treaty Series,* vol. 276, no. 3988, p. 36. Amendments to the Charter of the United Nations, which may be adopted by only a two-thirds majority of the one-nation-one-vote General Assembly, enter into force for all members upon ratification "in accordance with their respective constitutional processes by two thirds of the Members of the United Nations, including all the permanent members of the Security Council." U.N. Charter, arts. 108, 109 (2). The Covenant of the League of Nations "permitted amendment on ratification by states members of the Council plus a simple majority of all the Assembly members. Under this process several important amendments were successfully written into the League Covenant." Morley, p. 50.

526. Yemin, p. 44. See n. 260 above and accompanying text; and also Yemin, pp. 28, 207. Allowing amendments to enter into force only for ratifying states would theoretically have preserved a greater degree of state sovereignty, but would have undermined the uniformity and stability of the seabed mining regime without preventing widely ratified amendments from crystallizing into general customary law. Oxman, "The Ninth Session," p. 251.

Unless adopted by all states parties pursuant to a simplied consensus procedure, amendments to nonseabed provisions of the Convention may be adopted at a separate 10-year review conference and would enter into force – only for ratifying states – upon ratification "by two thirds of the States Parties or by 60 States Parties, whichever is greater." U.N. Convention on the Law of the Sea, art. 316 (1). "Such amendments shall not affect the enjoyment by other States Parties of their rights or the performance of their obligations." U.N. Convention on the Law of the Sea, art. 316 (1).

527. U.N. Convention on the Law of the Sea, art. 317 (1). See Oxman, "The Ninth Session," p. 253. Denunciation will not discharge a state from "the financial and contractual obligations which accrued while it was a Party . . . , nor shall the denunciation affect any right, obligation or legal situation of that State created through the execution of [the] Convention prior to its termination for the State." U.N. Convention on the Law of the Sea, art. 317 (2).

528. Yemin, pp. 18, 54. Cf. Vienna Convention on the Law of Treaties, arts. 56, 70, reprinted in *International Legal Materials,* 8 (1969): 699, 705. The U.N. Charter does not provide specifically for denunciation, but an interpretation adopted at the 1945 San Francisco Conference permits a member to withdraw from the United Nations in the event an amendment it has formally opposed enters into force or where an amendment it has formally supported fails to enter into force. Kelsen, *Principles of International Law,* p. 338.

The United States has exercised rights of denunciation with respect to the constitutive conventions of other international organizations. See Rachel Roat, "Promulgation and Enforcement of Minimum Standards for Foreign Flag Ships," *Brooklyn Journal of International Law* 6 (1980): 59 n. 16. In fact, the one-year denunciation period was originally included in the U.S. 1970 Draft Treaty. *Draft United Nations Convention* (U.N. Doc. A/AC.138/25), art. 77.

529. The Reagan administration asserted that the option of denunciation "is obviously not acceptable when dealing with major economic interests of countries which have invested significant capital in the development of deep seabed mining." Malone, "U.S. Participation," p. 62. See also Finlay, p. 145. While it may be true that a state generally would not resort to denunciation unless the amendment seriously threatened its national interest, it is most certainly true that a state ought not to reject an important treaty prior to any adverse – or otherwise – experience under its provisions, particularly when investments made by it under the treaty regime would remain unaffected by denunciation. See n. 527 above. Cf. Yemin, p. 213.

Concern has been expressed that the United States would be effectively foreclosed from denouncing the Convention, since customary international law at the time might preclude unilateral seabed mining operations. Bandow, p. 486. Cf. Hauser, pp. 129–30. But see Lord McNair, *The Law of Treaties* (Oxford: Clarendon, 1961), pp. 520–21, 531–33. The United States might seek to prevent any such loss of unilateral seabed mining rights by openly asserting their existence under customary international law, as it is now doing. An intention for the treaty provisions to be "replaced by subsequent custom" following denunciation is implicit in the use of a denunciation clause. *Rest. 3rd, Restatement of the Foreign Relations Law of the*

United States, §102, reporters' note 4. Even the loudest among the treaty critics within the Reagan administration admitted that the possibility of losing such rights by participating in the treaty regime was "mere speculation." Malone, "Who Needs the Sea Treaty?" p. 62.

530. Seabed Act, 30 U.S.C. §1441 (2). See text accompanying n. 479 above.

531. See Finlay, pp. 150–51; and Willsey, p. 531. Congress rejected language which would have required the terms of the Convention to be "substantially the same" as those of the Act, a requirement clearly not satisfied. Willsey, p. 531. See also statement of Senator Jacob Javits in *Congressional Record* 125 (December 14, 1979): 36071. The Seabed Act "recognizes that a treaty will provide for additional financial obligations and other new requirements not contained in this legislation, such as the transfer of technology on fair and reasonable terms and conditions." Statement of Representative Jonathan Bingham in *Congressional Record* 126 (June 9, 1980): 13685. See also n. 417 below.

532. U.N. Convention on the Law of the Sea, arts. 151 (10), 161 (8) (d), 162 (2) (n). See nn. 126–27, 227 above and accompanying text; and also U.N. Convention on the Law of the Sea, Annex III, art. 17 (1) (d). Such measures may be adopted by the Assembly "upon the recommendation of the Council," both organs acting by a two-thirds-majority vote. U.N. Convention on the Law of the Sea, arts. 159 (8), 160 (2) (1), 161 (8) (b), 162 (2) (n). The Assembly must either accept or reject the recommendations forwarded to it, without modification. See Hauser, pp. 47, 138.

533. Malone, "U.S. Participation," p. 61. White House fact sheet, in *U.S. Department of State Bulletin* 82 (March 1982): 55. "Discriminatory provisions allowing the Enterprise to produce minerals under conditions unrelated to practical economic forces will artificially stimulate . . . production, driving mineral prices down. Private investors, who must be concerned about market prices that determine the profitability of projects, will thus be at an unfair cost disadvantage." Malone, "Who Needs the Sea Treaty?" p. 62. See also statement of Theodore Kronmiller, September 15, 1982, in *Law of the Sea Negotiations*, U.S. Senate, p. 3.

534. Malone, "U.S. Policy and the Law of the Sea," p. 49. "Avoiding any potential monopolization of mining operations by the Enterprise . . . was recognized as central to U.S. efforts to protect American security needs and economic interest in commercial development." Malone, "Who Needs the Sea Treaty?" p. 50. U.S. opposition to monopolization has been consistent only with respect to the Enterprise however. See U.S. Senate, *The Third United Nations Law of the Sea Conference*, p. 54; and also n. 464 above and accompanying text.

535. See Finlay, pp. 151–52; and Breaux, "The Diminishing Prospects," pp. 257, 259. But see Chap. 6 n. 29.

536. Statement of Northcutt Ely, September 30, 1982, in *Congressional Digest* 62 (January 1983): 29–31. "That is like playing in a baseball game with the umpire being an active member of the other team and he has a right to change the rules as the game proceeds." Statement of James Wenzel in *Deep Seabed Mining*, U.S. House of Representatives, p. 217. Such objections fail to take into account the multinational structure of the Authority, which provides Western mining states with a large measure of control over decisions affecting their important interests. See nn. 318, 323–24, 328, 354, 360–62, 373, 378, 392–93, 409–10 above and accompanying text.

537. American Mining Congress, Declaration of Policy, September 26, 1976, reprinted in *Deep Seabed Mining*, U.S. House of Representatives, p. 58.

538. Hicks, p. 110.

539. Bernard H. Oxman, "The Third United Nations Conference on the Law of the Sea: The Seventh Session (1978)," *American Journal of International Law* 73 (1979): 39.

540. U.N. Convention on the Law of the Sea, arts. 148, 152. See n. 257 above and accompanying text. For example, the needs of developing states may be taken into consideration by the Authority in its allocation of seabed mining revenues. U.N. Convention on the Law of the Sea, arts. 160 (2) (f) (i), 162 (2) (o) (i). "Cynics have noted that whenever the Authority is bidden to act 'without discrimination,' these words are always qualified by the rider, 'special consideration must be given to the interests of developing countries.'" Watt, "The Need for an Agreement," p. 87.

541. See nn. 543–47, 553, 573 above and accompanying text. Cf. Hicks, p. 113. Rather, the effect of such provisions would be to permit the Authority to promote seabed mining by developing states under certain exceptional circumstances. See Geddes, p. 617.

542. See Bailey, p. 74.

543. Bandow, p. 481.

544. Remarks by Secretary of State Kissinger, September 1, 1976, in *U.S. Department of State Bulletin* 75 (1976): 398. See nn. 47–62, 65 above and accompanying text; and also Kissinger, "A Test of International Cooperation," p. 540. "In fact, in the absence of external subsidies, the newly created Enterprise would not be able to conduct mining operations." Charney, "The Law of the Deep Seabed Post UNCLOS III," p. 37. See Hauser, pp. 153–56.

545. Aldrich, "A System of Exploitation," p. 246. See also Hauser, p. 65.

546. See Brewer, "Deep Seabed Mining," p. 49; and also Chap. 3 n. 405 and accompanying text, and n. 192 above and accompanying text. Significantly, national subsidization of deep-sea mining operations has been opposed primarily by developed producer states, such as Australia and Canada — rather than by the United States, which might prefer to subsidize U.S. seabed mining consortia through tax incentives. See nn. 125, 163 above. Cf. n. 652 below. See also Sanger, p. 182. European states, on the other hand, are more inclined to support national subsidization of seabed mining activities. Two thirds of total West German investment in nodule mining operations, for example, was government-subsidized. U.N. Office for Ocean Affairs and the Law of the Sea, "Seminar on the Current Status of Developments," p. 61.

547. Faber and Brown, pp. 118–19. See also Brewer, "Deep Seabed Mining," p. 42.

548. Statement of Northcutt Ely, September 30, 1982, in *Congressional Digest* 62 (January 1983): 29–31.

549. See nn. 4, 6, 12, 25, 57, 433, and text accompanying nn. 3–6, 10, 12, 20, 62, 69, 72, 432–33 above; and also Hauser, pp. 65–66, 115. "It is not in the interest of our investors to make our technology immediately available to competitors worldwide. They hope to achieve the advantage in both time and profitability, and for that they take the risk of the investment." Statement of John E. Flipse in *Deep Seabed Mining*, U.S. House of Representatives, p. 338. Although such considerations, as well as the administrative burden of compliance, would represent somewhat of a competitive disadvantage for transferors, the benefit to transferees would be offset by the requirement of fair and reasonable compensation, as well as by practical difficulties involved in receiving transferred technology. Hauser, pp. 148–49, 154–55.

550. Although technology could be obtained by the Enterprise or by developing-state mining operators only "on fair and reasonable commercial terms and conditions," the third-party written assurance requirement may represent a potential burden for state-sponsored operators. See n. 3 above, and nn. 64, 67 above and accompanying text; and Hauser, pp. 114–15.

The cost of prospecting the additional mining has been set at less than $10 million, a small amount in relation to anticipated total investment costs of more than $2 billion. Sebenius, p. 97. See n. 436 above. Exploration of a reserved site for the Enterprise — a proposal rejected at UNCLOS III — would have cost each private mining firm as much as $50 million. Breaux, "The Diminishing Prospects," p. 284 n. 124. Special Commission 3 of the Preparatory Commission has taken steps to limit the amount of proprietary information required from state-sponsored operators in connection with their applications for plans of work, thereby reducing the prospecting expense to be incurred by them with respect to reserved areas as well as the concomitant competitive advantage to be gained by the Enterprise. Walkate, pp. 155–57. See Chap. 3 n. 347 and accompanying text.

551. Katz, p. 125. See also ibid., pp. 126–27.

552. See n. 117 above and accompanying text. This advantage would not be substantial, particularly if private mining operators are given domestic tax credits. The grace period was an integral part of a compromise reached in 1980 whereby developing states agreed that the Enterprise should be required to make equivalent revenue-sharing payments to the Authority. Hauser, pp. 135–36. The Enterprise is subject to taxation by states in which it conducts business, although it may seek to negotiate exemptions with the host state. U.N. Convention on the Law of the Sea, Annex IV, art. 13 (5). See Hauser, pp. 133–34.

553. *Report of the Co-Ordinators* (U.N. Doc. A/CONF.62/C.1/L.30), para. 29, in *Official Records*, UNCLOS III, vol. XVI, pp. 272–73. See Chap. 3 n. 373, and n. 9 above and accompanying text; and also Jagota, p. 317. Cf. U.N. Convention on the Law of the Sea, Annex III, art. 13 (1) (e). The Enterprise may be expected to encounter difficult — if not insurmountable — technical and political start-up problems.

554. Statement of Senator Edmund Muskie in *Congressional Record* 125 (December 14, 1979): 36077. See, e.g., Deep Seabed Miing Regulations, 15 CFR §971.204. The Reagan administration agreed to even more burdensome environmental restrictions upon access to Antarctic minerals in 1988. See Convention on the Regulation of Antarctic Mineral Resource Activities, art. 4 (2), (3), in *International Legal Materials* 27 (1988): 870–71.

555. Deep Seabed Mining Regulations, 15 CFR §§971.419 (a), 971.423, 971.604. See Seabed Act, 30 U.S.C. §1419 (b); and also n. 456 above and accompanying text. "Furthermore, each permit issued under the Act must require the permittee to monitor the environmental effects of commercial recovery activities in accordance with guidelines issued by the Administrator, and to submit information the Administrator finds necessary and appropriate to assess environmental effects and to develop and evaluate possible methods of mitigating adverse effects." Deep Seabed Mining Regulations, 15 CFR §971.600.

556. Seabed Act, 30 U.S.C. §1413 (a) (2) (C). Similar details must be included in each exploration plan. Ibid., §1413 (a) (2) (B).

557. Deep Seabed Mining Regulations, 15 CFR §971.804. See nn. 261, 504 above and accompanying text.

558. Seabed Act, 30 U.S.C. §1412 (c) (2), (3). See also Deep Seabed Mining Regulations, 15 CFR §§970.521, 971.205 (a).

559. Deep Seabed Mining Regulations, 15 CFR §§971.209, 971.408, 971.427. See Seabed Act, 30 U.S.C. §1412 (c) (5).

560. Statement of Marne Dubs, American Mining Congress, in *Deep Seabed Mining,* U.S. House of Representatives, p. 73. "[T]here would appear . . . to be no advantage, from an industry perspective, to assuming this onerous burden which, in view of the large proportion of the total capital investment it represents, could make an otherwise profitable project marginal or unacceptable." Ibid.

561. Laursen, p. 122.

562. Keith, p. 286. See text accompanying Chap. 2 n. 151, and Chap. 5 generally.

563. U.N. Convention on the Law of the Sea, art. 162 (2) (k). See also ibid., art. 170 (1); Annex IV, art. 12 (1), (2). It is simply not true that the Authority "is granted broad discretionary powers . . . in granting preferred status to its own mining arm or to applicants sponsored by developing countries." Nordquist, "The Law of the Sea Conference," p. 76. Strangely, the Reagan administration's April 13 set of amendments would have deleted the requirement that plans of work submitted by the Enterprise be evaluated "applying, *mutatis mutandis,* the procedures" applicable to prospective contractors. *Amendments* (U.N. Doc. A/CONF.62/L.121), proposed art. 162 (2) (z), in *Official Records,* UNCLOS III, vol. XVI, p. 228.

564. U.N. Convention on the Law of the Sea, Annex III, art. 8; Annex IV, art. 12 (1). See nn. 423, 426, 432–37 above and accompanying text; and also Hauser, pp. 132–33.

565. See Hollick, *U.S. Foreign Policy,* p. 342; and Chap. 3 n. 330 and text accompanying n. 363, and nn. 165–66, 212, 214, 486 above and accompanying text. "The crowning blow would come if the contractor were denied a production authorization because of lack of quota at the time that the priority provisions . . . were applied to give the Enterprise or one or more developing countries . . . a production authorization for the reserved area stemming from the contractor's own application." Finlay, p. 151. Because the basic priority applies only with respect to nonpioneer applicants and only when there are less reserved than nonreserved sites under exploitation, the priority is not an "absolute" one, as one Congressman has claimed. Cf. Breaux, "The Diminishing Prospects," p. 292. See n. 166 above and accompanying text; and Aldrich, "A System of Exploitation," p. 252.

Use of the priority "of course will depend on the ability of the Enterprise to obtain financing for such additional sites." Brewer, "Deep Seabed Mining," p. 66 n. 64. Its capacity in this respect is likely to be rather limited. See nn. 571, 574–75 below and accompanying text. Joint-venture arrangements with state-sponsored mining operators might permit the Enterprise to engage in more than one initial mining operation. See Aldrich, "A System of Exploitation," p. 252. Unless economic conditions improve significantly, however, the 20-year production limitation may well expire before the Enterprise is able to undertake seabed exploitation at a third site. See n. 161 above and accompanying text.

566. Statement of Theodore Kronmiller, September 15, 1982, in *Law of the Sea Negotiations,* U.S. Senate, p. 4. See nn. 275–77, 279–83, 459–65 above and accompanying text. One

Congressman has argued that exemption of the Enterprise from the antimonopoly provisions could give it "an enormous economic advantage." Breaux, "The Diminishing Prospects," p. 285. It remains to be seen whether the Enterprise will be successful enough to benefit from economies of scale, however. Similarly, exemption of the Enterprise from the standard enforcement penalties set forth in the Convention is unlikely to create a significant competitive imbalance, since the Enterprise has been rendered more directly subject to the control of the Council than are state-sponsored seabed mining operators. See n. 590 below and accompanying text.

567. U.N. Convention on the Law of the Sea, Annex III, art. 12 (1); Annex IV, art. 1 (2). See also n. 170 above, and n. 253 above and accompanying text.

568. U.N. Convention on the Law of the Sea, Annex IV, arts. 5 (8), 6 (c), (e), (f), (g), 7 (2). See n. 412 above, and n. 409 and accompanying text above; and also Hauser, p. 137. The Director-General must comply with rules and regulations enacted by the governing board, the Council, and the Assembly governing "the organization, management, appointment and dismissal of the staff of the Enterprise." U.N. Convention on the Law of the Sea, Annex IV, art. 6 (1). "The Governing Board directs the business operations of the Enterprise and possesses all the power necessary to fulfill the purposes of the Enterprise." R. Lee, "The Enterprise," p. 63. See U.N. Convention on the Law of the Sea, Annex IV, art. 12 (6).

569. See Breaux, "The Diminishing Prospects," p. 285. "The only certain way of preventing this possibility is by insisting at the outset that the international regime have no operating functions." American Bar Association, "Natural Resources of the Sea," p. 663. See n. 536 above and accompanying text; and also Dubow, p. 188.

570. See nn. 414–15, 445–48, 454, 542 above and accompanying text; and Hauser, pp. 141–42. "The Director-General and the staff of the Enterprise, in the discharge of their duties, shall not seek or receive instructions from any Government or from any other source.... The members of the Authority shall respect the international character of the Director-General and the staff of the Enterprise and shall refrain from all attempts to influence any of them in the discharge of their duties." U.N. Convention on the Law of the Sea, Annex IV, art. 7 (3). The Sea-Bed Disputes Chamber would have jurisdiction to determine any violation of these provisions, for which the Authority may be held liable. U.N. Convention on the Law of the Sea, art. 187 (b) (i); Annex III, art. 22. See nn. 278, 498 above and accompanying text.

571. See UNCLOS III, *Alternative Means of Financing the Enterprise: Preliminary Note by the Secretary-General* (U.N. Doc. A/CONF.62/C.1/L.17, September 3, 1976), para. 5, reprinted in *Official Records*, UNCLOS III, vol. VI, p. 156. Once it has become economically successful, the Enterprise will be able to finance its capital and operating expenditures from its own reserve funds, which will be kept separate from those of the Authority. See U.N. Convention on the Law of the Sea, Annex IV, arts. 10 (2), (3), 11 (4).

572. Breaux, "The Diminishing Prospect," p. 259.

573. Bandow, p. 491. Sentiment against taxation in general — and against foreign aid in particular — has been growing within the United States since the late 1960s.

The treaty provides that the Enterprise "shall be exempt from all direct taxation and from all customs duties on goods imported or exported for its official use." U.N. Convention on the Law of the Sea, art. 183 (1). Representative Breaux has stated that this provision "seems to be ideological," observing that U.S. private enterprise is subject to a corporate tax. Breaux, "The Diminishing Prospects," p. 283. See also Dubow, pp. 186–87. National taxation of international organizations would likely require reciprocity, however, and proponents of U.S. nationalism have traditionally opposed vesting such organizations with any power to tax nation-states. See Francis O. Wilcox and Carl M. Marcy, *Proposals for Changes in the United Nations* (Washington: Brookings Institution, 1955), p. 443. See also nn. 124–25 above and accompanying text.

574. Oxman, "The Third United Nations Conference," p. 21.

575. *Alternative Means of Financing* (U.N. Doc. A/CONF.62/C.1/L.17), para. 30, in *Official Records*, UNCLOS III, vol. VI, p. 159. "The Enterprise will, in one sense, be in a position of great commercial strength in that as a branch of the Authority it will have a privileged position with respect to access to the area and its resources.... On the other hand from the point of view of any lender it is bound to be viewed initially as a high-risk borrower." Ibid.,

para. 22, in *Official Records,* UNCLOS III, vol. VI, p. 159. See U.N. Convention on the Law of the Sea, Annex IV, art. 11 (1) (c), (2).

576. U.N. Convention on the Law of the Sea, Annex IV, art. 11 (3) (a), (b). See Gertler and Wayne, pp. 557-58; and also Hauser, p. 134. In order to provide potential states parties with adequate notice of their obligations, the Preparatory Commission will establish the total amount of initial funding before the Convention enters into force. U.N. Convention on the Law of the Sea, Annex IV, art. 11 (3) (a). See Gertler and Wayne, p. 540 n. 50. "Such a ceiling would tend to quiet fears about runaway expenditures by the Enterprise as it sets out to acquire the expensive technology being developed for deep seabed mining." W. Smith, p. 69. Interest-bearing loans are to have repayment priority. U.N. Convention on the Law of the Sea, Annex IV, art. 11 (3) (f).

The Reagan administration pointed out that the developed states which failed to vote in favor of the Convention at the final session of UNCLOS III account for "more than 65 percent of the contributions to the United Nations — and would be expected to provide comparable levels of funding to" the Authority. Malone, "Who Needs the Sea Treaty?" p. 56. "Severe doubts exist about . . . whether the parties to the Convention will be able without the U.S. and possibly other industrial nations, to meet the high financial obligations required by parties to the Convention." Kimball and Schneider, p. 72. See also Langevad, p. 233. However, loss of the U.S. contribution alone may not completely prevent the Enterprise from beginning operations, as the Reagan administration apparently expected it would. See Richardson, "The United States Posture," p. 512; and also Bailey, p. 74. Indeed, although concern within the Preparatory Commission regarding the amount of financial obligations to be assumed by states parties has in recent years led to widespread support for a scaled-back "nucleus" Enterprise which would operate on a substantially lower budget, the Commission has nonetheless proceeded with implementation of the parallel system. See Chap. 3 nn. 346, 364, 366 and accompanying text; and Nandan, p. 518. Any shortfall in initial funding is to be remedied by measures adopted by a consensus of the Assembly. U.N. Convention on the Law of the Sea, Annex IV, art. 11 (3) (c). See Hauser, p. 135.

577. U.N. Convention on the Law of the Sea, arts. 160 (2) (e), 171 (a). See n. 274 above and accompanying text; and *Alternative Means of Financing* (U.N. Doc. A/CONF.62/C.1/L.17), para. 6, in *Official Records,* UNCLOS III, vol. VI, p. 156. The administrative expenses of the Authority will have "first call" upon funds received by the Authority from other sources. U.N. Convention on the Law of the Sea, art. 173 (2). The U.S. share of the Authority's initial administrative budget was expected to be $5 million to $10 million. Statement of Theodore Kronmiller, September 15, 1982, in *Law of the Sea Negotiations,* U.S. Senate, p. 5.

Assessed contributions are a common method for funding such expenses of international organizations. See, e.g., Convention on the Regulation of Antarctic Mineral Resource Activities," art. 35 (5), in *International Legal Materials* 27 (1988): 885.

578. See *Alternative Means of Financing* (U.N. Doc. A/CONF.62/C.1/L.17), paras. 18, 19, in *Official Records,* UNCLOS III, vol. VI, p. 158. Loan guarantees have been freely employed domestically by the U.S. government to stabilize agriculture — the fiscal 1985 federal budget included approximately $630 million in such guarantees — as well as other industries. Internationally, the Reagan administration provided developing states with $30 billion in direct and indirect assistance annually. President Ronald W. Reagan, address before U.N. General Assembly, September 24, 1984, reprinted in *New York Times,* September 25, 1984, p. A10.

579. R. Lee, "The Enterprise," p. 68. See U.N. Convention on the Law of the Sea, art. 170 (4); and Jagota, p. 310.

580. See Hauser, p. 148. Kissinger "proposed on behalf of the U.S. Government that the United States would be prepared to agree to a means of financing the Enterprise in such a manner that the Enterprise could begin its mining operation either concurrently with the mining of state or private enterprises or within an agreed timespan that was practically concurrent." Remarks by Secretary of State Kissinger, September 1, 1976, in *U.S. Department of State Bulletin* 75 (1976): 398. Kissinger had expected states parties to provide "something in the nature of a billion dollars" in financial assistance. Aldrich, "A System of Exploitation," p. 246.

581. "U.S. Delegation Report," in *Law of the Sea,* U.S. Senate, 1976, p. 33. "It became

obvious that the only viable approach would be for states to make loans to the Authority."
R. Lee, "The Enterprise," p. 73 n. 34.

582. See Langevad, p. 237. "Administratively, there will be a time gap before the money
is made available to the Enterprise for use, since the whole procedure cannot begin until the
Assembly has met and the Council has been established, and has elected members of the
Governing Board." R. Lee, "The Enterprise," p. 66. Indeed, it has been pointed out that a successful Enterprise "would require the kind of management that may be unacceptable or
unavailable to an international bureaucracy." Brewer, "Deep Seabed Mining," p. 46.

583. U.N. Convention on the Law of the Sea, Annex IV, art. 11 (1) (a)-(b). Initial determinations with respect to transfers from the Authority are to be made by the Assembly pursuant to its residual authority. See n. 399 above and accompanying text. While, theoretically,
the Assembly could promote competitive imbalance by voting to retransfer a substantial
amount of funds to the Enterprise, effectively offsetting its revenue-sharing payments, the
number of states likely to oppose such use of the Authority's limited resources — including
developing producer states seeking compensation, seabed mining states seeking to preserve
a competitive balance, and states seeking to benefit directly from the equitable sharing of seabed mining revenues paid to the Authority — should be sufficient to block the necessary two-
thirds vote. Cf. Hauser, pp. 136, 142-43 nn. 60, 149.

Acting "upon the recommendation of the Governing Board," which it may accept or reject
but not modify, the Assembly may transfer excess funds from the Enterprise to the Authority,
in accordance with rules, regulations, and procedures over which the United States would
have an effective veto power in the Council. U.N. Convention on the Law of the Sea, art. 160
(2) (f) (ii); Annex IV, art. 10 (2). See Hauser, pp. 135, 139.

584. U.N. Convention on the Law of the Sea, arts. 160 (2) (f) (ii), (h), 161 (8) (c), (d),
162 (2) (o) (ii), (r), 172, 174 (3). See nn. 378-79 above and accompanying text; and also
Willsey, p. 521. The Preparatory Commission has decided that the Authority's decisionmaking process should be assisted by a Finance Committee. See Chap. 3 n. 357 and accompanying
text. Cf. U.N. Convention on the Law of the Sea, arts. 161 (8) (c), 162 (2) (y).

585. See nn. 480, 483-86 above and accompanying text.

586. See n. 116 above and accompanying text; and U.N. Convention on the Law of the
Sea, Annex III, arts. (4) (1), 13 (6) (o).

587. See n. 164 and text accompanying n. 477 above. "Some of our European friends are
very worried about the possibility of competitive bidding. They work on the theory that
American companies have all the money and will out bid them." Aldrich, "A System of Exploitation," p. 251.

588. See U.N. Convention on the Law of the Sea, Annex III, arts. 11 (2), 13 (14). Any
such financial incentives must "ensure that . . . contractors are not subsidized so as to be given
an artificial competitive advantage with respect to land-based miners" — a requirement applicable also to financial incentives employed in the revision of contracts and in the formation
of joint ventures. Ibid., art. 13 (1) (f).

589. See nn. 496, 501, 519 above and accompanying text.

590. U.N. Convention on the Law of the Sea, Annex IV, arts. 1 (3), 2 (1), (2), 12 (7). See
nn. 325, 402, 414-15 above and accompanying text; R. Lee, "The Enterprise," pp. 64-65; and
also W. Smith, pp. 60-61. Although it is true that developing states could muster the two-
thirds-majority vote necessary to control these directives, any tendency to base such votes
upon political rather than economic factors would seem to cast further doubt upon the ability
of the Council to manage the Enterprise in a manner which might enable it to become truly
competitive with state-sponsored operators. See n. 606 below and accompanying text.

591. W. Smith, p. 68.

592. See Weiss, p. 65; and Knight, "The Draft United Nations Conventions," pp. 529,
542.

593. U.N. Convention on the Law of the Sea, Annex III, art. 11 (1). See also ibid., art.
153 (3). When entering into joint ventures with respect to the conduct of mining activities in
reserved areas, the Enterprise may form such agreements with noncontractors otherwise
qualified to obtain contracts, but it must offer developing states parties "the opportunity of
effective participation." Ibid., Annex III, art. 9 (2). The rules, regulations, and procedures of
the Authority "may prescribe . . . substantive and procedural requirements and conditions

with respect to such contracts and joint ventures." Ibid., Annex III, art. 9 (3). For a discussion of possible forms of joint-venture arrangements, see Hauser, pp. 157-74.

594. U.N. Convention on the Law of the Sea, art. 153 (3); Annex III, art. 11 (1). See n. 496 and text accompanying n. 11 above. Cf. Breaux, "The Diminishing Prospects," p. 295. Private firms entering into such agreements with the Enterprise are liable for revenue-sharing payments "to the extent of their share in the joint ventures," although the amount of such payments could be reduced in order to provide a financial incentive for mining firms to enter into joint-venture agreements. U.N. Convention on the Law of the Sea, Annex III, art. 11 (3). Joint-venture agreements may also permit seabed mining firms to avoid application of the antimonopoly provisions in the unlikely event that they should threaten U.S. access, since reserved areas are not included in the limitation formulas. Hauser, p. 156. See nn. 459, 462 above.

In the absence of U.S. adherence to the Convention, U.S. seabed mining firms are prohibited from entering into joint ventures with the Enterprise without express authorization from the United States or a reciprocating state. Deep Seabed Mining Regulations, 15 CFR §971.103.

595. Malone, "U.S. Participation," p. 61. "The net effect is what amounts, as a practical matter, to a "unitary system of joint ventures" — for few, if any, rational investors could be expected to seek contracts except as joint ventures with the Enterprise." Darman, "Legislative Policy," p. 36051. See nn. 534, 598 above and below and accompanying text. But see n. 588 above.

596. U.N. Convention on the Law of the Sea, Annex III, arts. 11 (2), 13 (1) (d), 17 (1) (c) (iii). See n. 594 above, and n. 588 and accompanying text above. Significantly, in 1988, the Reagan administration agreed that the Antarctic Mineral Resources Commission must provide "opportunities for joint ventures or different forms of participation, up to a defined level, including procedures for offering such participation . . . by interested Parties . . . , in particular, developing countries" — a provision which, opponents of the 1982 Convention should agree, could lead to backdoor creation of an "Enterprise" in Antarctica. Convention on the Regulation of Antarctic Mineral Resource Activities, art. 41 (1) (d), in *International Legal Materials* 27 (1988): 888.

597. See Katz, p. 115.

598. Bailey, p. 76. The flexibility inherent in such joint-venture arrangements should permit private mining operators to avoid "most of the complexities, costs and inefficiencies which are embodied in the Convention." Pardo, "A Preliminary Appraisal," p. 500 n. 45. See nn. 11, 71 above and accompanying text.

599. See nn. 53, 58 above, and n. 168 above and accompanying text; and also Sebenius, p. 100.

600. See Post, "United Nations Involvement," p. 189; and *Costs of the Authority* (U.N. Doc. A/CONF.62/C.1/L.19), para. 42, in *Official Records*, UNCLOS III, vol. VII, p. 65.

601. Goldschneider, p. 261. Joint ventures have generally proven profitable, however. Barkenbus, p. 138.

602. See Aldrich, "A System of Exploitation," pp. 245-46.

603. Friedheim, p. 4. "Total" national self-sufficiency in such "strategic materials" was indeed identified in the U.S. Senate as a primary motivation for the 1980 Seabed Act. Statement of Senator Spark Matsunaga in *Congressional Record* 125 (December 14, 1979): 36064. See text accompanying n. 559 above.

604. Kissinger, address before American Bar Association, August 11, 1975, p. 357. See Brewer, "Deep Seabed Mining," pp. 64-65 n. 55; nn. 62, 83-84 above and accompanying text, and text accompanying nn. 38-41, 88, 544, 580 above; and also Laursen, p. 10.

605. Jagota, pp. 318-19. See n. 521 and text accompanying nn. 361, 542, 545 above; and also Arrow, "The Proposed Regime," p. 416. The success of the Enterprise will depend upon the degree of cooperation given to it by developed states, as well as upon its working relationship with the Authority and the competence of its Director-General. See W. Smith, p. 70; and n. 582 above and accompanying text.

606. R. Lee, "Deep Seabed Mining and Developing Countries," p. 217. "The early arguments of industrial states against the Enterprise, and the current arguments of its proponents for special favorable treatment as against both state and private mining operations,

reveal a curious consensus: the Enterprise is not expected to be as efficient as other operators." Bernard H. Oxman, "The Third United Nations Conference on the Law of the Sea: The 1976 New York Sessions," *American Journal of International Law* 71 (1977): 253.

607. Nordquist, "The Law of the Sea Conference," pp. 77–78.

608. Yost, pp. 677–78. See n. 42 above and accompanying text, and Chap. 5 nn. 223–26 and accompanying text. Although international control of seabed mining through the regulatory mechanism of the Authority may require some sacrifice of economic efficiency, the interests of many developing states can be represented only through such a political institution. See Barkenbus, p. 112. Indeed, for purposes of economic analysis, political power might be regarded as a commodity, with its own supply and demand schedules; it is thus somewhat hypocritical for the United States to argue, on economic grounds, that its economic demands for nodule minerals must be satisfied but that demands by Third World states for political power must not be. Such an argument cannot rest upon considerations of economic efficiency, since it is not clear that any lost economic efficiency would not be more than offset by an aggregate gain in political benefits. For an example of the simplistic conclusions which economic theorists may be expected to reach when it is wishfully assumed that most states "can be thought of as being indifferent with respect to seabed mining," see Federico Foders, "International Organizations and Ocean Use: The Case of Deep-Sea Mining," *Ocean Development and International Law* 20 (1989): 519–30.

609. Robertson and Vasatura, pp. 689–90 n. 75.

610. If state-sponsored operators should fail to attain economic viability under the Convention, the well-being of the Enterprise could require developing states to offer greater concessions to private contractors at the review conference in order to ensure continued assistance. See nn. 520–21 above and accompanying text.

611. Statement of President Reagan, January 29, 1982, in *U.S. Department of State Bulletin* 82 (March 1982): 54. "Most, if not all, of the adverse precedents which would be established by the . . . treaty could be avoided by achieving the six objectives set out by the President." Malone, "U.S. Participation," p. 62. See text accompanying Chap. 3 n. 280. The Reagan administration assigned greater importance to the Convention's precedential and ideological impact than had prior administrations. Sebenius, pp. 104, 107.

612. Malone, "Who Needs the Sea Treaty?" p. 46. See also James L. Malone, "The United States and the Law of the Sea After UNCLOS III," *Law and Contemporary Problems* 46 (Spring 1983): 31. Even prior to President Reagan's election, the United States officially viewed major changes in the norms or structures of international organizations as both unnecessary and impractical. See A.A. Fatouros, "The International Economic Order: Problems and Challenges for the United States," *Willamette Law Review* 17 (1980): 1107.

613. Statement of Representative John Breaux before Subcommittee on Arms Control, Oceans and International Operations and Environment, Committee on Foreign Relations, U.S. Senate, March 5, 1981, quoted in Larson, "The Reagan Administration," p. 301.

614. Statement of Representative Ronald Paul, April 28, 1982, reprinted in *Congressional Digest* 62 (January 1983): 25. See also ibid., p. 29. Notwithstanding the assertions of Representative Paul, "collective ownership," in the form of publicly held corporations, has played an important role in the economic development of the United States. See Christopher D. Stone, *Where the Law Ends* (New York: Harper & Row, 1975), pp. 19–23. In any event, it is not at all clear that the collective interest of the international community in deep-sea resources must have originated in the Convention. See Chap. 5.

615. Fuandez, p. 488. See also statement of Brian Hoyle in *Consensus and Confrontation*, ed. Van Dyke, p. 263. Cf. nn. 619–20 below and accompanying text. The World Bank also exercises a direct operational competence. See Leive, pp. 48–49; and also Post, "United Nations Involvement," pp. 181–82. Other international organizations empowered to allocate scarce resources include the IMF, the United Nations Development Program (UNDP), and the North Atlantic Treaty Organization (NATO). H. Jacobsen, p. 89.

616. Caron, "Reconciling Domestic Principles," p. 6. Hicks, p. 118. It has also been claimed that the Convention would establish a "world government system." "Summary of Detailed Ocean Mining Industry Comments," in *Congressional Record* 126 (June 9, 1980): 13682.

The precedential significance of the Convention can only be understood in the context of the continuing struggle between developed and developing states. See Saffo, p. 494; and also Barkenbus, p. 153. "The position of the developed countries on the access issue is based on theories of international equity and political expediency, while the posture of the developing States is grounded on the concept of a fundamental redistribution of global wealth." Breaux, "The Diminishing Prospects," p. 261.

617. Brokaw and Burke, p. 43. But see n. 700 below and accompanying text. Treaty critics have warned of widespread "assimilation into customary international law of principles harmful to the economic well-being of the West." Brokaw and Burke, pp. 51, 55.

618. Nasila S. Rembe, *Africa and the International Law of the Sea* (Germantown, Md.: Sijthoff & Noordhoff, 1980), p. 64. Fuandez, p. 493. See text accompanying Chap. 3 nn. 52, 87–88, and text accompanying n. 318 above.

619. See nn. 358, 390, 674 above and below, and nn. 296, 406, 629 above and below and accompanying text. "INTELSAT owns and operates the space segment of the global commercial communications satellite system consisting of satellites and related equipment required to maintain and operate the satellites." Colino, p. 107. See Doyle, "Permanent Arrangements," pp. 258–91. "The U.S. participant, Communications Satellite Corporation, is the operator of the overall system and has had little trouble in its relations with INTELSAT." Frank, "Jumping Ship," p. 132.

Similar to the seabed mining regime set forth in the 1982 Convention, article XIV (d) of the 1971 INTELSAT Agreement specifically provides for parallel access by state-sponsored satellite system operators, with INTELSAT being authorized to assure the compatibility of such other satellite systems. Colino, pp. 108–9. "Because applicants would have to undergo the INTELSAT coordination process, the status of the application, pending this process, would be uncertain and the applicant would be unlikely to receive financial backing until this process was completed." Colino, p. 143.

620. Although the United States had previously supported INTELSAT as the sole international satellite communications system, in 1983 the Reagan administration began working actively to alter INTELSAT's uniform pricing structure and to expedite applications by private U.S. firms for licenses to compete directly with INTELSAT in providing international satellite communications services. See Colino, pp. 103–56. "[I]n the pursuit of deregulatory and competitive policies, the Government of the United States is reversing its long-held views with respect to the role and place of INTELSAT in the international scene." Ibid., p. 153.

Following adoption of General Assembly Resolution 1721, INTELSAT was provisionally established in 1964, expressly in recognition "that communications by means of satellites should be available to the nations of the world as soon as practicable on a global and non-discriminatory basis." Agreement Establishing Interim Arrangements for a Global Communications Satellite System, preamble, quoted in ibid., p. 258. See nn. 654–56 below and accompanying text, and Chap. 5 n. 84; and Doyle, "Permanent Arrangements," p. 259. Agreements establishing INTELSAT on a permanent basis were adopted in 1971 by a vote of 73 to 0, with 4 abstentions. Doyle, "Permanent Arrangements," p. 273. By late 1985 INTELSAT was an organization of 110 states providing international telephone and television services to more than 170 states and territories with a 16-satellite system. Colino, p. 108. "This unique international organization is run on business principles as it exploits high technology but is essentially a non-profit cooperative." Colino, p. 107.

621. See Buzan, *Seabed Politics*, pp. 297–98; and also Post, "United Nations Involvement," pp. 183–84.

622. See Friedheim, 12, 23–24; and Gregg and Barkun, p. 404.

623. Friedheim, p. 17. See nn. 242, 315 above and accompanying text.

624. Breaux, "The Diminishing Prospects," pp. 261, 277, 281. Ambassador Richardson had, more positively, perceived an "opportunity . . . to establish a precedent which can serve as a blueprint for the development of future international institutions concerned with common resources." U.S. Senate, *The Third United Nations Law of the Sea Conference*, p. 115.

Certain areas have historically been uninhabitable and open to free access by persons of all nations. Honnold, pp. 847–48. "It is their location beyond the jurisdiction of any nation that gives them their special characteristics." Aldrich, "Law of the Sea," p. 57. See also Mary Victoria White, "The Common Heritage of Mankind: An Assessment," *Case Western Reserve Journal of International Law* 14 (1982): 535.

625. Sebenius, p. 104. See, e.g., statement of Representative John Breaux in *Congressional Record* 126 (June 9, 1980): 13695. Breaux has elsewhere described these common areas as "vital," and failure of the United States to cooperate with other states in their use might well jeopardize access for U.S. nationals. Breaux, "The Diminishing Prospects," p. 261. See Chap. 5 n. 444 et seq. and accompanying text. The earth's weather and climate may also be considered common resources. S. Brown et al., p. 1. Concern has been expressed that common-resource principles might be applied to international financial institutions, such as the IMF and the World Bank, as well as to international negotiations relating to food, commodities, the environment, and renewable energy. Brokaw and Burke, pp. 44, 55. Van Dyke and Teichmann, p. 526. Sebenius, p. 104.

626. See Shigeru Oda, "Fisheries Under the United Nations Convention on the Law of the Sea," *American Journal of International Law* 77 (1983): 755. Cf. U.N. Convention on the Law of the Sea, art. 133 (a). UNCLOS III delegates rejected any comprehensive regime for living high-seas resources at an early stage of the negotiations, making creation of such a regime unlikely for the foreseeable future. See text accompanying Chap. 2 nn. 99, 102, and text accompanying Chap. 3 n. 89; and also Barkenbus, p. 145.

627. McCormick, p. 112. See also Malone, "Law of the Sea and Oceans Policy," p. 50. Such an argument would seem to establish only its authors' low opinion of the negotiating ability of U.S. diplomats, unless it may be understood as a projection of personal weakness — by analogy, every loss of virginity would inevitably lead to promiscuity. In fact, the State Department has viewed this aspect of the seabed mining regime as unique and without significant precedential value. Breaux, "The Diminishing Prospects," p. 263. See also Van Dyke and Teichmann, p. 526.

628. Barnes, p. 11. See n. 14 above, and nn. 31, 64–65 above and accompanying text; and also Brewer, "Transfer of Mining Technology," pp. 4–5. Although the United States has successfully resisted attempts to maake multilateral codes of conduct legally binding upon all multinational corporations, developing states have been able to require technology transfer as a condition of access to natural resources within their own territorial jurisdiction. Statement of William J. Roche, Vice President, Texas Instruments, in R. Brown, p. 285. Cf. Van Dyke and Teichmann, p. 525. Representative Breaux has asserted, without explanation, that the Convention's technology transfer provisions "could affect the ability of U.S. citizens to export their technology to developing countries on freely negotiated commercial terms and conditions." Breaux, "The Diminishing Prospects," p. 262.

629. See Leive, p. 51. "The arrangements, though complex, have worked reasonably well in practice in terms of an effective transfer of the technology to members wishing to utilize it. . . . The applicability of this type of arrangement to other areas would depend, among other factors, on the nature of the technology . . . and the degree of its demand by member states." Ibid.

The United States has historically been willing to share its inventiveness with disadvantaged nations. See text accompanying nn. 42, 44 above. "We have . . . helped make discoveries of the high-yielding varieties of the Green Revolution available throughout the world." President Ronald W. Reagan, address before World Affairs Council, Philadelphia, October 15, 1981, reprinted in *U.S. Department of State Bulletin* 81 (December 1981): 17.

630. Hicks, p. 116. See also Breaux, "The Diminishing Prospects," p. 277; and Bandow, p. 486. The revenue-sharing provisions, along with those conferring limited supranational competence upon the Authority, could in fact do much to strengthen the international system. "Many people fear these twin developments, which are part of the new international economic order, because they are aware that in the United States the transition from a defective, non-working government under the Articles of Confederation, to a working, central government, under the Constitution rested on exactly those two things." C. Clyde Ferguson, "The New International Economic Order," *University of Illinois Law Forum* (1980): 699. See also Wilcox and Marcy, p. 443. Customary international law may in any event require the sharing of revenues produced from exploitation of seabed nodules. See Chap. 5 nn. 259–61 and accompanying text.

631. Knight, "The Draft United Nations Conventions," p. 549. See nn. 105–8 above and accompanying text. The U.S. government has also set many precedents in this area. See nn. 90, 94 above and accompanying text, and Chap. 6 nn. 39–40 and accompanying text.

632. Hicks, p. 113. See also Breaux, "The Diminishing Prospects," p. 269.

633. See nn. 188, 259 and text accompanying n. 224 above. U.S. import quotas — a major interference with "free-market" principles — could cause U.S. copper prices to rise as much as five cents per pound above world market prices. See Eduardo Lachica, "Copper Imports Hurt U.S. Firms, ITC Decides," *Wall Street Journal*, June 15, 1984, p. 2. In this respect at least, the U.S. position appears to be grounded in self-interest rather than adherence to free-market principles.

634. See nn. 299, 315 and text accompanying nn. 295–306, 308 above; Jens Evensen, "Banquet Address," in *Law of the Sea: Neglected Issues*, ed. John King Gamble, Jr. (Cambridge, Mass.: Ballinger, 1979), p. 534; and also Yemin, pp. 2, 206. Cf. speech by Ambassador James L. Malone, January 6, 1983, in *Environmental Policy and Law* 10 (1983): 97–98. Of course, the Convention also includes a massive delegation of power by the international community to coastal states in the form of expanded offshore jurisdiction. See Chap. 3 nn. 24–27, 62–64, 97 and accompanying text, n. 694 below and accompanying text, and Chap. 5 nn. 407–10, 417–20 and accompanying text. Private corporations have become accustomed to the exercise of even broader powers by the U.S. government. See nn. 261, 266, 268–69 above and accompanying text.

635. Malone, "Who Needs the Sea Treaty?" p. 54. See also Breaux, "The Diminishing Prospects," p. 281. In fact, of course, the United States and its Western allies would be able to prevent adoption of measures affecting their important interests. See nn. 333, 338, 340, 354, 361, 374, 397 and text accompanying nn. 324, 340, 359–79 above.

636. See nn. 316, 345–49, 404–8 above. The original IMCO council consisted of 18 member states: 6 from among the largest international shipping service states, 6 from among the largest maritime trading states, and 6 from among specially interested states. S. Brown et al., pp. 39–40. Amendments adopted in 1974 gave developing states greater representation at the expense of traditional maritime powers. S. Brown et al., pp. 39–40.

637. See nn. 355, 363, 369–72 above and accompanying text. The procedures may actually represent at least a partial reversal of the recent trend toward decreasing voting power for the United States and its allies.

638. Cf. Breaux, "The Diminishing Prospects," pp. 273, 275. See nn. 457, 460, 516, 525 above and accompanying text.

639. Breaux, "The Diminishing Prospects," p. 262.

640. In November 1990 the United States spurned efforts by the rest of the internaitonal community — including Australia, Japan, and Western European states — to set limits upon carbon dioxide emissions. See Michael Weisskopf, "Running Hot and Cold on the Issue of Global Warming," *Washington Post National Weekly Edition*, November 26–December 2, 1990, p. 32. Two years earlier, the U.N. General Assembly had adopted a resolution recognizing climate change as "a common concern of mankind, since climate is an essential condition which sustains life on earth." U.N. General Assembly, 43rd Session, Resolution 43/53, December 6, 1988. "Clearly, if climate change is a matter of 'common concern,' international regulation of it is legitimate." Frederic L. Kirgis, Jr., "Editorial Comment: Standing to Challenge Human Endeavors That Could Change the Climate," *American Journal of International Law* 84 (1990): 527.

Confrontationalist, single-issue negotiating tactics may also have disrupted plenipotentiary conferences on drugs, as well as the Uruguay Round of GATT talks. See United Nations, *Law of the Sea: Report of the Secretary-General* (U.N. Doc. A/44/650, November 1, 1989), p. 5 (para. 3); and Stuart Auerbach, "Trade Talks Collapse Over Farm Issue," *Washington Post*, December 8, 1990, pp. A1, A13. In 1991 the United States raised last minute objections to a carefully negotiated protocol to the 1959 Antarctic Treaty which was supported by the other 25 Consultative Parties and which would have imposed a 50-year moratorium on Antarctic mining activities. See "U.S. Opposes Antarctic Mining Ban Now," *New York Times International*, June 23, 1991, p. 3.

641. U.N. Convention on the Law of the Sea, art. 156 (3). See Chap. 3 nn. 38, 294 and accompanying text. "This allows them to represent the views of their people and request protection of their interests." Roberts and Vasatura, p. 700.

Four national liberation movements participated in UNCLOS III as observers: the African National Congress, the Pan African Congress, the Palestine Liberation Organization, and the South West African People's Organization. Kimball and Schneider, p. 69. "Other liberation movements also exist but have no UN status, for example the Eritrea liberation organisations and the Western Somali liberation front. . . . Other organisations claim to be liberation movements, for example the IRA, the Quebec separatists and various left and right wing organisations of violent inclination, but here again no status in international law is apparent or in prospect." Malcolm Shaw, "The International Status of National Liberation Movements," *Liverpool Law Review* 5 (1983): 32. See also David M. Galligan, "Wrapping Up the UNCLOS III 'Package': At Long Last the Final Clauses," *Virginia Journal of International Law* 20 (1980): 396–97.

642. U.N. Convention on the Law of the Sea, arts. 140 (1), 160 (2) (f) (i), 162 (2) (o) (i).

643. Malone, "U.S. Participation," p. 62. The United States would have deleted all references to such peoples. *Amendments* (U.N. Doc. A/CONF.62/L.121), proposed art. 140 (1), in *Official Records*, UNCLOS III, vol. XVI, p. 226 n. 29.

644. Malone, "The United States and the Law of the Sea After UNCLOS III," p. 29. It has been claimed that criticism of these provisions largely "stems from the belief that some States or peoples thereby will be doubly represented . . . and that non-State entities cannot be counted on to fulfill their international obligations." Galligan, p. 393. The raison d'être of national liberation movements is generally rooted in the failure of the titular governments to represent the interests of such peoples adequately, however, and the Convention provides for no specific rights or obligations with respect to any of the four movements represented at UNCLOS III.

645. Robertson and Vasatura, pp. 685–86. See Breaux, "The Diminishing Prospects," p. 296; and also Malone, "Who Needs the Sea Treaty?" p. 47.

646. U.N. Convention on the Law of the Sea, arts. 161 (8) (d), 162 (2) (o) (i). See n. 367 and text accompanying nn. 323, 368, 379 above. Indeed, all rights of participation conferred upon national liberation organizations by the Convention remain to be defined in the rules, regulations, and procedures of the Authority, again giving the United States effective control over the extent of such participation. U.N. Convention on the Law of the Sea, art. 156 (3).

647. See Shaw, p. 19; and Kelsen, *Principles of International Law*, pp. 158–61. "The question as to the status of national liberation movements must, therefore, be understood in the context of the rise of non-State participants upon the international scene and in the light of the developing requirements of the international community as demonstrated through practice." Shaw, p. 20. See also Helmut Freudenschuss, "Legal and Political Aspects of the Recognition of National Liberation Movements," *Millennium* 11 (1982): 115.

648. Shaw, p. 32.

649. Ibid., pp. 24–26. The U.N. General Assembly has granted observer status to several classes of national liberation movements since 1973. See Freudenschuss, pp. 116–18; and also Shaw, pp. 23–24. In 1974, "after noting that the Geneva Conference on the Reaffirmation and Development of Humanitarian Law, the World Population Conference and the World Food Conference had in fact already invited the PLO to their deliberations as an observer, the General Assembly . . . decided finally 'that the PLO is entitled to participate as an observer in the sessions and the work of all international conferences convened under the auspices of other organs.'" Freudenschuss, p. 118. See U.N. General Assembly, 29th Session, Resolution 3237, November 22, 1974. In 1980 the U.N. General Assembly recognized that such participation by the four liberation movements in question "helps to strengthen international peace and co-operation." U.N. General Assembly, 35th Session, Resolution 35/167, December 15, 1980.

In 1975 the Security Council voted to allow the PLO to participate in its Middle East debates, with "the same rights of participation as . . . a member-State." Shaw, pp. 25–26. "In other words, a precedent appears to have become established." Shaw, p. 26. "[T]here seems to be enough evidence to assume that a new category of recognition, to say the least, in the making." Freudenschuss, p. 118. See also Freudenschuss, p. 115. States nevertheless "remain—legally—free to decide whether or not to extend similar recognition . . . and they may apply . . . different criteria for different purposes. Thus, collective recognition does not spill over to a duty to recognize individually." Freudenschuss, p. 119. Such individual recognition is growing, however. See Freudenschuss, p. 124; and Shaw, pp. 27, 31.

650. Freudenschuss, p. 124. The fact that some existing liberation movements are

currently viewed as inimical to U.S. interests should not be considered dispositive: future such movements could be composed of more favorable elements, and their ability to participate in international organizations could prove beneficial.

651. In December 1988, following the utterance of carefully negotiated words by Chairman Yasser Arafat regarding terrorism and Israel's right to exist, the United States resumed diplomatic contacts with the PLO after a 13-year hiatus. See Don Oberdorfer, "George Shultz's Roller-Coaster Ride with Yasser Arafat," *Washington Post National Weekly Edition*, December 26, 1988–January 1, 1989, pp. 16–17.

The African National Congress (ANC) was legitimized by high-level meetings with the U.S. officials during 1987. See Robert I. Rotberg, "Shultz: Time to 'Deal In' the ANC?" *Christian Science Monitor*, January 16, 1987, p. 13. "One effect of the encounter seems certain, and irreversible. By officially receiving ANC President Oliver Tambo, the Reagan administration [was] extending a large measure of formal recognition to insurgents that both the South African authorities and its own more staunchly conservative members have long criticized as pro–Soviet 'terrorists.'" Ned Temko, "Shultz Meeting with ANC Chief Lends Validity to S. Africa Rebels," *Christian Science Monitor*, January 28, 1987, p. 7. International acceptance of the ANC increased dramatically following its legitimization by the South African government on February 2, 1990.

652. See, e.g., Breaux, "The Diminishing Prospects," p. 261. The Reagan administration has sought to provide government subsidies to encourage "essential" private enterprise for the commercial development of outer space, although such subsidies appear to be inconsistent with free-market principles. See Penny Pagano, "Man's First Steps on Moon Commemorated by Reagan," *Los Angeles Times*, July 21, 1984, part 1, p. 14.

653. See U.N. Department of Public Information, "Outer Space Conference to Focus on Potential Benefits to Mankind," *U.N. Monthly Chronicle* 19 (January 1982): 43. The establishment of industrial manufacturing operations in space has been anticipated. See Stephen Gorove, "Legal Aspects of the Space Shuttle," *International Lawyer* 13 (1979): 160. The potential for monopolization of space activities has come under criticism in international forums. See U.N. Department of Public Information, "UNISPACE '82: Getting a New Perspective on Earth (Vienna, 9–21 August)," *U.N. Chronicle* 19 (October 1982): 26–27.

654. U.N. General Assembly, 16th Session, Resolution 1721A, December 20, 1961. President Dwight Eisenhower had pressed an initiative based upon such principles in 1960. See Martin Menter, "Commercial Participation in Space Activities," *Journal of Space Law* 9 (1981): 55. President Kennedy had proposed similar guidelines in an address to the General Assembly earlier in 1961. Lincoln P. Bloomfield, "Outer Space and International Cooperation," in *The United Nations in the Balance: Accomplishments and Prospects*, ed. Norman J. Padelford and Leland M. Goodrich (New York: Praeger, 1965), p. 247. U.S. policy requiring assistance to be provided to developing states in order to assure equal access to satellite communications systems was codified in the Communications Satellite Act of 1962. 47 U.S.C. §701 (b). These policies and initiatives led directly to the formation of INTELSAT. See Colino, pp. 104–5.

655. U.N. General Assembly, 18th Session, Resolution 1962, December 13, 1963. "States shall regard astronauts as envoys of mankind in outer space." Ibid., para. 9.

656. Treaty on Principles Governing the Activities of States in the Exploration and Use of Outer Space, Including the Moon and Other Celestial Bodies, reprinted in *International Legal Materials* 6 (1967): 386–90. "In testifying before the Senate Committee on Foreign Relations in connection with the meaning attributed by the negotiators to the term 'province of all mankind' Ambassador Goldberg stated that the U.S. policy relating to the use of the space environment had been fixed in the 1958 National Aeronautics and Space Act, section 102 (a). It provided that 'activities in space should be devoted to peaceful purposes for the benefit of all mankind.' ... The negotiators reviewed with care the acceptance of the term 'province,' and, at the insistence of the United States, it was accepted as being the equivalent of 'benefit of all mankind.'" Christol, "The Common Heritage of Mankind Provision," pp. 449–50.

The Outer Space Treaty was ratified by the space powers before the first moon landing in 1969. Williams, p. 1176. "When the 1967 Treaty was being formulated, claims of sovereignty over outer space and celestial bodies were seen as the source of future conflicts which could disrupt international cooperation and lead to hostilities." Eilene Galloway, "Perspectives of Space Law," *Journal of Space Law* 9 (1981): 22–23.

657. Galloway, p. 25. Subsequent space treaties negotiated by COPUOS and its Legal Subcommittee have included the Rescue and Return Agreement of 1968, the Liability convention of 1973, the Registration Convention of 1976, and the Moon Treaty of 1979. See Stephen E. Doyle, "Significant Developments in Space Law: A Projection for the Next Decade," *Journal of Space Law* 9 (1981): 109.

658. Patricia Minola, "The Moon Treaty and the Law of the Sea," *San Diego Law Review* 18 (1981): 466–67. See also Kimball, "Implications of the Arrangements Made," p. 58.

659. Aldo Armando Cocca, "The Advances in International Law Through the Law of Outer Space," *Journal of Space Law* 9 (1981): 20. See also Manfred Lachs, "Some Reflections on the State of the Law of Outer Space," *Journal of Space Law* 9 (1981): 9. The Outer Space Treaty marked "the first time that States … have renounced in advance an area which they were in a condition to reach and effectively occupy." Williams, p. 1176.

660. See Agreement Governing the Activities of States on the Moon and Other Celestial Bodies, reprinted in *International Legal Materials* 18 (1979): 1434–41; and also K. Narayana Rao, "Editorial Comment: Common Heritage of Mankind and the Moon Treaty," *Indian Journal of International Law* 21 (1981): 275.

661. Hollick, *U.S. Foreign Policy*, p. 370. See, e.g., statement of Representative John Breaux in *Congressional Record* 126 (June 9, 1980): 13698. "The worth of the agreement, which had resulted from a careful assessment at the UN lasting from 1970 down to 1979, and which was considered sufficient to merit the unanimous approval of the General Assembly, has come under scrutiny in the United States." Carl Q. Christol, "The American Bar Association and the 1979 Moon Treaty: The Search for a Position," *Journal of Space Law* 9 (1981): 77. Particular — and familiar — concerns have been expressed regarding the potential ability of developing states to control the use of lunar resources. See Orlove, pp. 610–11. It has been claimed that the Moon Treaty would impose a moratorium on the development of celestial resources, prevent all such development by private enterprise, and serve only the interests of developing and Socialist states. Christol, "The American Bar Association and the 1979 Moon Treaty," p. 79. See also Menter, p. 59.

662. See Marian L. Nash, "Contemporary Practice of the United States Relating to International Law," *American Journal of International Law* 74 (1980): 423; and also Christol, "The Common Heritage of Mankind Provision," p. 469. The Soviet Union, not the United States, had vehemently opposed inclusion of the common heritage principle. At a 1974 COPUOS session, for example, a Soviet "expert" complained that the common heritage of mankind concept would lead to the erosion of the nation-state system and to the creation of a powerful supranational organization with wide-ranging jurisdiction. Williams, p. 1173. The United States did not share this fear; a resolution endorsing the Moon Treaty, ultimately adopted in the Special Political Committee of the General Assembly by consensus, was cosponsored by the United States and many of its Western allies. "Until July of 1979, the Soviet Union maintained strong opposition to the common heritage concept, and it was essentially because of this opposition that the Treaty was not concluded several years ago." Letter from Secretary of State Cyrus Vance to Senator Jacob Javits, November 28, 1979, reprinted in *Congressional Record* 125 (December 14, 1979): 36070.

663. Agreement Governing the Activities of States on the Moon and Other Celestial Bodies, art. 11 (1), (5), in *International Legal Materials* 18 (1979): 1438. See Minola, p. 468. The common heritage concept — expressly limited, however, to its meaning within the context of the Moon Treaty itself — was intended to operationalize the more general "province of all mankind" principle of the 1967 Outer Space Treaty. Cocca, p. 16.

The treaty imposes an obligation to negotiate the future regime in good faith. See K. Rao, p. 278. The regime is to promote orderly and safe development, "equitable sharing" of benefits, and expanded opportunities for use of the natural resources of the moon. See Menter, p. 60.

664. Letter from J. Brian Atwood, Assistant Secretary of State for Congressional Relations, to Senator Richard B. Stone, January 2, 1980, reprinted in Nash, p. 424. Critics of the Law of the Sea Convention have nevertheless argued that it would set "an adverse detailed and unavoidable model for the space resources exploitation regime." Statement of Representative John Breaux in *Congressional Record* 126 (June 9, 1980): 13699. See also

Christol, "The American Bar Association and the 1979 Moon Treaty," pp. 82–83. "It has been evident that the critics of the [common heritage of mankind] principle possess preferred policy outlooks. In general terms they are opposed to regulation of almost any kind, and particularly a regulatory process that is international in character." Christol, "The Common Heritage of Mankind Provision," pp. 475–77.

665. Letter from Secretary of State Cyrus Vance, November 28, 1979, in *Congressional Record* 125 (December 14, 1979): 36070. Cf. Agreement Governing the Activities of States on the Moon and Other Celestial Bodies, art. 11 (3), in *International Legal Materials* 18 (1979): 1438. "The lack of a moratorium on exploitation, in conjunction with a provision that allows property rights in resources after their extraction from the natural environment, makes the practical effect of the compromise exactly what the developed countries wanted. Exploitation will be allowed on a 'first come, first served' basis accompanied only by a duty to try to set up an international regime." Minola, pp. 468–69. See also Christol, "The Common Heritage of Mankind Provision," p. 470. It was nevertheless claimed—without elaboration—that U.S. firms would be restrained by a "*de facto* moratorium" pending agreement on the new regime while the Soviet Union, for some unstated reason, would be able to "move forward in the area of resource development at their own pace under the guise of scientific investigation, with no fear of significant competition from the West." Letter from Senator Frank Church to Secretary of State Cyrus Vance, October 30, 1979, reprinted in *Congressional Record* 125 (December 14, 1979): 36070.

666. Minola, p. 468. Unlike the situation with respect to the deep seabed, developed states would thus benefit from any impasse in the negotiations. Ibid., p. 469.

667. Menter, p. 64. See n. 784 below; and Williams, p. 1191. To the extent that such resources are inexhaustible, application of the *res communis* principle would call for a minimal regulatory regime—to prevent harmful disturbance of the lunar environment, for example—in contrast to that required for exploitation of deep ocean nodules and other exhaustible resources. See text accompanying Chap. 5 n. 293 et seq. "The gathering of a moon rock or of lunar substances for use is the substantial equivalent of the harvesting of a *living* resource in the ocean." Christol, "The Common Heritage of Mankind Provision," p. 471 (emphasis added). Cf. Chap. 5 nn. 44–45, 300, 347, 488 and accompanying text.

668. See S. Brown et al., p. 6; Gregg, p. 38; Martin A. Rothblatt, "ITU Regulation of Satellite Communication," *Stanford Journal of International Law* 18 (1982): 25; and also Glen O. Robinson, "Regulating International Airwaves: The 1979 WARC," *Virginia Journal of International Law* 21 (1980): 5. "The orbit-spectrum resource consists of the electromagnetic spectrum through which radio waves are transmitted and the orbits in space in which satellites and space platforms are placed. It has traditionally been viewed as a common resource that no single country may appropriate." Per Magnus Wijkman, "Managing the Global Commons," *International Organization* 36 (1982): 534. See U.N. Department of Public Information, "UNISPACE '82," p. 29; and also Soroos, p. 667.

669. Robinson, p. 5. See Soroos, p. 667. The ITU has been responsible for such international regulation for more than 50 years, with the full approval of the United States. Rothblatt, pp. 3–4, 6. New international legal norms have been promulgated at World Administrative Radio Conferences (WARCs), as well as at ITU plenipotentiary conferences. Robinson, pp. 6–7.

Orbiting satellites themselves are governed by the "province of all mankind" provisions of the 1967 Outer Space Treaty. Rothblatt, p. 20. See nn. 656–57 above and accompanying text. "Even though agreement has not been reached on where to draw the boundary between national airspace and outer space, the principle has been established in customary law that artificial earth-orbiting satellites move in outer space." Soroos, p. 669. See also Soroos, pp. 665, 668. In 1976 eight developing states nevertheless attempted to claim sovereignty over the geostationary orbital positions above their terrestrial jurisdiction. See Orlove, pp. 617–19.

670. Rothblatt, p. 5.

671. Ibid., p. 7. See n. 460 above. It has been estimated that the geosynchronous orbit can sustain a maximum of 180 to 1,800 satellites. Soroos, p. 667. "The 110 satellites operating in geosynchronous orbit in 1981 are significantly below its estimated carrying capacity, but the day of saturation may not be far off. . . . NASA anticipates a tenfold increase in demand for commercial satellite circuits between 1982 and 2000." Rothblatt, p. 668. See also

Lachs, p. 10. There are even fewer positions available in the geostationary orbit, which held more than 80 communications satellites in 1986. Doyle, "Regulating the Geostationary Orbit," p. 5. See Rothblatt, pp. 7-8; and U.N. Department of Public Information, "Lack of Space in Space: Geostationary Orbit," *UN Chronicle* 19 (July 1982): 58. "The attractiveness of this orbit is based on the absence of satellite motion. Static earth stations do not have to track satellites moving across the sky and that fact has significant economic and operational implications." Doyle, "Regulating the Geostationary Orbit," p. 5.

It has been pointed out that since they will exist "in perpetuity," the radio frequency spectrum and geostationary orbit are not, properly speaking, exhaustible. Doyle, "Regulating the Geostationary Orbit," p. 3. Doyle has nonetheless acknowledged that the frequency spectrum and geostationary orbit are "commonly shared natural phenomena like oceans, air or sunlight, the use of which requires common sense, cooperation and mutual accommodations." Doyle, "Regulating the Geostationary Orbit." Indeed, as the ITU has itself recognized, the frequency spectrum and geostationary orbit are limited in the sense that allocation of all available frequencies and orbital positions would effectively exhaust these resources. See n. 676 below and accompanying text. "Like competitors for other limited world resources, potential users of radio are bound to come into conflict." Gregg, p. 39. The limited nature of these common resources may affect their legal status under international law. See Chap. 5 n. 285 et seq. and accompanying text.

672. Robinson, pp. 8-9. See Gregg, pp. 44-45. Although the IFRB has traditionally allocated usage on a first-come, first-served basis, a 1971 WARC resolution provided that, in accordance with the principle of equitable access, "'registration should not provide any permanent priority' and that registrants should take 'all practicable measures' to help non-registrants exploit space systems." Rothblatt, pp. 8, 10-11.

673. Brokaw and Burke, p. 56. See Robinson, p. 18; and also Lachs, p. 9. "At the time of the 1979 General WARC it was observed that 90 percent of the radio spectrum was controlled by countries with only 10 percent of the world's population. Moreover, only a few countries, all developed, had the technological capacity to launch satellites into orbit." Soroos, p. 673. "The 'cheap seats' will all be taken by the time the developing nations are ready to utilize the geostationary orbit." Orlove, p. 628. See also Christol, "Telecommunications, Outer Space, and the New International Information Order," p. 357. Even some of the smallest states in the world have sought to use satellite technology to promote their economic development. Doyle, "Regulating the Geostationary Orbit," p. 2. "The proposed New World Information Order seeks wider distribution of more bandwidth and connectivity, particularly among the developing countries." Rothblatt, p. 19. See Soroos, pp. 673-74.

674. Breaux, "The Diminishing Prospects," p. 261. See also Rothblatt, pp. 20, 24-25. Such influence might include technology transfer. Soroos, pp. 675-76. Revenue sharing is also possible, although it would be "at odds with traditional policy." Robinson, pp. 40-43. In 1973, before the convening of UNCLOS III, developing states sought formal designation of the radio spectrum and the geostationary orbit as "limited natural resources," in the allocation of which developing states were to be given special consideration. See Doyle, "Regulating the Geostationary Orbit," p. 3. In fact, in a number of ways the ITU's system of frequency and orbital management may have set a precedent for the deep seabed mining regime. See nn. 296, 390 above and accompanying text. Cf. Rothblatt, p. 20 n. 89. Generally, ITU members are more likely to look to other space-related organizations as models for international telecommunications policy. U.N. Department of Public Information, "Lack of Space in Space," p. 58.

U.S. law has long embodied the policy of requiring telecommunications services to be made available to developing states. See Communications Satellite Act of 1962, 47 U.S.C. §701 (b); and also n. 654 above and accompanying text. Yet, despite this policy—and despite U.S. rejection of the 1982 Convention—the United States cast the only vote in opposition to a 1986 General Assembly resolution, adopted by a vote of 148 to 1 with 4 abstentions (Canada, Israel, Malawi, and the United Kingdom), which endorsed "an ongoing and continuous process" to establish a New International Information Order. U.N. General Assembly, 41st Session, Resolution 41/68A, December 3, 1986. See n. 809 below.

675. Robinson, p. 12. See Soroos, p. 672; and also Milton L. Smith, "The Space WARC Concludes," *American Journal of International Law* 83 (1989): 599. But see Doyle, "Regulating the Geostationary Orbit," pp. 19-20.

676. International Telecommunication Convention, Nairobi, 1982, art. 33 (2), quoted in Doyle, "Regulating the Geostationary Orbit," p. 3. See M. Smith, p. 599. See n. 673 above. Approximately half of the satellites in geostationary orbit are U.S. owned. Doyle, "Regulating the Geostationary Orbit," p. 5.

By 1986 telecommunications was generating more than $1 billion in annual business. Doyle, "Regulating the Geostationary Orbit," p. 1. The cost of a three-satellite direct-broadcast system approximates that of a seabed mining system. See Gregg, p. 40 n. 12.

677. M. Smith, pp. 596–99. See also Doyle, "Regulating the Geostationary Orbit," pp. 15–17. "The Space WARC achieved its objective by guaranteeing equitable access to the [geostationary orbit] and the frequency bands allocated to space services using it. The demands of developing countries have been met in a manner that is sufficiently flexible for acceptance by most developed countries." M. Smith, p. 599.

678. "The world shortages, becoming more acute in recent years, have given an impetus and urgency to the drive to tap the resources of Antarctica.... About a dozen potentially minable minerals—e.g., coal, copper, lead, gold and iron—are believed to lie in the trans-Atlantic mountains." Rahmatullah Khan, "Editorial Comment: Ocean Resources Development—India's Options," *Indian Journal of International Law* 22 (1982): 455. But see Harald Heimsoeth, "Antarctic Mineral Resources," *Environmental Policy and Law* 11 (1983): 61. Cf. Christopher C. Joyner, "Antarctica and the Law of the Sea: Rethinking the Current Legal Dilemmas," *San Diego Law Review* 18 (1981): 430. It has been claimed that the Antarctic continental shelf contains as much as 45 billion barrels of oil and 115 trillion cubic feet of natural gas. Christopher C. Joyner, "The Exclusive Economic Zone and Antarctica," *Virginia Journal of International Law* 21 (1981): 702–4. See also M.J. Peterson, "Antarctica: The Last Great Land Rush on Earth," *International Organization* 34 (1980): 386. The deep seabed off the Antarctic continent is also believed to contain an undetermined amount of nodule deposits. Christopher C. Joyner, "Antarctica and the Law of the Sea: An Introductory Overview," *Ocean Development and International Law Journal* 13 (1983): 282.

By the end of the 1980s no commercially attractive mineral deposits had been identified in the Antarctic region, however. M. Jacobsen, pp. 239–40. "The cold continent has been depicted as a genuine treasure vault. The findings of the scientists, however, have been much less encouraging." Heimsoeth, p. 61. The environmental and economic risks of Antarctic mineral exploitation would be high, in any event. Robert A. Jones, "Dividing the Pie: History in Making at South Pole," *Los Angeles Times*, January 24, 1985, part 1, p. 24.

679. See Joyner, "Rethinking the Current Legal Dilemmas," pp. 425–27. The 1980 Convention on the Conservation of Antarctic Marine Living Resources, which entered into force in 1982, was designed to prevent the depletion of krill stocks through novel ecosystem-based management techniques. See Malcolm J. Forster, "The Question of Antarctica," *Environmental Policy and Law* 14 (1985): 2. Recent studies have indicated that the annual metric tonnage of krill caught could safely surpass that of current worldwide fishery harvests without causing ecological harm. Joyner, "An Introductory Overview," p. 280. See also Joyner, "The Exclusive Economic Zone and Antarctica," p. 702.

The legal status of Antarctic waters will be determined by the regime ultimately established to govern use of the Antarctic continent. Joyner, "The Exclusive Economic Zone and Antarctica," p. 717. See n. 688 below and accompanying text; Peterson, pp. 396–97; Joyner, "An Introductory Overview," p. 279. The Antarctic Treaty preserved the legal status of the high seas within the Antarctic region, but some claimant states contend that they possess sovereign rights out to the northern boundaries of the treaty area, encompassing the area of greatest krill concentration; Argentina and Chile have proclaimed 200-mile offshore economic zones in conjunction with their terrestrial Antarctic claims. Peterson, p. 381. Joyner, "Rethinking the Current Legal Dilemmas," p. 424. See Oxman, "Antarctica and the New Law of the Sea," pp. 222–33. In the absence of a division of Antarctic territory, however, all such offshore claims remain "questionable and highly suspect." Joyner, "An Introductory Overview," p. 279.

680. See Roland Rich, "A Minerals Regime for Antarctica," *International and Comparative Law Quarterly* 31 (1982): 710–11; and generally Peterson, pp. 389–90.

681. Breaux, "The Diminishing Prospects," p. 261. Breaux has even claimed that the common heritage of mankind principle "does raise the spectre of war over Antarctic resources." Statement of Representative John Breaux in *Congressional Record* 126 (June 9, 1980): 13699.

The United States has favored an "open-access" regime for Antarctica. Peterson, p. 401. "The United States and other industrialized states have a strong economic interest in free appropriation, since these states have the procurement technology and capital needed for exploitation. Free appropriation of Antarctica's nonrenewable resources, however, could allow economically developed states to deplete valuable supplies before developing states can take a fair share. Antarctica's non-replenishable resources of oil and coal thus cannot be analogized to fish of the high seas; a better comparison is drawn to nonrenewable deep seabed resources, widely identified as belonging to all states and peoples in common and not available for private appropriation." Honnold, pp. 841–42. See Chap. 5 nn. 43–44, 353–57, 488–89 and accompanying text.

682. See Honnold, pp. 806–7; and also Kimball, "Implications of the Arrangements Made," p. 58. Like outer space and the deep seabed, Antarctica has been described as an area which "is, or may become, of value to mankind generally and for which there has developed a practice or general expectation of common access, use, or control." Honnold, p. 848.

683. See statement of Bill Hayden, Minister for Foreign Affairs, before 12th Antarctic Treaty Consultative Meeting, September 13, 1983, reprinted in *Australian Foreign Affairs Record* 54 (1983): 531. "Some who favor or oppose the Law of the Sea Convention because of the potential precedential effects of its deep seabed mining regime may understand too little about Antarctica and worry too much about the moon." Oxman, "Antarctica and the New Law of the Sea," pp. 245–46. See also White, p. 509.

684. Expeditions were sent by Spain, France, Great Britain, Germany, Belgium, Norway, Russia, and the United States. White, p. 511. Outposts established by these expeditions were generally abandoned long before effective occupation could be established. Honnold, pp. 809–10. Claims have been made by Great Britain, New Zealand, France, Australia, Norway, Chile, and Argentina. These states have claimed sovereign rights with respect to wedge-shaped sectors stretching northward from the South Pole to the Antarctic coast. See Honnold, pp. 812–13. "The territorial claims which began in 1908 on the basis of exploitation had solidified by 1946 to hard cases, well grounded in international law, on the basis of discovery, contiguity, occupation and inheritance. Given the climatic conditions, no court or arbitral tribunal is likely to insist on a perfect demonstration of title by these criteria." Khan, p. 456. See Rich, pp. 724–25; and also Peterson, pp. 392–93. If Antarctica were generally acknowledged to be a *terra nullius*, any state might indeed appropriate it through the exercise of effective occupation. Joyner, "Rethinking the Current Legal Dilemmas," p. 423. "The claims of Argentina, Britain, and Chile to the partially ice-free Palmer Peninsula, an area seemingly most accessible to mineral exploitation, conflict and overlap in substantial part. The other five claimant States do not recognize each other's claims to some unspecified extent, though Japan, the Soviet Union, and the United States historically have not legally acknowledged Antarctic claims by any State. Each of the latter has reserved all rights accruing through efforts made on its behalf, while at the same time each has refused to make any formal claims to sovereignty." Joyner, "Rethinking the Current Legal Dilemmas," p. 422. See also Peterson, pp. 391–92. The claims of Argentina and Chile are based on the papal bulls of the late 15th century. Peterson, p. 392. See Chap. 1 n. 8 and accompanying text.

Serious objections have been raised against a "division" of Antarctica. See Honnold, pp. 821–22; and also Joyner, "Rethinking the Current Legal Dilemmas," pp. 422–24 nn. 29, 30. The basis of the original claims has in fact been called into quesiton. See Honnold, pp. 807–8, 816–18. Apart from the diplomatic fallout and potential military conflict which could result, a "free-for-all" scramble for territorial control could immeasurably damage the fragile Antarctic ecosystem. See Joyner, "Rethinking the Current Legal Dilemmas," pp. 424–25.

685. The Consultative Parties include Argentina, Australia, Belgium, Brazil, Chile, China, France, Germany, Great Britain, India, Japan, New Zealand, Norway, Poland, South Africa, the Soviet Union, the United States, and Uruguay – states which together represent approximately two thirds of the world's population. Activities in Antarctica are governed by policies adopted by the Consultative Parties. See Honnold, pp. 832–34. The Consultative Parties take action by consensus. See n. 372 above. A review conference may be called after 1991 to adopt amendments to the Treaty for ratification by the Consultative Parties. Peterson, p. 391.

686. Antarctic Treaty, art. 4 (2), reprinted in *Cornell International Law Journal* 19 (1986): 304. See White, p. 514; and also Khan, pp. 454–55.

687. See Bruno Simma, "The Antarctic Treaty as a Treaty Providing for an 'Objective Regime,'" *Cornell International Law Journal* 19 (1986): 193, 203–5; and also Chap. 5 nn. 444–45 and accompanying text. "Not surprisingly, there are deep divisions among Antarctic Treaty Consultative Parties as to whether exploitation of oil, land-based minerals, and other resources should take place." Joyner, "Rethinking the Current Legal Dilemmas," p. 431. The Consultative Parties have nevertheless taken it upon themselves to draft and adopt supplementary conventions to govern exploitation of Antarctic resources. Because the Consultative Parties have carefully sought to accommodate the interests of the international community, these supplementary regimes have not been directly challenged by non–Treaty states. See Rich, p. 715.

688. "Internationalization (the exercise of common sovereignty through institutions allowing all states a share in decisions) and division (incorporation of the various parts of each area into national territories of individual states) are the opposites in the debate. Condominium (joint sovereignty exercised by a few states) and consortium (exercise of limited jurisdiction by a few states) lie somewhere in between." Peterson, p. 378.

Division is unlikely. Ibid., pp. 401, 403. See also Honnold, pp. 844–45. The interests of claimant states would have to be accommodated in any regime eventually to emerge, however. Rich, pp. 718–19. Such accommodation might be difficult were internationalization to occur. See Joyner, "Rethinking Current Legal Dilemmas," pp. 441–42; and also n. 690 below.

The sovereign rights inherent in a condominium regime would facilitate joint resource exploitation by participant states. Joyner, "Rethinking the Current Legal Dilemmas," p. 438. "Creating 'consortium Antarctica' would pose no legal problem within the Antarctic Treaty group because the Treaty prohibits the enlargement but not the diminution of claims. The problems would be political: agreement by claimants to yield part of their claim, agreement by nonclaimants to adopt an exclusive regime with few legal precedents, and agreement among all on the precise institutions to manage the resource regime." Peterson, p. 397. See also Honnold, p. 843.

The consortium alternative has been pursued by the Consultative Parties within the treaty framework. See Peterson, p. 379. Such an approach "attracts those who believe that national appropriation is impossible politically, want to settle the sovereignty question, but fear that internationalization would mean control by states unfriendly to their management preferences and/or insufficiently familiar with Antarctic conditions to provide sound management." Peterson, p. 396.

689. See R. Jones, pp. 24–25. The Treaty is viewed by many developing states as a remnant of a past colonial era when Western states sought to divide the world among themselves without regard to the interests of occupied nations. Ibid., p. 3. See also Joyner, "Rethinking the Current Legal Dilemmas," p. 421. "As Antarctic resources become the subject of strict world scrutiny, narrow forms of decisionmaking are certain to be challenged by new states seeking a greater role in world affairs, and doctrinal fictions of exclusive right are likely to give way to more contemporary principles of sharing and common right." Honnold, pp. 825–26. See Peterson, p. 398; and White, pp. 537–38. A number of developing states have advocated application of the common heritage of mankind principle to Antarctica. See R. Jones, p. 25. It may be regarded as significant that U.S. rejection of the Law of the Sea Convention did not, at least in this case, forestall efforts by developing states to protect their perceived interests in common resource areas. See Extract from the Closing Statement of the 7th Summit of Non-Aligned Nations at New Delhi, March 11, 1983, reprinted in *Environmental Policy and Law* 11 (1983): 54. See n. 690 below. Cf. Oxman, "Antarctica and the New Law of the Sea," pp. 238, 240.

690. See, e.g., U.N. General Assembly, 44th Session, Resolution 44/124A, December 15, 1989. See also U.N. Department of Public Information, "No Consensus on Antarctica," *UN Chronicle* 25 (March 1988): 77. Conceivably, the United Nations might place Antarctica under direct administrative control as a trusteeship pursuant to Chapter XII of the U.N. Charter. Honnold, pp. 853–54. The United States itself sought such a joint-management regime for Antarctica under U.N. auspices unsuccessfully in 1948, but such attempts to safeguard global common interests in Antarctica have been described as "noble, if perhaps politically naive." Joyner, "Rethinking the Current Legal Dilemmas," pp. 438–39. "In the absence of any principle linking developing countries to the Antarctic continent and its resources, it would seem that

the only linkage could come about through state practice, i.e., the persistent assertion of a right to participate in Antarctica's affairs, the formal protesting against exclusive management by the Consultative Parties and some capacity to exercise these perceived rights." Rich, p. 714. Absent such an assertion of interest – which the General Assembly resolutions of recent years arguably satisfy – members of the Antarctic Treaty regime were expected to be able to satisfy the demands of nonparty states by measures short of a universal system of joint management. Peterson, p. 402. Thus the Consultative Parties in 1981 agreed that the Antarctic mineral regime "should not prejudice the interests of all mankind in Antarctica." M. Jacobsen, p. 239. See Peterson, p. 402; and Shearer, p. 8. Indeed, acceptance of India and Brazil as Consultative Parties was intended to demonstrate the regime's ability to accommodate nonclaimant developing states. R. Jones, p. 25. See also Oxman, "Antarctica and the New Law of the Sea," p. 238.

Despite opposition by the Consultative Parties, the issue was first raised by the Malaysian delegation in 1983. Deporov, p. 37. See also R. Jones, p. 24. Following preparation of a report by the U.N. Secretary-General and elicitation of comments from governments, the General Assembly deferred consideration of the issue to the 39th and 40th sessions. "The comments reflect a widely held opinion that the mineral resources of Antarctica at least represent the common heritage of mankind." Forster, "The Question of Antarctica," p. 4. In an unopposed 1985 resolution, the General Assembly formally recognized "the interest of mankind in Antarctica." U.N. General Assembly, 40th Session, Resolution 40/156A, December 16, 1985. A companion resolution, adopted by a vote of 92 to 0, with 14 abstentions, affirmed "that any exploitation of the resources of Antarctica should ensure the maintenance of international peace and security in Antarctica, the protection of its environment, the non-appropriation and conservation of its resources and the international management and equitable sharing of the benefits of such exploitation." U.N. General Assembly, 40th Session, Resolution 40/156B, December 16, 1985. Such sharing of benefits would be inconsistent with a genuine condominium. Oxman, "Antarctica and the New Law of the Sea," pp. 244–45. See n. 688 above. In an even more strongly supported resolution in 1987, the General Assembly sought to ensure the participation of the entire international community as well as the U.N. Secretary-General in the negotiations on an Antarctic minerals regime. U.N. Department of Public Information, "No Consensus on Antarctica," p. 77.

691. The Antarctic Mineral Resources Convention was adopted on June 2, 1988, by 33 states, including the United States and the other Consultative Parties; it was to enter into force following ratification by 16 Consultative Parties, at least 5 of which must be developing states. M. Jacobsen, pp. 237–38 n. 8. By the end of 1990, New Zealand had decided not to ratify the 1988 Convention, which was not expected to enter into force in the foreseeable future. Telephone conversation with Claire Finley, Legal Division, New Zealand Ministry of Foreign Affairs, December 21, 1990. On November 16, 1990, President Bush signed into law the Antarctic Protection Act of 1990, imposing an indefinite moratorium upon U.S. commercial mining activities in Antarctica and calling for the negotiation of a multilateral ban on Antarctic mineral exploitation. See Public Law No. 101–594. Several weeks later the U.N. General Assembly adopted, by a vote of 98 to 0, with 7 abstentions, a resolution which urged all states "to support all efforts to ban prospecting and mining in and around Antarctica." U.N. General Assembly, 45th Session, Resolution 45/78A, December 12, 1990. But see n. 640 above.

692. See nn. 452, 596 above, and n. 300 above and accompanying text. "The agreement ... is an important environmental and resource management treaty which establishes the legal obligations necessary for considering and regulating commercial mineral resource activities in the Antarctic, should interest in them emerge in the future." Statement of U.S. Department of State, December 2, 1988, reprinted in *U.S. Department of State Bulletin* 89 (February 1989): 23.

693. See, e.g., nn. 654, 657, 664, 671, 682 above and accompanying text. The UNCLOS III seabed mining regime is "the unique result of the quid pro quo of those negotiations and stands as no generalizable precedent to be used in other settings." Jonathan I. Charney, "The United States and the Law of the Sea After UNCLOS III – The Impact of General International Law," *Law and Contemporary Problems* 46 (Spring 1983): 53. See also Oxman, "The Two Conferences," p. 139. It has also been pointed out that the Law of the Sea Conference would provide a procedurally flawed model for negotiation of other international issues: "The

negotiations simply took too long, were too costly, and produced too little to be used again in the near future." Charney, "The United States and the Law of the Sea After UNCLOS III," p. 53.

694. See Oxman, "The Two Conferences," p. 140. Arvid Pardo originally had intended the common heritage of mankind principle to prevent the enclosure of high-seas resources. "Extended national jurisdiction over ocean space leaves less area for international institutions to regulate and gives much more power to national states to make decisions about the most critical areas of ocean activities — resource development (except manganese nodules), pollution control, ocean transportation and scientific research." Robert L. Friedheim and Judith T. Kildow, "Report of the Ocean Policy Research Workshop," Occasional Paper No. 26 (Kingston, R.I.: Law of the Sea Institute, 1975), p. 12. See text accompanying Chap. 2 nn. 136–37.

695. Charney, "The United States and the Law of the Sea After UNCLOS III," p. 53. See Chap. 3 nn. 123, 410–14 and accompanying text.

696. See T. Alexander, p. 143; and e.g., nn. 31–34, 59, 64, 94, 104–8, 261, 268–69, 284, 295–306, 311, 316, 345–49, 358, 370–72, 390, 404–8, 628–29, 637–38, 649, 656, 659, 672 above and accompanying text, and Chap. 6 nn. 39–40 and accompanying text. "In reality, the [seabed] regime may actually function in a way comparable to a private enterprise system subject to normal government regulations. Consequently, it does not appear that this constitutes a substantial precedent that is adverse to U.S. interests." Charney, "The United States and the Law of the Sea After UNCLOS III," p. 53.

697. See Michel Virally, "Preface," in Yemin, p. xi. The number of areas in which such competences are deemed necessary has been increasing rapidly in recent years. "The trebling of the number of states in the world; the very rapid increase in the volume of transactions among these states; revolutionary changes in communication, transport and other technologies; much greater demands and pressures on man's physical environment; the emergence of trans- or multinational corporations as a major force in international economic relations; all of these point to the necessity for developing multilateral negotiating processes to deal with the new and unprecedented complexity and to respond to the new awareness of global environmental problems." Malcolm Fraser, "The Third World and the West," *Atlantic Community Quarterly* 20 (1982): 105. See also Richardson, "The Politics of the Law of the Sea," pp. 22–23.

In most states public goods are protected from unrestrained "private" exploitation. "Whether the good in question is the community's supply of fresh water, recreational parkland, or essential communications and transportation networks, the hope has proved ill-founded that, in the absence of incentives or sanctions imposed by the community, all users would spontaneously act in their long-term, enlightened self-interests and limit immediate consumption in order to conserve the long-term supply and quality of the resource." S. Brown et al., p. 17. See Chap. 5 nn. 288, 306, 340, 350–51 and accompanying text.

698. See nn. 516–20 above and accompanying text. By then, sufficient time may have passed to permit a more realistic evaluation of the operations of both the Authority and the seabed mining industry. See Barkenbus, p. 178. It is possible that the review conference may have an opportunity to amend the Convention before any nonpioneer operator was obliged to proceed under the original seabed mining regime set forth in the Convention. E. Brown, "From UNCLOS to Prep Com," p. 153. See also address by Leigh S. Ratiner before the Bar of the City of New York, May 25, 1982, reprinted in *Congressional Digest* 62 (January 1983): 24.

Treaty critics have predicted that U.S. negotiators would be unable to withstand "the inevitable foreign pressure" at the conference. See Breaux, p. 268. In light of the staunch resistance by the United States to the UNCLOS III package throughout the 1980s, one must question the validity of such assumptions.

699. See Barkenbus, pp. 172, 174; Harry, p. 11; and also Finlay, p. 138. See Chap. 2 n. 216 and accompanying text, and n. 617 above and accompanying text. In recent years the Group of 77 has enjoyed "minor and fragmentary" successes in other areas which "in no way constitute a serious movement toward a restructuring of the world economic order." Fatouros, p. 95.

700. It is not at all clear that such intertemporal linkages exist, except in the minds of

ideologues and unprofessional negotiators. See n. 627 above and accompanying text. "In fact, it is difficult to conceive of a way that a concession made at the Law of the Sea Conference could produce a binding effect or even strong precedential effect on a United States position at any other negotiation. . . . [T]he context and nature of the Law of the Sea negotiations, particularly the deep seabed negotiations, are such that substantive concessions made there can be readily distinguished from any other negotiation." Charney, "United States Interests," p. 42. The scope of Part XI has been carefully limited in order to permit the drawing of such distinctions. See Pardo, "Before and After," p. 103. Even one of the most vociferous treaty critics has acknowledged that the "reality" of such linkages "is dictated by . . . the willingness of the developed world to entertain and encourage" overly idealistic and frequently unrealistic proposals on the part of developing states. Breaux, "The Diminishing Prospects," p. 261.

    701. See Chap. 6 n. 19 and accompanying text.

    702. See Ratiner, "A Crossroads for American Foreign Policy," p. 1020. "We will stand as the emperor without clothes — for the entire world will see that it can do amazing and stupendous things without American money, leadership or technology." Ibid. Unless the U.S. position is grounded in a rational and consistent understanding of its national interest, the credibility of the United States may be expected to be diminished in other negotiating forums. Charney, "The United States and the Law of the Sea After UNCLOS III," p. 54. See also Richardson, "The United States Posture," pp. 511–12.

    703. See Buzan, *Seabed Politics*, pp. 278–79. "If an Enterprise 'so conceived and so dedicated cannot endure,' no nation will support erecting a new one. This verdict will be in before negotiations commence over a celestial resources regime." Kimball, "Implications of the Arrangements Made," p. 59. See also Pardo, "A Preliminary Appraisal," p. 501.

Chapter 5

    1. Deep Seabed Hard Mineral Resources Act, 30 U.S.C. §1401 (a) (12). See also statement of President Ronald W. Reagan, March 10, 1983, reprinted in *U.S. Department of State Bulletin* 83 (June 1983): 71. President Reagan evidently viewed the high seas as a refuge from regulatory excesses within territorial jurisdiction. Finn Laursen, *Superpower at Sea* (New York: Praeger, 1983), p. 148. See Chap. 3 n. 304. Ambassador Elliot Richardson had previously asserted that "[w]ith respect to seabed mining we are unaware of any [legal] restraints other than those that apply generally to the high seas and the exercise of high seas freedoms." Thomas A. Clingan, Jr., "Legal Problems Relating to the Extraction of Resources of the Deep Sea Other Than Manganese Nodules," in *Law of the Sea: Neglected Issues*, ed. John King Gamble, Jr. (Cambridge, Mass.: Ballinger, 1979), p. 76. The State Department had begun emphasizing this position in the mid-1970s. Steven J. Burton, "Freedom of the Seas: International Law Applicable to Deep Seabed Mining Claims," *Stanford Law Review* 29 (1977): 1169 n. 150. A former U.S. negotiator has stated that the possibility that deep-sea mining might not be a high-seas freedom was never seriously discussed within the Reagan administration. Leigh S. Ratiner, "Reciprocating State Arrangements: A Transition or an Alternative?" in *The 1982 Convention on the Law of the Sea*, ed. Albert W. Koers and Bernard H. Oxman (Honolulu: Law of the Sea Institute, 1984), p. 201. The revised *Restatement* of U.S. foreign relations law, while omitting deep-sea mining activities from its list of high-seas freedoms, nonetheless asserts that such activities are permissible under customary law. See nn. 33, 62 below.

    2. Theodore G. Kronmiller, *The Lawfulness of Deep Seabed Mining* (New York: Oceana Publications, 1980), vol. 1, pp. 514–17. See Benedeto Conforti, "Notes on the Unilateral Exploitation of the Deep Seabed," *Italian Yearbook of International Law* 4 (1978–79): 9. Individual nodules themselves are considered to have the corollary status of *res nullius*. Dennis W. Arrow, "The Proposed Regime for the Unilateral Exploitation of Deep Seabed Mineral Resources by the United States," *Harvard International Law Journal* 21 (1980): 354–56. See also statement of H. Gary Knight in *Deep Seabed Mining*, U.S. 95th Congress, 1st Session, House of Representatives, Committee on Merchant Marine and Fisheries, Subcommittee on Oceanography (Washington: U.S. Government Printing Office, 1977), p. 425.

    3. See statement of Ambassador Elliot L. Richardson, quoted in Clingan, "Legal

Problems," p. 76; and also Lilliana Torreh-Bayouth, "UNCLOS III: The Remaining Obstacles to Consensus on the Deepsea Mining Regime," *Texas International Law Journal* 16 (1981): 85. A U.S. Senator has erroneously intimated that the freedom to mine the deep seabed is a peremptory norm of international law by characterizing it as an "inalienable" right. Statement of Senator James McClure in *Congressional Record* 125 (December 14, 1979): 36045. See nn. 404–5 below and accompanying text.

4. Seabed Act, 30 U.S.C. §§1401 (a) (12), 1402. See Arrow, "The Proposed Regime," pp. 337–38; and also Kronmiller, vol. 1, pp. 338, 481. "Being unowned, in the absence of new agreements, the nodules are free for the taking, just as loose fish are in international waters." Robert A. Goldwin, "Locke and the Law of the Sea," *Commentary* 71 (June 1981): 48. U.S. seabed mining operators are subject to flag-state jurisdiction. See Ruth Lapidoth, *Freedom of Navigation with Special Reference to International Waterways in the Middle East* (Jerulsalem: Jerusalem Post, 1975), pp. 34, 37. "No pre-emptive claim is made as against the world; what is asserted is a universal right of mine and a right of each state to prevent its nationals from mining." Bernard H. Oxman, "The High Seas and the International Seabed Area," *Michigan Journal of International Law* 10 (1989): 534.

The Act embodied the official position of the United States toward deep seabed mining. Torreh-Bayouth, p. 87. "Ample evidence exists in the legislative history of the Act to demonstrate that the disclaimer of sovereignty and exclusive rights was made, not as a matter of political comity or expediency, but out of a sense of legal obligation." Jon Van Dyke and Christopher Yuen, "'Common Heritage' v. 'Freedom of the High Seas': Which Governs the Seabed?" *San Diego Law Review* 19 (1982): 540.

5. Torreh-Bayouth, p. 90. See text accompanying Chap. 1 nn. 96, 104. It was widely believed in 1958 that all recoverable marine resources were *res nullius*, and thus subject to appropriation by capture. Arrow, "The Proposed Regime," p. 356. See also Kronmiller, vol. 1, pp. 413–14.

6. See, e.g., article 24 of the High Seas Convention, which requires regulation of pollution "resulting from the exploitation and exploration of the seabed and subsoil." Convention on the High Seas, art. 24, reprinted in *United Nations Treaty Series*, vol. 450, no. 6465, p. 96. See also Oxman, "The High Seas and the International Seabed," pp. 528–30.

7. Convention on the High Seas, art. 2, in *United Nations Treaty Series*, vol. 450, no. 6465, p. 84. "There is no hierarchy of lawful uses of the sea such that, *a priori*, one use has precedence over any other." Kronmiller, vol. 1, p. 449.

8. Letter from John Norton Moore, Counselor on International Law, Department of State, to U.S. Senate, 1974, quoted in Roger A. Geddes, "The Future of United States Deep Seabed Mining: Still in the Hands of Congress," *San Diego Law Review* 19 (1982): 614 n. 9. See also Kronmiller, vol. 1, pp. 447–48. "The innumerable forms which such controversies may indeed take makes it impracticable to adopt any standard more explicit than that of reasonableness, determined by the familiar process of balancing the 'utility of the conduct' causing damage, and the 'gravity of the harm' to the injured party." Myres S. McDougal and Norbert A. Schlei, "The Hydrogen Bomb Tests in Perspective: Lawful Measures for Security," *Yale Law Journal* 64 (1955): 691. See n. 40 below and accompanying text.

It has been argued that the "balancing process" used to determine reasonableness is "inherent in the principle of the freedom of the seas." Kronmiller, vol. 1, pp. 9–10, 519. See also D.P. O'Connell, *The International Law of the Sea* (Oxford: Clarendon, 1982), vol. 1, p. 57. According to this frequently asserted view, it may be presumed that any high-seas activity is legal as long as it is not specifically prohibited and does not unreasonably interfere with other uses of the oceans. William C. Brewer, Jr., "Deep Seabed Mining: Can an Acceptable Regime Ever Be Found?" *Ocean Development and International Law Journal* 11 (1982): 31. See also Burton, p. 1173. "Thus, the fact that deep seabed mining is a new use which may entail adverse effects upon other uses does not render it unlawful" — the reasonableness standard would be applied to determine the legality of an ocean use both generally and specifically. Kronmiller, vol. 1, pp. 350, 454. This broad application of the reasonableness standard had been endorsed by several members of the International Law Commission in its deliberations prior to UNCLOS I. Myres S. McDougal and William T. Burke, *The Public Order of the Oceans* (New Haven, Conn.: Yale University Press, 1962), p. 760. But see n. 42 below and accompanying text.

9. Seabed Act, 30 U.S.C. §1428. See n. 57 below and accompanying text; and also Kent M. Keith, "Laws Affecting the Development of Ocean Resources in Hawaii," *University of Hawaii Law Review* 4 (1982): 283–84.

10. Agreement Concerning Interim Arrangements Relating to Polymetallic Nodules of the Deep Seabed, reprinted in *Environmental Policy and Law* 9 (1982): 134. Provisional Understanding Regarding Deep Seabed Matters, reprinted in *International Legal Materials* 23 (1984): 1354–65. See Chap. 3 nn. 385–88 and accompanying text and also Louis B. Sohn and Kristen Gustafson, *The Law of the Sea in a Nutshell* (St. Paul, Minn.: West, 1984), p. 177.

11. Kronmiller, vol. 1, p. 343. Kronmiller claims that the International Law Commission had believed deep seabed mining to be a freedom of the high seas. Ibid., vol. 1, p. 377.

12. The president of UNCLOS III, Tommy T.B. Koh, has threatened to seek an advisory opinion from the Court challenging any seabed mining efforts undertaken outside the new Convention. Address by Leigh S. Ratiner before Association of the Bar of the City of New York, May 25, 1982, reprinted in *Congressional Digest* 62 (January 1983): 26. Under article 96 of the U.N. Charter, the General Assembly may request such advisory opinions. See Hans Kelsen, *Principles of International Law* (New York: Rinehart, 1952), p. 395. "The result of such a case would be difficult to predict." Jonathan I. Charney, "The United States and the Law of the Sea After UNCLOS III – The Impact of General International Law," *Law and Contemporary Problems* 46 (Spring 1983): 50. Former Ambassador Richardson has predicted that a judgment against the United States would be "overwhelmingly likely." Elliot L. Richardson, "Law of the Sea: A Reassessment of U.S. Interests," *Mediterranean Quarterly* 1 (Spring 1990): 7.

13. See Myron Nordquist, "Customary Law Status of Deep Seabed Mining," in *United States Law of the Sea Policy: Options for the Future* (Oceans Policy Study Series, no. 6, November 1985), ed. Sharon K. Shutler (New York: Oceana Publications, 1985), p. 83. "The *travaux preparatoires* of the Geneva Convention can be cited in support of either position." Charles E. Biblowit, "Deep Seabed Mining: The United States and the United Nations Convention on the Law of the Sea," *St. John's Law Review* 58 (1984): 276.

14. See Torreh-Bayouth, pp. 86–87; Dennis W. Arrow, "The Customary Norm Process and the Deep Seabed," *Ocean Development and International Law Journal* 9 (1981): 20–21; and also Biblowit, p. 278. The case was decided by a vote of 7 to 5, and the reasoning of the majority has been criticized as having been "based on the highly contentious metaphysical proposition of the extreme positivist school that the law emanates from the free will of sovereign independent states. . . . [F]rom this premise they argued that restrictions on the independence of States cannot be presumed. Neither, it may be said, can the absence of restrictions; for we are not entitled to deduce the law applicable to a specific state of facts from the mere fact of sovereignty or independence." J.L. Brierly, "The *Lotus* Case," *Law Quarterly Review* 44 (1928): 155–56.

15. See, e.g., H. Gary Knight, "The Draft United Nations Conventions on the International Seabed Area: Background, Description, and Some Preliminary Thoughts," *San Diego Law Review* 8 (1971): 476. The 1974 claim of Deepsea Ventures asserting exclusive jurisdiction over a large portion of the seabed of the Pacific Ocean was based in part on the absence of a prohibitive rule. Francisco Orrego Vicuña, "National Laws on Seabed Exploitation: Problems of International Law," *Lawyer of the Americas* 13 (1981): 148. See n. 23 below and accompanying text; and also E.D. Brown, "Freedom of the High Seas Versus the Common Heritage of Mankind: Fundamental Principles in Conflict," *San Diego Law Review* 20 (1983): 560.

16. Craig W. Walker, "Jurisdictional Problems Created by Artificial Islands," *San Diego Law Review* 10 (1973): 657. Because the body of generally accepted customary law is limited, state practice frequently develops without any international normative authority and without any governing *opinio juris*. Biblowit, p. 278. See also Kelsen, *Principles of International Law*, p. 305. This has been the case with regard to space exploration. See Carl Q. Christol, "The Common Heritage of Mankind Provision in the 1979 Agreement Governing the Activities of States on the Moon and Other Celestial Bodies," *International Lawyer* 14 (1980): 448. It has been argued that the theory governs both the formulation of general norms of state conduct and their application in specific instances. See David A. Colson, "The United States, the Law of the Sea, and the Pacific," in *Consensus and Confrontation: The United States and the Law of the Sea Convention*, ed. Jon Van Dyke (Honolulu: Law of the Sea Institute, 1985), p. 41.

17. See Chap. 1 n. 13 and accompanying text. Cf. Kronmiller, vol. 1, p. 330. "With the exception of the regimes relating to shipways, pipelines or certain installations used in exploiting the live sedentary resources which are the only cases involving appropriations of the seabed, there are practically no precedents on which to base a customary law." René-Jean Dupuy, *The Law of the Sea* (Dobbs Ferry, N.Y.: Oceana Publications, 1974), p. 106. See also Arrow, "The Proposed Regime," p. 353.

18. See Walker, p. 658. The legal vacuum theory was implicitly rejected by a 1920 Committee of Jurists. Jon Gregory Jackson, "Deepsea Ventures: Exclusive Mining Rights to the Deep Seabed as a Freedom of the Sea," *Baylor Law Review* 28 (1976): 175. The "mischief" created by the *Lotus* case was remedied during the 1950s through the negotiation of comprehensive multilateral conventions. R.R. Baxter, "Multilateral Treaties as Evidence of Customary International Law," *British Year Book of International Law* 41 (1965–66): 289.

19. Colson, "The United States, the Law of the Sea, and the Pacific," p. 73.

20. Budislav Vukas, "The Impact of the Third United Nations Conference on the Law of the Sea on Customary Law," in *The New Law of the Sea*, ed. Christos L. Rozakis and Constantine A. Stephanou (New York: Elsevier Science Publishers B.V., 1983), pp. 45–46. See Burton, p. 1168; and nn. 47, 284 below and accompanying text.

21. See Walker, p. 658; American Bar Association, "Natural Resources of the Sea: Report of Section of Natural Resources Law," in *Status Report on the Law of the Sea Conference*, U.S. 93rd Congress, 1st Session, Senate, Committee on Interior and Insular Affairs, Subcommittee on Minerals, Materials, and Fuels (Washington: U.S. Government Printing Office, 1973), p. 638; and also Oxman, "The High Seas and the International Seabed," pp. 528, 534, 542.

22. See Chap. 1 nn. 3, 83–102, 106 and accompanying text, and nn. 24, 33 below and accompanying text; and also Arrow, "The Customary Norm Process," p. 13. "The most active of these writers . . . have at one time or another been retained by companies that hope to mine polymetallic nodules." Van Dyke and Yuen, p. 518.

23. See Burton, p. 1140. No claim was made to the seabed or the subsoil upon which the nodules rest. Ibid. Counsel for Deepsea Ventures has characterized the "right of capture" doctrine favored by the United States as "an unnecessarily restrictive view." L.F.E. Goldie, "A General International Law Doctrine for Seabed Regimes," *International Lawyer* 7 (1973): 816 n. 54. "Legally Deepsea's claim may be dispatched quickly on the grounds that it was not advanced by a party having capacity to assert it internationally." Burton, p. 1145. Goldie nevertheless "argues that the land mining in the early American West and the mining of *terra nulli* (ownerless land), as was done on the Spitzbergen archipelago, is evidence of basic and widespread mining concepts which should be adopted as the basis for a practice that could mature into customary international law." J. Jackson, p. 173. See L.F.E. Goldie, "Title and Use (and Usufruct) — An Ancient Distinction Too Oft Forgot," *American Journal of International Law* 79 (1985): 708–9. Another supporter of U.S. unilateral seabed mining rights has described Goldie's conclusions as "somewhat tenuous," however. Kronmiller, vol. 1, p. 200. In fact, the *res nullius* theory is "rarely applied today." *The American Law of Mining*, ed. Rocky Mountain Mineral Law Foundation (New York: Matthew Bender, 1982), vol. 1, p. 5. See n. 27 below. Even hard-mineral exploitation on federal lands is governed by a positive-law regulatory regime. See n. 313 below.

24. S.K. Eaton, Jr., and Janet Judy, "Seamounts and Guyots: A Unique Resource," *San Diego Law Review* 10 (1973): 617. See also Kronmiller, vol. 1, pp. 3–4. Three centuries ago, one of the requirements for occupation was that the *res* "must be susceptible of human ownership, and, without any breach of Natural Reason, it must be possible to exclude other men from the use of it." Johann Wolfgang Textor, *Synopsis of the Law of Nations*, trans. John Pawley Bate, orig. pub. 1680 (Washington: Carnegie Institution of Washington, 1916), p. 67. During the first half of the 20th century, however, effective occupation of a *terra nullius* could be established by a simple display of sovereign authority. See Burton, pp. 1157–59; H. Lauterpacht, "Sovereignty Over Submarine Areas," *British Year Book of International Law* 27 (1950): 416–17; and also Richard Young, "The Legal Status of Submarine Areas Beneath the High Seas," *American Journal of International Law* 45 (1951): 230. It has thus been argued that a private mining firm could acquire exclusive rights to a part of the deep seabed by establishing a form of constructive possession. "The taking of the first sample nodule should be viewed

as a symbolic taking of possession of all the nodules . . . since the enterprise has, from that moment on, a clear intention of taking all the others in the claimed area, plus the technological power to reduce them to its physical control by means of the knowledge it gains from the sample, and the procedures of collection and processing which that sample calls for." Goldie, "A General International Law Doctrine," p. 816. Six years later Goldie seemed to relax his requirements for constructive possession even further. Van Dyke and Yuen, p. 542.

25. See Burton, pp. 1154–55. Sedentary fisheries are located on the continental margin, in areas now encompassed by the territorial sea and continental shelf regimes. See Van Dyke and Yuen, p. 516. "Clearly there can be no claim of historic rights to manganese nodule tracts." Arrow, "The Proposed Regime," p. 358.

26. See Young, "The Legal Status of Submarine Areas," p. 230. The standards used to determine the requisite nature and extent of such occupation are inherently vague, arbitrary, and controversial, particularly in light of the discontinuous nature of nodule mining sites. Van Dyke and Yuen, pp. 542–43. See also Mina Mashayekhi, "The Present Legal Status of Deep Sea-Bed Mining," Journal of World Trade Law 19 (1985): 233. The issue of when possession becomes effective is also likely to produce disputes: "It would be difficult, if not impossible, to determine who pulled up the first nodule from a given area. If, on the other hand, the area must be 'actively worked,' how is this term to be defined?" Van Dyke and Yuen, p. 543.

27. See Arrow, "The Proposed Regime," pp. 361–62. "Europeans of the 19th century developed the doctrine of territorial acquisition by occupation of terra nullius to justify extending the European system of state sovereignty to the rest of the world, ostensibly to create international order, foster trade, and maintain international peace. . . . The Island of Palmas case recognized that the justification for this award of sovereignty rested primarily on the fulfillment of international obligations, such as protecting the inhabitants and the interests of other states, rather than on the acquisition of rights or mere exercise of rights or mere exercise of power in the territory." Burton, p. 1165.

The Spitzbergen case, frequently cited by advocates of the occupation theory, involved a situation, unlike the deep seabed, where the interested states had previously acknowledged the archipelago's res nullius status. Goldie, "A General International Law Doctrine," p. 808. See also Arrow, "The Customary Norm Process," p. 46 n. 82. At the time, it was acknowledged that resolution of the Spitzbergen issue could not serve as "a precedent of importance" even with respect to other unoccupied land areas. Fred K. Nielson, "Editorial Comment: The Solution of the Spitzbergen Question," American Journal of International Law 14 (1930): 234. "In fact, there were no exclusive rights to land or minerals on Spitzbergen prior to the treaty." Burton, p. 1156. See also Robert Lansing, "A Unique International Problem," American Journal of International Law 11 (1917): 769.

The only law of the sea cases cited in support of the doctrine of constructive occupation relate to wounded whales and wrecked ships. "Neither example of 'constructive possession' would support a claim to an expanse of seabed twice the size of Maryland." Van Dyke and Yuen, p. 542.

28. See Lauterpacht, "Sovereignty Over Submarine Areas," p. 421; Chap. 4 n. 138, Chap. 4 nn. 189–90 and accompanying text, and nn. 317–19 below and accompanying text; Arrow, "The Customary Norm Process," p. 13; and also Oxman, "The High Seas and the International Seabed," pp. 527–28. Unilateral appropriation of the deep seabed under the res nullius theory would conflict with the laying of submarine pipelines and cables, with scientific research on the ocean floor, and, potentially, with free use of the superjacent waters. Burton, p. 1166. "Even acceptance of the analogy to land does not decide the issue, because all unclaimed land has not automatically been treated as terra nullius." Van Dyke and Yuen, pp. 518–19. See also Burton, p. 1159.

29. Convention on the High Seas, arts. 1, 2, in United Nations Treaty Series, vol. 450, no. 6465, p. 82. See Van Dyke and Yuen, p. 519; Paul Lawrence Saffo, "The Common Heritage of Mankind: Has the General Assembly Created a Law to Govern Seabed Mining?" Tulane Law Review 53 (1979): 509–10; and also Kronmiller, vol. 1, p. 126. The initiative had been prompted by Malta's concern "that, because of a number of factors, including the increasing value of ocean space, a total division of ocean space among coastal states was ultimately unavoidable," unless action were taken. Arvid Pardo, "Before and After," Law and Contemporary Problems 46 (Spring 1983): 96.

30. United Nations, *Report of the Ad Hoc Committee to Study the Peaceful Uses of the Sea-Bed and the Ocean Floor Beyond the Limits of National Jurisdiction* (U.N. Doc. A/7230, 1968), p. 48 (Annex II, para. 40). "By definition, an area beyond national jurisdiction is one to which no national authority can accord such exclusive rights." George H. Aldrich, Acting Special Representative of the President for the Law of the Sea Conference, "Law of the Sea," *U.S. Department of State Bulletin* 81 (February 1981): 57.

31. Statement of the representative of the United States in *Provisional Verbatim Records*, U.N. General Assembly, 23rd Session, First Committee, 1590th meeting (U.N. Doc. A/AC.1/PV.1590, October 29, 1968), p. 7. U.N. General Assembly, 25th Session, Resolution 2749, December 17, 1970, para. 2. U.N. Convention on the Law of the Sea, art. 89. See Kronmiller, vol. 1, pp. 5–6, 516. At UNCLOS III no state asserted a right to claim exclusive *in situ* rights to seabed resources. Van Dyke and Yuen, p. 539. "The prohibition against appropriation and sovereignty no doubt is accepted universally today as customary law." Biblowit, p. 289.

32. The claim was quickly challenged – by Australia, Canada, and the United Kingdom, as well as the United States – and was not recognized by any other state. Burton, p. 1147. See Kronmiller, vol. 1, p. 348; and also Kathryn Surace-Smith, "United States Activity Outside of the Law of the Sea Convention: Deep Seabed Mining and Transit Passage," *Columbia Law Review* 84 (1984): 1045. U.S. support was withheld in express reliance upon the freedom of the seas principle. L.F.E. Goldie, "A Selection of Books Reflecting Perspectives in the Seabed Mining Debate: Part I," *International Lawyer* 15 (1981): 302. Arguments advanced in support of the claim "could just as easily apply to a particular trade route on the high seas. . . . Nations do not have a high seas duty to ensure that the operations of others are profitable." Van Dyke and Yuen, pp. 540–41. See also Burton, p. 1172 n. 161. Indeed, it would appear that the United States could not avail itself of this argument without conceding that the construction and outfitting of trawlers, together with the identification of the migratory patterns of certain high-seas fisheries, had vested exclusive fishing rights in their Soviet and Japanese owners.

33. See text accompanying Chap 3. n. 392, and n. 137 below. Committee II, which negotiated article 87 at UNCLOS III, never intended to encompass deep seabed mining. Kronmiller, vol. 1, p. 452. At least one knowledgeable commentator regards the 1958 language of article 2 as having "little if any significance" in light of article 87. Oxman, "The High Seas and the International Seabed," p. 529 n. 12. Even the revised *Restatement*, which employs the 1958 "reasonable regard" language rather than the 1982 "due regard" language, does not list seabed mining as one of the high-seas freedoms, although its list is, like that in article 87 of the Convention, nonexhaustive. American Law Institute, *Restatement 3rd, Restatement of the Foreign Relations Law of the United States*, §521 (2), (3). A separate section of the *Restatement* would nevertheless permit deep-sea mining activities to be conducted on a nonexclusive basis in conformity with the "reasonable regard" standard. *Rest. 3rd, Restatement of the Foreign Relations Law of the United States*, §523 (1) (b).

Within two years after formally notifying the United Nations of the coordinates of sites it had licensed to the four private consortia, the United States had achieved the preservation of exclusive rights to those sites pending the entry into force of the Convention and had concluded an agreement with the Soviet Union which resolved remaining site overlaps. See Chap. 3 nn. 330–31, 342 and accompanying text, and Chap. 4 n. 485. It has been argued, rather tenuously, that exclusive mining rights could thus be banked until seabed mining becomes profitable for these private firms, since "the obligation to pay due regard . . . protects the first user to the extent necessary for a reasonable and continued exercise of his activity." Günther Jaenicke, "Conflicts Between Mine Sites of Signatories and Non-Signatories of the Law of the Sea Convention," in *The Law of the Sea: What Lies Ahead?* ed. Thomas A. Clingan, Jr. (Honolulu: Law of the Sea Institute, 1988), p. 510. Jaenicke would vest "a quasi-possessory relationship to that mine site" in any seabed mining operator which "has already made a special technical effort and expended a substantial amount of money in the discovery and exploration of a mine site and has made known its intention to proceed with the exploration and development of that mine site." Jaenicke, p. 510. See also statement of David Colson in *Consensus and Confrontation*, ed. Van Dyke, p. 57. But see n. 179 below. Apart from the inherent unfairness and impracticability of basing the application of legal rules upon such economic and technological factors, these arguments are essentially indistinguishable from those advanced by Deepsea Ventures in support of its 1974 claim, which was rejected by the United States and other members of the international community. See nn. 31–32 above. "There is no precedent for such extended application of the principle." Oxman, "The High Seas and the International Seabed," p. 535.

The revised *Restatement*—which would permit unilateral seabed mining apparently only because of the adamancy of the U.S. position—nonetheless rejects the notion that conduct of deep-sea mining activities may suffice under the "reasonable regard"/noninterference doctrine to establish exclusive mining rights with respect to a specific seabed area: "A claim by a state or person that it has begun a mining activity and can, therefore, exclude others, must be limited in scope, area and duration to the extent strictly necessary." *Rest. 3rd, Restatement of the Foreign Relations Law of the United States*, §523, comment (b). See Myron H. Nordquist, "The Law of the Sea Conference and Deep Seabed Mining Legislation," in *Current Issues in the Law of the Sea*, ed. Christopher C. Joyner (Allentown, Penn.: Muhlenberg College, 1980), p. 80. It is unclear to what extent the "strictly necessary" language might be intended to prevent the intrusion of foreign mining operators into the claimed mining site rather than merely to prevent physical interference with actual mining operations. See William T. Burke, "Customary Law of the Sea: Advocacy or Disinterested Scholarship?" *Yale Journal of International Law* 14 (1989): 525–26. According to the reporters, a claim of 25,000 sq. km (the size of a small nodule mining site) would be precluded. *Rest. 3rd, Restatement of the Foreign Relations Law of the United States*, §523, reporters' note 2. See Chap. 4 n. 432; and also Oxman, "The High Seas and the International Seabed," p. 535.

34. Kronmiller, vol. 1, pp. 456–57.

35. Although nodules are technically a renewable resource, their rate of formation is so slow—approximately one millimeter every million years—that they may be regarded as essentially nonrenewable. See Introducion n. 5 and accompanying text, and n. 315 below and accompanying text. Furthermore, the composition of nodules varies geographically. See Introduction n. 4 and accompanying text; and also Clyde Sanger, *Ordering the Oceans: The Making of the Law of the Sea* (London: Zed Books, 1986), p. 160. Thus, the exclusive claim of a state to any of the limited number of viable seabed mining sites will necessarily diminish the resource base remaining for exploitation by others. Van Dyke and Yuen, p. 549.

36. See nn. 357, 368, 443 below and accompanying text. Indeed, because individual living marine resources—unlike marine minerals—have a limited lifespan, they may go to waste unless harvested.

37. Kronmiller, vol. 1, p. 458.

38. See Chap. 1 nn. 19–23 and accompanying text, nn. 261–62, 266–68, 310–11, 357–58, 368, 443 below and accompanying text; and also H. Gary Knight, "International Jurisdictional Issues Involving OTEC Installations," in *Ocean Thermal Energy Conversion*, ed. H. Gary Knight, J.D. Nyhart, and Robert E. Stein (Lexington, Mass.: D.C. Heath, 1977), p. 59.

39. See nn. 46, 293 below and accompanying text; and also Knight, "International Jurisdictional Issues," pp. 53–54. The reasonableness standard had been originally proposed in the 1950s as a means of reconciling the continental shelf doctrine and nuclear weapons testing with the exercise of traditional high-seas freedoms. See McDougal and Schlei, pp. 684–86.

40. Myres S. McDougal, "International Law and the Law of the Sea," in *The Law of the Sea: Offshore Boundaries and Zones*, ed. Lewis M. Alexander (Columbus: Ohio State University Press, 1967), p. 18. See Arrow, "The Proposed Regime," p. 359; and also Goldie, "A General International Law Doctrine," p. 824. Such interests may include "indicia of relative economic importance." McDougal and Burke, p. 741. Other relevant factors include "the significance of the interest sought to be protected by the state claiming exclusive access, the relationship between the authority claimed and the interest at stake, the types of activities affected, the intensity of their occurrence, the significance of such activities for the general community, the modality and degree of interference with affected uses, and the duration." McDougal and Burke, p. 765. "Free-market" theorists believe that quantification of such factors produces market prices. See D.W. Bowett, *The Law of the Sea* (Dobbs Ferry, N.Y.: Oceana Publications, 1967), pp. 49–50. Implementation of such a procedure would seem to be a practical impossibility, however, given imperfect information and pluralistic value systems among mortal beings. It has been claimed that the "major policy purpose" governing

high-seas usage is the "promotion of the most advantageous" use, as if it should be obvious to everyone which use is indeed "most advantageous." McDougal and Schlei, p. 657. U.S. dependence upon foreign sources of copper, nickel, and cobalt has been cited as an example of "policy factors" supporting the "reasonability" of seabed mining. Arrow, "The Proposed Regime," p. 360. Of course, the Group of 77 would undoubtedly argue that the impact of nodule mining upon producer states should weigh heavily in any calculation of reasonableness. See Chap. 4. nn. 180–87 and accompanying text. Indeed, the entire history of Committee I negotiations at UNCLOS III would seem to indicate that the "reasonableness" of deep seabed mining is not a simple issue.

41. Even proponents of the reasonableness standard have acknowledged that its inherent ambiguities "have given, and continue to give, to . . . decision makers, a very large discretion." McDougal and Schlei, p. 659. "Reasonable use, in the end, is simply a justification for the assertion of a unilateral claim to make a particular use of some portion of ocean space." Knight, "International Jurisdictional Issues," p. 55.

42. Bernard H. Oxman, "The Third United Nations Conference on the Law of the Sea: The 1976 New York Sessions," *American Journal of International Law* 71 (1977): 258. "There is an independent duty *to exercise* high seas freedoms with reasonable regard for the exercise of the freedom of the high seas by other states. To convert a duty of reasonable regard into a limiting as well as expansive concept of a right to *do anything* reasonable is to substitute subjective judgment for law." Ibid., pp. 258–59 (emphasis added). See also O'Connell, vol. 1, p. 58.

43. McDougal and Burke, pp. 48–49 n. 125.

44. See John M. Junker, "The Structure of the Fourth Amendment: The Scope of the Protection," *Journal of Criminal Law and Criminology* 79 (1989): 1118–24, 1166–69, 1178–84; and also Silas J. Wasserstrom and Louis Michael Seidman, "The Fourth Amendment as Constitutional Theory," *Georgetown Law Journal* 77 (1988): 44–50, 111.

45. Kelsen, *Principles of International Law.* Indeed, it might be asked whether a system of laws would even be necessary under such an approach. If all relevant data could be quantified and fed into a computer programmed to calculate reasonableness, conflicts on the high seas — and elsewhere — might be resolved peacefully without resort to legislation, judicial proceedings, or principles of morality. The world awaits such a utopian development.

46. Eaton and Judy, p. 619. It has been pointed out that the reasonableness standard has demonstrated "an uncanny ability to justify, in pseudo-legal terms, whatever course of behavior the United States government deems expedient to its national interest under particular historical conditions." Francis A. Boyle, "The Irrelevance of International Law: The Schism Between International Law and International Politics," *California Western International Law Journal* 10 (1980): 213.

At UNCLOS I, the United States insisted upon inclusion of the reasonableness standard instead of a stricter test of permissible high-seas uses after the International Law Commission had emphasized that nuclear weapons testing might be prohibited under the latter standard. Burton, p. 1172 n. 161. See *Yearbook of the International Law Commission, 1956,* vol. 1, pp. 11–13, 32; and also Kronmiller, vol. 1, pp. 389–90. Defenders of such tests, emphasizing their overriding value for the preservation of Western culture, and fully discounting the value of resulting environmental damage, not surprisingly concluded "that the tests are reasonable and hence lawful." McDougal and Schlei, pp. 686–90, 692–95. Thus, under this approach, "technical prescriptions" such as "freedom of the seas" become "highly flexible policy preferences invoked by decision-makers to record or justify whatever compromise or adjustment of competing claims they may reach in any particular controversy." McDougal and Schlei, pp. 659–60. This is not law — it is after-the-fact rationalization of political decisions.

47. See nn. 20, 284 above and below and accompanying text; and also Don C. Piper, "On Changing or Rejecting the International Legal Order," *International Lawyer* 12 (1978): 294. "As new problems arise demanding new legal remedies, the rule which is most rational and just is the one most likely, perhaps, to be accepted, but it will be its acceptance by the nations and not the justice of the rule that gives it the binding character of law." J. Jackson, p. 182. International consent is particularly important for the legitimization of new high-seas freedoms. Van Dyke and Yuen, p. 513. See also Louis B. Sohn, "'Generally Accepted' International Rules," *Washington Law Review* 61 (1986): 1073–74.

48. See William T. Burke, *Ocean Sciences, Technology, and the Future International Law of the Sea* (Columbus: Ohio State University Press, 1966), p. 64; and Lauterpacht, "Sovereignty Over Submarine Areas," p. 395.

49. See generally O'Connell, vol. 1, pp. 38–43. Objections must be accompanied by *opinio juris*. Burton, p. 1148 n. 51. See n. 157 below and accompanying text. Where claims are met by counterclaims, customary law will remain uncertain until one practice becomes generally accepted. See Mark W. Janis, *Sea Power and the Law of the Sea* (Lexington, Mass.: D.C. Heath, 1976), p. 76.

50. R.Y. Jennings, *The Acquisition of Territory in International Law* (New York: Oceana Publications, 1963), p. 39.

51. Robert W. Tucker, "The Principle of Effectiveness in International Law, in *Law and Politics in the World Community*, ed. George A. Lipsky (University of California Press, 1953), p. 42. See n. 62 below.

52. See Chap. 1 n. 31 et seq. and accompanying text. The Truman Proclamation was codified in the Convention on the Continental Shelf just 13 years later at UNCLOS I. See Chap. 1 nn. 35, 51–53 and accompanying text. In contrast, a companion proclamation relating to offshore fisheries was quickly retracted in the face of adverse foreign reaction. See Chap. 1 n. 30 and accompanying text.

53. Kronmiller, vol. 1, p. 103. See also S.H. Amin, "The Regime of the Sea-Bed and Ocean Floor: A Legal Analysis," *Juridical Review* (1983), p. 52. In 1980 the Group of 77 voiced its opposition to unilateral seabed mining in the wake of passage of the U.S. Seabed Act. Elliot L. Richardson, "Law in the Making: A Universal Regime for Deep Seabed Mining?" *New York State Bar Journal* 53 (1981): 441. The Soviet Union subsequently protested U.S. efforts to promote deep-sea nodule mining outside the UNCLOS regime. See I. Yakovlev, "The World Ocean and International Law," *International Affairs* (Moscow) (1983, no. 8), p. 76. In response to adoption of the Soviet Union's own national seabed mining legislation, the Group of 77 made clear its opposition to all national mining legislation inconsistent with the UNCLOS III regime. "USSR Mining Decree," *Environmental Policy and Law* 9 (1982): 96.

At the 1982 signing ceremony in Jamaica, the President of the Conference noted that "speakers from every regional and interest group expressed the view that the doctrine of the freedom of the high seas can provide no legal basis for the grant by any State of exclusive title to a specific mine site." Tommy T.B. Koh, quoted in U.N. Department of Public Information, "Unique Ceremony Marks End to Long Sea Law Conference," *UN Chronicle* 20 (February 1983): 5. See also Lee Kimball, "Turning Points in the Future of Deep Seabed Mining," *Ocean Development and International Law* 17 (1986): 371. Indeed, the claim that deep-sea nodule mining is a high-seas freedom has been maintained only by a small number of advanced developed states. R.P. Anand, "UN Convention on the Law of the Sea and the United States," *Indian Journal of International Law* 24 (1984): 180.

54. Torreh-Bayouth, p. 90. See also Patricia Minola, "The Moon Treaty and the Law of the Sea," *San Diego Law Review* 18 (1981): 460.

55. Aldrich, "Law of the Sea," p. 57. "There is no way that the investor can be guaranteed such a right except in accordance with a widely supported international mechanism." Richardson, "A Reassessment of U.S. Interests," pp. 7–8.

56. H. Gary Knight, "The Deep Seabed Hard Mineral Resources Act – A Negative View," *San Diego Law Review* 10 (1973): 455. See also Biggs, p. 244; and J. Jackson, p. 178. "The inadequacy of the rule of non-interference is seen when it is used to justify acts imperiling the conservation of limited resources such as fisheries or the disciplined exploitation of submarine resources of untold value." Herbert W. Briggs, "Editorial Comment: Jurisdiction Over the Sea Bed and Subsoil Beyond Territorial Waters," *American Journal of International Law* 45 (1951): 338.

57. Seabed Act, 30 U.S.C. §1428 (a) (2). See Surace-Smith, p. 1048.

58. See Mashayekhi, p. 243.

59. See John S. Bailey, "The Future of the Exploitation of the Resources of the Deep Seabed and Subsoil," *Law and Contemporary Problems* 46 (Spring 1983): 72; and also Elliot L. Richardson, "Superpowers Need Law: A Response to the United States Rejection of the Law of the Sea Treaty," *George Washington Journal of International Law and Economics* 17 (1982): 12. It has elsewhere been observed that an RSA regime in which all potential seabed mining

states participated "might be workable" if the Authority were never established and if the legitimacy of such a regime were upheld by the International Court of Justice. Ted L. McDorman, "The 1982 Law of the Sea Convention: The First Year," *Journal of Maritime Law and Commerce* 15 (1984): 220.

60. E.D. Brown, "Seabed Mining: From UNCLOS to Prep Com," *Marine Policy* 8 (1984): 163. See Chap. 3 n. 385 and accompanying text. The Agreement applied only to claims filed with these governments by March 12, 1982, and expressly reserved the rights of its parties with respect to the Law of the Sea Convention, but was regarded as an initial step toward the establishment of an RSA regime. Susan M. Banks, "Protection of Investment in Deep Seabed Mining: Does the United States Have a Viable Alternative to Participating in UNCLOS?" *Boston University International Law Journal* 2 (1983): 287. See McDorman, "The First Year," p. 221.

61. See Chap. 3 nn. 376, 388. As of November 1984, only France, West Germany, and the United Kingdom had been designated reciprocating states. Conversation with Maureen Walker, Office of Oceans Law and Policy, Department of State, November 26, 27, 1984. The Director of the State Departent's Office of Oceans Law and Policy has acknowledged that the Provisional Understanding "is not a mutual recognition of rights agreement," but rather "a noninterference agreement." Statement of Brian Hoyle in *Consensus and Confrontation*, ed. Van Dyke, p. 250. See also Sanger, p. 5; and Mashayekhi, p. 247. "Because the . . . Understanding . . . provides for denunciation effective after 180 days' notice, it is unlikely to provide long-term stability to mining operators from countries that do not ultimately protect their claims under a more universally-agreed treaty." *Oceans Policy News* (August 1984), p. 1.

62. As signatories to the Convention, Belgium, France, Italy, Japan, and the Netherlands are "obliged to refrain from acts which would defeat the object and purpose of a treaty . . . until it shall have made its intention clear not to become a party to the treaty." Vienna Convention on the Law of Treaties, art. 18 (a), reprinted in *International Legal Materials* 8 (1969): 686. See Paul V. McDade, "The Interim Obligation Between Signature and Ratification of a Treaty," *Netherlands International Law Review* 32 (1985): 20; and also *Rest. 3rd, Restatement of the Foreign Relations Law of the United States*, §312 (3). Signatory states are also prohibited under paragraphs 1 and 3 of article 137 from recognizing rights to minerals from the deep seabed unless they are exploited under the Convention regime. James K. Sebenius, *Negotiating the Law of the Sea* (Cambridge, Mass.: Harvard University Press, 1984), p. 102. See nn. 132–34 below and accompanying text; and also Bailey, pp. 72–73. "Thus, a license or permit issued by any State outside the . . . Convention, once the Convention enters into force for a substantial number of countries, would be of questionable legal validity." Anand, "UN Convention on the Law of the Sea and the United States," p. 181. Pioneer registrants arguably have an even greater duty to abide by such provisions of the Convention pending its entry into force for them. See McDade, p. 30.

Significantly, prohibitions on any claim or recognition of "sovereignty or sovereign rights over any part of the . . . Area or its resources" and on acquisition of "any right, title, or interest in the . . . Area or its resources except as provided in this Convention" were originally included in the 1970 U.S. Draft Treaty. United Nations, *Draft United Nations Convention on the International Seabed Area* (U.N. Doc. A/AC.138/25, August 3, 1970), art. 2. While the revised *Restatement* would prohibit appropriation of the deep seabed, it expressly authorizes, "unless prohibited by international agreement, . . . activities of exploration for and exploitation of the mineral resources of that area." *Rest. 3rd, Restatement of the Foreign Relations Law of the United States*, §523 (1) (b). Although the *Restatement* asserts that this authorization "corresponds to" article 137 and other principles set forth in section 2 of Part XI, it is plainly inconsistent with article 137 (1). Cf. *Rest. 3rd, Restatement of the Foreign Relations Law of the United States*, §523, source note, reporters' note 1; and nn. 132, 217 below and accompanying text. See also Burke, "Customary Law of the Sea," p. 526.

63. See Kronmiller, vol. 1, pp. 343–44. The Provisional Understanding is, by its own terms, "without prejudice to, nor does it effect, the positions of the Parties, or any obligations assumed by any of the Parties, in respect of the United Nations Convention on the Law of the Sea." Provisional Understanding, para. 15, in *International Legal Materials* 23 (1982): 1357. See also Mashayekhi, p. 247. The denunciation clause thus assumes particular importance. See n. 61 above. "In respect to the legal status of the Provisional Understanding there

seems to be little doubt that it is consistent with the U.N. Convention." David L. Larson, "Deep Seabed Mining: A Definition of the Problem," *Ocean Development and International Law* 17 (1983): 282.

The seven seabed mining states which joined the United States in the Provisional Understanding also issued a joint statement to the Commission indicating that the Understanding "[i]n no way" is "an alternative to . . . the United Nations Convention on the Law of the Sea. It is essentially concerned with conflict resolution." Preparatory Commission, *Statement by the Chairman of the Delegation of the Netherlands on Behalf of the Delegations of Belgium, France, Germany, Federal Republic of, Italy, Japan and the United Kingdom of Great Britain and Northern Ireland Delivered on 14 August 1984* (U.N. Doc. LOS/PCN/52, August 24, 1984), p. 1. See also Larson, "A Definition of the Problem," p. 281. In the same month, in fact, Japan and France submitted applications to the Preparatory Commission for registration of their own pioneer applicants. See Chap. 3 n. 313 and accompanying text. France, Japan, and the Netherlands resisted U.S. pressure for the kind of mutual recognition of claims which would have constituted a genuine RSA regime. See McDorman, "The First Year," p. 221; and also J.M. Broadus and Porter Hoagland, III, "Conflict Resolution in the Assignment of Area Entitlements for Seabed Mining," *San Diego Law Review* 21 (1984): 562.

Two of the private consortia have each applied for two mine sites under the national mining laws, apparently in an effort to be able to submit a site to the Authority for banking under the Convention. Larson, "A Definition of the Problem," p. 281. See n. 68 below and accompanying text. It is unclear what the consortia would do if the Convention were ratified by some but not all of the states whose nationals comprise the consortia membership. William B. Jones, "Risk Assessment: Corporate Ventures in Deep Seabed Mining Outside the Framework of the U.N. Convention on the Law of the Sea," *Ocean Development and International Law* 16 (1986): 348. See Chap. 4 n. 485 and accompanying text; and also Tullio Treves, "Seabed Mining and the United Nations Law of the Sea Convention," *Italian Yearbook of International Law* 5 (1980–81): 51.

64. Hasjim Djalal, "Law of the Sea Conference: Other Alternatives for Seabed Mining?" *New York Law School Journal of International and Comparative Law* 3 (1981): 48. See Chap. 3 nn. 390, 393–95 and accompanying text, and n. 53 above and accompanying text. At the 1982 signing ceremony, Peru, speaking on behalf of the Group of 77, asserted that any such regime would be actively opposed by the international community. U.N. Department of Public Information, "Unique Ceremony," p. 6. The Group of 77 later protested that the Provisional Understanding went "beyond the resolution of conflicts arising from overlapping claims, by including provisions regarding exploration and exploitation of the sea-bed resources," and that "such agreements are contrary to the letter and spirit of the . . . Convention and have no legal validity." Preparatory Commission, *Statement by the Chairman of the Group of 77 Delivered on 13 August 1984* (U.N. Doc. LOS/PCN/48, August 16, 1984), pp. 1–2. The Eastern-bloc states issued a supporting statement on the same day. See Preparatory Commission, *Statement by the Chairman of the Group of East European Socialist Countries Delivered on 13 August 1984* (U.N. Doc. LOS/PCN/49, August 17, 1984), p. 1; and also Larson, "A Definition of the Problem," p. 281.

65. Seabed Act, 30 U.S.C. §§1401 (b), 1441 (3) (B). See Chap. 3 n. 253; n. 244 below and accompanying text; and also W. Jones, "Risk Assessment," p. 342. "Nowhere in the legislative history of this bill is there any indication that this bill is designed as a permanent substitute for a Law of the Sea treaty." Statement of Senator Claiborne Pell in *Congressional Record* 125 (December 14, 1979): 36071. But see statement of Senator Henry M. Jackson in *Congressional Record* 125 (December 14, 1979): 36075.

The 1982 Interim Agreement implicitly recognized the interim character of unilateral legislation in anticipation of the Convention's entry into force. Broadus and Hoagland, p. 573. See Agreement Concerning Interim Arrangements, paras. 2, 4 (4), in *Environmental Policy and Law* 9 (1982): 134; and also Chap. 3 n. 315. Germany and the United Kingdom have made clear that their legislation is temporary in nature and that their unilaterally issued licenses would be repealed upon general acceptance of an international regime. U.N. Department of Public Information, "Sea-Bed Commission Condemns Issuing of Licenses for Exploration of International Area," *UN Chronicle* 23 (August 1986): 107. See statement of the representative of the Federal Republic of Germany in *Provisional Summary Record*, UNCLOS III, 110th plenary meeting (U.N. Doc. A/CONF.62/SR.110, March 20, 1979), reprinted in Kronmiller, vol. 2, p. 47; and also "U.K.: Deep Sea Mining (Temporary Provisions) Bill," *Environmental Policy and Law* 7 (February 1981): 41–42. The title to the Japanese seabed mining act indicated that it, too, was intended to be "interim." See Moritaka Hayashi, "Japan and Deep Seabed Mining," *Ocean Development and International Law* 17 (1986): 362. The "temporal scope" of the laws could nevertheless be extended to satisfy the requirements of an RSA. McDade, p. 32.

66. Donald Cameron Watt, "The Law of the Sea Conference and the Deep Sea Mining Issue: The Need for an Agreement," *International Affairs* (London) 58 (1981–82): 93. See Tommy T.B. Koh, "Deep Seabed Resources Are the Common Heritage of Mankind," in *Consensus and Confrontation*, ed. Van Dyke, p. 231; and also Broadus and Hoagland, p. 554. The ability of private consortia to raise the capital necessary to begin exploitation under the existing RSA framework has been described as "doubtful." Larson, "A Definition of the Problem," p. 291. See also W. Jones, "Risk Assessment," p. 349. The General Accounting Office has also reported that underwriters would not finance private seabed mining operators under the RSA, requiring instead "a satisfactory Law of the Sea Treaty." Jesper Grolin, "The Future of the Law of the Sea: Consequences of a Non-Treaty or Non-Universal Treaty Situation," *Ocean Development and International Law* 13 (1983): 22.

67. See Wolfgang Hauser, *The Legal Regime for Deep Seabed Mining Under the Law of the Sea Convention*, trans. Frances Bunce Dielmann (Deventer, Netherlands: Kluwer, 1983), p. 28; and also E.J. Langevad, "Exploitation of the Mineral Resources of the Oceans as Affected by the Provisions of the Convention on the Law of the Sea," *Natural Resources Forum* 7 (1983): 234. "To reach this conclusion, it is not even necessary to imagine such radical scenarios as armed attacks on mining or transport ships. Such actions are quite improbable, but also not necessary in frightening off possible investors." Hauser, pp. 28–29. In fact, in the late 1980s, few developing states were willing to consider economic or military sanctions against states conducting unilateral mining operations, most preferring to oppose such activities by diplomatic methods. David L. Larson, "When Will the UN Convention on the Law of the Sea Come Into Effect?" *Ocean Development and International Law* 20 (1989): 181. A state may nevertheless legitimately act against a state pilfering common sea resources, to protect both its own interest and that of the international community. See Conforti, p. 16. "Every large American corporation which might conceivably be involved in deep seabed mining has an overseas component. These would be vulnerable to attack." W. Jones, "Risk Assessment," pp. 344–45. Military efforts to guarantee security under such circumstances in the past have proven unreliable as well as costly. See W. Jones, "Risk Assessment," p. 350.

68. Leigh S. Ratiner, "The Law of the Sea: A Crossroads for American Foreign Policy," *Foreign Affairs* 60 (1982): 1020. A NOAA official has acknowledged that U.S. firms "may be" piggybacking on the Convention signatures of other Western mining states. Conversation with Joseph P. Flanagan, November 26, 1984. "[I]t remains very difficult to secure reliable information on the detailed thinking of the mineral companies; nor is there any guarantee that publicly stated positions are anything more than part of the lobbying process designed to persuade the Government to press for commercially more advantageous terms." E.D. Brown, "The United Nations Convention on the Law of the Sea 1982: The British Government's Dilemma," *Current Legal Problems* 37 (1984): 280. Standard Oil of Ohio, the only U.S. partner in the Kennecott Consortium, has been controlled by British Petroleum, which moved to acquire complete ownership of Standard Oil of Ohio in 1987. See Broadus and Hoagland, p. 558; and Clemens P. Work, "Crumpets in Cleveland," *U.S. News and World Report*, April 6, 1987, p. 41. Canada and Japan, the other states with national firms participating in the Kennecott Consortium, have both signed the Convention. "The Kennecott Group's members apparently have been motivated by the prospect of eventual seabed mining profits, rather than, as may have been the case in the Lockheed Group, by more immediate revenues from the sale of research and development services." Broadus and Hoagland, p. 559. It has been foreseen that at least some of the consortia may simply abandon their seabed mining efforts. A.R.H. Schneider, "UNCLOS III Revisited—Recent Events in the Law of the Sea," *Environmental Policy and Law* 9 (1982): 109. See also Lance N. Antrim and James K. Sebenius, "Incentives for Ocean Mining Under the Convention," in *Law of the Sea: U.S. Policy Dilemma*, ed. Bernard H. Oxman with Charles L.O. Buderi and David D. Caron (San Francisco: ICS, 1983), p. 95.

69. Brewer, "Deep Seabed Mining," p. 33. See Chap. 2 nn. 86–88 and accompanying text. As early as 1960, developing states attributed a "limited" legislative competence to the General Assembly, whereby its resolutions might supersede rules of international law previously established through the practice of the dominant colonial states. E. Brown, "Fundamental Principles in Conflict," pp. 551–52. Mark E. Ellis, "The New International Economic Order and General Assembly Resolutions: The Debate Over the Legal Effects of General Assembly Resolutions Revisited," *California Western International Law Journal* 15 (1985): 655.

70. See Kronmiller, vol. 1, p. 334. But see nn. 242–43 below and accompanying text. The United States has not rejected the Declaration – or even the common heritage of mankind principle – outright. Edward L. Miles, "Preparing for UNCLOS IV?" *Ocean Development and International Law* 19 (1988): 428. See n. 88 below.

71. See Biblowit, p. 285; Van Dyke and Yuen, pp. 538–41; and also Ellis, pp. 662, 665. "The persuasive force of Assembly resolutions can indeed be very considerable – but this is a different thing." South West Africa, Second Phase, Judgment, *I.C.J. Reports 1966*, pp. 50–51.

72. See Piper, p. 295; and also Orrego Vicuña, "National Laws on Seabed Exploitation," p. 145. The view that such resolutions can themselves be binding "is overwhelmingly rejected by both traditional and progressive legal scholars on theoretical and empirical grounds." Ellis, p. 685.

73. Christopher C. Joyner, "U.N. General Assembly Resolutions and International Law: Rethinking the Contemporary Dynamics of Norm-Creation," *California Western International Law Journal* 11 (1981): 463. See also Van Dyke and Yuen, p. 524. "Declarations carry more 'weight' than ordinary resolutions." Ellis, p. 665. See also *Rest. 3rd, Restatement of the Foreign Relations Law of the United States*, §102, reporters' note 2. "A unanimous vote is at least presumptive evidence that the resolution expresses the *opinio juris* of the nations involved. If the nations voting represent all geographical areas and all economic systems, the presumption is virtually uncontestable." Minola, p. 463. See *Rest. 3rd, Restatement of the Foreign Relations Law of the United States*, §103, comment (c); and also Sohn, "'Generally Accepted' International Rules," p. 1078.

Resolution 1721, adopted by the General Assembly in 1961, was acknowledged by the United States to have enunciated principles of international law applicable to outer space. Van Dyke and Yuen, pp. 525–26. See Chap. 4 nn. 654–55 and accompanying text. The General Assembly has adopted other widely cited "lawmaking" declarations, including the Universal Declaration of Human Rights, the Declaration on Principles of International Law Concerning Friendly Relations and Co-operation Among States, the Resolution on Apartheid as an International Crime, and the Convention on the Prevention and Punishment of the Crime of Genocide.

74. Edward Yemin, *Legislative Powers in the United Nations and Specialized Agencies* (Leyden: A.W. Sijthoff, 1969), p. 24. See *Rest. 3rd, Restatement of the Foreign Relations Law of the United States*, §§102, reporters' note 2, 103, reporters' note 2; and also E. Brown, "Fundamental Principles in Conflict," pp. 539–40. "It is in this sense that – no matter how premature or unrealistic the inclusion of the phrase in a General Assembly resolution may have been – the United Nations may have taken a landmark decision by declaring the seabed and the ocean floor beyond the limits of national jurisdiction to be the common heritage of mankind." Gorove, "The Concept of 'Common Heritage of Mankind,'" p. 402.

75. See Kronmiller, vol. 1, p. 283. Piper, p. 305; and also Knight, "A Negative View," p. 458. At the very least, such resolutions may provide an important source of *opinio juris*. Joyner, "U.N. General Assembly Resolutions and International Law," p. 478. See Jennings, p. 84; and also Burton, p. 1150 n. 52. Arbitral and judicial decisions have affirmed that such resolutions do have legal significance. Van Dyke and Yuen, p. 526. See also Ellis, pp. 682–83, 704. Absent a formal legislative competence, the determination of the legal effect of such instruments will nevertheless remain problematic. See Jerzy Sztucki, *Jus Cogens and the Vienna Convention on the Law of Treaties* (New York: Springer, 1974), p. 167. "It may safely be inferred that the consensus of opinion amongst publicists is to attribute differing legal value to different types of resolutions, taking into account the circumstances in which they were adopted and the language in which they are couched." S.K. Agrawala, "The Role of General Assembly Resolutions as Trend-Setters of State Practice," *Indian Journal of International Law* 21 (1981): 515.

76. Jennings, p. 85.

77. Van Dyke and Yuen, p. 515 n. 89. See Chap. 1 nn. 148, 157 and accompanying text, Chap. 2 n. 61 et seq. and accompanying text, and nn. 29–31 above and accompanying text.

78. Kronmiller, vol. 1, pp. 251–52, 260. See also Torreh-Bayouth, p. 92 n. 74. "[T]he reservations and exceptions that . . . countries usually express even when voting in favour of declarations of this kind normally entail no serious intention on their part to be bound to the declared principles." Conforti, p. 7.

79. Arrow, "The Customary Norm Process," p. 30. See Chap. 2 nn. 28, 43 and accompanying text; Knight, "The Draft United Nations Conventions," p. 493; and also "Discussion," in *Consensus and Confrontation,* ed. Van Dyke, pp. 257–58. There was, for example, disagreement over whether the principle mandated the exercise of exclusive jurisdiction over the deep seabed by an international agency. Stephen Gorove, "The Concept of 'Common Heritage of Mankind': A Political, Moral, or Legal Innovation," *San Diego Law Review* 9 (1972): 401. "The legislative history of the declaration indicates that there was no explicit meaning given to the phrase, common heritage of mankind." Rudolph Preston Arnold, "The Common Heritage of Mankind as a Legal Concept," *International Lawyer* 9 (1975): 153. See also Van Dyke and Yuen, p. 522.

80. Van Dyke and Yuen, p. 529. Richardson, "Law in the Making," p. 445.

81. Roy S. Lee, "Machinery for Seabed Mining: Some General Issues Before the Geneva Session of the Third United Nations Conference on the Law of the Sea," in *Law of the Sea: Caracas and Beyond,* ed. Francis T. Christy et al. (Cambridge, Mass.: Ballinger, 1975), p. 119. The President of UNCLOS I had employed the phrase in reference to the seabed and its resources, however. Koh, "Deep Seabed Resources," p. 228. See also Chap. 1 n. 129.

82. See Chap. 4 n. 663 and accompanying text. General Assembly resolutions in 1958, 1959, and 1961 referred to the "common interest of mankind" in outer space, and the 1967 Outer Space Treaty proclaimed that area to be the "province of all mankind." Christol, "The Common Heritage of Mankind Provision," pp. 449, 480. See Chap. 4 nn. 654–56 and accompanying text. The concept contained in the Declaration of Principles was influenced by these previous invocations. Christol, "The Common Heritage of Mankind Provision," p. 450. See Chap. 4 nn. 658–59 and accompanying text. It has, however, been noted that the concept set forth in the Moon Treaty "does not even vaguely resemble the definition of the term advocated by the developing countries at UNCLOS III." Torreh-Bayouth, p. 99 n. 100. Indeed, there would seem to be no reason why the same meaning should necessarily be attributed to the principle in each context of its use by the international community.

83. Gorove, "The Concept of 'Common Heritage of Mankind,'" pp. 393, 398. See also Arnold, p. 154. The Authority would thus serve as the institutional machinery through which the international community allocates rights to common deep-sea mineral resources. Surace-Smith, p. 1039. The term "heritage" is intended to imply a responsibility on the part of the present generation to manage the resources prudently on behalf of future generations. Biblowit, p. 294.

84. H. Gary Knight, "Foreword: Law of the Sea Negotiations 1971–1972 — From Internationalism to Nationalism," *San Diego Law Review* 9 (1972): 384. See also Djalal, p. 44. "Considered alone, the phrase would appear to be entirely devoid of meaning, yet the nature of its use in subsequent resolutions and law of the sea rhetoric since 1970 would suggest that it has become a term of art." Saffo, pp. 513–14. The principle may also require that the area be used only for peaceful purposes and that seabed exploitation occur in an environmentally sensitive manner, "limiting both depletion and pollution." George Kent, "Fisheries and the Law of the Sea: A Common Heritage Approach," *Ocean Management* 4 (1978): 14–15.

Arvid Pardo has indicated that his concept of the common heritage included "five basic characteristics": non-appropriation, international management, "active" sharing of benefits, use for peaceful purposes, and protection of the interests of future generations. Pardo, "Before and After," p. 96. Arvid Pardo, "Foreword," *San Diego Law Review* 14 (1977): 516. See also Arvid Pardo and Elisabeth Mann Borgese, "The New International Economic Order and the Law of the Sea," Occasional Paper No. 5 (Malta: International Ocean Institute, 1976), pp. 14–15.

85. Hamilton S. Amerasinghe, "The Third World and the Seabed," in *Pacem in Maribus,* ed. Elisabeth Mann Borgese (New York: Dodd, Mead, 1972), p. 245. The Soviet Union had

taken a similarly broad view of the common heritage of mankind principle. Christopher C. Joyner, "Towards a Legal Regime for the International Seabed: The Soviet Union's Evolving Perspective," *Virginia Journal of International Law* 15 (1975): 875. An advocate of free-market economics has claimed, without supporting analysis, that under the common heritage of mankind "the interests of the ruling establishment in a few developing countries are to take precedence over the fundamental rights of all individuals, including those of citizens of developing countries." Doug Bandow, "UNCLOS III: A Flawed Treaty," *San Diego Law Review* 19 (1982): 479–80.

86. Torreh-Bayouth, p. 85. See Conforti, p. 8; and also Kronmiller, vol. 1, pp. 263–64. Developed states have generally argued that they may only be obligated to share revenues derived from nodule exploitation. Torreh-Bayouth, p. 98. See nn. 229–31 below and accompanying text; and also Saffo, p. 514. Prior to the adoption of the Declaration of Principles in 1970, the application of slogans such as "common legacy," "common patrimony," and "common heritage" to parts of the high seas had not been viewed as inconsistent with free use of ocean resources. Biblowit, p. 295. Representatives of the U.S. seabed mining industry thus believe that the common heritage of mankind principle "cannot reasonably be considered to be in opposition to the free-enterprise system of economic and political relationships." Christol, "The Common Heritage of Mankind Provision," p. 454. One industry lawyer has even argued that the principle would permit unilateral seabed mining free from the "restraints of accommodation" with respect to other ocean users which normally inhere in the freedom of the seas principle. Northcutt Ely, "Commentary," in *The Law of the Sea: U.S. Interests and Alternatives,* ed. Ryan C. Amacher and Richard James Sweeney (Washington: American Enterprise Institute for Public Policy Research, 1976), p. 150.

87. See Jack N. Barkenbus, *Deep Seabed Resources: Politics and Technology* (New York: Free Press, 1979), pp. 41–42. "The fact that the phrase appears in the operative part of the declaration, and not in the preamble, suggests an intention that it should be a legal concept." Arnold, p. 153. The French and Spanish translations of the Declaration, moreover, employ language with legal connotations (specifically, property rights). J. Jackson, p. 184. But see Biblowit, p. 294. The strongest legal interpretation of the principle "would require that without the agreement of all joint owners, the states of the world, no individual state could exercise its individual right to the property held jointly with the other states of the world." Arnold, p. 155.

88. Gorove, "The Concept of 'Common Heritage of Mankind,'" p. 402. See also Barkenbus, p. 41. The ambiguity of the concept may have hindered its acceptance as a principle of law. Gorove, "The Concept of 'Common Heritage of Mankind,'" p. 394. "By its inclusion, the phrase was quite clearly intended by all parties to be more than a mere label, yet at present, the concept is something less than a fully defined principle. It can perhaps be considered as a principle having a latent value to be assigned by subsequent practice, essentially the view taken by the United States." Saffo, p. 514. See also Torreh-Bayouth, p. 81. The Reagan administration complained, however, that the concept had become "a guise under which the seabed's wealth was to be governed, regulated, and allocated under the false assumption that every nation has an undivided property interest in the deep seabed." James L. Malone, "The United States and the Law of the Sea After UNCLOS III," *Law and Contemporary Problems* 46 (Spring 1983): 32.

89. U.N. General Assembly, Resolution 2749, para. 7. See also Chap. 2 n. 53 and accompanying text. This principle had become closely associated with the common heritage of mankind principle during the pre-conference negotiations. See n. 225 below; and also Arrow, "The Customary Norm Process," p. 22. Indeed, as early as 1950 the International Law Commission had endorsed exploitation of submarine resources "for the benefit of all mankind." *Yearbook of the International Law Commission, 1950,* vol. 2, p. 384 (paras. 193, 198).

90. Kronmiller, vol. 1, pp. 257–58, 281, 516–17; and also Treves, "Seabed Mining and the United Nations Law of the Sea Convention," p. 27. It has been observed that the benefits of unregulated private mining efforts are more likely to accrue to private companies than to mankind as a whole. Djalal, p. 45.

91. See Arnold, p. 157.

92. "Although this provision is merely preambular, distinguished jurists have cited it as authority for the proposition that customary law does not authorize seabed mining." Biblowit,

p. 289. Western states have argued, however, that such substantive rules are not necessary, since seabed mining may be undertaken as a freedom of the high seas. Ibid. See also Kronmiller, vol. 1, pp. 256–57. They have cited paragraph 5 of the Declaration in support of their position. Kronmiller, vol. 1, p. 277. See text accompanying Chap. 2 n. 88.

93. U.N. General Assembly, Resolution 2749, para. 4. See also ibid., paras. 3, 14.

94. Ibid., para. 9. See also Kronmiller, vol. 1, p. 281. Paragraph 9 also specifies broad guidelines which are to govern the international regime. See text accompanying Chap 2 n. 88; and also Chap. 2 n. 85 and accompanying text. It was recognized at the time that opposition by a number of states might prevent the future establishment of such a regime. Van Dyke and Yuen, p. 537. "By specifying that the new regime must be of a universal character, this principle makes clear that the regime cannot simply be an agreement among a few States in a region, or in one group or another." Statement of the representative of the United States in *Official Records*, U.N. General Assembly, 25th Session, First Committee, 1799th meeting (U.N. Doc. A/C.1/PV.1799, December 15, 1970), p. 3.

It has been argued that the 1982 Law of the Sea Convention may not meet the requirements of paragraph 9 because it was adopted by vote rather than by consensus. Biblowit, p. 298. The degree of support for the new Convention will ultimately be determined by the number of ratifications it receives, however. The United States itself has clearly failed to implement an RSA which satisfies the requirements of paragraph 9. See n. 156 below.

95. See Brewer, "Deep Seabed Mining," p. 33; and also Kronmiller, vol. 1, p. 2. "The strongest argument for reading a prohibition on mining outside the U.N. Convention into the Declaration of Principles derives from reading together the 3rd, 4th, 9th, and 14th paragraphs of the Declaration." Biblowit, p. 296.

96. Hauser, p. 27. See n. 92 above; Richardson, "Law in the Making," p. 410; and Kronmiller, vol. 1, pp. 269, 272. "The Declaration does not expressly forbid or authorize the exploration for and the exploitation of the mineral resources of the deep seabed pending the establishment of a generally accepted international treaty of a universal character." Sohn and Gustafson, p. 175. In fact, the ambiguous language was drafted in an effort to reconcile the significant differences of opinion which existed on the issue. See text accompanying Chap. 2 nn. 28, 86.

Even one of the strongest defenders of U.S. unilateral seabed mining rights has acknowledged that, under the terms of the Declaration, "present activities should arguably be compatible" with the future international regime. Kronmiller, vol. 1, p. 281. It has been argued that developed states, having opposed the Moratorium Resolution only the year before, would not have voted for a declaration which prohibited deep-sea mining. Goldie, "A General International Law Doctrine," p. 818. See also Kronmiller, vol. 1, p. 257. Much of the opposition to the Moratorium Resolution was designed to enhance the bargaining power of maritime states with respect to the negotiation of rights of passage through international straits, however. Virginia A. Pruitt, "Unilateral Deep Seabed Mining and Environmental Standards: A Risky Venture," *Brooklyn Journal of International Law* 8 (1982): 346 n. 7. See Chap. 2 nn. 1, 64, 79 and accompanying text.

97. See Chap. 2 n. 91 and accompanying text; and also Robert F. Pietrowski, Jr., "Hard Minerals on the Deep Ocean Floor: Implications for American Law and Policy," *William and Mary Law Review* 19 (1977): 56.

98. Kronmiller, vol. 1, pp. 285, 308.

99. See Chap. 2 nn. 89, 131, 233 and accompanying text. Altogether, more than 115 states, representing "a remarkable proportion of the international community," have indicated support for the binding legal effect of the principles set forth in the Declaration; the developed states supporting the U.S. position comprise "a group of large, densely populated, highly industrialized, wealthy States which together contribute nearly half of the United Nations budget." E. Brown, "Fundamental Principles in Conflict," pp. 554–55. See also Arrow, "The Proposed Regime," p. 379. The Soviet Union expressed support for the legal position of the Group of 77, as have other Eastern European states. Kronmiller, vol. 1, pp. 207, 325–26. Australia also indicated in 1974 that it viewed rights to the deep seabed and its resources as vested in the entire international community. Kronmiller, vol. 1, p. 348 n. 627. Both of these states had previously refused to attribute legal effect to the Declaration. Arrow, "The Proposed Regime," p. 376.

100. See Kronmiller, vol. 1, pp. 234–35, 239, 313; Gonzalo Biggs, "Deep Seabed Mining and Unilateral Legislation," *Ocean Development and International Law Journal* 8 (1980): 239; Chap. 2 n. 223 and accompanying text, and Chap. 3 nn. 394–95 and accompanying text; and also E. Brown, "Fundamental Principles in Conflict," pp. 546–47.

101. See, e.g., Kronmiller, vol. 1, p. 285. "Thus, what began in 1970 as simply a non-binding pronouncement by the General Assembly has evolved in only a decade to what many commentators perceive to be an incipient legal norm or a new principle of law." Joyner, "U.N. General Assembly Resolutions and International Law," p. 476. See also nn. 73–74 above and accompanying text.

102. H. Gary Knight, quoted in Barkenbus, p. 36. See Sohn, "'Generally Accepted' International Rules," p. 1079: and L.H.J. Legault, "The Freedom of the Seas: A License to Pollute?" *University of Toronto Law Journal* 21 (1971): 220–21. From the standpoint of political force, "one could say that the Declaration is perhaps even more binding than treaties." R. Lee, "Machinery for Seabed Mining," p. 120. See also John Temple Swing, "Address," in *Current Issues in the Law of the Sea*, ed. Joyner, p. 117; and J. Jackson, p. 183. The U.N. Office of Legal Affairs has stated that, "in so far as the expectation is gradually justified by State practice, a declaration may by custom become recognized as laying down rules binding upon states." Minola, p. 464. See also Van Dyke and Yuen, pp. 524–25.

103. Arrow, "The Proposed Regime," p. 405. Until it is fully defined, "the resolution enjoys a 'twilight existence.' It is probably more than merely recommendatory, but it is not yet a legally binding norm." Ellis, p. 702.

104. U.N. Convention on the Law of the Sea, preamble. "The acceptance of the principles of the Declaration in the Convention (Part XI, Section 2), with no opposition, should dissipate all suspicions in relation to the general character of these principles." Vukas, p. 46. See also Nordquist, "Customary Law Status," p. 79; and Kohn, "Deep Seabed Resources," p. 230.

105. Luke T. Lee, "The Law of the Sea Convention and Third States," *American Journal of International Law* 77 (1983): 543. See Vienna Convention on the Law of Treaties, arts. 34–36, in *International Legal Materials* 8 (1969): 693. But see McDade, pp. 25–27. Under article 35 of the Vienna Convention, a third state could only be bound by a provision prohibiting unilateral seabed mining if it "expressly accepts that obligation in writing." *International Legal Materials* 8 (1969): 693. The applicability of such a requirement under customary international law is doubtful, however, insofar as it represents progressive development of the law. See n. 392 below. The International Court of Justice nevertheless has taken the view that conventional norms may acquire binding force for nonparty states through subsequent state practice and *opinio juris*. Jonathan I. Charney, "International Agreements and the Development of Customary International Law," *Washington Law Review* 61 (1986): 971–72. See North Sea Continental Shelf, Judgment, *I.C.J. Reports 1969*, p. 25 (paras. 26, 27); and nn. 188–92 below and accompanying text.

106. See n. 223 below and accompanying text. "The Vienna Convention on the Law of Treaties makes the intent of parties the controlling factor." Surace-Smith, p. 1057. At the Montego Bay signing ceremony in December 1982, a spokesman for the Group of 77, representing nearly three fourths of the 167 delegations present, expressly denied the applicability of "the new rules and rights established by the convention" to any state not becoming a party to it, a view seconded by delegates representing several other states, including the Soviet Union (on behalf of the Soviet bloc) and Canada. See L. Lee, pp. 547–48; and Surace-Smith, p. 1057. However, an intention to apply certain select provisions of the Convention to third states might permit such an application.

107. Kelsen, *Principles of International Law*, pp. 347–48. See Arrow, "The Proposed Regime," pp. 405–6; and also Van Dyke and Yuen, p. 550. The U.N. Charter is itself a convention which purports to bind nonmember states and which has come to be accepted as general international law "to such an extent that a number of rules contained in the Charter have acquired a status independent of it." Military and Paramilitary Activities in and Against Nicaragua (Nicaragua v. United States of America), Merits, Judgment, *I.C.J. Reports 1986*, p. 97 (para. 181). See also Kelsen, *Principles of International Law*, p. vii. "There is indeed a remarkable difference between treaties concluded by many states – multilateral treaties – by which general norms are created, regulating the mutual behavior of the contracting states, as

the Covenant of the League of Nations or the Charter of the United Nations, and treaties concluded by only two states—bilateral treaties—by which an individual norm is created, establishing only one obligation of one state and one right of the other." Kelsen, *Principles of International Law*, p. 320. See also Quincy Wright, "Conflicts Between International Law and Treaties," *American Journal of International Law* 11 (1917): 572–74; and L. Lee, pp. 564–65. The "objective regime" thus established may have binding effect *erga omnes*. Biblowit, pp. 279–80. See also Ian Sinclair, *The Vienna Convention on the Law of Treaties*, 2nd ed. (Manchester: Manchester University Press, 1984), p. 104. Narrow interpretations of the doctrine of objective regimes cannot account for such instances of its application as the identification of high-seas pollution as an international crime, *erga omnes*. See n. 406 below; and Stephen Vasciannie, "Part XI of the Law of the Sea Convention and Third States: Some General Observations," *Cambridge Law Journal* 48 (1989): 90. While it may not be possible, by processes of logical deduction, to prove the existence of objective regimes directly, such processes may achieve the result by disproving the negative: the application of two or more inconsistent normative rules simply cannot be simultaneously valid with respect to a single object such as the sea. See n. 383 below and accompanying text.

"Strictly speaking, the process by which the provisions of Part XI *qua* treaty rules became potential rules of customary law could not have begun prior to the adoption of the Convention in 1982." Vasciannie, p. 88. However, the 1969 Vienna Convention was itself generally applied before it officially entered into force in 1980, a process which has been emulated with respect to the Law of the Sea Convention. See Louis B. Sohn, "The Law of the Sea: Customary International Law Developments," *American University Law Review* 34 (1985): 276. Existing "restrictions on the law-creating process were self-made, and they can be changed by the very method that established them in the first place.... There is no rule of international law preventing all states meeting in a conference ... to agree that henceforth they will use certain methods of law-creation and will consider themselves bound by the rules established through that method." Sohn, "'Generally Accepted' International Rules," pp. 1079–80. It must therefore be considered particularly significant that UNCLOS III was the largest such conference in history, with more than 150 states participating. Anand, "UN Convention on the Law of the Sea and the United States," p. 153. See also Biblowit, p. 280.

108. *Rest. 3rd, Restatement of the Foreign Relations Law of the United States*, §102 (3). See ibid., §102, reporters' note 2; and also Ted L. Stein, "The Approach of the Different Drummer: The Principle of the Persistent Objector in International Law," *Harvard International Law Journal* 26 (1985): 472. Such treaties themselves constitute state practice. Charney, "International Agreements," p. 974. The revised *Restatement* notes that multilateral conventions have had an important formative effect in areas such as human rights and the law of treaties. *Rest. 3rd, Restatement of the Foreign Relations Law of the United States*, §102, comment (f), reporters' note 4; Introductory Note to Part III. See also Charney, "International Agreements," p. 975. "In its treatment of the law of the sea the Restatement assigns international treaty negotiations a greater role in the formation of customary law than in any other substantive area of international law." Charney, "International Agreements," p. 989. The *Restatement* is recognized in the United States as authoritative evidence of international law. Charney, "International Agreements," pp. 972–73 n. 8.

109. Convention on the High Seas, preamble, in *United Nations Treaty Series*, vol. 450, no. 6465, p. 82. See Kronmiller, vol. 1, p. 384. "If there is general consensus that such a statement is not self-serving, but reflects the fact that the convention codifies prior practice, making the rules clearer and more precise, the rules contained in that convention are then considered as binding not only on states that are parties ... but also on all other states." Sohn, "'Generally Accepted' International Rules," p. 1075. See also Baxter, p. 286. The other three 1958 conventions expressed no such preambular intention, and their effectiveness has in any event been undermined by technological developments and the emergence of new developing states. Remarks by Cecil J. Olmstead in *Proceedings of the 76th Annual Meeting*, American Society of International Law (1984), p. 111. See Chap. 1 nn. 62, 64–73, 101–2 and accompanying text; and also Baxter, p. 299. It has been pointed out that the amount of "codification" in the 1982 Convention is likely to be far outweighed by the amount of "progressive development," since most of the Convention is a result of diplomatic bargaining rather than studied analysis of the state of existing law. Vukas, p. 36. See n. 157 below.

111. Statement of former Secretary of State Dean Rusk in *Law of the Sea Resolution*, U.S. 93rd Congress, 1st Session, House of Representatives, Committee on Foreign Affairs, Subcommittee on International Organizations and Movements (Washington: U.S. Government Printing Office, 1973), p. 37. See n. 94 above and accompanying text. "The meaning of 'generally agreed upon' is not clear, but would seem to require a very broad consensus." Sohn and Gustafson, p. 179. The phrase would seem to be the effective equivalent of "general practice accepted as law," which is specified as one source of international law in article 38 (1) (b) of the Statute of the International Court of Justice. Universal adherence is not required by the Declaration of Principles, which calls for a treaty of "a universal character" – a requirement which would seem to have been met by the intended universal and comprehensive scope of the subject matter of the Convention. See Vukas, p. 47. In the early part of this century, apart from "a vague impression among some writers," customary law did require universality: states could not be bound against their will. Ronald F. Roxburgh, *International Conventions and Third States* (New York: Longman, Green, 1917), p. 103. See also Fisheries Case, Judgment of December 18th, 1951, *I.C.J. Reports*, 1951, p. 131. Today, however, "the very generality of a rule of customary international law is held to make it binding on all states without distinction." Prosper Weil, "Towards Relative Normativity in International Law?" *American Journal of International Law* 77 (1983): 436. In the North Sea Continental Shelf case the International Court of Justice "clearly implied that much less must be understood for 'general' than for 'universal.'" H. Meijers, "How Is International Law Made? – The Stages of Growth of International Law and the Use of Its Customary Rules," *Netherlands Yearbook of International Law* 9 (1978): 15.

112. Van Dyke and Yuen, p. 537. See Levan Alexidze, "Legal Nature of *Jus Cogens* in Contemporary International Law," *Recueil des Cours* 172 (1981): 247; and Christos L. Rozakis, "The Greek-Turkish Dispute Over the Aegean Continental Shelf," Occasional Paper No. 28 (Kingston, R.I.: Law of the Sea Institute, 1975), p. 5. It has been observed that opposition by "a number of important states" would prevent general acceptance of the Convention. Sohn and Gustafson, p. 179. Sohn has elsewhere stated that opposition by "a few states" would not be sufficient to hinder development of general international law. Sohn, "'Generally Accepted' International Rules," p. 1074.

The UNCLOS III delegations worked long and hard to accommodate the interests of all states in Part XI of the Convention in order to ensure widespread participation. See Chap. 3. President Reagan was quick to point out that the large number of states which voted to adopt the Convention at the final session of the Conference actually represented less than 40 percent of the world's gross national product. Statement of President Ronald W. Reagan, July 9, 1982, reprinted in *U.S. Department of State Bulletin* 82 (August 1982): 71. The 159 states which have since signed the Convention are "broadly representative of the international community," however. Bernardo Zuleta, "The Law of the Sea After Montego Bay," *San Diego Law Review* 20 (1983): 481. See n. 113 below and accompanying text.

113. Daniel S. Cheever, "The Politics of the UN Convention on the Law of the Sea," *Journal of International Affairs* 37 (1984): 247. See Chap. 3 nn. 318–19, 376–77 and accompanying text. "The United States position of opposition to the treaty is partly responsible for the small number of ratifications it has received in the . . . years since its conclusion." William T. Burke, "The Law of the Sea Treaty, Customary Law, and the United States," *Water Log* 5 (April–June 1985): 2. See also E. Brown, "From UNCLOS to Prep Com," p. 152. The dearth of ratifications may also be attributable to a belief that many of its beneficial provisions have already "been so warmly received into customary international law that ratification of the treaty or accession to it would be supererogatory." Baxter, p. 299. It is expected that ratification will become easier for many states once the Convention has entered into force. Norman A. Wulf, "Comment," *Law and Contemporary Problems* 46 (Spring 1983): 156. Generally, about 60 percent of the signatories of a multilateral convention may be expected to ratify, but many states may want to await the results of the work of the Preparatory Commission before ratifying the Law of the Sea Convention. Renate Platzoeder, "Who Will Ratify the Convention?" in *The 1982 Convention*, ed. Koers and Oxman, pp. 662–63. "Signature does not guarantee that the Convention will come into force or that all the signing states will become parties. It does, however, create a substantial possibility that the Convention will enter into force." Charney, "The United States and the Law of the Sea After UNCLOS III," p. 37. See also John King

Gamble, Jr., "Assessing the Reality of the Deep Seabed Regime," *San Diego Law Review* 22 (1985): 790.
114. Larson, "When Will the UN Convention on the Law of the Sea Come Into Effect?" p. 179. See also Manjula R. Shyam, "Deep Seabed Mining: An Indian Perspective," *Ocean Development and International Law* 17 (1986): 326.
115. Wulf, "Comment," pp. 158–59. "Many Caribbean countries will ratify as a measure of support for Jamaica, the future site of the proposed Seabed Authority, while others will find too few national interests to be anything other than massively indifferent." Ibid., p. 159.
116. Wulf, "Comment," p. 159. See also Broadus and Hoagland, pp. 569–70. Ratification by "almost all the Pacific states" is expected. Statement of Rabbie Namaliu in *Consensus and Confrontation*, ed. Van Dyke, p. 34.
117. United Nations, *Law of the Sea: Report of the Secretary-General* (U.N. Doc. A/41/742, October 28, 1986), p. 5. Many of the early ratifications have been provided by African states, despite predictions that African participation in the Convention might be difficult to achieve. See Chap. 3 nn. 378–80 and accompanying text; and Wulf, "Comment," pp. 160–61. "Africa has almost enough votes to singlehandedly bring the Treaty into force." Wulf, "Comment," p. 160.
118. Pietrowski, "Hard Minerals," p. 69 n. 129. See Chap. 3 n. 375 and accompanying text. Soviet-bloc states abstained in the 1982 vote on the Convention in protest of a relatively minor procedural preference accorded Western mining states at the final UNCLOS III session, but have since signed the Convention. See Ratiner, "A Crossroads for American Foreign Policy," p. 1012. It was believed that the Soviet Union would prefer that the Convention enter into force, but that it was unlikely to ratify if Western European states did not. Wulf, "Comment," p. 159.
119. See Chap. 3 nn. 305, 309, 319, 388–89 and accompanying text; Kimball, "Turning Points," p. 375; and Wulf, "Comment," pp. 157–58. Both Japan and France have demanded substantial improvements in the seabed mining regime. John Tagliabue, "Bonn Cabinet Said to Oppose Sea Treaty," *New York Times*, November 24, 1984, p. 3. See n. 120 below. Entry of the Convention into force before satisfactory resolution of Commission negotiations could jeopardize developed-state participation. Kimball, "Turning Points," p. 387.
120. See Banks, p. 293. All of these developed states have carefully preserved their ability to proceed with seabed mining under the Convention, despite their participation with the United States in the Provisional Understanding. See Chap. 3 n. 388 and accompanying text, and nn. 61, 63 above and accompanying text. "To the extent that the parties to the August 3 agreement set in motion a process to achieve consistent application requirements and operating standards and other measures to implement the agreement that are compatible, comparable and equally effective, their work will actually contribute to the regulation-writing efforts of the Preparatory Commission." *Oceans Policy News* (August 1984), p. 2.
It is believed that Japan is the developed state most likely to proceed with deep-sea mining, "possibly for strategic reasons." *Ocean Science News*, May 7, 1984, p. 2. See Hayashi, "Japan and Deep Seabed Mining," pp. 353–54; and also Takeo Iguchi, "Japan and the New Law of the Sea: Facing the Challenge of Deep Seabed Mining," *Virginia Journal of International Law* 27 (1987): 531–32. Despite close economic ties with the United States, Japan is also "especially ... interested in maintaining friendly relations" with developing states which supply it with large quantities of raw materials. Broadus and Hoagland, p. 567. See also Iguchi, pp. 527, 537, 541, 544. Even so, Japanese ratification is not assured in the absence of a favorable outcome to negotiations within the Preparatory Commission. Iguchi, pp. 529–30, 541.
France's jurisdiction over the exclusive economic zones of its many island possessions, giving it the third largest such zone in the world, would be solidified by the Law of the Sea Convention. Anand, "UN Convention on the Law of the Sea and the United States," p. 177. It has been reported that Canada is not only expected to ratify, but is working actively to ensure the Convention's rapid entry into force. McDorman, "The First Year," p. 212. See also Iguchi, p. 548. The Italian national seabed mining bill called upon the government to work within the Preparatory Commission to improve the seabed mining provisions of the Convention, so that "our country can become a party to it." *Oceans Policy News* (October/November 1984), p. 3.

121. See Chap. 3 nn. 318–19, 336 and accompanying text. The British government has stated its preference for a generally accepted conventional regime. Watt, "The Need for an Agreement," p. 83. "In every field Britain is one of the most vulnerable of countries to the absence of a generally accepted regime – as three lost 'Cod Wars' have shown." Watt, "The Need for an Agreement," p. 93.

Germany, which has also decided to work within the Commission to improve the seabed mining regime, may have a special incentive to ratify the Convention so that the new Law of the Sea Tribunal will be located in Hamburg. See Chap. 3 n. 349. France, in particular, has urged German ratification. Tagliabue, p. 3.

A Department of Commerce study has concluded that U.S. opposition to the Convention cannot succeed without the support of Germany and the United Kingdom. Schneider, p. 109. See n. 124 below. The decision of both states regarding participation in the Convention may ultimately be determined by the position adopted by the EEC membership as a whole. See n. 122 below; and also Sanger, p. 5.

122. A majority of the 12 EEC member states must join the Convention before it is eligible to participate. U.N. Convention on the Law of the Sea, Annex IX, art. 3 (1). A similar requirement with respect to signature had encouraged eight of the (then) 10 Community member states to sign the Convention. "Oddfellows," *Economist* 293 (December 22, 1984): 30. See Holger Rotkirch, "The Future of the Convention," in *The 1982 Convention*, ed. Koers and Oxman, p. 681; and also William Riphagen, "The UN Convention on the Law of the Sea and the Treaty of Rome," in *The 1982 Convention*, ed. Koers and Oxman, p. 644. Moreover, the EEC's ultimate decision on whether or not to confirm the treaty will be a group decision, since "uncoordinated ratification by Member States would violate Community competences." "Questions in the European Parliament," *Environmental Policy and Law* 15 (1985): 96.

As an international organization, the EEC may confirm or accede to the Convention "in respect of matters relating to which competence has been transferred to it by its member States which are parties." U.N. Convention on the Law of the Sea, Annex IX, art. 4 (1). As in the case of signature, confirmation must be authorized by a unanimous vote of the EEC Council of Ministers. Tullio Treves, "The EEC and the Law of the Sea: How Close to One Voice?" *Ocean Development and International Law Journal* 12 (1983): 184–85. "Although the extent to which responsibility [is] divided between the Community and its member states varie[s] from topic to topic, the Community ha[s] competence in a number of areas which the member states no longer possess." Michael Hardy, "The Law of the Sea and the Prospects for Deep Seabed Mining: The Position of the European Community," *Ocean Development and International Law* 17 (1986): 312. See also Treves, "The EEC and the Law of the Sea," p. 174. The European Parliament has proposed the establishment of "a Community deep-sea mining regime . . . compatible with and complementary to that proposed in the . . . Convention." European Parliament, Resolution C101/65–68, June 1, 1981, reprinted in *Environmental Policy and Law* 7 (1981): 184. In a declaration made at the time of its signing, the EEC indicated that its ultimate decision on confirmation "will be taken in light of the results of the efforts made to attain a universally acceptable Convention," indicating that negotiations in the Preparatory Commission were of primary importance. "Questions in the European Parliament," p. 96. See also Hardy, p. 314. Significantly, both West Germany and the United Kingdom permitted EEC signature, despite their own refusal to sign the Convention. See Tagliabue, p. 3; and *Oceans Policy News* (January/February 1985), p. 5.

123. See Chap. 3 n. 319 and accompanying text. "The main problem is that, if the major powers do not ratify the treaty, the cost of financing the Enterprise could be a heavy burden on those countries which do ratify it." Langevad, p. 233. The mid-sized developed states have been particularly concerned about the potential extent of their financial obligations, and even the Soviet Union indicated support for cost-saving efforts. Kimball, "Turning Points," pp. 380–81. Because states assume no financial obligations until the Convention enters into force for them, they may be in no hurry to achieve 60 ratifications, but it has been predicted that several of these states will ultimately decide to ratify. Iguchi, p. 548.

124. Statement of John Norton Moore in *Law of the Sea Resolution*, U.S. House of Representatives, p. 27. See Chap. 3 nn. 286, 318, and n. 113 above. "States may decide that without U.S. involvement the . . . Authority will be unable to function or will be subject to direct conflict with the United States and other seabed mining interests and thus effectively

become an expensive, useless institution. . . . U.S. action may encourage states which did not achieve all their desired goals at UNCLOS III to reject the . . . Convention and remain outside the treaty regime." McDorman, "The First Year," p. 228. Rejection of the Convention by Western seabed mining states could be expected to undermine its support among developing states as well. Grolin, p. 24.

A former U.S. ambassador to UNCLOS III has predicted that "entry into force of a widely accepted" Convention is "probable." Elliot L. Richardson, "The United States Posture Toward the Law of the Sea Convention: Awkward But Not Irreparable," *San Diego Law Review* 20 (1983): 508–9 n. 14. See also Kimball, "Turning Points," p. 381. For most states, any national costs or benefits associated with Part XI are substantially outweighed by the net benefits conferred by the rest of the Convention, including offshore fisheries jurisdiction and navigational rights. See McDorman, "The First Year," p. 229; and also Platzoeder, pp. 666–67.

125. "[N]o single nation, not even one as powerful as the United States, can block the creation of international law." Van Dyke and Yuen, p. 537. See Ann L. Hollick, *U.S. Foreign Policy and the Law of the Sea* (Princeton, N.J.: Princeton University Press, 1981), p. 373; and also Alexidze, p. 246. U.S. opposition was thus unable to prevent the expansion of the territorial sea to a 12-mile limit during this century. See text accompanying Chap. 1 nn. 25–26, 36–43, 55–56, 58, 61, 78–83 and nn. 187, 251 below and accompanying text.

126. Malone, "The United States and the Law of the Sea After UNCLOS III," p. 33. "It is clear that provisions found within international agreements that require the creation of international machinery, such as international organizations or dispute settlement forums, are not suitable for incorporation as new norms of general international law." Charney, "The United States and the Law of the Sea After UNCLOS III," p. 38. See also Vukas, p. 49. This view has been endorsed in the revised *Restatement*. See *Rest. 3rd, Restatement of the Foreign Relations Law of the United States*, Introductory Note to Part V; §523, reporters' note 1.

127. Richardson, "Superpowers Need Law," p. 13 n. 69. See n. 107 above and accompanying text; and also Caminos and Molitor, p. 882. In a 1949 case involving diplomatic privileges, the Court stated its "opinion . . . that fifty States, representing the vast majority of the members of the international community, had the power, in conformity with international law, to bring into being an entity possessing objective international personality, and not merely personality recognized by them alone, together with the capacity to bring international claims." Reparation for Injuries Suffered in the Service of the United Nations, Advisory Opinion, *I.C.J. Reports 1949*, p. 185 (para. 53). "This statement . . . clearly indicates that a multilateral treaty like the Charter of the United Nations may be binding also on third-party states." Hombro, p. 250. See Vasciannie, pp. 91–92.

128. U.N. Convention on the Law of the Sea, art. 176.

129. Charney, "The United States and the Law of the Sea After UNCLOS III," pp. 49–50. Deep seabed mining licenses issued unilaterally by the United States in violation of the norm of multilateral regulation would therefore be subject to legal preemption by the international community, if not by the Authority itself. See David D. Caron, "Municipal Legislation for Exploitation of the Deep Seabed," *Ocean Development and International Law Journal* 8 (1980): 283–84.

130. The provisions of Section 2 are generalizable. See Constantine A. Stephanou, "A European Perception of the Attitude of the United States at the Final Stage of UNCLOS III with Respect to the Exploitation of the Deep Sea-Bed," in *The New Law of the Sea*, ed. Rozakis and Stephanou, pp. 270–73. Section 1, entitled "General Provisions," would also appear to be generalizable. Together, these two sections encompass articles 133–49.

131. See Betzy Ellingsen Tunold, "The Deep Sea-Bed Regime: Innovation or Perpetuation of the Status Quo?" in *The 1982 Convention*, ed. Koers and Oxman, pp. 113–14.

132. U.N. Convention on the Law of the Sea, arts. 137, 153 (1). "Title to minerals shall pass upon recovery in accordance with this Convention." U.N. Convention on the Law of the Sea, Annex III, art. 1.

133. U.N. Convention on the Law of the Sea, art. 137 (3). See Ratiner, "A Crossroads for American Foreign Policy," p. 1017 n. 3. "The provision is designed to ensure that a mini-treaty, inconsistent with the . . . Convention is challengeable by parties to the . . . Convention and the . . . Authority." McDorman, "The First Year," p. 219. Paragraph 1 of article 137 prohibits recognition of any appropriation of the Area or deep-sea minerals. See n. 62 above.

Because these provisions are set forth in the Convention as generally applicable "principles," once the Convention has achieved general acceptance they would no longer be regarded as merely contractual in nature. See nn. 105-7 above and accompanying text; and also Oxman, "The High Seas and the International Seabed," p. 540.

134. In addition to article 137, such terms are employed, *inter alia*, in article 138 (concerning the "general conduct of states in relation to the Area"), article 141 (concerning use of the Area "exclusively for peaceful purposes"), and article 273 (concerning international cooperation to promote technology transfer). Use of such terms is evidence of an intent for these provisions to bind nonparties. L. Lee, p. 545. See also Hauser, pp. 129-30.

135. Such provisions include article 139 (concerning enforcement of compliance with Part XI), article 143 (concerning marine scientific research in the Area), article 144 (concerning promotion of technology transfer), article 153 (concerning the structure of the parallel system), article 156 (2) (concerning membership in the Authority), article 157 (1) (concerning the nature of the Authority), and article 187 (concerning the jurisdiction of the Sea-Bed Disputes Chamber). Many of these provisions concern participation in the institutional machinery established by the Convention.

Use of the term "States Parties" tends to refute the contention that "reference to 'all States' or 'every State' reflects the assumption that all states would be parties to the Convention." Cf. L. Lee, p. 548. The preamble indicates that the Convention as a whole is to be an agreement among "States Parties," but also expresses an intention to establish "a legal order for the seas" pursuant to the provisions of the 1970 Declaration of Principles. See n. 104 above and accompanying text. In fact, the UNCLOS III delegates were aware that use of terms other than "states parties" implied an intent to create a general obligation, and employed the latter term specifically with an intent to limit the application of certain of the Convention's provisions. See n. 444 below.

136. U.N. Convention on the Law of the Sea, arts. 1 (3), 140. See Chap. 4 n. 264 and text accompanying n. 279.

137. Freedoms of the high seas are to be exercised "with due regard for the interests of other States in their exercise of the freedom of the high seas, *and also* with due regard for the rights under this Convention with respect to activities in the Area." U.N. Convention on the Law of the Sea, art. 87 (2) (emphasis added). See also ibid., art. 147 (1), (3); and Oxman, "The High Seas and the International Seabed," p. 532 n. 20. Article 125 makes a similar distinction.

138. Sohn, "Customary International Law Developments," p. 280. See also Richardson, "A Reassessment of U.S. Interests," p. 7. It has been pointed out that "several coastal state rights and flag state rights" also appear to depend upon "the institutional framework" created by the Convention. Riphagen, p. 643. The United States is apparently willing to accept these other institutional provisions as binding law. See Chap. 3 n. 416 and accompanying text.

139. See McDorman, "The First Year," p. 215; and U.N. Department of Public Information, "Unique Ceremony," p. 6. In 1981, prior to its final decision to oppose the Convention, the U.S. ambassador to UNCLOS III acknowledged that "[i]t has always been well understood at the ... Conference that a successful treaty must be based on a package deal." James L. Malone, quoted in Caminos and Molitor, p. 876 n. 30. An assistant legal adviser in the U.S. State Department has nevertheless more recently claimed that the negotiations had not initially been predicated upon "a grand package deal" in which the United States agreed to any rights to unrestrained access to deep-sea nodules in exchange for secure rights of navigation through narrow straits. Statement of David A. Colson in *Consensus and Confrontation*, ed. Van Dyke, p. 24. But see nn. 145-46 below and accompanying text. This position was "cautiously" endorsed in a tentative draft of the revised *Restatement*. Charney, "International Agreements," pp. 985-86.

Apparently, the U.S. decision to rely on customary international law to protect its nonseabed interests was made without an empirical analysis of state practice in relation to the provisions of the Convention. Stein, p. 465 n. 24. In fact, the United States may actually regard the nonseabed portions of the Convention, "not as binding law ... but as a guide for conduct and as the measure of the rights of others that we will recognize." Colson, "The United States, the Law of the Sea, and the Pacific," p. 45.

140. D.H.N. Johnson, "The Conclusions of International Conferences," *British Year Book of International Law* 35 (1959): 31.

141. Carl August Fleischer, "Significance of the Convention: Second Committee Issues," in *The 1982 Convention,* ed. Koers and Oxman, pp. 62–63. See statement of Anatoli Kolodkin in *The 1982 Convention,* ed. Koers and Oxman, p. 101; and Vasciannie, pp. 93–97. A less comprehensive application of the package deal concept would only prevent a state from claiming the benefit of treaty provisions without complying with provisions which impose corresponding duties. Fleischer, p. 63. In any event, it remains significant "that the vast majority of states that participated in the Conference, including the United States, consented in some fashion to the idea of the package deal." Caminos and Molitor, p. 878.

142. McDorman, "The First Year," p. 225. See n. 157 below. But see n. 139 above.

143. Cf. *Rest. 3rd, Restatement of the Foreign Relations Law of the United States,* §102, reporters' note 2. See nn. 108, 162–65 above and below and accompanying text; and Surace-Smith, p. 1056. "The participating states reached nearly all of their decisions by consensus and, believing that 'the problems of ocean space are clearly interrelated,' they treated the conference's informal negotiating texts as a provisional package of indivisible constituent compromises; consequently, the traditional rules may not apply to the 1982 Convention." Caminos and Molitor, p. 873. See text accompanying n. 153 below.

144. See Vasciannie, p. 97.

145. Hollick, *U.S. Foreign Policy,* pp. 236–37. See Chap. 2 nn. 1, 79 and accompanying text. Such linkage was advocated by the Defense Department, which was anxious to assure navigational access for U.S. naval forces. See Chap. 2 nn. 91, 202, 204. The change in U.S. negotiating strategy resulting from such a deal would have explained the U.S. decision to support the 1970 Declaration of Principles, despite its opposition to the Moratorium Resolution one year earlier. See n. 96 above.

146. Richardson, "Superpowers Need Law," p. 9. The U.S. navigation-for-seabed-mining agreement "was part of the larger 'package deal' under U.N. General Assembly Resolution 3067, which set forth the goal of a single, comprehensive convention on the law of the sea, under which such trade-offs, concessions, and compromises would be negotiated." Larson, "A Definition of the Problem," p. 272. It remains a matter of some dispute whether rights of transit through international straits could have been (or can be) obtained without concessions on the regime to govern deep-sea mining. Cheever, "The Politics of the UN Convention," p. 250.

147. Brewer, "Deep Seabed Mining," p. 37. See Sebenius, pp. 92, 108. In practice, the "package deal" concept became "the basis of much of the negotiations." E. Brown, "The British Government's Dilemma," p. 261. See text accompanying Chap. 2 n. 193.

148. Wayne R. Smith, "Law of the Sea Treaty: Report on the Enterprise," *New York Law School Journal of International and Comparative Law* 3 (1981): 54. "It was an implicit argument always that the developing countries would get the resources and the major powers would get the freedom of navigation. Now we find some developed countries demanding both, and one definitely demanding both on its own terms, so I do not think you can analyze this Convention and its legal import divorced from that legislative history." Statement of Alan Beesley in *Law and Contemporary Problems* 46 (Spring 1983): 143. See also statement of Edward Miles in *The Law of the Sea and Ocean Industry,* ed. Douglas M. Johnston and Norman G. Letalik (Honolulu: Law of the Sea Institute, 1983), p. 121.

149. L. Lee, p. 546. The position was reaffirmed several months later in a formal declaration. See *Declaration of the Group of 77* (U.N. Doc. LOS/PCN/5, April 11, 1983).

150. U.N. Department of Public Information, "Unique Ceremony," p. 6. See text accompanying Chap. 3 nn. 394–95; and Caminos and Molitor, pp. 877–78. "Although it is true that various chapters, mini-packages, were negotiated, *none of these mini-packages was regarded as having been concluded until all the other mini-packages were concluded.*" Statement of Tommy T.B. Koh in *Consensus and Confrontation,* ed. Van Dyke, p. 60 (emphasis in original).

151. U.N. Convention on the Law of the Sea, art. 309. See North Sea Continental Shelf, Judgment, *I.C.J. Reports 1969,* pp. 38–39 (para. 63); n. 153 below, and n. 78 above and accompanying text; and also Kronmiller, vol. 1, p. 385. "It would be paradoxical in the extreme if a nonparty were to be regarded as bound unqualifiedly by the obligations of the Conventions, while a party might limit its duties by the entry of reservations." Baxter, p. 285.

This prohibition had been sought by the United States and, like the package deal concept which it reinforced, "was implicit from the beginning." Statement of Alan Beesley in *Law and Contemporary Problems* 46 (Spring 1983): 143. See also statement of Bernard H. Oxman in *Consensus and Confrontaton*, ed. Van Dyke, pp. 62–63.

152. U.N. Convention on the Law of the Sea, art. 155 (2), 311 (6). See nn. 132–37 above and accompanying text.

153. U.N. Convention on the Law of the Sea, art. 311 (3). The Convention further requires that the review conference "ensure the maintenance of" many of the principles set forth in Part XI. Ibid., art. 155 (2). See Chap. 4 n. 517 and accompanying text; and also Riphagen, p. 648.

The package deal concept is also implicit in the preamble of the Convention, which states that "the problems of ocean space are closely interrelated and need to be considered as a whole." See E. Brown, "The British Government's Dlemma," pp. 260–61. The preamble also states that the Convention is a product of "the desire to settle . . . all issues relating to the law of the sea." U.N. Convention on the Law of the Sea, preamble.

154. Such failure might result from a lack of intention to confer rights and obligations upon nonparties, from a finding that the Convention is not, in fact, an integrated "package deal," from a lack of recognition under international law of the procedure whereby such a comprehensive treaty might become binding as general law, or simply from failure of the Convention to attract sufficiently widespread support.

155. See Stein, p. 458; and also L. Lee, p. 562. In addition to convention texts, evidence of state practice may be found in "every written document, every record or act or spoken word, which presents an authentic picture of the practice of states in their international dealings." Jerome Morenoff, *World Peace Through Space Law* (Charlottesville, Va.: Michie, 1967), p. 169. See Conforti, p. 7; and also Rozakis, "The Greek-Turkish Dispute," p. 4.

156. Morenoff, p. 170. See Rozakis, "The Greek-Turkish Dispute," p. 5; and also Ellis, pp. 670–71. "In order to deduce the existence of customary rules, the Court deems it sufficient that the conduct of States should, in general, be consistent with such rules, and that instances of State conduct inconsistent with a given rule should generally have been treated as breaches of that rule, not as indications of the recognition of a new rule." Military and Paramilitary Activities in and Against Nicaragua (Nicaragua v. United States of America), Merits, Judgment, *I.C.J. Reports 1986*, p. 98 (para. 186). See also Baxter, p. 278.

Few rules of international law are universally supported. O'Connell, vol. 1, p. 39. Indeed, logically, the application of a norm of international law in a given case cannot depend upon its recognition by the state in question, if an effective system of international law is to be maintained. See Kelsen, *Principles of International Law*, p. 433; and also nn. 181, 249–51 below and accompanying text. "The withholding of consent by one State has never been decisive in the translation of *lex ferenda* into *lex lata*, for State practice is not a matter of counting heads but of juristic evaluation of the factors that tend to legitimize action by individual States." O'Connell, vol. 1, p. 39. The states adhering to the practice should be representative of the various political and legal systems existing in the international community. Rozakis, "The Greek-Turkish Dispute," p. 5. See n. 112 above and accompanying text. "[T]he number of States participating is more important than the frequency or duration of the practice. Even a practice followed by a few States, on a few occasions and for a short period of time, can create a customary rule, provided that there is no practice which conflicts with the rule, and provided that other things are equal; but other things are seldom completely equal." Michael Akehurst, "Custom as a Source of International Law," *British Year Book of International Law* 47 (1974–75): 53.

157. Arrow, "The Customary Norm Process," p. 3. See North Sea Continental Shelf, Judgment, *I.C.J. Reports 1969*, p. 44 (para. 77); and also Kelsen, *Principles of International Law*, p. 307. "A practice initially followed by states as a matter of courtesy or habit may become law when states generally come to believe that they are under a legal obligation to comply with it. It is often difficult to determine when that transformation into law has taken place. Explicit evidence of a sense of legal obligation (e.g., by official statements) is not necessary; *opinio juris* may be inferred from acts or omissions." *Rest. 3rd, Restatement of the Foreign Relations Law of the United States*, §102, comment (c). Negotiating positions adopted during a multilateral law-making conference will not qualify as *opinio juris* if they merely reflect bargaining tactics; the votes of individual states on particular provisions may not

necessarily be an accurate reflection of their attitudes with respect to international law on the subject, even when approved by consensus. Baxter, p. 293. Theodor Schweisfurth, "The Influence of the Third United Nations Conference on the Law of the Sea on International Customary Law," *Zeitschrift für ausländisches öffentliches Recht und Völkerrecht*, 43 (1983): 578. Agrawala, p. 532. See n. 163 below.

158. Sinclair, p. 22. In the North Sea Continental Shelf case, the International Court of Justice was prepared to discount this element in the face of widespread practice recognized as obligatory. North Sea Continental Shelf, Judgment, *I.C.J. Reports 1969*, p. 43 (para. 74). See also Arrow, "The Customary Norm Process," p. 3. "All states which fall within the potential reach of the nascent rule must get the opportunity to protest against its emergence.... It will depend on the nature of the acts by which the will of states is explicitly expressed how much time must elapse before it can be presumed that all those with an interest in the matter and who are reasonably alert, have received the message. When the necessary time between the sending and the receiving of the message has elapsed, a reasonable period will still be required in order to make a reaction possible. In this age of fast communication and of frequent, even permanent, conferences, the total minimum necessary time can be much shorter than in the days of the horse and cart." Meijers, pp. 23–24. See also Morenoff, pp. 171–73.

159. Arrow, "The Proposed Regime," p. 371. "The premise ... is easy to illustrate: should, for example, a less developed state have as much input into the law of space exploration as the United States or the Soviet Union? Should landlocked states have an equal voice with coastal states concerning, for example, the delimitation of the continental shelf?" Ibid. These two examples are distinguishable, however. See nn. 170–98 below and accompanying text. In any event, the practice need not be supported by all specially affected states. See n. 186 below and accompanying text.

160. Separate opinion of Judge Ammoun, North Sea Continental Shelf, Judgment, *I.C.J. Reports 1969*, p. 104. Analytically, this consent is expressed through *opinio juris*. Christos L. Rozakis, *The Concept of Jus Cogens in the Law of Treaties* (New York: North-Holland, 1976), p. 59. "Expressions of consent by States are quite varied; they may be overt, vocal, time certain, and otherwise clear in all respects; but they also may be no more than a quiet acceptance of a state of things, or quite ambiguous." David A. Colson, "How Persistent Must the Persistent Objector Be?" *Washington Law Review* 61 (1986): 962. See also Akehurst, p. 53.

161. See nn. 224, 251 below and accompanying text.

162. North Sea Continental Shelf, Judgment, *I.C.J. Reports 1969*, p. 41 (para. 71). "There is no doubt that this process is a perfectly possible one and does from time to time occur: it constitutes indeed one of the recognized methods by which new rules of customary international law may be formed. At the same time this result is not lightly to be regarded as having been attained." Ibid., p. 41 (para. 71). See Sztucki, *Jus Cogens*, p. 75; and also Colson, "The United States, the Law of the Sea, and the Pacific," pp. 71–72. The Court followed the rule set forth in article 38 of the Vienna Convention on the Law of Treaties: "Nothing ... precludes a rule set forth in a treaty from becoming binding on a third state as a customary rule of international law, recognized as such." See Schweisfurth, p. 573; and also Vukas, p. 38. The sources of the norm remain distinct, however. See Caminos and Molitor, p. 889.

At a minimum, acceptance of a particular provision through ratification of the convention would constitute evidence of state practice. See *Rest. 3rd, Restatement of the Foreign Relations Law of the United States*, §102, reporters' note 2; and also Charney, "International Agreements," pp. 978–79. "If fifty States are parties to a treaty that represents itself as reflecting customary international law, the treaty has the same persuasive force as would evidence of the State practice of fifty individual States. Moreover, since the treaty speaks with one voice rather than fifty, it is much clearer and more direct evidence of the state of the law than the conflicting, ambiguous and multi-temporal evidence that might be amassed through an examination of the practice of each of the individual States." Baxter, pp. 277–78. A negotiated agreement such as the Law of the Sea Convention which has not yet entered into force has greater legal significance than formal resolutions such as the Declaration of Principles, but less than acts of states in specific instances. Charney, "International Agreements," p. 994. See also Baxter, p. 292.

163. Fisheries Jurisdiction (United Kingdom v. Iceland), Merits, Judgment, *I.C.J. Reports 1974*, p. 23. See O'Connell, vol. 1, p. 34. It was thus recognized that the written

370 Notes – Chapter 5

product of multilateral negotiations may promote uniformity of state practice, and that official statements made in relation to such negotiations may provide a source of *opinio juris.* Sørensen, p. 147. See also Sinclair, pp. 255–56. Previously adopted resolutions may also be considered as part of a continuous process of normative development. See Agrawala, p. 532. 'The Court's finding of the existence in contemporary international law of a rule to be applied based on resolutions affirmed at international conferences, as evidence of a 'consensus revealed,' must be regarded as a rather interesting development, which can be of relevance also in regard to the text established by UNCLOS III." International Law Association, "The Exclusive Economic Zone: First (Preliminary) Report of the Committee" in *Report of the Sixtieth Conference* (London: International Law Association, 1983), p. 304.

164. Continental Shelf (Tunisia v. Libyan Arab Jamahiriya), *I.C.J. Reports 1982,* p. 38. In 1985 the International Court reiterated as "axiomatic that the material of customary international laws is to be looked for primarily in the actual practice and *opinio juris* of States, even though multilateral conventions may have an important role to play." Continental Shelf (Malta v. Libyan Arab Jamahiriya), *I.C.J. Reports 1985,* p. 13. The modern trend is nevertheless to look to such multilateral conferences, rather than to bilateral diplomacy, for evidence of state practice and *opinio juris.* Sohn, "Customary International Law Developments," pp. 273–74. See also Stein, p. 464.

165. In a 1984 case, the International Court stated that provisions of the 1982 Convention which were adopted by consensus may "be regarded as consonant at present with general international law on the question." Delimitation of the Maritime Boundary in the Gulf of Maine Area, Judgment, *I.C.J. Reports 1984,* p. 294 (para. 94). See also *Rest. 3rd, Restatement of the Foreign Relations Law of the United States,* §102, reporters' note 2. "It seems, therefore, that once a consensus is reached at an international conference, a rule of customary international law can emerge without having to wait for the signature of the convention. Once a convention is signed by a vast majority of the international community, its stature as customary international law is thereby strengthened, as such signatures are a clear evidence of an *opinio juris* that the convention contains generally acceptable principles." Sohn, "Customary International Law Developments," pp. 278–79. See Charney, 'The United States and the Law of the Sea After UNCLOS III," pp. 38–39.

166. North Sea Continental Shelf, Judgment, *I.C.J. Reports 1969,* p. 42 (para. 72). See Rozakis, *The Concept of Jus Cogens,* p. 61. Such norm-creating provisions are more likely to be found in conventions dealing with human rights, for example, than in those relating to taxation and international trade. Charney, "International Agreements," p. 282.

167. Stein, p. 465.

168. E. Brown, "Fundamental Principles in Conflict," p. 536. Arrow, "The Customary Norm Process," p. 10; and nn. 14, 129 above and accompanying text. "To establish a customary rule that prohibits ocean mining, it would be necessary to show an abstention from ocean mining activity and a belief on the part of the abstaining states that the abstention is obligatory." Robert F. Pietrowski, Jr., "International Law Applicable to Deep Sea Mining," *Congressional Record* 125 (December 14, 1979): 36067.

169. See Vukas, p. 48. 'The rule that the effect of an act is to be determined by the law of the time when it was done, not of the law of the time when the claim is made, is elementary and important." Jennings, p. 28.

170. North Sea Continental Shelf, Judgment, *I.C.J. Reports 1969,* p. 42 (para. 73). See n. 159 above.

171. Kronmiller, vol. 1, pp. 294–95, 337. See also Arrow, "The Proposed Regime," p. 379. As a corollary of this doctrine, states opposing unilateral seabed mining are perceived as lacking the capacity to undertake mining activities themselves. See Chap. 4 n. 47 and accompanying text, and n. 174 below and accompanying text; and also Arrow, "The Customary Norm Process," p. 24. Preferential status for such states has been favored on the basis of their superior acquaintance with the subject activities. C. John Colombos, *The International Law of the Sea,* 4th revised ed. (London: Longmans, Green, 1959), p. 7.

172. E.W. Seabrook Hull, 'The International Law of the Sea: A Case for a Customary Approach," Occasional Paper No. 30 (Kingston, R.I.: Law of the Sea Institute, 1976), p. 8. Manganese is necessary for steel production; cobalt for manufacture of jet engines. Marne A. Dubs, "Industrial Interests and the Deep Seabed," in *Current Issues in the Law of the Sea,* ed.

Joyner, p. 143. "[N]ickel and copper are commodities so basic to the continuous functioning of our society as we know it that it would be difficult to describe the state of affairs which would exist in our society and in other similarly situated societies were those commodities to be in short supply or obtainable only at substantially higher prices." Statement of Leigh S. Ratiner in *Status Report*, U.S. Senate, p. 31.

In the 1980 Seabed Act, Congress cited the importance of seabed nodules as an alternative source of supply of minerals for which the United States has become dependent upon imports from other states. 30 U.S.C. §1401 (a) (1)–(6). See also Keith, p. 281. Many of the fears regarding security of access to such minerals are traceable to the 1973 OPEC oil embargo. See Chap. 2 n. 206 and accompanying text; and also Stanley H. Dempsey, "Undersea Minerals," in *Status Report*, U.S. Senate, pp. 365, 368. In the mid-1970s the United States imported more than 70 percent of its annual nickel consumption and virtually all of its manganese and cobalt consumption. U.S. Department of Commerce, *U.S. Ocean Policy in the 1970s: Status and Issues* (Washington: U.S. Government Printing Office, 1978), part 6, p. 25. It has been predicted that seabed mining would fulfill "[a] significant portion of total U.S. demand and U.S. import demand," although such predictions were based upon overly optimistic projections of nodule mining activities. David B. Johnson and Dennis E. Logue, "U.S. Economic Interests in Law of the Sea Issues," in *The Law of the Sea*, ed. Amacher and Sweeney, p. 45. "Five nations control virtually all of the free world's manganese reserves; two nations control two-thirds of the free world's nickel resources; and five nations control virtually all of the free world's cobalt supplies. This, coupled with a lack of domestic supply, increases vulnerability to both price fixing and curtailment." Geddes, p. 621. Wars have been fought over access to such minerals "from the beginning of human civilization." John L. Mero, "A Legal Regime for Deep Sea Mining," *San Diego Law Review* 7 (1970): 488.

173. Dempsey, p. 365. See also Geddes, p. 620. The Treasury Department has indicated that nodules recovered by U.S. nationals would be considered to be of domestic origin. Hull, p. 8. In the early 1970s the U.S. trade deficit for the four nodule minerals was almost $600 million annually. American Bar Association, "Natural Resources of the Sea," p. 674. By the early 1980s the deficit amounted to more than $1 billion annually, and was expected to increase to $6 or $7 billion by the year 2000 in the absence of deep-sea nodule recovery operations, which could make the United States a net exporter of nodule minerals. Geddes, p. 622. See also Arrow, "The Proposed Regime," p. 342. But see n. 176 below. Deep seabed mining could also be expected to save the United States several million dollars per year through a more efficient allocation of resources generally. Johnson and Logue, p. 54.

174. Dempsey, pp. 365, 367–68. See Chap. 4 nn. 47–49 and accompanying text.

175. See Chap. 4 nn. 48, 52–54 and accompanying text. "The level of technology required is not indicated to be of a level unattainable by most nations of the world." Mero, "A Legal Regime for Deep Sea Mining," p. 490. This fact was recognized by the United States as early as 1969. See Chap. 4 n. 47. Several years later it was acknowledged that the U.S. lead in technology was being eroded. David W. Proudfoot, "Guarding the Treasures of the Deep: The Deep Seabed Hard Mineral Resources Act," *Harvard Journal on Legislation* 10 (1973): 603. More recently, development of seabed mining technology by private firms has been retarded by a depressed market for nodule minerals. See n. 179 below and accompanying text. It would seem somewhat anomalous for the United States to claim a special interest in deep-sea mining on the basis of its technological capacity when its own nationals have failed to utilize that advantage. Application of a "specially affected" criterion in this context would undermine the state practice requirement prematurely and arbitrarily by excluding much of the international community from influencing the development of international law. See nn. 169, 191–93 above and below and accompanying text.

176. The expected improvement in the U.S. balance of payments position has been described as "a pleasant side effect." Johnson and Logue, p. 46. Furthermore, it is doubtful that the U.S. interest in an improved balance of payments position is significantly greater than that of other states, if balance of payments is considered as a percentage of gross national product or of population. Projected growth estimates for U.S. demand have generally been lower than those for the rest of the international community. Johnson and Logue, p. 41.

In 1976 the United States imported nearly two thirds of its national fish consumption, adding nearly $2 billion to the balance of payments deficit. U.S. Department of Commerce,

U.S. Ocean Policy in the 1970s, part 2, p. 2. Yet this "special" interest was not sufficient to prevent the expansion of offshore fishing zones which denied U.S. fishermen access to productive foreign fisheries. See Chap. 1 nn. 32–35, 55–57, 61, 68 and accompanying text.

177. In light of the direct economic interest of producer states in mining operations, it is difficult, if not impossible, to credit the United States with a particularly "special" interest in the effect of nodule mining upon international trade flows. See Chap. 4 nn. 172–87 and accompanying text. Moreover, the EEC, representing an aggregate gross national product and population approximately equal to that of the United States, is also "100 percent dependent" upon mineral imports. Hardy, p. 313. The same may be said of many developing states. India, for example, imports all of its nickel and cobalt and more than two thirds of its copper, and demand is expected to increase rapidly with economic growth. Shyam, pp. 327, 340. Even one of the strongest proponents of U.S. pursuit of its national economic interests on the world's oceans has noted that "every state has comparable interests." Comment by Myres S. McDougal, in *The Law of the Sea*, ed. Amacher and Sweeney, p. 159. The United States has itself argued that seabed mining would benefit all states. See *United States of America: Working Paper on the Economic Effects of Deep Sea-Bed Exploitation* (U.N. Doc. A/CONF.62/C.1/L.5, August 8, 1974), reprinted in *Official Records*, UNCLOS III, vol. XVI, p. 166.

178. See Chap. 6 nn. 18–20 and accompanying text. "In one way this dependence is a force for peace and stabilization as it makes all nations more dependent on one another." Mero, "A Legal Regime for Deep Sea Mining," pp. 488–89.

During the 1970s, the United States recognized that the OPEC experience could not simply be translated into a general threat of economic upheaval: "The interest in developing a new source for these metals is not because of threats of shortages, cartel actions, or embargoes. Rather, the interest is in diversifying the sources of these important metals so that the United States will not depend on only one or a few sources for any of them." U.S. Department of Commerce, *U.S. Ocean Policy in the 1970s*, part 6, p. 26. In fact, the OPEC oil embargo may have actually resulted in reduced demand, increased supply, and lower prices for nodule minerals. Joseph Haggin, "Marine Mining to Improve Its Organization, Direction, Financing," *Chemical and Engineering News* (November 18, 1985), p. 64.

179. See Chap. 3 nn. 401–2 and accompanying text; and also Hardy, p. 315. "Furthermore, while increasing consumption of current, conventional resources signals greater promise for seabed materials, it also triggers economic mechanisms that will expand onshore resources (through price effects and discoveries), while moderating consumption (through higher costs, conservation, recycling, and substitution)." James M. Broadus, "Seabed Materials," *Science* 235 (February 20, 1987): 859. In a seemingly desperate attempt to establish the existence of a customary norm validating unilateral mining operations, a State Department official has even argued that "a great deal of money being spent" in preparation for future nodule exploitation is sufficient to constitute state practice. Statement of David A. Colson in *Consensus and Confrontation*, ed. Van Dyke, p. 57. "International law does not necessarily function to insure the economic viability of any particular enterprise. Thus, whether deep seabed mining would proceed or not cannot be made a criteria [sic] for what the law is or ought to be here." Burton, p. 1140 n. 17. See nn. 182–83, 188–98 below and accompanying text.

180. See Chap. 3 nn. 407–12 and accompanying text; and Jaenicke, "Conflicts Between Mine Sites," pp. 514–15 n. 1.

181. See W. Frank Newton, "Inexhaustibility as a Law of the Sea Determinant," *Texas International Law Journal* 16 (1981): 428 n. 346. International law, like national law, developed from a conception of the subjects of the law — states and individuals, respectively — as isolated and mathematically equal entities. See P.H. Kooijmans, *The Doctrine of the Legal Equality of States* (Leyden: A.W. Sijthoff, 1964), pp. 79–80. "The doctrine of legal equality ... arose out of the historical circumstances surrounding the Peace of Westphalia.... It means that, whenever a question arises to be settled by the consent of the Family of Nations, every state is entitled to one vote, and to one vote only; and that the vote of the state politically weak carries as much weight as the vote of the state politically strong." Roxburgh, pp. 100–1. "Every acquiescence in an assumption of rights on the ground of a position of greater political power or on other grounds means acquiescence in infringement of the personality of the state; and this means a negation of its sovereignty." Kooijmans, *The*

Doctrine of the Legal Equality of States, p. 144. See also S.W. Armstrong, "The Doctrine of the Equality of Nations in International Law and the Relation of the Doctrine to the Treaty of Versailles," *American Journal of International Law* 14 (1920): 542. Indeed, on moral grounds, it is difficult to argue that international law should accord preferential treatment to the more powerful states, rather than the opposite. "Notions like 'Great Power' or 'hegemony' are not juridical notions and will never be so without loss to their significance to politics." Kooijmans, *The Doctrine of the Legal Equality of States*, p. 121. See T.J. Lawrence, *The Principles of International Law*, 2nd ed. (Boston: D.C. Heath, 1898), p. 242.

The principle of sovereign equality is reaffirmed in the preamble of the U.N. Charter, as well as in article 2, which requires all member states to "act in accordance with" that principle; although five militarily powerful states have been granted permanent seats on the U.N. Security Council, their permanent membership is associated with a duty (rarely fulfilled) to ensure international peace and security rather than any right of preferential access to natural resources. According "specially affected" status to states deemed financially able to develop seabed mining technology would be analogous to according preferential voting rights to property-holding citizens. Thus it should not be surprising that UNCLOS III delegates rejected efforts by the Reagan administration to secure control of the Council of the Authority for Japan, West Germany, the United Kingdom, and the United States: "That goal is simply not achievable in this day and age." Statement of Tommy T.B. Koh in *Consensus and Confrontation*, ed. Van Dyke, p. 273. "The acceptance of [U.S.] justifications as a basis for ignoring the sovereign equality of states — however valid they may be for purposes of national legislation — could precipitate a serious crisis in international law." Biggs, "Deep Seabed Mining and Unilateral Legislation," p. 248. See Chap. 6 nn. 123–25 and accompanying text; and also Kooijmans, *The Doctrine of the Legal Equality of States*, p. 113. At the very least, any favoritism shown to the more powerful states under international law could be expected to produce subsequent demands by developing states for remedial preferences.

182. North Sea Continental Shelf, Judgment, *I.C.J. Reports 1969*, p. 42 (para. 73). "[T]he doctrine of the continental shelf and its treatment in the International Court of Justice have consistently been based on geographic and geological factors viewed in the context of the land boundaries of states." Jonathan I. Charney, "The Delimitation of Lateral Seaward Boundaries Between States in a Domestic Context," *American Journal of International Law* 75 (1981): 67. Thus a coastal state might be considered to be specially interested for purposes of offshore lateral boundary delimitation (in contrast to landlocked states unaffected by the rule), but only to be "equally interested" with respect to space exploration activities in which it was involved. See n. 159 above. It is therefore incorrect to equate "specially affected" with "vitally interested," as some proponents of U.S. unilateral mining have done. See, e.g., Kronmiller, vol. 1, p. 295.

183. Continental Shelf (Tunisia/Libyan Arab Jamahiriya), Judgment, *I.C.J. Reports 1982*, p. 77 (para. 107). "They are virtually extraneous factors since they are variables which unpredictable national fortune or calamity, as the case may be, might at any time cause to tilt the scale one way or the other. A country might be poor today and become rich tomorrow as a result of an event such as the discovery of a valuable economic resource." Ibid., pp. 77–78 (para. 107). Juridical use of such factors has been similarly rejected within the United States itself. Charney, "The Delimitation of Lateral Seaward Boundaries," p. 53.

184. Fisheries Jurisdiction (United Kingdom v. Iceland), Merits, Judgment, *I.C.J. Reports 1974*, p. 30 (para. 70). In addition to state practice and diplomatic correspondence between the contending states, the Court cited the recognition of such "preferential rights of coastal States" at the UNCLOS II negotiations in 1960. Ibid. (para. 69). See F.V. Garcia-Amador, *The Exploitation and Conservation of the Resources of the Sea*, 2nd ed. (Leyden: A.W. Sijthoff, 1963), p. 3. "This principle rests on the presumption that certain states gain much greater net socioeconomic benefit than others from the enjoyment of prior rights over the resource, and would suffer some inequity if equality of treatment were strictly applied." Douglas M. Johnston, *The International Law of Fisheries* (New Haven, Conn.: Yale University Press, 1965), p. 352. See also O'Connell, vol. 1, pp. 540–41.

185. Fisheries Case, Judgment of December 18th, 1951, *I.C.J. Reports 1951*, p. 133. "Such rights, founded on the vital needs of the population and attested by very ancient and peaceful usage may legitimately be taken into account in drawing a [jurisdictional] line." Ibid., p. 142.

186. Meijers, pp. 7, 16. See also Rozakis, "The Greek-Turkish Dispute," p. 5. "The states which together make up all the potential subjects of the rule do not have to be the same states as those constituting the necessary minimum for the creation of the rule." Meijers, p. 20.

187. See n. 125 above. "The United States [was] virtually isolated" in its view that the three-mile limit remained binding under customary international law. Charney, "The United States and the Law of the Sea After UNCLOS III," p. 45. U.S. efforts to impose sanctions against states enforcing their extended territorial jurisdiction against U.S. ships proved ineffectual. Grolin, p. 15.

188. The concept may have first been put forward by Grotius. Thomas Wemyss Fulton, *The Sovereignty of the Sea* (Edinburgh: William Blackwood and Sons, 1911), pp. 156–57 n. 1. The principle is popularly attributed to the Dutch jurist Cornelius van Bynkershoek, who published *De Dominio Maris* in 1702. George P. Smith, III, *Restricting the Concept of Free Seas* (Huntington, N.Y.: Robert E. Krieger, 1980), p. 21. "It is probable that in the 18th century, the three-mile limit did represent the approximate distance that a cannon could be fired off-shore." A.E. Gotlieb, "The Impact of Technology on the Development of Contemporary International Law," *Recueil des Cours* 170 (1981): 157.

189. See Fulton, pp. 684–87. "The trouble with [the cannon-shot] rule was that it was applicable only in those areas where guns had actually been emplaced on the coast. In addition, it did not produce territorial seas of uniform breadth because the range of cannons varied greatly." Tommy T.B. Koh, "Negotiating a New World Order for the Sea," *Virginia Journal of International Law* 24 (1984): 763. Some authorities tried unsuccessfully to promote the idea that the outer limit of the territorial sea should be determined by "the range of artillery at any particular period," and others, presaging the exploitability clause of the 20th century, defined the limit as three miles "*or the range of cannon.*" Fulton, pp. 688–89 (emphasis in original). Today, of course, intercontinental missiles would render the rule meaningless.

190. See Chap. 1 nn. 83–106 and accompanying text. Use of economically based criteria to determine exploitability would have created even greater legal difficulties. O'Connell, vol. 1, p. 493. See nn. 192–98 below and accompanying text.

Concern had been expressed that the clause would have led to a division of the seabed by coastal states out to the median line. See Gotlieb, p. 168. Alternatively, if the exploitability test had been construed to apply to the technology available to coastal states individually—rather than to the best technology available anywhere at any particular time—a state with adequate seabed mining technology would have been able to claim exclusive rights clear up to the outer edge of the territorial seas of opposite states having no such technology.

191. Pietrowski, "International Law," p. 36066. See also Alexandra M. Post, *Deepsea Mining and the Law of the Sea* (Boston: Martinus Nijhoff, 1983), pp. 5, 76–77.

192. Convention on the High Seas, arts. 2–4, in *United Nations Treaty Series*, vol. 450, no. 6465, pp. 82, 84. The U.S. Seabed Act of 1980 itself encourages negotiation of a treaty which "provides assured and nondiscriminatory access to the hard mineral resources of the deep seabed for *all* nations." Seabed Act, 30 U.S.C. §1402 (b) (1) (emphasis added). "[T]he attempt to reach worldwide agreement on a modern and pervasive law of the sea is itself demonstrative of a recognition that finite ocean resources must be managed with reference to the interests of all nations." Mary Julie Ann Brodd, "A 'Common Heritage' Approach to Fisheries Through Regional Control," *New York University Journal of International Law & Politics* 10 (1977): 189.

193. See n. 175 above and accompanying text. "There is nothing unique about the capacity to engage in deep sea-bed mining. It is an exercise which can be duplicated by anyone willing to spend a certain amount of money on technological development and engineering." Elliot L. Richardson, "The Case for the Convention," in *The 1982 Convention*, ed. Koers and Oxman, p. 6. See also statement of H. Gary Knight in *Deep Seabed Mining*, U.S. House of Representatives, p. 431. Similarly, it has been observed that even the poorest coastal states may obtain financial and technological assistance from other states to conduct fishing within their exclusive econmic zones, and that therefore "any least devleoped country may be considered always to have the *capacity* to harvest the total allowable catch." Shigeru Oda, "Fisheries Under the United Nations Convention on the Law of the Sea," *American Journal of International Law* 77 (1983): 744 (emphasis in original). A decision not to devote scarce national resources to seabed mining operations should not be confused with a genuine lack of

capacity to do so, although such confusion may seem unavoidable where states decline to undertake their own nodule mining operations.

194. Bernice R. Kleid, "Synopsis: Recent Developments in the Law of the Sea 1980–1981," *San Diego Law Review* 19 (1982): 657). See Chap. 3 nn. 291, 340, 406 and accompanying text. India first announced its decision to proceed with deep-sea mining in 1981. See Brewer, "Deep Seabed Mining," p. 4 n. 53. "Plans are to purchase more vessels from France, Denmark, and the Federal Republic of Germany." Post, *Deepsea Mining*, p. 23. See also "Taking a Dive," *Economist* 278 (March 21, 1981): 85.

195. See Chap. 3 nn. 291, 331, 375, 403–4 and accompanying text, and Chap. 4 n. 480. "The number of states that have reasonable expectations of being involved in at least some aspects of deep seabed mining (perhaps by the term of the century) is by no means small, and by no means limited to currently highly industrialized states." Oxman, "The 1976 New York Sessions," p. 252. See also remarks by Cecil J. Olmstead in *Proceedings of the 76th Annual Meeting*, American Society of International Law (1982), p. 119.

196. See Post, *Deepsea Mining*, p. 3. "The high seas are open and free to all states, whether coastal or land-locked." *Rest. 3rd, Restatement of the Foreign Relations Law of the United States*, §521 (1).

197. Convention on the High Seas, art. 2, in *United Nations Treaty Series*, vol. 450, no. 6465, p. 84. U.N. Convention on the Law of the Sea, arts. 124–32, 141, 150 (g). The principle of equal rights of access for such states was also included in the 1970 Declaration of Principles. See text accompanying n. 89 above. UNCLOS III participants "agreed that landlocked nations should share in the exploitation of offshore resources."

198. See Jennings, p. 26. For an example of the simplistic conclusions which economic theorists may be expected to reach when it is wishfully assumed that most states "can be thought of as being indifferent with respect to seabed mining," see Federico Foders, "International Organizations and Ocean Use: The Case of Deep-Sea Mining," *Ocean Development and International Law* 20 (1989): 519–30.

Application of a "most vitally interested" criterion in cases of common resources such as the high seas, where each state has a shared, arguably equal interest, would thus reward states which cultivate import dependencies or which expend huge sums on exotic and economically unjustifiable exploitation technology in order to secure preferential rights to finite common resources. "Some political objectives may be achieved but, from a resource conservation point of view, mankind as a whole will be the loser." Langevad, p. 236.

199. See, e.g., statement of John Quigley in *Deep Seabed Mining*, U.S. House of Representatives, pp. 184–85; Rotkirch, p. 683; and Yakovlev, pp. 80–81.

200. See, e.g., E. Brown, "Fundamental Principles in Conflict," p. 554. See also nn. 2, 15 above and accompanying text. It has also been argued that unilateral mining may occur, but only subject to certain principles embodied in the Convention. Conforti, pp. 3–19.

201. See separate opinion of Judge Ammoun, North Sea Continental Shelf, Judgment, *I.C.J. Reports 1969*, pp. 123–24.

202. Charney, "International Agreements," p. 283. See nn. 91–97, 166 above and accompanying text; and Leslie N. MacRae, "Customary International Law and the United Nations Law of the Sea Treaty," *California Western International Law Journal* 13 (1983): 221. This rule of multilateral regulation has elsewhere been described as "the principle of joint management." Stephanou, p. 269. It is thus significant that the Convention does establish an Authority to allocate exploitation rights with respect to deep-sea resources. McDade, p. 29. See n. 83 above; and also Richardson, "Law in the Making," p. 411. "While a universal convention may not be the only way in which such things can be elaborated, some form of negotiated solution seems indispensable." Brewer, "Deep Seabed Mining," pp. 36–37.

It has been argued that "mining-subject-to-Convention-regulation" cannot be a generalizable norm of international law because a nonparty undertaking seabed mining "would have a lesser legal position with respect to the regulatory regime set up by the Convention than would the nations party to it." Anthony D'Amato, "Editorial Comment: An Alternative to the Law of the Sea Convention," *American Journal of International Law* 77 (1983): 282. Under such a norm, however, mining is simply not permitted outside the framework of a generally accepted multilateral convention — not necessarily the particular regulatory regime set forth in the 1982 Convention, although the general framework of that regime did achieve widespread support at UNCLOS III – and the generalizable norm therefore might better be characterized as "mining-subject-to-multilateral-regulation." See nn. 129–32 above and accompanying text.

Unlike the norm of seabed-mining-subject-to-multilateral-regulation, the principle of equidistance was found by the International Court of Justice, in the North Sea Continental Shelf case, to have been proposed "with considerable hesitation, somewhat on an experimental basis." North Sea Continental Shelf, Judgment, *I.C.J. Reports 1969*, p. 38 (para. 62). The Court also cited "very considerable, still unresolved controversies as to the exact meaning and scope of" the principle, similar to the difficulties posed by the common heritage of mankind concept – but not by the requirement of a generally accepted multilateral regime for the regulation of nodule mining. Cf. North Sea Continental Shelf, Judgment, *I.C.J. Reports 1969*, p. 42 (para. 72). Unlike the rule of multilateral regulation, the equidistance principle was found to be conditional and subject to specific reservation. North Sea Continental Shelf, Judgment, *I.C.J. Reports 1969*, pp. 41–42 (para. 72). Nor are the basic provisions of Section 2 of Part XI, which set forth the requirement of multilateral regulation, subject to amendment. See n. 152 above and accompanying text.

203. North Sea Continental Shelf, Judgment, *I.C.J. Reports 1969*, p. 37 (para. 60). See also Arrow, "The Proposed Regime," p. 378.

204. Unilateral licensing of seabed mining operations under domestic legislation implementing such a regime provides "something that may be called state practices." Statement of Marne Dubs in *Law and Contemporary Problems* 46 (Spring 1983): 90. This practice is not necessarily inconsistent with a norm requiring multilateral regulation, however, since the Seabed Act is designed to implement U.S. seabed mining operations provisionally pending acceptance of a multilateral regulatory regime – a procedure also embodied in Resolution II. See Chap. 3 nn. 253, 290–91, 342–43 and accompanying text, and n. 232 below and accompanying text; and also Kronmiller, vol. 1, p. 452.

It has been argued that the practice established by the four private consortia named as pioneer investors should have special legal significance because of the status of Resolution II. Jaenicke, "Conflicts Between Mine Sites," pp. 508, 511. Yet, under Resolution II, all developing states are also permitted to qualify as pioneer registrants. See Chap. 3 n. 343, and n. 195 above and accompanying text. Moreover, Jaenicke fails to explain how exclusive rights could accrue to pioneer investors who have failed to apply to the Preparatory Commission for registration or to comply with other requirements of Resolution II.

205. Statement of John S. Bailey in *Law and Contemporary Problems* 46 (Spring 1983): 90. See Chap. 3 nn. 384–89, 397–406 and accompanying text, and n. 179 above. The uncertainty engendered by such forbearance may have delayed the formation of a positive rule of law on the subject. Stephanou, p. 270. It has been noted that "the U.S. legal position . . . would appear considerably stronger if the U.S. (and preferably, other mining countries as well) had actually explored and mined the seabed for some time under a framework of domestic law." Richard G. Darman, "Legislative Policy for Deepsea Mining – in the Context of International Law of the Sea Negotiations," *Congressional Record* 125 (December 14, 1979): 36050. The fact that more than half the states in the world consider themselves sufficiently interested in deep-sea mining to participate in the ongoing sessions of the Preparatory Commission must be considered important evidence of state practice, as must the potentially inconsistent issuance of unilateral exploration licenses by the United States, West Germany, and the United Kingdom. The weight of official pronouncements by states participating in the Preparatory Commission would certainly favor the crystallization of a norm of multilateral regulation, the unilateral assertions of the United States notwithstanding. See nn. 209–12 below and accompanying text. It is perhaps not surprising, therefore, that at least one proponent of U.S. unilateral mining efforts has characterized the evidentiary value of such verbal practice as "weak." Kronmiller, vol. 1, p. 354. But see Sinclair, p. 254.

206. D'Amato, "Editorial Comment," p. 284. See also Vukas, p. 33. To the extent that future seabed mining practice occurs "in application of the Convention" *per se*, the North Sea Continental Shelf case might seem to require that "no inference could legitimately be drawn as to the existence of a rule of customary international law"; yet in that same case the International Court recognized the customary status of the continental shelf doctrine codified in the 1958 Continental Shelf Convention, despite the substantial identity of the Convention's

with the 1945 Truman Proclamation. North Sea Continental Shelf, Judgment, *I.C.J. Reports 1969*, pp. 43 (para. 76), 32–33 (para. 47). See statement of John Lammers in *The 1982 Convention*, ed. Koers and Oxman, p. 102; and also Hollick, *U.S. Foreign Policy*, pp. 372–73. Moreover, the better view would appear to be that participation in a treaty regime does constitute state practice. Meijers, p. 15. See n. 155 above. If the Convention were "ratified not only by the States that voted in its favor but also by all maritime powers that abstained at the time of the vote, we would not hesitate to consider that the principle of joint management of the resources of the Area and the rule of the inadmissibility of the exercise of sovereign rights which stems therefrom are part of customary law." Stephanou, p. 273. See also Grolin, p. 22. Indeed, "if participation in the multilateral agreement . . . is truly widespread. . . , state practice dehors the agreement will be unlikely." Charney, "International Agreements," p. 981.

207. Kronmiller, vol. 1, p. 352. See Chap. 3 nn. 397–402 and accompanying text, and nn. 157, 191–98 above and accompanying text.

208. See nn. 78, 85–86, 91–96 above and accompanying text.

209. See nn. 105, 108 above. "The fact that representatives of all UN member states met officially year after year in the Conference and worked out compromises in the form of articles is not devoid of legal significance, even when the guiding principles are the consensus principle and the package deal." Jens Evensen, "Keynote Address," in *The 1982 Convention*, ed. Koers and Oxman, p. xxxii. See Baxter, p. 287; and also Charney, "International Agreements," p. 992. Because of the package deal principle, however, the articles may be regarded as "not necessarily evidence of *opinio juris* at all but of elements of a bargain." Schweisfurth, p. 577. See nn. 145, 148, 157 above and accompanying text; and Caminos and Molitor, p. 885. Officially, the various drafts of the negotiating text were "purely a procedural device," which "do not necessarily reflect the views of States as to the nature of existing or developing law." Kronmiller, vol. 1, pp. 450–51. See also Saffo, p. 495. It has also been argued that the consensus procedure "require[d] a high degree of compromise," and that the Convention is therefore "fraught with ambiguous phrases subject to differing interpretations." Colson, "The United States, the Law of the Sea, and the Pacific," p. 73. These difficulties would seem to inhere in the formative process of customary international law, however, and the International Court of Justice has cited the achievement of negotiated consensus as evidence of *opinio juris*. See nn. 162, 165 above. Many, if not most states did express the opinion that deep-sea minerals are the shared property of the entire international community. Statement of John Quigley in *Deep Seabed Mining*, U.S. House of Representatives, p. 184.

210. UNCLOS III, *Closing Statement by President Koh* (U.N. Doc. A/CONF.62/17, April 30, 1982), quoted in Burke, "Customary Law of the Sea," p. 526 n. 64. See n. 62 above, and nn. 132–33 above and accompanying text.

211. See Chap. 3 nn. 296, 298–99, 319 and accompanying text. "The vast majority of states in the world support Part XI as the *only* auspices and rules under which deep seabed mining may occur." Gamble, p. 789 (emphasis in original). See also statement of John Quigley in *Deep Seabed Mining*, U.S. House of Representatives, p. 182. It must be regarded as particularly significant that the EEC membership believes that deep-sea mining must be regulated by an international Authority. Treves, "The EEC and the Law of the Sea," p. 178. Although the Soviet Union originally adhered to the freedom of the seas position, it subsequently opposed unilateral mining as illegal under international law. See n. 85 above. Despite the failure to adopt the Convention by consensus, the final vote on the entire "package" did facilitate the identification of *opinio juris* with respect to the deep-sea mining regime set forth in the Convention. Schweisfurth, p. 578. See also Yakovlev, p. 81. Acceptance of multilateral regulation by the states specified in Resolution II as potential sponsors of pioneer investors may be of particular importance to the formation of a customary norm. See n. 214 below and accompanying text; and also Jaenicke, "Conflicts Between Mine Sites," p. 511.

212. See Chap. 3 nn. 390–95 and accompanying text, and n. 64 above. The declarations appear to constitute the kind of "authoritative utterances" of *opinio juris* rarely found in traditional diplomatic practice. Cf. Roxburgh, p. 80. The declaration of August 30, 1985, expressly recalled "Article 137 of the Convention which proclaims that no State or natural or juridical person shall claim, acquire or exercise rights with regard to the minerals recovered from the Area except in accordance with Part XI of the Convention." Preparatory Commission, *Declaration Adopted by the Preparatory Commission on 30 August 1985* (U.N. Doc.

LOS/PCN/72, September 2, 1985). See nn. 62, 132–34 above and accompanying text. In the opinion of many commentators, such a "declaration of 'nonrecognition' is sufficient to prevent acts illegal in origin from having law-creating effects." Tucker, p. 331 n. 34.

213. Although the unilateral legislation of Western seabed mining states which signed the Convention is generally consistent with the common heritage of mankind principle and the requirements of the Convention, the national seabed mining acts affirmed mining as a freedom of the high seas and were designed to be simultaneously consistent with an alternative RSA regime pending final decisions on whether to ratify the Convention – a position which is at least potentially inconsistent with article 137. Mashayekhi, p. 243. McDade, pp. 32–33. See nn. 58, 62 above; and also Kronmiller, vol. 1, p. 329. But see "USSR Mining Decree," *Environmental Policy and Law* 9 (1982): 96. The United States and its allies maintained this position throughout the UNCLOS III negotiations. Brewer, "Deep Seabed Mining," p. 35. See n. 86 above and accompanying text. Again, however, it is unclear to what extent the position represents a bargaining strategy. See n. 157 above.

214. Nordquist, "Customary Law Status," pp. 79–80. See Chap. 3 nn. 319, 388 and n. 62 above and accompanying text; and also Biblowit, p. 297; and Gamble, p. 790. All but 3 of the 11 states named in Resolution II as potential sponsors of pioneer investors signed the Convention. The importance of signature is evident in the U.S. decision not to sign the Convention, a decision prompted by a National Security Council conclusion that signing the Convention would have "significant effects" on the legal position of the United States. David L. Larson, "The Reagan Administration and the Law of the Sea," *Ocean Development and International Law Journal* 11 (1982): 311–12. Indeed, all of the states parties to the 1984 Provisional Understanding except the United States have gone to some length to reassure the international community that the Understanding was not inconsistent with the Convention regime. *Oceans Policy News* (September 1984), p. 2. See nn. 63, 120 above; and Larson, "When Will the UN Convention on the Law of the Sea Come Into Effect?" p. 180.

It must nonetheless be regarded as significant that the practice of forbearance by the private U.S.-licensed consortia was not accompanied by *opinio juris*. See n. 205 above. Indeed, forbearance due to unfavorable economic conditions would be no more evidence of *opinio juris* than the desire of navigators to avoid radioactive contamination was evidence of a duty to avoid the nuclear weapons testing zones formerly established on the high seas of the Pacific Ocean, particularly in light of persistent U.S. assertions of high-seas rights. See Note, "Exclusion of Ships from Nonterritorial Weapons Testing Zones," *Harvard Law Review* 99 (1986): 1051. However, any norm not reinforced by state practice may dissipate over time. Statement of John Quigley in *Deep Seabed Mining*, U.S. House of Representatives, p. 189. See n. 249 below and accompanying text.

215. Western seabed mining states might decide not to ratify an unacceptable Convention, freeing them from their existing as signatories to act in a manner compatible with its provisions. Tagliabue, p. 3. See Chap. 3 n. 306 and accompanying text, and nn. 119, 121 above and accompanying text. Yet failure to participate in the Convention for political or economic reasons, or as a result of pressure by the United States, would not indicate a belief that mining could proceed *dehors* a multilateral regulatory regime. See n. 124 above and accompanying text.

216. Jonathan I. Charney, "The Persistent Objector Rule and the Development of Customary International Law," *British Year Book of International Law* 56 (1985): 2. Nordquist, "Customary Law Status," p. 83. See also Kronmiller, vol. 1, pp. 224, 517. "[W]hile a customary rule may indeed be formed on the basis of consent that, though general, does not have to be universal, the scope of the normativity attributable to it once formed will likewise be, though general, not necessarily universal." Weil, p. 434. Persistent objection must occur from the inception of the rule, and must continue during its subsequent application. Meijers, p. 22. Lauterpacht, "Sovereignty Over Submarine Areas," p. 406. See also Akehurst, p. 53; and Arrow, "The Proposed Regime," p. 373.

The International Court of Justice has not specified precisely what positive actions a state must take in order to preserve its status as a persistent objector. Stein, p. 478. It is uncertain, for example, whether the United States would actually have to undertake unilateral seabed mining operations in order to preserve its rights. Colson, "How Persistent Must the Persistent Objector Be?" pp. 963–64. Arguably, it should not be required to do so, although it would

certainly be required to protest efforts to develop a regime of mandatory multilateral regulation. Roxburgh, p. 94. It has been asserted that the United States "probably" has protested
sufficiently to be able to claim persistent objector status with respect to seabed mining. Arrow,
"The Proposed Regime," p. 384. In the absence of further objection, however, acquiescence
may be inferred without a positive act of recognition. See O'Connell, vol. 1, pp. 42–43.

Because it permits states to avoid application of rules of international law, the persistent
objector rule has been described as an "example of the viability of absolute sovereignty." Brian
Smith, "Innocent Passage as a Rule of Decision: Navigation v. Environmental Protection,"
Columbia Journal of Transnational Law 21 (1982): 59. "This means that states are at once the
creators and the addressees of the norms of international law and that there can be no question
today, any more than yesterday, of some 'international democracy' in which a majority or
representative proportion of states is considered to speak in the name of all and thus be entitled to impose its will on other states." Weil, p. 420. See also Arrow, "The Customary Norm
Process," pp. 4–6. In this regard, state sovereignty is limited by the doctrine of *jus cogens*,
whereby persistent objections are rendered ineffectual with respect to peremptory norms,
from which no derogation is permitted. Stein, pp. 480–81. See n. 393 below. The Group of
77 has, in fact, argued that Part XI represents *jus cogens*. See Larson, "A Definition of the
Problem," p. 275. This position was disputed at UNCLOS III, however, and the Convention
only prohibits "States Parties" from being "party to any agreement in derogation" from the
principle of the common heritage of mankind, which remains undefined. U.N. Convention
on the Law of the Sea, arts. 136, 311 (6).

217. *Rest. 3rd, Restatement of the Foreign Relations Law of the United States*, §§102,
comment (i), 523, comment (e), reporters' note 1. See n. 62 above. But see n. 251 below.

218. See Chap. 2 nn. 89, 182 and accompanying text.

219. Statement of the representative of the United States in *Provisional Verbatim
Records*, U.N. General Assembly, First Committee (U.N. Doc. A/AC.1/PV.1590), p. 10.

220. See Kronmiller, vol. 1, pp. 230, 232. The United States nonetheless considered itself
obligated "to give good faith consideration to the Resolution in determining its policies." Letter
from John R. Stevenson, Legal Adviser, Department of State, to Senator Lee Metcalf, January
16, 1970, reprinted in *International Legal Materials* 9 (1970): 832. See Chap. 2 n. 64 and accompanying text. It is unclear to what extent the initial U.S. position toward the Moratorium
Resolution was intended to represent a bargaining position. See nn. 96, 145 above. The nonbinding effect of the Moratorium Resolution was reasserted during the Ford and Carter administrations. Arrow, "The Proposed Regime," pp. 374–75. See also Arrow, "The Customary
Norm Process," p. 35.

221. Statement of the representative of the United States, August 24, 1979, in *Official
Records*, UNCLOS III, vol. XII, p. 14 (para. 33). As of 1978, the United States did not regard
title to deep-sea mineral resources as being vested in the international community. U.S. 95th
Congress, 2nd Session, Senate, Committee on Commerce, Science, and Transportation,
Report: The Third United Nations Law of the Sea Conference (Washington: U.S. Government
Printing Office, 1978), p. 12.

222. Seabed Act, 30 U.S.C. §1402 (a) (1). Again, this position seems to reflect a further
expansion of the claimed prerogative to determine unilaterally which international legal
norms are "recognized by the United States," and goes considerably beyond the traditional formulation of the persistent objector doctrine. See n. 216 above. It would be difficult to find
much utility in any system of law which permits its subjects to observe only those rules which
they themselves have expressly recognized. See nn. 45–47, 198 above and accompanying text.

223. O'Connell, vol. 1, p. 41. See Chap. 3 n. 295 and accompanying text.

224. See Chap. 2 n. 192 et seq. and accompanying text, and Chap. 3 n. 253 and accompanying text. A state's lack of objection, "a function of its stance and the negotiations and
tradeoffs into which it entered," may be construed as acquiescence—i.e., as a waiver of the
right to claim persistent objector status with respect to customary norms emerging from the
UNCLOS III negotiations. Burke, "Customary Law of the Sea," p. 511 n. 13. "The U.S. could
hardly charge that it had insufficient notice of what the Convention would look like—
particularly since its fundamental structure was embodied in a working paper submitted by
the U.S. in 1970 to the Conference." Nordquist, "Customary Law Status," p. 79.

225. Statement of the representative of the United States in *Summary Records*, U.N.

General Assembly, Ad Hoc Committee, Legal Working Group, 2nd meeting (U.N. Doc. A/AC.135/WG.1/SR.2, June 20, 1968), p. 4. See text accompanying Chap. 1 nn. 112, 132. "Similar statements were made by Ambassador Goldberg before the First Committee of the General Assembly in 1967; by President Johnson in his transmittal to Congress of the second annual report of the National Council of Marine Resources and Engineering Development of March 11, 1968; . . . and by Ambassador Wiggins, before the First Committee of the General Assembly, on November 6, 1968." Biggs, "Deep Seabed Mining and Unilateral Legislation," p. 239. The Draft Declaration of Agreed Principles submitted to the Ad Hoc Committee by the United States and other developed states in 1968 proposed that "use of this area . . . be carried on for the benefit and in the interests of all mankind, taking into account the special needs of the developing countries." *Report of the Ad Hoc Committee* (U.N. Doc. A/7230), p. 19 (para. 88). The United States indicated that these draft principles represented "a minimum balanced statement . . . consistent with our own views." Statement of the representative of the United States in *Provisional Verbatim Records*, U.N. General Assembly, First Committee (U.N. Doc. A/AC.1/PV.1590), p. 12. The United States subsequently supported General Assembly Resolution 2467A, which contained virtually identical language.

226. Testimony of Ambassador Elliot L. Richardson, May, 1977, quoted by John Quigley, in *Deep Seabed Mining*, U.S. House of Representatives, p. 184. See also n. 192 above and accompanying text. The Carter administration also acknowledged that the deep seabed is "internationally owned." Statement of Secretary of Commerce Juanita Kreps in *Deep Seabed Mining*, U.S. House of Representatives, p. 484.

227. See Barkenbus, p. 41; and n. 88 above and accompanying text. The 1982 Green Book would have changed the definition of the resources to which article 137 would apply, providing the Authority with jurisdiction over only those nodules "for which rules, regulations and procedures have been adopted." UNCLOS III, *The U.S. Proposals for Amendments to the Draft Convention on the Law of the Sea* (U.N. Doc. A/CONF.62/WG.21/Informal Paper 18, 1982), proposed art. 133 (1). The only other substantive change in articles 133–49 related to the treatment of national liberation movements. See Chap. 4 n. 643 and accompanying text. The April 13 proposal would only have eliminated the requirement that nodule mining occur pursuant to the terms of the Convention (as opposed to some other generally accepted multilateral regulatory framework), while leaving the prohibition on unilateral nodule appropriation intact. UNCLOS III, *Belgium, France, Federal Republic of Germany, Italy, Japan, United Kingdom of Great Britain and Northern Ireland and United States of America: Amendments* (U.N. Doc. A/CONF.62/L.121, April 13, 1982), proposed art. 137 (2), reprinted in *Official Records*, UNCLOS III, vol. XVI, p. 226.

228. See Van Dyke and Yuen, p. 534; and also E. Brown, "Fundamental Principles in Conflict," p. 558. In 1976 Secretary of State Kissinger acknowledged "that the world community should share in the benefits of deep seabed exploitation." Henry Kissinger, "The Law of the Sea: A Test of International Cooperation," *U.S. Department of State Bulletin* 74 (1976): 539. See also n. 295 below and accompanying text.

229. Statement of Elliot L. Richardson in *The Third United Nations Law of the Sea Conference*, U.S. Senate, p. 120. But see statement of Elliot L. Richardson in *Deep Seabed Mining*, U.S. House of Representatives, p. 459. See also President Richard M. Nixon, "United States Policy for the Seabed," in *U.S. Department of State Bulletin* 62 (1970): 737. Even treaty opponents have recognized "that there are some common heritage principles that need to be addressed." Statement of Representative John Breaux in *The 1982 Convention*, ed. Koers and Oxman, p. 16. At the same time, however, the United States has officially maintained that the common heritage principle remains to be defined in a convention. Kronmiller, vol. 1, p. 6 n. 2.

230. Draft Declaration of Agreed Principles, quoted in U.S. Congressional Research Service (George A. Doumani), *Exploiting the Resources of the Seabed* (Washington: U.S. Government Printing Office, 1971), p. 69. The dedication sought was consistent with U.S. policy. See n. 225 above.

231. Arrow, "The Proposed Regime," p. 393. See Chap. 4 nn. 119–20 and accompanying text. The small royalty is payable only for a limited period, and no provision is made for its disbursement. E. Brown, "Fundamental Principles in Conflict," p. 559. See also D'Amato, "Editorial Comment," p. 283. "One could not pretend that reserving a small, unilaterally

decided, portion of the proceeds for developing countries amounted to fulfilling the obligation of exploiting the resources under the regime to be established." Statement of the Chairman of the Group of 77, September 15, 1978, para. 23, in *Official Records*, UNCLOS III, vol. X, p. 103.

"The United States has repeatedly emphasized its commitment to the principle of some type of revenue sharing from deep ocean mining." Statement of Helen Junz, Deputy Assistant Secretary of the Treasury for Energy, Raw Materials and Oceans Policy, in *Deep Seabed Mining*, U.S. House of Representatives, pp. 526–27. The Treasury Department also endorsed revenue sharing as a way "to bolster the principle of the 'common heritage of mankind,'" and "to demonstrate the United States' commitment to this principle." Letter from Henry C. Stockwell, Jr., Deputy General Counsel, to Representative Morris K. Udall, April 30, 1979, reprinted in Kronmiller, vol. 3, p. 109. "Similar legislation reflecting the same acceptance of this norm has been adopted by other states. Such a generalized acceptance of this practice among interested states, when coupled with the Convention's stipulations, provides strong support for the existence of an international rule of law requiring some form of revenue sharing." Surace-Smith, p. 1047. It has nevertheless been asserted that acquiescence to such a norm would only bind the United States to implement a system of revenue sharing, and not to comply with other obligations which may inhere in the common heritage of mankind principle. Arrow, "The Proposed Regime," p. 394.

232. See n. 65 above and accompanying text, and n. 244 below. The interim nature of the legislation may render it difficult to implement a viable alternative regime within the framework of the Act. See Banks, p. 287.

233. Seabed Act, 30 U.S.C. §1402 (c) (1) (D). See Michael R. Molitor, "The U.S. Deep Seabed Mining Regulations: The Legal Basis for an Alternative Regime," *San Diego Law Review* 19 (1982): 603; and also Hauser, p. 30; and Van Dyke and Yuen, p. 499 n. 21. According to one treaty opponent, who may have been concerned about the impact of the delay upon the status of the United States as a persistent objector, the provision was merely "the practical recognition of the fact that 'commercial recovery' . . . [could not] take place, for technological and financial reasons, prior to 1988." Statement of Representative John Breaux in *Congressional Record* 126 (June 9, 1980): 13695. The self-imposed moratorium has been described elsewhere as "confusing" in light of the U.S. position that deep seabed mining is a freedom of the high seas. Molitor, p. 610. Significantly, most of the seabed mining acts enacted by other seabed mining states also prohibited nodule exploitation prior to 1988, as called for in the 1984 Provisional Understanding. See Chap. 3 n. 386 and n. 57 above and accompanying text; and also Keith, pp. 289–90.

234. See nn. 139–48 above and accompanying text. To protect this package deal, the United States in 1980 successfully argued that reservations to the Convention should be prohibited. Caminos and Molitor, p. 875. See n. 151 above.

235. See nn. 62, 214 above and accompanying text. "If mere attendance at an international conference could produce binding effects, no State would be willing to take part in any conference, the concrete results and implications of which are unknown." Separate opinion of Judge Nervo, North Sea Continental Shelf, Judgment, *I.C.J. Reports 1969*, p. 95. Failure of the United States to object to the emergence of particular norms during the Conference might therefore be accorded less weight in determining U.S. acquiescence than, for example, its vote in favor of the Declaration of Principles. See nn. 162–65, 241, 244 above and below and accompanying text. But see Torreh-Bayouth, p. 112.

236. Colombian-Peruvian Asylum Case, Judgment of November 20th, 1950, *I.C.J. Reports 1950*, p. 278. See also Stein, p. 478 n. 62. Clearly, in its opposition to the crystallization of a new norm affecting any existing rights to mine the deep seabed, the United States has gone far beyond simply electing not to participate in the Law of the Sea Convention. See text accompanying n. 223 above.

237. See North Sea Continental Shelf, Judgment, *I.C.J. Reports 1969*, p. 26 (paras. 31, 32).

238. See Arrow, "The Proposed Regime," pp. 386–87; and nn. 91–104 above and accompanying text. It has been argued that the United States has acted in good faith by including revenue-sharing provisions in its Seabed Act. Arrow, "The Proposed Regime," p. 387 n. 337. See also Conforti, p. 9. U.S. participation in the UNCLOS III negotiations and its delaying

of unilateral exploitation until 1988 may provide additional evidence of good faith on the part of the United States. Proudfoot, p. 617. The U.S. position with respect to the legality of unilateral seabed mining has changed since 1970 in a subtle but important way, however, and this shift has occasioned charges of bad faith. See text accompanying Chap. 3 n. 266 and nn. 221, 246–47 above and below and accompanying text. "It should be recalled that it was the United States that initiated a number of important provisions in the regime. . . . But today the United States is opposed to such regulation and contrasts it with the freedom of the high seas." Yakovlev, p. 81.

239. Van Dyke and Yuen, p. 527. See Stephanou, p. 269; and also Zuleta, p. 482. "It is difficult to understand how a country could have approved the Declaration of Principles and be working in the Conference on the preparation of an international regime, and at the same time support initiatives that clearly conflict with the legal framework and the negotiating effort." Orrego Vicuña, "National Laws on Seabed Exploitation," p. 153.

240. Keith, p. 281. See Antrim and Sebenius, p. 80; Conforti, p. 14; and also Surace-Smith, pp. 1045–47. One of the reasons prompting the United States to decline to support the 1974 Deepsea Ventures claim was its recognition that the law of the sea should be developed through UNCLOS III rather than through unilateral claims. Orrego Vicuña, p. 149.

241. U.N. General Assembly, Resolution 2749, paras. 9, 14. See nn. 94, 111–12 above and accompanying text. Unilateral seabed mining operations conducted under national legislation "would fall far short of fulfilling the letter and spirit of the 1970 Declaration." Grolin, p. 21.

242. See Kronmiller, vol. 1, p. 5. "The United States . . . consistently took the position in adopted policy that these resources were to be developed under the auspices of international control." MacRae, p. 220. In 1968 the competence of the U.N. General Assembly to develop a set of principles to govern deep-sea nodule mining was acknowledged by President Johnson and by the U.S. delegate to the Ad Hoc Committee. Biggs, "Deep Seabed Mining and Unilateral Legislation," p. 238. In doing so, they were no doubt influenced by the series of General Assembly resolutions which had succeeded in establishing a set of legal principles to govern the use of outer space. See Chap. 4 nn. 654–56 and accompanying text. According to U.S. Ambassador Arthur J. Goldberg, Resolution 1721, adopted unanimously on December 20, 1961, "set forth the essential legal principles applicable to outer space." Letter of May 9, 1966, reprinted in Charles I. Bevans "Contemporary Practice of the United States: Exploration of the Moon and Other Celestial Bodies," *American Journal of International Law* 60 (1966): 835. See n. 73 above. As early as 1947 "the United States delegation contended that U.N. 'recommendations' have 'virtually the force of law'." Ellis, p. 680. More recently, the U.S. position has emphasized the value of unanimously adopted resolution as evidence of customary international law. Ellis, pp. 680–81. See letter from Stephen M. Schwebel, Deputy Legal Adviser, Department of State, to Marcus G. Raskin, Co-Director, Institute for Policy Studies, April 25, 1975, quoted in Eleanor C. McDowell, "Contemporary Practice of the United States Relating to International Law," *American Journal of International Law* 69 (1975): 863. "The United States Government's acknowledgment that resolutions may have some legal significance is important because the United States has traditionally represented the viewpoints of other developed nations on this issue. It is also important because historically, the United States has been the developed nation with one of the most conservative stands." Ellis, p. 681.

Clearly, it is possible that votes in favor of such resolutions may not represent *opinio juris*. See n. 157 above and accompanying text. But see n. 162 above. To qualify for persistent objector status, however, a protest need not be accompanied by a belief in the invalidity of the usage in question; conversely, acquiescence does not require a belief in the validity of the practice. See n. 216 above; and also Colson, "How Persistent Must the Persistent Objector Be?" pp. 958–59. If the Declaration was not in itself binding, it may nevertheless have acquired the force of law through subsequent state practice if the United States continued to acquiesce. See n. 247 below and accompanying text.

243. Barkenbus, p. 42. See also statement of H. Gary Knight in *Deep Seabed Mining*, U.S. House of Representatives, p. 438. The United States acquiesced in the principle of multilateral regulation, having agreed in 1970 that the seabed mining regime could not result from an agreement among a small number of states. Statement of the representative of the United States in *Provisional Verbatim Records*, U.N. General Assembly, First Committee

Notes — Chapter 5 383

(U.N. Doc. A/C.1/PV.1799), p. 3. See nn. 94, 242 above. It is thus inaccurate to claim that "the United States consistently has repudiated attempts to alter existing international law." Pietrowski, "Hard Minerals," p. 69.

244. Nixon, "United States Policy for the Seabed," p. 738. See Knight, "The Draft United Nations Conventions," p. 495; and also Kronmiller, vol. 1, p. 271. Such statements by a head of state "must be held to constitute an engagement of the state, having regard to their intention and to the circumstances in which they were made." Nuclear Tests (Australia v. France), Judgment of 20 December 1974, I.C.J. Reports 1974, p. 269.

The U.S. vote for the Declaration of Principles must be evaluated in light of this policy, which upheld unilateral deep-sea mining only as an interim right, pending implementation of the kind of grandfather rights subsequently set forth in the system of preparatory investment protection. It was clearly contemplated "that there shall be no rights with respect to the area and its resources incompatible with the regime [to be established] or the principles of this declaration." Statement of the representative of the United States in *Provisional Verbatim Records*, U.N. General Assembly, First Committee (U.N. Doc. A/C.1/PV.1799), p. 3 (para. 23). "The point is that it would be an international regime and we will have to work out in the treaty the extent to which national agencies and national laws are relevant." Statement of John R. Stevenson, Legal Adviser, Department of State, at White House news conference, May 23, 1970, reprinted in *International Legal Materials* 9 (1970): 815.

245. Note presented to U.N. Secretary-General from the United States, June 12, 1970, reprinted in *International Legal Materials* 9 (1970): 834. See n. 65 above and accompanying text.

246. Van Dyke and Yuen, p. 497. See Chap. 2 n. 205 and accompanying text, Chap. 3 nn. 165, 253, 281, 295, 384 and accompanying text, and nn. 65, 221–23, 242 above and accompanying text; and also Arrow, "The Customary Norm Process," p. 29. "Although it continues to give verbal adherence to the principle of the common heritage of mankind, when the United States is pressed it appears to the rest of the world that this support is more directly refuted by more forceful adherence to the high-seas or *res nullius* principles." Barkenbus, p. 40.

247. See nn. 93–94, 162, 202, 241–42, 245 above and accompanying text; and also Oxman, "The High Seas and the International Seabed," p. 540. State practice consistent with the procedural requirements of the Declaration took the form of the Ad Hoc and Seabed Committee negotiations and the near-universal participation in UNCLOS III.

248. See Charney, "The Persistent Objector Rule," pp. 16, 18, 21; and Stein, pp. 457–59.

249. Stein, p. 479. See Charney, "The Persistent Objector Rule," pp. 21–24. The isolation encountered by persistent objector states generally will produce political costs which tend to increase exponentially with the saliency of the issue. Colson, "How Persistent Must the Persistent Objector Be?" p. 967. "It is clear that one State by protesting cannot stand out in perpetuity against the *opinio juris* of the majority of States, and what the protesting State would be seeking to protect would no longer be rights but only interests." O'Connell, vol. 1, p. 44.

The emergence of new multilateral rule-making procedures in recent years has led to the prediction of a possible resurgence of the persistent objector rules as a type of parliamentary maneuver: "A vote, an explanation of position, a refusal to ratify — these are the steps available to the objector today that were not available to the states of the classical era." Stein, p. 467. Yet the rule's lack of resurgence in such an environment argues against its validity; the rule has in this respect been compared to the special status claimed by "specially affected" states. Charney, "The Persistent Objector Rule," p. 23. See nn. 170–98 above and accompanying text.

250. Charney, "The Persistent Objector Rule," p. 11. "The paucity of empirical referents for the persistent objector principle is striking." Stein, p. 459. See also Charney, "The Persistent Objector Rule," p. 21. Despite persistent Soviet objections to the erosion of the doctrine of sovereign immunity, other states enforced restrictive new rules against Soviet diplomatic personnel and foreign agencies. Stein, pp. 460–62, 474. Similarly, the objections of South Africa and Rhodesia against the coalescence of new human rights norms failed to preserve the legitimacy of their systems of apartheid. Charney, "The Persistent Objector Rule," p. 15. See also Stein, pp. 463, 474. "Indeed, as in the other cases . . . , the principle of the persistent objector does not seem to have created sufficient intellectual embarrassment so as to require an effort to explain it away." Stein, p. 463.

251. See Charney, "The Persistent Objector Rule," p. 23; and B. Martin Tsamenyi, "The South Pacific States, the USA and Sovereignty Over Highly Migratory Species," *Marine Policy* 10 (1986): 32–36, 41. During 1986 the United States in effect admitted the legal inadequacy of its position that highly migratory species are not subject to coastal-state jurisdiction within the exclusive economic zone, agreeing to pay 16 South Pacific states $12 million annually to secure access to tuna stocks within their 200-mile offshore zones. *Oceans Policy News* (October/November 1986), p. 1. See *Oceans Policy News* (April 1987), p. 1; and *Oceans Policy News* (March 1988), p. 8. The South Pacific states had "warned that if the U.S. would not agree to adequate compensation for fishing rights, the island nations would look elsewhere." *Oceans Policy News* (August 1986), p. 6. Other coastal states have recognized coastal-state jurisdiction over highly migratory species within the exclusive economic zone, as reflected in the 1982 Convention. Colson, "How Persistent Must the Persistent Objector Be?" p. 967. See U.N. Convention on the Law of the Sea, arts. 56 (1) (a), 61 (1), 62 (1), 64 (2); and Burke, "The Law of the Sea Treaty, Customary Law, and the United States," p. 4. In 1990 Congress itself moved to repudiate the administration's position by including highly migratory species in U.S. exclusive fisheries jurisdiction. See *Oceans Policy News* (June 1990), p. 8. International law "may have some marginal value in this effort but, as in the case of the 12-mile territorial sea and the 200-mile exclusive economic zone, the law will settle and the US can be expected to abandon its positions if they do not prevail." Charney, "The Persistent Objector Rule," p. 24. Indeed, the United States dropped its ineffectual objections to extensions of territorial waters beyond three miles from shore in 1983. See Chap. 1 nn. 10, 21, 25, 80, 82, 136 and accompanying text, and n. 187 above and accompanying text; and also Stein, p. 462. The revised *Restatement*, which reflects the U.S. position, invokes the persistent objector rule on behalf of the United States only with respect to deep seabed mining. Stein, p. 474. See *Rest. 3rd, Restatement of the Foreign Relations Law of the United States*, §523, comment (e), reporters' note 1.

252. Hugo Grotius, *The Freedom of the Seas*, trans. Ralph van Deman Magoffin (New York: Oxford University Press, 1916), pp. 5, 53–54. Grotius observed that "in the legal phraseology of the Law of Nations" the sea had also been referred to as *res nullius* or *res publica* (public property), and that the Romans had called the sea "sometimes common, sometimes public." Ibid., pp. 22, 29 (the translator employed the term "*res nullius*" where Grotius had not himself used the words – cf. original Latin text, ibid., p. 29). See n. 328 below. Grotius's endorsement of the application of the *res communis* principle is thus particularly significant.

Grotius was strongly influenced by principles of classical Roman law. Lawrence, pp. 49–51. See Mark W. Janis, "The Seas and International Law: Rules and Rulers," *St. John's Law Review* 58 (1984): 308; and also H. Lauterpacht, "The Grotian Tradition in International Law," *British Year Book of International Law* 23 (1946): 24. Laws formed through the customary practice of states were regarded by Grotius as subordinate to the law of nature, which God himself could not alter. Ernest Nys, "The Development and Formation of International Law," *American Journal of International Law* 6 (1912): 2. See n. 262 below and also Charles S. Edwards, *Hugo Grotius: The Miracle of Holland* (Chicago: Nelson-Hall, 1981), pp. 104–5.

253. Author's translation. The Latin phrase may also be translated as "property of the community" (of nations), or, more generally, as "common thing."

254. Grotius, pp. 22–23.

255. Ibid., pp. 23–25. "[T]wo categories were made of the things which had been wrested away from early ownership in common. For some things were public, that is, were the property of the people (which is the real meaning of that expression), while other things were private, that is, were the property of individuals." Ibid., p. 26.

John Locke's reasoning was premised on the existence, originally, of a similar "state of nature." John Locke, *Second Treatise of Civil Government*, ed. Lester de Koster, orig. pub. 1690 (Grand Rapids, Mich.: William B. Eardmans, 1978), pp. 17, 26–30.

256. D.F. Scheltens, "Grotius' Doctrine of the Social Contract," *Netherlands International Law Review* 30 (1983): 59.

257. Grotius, pp. 12–13, 29. Conceptually, insofar as it applies to fish and other common marine resources subject to capture, the *res nullius* doctrine may be regarded as subordinate to the governing *res communis* principle applicable to the sea as a whole.

258. "There appears to be nothing truer than what our learned jurists have enunciated, namely, that since the sea is just as insusceptible of physical appropriation as the air, it cannot be attached to the possessions of any nation." Ibid., p. 39. Conceptually, appropriation — by occupation or otherwise — may be viewed as an exhaustion, vis-à-vis other states, of ocean uses. Cf. ibid., p. 31.

259. Ibid., p. 27. See also Chap. 1 n. 9 and accompanying text; and Grotius, p. 34.

260. Grotius, p. 30. "This qualification is deservedly recognized." Ibid. Occupation and possession of the sea itself "could hardly happen without hindrance to the general use." Ibid., p. 31. The dual tenets of inexhaustibility and ineffectiveness of occupation, upon which the Grotian freedom of the seas doctrine was based, are thus properly understood as complementary and mutually reinforcing limitations upon the exercise of territorial sovereignty — "really two facets of a single argument." Newton, "Inexhaustibility," p. 394.

261. Grotius, p. 43. See also Percy Thomas Fenn, Jr., *The Right of Fishery in Territorial Waters* (Cambridge, Mass.: Harvard University Press, 1926), p. 157.

262. Grotius, p. 2. In contrast to the unchanging law of nature, such regulation is to be established consensually by custom or convention, taking into account the extent to which states have reached agreement. Edwards, *Hugo Grotius*, pp. 110–11. See n. 252 above. Grotius was able to allow for operation of a subordinate "law of nations" within his natural-law scheme "by explaining that natural law is unchangeable only with regard to matters which it expressly prohibits and enjoins, but not with respect to those which it merely permits." Lauterpacht, "The Grotian Tradition," p. 11. In the more modern terminology of H.L.A. Hart, *res communis* is a "primary norm," which may be implemented by a "secondary" regulatory regime. See Francesco Francioni, "Legal Aspects of Mineral Exploitation in Antarctica," *Cornell International Law Journal* 19 (1986): 182.

263. See MacRae, p. 185. "For every one admits that if a great many persons hunt on the land or fish in a river, the forest is easily exhausted of wild animals and the river of fish, but such a contingency is impossible in the case of the sea." Vasquez, quoted in Grotius, p. 57. Although a Spaniard, Vasquez opposed the claims of Spain to maritime dominion. Pitman B. Potter, *The Freedom of the Seas in History, Law, and Politics* (New York: Longmans, Green, 1924), p. 51.

264. See Fenn, *The Right of Fishery*, pp. 150–51; and also D. Johnston, p. 164.

265. Gentili cited Andrea Alciati in support of the legal status of the sea as "the common property of all." Alberico Gentili, *De Jure Belli Libri Tres*, trans. John C. Rolfe, orig. pub. 1612 (Oxford: Clarendon, 1933), vol. 2, p. 24. The sea "is by nature open to all men and its use is common to all, like that of the air. It cannot therefore be shut off by any one." Ibid., p. 90. Gentili nevertheless supported a limited form of offshore sovereign jurisdiction. "After Gentilis, it is literally correct to speak of territorial waters in international law." Percy Thomas Fenn, Jr., "Origins of the Theory of Territorial Waters," *American Journal of International Law* 20 (1926): 478. See also Alison Reppy, "The Grotian Doctrine of the Freedom of the Seas Reappraised," *Fordham Law Review* 19 (1950): 276; and C.H. Alexandrowicz, "Freitas *Versus* Grotius," *British Year Book of International Law*, 35 (1959): 180–81.

266. William Welwood, *An Abridgement of All Sea-Lawes*, orig. pub. 1613 (New York: Da Capo, 1971), pp. 57, 71–72. Welwood was the first jurist to articulate an exclusive usufructory right of coastal states to fisheries in their offshore waters based upon the risk of exhaustion from "promiscuous use." Fulton, p. 355. "To Grotius' statement that it was worse to prohibit promiscuous fishing than to forbid navigation, Welwood justly replied that if the free use of the sea is interfered with for any purpose, it ought to be chiefly for the sake of the fishings, if the fishes become exhausted and scarce," as they had become off Scotland. Fulton, p. 354. The "libertie onely to saile on Seas," on the other hand, was "a thing farre off from all controversie, at least upon the Ocean." Welwood, pp. 61–62.

267. John Selden, *Of the Dominion, Or, Ownership of the Sea*, trans. Marchamont Nedham, orig. pub. 1652 (New York: Arno, 1972), pp. 142–43. "Yea, the plenty of such seas is lessened every hour, no otherwise than that of Mines of Metal, Quarries of Stone, or of Gardens, when their Treasures and Fruits are taken away." Ibid., p. 141. See also Fulton, p. 372; and Newton, "Inexhaustibility," p. 390.

At one point, Selden, straining to justify Charles I's pretension to expansive sovereignty over the "British Seas," sought to argue that proof of "inexhaustible abundance" would in itself be "of very little weight. . . . Suppose it be inexhaustible, so that he which shall appropriate it to himself, can receive no damage by other men's using it, what more prejudice is this to the right of Ownership or Dominion, than it is to the Owner of a Fire or Candle, that another man's should be lighted by his? Is he therefore less Master of his own Fire or Candle?" Selden, p. 141. Elsewhere, however, Selden acknowledged "that it in no wise diminisheth from any man's Right or Power, to permit another to participate of what is his, when he himself loseth nothing thereof." Selden, p. 124.

By 1625, several years prior to the publication of *Mare Clausum*, Grotius had revised his theory to reflect the widespread exercise of coastal-state control over use of such offshore waters. "The real problem, he admitted, was to determine how far such coastal prerogatives should extend." Seyom Brown et al., *Regimes for the Ocean, Outer Space, and Weather* (Washington: Brookings Institution, 1977), p. 31. In essence, there appears to have been little disagreement between Grotius and Selden regarding the applicable law where conditions of exhaustibility existed; rather, disagreement centered on the factual question of whether and to what extent conditions of exhaustibility actually existed.

268. Fulton, pp. 546–47. Gerard Malynes was another 17th-century British proponent of this view. Ibid., p. 358. Their disagreement with Grotius was more one of fact than of law, however. See n. 267 above. The contention that the high seas are *res nullius* rather than *res communis* was never accepted. Newton, "Inexhaustibility," pp. 402–3. "For as to the Sovereignty of the vast Ocean, no Man can pretend to it unless he was Lord of the Universe." Alexander Justice, *A General Treatise of the Dominion of the Sea*, 2nd ed. (London: D. Leach, 1709), p. 2.

269. See, e.g., Goldwin, "Locke and the Law of the Sea"; and also Bandow, p. 478. "Such a right to economic freedom arises from the natural right of self-ownership and the necessary corollary right to transfer and trade the fruits of one's own labor." Bandow, p. 480. See Locke, p. 20. See n. 270 below.

270. Locke, pp. 27–28. "So that, in effect, there was never less left for others because of one's enclosure for himself. For he that leaves as much as another can make use of does as good as take nothing at all." Ibid., p. 29. In the pre-industrial era, Locke evidently regarded the supply of land as inexhaustible under natural conditions. See ibid., pp. 29–30. "One of the obstacles to understanding Locke's theory of natural individuation is the predisposition to read 'property' as a term comprising unconditional rights over land and so to equate it with 'private property.' . . . Locke's tenant in common has a use right in his improved land, conditional on his continuing strict use and on his due use of the products." James Tully, *A Discourse on Property: John Locke and His Adversaries*, Cambridge: Cambridge University Press, 1980), p. 124. "Part XI of the 1982 Convention is much closer to pure Lockean theories of property than its ideological opponents seem prepared to admit." Oxman, "The High Seas and the International Seabed," p. 541. "For where there is an authority, a power on earth from which relief can be had by appeal, there the continuance of the state of war is excluded, and the controversy is decided by that power." Locke, p. 23.

271. See Fulton, pp. 559–60. "[S]ince no one can acquire ownership in property absolutely natural, subject to unlimited use, no nation either is allowed to bring under its ownership the open sea, even if that were possible, nor can it acquire ownership of it without contravention of natural law." Christian Wolff, *Jus Gentium Methodo Scientifica Pertractatum*, trans. Joseph H. Drake, orig. pub. 1764 (Oxford: Clarendon, 1934), vol. 2, p. 69. "For the use in the parts of the sea . . . near the shores, which consists in fishing and collection of things produced in the sea, and not in navigation alone, is not inexhaustible, nor is the use which consists in navigation always innocent. . . . Since, therefore, for the sake of advantage nations have occupied portions of the earth, just as individuals have occupied farms, for the same reason it cannot be doubted that nations dwelling on the shores of the sea can occupy portions of the sea, so far as they can protect their ownership over the same." Wolff, vol. 2, p. 72.

272. Samuel Pufendorf, *De Jure Naturae et Gentium Libri Octo*, trans. C.H. Oldfather and W.A. Oldfather, orig. pub. 1688 (Oxford: Clarendon, 1934), vol. 2, pp. 562, 566. See Newton, "Inexhaustibility," pp. 399–400. Inexhaustibility was cited as "[t]he moral reason why ownership is not suitable to the sea" in general. Pufendorf, vol. 2, p. 561. Yet Pufendorf "held with Selden and Welwood that fisheries in the sea might be exhausted by promiscuous use. . . . On this ground, the right of exclusive fishery, and also for the security and defence

of the state, a nation was justified in claiming dominion in the neighboring sea." Fulton, p. 551. See Pufendorf, vol. 2, p. 567, and see also pp. 559, 565–66.

273. M.D. Vattel, *The Law of Nations*, trans. Joseph Chitty, orig. pub. 1758 (Northampton, Mass.: Thomas M. Pomroy, 1805), p. 187. See also Vattel, p. 175; and Kronmiller, vol. 1, p. 139. "Of concern to Vattel appeared to be, quite simply, the exhaustibility of the resource." Kronmiller, vol. 1, pp. 134–35. Consistent with the governing *res communis* regime, Vattel viewed exhaustibility of ocean uses, as well as state security, as justification for an allocation of offshore jurisdiction to coastal states. Newton, "Inexhaustibility," pp. 405–6. See also MacRae, pp. 196, 206.

274. "As each one may find, in a common participation, enough to satisfy his wants, to seize on such things, and exclude others from their use, would be to deprive them, without reason, of the benefits conferred upon all mankind, by the great author of nature." M.D.A. Azuni, *The Maritime Law of Europe*, trans. William Johnson (New York: I. Riley, 1806), vol. 1, p. 7.

Sir Cecil Hurst acknowledged that the high seas were *res communis* and emphasized that the laying of permanent submarine cables on the seabed did not give rise to any rights of appropriation since such cables could be laid on top of one another. Kronmiller, vol. 1, p. 141. See O'Connell, vol. 1, p. 456.

275. See Post, *Deepsea Mining*, p. 99. Gidel, who regarded the sea as *res communis*, distinguished between the essentially inexhaustible use of the high seas for navigation and the depletion of marine resources resulting from unregulated exploitation of the seas as a source of wealth. R.P. Anand, *Origin and Development of the Law of the Sea* (Boston: Martinus Nijhoff, 1983), p. 233. See also D. Johnston, p. 308. "As long as those resources were inexhaustible or were not liable to waste, damage or extermination as a result of the means and methods used to exploit them, any form of regulation would have been bound to be unjustified." Garcia-Amador, p. 212.

276. Myres McDougal, "Commentary," in *The Law of the Sea*, ed. Amacher and Sweeney, pp. 157–58.

277. See text accompanying Chap. 1 nn. 139–40. Although Pardo has stated that, at the time, he believed the deep seabed to be *res nullius*, his 1967 initiative was nevertheless equally consistent with an effort to achieve a particular positive-law allocation of rights to the use — and benefit — of an exhaustible marine resource. See Post, *Deepsea Mining*, p. 100.

278. Goldie, "Title and Use (and Usufruct)," pp. 691–92.

279. Nordquist, "The Law of the Sea Conference," p. 80. Nordquist, adhering to the traditional U.S. position, would permit exclusive use, subject to a duty to pay reasonable (or "due") regard to rights of other high-seas users. Ibid. See n. 8 above and accompanying text; and Jaenicke, "Conflicts Between Mine Sites," pp. 509–10. Use of such a reasonableness standard contributes little to an analysis of the applicable legal regime, however. See nn. 39–47 above and accompanying text. The exhaustibility criterion provides a more concrete and historically relevant basis upon which rights to ocean use may be allocated by positive law when necessary: "If one assumes that change is necessary only when a particular use is exhaustible, then the freedom of such uses as navigation and communication need not be affected even though exploitation of exhaustible living and nonliving resources requires a modification of the traditional norm." Newton, "Inexhaustibility," p. 431.

280. "A State which arbitrarily and without good reason, in rigid reliance upon the principle of the freedom of the seas, declines to play its part in measures reasonably necessary for the preservation of valuable, or often essential, resources from waste and exploitation, abuses a right conferred upon it by international law." *Yearbook of the International Law Commission, 1953*, vol. 2, pp. 218–19. For a discussion of the abuse of rights doctrine, see Bowett, p. 62. "The whole concept of conservation of high seas resources is capable of explanation as a means of limiting a possible abuse of rights." Bowett, p. 44.

281. See Oxman, "The High Seas and the International Seabed," p. 527, and Burton, p. 1175. Under ancient Roman law, the seabed was included with surface waters under a *res communis* regime. Eldon H. Reiley, "Introduction to a Tempest: The Legal, Technological and Political Dimensions of the 1984 Law of the Sea Conference in San Francisco," *University of San Francisco Law Review* 18 (1984): 421–22. This juridical unity was unquestioned for centuries, but with no agreement on a comprehensive legal theory the idea of distinct regimes for

the surface and the seabed emerged in the 1920s "based on the dissimilar uses of the respective areas." Brad Shingleton, "UNCLOS III and the Struggle for Law: The Elusive Customary Law of Seabed Mining," *Ocean Development and International Law Journal* 13 (1983): 38–39. See also Saffo, p. 505. As the discovery of economically valuable mineral resources gave rise to national demands for exclusive seabed exploitation rights during the 20th century, claims to such rights were supported by the doctrine of *res nullius* or, alternatively, by the continental shelf doctrine and the exploitability clause. Pardo and Borgese, p. 74. See Chap. 1 nn. 3, 84–101, 106 and accompanying text, and n. 22 above and accompanying text. Unable to identify the theoretical justification for these national claims, traditional supporters of *res communis* protested ineffectually that recognition of sovereign rights to the seabed and its resources would interfere with the exercise of high-seas freedoms in superjacent waters. Burton, pp. 1152–53 n. 62. Ultimately, the supposed "obvious difference in legal status" between the deep seabed and superjacent waters led to the emergence of the common heritage of mankind concept to govern exploitation of the former area under a separate legal regime. Cf. Group of Legal Experts, *Letter Dated 23 April 1979*, reprinted in Kronmiller, vol. 2, pp. 62–63. "Although the scientists have seen the sea as a biological whole for a long time, policy makers and lawyers still attempt to place their policies and legal drafts into neat compartments." Edgar Gold, *Maritime Transport* (Lexington, Mass.: D.C. Heath, 1981), p. xxi.

Article 1 of the 1958 High Seas Convention defines "high seas" residually, and article 26 expressly permits the laying of submarine cables and pipelines "on the bed of the high seas." Convention on the High Seas, arts. 1, 26 (1), in *United Nations Treaty Series*, vol. 450, no. 6465, pp. 82, 96. See n. 29 above and accompanying text; and also Arrow, "The Proposed Regime," p. 366. The newly emergent exclusive economic zone concept encompasses both the seabed and its superjacent waters, and the UNCLOS III Convention functionally preserves the juridical unity of all parts of the sea, including the seabed. In 1989, the U.N. General Assembly declared with virtual unanimity that the Convention "encompasses all uses and resources of the sea." U.N. General Assembly, 44th Session, Resolution 44/26, November 20, 1989. See Chap. 3 nn. 394–95 and accompanying text. Treatment of all parts of the sea as a juridical unity is ultimately justified – if not necessitated – by a functional analysis based on use. See Don Walsh, "Comment," *Law and Contemporary Problems* 46 (Spring 1983): 167; and also Newton, "Inexhaustibility," p. 370.

282. See nn. 262, 385, 401–2, 405 above and below and accompanying text.

283. See Kelsen, *Principles of International Law*, p. 191. "When it became necessary to subject to legal regulation matters requiring rapid, purposive solutions, states resorted to the contractual mode of law-making." Yemin, p. 1. This process of allocation of scarce ocean uses has been described as "organized community prescription." McDougal and Burke, p. 750. International regulation may be achieved through the observance of stronger international norms, the extension of exclusive national jurisdiction, or the creation of multilateral institutions to administer an allocation of ocean uses. S. Brown et al., p. 9. The benefits of such regulation should exceed its costs, although precise calculation of such costs and benefits is generally difficult, if not impossible. See Wilfried Prewo, "Ocean Fishing: Economic Efficiency and the Law of the Sea," *Texas International Law Journal* 15 (1980): 262. Internationalists have been disappointed that the limited efforts to achieve such a comprehensive regulatory regime have been rejected in favor of an increasing allocation of exclusive zones of national jurisdiction. See Chap. 2 nn. 99–102 and accompanying text; and also Pardo, "Foreword," pp. 515–19. Yet there is widespread support for the view that limited coastal-state management is likely to prove more effective than control by an international organization. P. Chandrasekhara Rao, "Editorial Comment: The UN Convention on the Law of the Sea: Some Reflections," *Indian Journal of International Law* 22 (1982): 465. See generally Ross Eckert, *Enclosure of Ocean Resources* (Stanford: Hoover Institution Press, 1979).

284. See Azuni, vol. 1, p. 15. This consent procedure has been described as "the basic, constitutive process by which states establish the allocation of competence." McDougal and Burke, p. 794. Under ancient Roman law, disposition of common property required the consent of *all* joint owners, a requirement foreign to traditional principles of Anglo-American common law. Arnold, pp. 155–56, 158. Since the Protestant Reformation, the international community has developed a decisionmaking standard of *general* acceptance for the development of international norms. See nn. 155–56 above and accompanying text; and also Azuni,

vol. 1, p. 15. Although diplomatic mechanisms allow the political and economic power of individual states to affect the development of specific regulatory regimes, the principle of sovereign equality ensures that each state retain a legal interest in the outcome. See n. 181 above. "There is ample support in municipal law for the premise that those with co-equal rights in a thing cannot be deprived of those rights without consent." Oxman, "The High Seas and the International Seabed," p. 540. For an unconvincing contention that no common-property rights attach to "a universal common," see Goldwin, "Locke and the Law of the Sea," p. 48.

285. Potter, p. 77. Cf. Selden, p. 22. Like Grotius, Selden acknowledged the primacy of an overriding natural-law regime, but he was more ready to recognize the validity of positive-law regulatory regimes seemingly in contradiction to it. Potter, p. 76. This readiness may be attributed to Selden's greater appreciation of the practical exhaustibility of ocean resources. MacRae, p. 187 n. 33. See nn. 267–68 above and accompanying text.

286. See O'Connell, vol. 1, p. 59. "The *res communis* may not be subjected to the sovereignty of any state, general acquiescence apart." Ian Brownlie, *Principles of Public International Law*, 3rd ed. (Oxford: Clarendon, 1979), p. 181. See also Garcia-Amador, p. 3. In the 1951 Fisheries case, for example, the International Court of Justice noted that "delimitation of sea areas always has an international aspect; it cannot be dependent merely upon the will of the coastal State as expressed in its municipal law." Fisheries case, Judgment of December 18, 1951, *I.C.J. Reports 1951*, p. 132. "A community, as well as every proprietor, has the right of alienating and mortgaging its property, but the present members have never a right to lose sight of the design of these common goods, nor to dispose of them otherwise than for the advantage of the body, or in the case of necessity." Vattel, p. 174.

287. See Martin A. Belsky, "Management of Large Marine Ecosystems: Developing a New Rule of Customary International Law," *San Diego Law Review* 22 (1985): 751–52; and also Mashayekhi, p. 234.

As used herein, exhaustibility refers to exhaustibility of resource use rather than to exhaustibility of the resource itself; conversely, depletion of a resource may not necessarily imply exhaustibility if the resource is renewable. See nn. 310, 443 below and accompanying text.

288. Pardo, "Foreword," p. 515. "The principle of freedom of the seas was explicitly based on the assumption that the living resources of the seas were inexhaustible and that the oceans were sufficiently vast to accommodate all navigational uses without need for regulation. Implicitly, it was assumed that man could not seriously impair the quality of the marine environment and that the oceans were so vast and their uses so limited that serious conflicts of use were impossible." Pardo and Borgese, p. 6.

289. Colombos, pp. 60–61.

290. Garcia-Amador, p. 2.

291. Gerald Fitzmaurice, "The General Principles of International Law," *Recueil des Cours* 92 (1957): 150–51, 160. "The only category of territory which undoubtedly consists of *res communis* is the high seas." Ibid., p. 151.

292. McDougal, "International Law and the Law of the Sea," p. 15. See text accompanying n. 261 above; and also McDougal and Burke, pp. 748–49. Elsewhere, McDougal emphasized the importance of the international regulation of surface navigation, noting that "it is no longer true to say that complete freedom of navigation and fishing cannot possibly interfere with free use of the seas by others." McDougal and Schlei, p. 662.

Normally, the exclusion of competing ocean users by surface vessels will be "virtually momentary." E. Brown, "Fundamental Principles in Conflict," p. 535. Yet economic analysis has demonstrated that the "costs associated with congestion" can necessitate international regulation in the absence of a market mechanism to allocate use of surface space. Francis T. Christy, Jr., "Marine Resources and the Freedom of the Seas," *Natural Resources Journal* 8 (1968): 430. See also Knight, "International Jurisdictional Issues," p. 60.

293. *Yearbook of the International Law Commission, 1956*, vol. 1, pp. 11–13. See Chap. 1 nn. 19–23 and accompanying text. The language was ultimately omitted from the Convention at the insistence of the United States, which prevailed upon other states to include instead the "reasonableness" standard. Burton, pp. 1171–72 n. 161. The reasonableness standard was employed largely out of concern that language such as the "adversely affect" provision would have interfered with the testing of atomic weapons on the high seas. *Yearbook of the Interna-*

tional Law Commission, 1956, vol. 1, pp. 11–13, 32, 60. The latter language was nevertheless retained in the comment to article 2 of the Convention.

294. See, e.g., statement of the representative of Norway in *Summary Records*, U.N. General Assembly, Ad Hoc Committee, Legal Working Group, 6th meeting (U.N. Doc. A/AC.135/WG.1/SR.6, June 26, 1968), p. 10. See also Arrow, "The Customary Norm Process," p. 48 n. 111.

295. See nn. 226–31 above and accompanying text. Even one of the more vociferous proponents of U.S. unilateral seabed mining has acknowledged the "coexistent, indeed transcendent, rights of the world community" in such minerals. Goldie, "Title and Use (and Usufruct)," pp. 713–14.

Defenders of the U.S. position have characterized unilateral deep-sea mining as "another variant of *res communis."* Reiley, p. 423. See nn. 1–2 above and accompanying text; and also Brewer, "Deep Seabed Mining," p. 32. It has even been mistakenly argued that, because a *res communis* regime supposedly would require the consent of the international community even for use of inexhaustible resources, the freedom of the seas doctrine was evidence that the concept had "been rejected with reference to the high seas." Burton, pp. 1174–75 n. 172.

296. See nn. 1–8 above and accompanying text. "Under this theory, the freedom of the seas would be a generic principle analogous in a way to that of economic free enterprise in accordance to which every new activity would be permitted unless specifically prohibited." Biggs, "Deep Seabed Mining and Unilateral Legislation," p. 228.

297. Biggs, "Deep Seabed Mining and Unilateral Legislation," p. 228. See nn. 292, 350 above and below and accompanying text; and also Lauterpacht, "Sovereignty Over Submarine Resources," p. 407.

298. Conforti, pp. 9–10. See Oxman, "The High Seas and the International Seabed," p. 536. During the deliberations of the International Law Commission on the language of the High Seas Convention, Georges Scelle noted that existing and emerging zones of coastal-state jurisdiction over traditional high-seas areas "did not constitute exceptions to the principle of the freedom of the high seas," but rather "constituted limitations of or restrictions on the absolute freedom of the high seas"; other Commission members agreed that free use and limitations thereon were understood to be dual aspects of a single unified regime rather than independent and mutually exclusive legal doctrines. *Yearbook of the International Law Commission, 1955*, vol. 1, p. 59.

299. Newton, "Inexhaustibility," p. 416. Newton nevertheless fails to provide a satisfactory theoretical answer to the question of "how the concept of freedom of the seas is to be accommodated with the concept of special coastal state jurisdiction and with international administration of exhaustible uses." Ibid., p. 431. See also Christy, "Marine Resources and the Freedom of the Seas," pp. 432–33. The solution is to be found in the *res communis* principle, with respect to which both concepts are subordinate. See n. 262 above and accompanying text.

300. See Amin, p. 65; and also Briggs, p. 338.

301. *Report of the International Law Commission to the General Assembly* (U.N. Doc. A/3159, 1956), commentary to art. 27, reprinted in *Yearbook of the International Law Commission, 1956*, vol. 2, p. 278. "Hence, the law of the high seas contains certain rules ... designed, not to limit or restrict the freedom of the high seas, but to safeguard its exercise in the interests of the entire international community." Ibid. See n. 293 above and accompanying text; Garcia-Amador, p. 212; and also Lauterpacht, "Sovereignty Over Submarine Areas," p. 398. Freedom of the seas is thus absolute only insofar as its exercise does not impair free use by others. See Emanuel Margolis, "The Hydrogen Bomb Experiments and International Law," *Yale Law Journal* 64 (1955): 634. See also Legault, pp. 217–18. Otherwise, free access must be restricted, "and there must be provisions under which a user can acquire a right to exclude others from participating directly in the same use." Christy, "Marine Resources and the Freedom of the Seas," p. 433.

302. When an open resource is sufficiently abundant for unlimited use, it will lack any market value and will consequently be available for use by all. Post, *Deepsea Mining*, p. 92. See also Koh, "Negotiating a New World Order," p. 762.

303. S. Brown et al., pp. 23, 109. See United Nations, *Law of the Sea: Report of the Secretary-General* (U.N. Doc. A/44/650, November 1, 1989), p. 9 (para. 17); and also

Newton, "Inexhaustibility," p. 410. "To a very real degree seabed resource exploitation, long thought of as completely compatible with surface navigation, shows itself to be incompatible in part. Fishermen with their trawls occasionally sever submarine cables. The discharge of effluents by merchant vessels causes so much pollution that sport fishermen and bathers are precluded from fully using the ocean. This representative list is exemplary but not exhaustive, and its thrust is magnified as more uses and users of ocean space appear." W. Frank Newton, "Seabed Resources: The Problems of Adolescence," San Diego Law Review 8 (1971): 557. While it is theoretically possible to implement an optimal regulatory scheme which would maximize the aggregate value of competing ocean uses, such a solution would necessarily imply the restriction of particular ocean uses. Kenneth W. Clarkson, "International Law, U.S. Seabeds Policy and Ocean Resource Development," Journal of Law and Economics 17 (1974): 125. Free-rider problems have been associated with voluntary self-regulation efforts among producers. Timothy H. Hennessey, "Multiple Uses of International Marine Resources: Theoretical Considerations," in The Law of the Sea and Ocean Industry, ed. Johnston and Letalik, p. 38. "Management of the commons requires, in Garrett Hardin's words, a 'system of mutual coercion mutually agreed upon.'" Per Magnus Wijkman, "Managing the Global Commons," International Organization 36 (1982): 520. The increasing obsolescence of "the common property system" of unrestricted access was recognized by the Stratton Commission in 1969. U.S. Commission on Marine Science, Engineering, and Resources, Our Nation and the Sea: A Plan for National Action (Washington: U.S. Government Printing Office, 1969), p. 86. See n. 365 below and accompanying text.

304. Prewo, pp. 264, 267–68. See also Francis T. Christy, Jr., "The Distribution of the Sea's Wealth in Fisheries," in Law of the Sea, Offshore Boundaries and Zones, ed. Alexander, pp. 112–13. The optimal yield "is reached when the marginal benefits of a management system that controls excess effort and associated problems equal the marginal costs of implementing the system." E.A. Keen, "Common Property in Fisheries: Is Sole Ownership an Option?" Marine Policy 7 (1983): 199.

305. "This extreme case leads to the most rapid rate of resource use and, consequently, represents the least conservative approach." Prewo, p. 265. The United States has itself recognized that a regime of unrestricted access will lead to the pursuit of "immediate profit" at the expense of environmental protection and marine resource preservation. U.S. Department of Commerce, U.S. Ocean Policy in the 1970s, part 6, p. 1.

306. See Christy, "The Distribution of the Sea's Wealth," pp. 113, 118–20; S. Brown et al., p. 110. It has been claimed, without substantiation, that the capture of economic rent is only important in the case of so-called "common-pool" resources. Richard James Sweeney, Robert D. Tollison, and Thomas D. Willett, "Market Failure, the Common-Pool Problem, and Ocean Resource Exploitation," Journal of Law and Economics 17 (1974): 181 n. 4. See nn. 312, 316 below and accompanying text.

307. Clarkson, p. 126. An efficient regulatory regime will provide for registration, transfer, and enforcement of exploitation rights. Ross D. Eckert, "Exploitation of Deep Ocean Minerals: Regulatory Mechanisms and United States Policy," Journal of Law and Economics 17 (1974): 165. In addition, regulatory mechanisms may include quotas, subsidies, revenue sharing, and court-enforced liability. See Wijkman, p. 519. "Which form of management and ownership is more efficient depends on the relative costs and benefits of extending property rights in each particular case." Wijkman, p. 513.

308. Eckert, "Exploitation of Deep Ocean Minerals," pp. 152–53. Without perfect information, however, identification of the optimal regulatory regime may be impossible, even if the political will for implementation exists within the international community. Clarkson, p. 133. See also Grolin, p. 61. The "process of constitutive decision in which all states participate" has nevertheless been recognized as "the best assurance that the great potential gains of cooperative activity on [a] common resource can be achieved and appropriately shared." McDougal and Burke, p. 748. In fact, market forces are themselves unable to operate effectively in the absence of appropriate regulatory controls. Carlyle L. Mitchell, "Commentary," in The Law of the Sea and Ocean Industry, ed. Johnston and Letalik, p. 575. Failure to implement a regulatory regime to stem the depletion of limited ocean resources may be expected to produce mounting international conflict as large numbers of resource users compete for increasingly scarce common resources. The accelerating rate of depletion which results has

been termed "the tragedy of the commons." See Garrett Hardin, "The Tragedy of the Commons," *Science* 162 (December 13, 1968): 1243–48; and also Hennessey, p. 35.

309. See Mitchell, p. 575; and Chap. 4 nn. 189–92 and accompanying text. Depletion of ocean resources has been compared to depreciation of capital assets. See Prewo, p. 263.

310. D. Johnston, p. 3. See also Sweeney, Tollison, and Willett, p. 183. "Excess effort has come to be recognized as the major and most intractable problem in marine fishery resources management. . . . An equally broad consensus exists that the problem rests in the common property, or open access, aspect of fishery resources." Keen, p. 197.

311. Christy, "Marine Resources and the Freedom of the Seas," p. 429. See also Prewo, p. 265. Competition for resources among individual states is governed by the same economic forces. See David A. Kay, "Operational Aspects of Managing the Oceans," *Columbia Journal of World Business* 10 (1975): 32. Prior to the development of the 200-mile exclusive economic zone, it was estimated that unrestricted access to high-seas fisheries was causing several billion dollars in economic waste annually. Christy, "Marine Resources and the Freedom of the Seas," p. 429.

312. Sweeney, Tollison, and Willett, pp. 182–83. Use of the sea as a sink for disposal of waste materials has created similar problems. See Kent, p. 10. "These common-pool features of resource exploitation may create divergences between private and social costs and benefits, and in the face of resulting 'market failure,' some form of collective agreement among producers or government action may be called for in order to achieve a fully efficient economic outcome." Sweeney, Tollison, and Willett, p. 180. See also Eckert, "Exploitation of Deep Ocean Minerals," pp. 159–61.

Reduced to its essentials, the common-pool argument establishes only that enforcement costs of allocated rights may be prohibitive with respect to certain ocean resources – exclusive exploitation rights "can conceptually be defined, but are uneconomic to implement." Sweeney, Tollison, and Willett, p. 188. See also Wijkman, p. 518. The argument overlooks the significance of the exhaustibility criterion. See n. 316 below and accompanying text.

313. See Mitchell, p. 574; and Hardin, pp. 1244–48. "Whether the good in question is the community's supply of fresh water, recreational parkland, or essential communications and transportation networks, the hope has proved ill-founded that, in the absence of incentives or sanctions imposed by the community, all users would spontaneously act in their long-term, enlightened self-interests and limit immediate consumption in order to convene the long-term supply and quality of the resource." S. Brown et al., p. 17. See Chap. 4 nn. 656–57, 669–72, 694 and accompanying text. Even the air has been subjected to pollution control regulation. See Chap. 4 n. 640; and S. Brown et al., p. 17 n. 2.

The traditional English commons was based on a system of cooperative agriculture. "To the eyes of some historians the cooperative element in the village community appears so strong that they describe it as an agrarian communism; but . . . it is almost as difficult to prove a true agrarian communism as it is to find the modern notion of individual private property in land." Theodore F.T. Plucknett, *A Concise History of the Common Law* (Rochester, N.Y.: Lawyers Co-Operative, 1929), p. 82.

Exclusive rights to open land in the western United States were leased to individual users under President Theodore Roosevelt after growing demand for limited space had resulted in range wars among unrestricted users. Christy, "The Distribution of the Sea's Wealth," pp. 111–12. "Divisions of this sort tend to be the first alternative considered when a common resource becomes scarce." S. Brown et al., p. 7. Even the minimal regulatory regime applicable to hard-mineral exploitation on federal lands was established by an act of positive regulation – the General Mining Law of 1872. See John Lancaster, "The 'Great Terrain Robbery,'" *Washington Post National Weekly Edition*, September 10–16, 1990, p. 6.

314. Barkenbus, p. 139. See U.S. Department of Commerce, *U.S. Ocean Policy in the 1970s*, part 6, p. 24. It was initially estimated that there were 400–500 "prime first-generation" nodule mining sites. Johnson and Logue, p. 40. By the mid-1970s the mining industry had admitted that previously published figures had been overly optimistic. Burton, p. 1144. See Chap. 4 n. 460; and also Robert A. Goldwin, "Common Sense v. The Common Heritage,'" in *Law of the Sea: U.S. Policy Dilemma*, ed. Oxman et al., pp. 63–64.

315. Van Dyke and Yuen, pp. 509–10. See also Gotlieb, p. 160. Like oil and gas deposits, seabed nodules take millions of years to form – a biochemical process which has been

described as "one of the slowest chemical reactions in nature." Eckert, "Exploitation of Deep Ocean Minerals," p. 145. See n. 35 above and accompanying text; and also Eckert, "Exploitation of Deep Ocean Minerals," p. 161. "Nodules can flourish only in areas where bottom disturbances are exceptionally rare and where sediments are growing at a rate equal to or less than the rate of growth of nodules." Eckert, "Exploitation of Deep Ocean Minerals," p. 145. Nodules vary greatly in their composition as well as their concentration. See National Oceanic and Atmospheric Administration, *Report to Congress*, p. 2. In recent years it has been estimated that there are less than two dozen viable mining sites. Goldwin, "Common Sense v. The Common Heritage,'" p. 71. One seabed mining operator has indicated that there are "probably" 3 to 10 suitable sites. J.P. Lenoble, "Polymetallic Nodule Resources and Reserves in the North Pacific from the Data Collected by AFERNOD," *Ocean Management* 7 (1981): 20. The U.S. consortia themselves were unable to avoid overlapping claims to the few available prime mining sites, necessitating a substantial reduction in the size of the exploration areas submitted in their license applications. See Chap. 3 nn. 310–11, 314–15, 396–97 and accompanying text; and also Broadus and Hoagland, p. 561 n. 98.

316. See David B. Brooks, "Deep Sea Manganese Nodules: From Scientific Phenomenon to World Resource," *Natural Resources Journal* 8 (1968): 411, 414. Unlike fish and oil, nodules are a fixed marine resource which may be exploited under a system of allocated exclusive rights, the exploitation of which, it has been mistakenly argued, may be carried out without affecting the rights of other users. See Sweeney, Tollison, and Willett, p. 185. The argument ignores the exhaustibility problem: manganese nodules "are not, inherently, common-pool resources, and yet, like the gravel around the Black Sea, or like a communal forest, the resources would be used uneconomically if they were treated as a commons." Roland McKean, "Commentary," in *The Law of the Sea*, ed. Amacher and Sweeney, p. 110. Arguably, nodule exploitation poses no less of a "common-pool" problem than does oil exploitation, since claim jumping is likely to deprive the operator of a proper return on investment. Cf. Sweeney, Tollison, and Willett, pp. 186–89.

317. See nn. 309, 365 above and below; and also Broadus, p. 858. "The object is to recover more of the investment in any time period by taking advantage of the low cost of additional units of output. Such action can be carried so far that markets are totally disrupted and the resource is quite incompletely recovered." D. Brooks, pp. 409–10. See also Keen, p. 198. Overcapitalization has been encouraged by national security concerns, which have led even Western states to subsidize seabed mining operations. See n. 172 above and accompanying text; and also Oxman, "The 1976 New York Sessions," p. 258.

318. See Saffo, pp. 496–97. "Apart from exclusive rights there is no way to insure that the returns from exploration accrue to the discoverer, hence no way to attract capital to the exploration effort nor any way to prevent the common property dilemma that bedevils fishing." D. Brooks, pp. 413–14.

It has been argued that high entry costs and low transaction and enforcement costs would prevent claim-jumping among "private, profit-motivated" seabed mining operators; "all that is actually required is tacit agreement not to claim-jump." Sweeney, Tollison, and Willett, pp. 185, 192. See also Eckert, "Exploitation of Deep Ocean Minerals," pp. 163, 173–74. Yet the motivation for deep-sea mining is not so simple, and the overlapping claims asserted by pioneer seabed mining operators tend to disprove such hopeful libertarian theories. Similarly, the urgent diplomatic efforts of the United States on behalf of unilateral deep-sea mining by an unproven seabed mining industry belie the assertion by former Ambassador Malone that unexploited nodule deposits are "without value." Cf. Ambassador James Malone, speech at Montego Bay, Jamaica, January 6, 1983, reprinted in *Environmental Policy and Law* 10 (1983): p. 97.

319. See n. 410 et seq. below and accompanying text. Even the 1982 Interim Agreement negotiated by the United States called upon adherents "to insure that adequate areas containing polymetallic nodules remain available for operations by other states and entities in conformity with international law." Agreement Concerning Interim Arrangements, reprinted in *Environmental Policy and Law* 9 (1982): 134.

320. See Chap. 3 n. 409. Deep-sea sulfide deposits have been recognized by the United Nations as "the first known renewable mineral resource." "A New Ocean Treasure?" *Environmental Policy and Law* 10 (1983): 57. If the sulfides can be commercially exploited

without depleting the supply of available deposits, they may be characterized as truly inexhaustible. Under such conditions a regime of unrestricted access should continue to apply.

321. Grotius, p. 23 et seq. See also Selden, p. 18.

322. S. Brown et al., p. 5.

323. See Azuni, vol. 1, pp. 20, 171–72; and also Selden, p. 61. The area of the Mediterranean surrounding Crete had been dominated by the Minoans, and the southern Levant by the Egyptians. D. Johnston, p. 70. See also Reppy, pp. 247–49 nn. 17, 22.

324. Reppy, p. 303. See also Garcia-Amador, pp. 13–14; and Newton, "Inexhaustibility," p. 373. "In other words, did the rulers and peoples of the Near East, the Levant, and primitive Greece, consciously conceive of their actions in terms of maritime dominion, or did they merely obtain and exercise what we should call maritime dominion, without themselves recognizing it as such? Beyond this comes the question whether such attempts to secure and maintain dominion were regarded by other peoples as proper, legitimate, legal, or not." Potter, p. 14. See also Lawrence, pp. 29–30. Because no conceptual system of international law had yet developed, the answers to such questions point to an exercise of little more than *de facto* "dominion." Potter, p. 15. Even Selden, while proffering myriad examples of maritime dominion, acknowledge that "some of the most eminent" lawyers of antiquity were of the opinion that the sea was "perpetually and necessarily common to all men." Selden, p. 150.

325. Maritime control was exercised by Athens, Lacedaemon, Thebes, and Macedonia. Azuni, vol. 1, pp. 39–40. See also Reppy, p. 248 n. 20. Ancient Greek jurisprudence still lacked any concept of property or dominion, however. Newton, "Inexhaustibility," p. 378. See also Percy Thomas Fenn, Jr., "Justinian and the Freedom 'of the Sea," *American Journal of International Law* 19 (1925): 718. Thus, for example, the Athenians sought to safeguard the ships of other Greek city-states against pirates and against Spartan and Persian attack, yet Athens "did not lay down any formal rules of legal right respecting sea dominion, but regarded [it] as, in the main, a matter of military and commercial power." Potter, pp. 23–25. See also Anand, *Origin and Development*, p. 11; and Fenn, *The Right of Fishery*, p. 7.

326. See Albert Hyma, *A Short History of Europe, 1500–1815* (New York: F.S. Crofts, 1929), p. 38. In addition to the eastern Mediterranean, Alexander also exerted control over the Red Sea and part of the Indian Ocean. Justice, p. 11. Reppy, p. 248 n. 21. By pursuing the ideal of a single world state, Alexander may have helped pave the way for subsequent development of a functional system of international law. See J. Westbury-Jones, *Roman and Christian Imperialism* (London: MacMillan, 1939), pp. 67–68.

327. Gordon W. Paulsen, "An Historical Overview of the Development of Uniformity in International Maritime Law," *Tulane Law Review* 57 (1983): 1068. See also Anand, *Origin and Development*, p. 11. Although the Rhodians "established a maritime empire more real and absolute than that of any other nation of antiquity," they worked to assure freedom of navigation through a system of public and private laws. Azuni, vol. 1, p. 32.

328. Fenn, *The Right of Fishery*, p. 28. Arvid Pardo, "The Law of the Sea: Its Past and Its Future," *Oregon Law Review* 63 (1984): 7. Because of the status of Marcianus as a jurisconsult, this classification amounted to an official pronouncement; there is also evidence that the sea had previously been regarded as *res communis*. See Fenn, "Justinian and the Freedom of the Sea," p. 716. Ulpian, writing at the end of the second century, agreed with Marcianus. Fenn, *The Right of Fishery*, p. 28.

The confusion between the *jus naturale* and the *jus gentium* which emerged in later Roman law has been attributed to Gaius. See Fenn, "Justinian and the Freedom of the Sea," p. 727 n. 59. "The Roman legal professors thought that this law of nations [*jus gentium*] was to be identified with the law of nature of the Stoics. In this idea they were wrong, because the law of nations was nothing more than a common set of rules for commerce, while the law of nature was something ethical, spiritual and eternal." Westbury-Jones, p. 150. Properly understood, things classified as *res communis* "are common by the *ius naturale*, but their use is subject to the *ius gentium*." Fenn, "Justinian and the Freedom of the Sea," p. 720. See nn. 252, 262 above and accompanying text. Believing the *res communis* regime to be part of the *jus gentium* rather than the *jus naturale*, some jurists have asserted that the sea could be subject to appropriation under the positive law of nations, and that the sea had in fact been so appropriated by Rome. See, e.g., Alexandrowicz, p. 172. Others have argued that the sea was *res publicae* – a part of national law applicable only to Roman citizens – and that foreigners therefore did not enjoy equal rights of fishing and navigation. See Potter, pp. 31–32; and also Gold, p. 13. Such distinctions based upon citizenship found no support among Roman jurists, however. Fenn, "Justinian and the Freedom of the Sea," p. 725. The ambiguity nevertheless persisted and ultimately fueled the dispute between Grotius and Selden, both of whom sought to justify their positions by citation to classical Roman law. MacRae, p. 185. See Grotius, pp. 34–35; Selden, pp. 81, 89–90, 151–52, 206, 209, 247; and also Newton, "Inexhaustibility," pp. 376–77, 381–82; and Fenn, *The Right of Fishery*, pp. 20–21. Because a true system of international law in the modern sense was still lacking, Roman jurists were never faced squarely with the need to define the precise legal character of the sea. Lawrence, pp. 31–32. See also Fenn, *The Right of Fishery*, pp. 27–28; and Potter, p. 35.

329. Post, *Deepsea Mining*, p. 89. See also Welwood, p. 64. In classifying the sea as *res communis*, Gaius recognized that it was common by the law of nature but that its use may be regulated by the positive-law *jus gentium*. Fenn, "Justinian and the Freedom of the Sea," p. 720. This is consistent with the Grotian *res communis* regime. See n. 262 above and accompanying text. The sea was also identified as *res communis* by the jurists who produced Justinian's official compilation of Roman laws in the sixth century A.D. Fenn, *The Right of Fishery*, pp. 12, 17. See also Newton, "Inexhaustibility," p. 380.

The importance of the inexhaustibility of common resources had been noted in the first century B.C. by Cicero: "Whatever can be granted without inconvenience should be bestowed even upon a stranger." Gentili, vol. 2, p. 91. According to Ulpian, Roman law permitted the building of structures on the seashore "if no man sustain damage," and Celsus agreed that piles could not be "cast into" the sea if "the use of the shore or Sea may be that means become the worse." Selden, pp. 87–88. See also Grotius, p. 31. Such interference with other users of the sea and its resources was rare at the time, however. See Clarkson, p. 119. Because they could be exploited without interference with other users, fish, sponges, and other self-replenishing movables were treated individually as *res nullius* — a classification consistent with the governing *res communis* regime. See nn. 4, 260–62 above and accompanying text.

330. Fenn, "Justinian and the Freedom of the Sea," p. 723. "The works of Cicero, Seneca, Paulus and Ovid express this belief as being a matter of common opinion shared by all men." Fenn, *The Right of Fishery*, pp. 5–6. See Fenn, "Justinian and the Freedom of the Sea," pp. 726–27. There is no record of any claim to maritime dominion in Roman law. Fenn, *The Right of Fishery*, p. 5. As Selden acknowledged, even after the fall of the western Empire, "the opinion of Ulpianus for a perpetual community of the Sea was so entertained as authentic by the Lawyers of the Eastern Empire, that there was no Law in force among them whereby an adjacent Sea might be made appropriate, or any man be debarred the liberty of Fishing by the Owner of such Lands as bordered thereon." Selden, pp. 95–96.

331. R.P. Anand, "Maritime Practice in South-East Asia Until 1600 A.D. and the Modern Law of the Sea," *International and Comparative Law Quarterly* 30 (1981): 447. "The influence of the Asian practices of inter-state conduct on the early development of European international law by classical jurists has been lost in the pages of political history written during the colonial period, or ignored because of later political developments when practically all the Asian states lost their international personality and identity." Anand, *Origin and Development*, pp. 5–6.

332. Vattel, p. 189. These claims have been asserted for more than 2,000 years. Arrow, "The Customary Norm Process," p. 14. Over the centuries Australia, Malaysia, Ireland, Tunisia, Panama, and Venezuela have also claimed exclusive rights to sedentary fisheries in shallow offshore waters where unrestricted access might have resulted in the exhaustion of valuable renewable resources. See E. Brown, "Fundamental Principles in Conflict," p. 527. The threat of exhaustion thus gave rise to a positive allocation of exclusive rights of exploitation based upon contiguity, effective occupation, and prolonged usage. See generally Cecil J.B. Hurst, "Whose Is the Bed of the Sea?" *British Year Book of International Law* 4 (1923–24): 34–43. The exclusive right to exploit these sedentary fisheries is in the nature of a usufruct, involving no claim to the seabed or to the superjacent water column. E. Brown, "Fundamental Principles in Conflict," p. 529.

The doctrinal source of the usufruct has remained uncertain among international jurists who have failed to appreciate the significance of Vattel's reference to exhaustibility. See O'Connell, vol. 1, pp. 452–53; and also E. Brown, "Fundamental Principles in Conflict," p.

526. Some jurists believed that the seabed areas where the sedentary fisheries were located were simply *res nullius*, subject to sovereignty by effective occupation, a view which Great Britain had officially rejected by the early 19th century. O'Connell, vol. 1, p. 516. See Kronmiller, vol. 1, pp. 139, 142, 168–70; and also Arrow, "The Proposed Regime," p. 356. Disregarding the contiguity and long usage requirements, supporters of unilateral deep-sea nodule mining have nevertheless cited the ancient pearl and sponge fisheries in support of their contention that seabed mining is a high-seas freedom. Burton, p. 1155 n. 73. See, e.g., Pietrowski, "International Law," pp. 36065–66; and also n. 25 above and accompanying text. All harvestable sedentary fisheries are, of course, now governed by the territorial sea and continental shelf regimes. See Gotlieb, p. 159.

333. Lawrence, pp. 35, 39. See also Robert Ward, *An Enquiry Into the Foundation and History of the Law of Nations in Europe from the Time of the Greeks and Romans to the Age of Grotius* (London: A. Strahan and W. Woodfall, 1795), vol. 1, p. 201. Germanic tribes overran the Empire in the fifth century, destroying not only its military power and political authority but much of Roman culture as well. See Ward, vol. 1, pp. 212–13. Elements of the Roman vulgar law were assimilated into the relatively primitive legal systems of the new European nations. See Edgar Bodenheimer, "The Influence of Roman Law on Early Medieval Culture," *Hastings International and Comparative Law Review* 3 (1979): 9, 15–16. Charlemagne tried unsuccessfully to reestablish the full authority of classical Roman law. John Ayliffe, *A New Pandect of the Roman Civil Law* (London: Thomas Osborne, 1734), p. xxxvii. Classical Roman law survived in the eastern Empire until the ninth century. Ayliffe, p. xxxiii.

334. Fenn, *The Right of Fishery*, p. 37. The Digest of Justinian received particular attention from legal scholars. Bodenheimer, p. 24. See also Ayliffe, p. xliii.

335. Otto Gierke, *Political Theories of the Middle Age*, trans. Frederic William Maitland (Cambridge: Cambridge University Press, 1938), pp. 75–76. See also Kooijmans, *The Doctrine of the Legal Equality of States*, pp. 45–46. "The church represented a tie to the civilized, unitary past and survived the empire as the one institution seemingly capable of promoting universal values and of imposing some measure of order throughout the politically shattered West." Edwards, *Hugo Grotius*, pp. 75–76. See also Westbury-Jones, pp. 27, 30. As Christianity spread throughout the known world, international bonds were formed which would lead ultimately to the emergence of the first true system of international law. Westbury-Jones, p. 200. See also James Brown Scott, *The Spanish Origin of International Law* (Washington: Georgetown University School of Foreign Service, 1928), p. 103.

336. See Lapidoth, p. 13. The laws of the Germanic tribes which had divided the Roman Empire among themselves made no mention of any right of free navigation — or of any restriction thereon. Fenn, *The Right of Fishery*, p. 50. "For centuries the sea was to remain an area of 'no-law,' by general juridical consent." D. Johnston, p. 159. As Selden put it, "the Dominion of the Sea ... returned unto the Natives." Selden, p. 248. Yet the *Basilicus* of the ninth century upheld the freedom of the seas doctrine. Fenn, *The Right of Fishery*, pp. 30–31. Even in the 14th century, the sea was identified as a common resource by Ange de Ubaldis. Ernest Nys, *Les Origines du Droit International* (Brussels: Alfred Castaigne, 1894), p. 381. "Legislation in the sense of lawmaking was foreign to mediaeval thought. Law was interpreted, not created; discovered, not made. ... When the sea was declared to be by the law of nature incapable of becoming the object of private property, the matter was closed." Percy Thomas Fenn, Jr., "Origins of the Theory of Territorial Waters," *American Journal of International Law* 20 (1926): 466. Freedom of the seas continued to be practiced in the Indian Ocean until Portugal achieved domination in the late 15th century. Anand, *Origin and Development*, pp. 19–20, 33–34.

337. Arrow, "The Customary Norm Process," p. 11. See Fulton, pp. 59–61, 83–84, 175–76. Although feudal law related only to ownership of land, and made no provision for use of the sea, the practice gradually emerged among coastal landowners of excluding others from waters along the shore. D. Johnston, pp. 159–60. The practice was sanctioned by medieval monarchs in whom "jurisdiction" — but not dominion or sovereignty — was said to reside, such waters being considered *res publicae* ("public" property). Clarkson, pp. 119–20. See Fenn, *The Right of Fishery*, pp. 52, 143–44, 156–57. There were, however, conflicting theories regarding the source and scope of this new form of offshore jurisdiction. Fenn, *The Right of Fishery*, p. 45.

The first recorded departure from the traditional freedom of fishing dates from the early 10th century. Garcia-Amador, p. 14. See also D. Johnston, p. 74. Near the end of the 14th century Baldus Ubaldus included exclusive offshore jurisdiction in the regalia of medieval monarchs. Fenn, *The Right of Fishery*, pp. 77–79, 105–10. "The doctrine of the sovereignty of the seas, as a general concept, really dates from the publication of Bidin's treatise on sovereignty in 1582, when he wrongly ascribed to Baldus the idea that governmental power was exercisable over shipping within sixty miles from the coast." O'Connell, vol. 1, pp. 2–3. Yet the freedom of fishing was reiterated in the *Sachsenspiegal*, "one of the most important of the feudal law books." Fenn, *The Right of Fishery*, p. 75. In any event, monarchical claims to full dominion were still barred by the natural-law *res communis* principle which remained applicable to the sea as a whole, and the exclusive *fructus* was limited to fisheries in coastal waters close to the shore. Fenn, *The Right of Fishery*, pp. 67, 77–78. See Fulton, p. 142.

338. Fenn, *The Right of Fishery*, pp. 91–92. See also Fulton, p. 66. Other early English jurists agreed. Fulton, p. 539. Bracton did, however, assert the king's right to control fishing in the waters surrounding Britain. Potter, p. 45.

339. An ordinance enacted in 1201 at the behest of King John, who was "intoxicated with pride" following a naval victory over the French, required foreign ships to lower their sails in salute upon the British navy's demand. Azuni, vol. 1, pp. 125–26. Fulton, pp. 6–7. "A great deal was made later of . . . the ordinance of John, as proving that the Angevin or Plantagenet kings possessed the sovereignty of the sea; but beyond the jurisdiction in question, which doubtless was exercised in the Straits of Dover and perhaps in the Channel when the coasts on each side were in possession of the crown, there is a lack of evidence to prove that any claim of the kind was made." Fulton, p. 8. The specific requirements of the salute remained uncertain, and there is little evidence of its enforcement prior to the 15th century. Fulton, pp. 43, 206. Although Elizabeth I, one of the first monarchs to enunciate a freedom of the seas doctrine, had expressly rejected Lord Plowden's argument that British jurisdiction extended as far as the middle of the surrounding seas, sovereignty over the full extent of such seas was asserted under James I, who was particularly concerned with the seeming assertion by Grotius of a Dutch right to fish freely in the waters off Scotland, where foreigners had been required to obtain permission in order to exploit valuable coastal fisheries. Fulton, pp. 9–10, 57, 83–86, 118, 293, 361–63. See Selden, pp. 443–44; and also Louise Fargo Brown, *The Freedom of the Seas* (New York: E.P. Dutton, 1919), p. 19. At the request of Charles I, Sir John Burroughs searched the records in the Tower and discovered the 13th-century ordinance, which was interpreted as historical proof of England's absolute sovereignty over the British Seas. Fulton, pp. 10–11. "No doubt more was read into its use of the word *possessio* in connection with the sea than the medieval draftsman intended." O'Connell, vol. 1, p. 5. England nonetheless sought to expand the scope of its maritime jurisdiction by imposing a license requirement upon foreign fishermen and exacting a tribute from vessels desiring the protection of the royal navy, as well as by enforcing the flag salute. Fulton, pp. 283–85, 289–90, 292–93, 309–11, 381, 478. Yet even Charles was unable to enforce the "extraordinary pretensions" set forth by Selden in *Mare Clausum*. Fulton, p. 11. See Selden, p. 181 et seq. "John Selden was to demonstrate what a powerful weapon the appeal to custom and usage could be, in the hands of a clever man, to justify a claim not only contrary to the received law, but at variance with the facts." Fenn, *The Right of Fishery*, p. 95. See also Fulton, pp. 26, 372–73. Britain's claim to dominion was asserted in less absolutist terms by Charles II, and the flag salute was officially ended in the 18th century. Fulton, pp. 14–15.

Selden argued that the British claim was justified by immemorial prescription. Lawrence, p. 170. The sporadic and ineffective enforcement of the claim called into question the validity of this justification, however. Fulton, pp. 20–21, 32–33. See n. 343 below and accompanying text. Significantly, Britain asserted no jurisdiction over disputes occurring within the British Seas among foreign flag vessels. Fulton, p. 525. Moreover, the boundaries of these Seas were never definitely fixed, the area of British sovereignty being vaguely and variously described as including the Channel, the North Sea, the Bay of Biscay, and some portions of the oceans west of Ireland and north of Scotland. Edward Hall, *A Treatise on International Law* (Oxford: Clarendon, 1924), p. 179. See also Fenn, *A Right of Fishery*, p. 194. "It is remarkable that . . . no authoritative definition was ever given of the extent of sea included in the term." Fulton, p. 15. See also L. Brown, pp. 18–19.

340. In 1177, in gratitude for Venetian military assistance, Pope Alexander III instituted a ring ceremony whereby the sea was symbolically "subjected to" Venice, "as the wife to her husband." Reppy, p. 250 n. 25. See also Azuni, vol. 1, pp. 77–79. In 1263 Venice began levying a toll on ships navigating the upper Adriatic Sea, and the protests of Ancona were rejected by Pope Gregory X on condition that the Venetians defend navigation in the region against attack by Saracens and pirates. Reppy, p. 250 n. 25; and Selden, pp. 102–4. See Azuni, vol. 1, pp. 79–80; and also Justice, pp. 25–27. The practice was maintained by force and reinforced by treaties, and it lasted until Napoleon's conquest of Venice in 1795. See Fulton, pp. 3–4, 567. Italian jurists of the post-glossator period did not ascribe the tribute to a formal claim of territorial sovereignty of the type exercised over gulfs and bays, but rather to "imperium" or to jurisdiction acquired by prescription and exercised to a distance of 60 to 100 miles for the purpose of safeguarding commercial navigation against hostile attack. Fenn, "Origins of the Theory," pp. 468–69, 478. See D. Johnston, p. 164; and also Fulton, pp. 539–43. "The claim to ownership of the sea itself seems to have been a later development, and was not readily accepted, if indeed it ever secured general consent." Fenn, "Origins of the Theory," p. 469.

Genoa claimed similar prescriptive rights in the Ligurian Sea, levying tolls on navigation until its naval power declined in the 16th century. See Newton, "Inexhaustibility," p. 382; Fulton, p. 4; and Reppy, pp. 250–51 n. 26. The earliest tolls were collected by Pisa and Tuscany in the Tyrrhenian Sea. Newton, "Inexhaustibility," p. 382. In exchange for toll revenues, these city-states generally were required to police their seas against pirates. Margolis, p. 632.

Denmark excluded foreign vessels from the Atlantic Ocean between Norway and Iceland, and exacted "sound dues" from shipping traffic in the Baltic Sea. See Newton, "Inexhaustibility," p. 383; and also Fulton, p. 108. "As early as the fourteenth century Denmark was levying tolls on vessels passing the straits, on the ground that she kept those waters free from pirates, and later that she maintained lighthouses to mark the passage." L. Brown, p. 24. See Charles E. Hill, *The Danish Sound Dues and the Command of the Baltic* (Durham, N.C.: Duke University Press, 1926), pp. 8, 10–11, 14; and also Justice, p. 30. Significantly, Denmark's exclusive claims to the northern Atlantic were prompted by the need to regulate competition for limited ocean fisheries. See Fulton, p. 339; and n. 337 above and accompanying text. Norway also claimed rights of access in these waters, and in the 17th century Sweden acquired control over the Bothnian Gulf. Newton, "Inexhaustibility," p. 383.

341. See n. 464 et seq. below and accompanying text. With a papal grant of exclusive navigation rights, Portugal explored the African coast in the late 15th century, reaching India in 1498. See Newton, "Inexhaustibility," p. 384. Portugal's claims were challenged by Spain as well as by other European maritime states. See Christopher Bell, *Portugal and the Quest for the Indies* (New York: Harper and Row, 1974), pp. 100–1, 133; and also Bailey W. Diffie and George D. Winius, *Foundations of the Portuguese Empire, 1415–1580* (Minneapolis: University of Minnesota Press, 1977), pp. 151–52. Portuguese control over European access to the Indian Ocean developed quickly, replacing the freedom of navigation which had been generally observed. See Hyma, pp. 25–26; and Anand, "Maritime Practice in South-East Asia," pp. 444–48. See n. 369 above. Spain was soon shipping large amounts of silver and gold from its own New World colonies, and territorial disputes between the two Iberian kingdoms produced several boundary adjustments. See nn. 6, 383 above and below and accompanying text.

342. Fulton, p. 541.

343. See ibid., p. 3 et seq. Many jurists attributed the exclusive claims to immemorial prescription, in derogation from the general rule of freedom of the seas. Fenn, *The Right of Fishery*, pp. 116, 143–44. Yet the popular theory of acquisitive prescription was attacked by prominent jurists who recognized that its application was inconsistent with a natural-law *res communis* regime, including Vasquez, Gentili, and Grotius. Justice, pp. 7–8. Gentili, vol. 2, pp. 91–92. Grotius, pp. 49–50. See n. 406 below and accompanying text.

Selden cited "a *Libel* published of old, or a *Bill of complaint* instituted" in the time of Edward I as evidence that "very many foreign Nations" had acknowledged English dominion in the sea. Selden, p. 398. Significantly, however, Britain's right to exact a salute, or otherwise to exercise sovereignty in the "British Seas," was challenged by the French and the Dutch, neither of whom ever acknowledged Britain's claim. Fulton, pp. 112, 212, 276. Vattel, p. 191. Azuni, vol. 1, pp. 119, 129–30. See also L. Brown, p. 19. "Danes and Swedes are on record as having refused it, and international incidents arising from the refusal of French captains occurred until England relinquished her claim in the Napoleonic wars.... After heroic struggles in the first Dutch war, Holland yielded the point but not the principle.... The Dutch always maintained that it was done as an act of courtesy, not as a recognition of British sovereignty over any part of the sea. Dutch ambassadors emphasized the point by offering to salute in any part of the world." L. Brown, pp. 22–23. See also Fulton, pp. 396, 436, 479–81, 495. The licensing requirement was never enforced against French or Spanish fishermen. Fulton, p. 150. France's brief effort to enforce its own century claim of sovereignty in the Channel ended when the French edict, sanctioned by the Pope but challenged by German merchants, "was declar'd void and of no effect in Law, in a Notable Case . . . before an Assembly of the States . . . held at *Tours* by *Henry* IV." Justice, p. 31. See also Fulton, p. 212.

The claim of Venice to sovereignty in the Adriatic has been described as "the most successful in terms of recognition by other states." Newton, "Inexhaustibility," p. 382. See Garcia-Amador, p. 15. In fact, Venetian claims to sovereignty, like those of ancient Mediterranean city-states, were founded upon military force rather than rule of law. Vattel, p. 191. See n. 340 above. Venice did not even find it necessary to attempt to justify its claims in legal terms until the 16th century, when it sought to invoke immemorial prescription. Fenn, *The Right of Fishery*, pp. 89–90, 224. See also Fulton, p. 351. Bologna and Ancona brought unsuccessful military challenges to Venetian naval supremacy. Hall, p. 180. The French ambassador, on at least one occasion, formally protested that the Adriatic Sea "ought rather to be common to all nations." Azuni, vol. 1, p. 81 n. 94. Spain also opposed Venice's claims, as did the Hanseatic League. See Justice, pp. 9, 32; and also Fulton, p. 107. "The situation seems to be that while the Venetians had the naval power to assert their claims, foreign nations could only accept them under protest; and that these claims declined in cogency with the decline of the power of Venice. The same general situation was true in the case of Genoa." Fenn, *The Right of Fishery*, p. 57. See also Reppy, p. 250. Unlike Venice, however, Genoa was unable to control Spanish navigation. Georges Pages, *The Thirty Years War 1618–1648*, trans. David Maland and John Hooper (New York: Harper and Row, 1970), p. 23.

The exercise of exclusive rights in the northern Atlantic by Denmark and Norway caused "a great many disputes" with Britain, which sought access to the fisheries of those waters, and later with Holland, which was supported by Britain, France, and Sweden. Fulton, pp. 108, 529. See L. Brown, pp. 23–25. Danish claims to sovereignty in the Baltic were opposed by Holland, Sweden, and the Hanse towns. Hall, p. 183. See also Lester Bernhardt Orfield, *The Growth of Scandinavian Law* (Philadelphia: University of Pennsylvania Press for Temple University Publications, 1953), p. 4. Efforts to assert such sovereignty "formed, indeed, one of the causes of the war by Gustavus Adolphus against Germany." Fulton, p. 377. The sound dues, Denmark's "last vestige of dominion over the Baltic," were abolished in the 19th century as a result of diplomatic pressure by Russia, Brandenburg-Prussia, and the United States. Hill, pp. vii–viii.

Portuguese claims to sovereignty were contested by Spain, even after treaties of demarcation had been confirmed by papal bulls. See Azuni, vol. 1, pp. 100–1; and L. Brown, p. 7. Waters off the west coast of Africa were plied regularly by French and other traders during the 16th century, despite Portuguese efforts to monopolize trade routes in the southern Atlantic. Garrett Mattingly, *Renaissance Diplomacy* (Boston: Houghton Mifflin, 1955), p. 182. "Foreign interlopers paid as little attention to the papal demarcation line of 1493 as they had paid to previous bulls granting Portugal exclusive rights south of Cape Bajador." Mattingly, p. 182. Significantly, no effort was made to restrict navigation or fishing by native populations in waters off the coasts of Africa and Asia. Grotius, pp. 60, 68. See also James A. Williamson, *Hawkins of Plymouth*, 2nd ed. (New York: Barnes and Noble, 1969), p. 6. Holland challenged the claims of Portugal, and by the middle of the 17th century most Portuguese trading stations in the Indian Ocean had been seized by the Dutch East India Company. Lyle N. McAlister, *Spain and Portugal in the New World 1492–1700* (Minneapolis: University of Minnesota Press, 1984), p. 300. The French and the British followed. See *European Treaties Bearing on the History of the United States and Its Dependencies to 1648*, ed. Frances Gardiner Davenport (Washington, D.C.: Carnegie Institution of Washington, 1917), p. 7.

Spain's exclusive claims in the western Atlantic were opposed by France, Britain, and Holland, each of which sponsored expeditions to the New World. Azuni, vol. 1, p. 94. "Not only did the French corsairs plague the fleets and oversea settlements of Spain, but, as indicated by the voyages of Jacques Cartier, they were bent on establishing themselves on the mainland of America." *European Treaties*, ed. Davenport, p. 3. In 1496 Henry VII of England granted John Cabot and his sons "full and free authoritie, leave, and Power, to sayle to all Partes, Countryes, and Seas, of the East, of the West, and of the North, under our banners and ensignes . . . to seeke out, discover, and finde, whatever Iles, Countryes, Regions, or Provinces . . . before this time have been unknowen to all Christians." William Wood, *Elizabethan Sea-Dogs* (New Haven, Conn.: Yale University Press, 1920), pp. 4–5. Although such official expeditions were largely abandoned by the English after 1505, they paved the way for exploitation of the Newfoundland fishery by French, Portuguese, "and almost certainly English" mariners in the early 16th century — "the first transatlantic commerce that was not under the control of the Spanish monarchy." David B. Quinn, *England and the Discovery of America, 1481–1620* (New York: Alfred A. Knopf, 1974), pp. 130–31. When Queen Elizabeth was confronted with Spanish protests against renewed incursions by English privateers, she responded that the sea was by its nature common and open to free use by all, and the raids continued. Azuni, vol. 1, pp. 130–31. Fulton, pp. 105–8, 338. See nn. 429–31 below and accompanying text. By the turn of the century, the Dutch had also begun to challenge Spain in the East Indies. Reppy, p. 255. See n. 433 below and accompanying text. "Thus by the middle of the seventeenth century the two Iberian powers were compelled to admit other nations to trade . . . in those overseas regions which they had hoped to monopolize." *European Treaties*, ed. Davenport, p. 7.

344. Medieval Europe formed "an harmoniously articulated Universal Community whose structure from top to bottom was of the federalistic kind." Gierke, p. 95. See ibid., pp. 3–4, 18–19; and also Kooijmans, *The Doctrine of the Legal Equality of States*, pp. 46–47. This community was wholly Christian. See P.H. Kooijmans, "Protestantism and the Development of International Law," *Recueil des Cours* 152 (1976): 92. "The Roman Catholic Church was by far the mightiest institution of the Middle Ages. Not only did it aim to control the spiritual destiny of every man, woman, and child in central and western Europe, but through the operation of its courts, its seven sacraments, and the exercise of its enormous temporal power, it surpassed the national and feudal governments in wealth and social prestige." Hyma, pp. 49–50. See also William Holdsworth, *A History of English Law* (London: Methisen, 1956), vol. 1, pp. 581–82.

The supreme authority of the Pope was acknowledged — and frequently invoked — in medieval treaties. Ward, vol. 2, p. 54. See Fenn, *The Right of Fishery*, p. 61. "[N]o feature in the history of Europe is more striking than that vast and frequent assemblage of all the Sovereigns of Christianity, or their Representatives, in what were called the Ecumenical Councils. In these, many things were settled exclusive of mere points of faith; more particularly the precedency of Nations, the rank and power of Sovereigns, and not infrequently their right to their Thrones themselves; points which it is palpably the province of the Law of Nations to determine." Ward, vol. 1, pp. 142–43. As head of the Church and purveyor of the canon law, the Pope functioned as "a sovereign legislator whose authority it was heresy to question." Holdsworth, vol. 1, pp. 582–83. Thus, the Pope had the authority to allocate exclusive rights to *res communis* resources threatened with exhaustion. See Fenn, *The Right of Fishery*, pp. 43–44. Cf. Lawrence, p. 30. See also n. 284 above and accompanying text. As Selden put it, the "Imperative Law of Nations" was observed "in obedience to the *Pope's* Authority and command." Selden, p. 15 (emphasis in original).

Theoretically, the temporal and spiritual authority of imperial Christendom was divided between the Holy Roman Emperor and the Pope, respectively. The actual power exercised by the Emperor failed to match his temporal responsibilities, and his authority was rejected by many of the newly emergent states of Europe — including England, France, and Spain — which were prepared to assert their national sovereignty in opposition to the established system of medieval feudalism but were nevertheless generally willing to continue to respect the authority of the Pope. Kooijmans, *The Doctrine of the Legal Equality of States*, pp. 47–49. Leo Gross, "The Peace of Westphalia, 1648–1948," *American Journal of International Law* 42 (1948): 30. See also Gentili, vol. 2, pp. 112–13; and Fenn, *The Right of Fishery*, pp. 99–100. "From Gregory VII onwards the Popes and their supporters are unanimous in holding that, so far as the substance is concerned, the Temporal as well as the Spiritual Power belongs to the Chair and Peter, and that the separation which is commanded by divine law affects only the Administration, not the Substance." Gierke, pp. 107–8 n. 13. See Holdsworth, vol. 1, pp. 581–82. Although papal authority in the east was lost to the Greek Orthodox Church in the 11th century, the Pope's control over Western Europe remained largely intact several hundred years later when excommunication still remained a potent sanction. See Ward, vol. 2, p. 60; and also Hyma, p. 51. By the 16th century, the temporal authority of the Pope had begun to decline. Newton, "Inexhaustibility," p. 384. See nn. 436–38 below and accompanying text; and Grotius, p. 16.

345. See nn. 436–40 below and accompanying text; and Ward, vol. 2, pp. 77, 334–35. "By 1560 most of the basic tenets of Protestantism had been asserted, and the changing circumstances dispelled once and for all any lingering notion that Christendom was a monolithic unity directed from Rome." Edwards, *Hugo Grotius*, p. 82.

346. O'Connell, vol. 1, p. 13. See, e.g., Justice, pp. 3–4. But see nn. 267–72 above and accompanying text. Portugal had relied on such a theory, as well as on papal donation, to support its claims to sovereignty. J. Williamson, p. 5. Medieval jurists even theorized that the claims of Venice and Denmark were justified by those states' exercise of territorial sovereignty over shores surrounding the sea. H.A. Smith, *The Law and Custom of the Sea* (London: Stevens and Sons, 1950), p. 43. Fulton, pp. 15–16. In fact, most such maritime claims were motivated by a desire to protect commercial navigation against pirate attack and to secure trading monopolies. H. Smith, pp. 43–44. See also Morenoff, p. 140 n. 8.

347. See Georges Pages, "The War as a Dividing Point Between Medieval and Modern Times," in *The Thirty Years' War*, ed. Theodore K. Rabb (Lexington, Ky.: D.C. Heath, 1972), pp. 33, 40; Richard A. Falk, "Introduction: The Grotian Quest," in Edwards, *Hugo Grotius*, pp. xv–xvi; Gross, pp. 28–29; and Lauterpacht, "The Grotian Tradition," pp. 16–17. Protestant theology recognized no single authority competent to allocate international rights and responsibilities, as the Pope had done in the Middle Ages. See P.H. Kooijmans, "How to Handle the Grotian Heritage," *Netherlands International Law Review* 30 (1983): 90. Absent such authority, international law was now made, adjudicated, and enforced by sovereign states, acting individually and collectively. See Janis, "The Seas and International Law," pp. 307, 309–10; and also Lawrence, pp. 34–35. "Each state was absolutely independent of any external human authority, and as a corollary all were equal before the law which Nature and common consent imposed. This is the fundamental doctrine of modern International Law." Lawrence, p. 48.

348. Kooijmans, "How to Handle the Grotian Heritage," p. 90. "The conscience of Christian rulers was expected, not totally without reason, to substitute for an enforcement mechanism and to allow an international order that lacked central institutions still to avoid . . . relapse into barbarism." Richard A. Falk, "The Quest for World Order: The Legacy of Optimism Re-Examined," *Dalhousie Law Journal* 9 (1984): 140. The concept of an international order was maintained by early Protestant reformers, including John Calvin, who attributed all secular authority to the supreme sovereignty of God. Kooijmans, "Protestantism and the Development of International Law," pp. 96–98. This concern for world order eroded with the spread of colonialism — and of the atomistic philosophy of Thomas Hobbes, which ultimately developed into the extreme version of positivism which subordinated the authority of the international community to the sovereignty of individual nation-states. Kooijmans, *The Doctrine of the Legal Equality of States*, pp. 75, 91–92. See also Anand, *Origin and Development*, p. 135. Ironically, the doctrine of unlimited state sovereignty has been rejected by modern Protestantism. Kooijmans, "Protestantism and the Development of International Law," p. 112.

349. See O'Connell, vol. 1, p. 30. See nn. 160, 284 above and accompanying text. "When a number of equal and independent states no longer own, even in theory, a common superior, the most obvious mode of escape from utter lawlessness in their mutual dealings seems . . . to be the regulation of their conduct towards one another by rules to which all have assented." Lawrence, p. 39. The conceptual and doctrinal similarity of 17th-century juristic thought to the UNCLOS III deliberations has been noted. Sanger, p. 11.

350. D. Johnston, p. 315. "The emerging notion was that use of the high seas would remain open to all states but the right of access would be subordinated to the general interest of all nations as regulated by the international community." Orrego Vicuña, "National Laws

on Seabed Exploitation," p. 140. The freedom of the seas had become "axiomatic" doctrine by the 19th century. Margolis, p. 633. See Fulton, pp. 523, 554–55; and also Lapidoth, pp. 15–16. "The larger claims disappeared, and those only continued at last to be recognised which affected waters the possesion of which was supposed to be necessary to the safety of a state, or which were thought to be within its power to command." Hall, p. 189. See also Colombos, pp. 55–56.

The unrestricted use of high-seas areas was reinforced by convention and judicial decision, as well as state practice. See MacRae, p. 205; and also Newton, "Inexhaustibility," p. 429. In asserting its right to fish in waters off Newfoundland, the United States formally declared that the sea by its nature was not susceptible to appropriation. L. Brown, pp. 106–7. The freedom of the seas doctrine was formally recognized in the 1856 Treaty of Paris. Hollick, *U.S. Foreign Policy*, p. 5.

351. See Hollick, *U.S. Foreign Policy*, p. 389. To ensure minimal interference by belligerent states with neutral shipping, rules were developed to cover the capture of contraband, the right of visit and search, and the imposition of maritime blockades. Rules were also developed for the promotion of navigational safety during peacetime. See Anand, *Origin and Development*, p. 151. These rules "may be said to form a 'common law of the sea, adopted by common consent of States.'" Colombos, p. 291. By the middle of the 20th century, the capacity of the sea to absorb vessel-source pollution had become a matter of international concern, giving rise to further navigational restrictions. Gotlieb, p. 193. See n. 374 below and accompanying text.

"In the sixteenth and seventeenth centuries the struggle for freedom of the seas was essentially a struggle for freedom of commerce." L. Brown, p. 27. See Azuni, vol. 1, pp. 113–15. Yet, in the late 18th century, the United States was still seeking universal acceptance of the principle of unrestricted navigation, which John Adams regarded as a natural condition. See L. Brown, p. 85. During the 19th century, Britain used its position as the world's dominant maritime power to enforce freedom of navigation on the high seas. See Anand, *Origin and Development*, pp. 126, 129–30, 133–34; and also Reppy, p. 244.

352. Koh, "Negotiating a New World Order," p. 762. See Fulton, p. 21; and also Gary M. Shinaver, "Fishery Conservation: Is the Categorical Exclusion of Foreign Fleets the Next Step?" *California Western International Law Journal* 12 (1982): 159. Grotius acknowledged that the assertion of exclusive jurisdiction over such limited offshore areas might be permissible. Potter, p. 68. See n. 267 above. Cf. Grotius, p. 31. Gentili, building upon the works of earlier jurists, had formulated the original concept of the territorial sea. Reppy, p. 278. See n. 265 above. "The unsettled questions in the eighteenth century were how far from shore the territorial sea extended, the nature of any right of passage through straits, and the nature of any right of passage through another state's territorial seas." Newton, "Inexhaustibility," p. 408. See Fenn, *The Right of Fishery*, p. 222. Selden had, in effect, advocated recognition of a territorial sea limited only by the capacity of maritime states to maintain effective control through the exercise of naval power. See Potter, p. 91. "Not only did the extravagances of its friends lay the doctrine of maritime dominion open to ridicule, but they also warned nations what serious inconveniences would flow from acquiescence in such a doctrine." Potter, pp. 89–90. See also Fenn, *The Right of Fishery*, pp. 168–69. The widespread acceptance of Bynkershoek's cannon-shot rule during most of the 18th century was thus based in part upon a recognition of technological limitations to effective maritime occupation, as well as upon considerations of exhaustibility and coastal-state security. Potter, p. 91. See n. 188 above and accompanying text; C. Wolff, vol. 2, p. 184; and Fulton, pp. 473, 549, 557, 565–66. The three-mile limit was suggested by Galiani in 1782 as the approximate distance of a contemporary cannon shot, and became widely accepted in the early 19th century with the backing of Great Britain and the United States. See Fulton, pp. 579–80, 658, 661, 663–64; and also Gold, p. 149. The Iberian and Scandinavian states had adopted territorial zones of slightly greater width several decades earlier. Fulton, pp. 552, 568–69, 664, 667, 669, 681.

353. Vattel, p. 189. See nn. 267–68, 337 above and accompanying text; Pufendorf, vol. 2, p. 562; and also C. Wolff, vol. 2, p. 72. In the 17th century, foreign fishermen depleted stocks by destroying the "spawn and brood" of fish on the British coast. Fulton, pp. 604–5. "By 1700, the idea was current that fisheries are exhaustible, and so were not, like navigation, susceptible of common use." O'Connell, vol. 1, p. 510. See Fulton, pp. 711–13, 738–39. The

exclusive jurisdiction exercised by coastal states within their territorial waters was thus described by Paul Fauchille as a "right of conservation." O'Connell, vol. 1, p. 73.

Various resource management schemes are available to protect exhaustible marine resources within territorial waters. See Hennessey, p. 41.

354. The adjacent sea "furnishes a means of protection to maritime countries, and therefore it is to the advantage of the inhabitants that no one should be allowed to remain there with armed ships." C. Wolff, vol. 2, p. 72. An allocation of offshore jurisdiction for this purpose is conceptually consistent with a natural-law *res communis* regime, since use of specific offshore areas for legitimate military purposes related to national defense may be viewed as an exhaustible use insofar as military operations by foreign vessels are necessarily excluded. See n. 296 above and accompanying text. Protection of peaceful navigation was thus cited as justification for the assertion of a 300-mile security zone in the Western hemisphere during World War II. Reppy, p. 283. See Chap. 1 nn. 24, 35 and accompanying text. The Truman Proclamation cited "self-protection" as one justification for the continental shelf doctrine proclaimed by the United States several years later. Presidential Proclamation No. 2667, reprinted in Hollick, *U.S. Foreign Policy*, p. 391. "A nation may appropriate things, where the free and common use of them would be prejudicial or dangerous." Vattel, p. 189. See also Fulton, p. 601.

355. Although unlimited territorial sovereignty was initially claimed, by the 19th century it had become clear that the ability of foreign vessels to navigate territorial waters peacefully could not be left to the discretion of coastal states. See B. Smith, pp. 53, 55, 83. "The problem of rationalizing a right of passage through the territorial sea found a quick solution in the disengagement of the concepts of territory and jurisdiction.... Territory ceased to be regarded as something owned, and came to be regarded as a spatial area within which the faculties of sovereignty could be exercised." O'Connell, vol. 1, pp. 61–62. See also Fenn, *The Right of Fishery*, pp. 130, 134; and Potter, p. 89.

The jurisdictional claims of the Middle Ages from which the territorial sea evolved had not traditionally been justified as an exercise of territorial sovereignty. Fenn, "Origins of the Theory," pp. 467–68. See nn. 337, 339–40 above; and also Fulton, p. 538. These medieval claims had, in fact, been asserted largely for the purpose of preventing foreign vessels from engaging in harmful activities. Pufendorf, vol. 2, p. 564. Gentili, who recognized the *res communis* character of the sea, was careful to limit his concept of the territorial sea by a right of innocent passage based upon both natural law and contemporary state practice. Reppy, p. 278. See n. 265 above; and Selden, pp. 124–25. As the doctrine of the territorial sea became current among newly independent European states anxious to maximize their national authority, the importance of innocent passage as a limitation apparently became obscured, only to be rediscovered when the juridical defects of absolute sovereignty became evident. "The reason is clear, because, inasmuch as one and the same thing is susceptible by nature to different uses, the nations seem on the one hand to have apportioned among themselves that use which cannot be maintained conveniently apart from private ownership; but on the other hand to have reserved that use through the exercise of which the condition of the owner would not be impaired." Grotius, p. 44. The underlying *res communis* character of territorial waters was recognized by Albert La Pradelle and others in the late 19th and early 20th centuries. O'Connell, vol. 1, pp. 266–74. The precise juridical nature of the territorial sea nevertheless remained unresolved. B. Smith, p. 56.

356. Anand, "UN Convention on the Law of the Sea and the United States," p. 156. See Chap. 1 nn. 14–43 and accompanying text; and also Sanger, p. 13. Derived as it was from the cannon-shot rule, the three-mile territorial sea was "in reality a product of the maritime wars in the latter part of the eighteenth and the beginning of the nineteenth century, and its application to the right of fishing is accidental and arbitrary." Fulton, pp. 694, 698. Coastal states therefore justified claims to expanded offshore jurisdiction as necessary conservation measures. Fulton, pp. 22, 646, 664. See n. 353 above; and also Hennessey, p. 41. Other claims to contiguous zones of limited functional jurisdiction were designed to permit coastal states to exercise their police power beyond the territorial sea with minimal interference to high-seas navigation. See n. 364 below and accompanying text. Latin American states expanding their offshore jurisdiction during the second half of the 20th century cited the erosion of the Grotian inexhaustibility premise as justification for their claims. See D. Johnston, p. 311; and Chap.

1 n. 52 and accompanying text. There has been a "realization that ocean resources need to be authoritatively allocated. There are simply not enough fish to go around, nor enough oil, nor gas. And there is a need for effective ocean management if problems such as pollution and ocean safety are to be mastered." Janis, *Sea Power*, p. 89. Multilateral codificatory efforts produced conventions regulating navigation, maritime safety, piracy, fisheries, the slave trade, and the laying of submarine cables. See Colombos, p. 61.

357. Convention on Fishing and Conservation of the Living Resources of the High Seas, preamble, and arts. 1, 3, 4 (1), 6 (3) (4), 8 (1), reprinted in *United Nations Treaty Series*, vol. 559, no. 8164, pp. 286, 288, 290, 292. This same "necessity" to conserve limited ocean resources had been identified three decades earlier by the League of Nations Committee of Experts, which recommended negotiation of "a new jurisprudence" based on recognition that the sea, "being the uncontrolled property of all, belongs to nobody." League of Nations Committee of Experts, "Report on the Exploitation of the Products of the Sea," p. 236. See Chap. 1 nn. 19–23; and also Kronmiller, vol. 1, pp. 363–64. In 1953 the International Law Commission had called for creation of a more comprehensive regulatory regime, recommending that states be required to accept as binding "any system of regulation of fisheries in any area of the high seas which an international authority . . . shall prescribe as being essential for the purpose of protecting the fishing resources of that area against waste or extermination." International Law Commission, *Report of the International Law Commission to the General Assembly* (U.N. Doc. A/2456, September 1953), reprinted in *Yearbook of the International Law Commission, 1953*, vol. 2, p. 218 (paras. 97–98). In the second of the 1945 proclamations, the United States had sought to establish "explicitly bounded conservation zones" in offshore high-seas waters. Presidential Proclamation No. 2668, September 28, 1945, reprinted in Hollick, *U.S. Foreign Policy*, p. 393. See Chap. 1 n. 30; and Belsky, p. 755. "The theoretical basis of freedom of fishing . . . argued by Grotius and followed by general opinion, had become unsound. The inexhaustibility of fisheries proved to be an illusion." Separate opinion of Judge de Castro, Fisheries Jurisdiction (United Kingdom v. Iceland), Merits, Judgment, *I.C.J. Reports 1974*, p. 83. See Chap. 1 nn. 67–69 and accompanying text, and Chap. 2 nn. 168–69 and accompanying text.

By the middle of the 20th century, a great number of bilateral and multilateral conservation treaties had been concluded with respect to various fisheries, in derogation from the freedom of the seas doctrine. See Newton, "Inexhaustibility," p. 410. The Convention on Fur Seals in the North Pacific, concluded between Canada, Japan, Russia, and the United States in 1911 and considered one of the most effective conservation agreements, provided for compensation of states which were excluded from the fur seal harvest by the Convention's allocation of exclusive rights. Prewo, p. 282. See Christy, "Marine Resources and the Freedom of the Seas," p. 431. Unilateral U.S. claims of exclusive rights to these fur seal stocks had been declared unlawful by an international arbitration panel. Note, "Power of a State to Extend Its Boundary Beyond the Three-Mile Limit," *Columbia Law Review* 39 (1939): 325. The International Whaling Commission (IWC) began setting quotas in 1949, although their effectiveness was limited by the absence of an enforcement mechanism. See n. 443 below. "[I]t would appear that, although all those who carry on this trade realise the harm they are doing, each is unwilling to restrict his activities for the benefit of the others, and they endeavor to kill as many whales as they can, realising that the total extinction of the species is approaching." League of Nations Committee of Experts, "Report on the Exploitation of the Products of the Sea," pp. 235–36. See nn. 310–11 above and accompanying text.

358. Fisheries Jurisdiction (United Kindgom v. Iceland), Merits, Judgment, *I.C.J. Reports 1974*, p. 3 et seq. See O'Connell, vol. 1, pp. 539–40, 542; and *Rest. 3rd, Restatement of the Foreign Relations Law of the United States*, §521, reporters' note 3. A separate opinion noted that the high seas "are regarded as *res omnium communis,*" their use belonging "equally to all peoples," and that "appropriation of an exclusive fisheries zone in an area hitherto considered as part of the free seas is equivalent to deprivation of other peoples of their rights." Separate opinion of Judge de Castro, Fisheries Jurisdiction (United Kingdom v. Iceland), Merits, Judgment, *I.C.J. Reports 1974*, p. 97.

359. See Burton, p. 1166; and also Young, "The Legal Status of Submarine Areas," p. 226. As it developed through state practice, the drafting efforts of the International Law Commission and the negotiations at UNCLOS I, the continental shelf was a novel legal doctrine based

upon practical economic and political considerations. See Kronmiller, vol. 1, p. 185. "The end result is that all states enjoy only some freedoms of the high seas on the seabed of the continental shelf, and that the coastal state enjoys exclusive rights with respect to the exploration and exploitation of the natural resources of the continental shelf, subject to certain obligations derived from the high seas regime." Oxman, "The High Seas and the International Seabed," pp. 529-30.

Article 2 of the 1958 Convention on the Continental Shelf granted coastal states only limited rights to manage and exploit shelf resources. See Chap. 1 nn. 46, 51-52, and accompanying text; and also Arrow, "The Customary Norm Process," pp. 18-19. Article 4 of the Convention expressly preserved the right to lay submarine cables on the continental shelf, a right inconsistent with genuine coastal-state sovereignty. See Garcia-Amador, p. 130. Judicial opinion has properly recognized that the continental shelf doctrine, which originated with the Truman Proclamation, was crystallized in the Continental Shelf Convention. See Pietrowski, "Hard Minerals," pp. 49-50. It has been observed that use of the phrase *"ipso facto* and *ab initio"* in the North Sea Continental Shelf case to describe coastal-state rights to shelf resources was not intended to indicate that such rights existed prior to 1945, but rather was "employed by the Court in 1969 simply to strengthen the regime of the continental shelf which had not yet achieved a firm status in international law." Dissenting opinion of Judge Oda, Continental Shelf (Tunisia/Libyan Arab Jamahiriya), Judgment, *I.C.J. Reports 1982,* p. 191. Cf. North Sea Continental Shelf, Judgment, 1969, p. 29 (para. 39). See also Briggs, p. 342. Early coastal-state claims to shelf resources were justified as necessary to avoid the conflict and waste which would otherwise result from unrestrained international competition for limited mineral resources. Reppy, p. 281. See Hennessey, p. 42. "The first of the elements historically controlling the basic nature of the law of the sea, whether the use is exhaustible, is a very real, albeit implicit, part of the Truman Proclamation." Newton, "Inexhaustibility," pp. 420-21.

Prior to the adoption of the Convention, there had been a great deal of concern that unilateral assertion of exclusive rights to shelf resources might be inconsistent with the *res communis* principle. See, e.g., *Yearbook of the International Law Commission, 1950,* vol. 1, p. 223 (para. 77), 227 (para. 8a). See also Kronmiller, vol. 1, pp. 172-75. The Commission nevertheless endorsed an allocation of limited exclusive jurisdiction to coastal states — including those lacking a geological shelf — but only after concluding that joint development of shelf resources by the international community "would meet with insurmountable practical difficulties, and . . . would not ensure . . . effective exploitation." International Law Commission, *Report of the International Law Commission to the General Assembly* (U.N. Doc. A/1316, July 1950) — Annex: Draft Articles on the Continental Shelf and Related Subjects, reprinted in *Yearbook of the International Law Commission, 1950,* vol. 2, p. 384. See also Garcia-Amador, p. 133; and O'Connell, vol. 1, p. 476. Still, at UNCLOS I West Germany emphasized that the "common property" nature of the continental shelf required the consent of the international community for an allocation of exploitation rights, and Lebanon too noted that this portion of the seabed was *res communis.* Kronmiller, vol. 1, pp. 181, 346-47 n. 625. Such statements were consistent with prior case law, which had recognized the shelf — along with the remaining area beyond the territorial sea — as *res communis.* Post, *Deepsea Mining,* p. 99.

360. See nn. 300-301 above and accompanying text. In acknowledgment of the increasing danger and complexity of modern navigation, and the decreasing capacity of the sea to accommodate such expanding uses, the Convention specified duties of states with regard to the promotion of maritime safety and the prevention of marine pollution. Convention on the High Seas, arts. 10, 24, 25, in *United Nations Treaty Series,* vol. 450, no. 6465, pp. 86, 88, 96. See nn. 351, 361 above and below and accompanying text. Air traffic over the high seas was regulated by ICAO pursuant to article 12 of the Chicago Convention. See Morenoff, p. 147 n. 29. Freedom of fishing was made subject to the general duty of conservation set forth in the Fisheries Convention, as well as to the limitations imposed by coastal states within their offshore fisheries zones. See nn. 357-58 above and accompanying text. The slight interference with other ocean uses posed by the laying of submarine cables and pipelines, on the other hand, gave rise to only minimal regulation. "Cables cross each other freely; pipelines can cross with appropriate engineering deisgn. They use only a tiny portion of the ocean bottom. Cable-laying is entirely different than the exploitation of mineral resources, which are exhaustible.

To the ILC, the nature of the activity was controlling, not the region in which it occurred." Van Dyke and Yuen, p. 508. See Kronmiller, vol. 1, pp. 484–87; and also Hurst, p. 42. Although the International Law Commission referred to "the freedom to . . . exploit the subsoil of the high seas," it nonetheless concluded that such exploitation would require "special regulation." See Chap. 1 n. 96 and accompanying text. It has been speculated that the Commission might have recommended a regulatory regime consistent with the common heritage of mankind principle had deep-sea mining been viable in the 1950s. Van Dyke and Yuen, p. 505.

361. Convention on the High Seas, arts. 24, 25 (1), in *United Nations Treaty Series*, vol. 450, no. 6465, p. 96. In the late 17th century, Pufendorf had been able to assert with confidence that use of the sea for navigation, bathing, and drawing water was "in truth inexhaustible." Pufendorf, vol. 2, p. 562. By 1926 the advent of engine-powered vessels had prompted the United States to convene an international conference of leading maritime states to address the problem of pollution resulting from the discharge of oil into the sea. See Colombos, pp. 371–72. The harmful effects of vessel-source pollution were addressed by the International Law Commission in 1937, and in 1954 maritime states adopted the International Convention for the Prevention of Oil Pollution, which prohibited oil discharges within a 50-mile-wide coastal zone. See Margolis, p. 643; and Gold, pp. 284–85. With the increase in the size of oil tankers in the postwar period, there was growing concern regarding the environmental threat posed by accidental oil spills, particularly in increasingly crowded coastal shipping lanes. See Chap. 1 n. 65 and accompanying text. The International Maritime Organization has assumed primary legislative responsibility for prevention of vessel-source pollution. See n. 374 below and accompanying text.

The dumping of radioactive waste had posed a particularly serious threat of long-term contamination to the marine food chain. See McDougal and Burke, p. 853. "[I]t was quite obvious that freedom of the high seas could not be enjoyed if either the water or the air were contaminated by radioactivity or if fish were poisoned by radio-active waste dumped in the sea." Statement of Jaroslav Zourek in *Yearbook of the International Law Commission, 1956*, vol. 1, p. 60 (para. 8). Pursuant to regulatory mechanisms established by the 1972 Convention on the Preservation of Marine Pollution by Dumping of Waste and Other Matter (the Dumping Convention), a moratorium has therefore been imposed upon the dumping and seabed disposal of radioactive wastes. See *Oceans Policy News* (March 1984), p. 3; and *Oceans Policy News* (December 1986), p. 6. Similar prohibitions have also been implemented regionally. See *Oceans Policy News* (February 1986), p. 8. The Dumping Convention also prohibited disposal—within territorial waters as well as upon the high seas—of pesticides, plastics, and biochemical weapons, and required permits for disposal at sea of other materials which could harm the marine environment or interfere with navigation. Gershon D. Greenblatt, James Robert Miller, and Alfred J. Waldchen, "Recent Developments in the Law of the Sea IV: A Synopsis," *San Diego Law Review* 10 (1973): 584–85. See also Terry L. Leitzell, "The Ocean Dumping Convention—A Hopeful Beginning," *San Diego Law Review* 10 (1973): 506. "Whatever one state does in the oceans in respect of radioactive waste is of direct concern to all other states, and this includes deposits within its own territorial sea." McDougal and Burke, p. 854. See David G. Spak, "The Need for a Ban on All Radioactive Waste Disposal in the Ocean," *Northwestern Journal of International Law and Business* 7 (1986): 803–32. It has recently been urged that oceanic waste disposal should, to the greatest extent possible, be based upon site- and pollutant-specific "assimilative capacity"—a standard conceptually consistent with the exhaustibility criterion of the governing *res communis* principle. See Alan B. Sielen, "Sea Changes? Ocean Dumping and International Regulation," *Georgetown International Environmental Law Review* 1 (1988): 1–32.

362. Although the United States had conducted atomic weapons tests in the South Pacific since the 1940s, detonation of the first hydrogen bomb in 1954 resulted in the radioactive contamination of scores of Japanese fishermen, Marshalese natives, and U.S. test personnel. Note, "Exclusion of Ships," p. 1044. See also Margolis, pp. 629–32. The test caused closure of a 400,000-square-mile area of the high seas for 57 days, and the United States closed several hundred other testing areas in the South Pacific, some for a period of years. McDougal and Schlei, pp. 651–52, 683–84. Other nuclear-power states also temporarily closed large areas of the sea in order to conduct weapons tests. See Morenoff, p. 141 n. 11. "Nuclear weapons testing

necessarily displaces free movement in the air and sea for thousands of square miles in the vicinity, and this activity has understandably occasioned much controversy about limits on free navigation." McDougal and Burke, pp. 771–72. See also Margolis, p. 636; and Kronmiller, vol. 1, p. 406. Nuclear testing also interferes with freedom of fishing by causing the contamination and destruction of fishery stocks. D. Johnston, p. 15. Significantly, defenders of such testing attempted to argue that it causes little or no interference with high-seas fishing or navigation, that it is a legitimate use of the sea for security purposes, and that the testing had been implicitly authorized by the U.N. Security Council. McDougal and Schlei, pp. 683–84, 686–90, 692–94, 695–708.

Despite international protests by Japan and others, the United States maintained that nuclear weapons testing was permissible as a "reasonable" use of the high seas and that no compensation was owing to displaced ocean users, although it did pay $2 million to Japan as an expression of "concern and regret" following the 1954 incident. Note, "Exclusion of Ships," pp. 1044–46. At the urging of nuclear power states, the International Law Commission avoided consideration of the "extremely delicate problem" posed by such testing, noting "that it was outside the Commission's terms of reference to prohibit such experiments and premature to take any stands on questions which were under active consideration by other United Nations bodies," and adopted instead the vague reasonableness standard. *Yearbook of the International Law Commission, 1955,* vol. 1, p. 263 (para. 50). *Yearbook of the International Law Commission, 1956,* vol. 1, p. 60 (para. 12). See nn. 39, 46, 293 above; and also Burton, p. 1172 n. 161. "Clearly there was a difference of view, both in the *I.L.C.* and at the 1958 Conference, as to whether any *de facto* closure of a temporary though geographically extensive nature, or only a juridical claim to exclusion of foreign vessels, constituted an infringement of high seas freedoms." Kronmiller, vol. 1, p. 499. See also Kronmiller, vol. 1, pp. 398–404. UNCLOS I delegates nonetheless adopted a resolution recognizing the "serious and genuine apprehension" of many states that the nuclear weapons tests infringed upon the freedom of the seas and referring the matter to the U.N. General Assembly "for appropriate action." McDougal and Burke, p. 762. Significantly, the 1963 Test Ban Treaty prohibited nuclear weapons testing within territorial waters as well as upon — or under — the high seas. Treaty Banning Nuclear Weapons Tests in the Atmosphere, in Outer Space and Under Water, art. I (1) (a), reprinted in *United Nations Treaty Series,* vol. 480, no. 6964, p. 45.

The United States has continued to conduct nonnuclear missile tests in the South Pacific, effectively excluding vessels from "caution areas" several thousand square miles large. Note, "Exclusion of Ships," pp. 1046–49. See also McDougal and Schlei, pp. 679–80. Exclusion from such areas has been achieved by "voluntary compliance" with widely publicized warnings, rather than by any U.S. claim of right; the *opinio juris* necessary for formation of a rule of customary international law is lacking where avoidance results from an apprehension of physical danger rather than from acknowledgment of a legal duty. Note, "Exclusion of Ships," pp. 1051, 1055, 1057–58. The forcible exclusion of a Greenpeace vessel from a high-seas testing area in 1989 raised the question of whether the United States is now unilaterally asserting a claim of exclusive rights to such areas, an assertion which would seem to contravene the *res communis* principle in the absence of international consent. See Mike Clary, "Navy Foils Greenpeace Ship to Fire Trident 2," *Los Angeles Times,* December 5, 1989, p. A4.

363. See S. Brown et al., p. 14. "[T]he complications of modern life, with greater diversity in the uses of resources and the interactions of people, have made the traditional, nonfunctional approach to the law of the sea somewhat otherworldly." D. Johnston, p. xvi. Submarine cables and pipelines, for example, have frequently become entangled with anchoring and fishing equipment, necessitating conventional regulation to minimize such interference. See Kronmiller, vol. 1, p. 482. Similarly, the emplacement on the continental shelf of semipermanent artificial islands and installations for the exploitation of offshore oil deposits necessarily interferes with the exercise of high-seas freedoms in surrounding waters. Lauterpacht, "Sovereignty Over Submarine Areas," pp. 409–10. Cf. Convention on the Continental Shelf, art. 5 (1), reprinted in *United Nations Treaty Series,* vol. 499, no. 7302, p. 314. See also Kronmiller, vol. 1, pp. 420, 428.

364. See Convention on the Territorial Sea and the Contiguous Zone, art. 24, reprinted in *United Nations Treaty Series,* vol. 516, no. 7477, pp. 220–22; and Fulton, pp. 594–95.

Unilateral extensions of territorial jurisdiction had unnecessarily interfered with the exercise of high-seas freedoms by maritime states. See Chap. 1 n. 29 and accompanying text.

365. U.S. Commission on Marine Science, Engineering, and Resources, *Our Nation and the Sea*, p. 19. The Commission pointed out that a system of unrestricted access "is no obstacle to economic development if resources are abundant, technology simple, and investment minimal. But it is not appropriate for large-scale industrial activities in a highly technological, mobile, and capital intensive economy, and it is slowly yielding to arrangements to assign resource development rights." Ibid., p. 86. See also ibid., pp. 88–90. "Without regulation, this 'open access' characteristic has tended in the past to lead to overcapitalization and resultant dissipation of economic rent." Charles N. Brower, Acting Legal Adviser, Department of State, Letter to Senator Henry M. Jackson Dated July 26, 1973, reprinted in *Status Report*, U.S. Senate, p. 404. See nn. 305–6, 309–11 above and accompanying text.

366. See Burton, p. 1167 n. 141. The detail of the 1982 Convention is itself a reflection of the increasing use of the oceans and the resulting inadequacy of the broad rule of unrestricted access. Janis, *Sea Power*, p. 92. See also Clarkson, p. 118. It has been noted that, in negotiating the various jurisdictional allocations set forth in the Convention, UNCLOS III delegates employed arguments sounding "rather like natural law." Newton, "Inexhaustibility," p. 387 n. 103. "Grotius, of course, wrote . . . long before the advent of nuclear ships, super-tankers, radioactive wastes, pesticides, nerve gas, and coastal cities with populations numbered in millions pouring their industrial and domestic effluents into the sea – not to mention vast fishing fleets accompanied by factory ships. The sea now can be exhausted by 'promiscuous use,' and the freedom of the seas cannot be distorted to permit such a result." Legault, p. 221.

367. Elisabeth Mann Borgese, "A Constitution for the Oceans," in *The Fate of the Oceans*, ed. John J. Logue (Villanova, Penn.: Villanova University Press, 1972), p. 7. There has been increasing recognition that sound management requires advance planning in anticipation of expanded ocean uses. See generally Belsky, pp. 733–63.

368. U.N. Convention on the Law of the Sea, arts. 61 (2) (3) (4), 64 (1), 117, 118, 119 (1). Similarly, the 1980 Convention on the Conservation of Antarctic Marine Living Resources sought to prevent depletion of krill stocks in the Antarctic Ocean through the institution of ecosystem-based conservation and management techniques. See Chap. 4 n. 679. In 1989 the U.N. General Assembly, at the urging of the United States, adopted by consensus a resolution recommending a moratorium on high-seas use of large-scale driftnets by mid-1992 unless effective, scientifically based conservation and management measures were implemented jointly by interested states. William T. Burke, "Driftnets and Nodules: Where Goes the United States?" *Ocean Development and International Law* 20 (1989): 238. See U.N. General Assembly, 44th Session, Resolution 44/225, December 22, 1989; and also *Oceans Policy News* (January 1990), p. 2. The incongruity of ideological opposition to deep seabed mining regulation in light of U.S. support for such severe restrictions on free access to other limited marine resources has been duly noted. Burke, "Driftnets and Nodules," pp. 237–40.

By 1976 the capacity of the world fishing fleet exceeded the annual catch by more than a 2 to 1 ratio. Pardo and Borgese, p. 71. Some UNCLOS III delegates even advocated complete elimination of the freedom of fishing from article 87. Oda, "Fisheries Under the United Nations Convention on the Law of the Sea," p. 741. The extension of exclusive fisheries jurisdiction to 200 miles by the United States in 1976 was prompted mainly by concern that offshore fisheries were, in fact, being exhausted. Brewer, "Deep Seabed Mining," p. 30. See Chap. 3 n. 123, and n. 311 above; and also Shinaver, p. 180.

369. U.N. Convention on the Law of the Sea, arts. 62 (2), 69, 70. See also ibid., art. 63. The 200-mile limit fails to encompass highly migratory and high-seas fisheries, and the division of the zone into nationally administered segments renders uniform species management schemes generally unworkable. Prewo, pp. 270–71, 284. See also Keen, p. 208. Supranational regulation of high-seas fisheries may yet prove necessary. See n. 357 above.

370. U.N. Convention on the Law of the Sea, arts. 56 (1) (b), 60, 80, 86, 87 (1), 246. See *Rest. 3rd, Restatement of the Foreign Relations Law of the United States*, §521, comment (g), reporters' note 5. It has been pointed out that, "theoretically," marine scientific research "should in no sense diminish the worth of" the sea to other users. Hennessey, p. 39. The conduct of such research by foreign states in zones of coastal-state jurisdiction nevertheless has

given rise to security concerns of the type which had been cited in support of the territorial sea. See n. 354 above. "The problem comes when the research vessels of one, often distant and much larger, state seeks to conduct research on the continental slope or shelf of another, perhaps smaller state. . . . The fear is that the 'research vessel' may be looking for exploitable resources or that it may have military business." Hull, p. 14. To accommodate this concern, the Convention subjects scientific research within the exclusive economic zone and on the continental shelf to a coastal-state consent regime. See Chap. 6 n. 78 et seq. and accompanying text.

371. It has been argued that the process of ocean thermal energy conversion (OTEC) is an inexhaustible use of the sea — ocean thermal energy itself being *res nullius* — and that such energy production may be implicitly recognized as a high-seas freedom in the Convention. Burton, p. 1174. See Knight, "International Jurisdictional Issues," p. 62. Cf. U.N. Convention on the Law of the Sea, art. 56 (1) (a). "The sticking point in this process relates to the matter of exclusivity with respect to OTEC deployment. If the claim were simply limited to the right to use a particular area of the ocean for an OTEC installation, without asserting an exclusive right to a broader area, deployment is likely to meet with little protest (provided the proper justifications are made for the claim). On the other hand, if a significantly large area of the ocean is prohibited to other OTEC devices by the claim (because of the drain on heat caused by two devices in too-close proximity to each other) or to other uses in general, then the issue has escalated to a new plane." Knight, "International Jurisdictional Issues," p. 57. Beyond such considerations, however, under the governing *res communis* principle, it may be regarded as dispositive that the number of available OTEC sites, like nodule mining sites, appears to be limited. See Knight, "International Jurisdictional Issues," p. 59; and nn. 314–15 above and accompanying text. Although the Convention does not expressly provide for the establishment of safety zones around installations in high-seas areas beyond the continental shelf, the omission may be explained by a present lack of production capability in such areas.

372. See n. 360 above. Because of the threat of pollution, UNCLOS III delegates permitted a greater degree of coastal-state regulation of pipelines. See U.N. Convention on the Law of the Sea, art. 79 (2), (3).

373. Van Dyke and Yuen, p. 509 n. 66. See n. 361 above and accompanying text. "Environmental protection at sea is frequently addressed on a global basis, as evidenced not only by Part XII of the 1982 Convention and various environmental treaties, but by the establishment of the U.N. Environmental Programme and the expanded composition of the Marine Environment Protection Committee of the International Maritime Organisation." Oxman, "The High Seas and the International Seabed," p. 537.

374. U.N. Convention on the Law of the Sea, arts. 94 (3) (4) (5), 211 (1) (2), 216 (1), 217–20. See Chap. 4 nn. 291–92 and accompanying text. Because states are required to comply with "generally accepted" international regulatory standards, maritime states may be bound by the terms of conventions to which they are not parties and by regulations adopted by competent international organizations. See Chap. 4 n. 291 and accompanying text. For a discussion of the environmental provisions of the 1982 Convention, see O'Connell, vol. 2, pp. 988–95; and Chap. 6 n. 100 et seq. and accompanying text. "The regulatory functions of IMO in the technical fields of navigation and vessel-source pollution are confirmed, and in some cases strengthened, by the provisions of the Convention." Mario Valenzuela, "IMO: Public International Law and Regulation," in *The Law of the Sea and Ocean Industry*, ed. Johnston and Letalik, p. 150. IMO's competence was expanded in 1982 to include prevention and control of pollution, as well as maritime safety and efficiency. Valenzuela, p. 141. IMO (as IMCO) had already drafted a variety of recommendations and regulations on subjects related to commercial shipping, including the International Convention of Civil Liability for Oil Pollution and several other antipollution conventions following the wrecking of the *Torrey Canyon* off the coast of England in 1967. See Sanger, pp. 100–2; and also UNCLOS III, *The Activities of the Inter-Governmental Maritime Consultative Organization in Relation to Shipping and Related Maritime Matters* (U.N. Doc. A/CONF.62/27, June 10, 1974), para. 63, reprinted in *Official Records*, UNCLOS III, vol. III, p. 47. IMO has undertaken the regulation of tanker construction, and the rate of acceptance of its conventions has increased substantially since the UNCLOS III Convention was opened for signature. *Law of the Sea* (U.N. Doc. A/41/742), pp. 10, 15–16.

375. U.N. Convention on the Law of the Sea, arts. 2 (3), 17–32, 34–45. See Chap. 3 n. 23 and accompanying text. Similar limitations have been made applicable within the newly created archipelagic waters. U.N. Convention on the Law of the Sea, arts. 49 (3), 51–54.

376. Statement of John R. Stevenson, Legal Adviser, U.S. Department of State, before Subcommittee on International Organizations and Movements House Committee on Foreign Affairs, April 10, 1972, reprinted in *U.S. Department of State Bulletin* 66 (1972): 676. See also text accompanying Chap. 2 n. 123.

377. U.N. Convention on the Law of the Sea, art. 76 (1). See nn. 359, 368–71 above and accompanying text. "While the freedoms in article 87 are at least *inter alia* freedoms, the high seas freedoms exercisable within the EEZ are specifically enumerated and, in addition, blurred as to their contours." Stefan A. Riesenfeld, "Comment," *Law and Contemporary Problems* 46 (Spring 1983): 12. See also Gotlieb, p. 198.

378. See Oxman, "The High Seas and the International Seabed," pp. 539–40. "The essence of this concept is that the areas of the ocean, its bed, and resources beyond national jurisdiction are to be jointly controlled, developed, and preserved by all nations, and not unilaterally claimed or exploited." Shingleton, p. 34. See nn. 83–84 above and accompanying text. It has been pointed out that the common heritage of mankind principle is based on "the likelihood of irreparable damage" to "non-rejuvenating" nodule resources. W. Jones, "Risk Assessment," p. 349. See nn. 34–35, 314–19 above and accompanying text. It has also been noted that deep-sea mining may adversely affect other lawful uses of the sea. Kronmiller, vol. 1, p. 8. Although the environmental effects of nodule mining remain largely uncertain, it has been predicted that local populations of dependent organisms would be depleted for hundreds of years, causing an unknown impact on the marine food chain. Peter A. Jumers, "Limits in Predicting and Detecting Benthic Community Responses to Manganese Nodule Mining," *Marine Mining* 3 (1981): 213–29. See Chap. 3 n. 372 and accompanying text, and n. 305 above and accompanying text.

Developing states have correctly observed that application of the *res communis* principle itself prohibits nodule exploitation outside a generally accepted regulatory regime. See Richardson, "Law in the Making," p. 409. Indeed, U.S. enactment of the 1980 Seabed Act confirms the necessity for a regulatory regime and belies official U.S. support for a true system of unrestricted access. See nn. 55–56, 59, 65–67 above and accompanying text. The United States has in fact acknowledged the *res communis* status of deep seabed resources but has misapplied the legal content of that principle. See nn. 2, 227–31 above and accompanying text; and also Torreh-Bayouth, p. 98.

379. See nn. 62, 91–97, 132–34, 136–37 above and accompanying text. The revenue-sharing provisions, the decisionmaking procedures, and the system of parallel access thus represent legitimate efforts to preserve the interests of the international community as part of a compromise system of exploitation, where that community was unable to reach agreement upon a system of either national or international access. See Oxman, "The 1976 New York Sessions," p. 253. There was widespread recognition at UNCLOS III of the need "to avoid a 'free-for-all' situation so that no state or entity obtained a disproportionate share of the resources or was able to hoard mine sites and manipulate price levels to its own advantage." McDade, p. 36. Significantly, prospecting activities "may be conducted simultaneously by more than one prospector in the same area," permitting recovery of "a reasonable quantity of minerals to be used for testing," subject to only minimal regulatory control. U.N. Convention on the Law of the Sea, Annex III, art. 2 (1) (c), (2). See Chap. 4 n. 514.

The preamble to the 1970 Declaration of Principles emphasized that it was "essential" for the international community to remedy the lack of "substantive rules for regulating the exploitation of" deep-sea mineral resources. In the U.N. debates which accompanied the votes on the Declaration and the 1969 Moratorium Résolution, delegates generally endorsed the applicability of the *res communis* principle to nodule resources. W. Smith, p. 51 n. 9.

380. Alfred Verdross, "*Jus Dispositivum* and *Jus Cogens* in International Law," *American Journal of International Law* 60 (1966): 55. See International Law Commission, *Report on the Law of Treaties* (U.N. Doc. A/CN.4/63, March 24, 1953), reprinted in *Yearbook of the International Law Commission, 1953*, vol. 2, p. 154; and also Sztucki, *Jus Cogens*, pp. 180–82. "[N]ormally the rules of general international law have the character of *jus dispositivum*. This means that they are not imperative but of a yielding nature. They must

be applied only if the individual states have not agreed otherwise *inter se.*" Verdross, "*Jus Dispositivum* and *Jus Cogens*," p. 58. See International Law Commission, *Law of Treaties: Third Report* (U.N. Doc. A/CN.4/115, March 18, 1958), reprinted in *Yearbook of the International Law Commission, 1958*, vol. 2, p. 40 (para. 76); and also Alfred Verdross, "Forbidden Treaties in International Law," *American Journal of International Law* 31 (1937): 571 (emphasis in original). Rules of *jus cogens*, in contrast, consist of compulsory norms of a juridically superior character, upon which the validity of inferior norms of *jus dispositivum* depends. See Rozakis, *The Concept of Jus Cogens*, p. 20.

381. Vienna Convention on the Law of Treaties, in *International Legal Materials* 8 (1969): 699. For a discussion of this definition, see Weil, pp. 425–27. Article 53 was adopted by a vote of 87 to 8, with 12 abstentions, and the objections voiced were directed to the practical content, rather than to the theoretical existence of peremptory norms. Rozakis, *The Concept of Jus Cogens*, pp. 51–52. See n. 394 et seq. below and accompanying text; and also Alexidze, p. 231. Article 64 of the Convention provides that with the emergence of a new peremptory norm "any existing treaty which is in conflict with that norm becomes void and terminates." *International Legal Materials* 8 (1969): 703. Articles 65 and 66 specify the institutional procedures whereby parties to an invalid convention may invoke *jus cogens. International Legal Materials* 8 (1969): 703–4. See Sinclair, p. 41. The lack of any procedure for impeachment of such treaties by third parties has been criticized: "The whole machinery of invalidation is left in the hands of the parties to an illegal treaty, that is to say, to those States who, knowingly most of the time, violated a peremptory norm of international law.... The international community, unable to use the Convention's machinery, has as its only weapon the usual diplomatic pressures and the traditional channels of the decentralized international legal system." Rozakis, *The Concept of Jus Cogens*, pp. 191–92.

382. "It may be said that no other general principle of law is so universally accepted as this one." Verdross, "*Jus Dispositivum* and *Jus Cogens*," p. 61. See Gordon A. Christenson, "The World Court and *Jus Cogens*," *American Journal of International Law* 81 (1987): 95; and also Alexidze, p. 229. Contracts violative of peremptory norms were routinely annulled by the courts of ancient Rome, and such norms continue to be enforced, particularly with continental code states such as France, Germany, Italy, and Spain. See Alexidze, pp. 234–35. The analogy to municipal law "informed the reasoning of Lauterpacht, McNair, and Mosler in their public order justification of *jus cogens* in international law." Gordon A. Christenson, "*Jus Cogens*: Guarding Interests Fundamental to International Society," *Virginia Journal of International Law* 28 (1988): 598. "The international law of international agreements, indeed, is derived in substantial part from general principles common to the contract laws of state legal systems." *Rest. 3rd, Restatement of the Foreign Relations Law of the United States*, Introductory Note to Part III. See Sinclair, *The Vienna Convention*, p. 203. "The application in international law of the general principle according to which treaties *contra bonos mores* are void, is not free from difficulties. These difficulties are the consequence of the fact that the ethics of the international community are much less developed than the ethics of national communities; further, the international community embraces different juridical systems, built upon different moral conceptions." Verdross, "Forbidden Treaties," p. 573. It has thus been argued, unconvincingly, that the municipal law analogy fails because there is no organized, overriding international public order capable of enacting and enforcing peremptory norms. Christenson, "Guarding Interests," pp. 598–601. Elsewhere, Christenson appears to acknowledge that the real problem with the "public order concept" is the system of positive law from which it is supposed to arise: "peremptory norms can reinforce any political ordering system with any content determined by those powerful enough to make and implement effective decisions, even paternalistic ones." Christenson, "Guarding Interests," p. 634. See n. 384 below; and also Verdross, "Forbidden Treaties," p. 576.

383. See Kelsen, *Principles of International Law*, p. 190. Although Kelsen asserted that "no conflict is possible" between two treaties concluded by entirely different parties, he acknowledged that "[i]n the application to concrete cases, particular conventional (or particular customary) law precedes general customary law." Ibid., pp. 305, 361. It seems clear that to the extent that such particular norms become predominant — theoretically, each state could agree to observe a different rule with each other state — a rule of *general* international law may cease to exist as such; it is not possible for such competing positive-law allocation

regimes to be simultaneously valid with respect to the use of a single, indivisible natural resource (or any part thereof). See n. 401 below and accompanying text.

384. See Sztucki, *Jus Cogens*, pp. 60–62. The Vienna Convention "closely connects international *jus cogens* with the consensual character of general international law, excluding any possibility of identifying it with 'natural law standing above the will of States.'" Alexidze, pp. 253–54. Significantly, the International Law Commission had sought to draft a *jus cogens* concept which did not depend upon international acceptance and recognition; but the Conference delegates, perceiving a threat to the theory of unlimited state sovereignty, modified the language of article 53 to ensure greater state control over the validity of treaties. See Sztucki, *Jus Cogens*, pp. 97–98; and also Christenson, "Guarding Interests," p. 594. "Otherwise, they would have to accept that peremptory norms which exist independently of the will of States had been established either by a supra-national authority or by law of nature." Sztucki, *Jus Cogens*, p. 98. See also Mary Ellen Turpel and Philippe Sands, "Peremptory International Law and Sovereignty: Some Questions," *Connecticut Journal of International Law* 3 (1988): 365, 367–68. Soviet-bloc states in particular were adamantly opposed to any implication of a natural-law basis for peremptory norms. Sinclair, p. 221. Somewhat contradictorily, there was nevertheless apparent agreement among the delegates that—to some undefined degree—the capacity of states to determine the substantive content of their treaty relations is necessarily limited. See Sztucki, *Jus Cogens*, p. 156. The positive-law character of the *jus cogens* doctrine is confirmed in other articles of the Convention. See n. 381 above; and Rozakis, *The Concept of Jus Cogens*, p. 114.

385. Sztucki, *Jus Cogens*, p. 64. See n. 328 above; and also Christenson, "Guarding Interests," pp. 632–33. "In a number, if not a majority, of writings supporting the category of an international *jus cogens*, this category is understood—explicitly or implicitly—as one of natural rather than positive law. True enough, the classical natural law school did not use the term *jus cogens* but spoke of 'necessary norms.' However, the essential meaning may well be understood and interpreted in both cases in the same way." Sztucki, *Jus Cogens*, p. 59. See Gierke, pp. 75–76, 84–85; Verdross, "*Jus Dispositivum* and *Jus Cogens*," pp. 55–56; and also Mark W. Janis, "The Nature of *Jus Cogens*," *Connecticut Journal of International Law* 3 (1988): 361–62. The existence of peremptory norms is thus "independent of and antecedent to the perception of their existence." Onuf and Birney, p. 188. "The law of nature has been rightly exposed to the charge of vagueness and arbitrariness. But the uncertainty of the 'higher law' is preferable to the arbitrariness and insolence of naked force." Lauterpacht, "The Grotian Tradition," p. 24. Indeed, the modern concept of *jus cogens* was formulated largely in response to the positivist excesses of Nazism in the first half of the 20th century. Janis, "The Nature of *Jus Cogens*," p. 361. It has been pointed out that, ultimately, the validity even of modern international law must depend on the existence of fundamental normative assumptions which cannot be accounted for by a system of positive law. See Tucker, pp. 31–32; Kelsen, *Principles of International Law*, pp. 314–15; and Christenson, "Guarding Interests," pp. 587, 628–29.

386. Nys, "The Development and Formation of International Law," p. 2. Natural law "is always stronger than the municipal law." Grotius, p. 47. "For, since the law of nature arises out of Divine Providence, it is immutable." Grotius, p. 53. See Edwards, *Hugo Grotius*, p. 105; and also n. 252 above and accompanying text.

387. See Christenson, "Guarding Interests," p. 603 n. 56. "Whence, as this law is immutable and the obligations that arise from it necessary and indispensable; nations can neither make any changes in it by their conventions, dispense with it themselves, nor reciprocally, with respect to each other." Vattel, p. 51.

388. As expressed by Cicero, the Roman conception of natural law was functionally equivalent to *jus cogens*: "Neither the Senate nor the people can absolve us from our obligation to obey this law. . . . It will not lay down one rule at Rome and another at Athens, nor will it be one rule today and another tomorrow. But there will be one law, eternal and unchangeable, binding at all times upon all peoples." *The Commonwealth*, book III, chap. 22, quoted in Edwards, *Hugo Grotius*, p. 33. This normative hierarchy was ultimately carried forward into the Middle Ages by the glossators. See nn. 333–34 above and accompanying text; and also Gierke, pp. 75–76, 84. "For the law of nature has a reason within itself, nor are the acts of nations a rule of law, but rather is the law a rule for the acts." C. Wolff, vol. 2, p. 70.

Even Selden regarded natural law as "obligatory," "universal," and "unchangeable." Selden, pp. 12–13. See also *Justice*, p. 7. The law "may leave a vast number of points at the absolute discretion, or even caprice, of Convention; it by no means goes so far as to say, that any thing which the heart of man can devise shall be legal." Ward, vol. 2, p. 237.

389. Alexidze, pp. 228–29.

390. Ibid., p. 230. See Sztucki, *Jus Cogens*, pp. 54, 58, 62, 75–76; and also Verdross, "*Jus Dispositivum* and *Jus Cogens*," pp. 56–57. Although no international tribunal has ever rested a decision squarely upon the principle of *jus cogens*, the Permanent Court of International Justice appeared to rely implicitly on peremptory norms in reaching more than a dozen of its decisions. Sztucki, *Jus Cogens*, pp. 12–16, 25–26. Similar implicit references have been identified in decisions of the International Court of Justice. Christenson, "Guarding Interests," pp. 605–8, 619–21. See also Alexidze, p. 228.

391. Rozakis, *The Concept of Jus Cogens*, p. 2. See Christenson, "Guarding Interests," p. 607; and also Sztucki, *Jus Cogens*, pp. 26–54, 69, 94, 101–2. Even the statements of delegates at the Vienna Conference and in the U.N. General Assembly were largely ambiguous. Giorgio Gaja, "*Jus Cogens* Beyond the Vienna Convention," *Recueil des Cours* 172 (1981): 287. Of course, the lack of state practice may simply indicate that compliance with peremptory norms is so widespread that instances of their violation are exceedingly rare.

392. Article 53 not only provides that a peremptory norm may only be modified "by a subsequent norm of general international law having the same character," but also states clearly that the definition of peremptory norms set forth therein applies only "[f]or the purposes of the present Convention." Vienna Convention, art. 53, in *International Legal Materials* 8 (1969): 698–99. The formulation of *jus cogens* contained in article 53 must thus be understood as an example of the "progressive development" of law mentioned in the preamble of the Convention, rather than as a codification of existing law. Cf. *International Legal Materials* 8 (1969): 680. See also Sinclair, pp. 17–18; and Rozakis, *The Concept of Jus Cogens*, p. 5. In this regard, it is significant that the United States, which has not ratified the Vienna Convention, only applies those of its provisions which codify customary international law. Sohn, "Customary International Law Developments," p. 276 n. 24. Cf. *Rest. 3rd, Restatement of the Foreign Relations Law of the United States*, Introductory Note to Part III.

The lack of acceptance and recognition of article 53 — or of other individual peremptory norms — as *jus cogens* is compounded by the failure of the Vienna Convention to achieve general acceptance. See Sztucki, *Jus Cogens*, pp. 106–7, 158–59. The Convention did not enter into force until 1980, more than 10 years after it had been opened for signature, and by 1986 it had been ratified by only 52 states — one third of the international community. As in the case of the Law of the Sea Convention, the legal status of treaty provisions representing progressive development of the law remains uncertain in the absence of general acceptance, particularly in light of article 4, which provides that "the Convention applies only to treaties . . . concluded by States after the entry into force of the present Convention with regard to such States." *International Legal Materials* 8 (1969): 682. See *Rest. 3rd, Restatement of the Foreign Relations Law of the United States*, Introductory Note to Part III; and also McDade, pp. 26–27. The failure of article 53 to achieve general acceptance even as customary law — attributable to the unwillingness of states to concede the existence of undefined limitations to their sovereign treaty-making powers — "makes it impossible to maintain that the provisions in the Convention relating to peremptory norms correspond to the existing law on the subject." Gaja, p. 279. See also Sztucki, *Jus Cogens*, p. 107.

If the *jus cogens* principle were not itself peremptory, the concept would lack practical legal significance, since states parties to a derogating convention would be free to avoid *jus cogens* by mutual consent — consent which might arguably be implied from the conclusion of any agreement in derogation of a peremptory norm. Unilateral derogation through the process of prescription would be even more problematic, since the derogating act would necessarily constitute a derogation from both the specific peremptory norm in question and the principle of *jus cogens* itself. See n. 393 below and accompanying text. The revised *Restatement*, while indicating that norms recognized as peremptory by the international community are *jus cogens*, does not define *jus cogens* exclusively in terms of such recognition, as does article 53. *Rest. 3rd, Restatement of the Foreign Relations Law of the United States*, §102, comment (k). Confusedly, the reporters, while describing *jus cogens* as "a principle of customary law (albeit of higher status)," assert that comment (k) "adopts" the definition set forth in article 53. *Rest. 3rd, Restatement of the Foreign Relations Law of the United States*, §102, reporters' note 6.

393. See *Rest. 3rd, Restatement of the Foreign Relations Law of the United States*, §338, comment (c). The *Restatement* recognizes that peremptory norms "prevail over and invalidate international agreements *and other rules of international law* in conflict with them." Ibid., §102, comment (k) (emphasis added). With respect to derogation from peremptory norms, acquiescence to unilateral prescriptive acts may properly be understood as the functional equivalent of impermissible consensual agreements. See Sztucki, *Jus Cogens*, pp. 67–68, 111. It would be undeniably incongruous if individual states were permitted to do on their own what they were not permitted to do in concert with other states. "The need to remove the cause of a possible breach of an obligation imposed by a peremptory norm appears to be applicable in the same way to treaties and to unilateral acts." Gaja, p. 295. Grotius recognized the invalidity of custom and prescription violative of natural law, pointing out that "it is impossible to acquire by usucaption or prescription things which cannot become property." Grotius, pp. 47, 52. See n. 402 below and accompanying text; and also MacRae, p. 190. Many other jurists have failed to appreciate this point. See Sztucki, *Jus Cogens*, p. 69.

The invalidity of unilateral acts in derogation of peremptory norms is consistent with the language of article 53 and is implied by article 45. See Rozakis, *The Concept of Jus Cogens*, p. 126; and also Sinclair, p. 224. "Current doctrine classifies as peremptory any particular ordering norm that mandatorily disables a State's international law-making capacity." Christenson, "Guarding Interests," p. 594. See also Brownlie, *Principles of International Law*, p. 514. The consequent difficulty of creating new peremptory norms by positive-law processes has been noted. See Onuf and Birney, p. 192. In the event of the emergence of a new peremptory norm, the persistent objector rule would not apply. Weil, p. 430.

394. See Sztucki, *Jus Cogens*, pp. 3, 119. Most states were concerned that the undefined *jus cogens* principle posed a threat to the doctrine of national sovereignty upon which the positivist character of modern international law has rested. See n. 384 above. The United States, for example, complained that *jus cogens* "is undeniable as an abstract proposition but is so lacking in legal content that there is no way of judging its effects." *Analytical Compilation of Comments and Observations Made in 1966 and 1967 with Respect to the Final Draft Articles on the Law of Treaties: Working Paper Prepared by the Secretariat* (U.N. Doc. A/CONF.39/5 February 10, 1968), p. 324. The International Law Commission had declined to make a formal recommendation with respect to the content of *jus cogens* norms, deferring the issue to subsequent international jurisprudence and state practice. Christenson, "Guarding Interests," pp. 604–5. There was nevertheless a consensus among Commission members that most international norms are not peremptory in nature. Verdross, *"Jus Dispositivum* and *Jus Cogens,"* p. 58. Discussions within the Commission also yielded informal agreement on the content of several peremptory norms. See nn. 397–99 below and accompanying text.

395. See Sztucki, *Jus Cogens*, pp. 58–59, 77, 81–84; and also Christenson, "Guarding Interests," p. 646. Among the few judicial decisions to invoke *jus cogens*, a U.S. military tribunal ruled in 1948 that any international agreement providing for the employment of prisoners of war in violation of the 1929 Geneva Convention would have been void as *contra bonos mores.* Sztucki, *Jus Cogens*, pp. 23–24. It is somewhat easier to identify norms from which derogation is permitted. Diplomatic immunities, for example, are frequently cited as an example of *jus dispositivum.* See Sztucki, *Jus Cogens*, p. 70 n. 277; and also Alexidze, p. 253.

396. An agreement derogating from the rule that treaty obligations must be observed "poses a logical conundrum, for its validity would appear to depend on the very norm which it purports to abolish." Sinclair, p. 207. Cf. Kelsen, *Principles of International Law*, pp. 304, 314. On this point, Grotius and Kelsen were in fundamental agreement, although from methodologically different perspectives. See Lauterpacht, "The Grotian Tradition," p. 22. Selden, too, agreed that, in contrast to the *"Permissive"* branch of natural law (i.e., positive law), "the *Universal Obligatory Law* . . . provides for the due observations of Compacts and Covenants." Selden, p. 25. "A rule such as *pacta sunt servanda* is natural to the international community of states because there would be no such community without such a rule." Janis, "The Nature of *Jus Cogens,"* p. 363. See also Burton, pp. 1162–63.

Derogation from the international legal personality of states or the principle of *res inter alios acta* also has been described as "a virtual logical impossibility." Sztucki, *Jus Cogens*, pp.

72, 80. See also Alexidze, p. 262. The principle of *jus cogens* must itself be peremptory. See n. 392 above and accompanying text. Other rules governing the validity and legal effect of international conventions may also be regarded as peremptory. See Verdross, "Forbidden Treaties," p. 572.

397. See, e.g., Christenson, "Guarding Interests," p. 586; and also Alexidze, pp. 262–63. See also *Rest. 3rd, Restatement of the Foreign Relations Law of the United States*, §102, comment (o). One commentator has incorrectly limited the entire content of *jus cogens* to such "moral or humanitarian" norms. Cf. Arrow, "The Proposed Regime," p. 382. "Most of the cases in this class are cases where the position of the individual is involved, and where the rules contravened are rules instituted for the protection of the individual." *Third Report* (U.N. Doc. A/CN.4/115), in *Yearbook of the International Law Commission, 1958*, vol. 2, p. 40 (para. 76). During the 20th century there has been a growing acceptance of the principle that a state may not violate these norms, even with respect to its own nationals. In 1987 the Inter-American Commission on Human Rights, applying a "shock the conscience" test, found that the United States had violated a peremptory norm prohibiting state execution of juveniles; yet the United States may join South Africa and the Soviet Union in refusing to accept domestic application of international *jus cogens* principles, the Nuremberg prosecutions notwithstanding. Christenson, "Guarding Interests," pp. 621–24, 638. See George A. Finch, "The Nuremberg Trial and International Law," *American Journal of International Law* 41 (1947): 22.

398. Sztucki, *Jus Cogens*, p. 102 n. 409. *Rest. 3rd, Restatement of the Foreign Relations Law of the United States*, §§102, comment (k), 331 (2) (a), 905, comment (g). See *Yearbook of the International Law Commission, 1966*, vol. 2, p. 247; and Christenson, "The World Court and *Jus Cogens*," pp. 93–94. Thus, states may not conclude alliances for the purpose of waging wars of aggression. *Third Report* (U.N. Doc. A/CN.4/115), in *Yearbook of the International Law Commission, 1958*, vol. 2, p. 40 (para. 76). In contrast, states may lawfully derogate from the principle of nonintervention, as has been done in the case of the European Convention on Human Rights. Sztucki, *Jus Cogens*, p. 50. Similarly, states may agree to avoid application, *inter se*, of the principles of territorial integrity and sovereign equality of states. Alexidze, p. 260.

399. "It is arguable that . . . because they purport to affect the rights of third States, such treaties are not only unenforceable against such States, but are also in themselves void on account of the fact that their object is illegal — such illegality consisting in the attempt to interfere with the rights of a third State in disregard of rules of international law." *Report on the Law of Treaties* (U.N. Doc. A/CN.4/63), in *Yearbook of the International Law Commission, 1953*, vol. 2, p. 154. See Vienna Convention on the Law of Treaties, art. 34, in *International Legal Materials* 8 (1969): 693; and also Kelsen, *Principles of International Law*, p. 345 et seq. International practice tends to support this view. Sztucki, *Jus Cogens*, p. 189. Yet the Commission's *Third Report* distinguished rules of *jus cogens* from rules of *jus dispositivum* which affect the interests of third parties. See *Yearbook of the International Law Commission, 1958*, vol. 2, p. 40 (para. 76). The distinction seems theoretically sound in light of the capacity of third states to consent to infringement of their positive-law rights, whereas states would lack the capacity to bind their nationals in contravention of natural-law peremptory norms. Cf. Vienna Convention on the Law of Treaties, art. 35, in *International Legal Materials* 8 (1969): 693. See nn. 385, 393 above and accompanying text; and also Sztucki, *Jus Cogens*, pp. 70–72.

Some agreements may unavoidably interfere with rights of third parties, as when upstream states deprive downstream states of water by diverting the path of a river, and such agreements would necessarily be invalid in the absence of third-party consent. In such cases, the applicable norm (preserving the riparian rights of downstream users) has the force of *jus cogens* only vis-à-vis the (upstream) states parties, and might therefore be termed quasi-peremptory. The principles of nonintervention and territorial sovereignty may in this sense also be regarded as quasi-peremptory, agreements in derogation of such principles being invalid in the absence of consent by the affected state. "Generally, it may be said that treaties affecting the rights of third States *may be* but *need not necessarily be* illegal — e.g., when it follows from the texts of treaties or from the conduct of the parties thereto that they rely on an eventually forthcoming consent of third States whose rights are affected." Sztucki, *Jus Cogens*, p. 170 (emphasis in original).

400. "The criterion for these rules consists in the fact that they do not exist to satisfy the needs of the individual states but the higher interest of the whole international community. Hence these rules are absolute." Verdross, *"Jus Dispositivum* and *Jus Cogens,"* p. 58. See also Gaja, p. 281. Thus the Commission mentioned international crimes — defined in article 19 of the Draft Articles on State Responsibility as "breach by a State of an international obligation so essential for the protection of fundamental interests of the international community that its breach is recognized as a crime by that community as a whole" — as acts which cannot be legitimized even *inter se* by agreement, citing massive pollution of the human environment as an example. Sinclair, pp. 213, 215. See Christenson, "Guarding Interests," p. 618 n. 139; and also Gaja, pp. 299–300. The International Court of Justice has recognized that such "obligations *erga omnes"* must by their very nature involve the rights of all states. Barcelona Traction, Light and Power Company, Limited, Judgment, *I.C.J. Reports 1970,* p. 32 (para. 33). See Gaja, pp. 280–81. "The term objective illegality means the objective recognition of an illegality, as such, which can, therefore, be invoked with a view to its extinction by all members of the international community regardless of whether there is a particular damage sustained by the invoking State or States." Rozakis, *The Concept of Jus Cogens,* p. 24. See Sinclair, pp. 104–5. The United States evidently views the obligation of states to prevent pollution of the "common" marine environment as *erga omnes* in nature. *Rest. 3rd, Restatement of the Foreign Relations Law of the United States,* §604, comment (a).

401. See Gerald Fitzmaurice, "The Law and Procedure of the International Court of Justice: General Principles and Substantive Law," *British Year Book of International Law* 27 (1950): 8–9. Elsewhere, Fitzmaurice emphasized that the *erga omnes* validity of such a regime depends upon the participation of all interested states. Bruno Simma, "The Antarctic Treaty as a Treaty Providing for an 'Objective Regime,'" *Cornell International Law Journal* 19 (1986): 193. See n. 400 above; and also *Rest. 3rd, Restatement of the Foreign Relations Law of the United States,* §324, comment (d). "If the totality of these areas is enlarged or diminished the general interest is *directly* affected." Meijers, p. 22 n. 50 (emphasis in original).

402. Grotius, pp. 15–16, 50. Grotius's observations were premised upon conditions of inexhaustibility. See n. 267 above, and nn. 257–58 above and accompanying text. Since the *res communis* principle *requires* international regulation under conditions of exhaustibility, acquiescence to prescriptive unilateral acts under such conditions would be a legitimate method of developing a customary regulatory regime. See nn. 262, 270, 279–80, 283–86, 301, 303, 307–10 above and accompanying text.

403. See, e.g., Rozakis, *The Concept of Jus Cogens,* p. 15; Tucker, p. 40; and Riesenfeld, "Comment," p. 13 n. 13. This view was shared by at least one delegation at the Vienna Conference. See Sztucki, *Jus Cogens,* p. 120.

404. See, e.g., Sztucki, *Jus Cogens,* pp. 20, 81; Brownlie, *Principles of Public International Law,* p. 514; C. Wolff, vol. 2, pp. 7, 69; and also Vattel, p. 188. "[T]here would seem to be no reason why two States shall not agree that, *as between themselves,* the width of territorial waters should be fifty miles." *Report on the Law of Treaties* (U.N. Doc. A/CN.4/63), in *Yearbook of the International Law Commission, 1953,* vol. 2, p. 154 (emphasis in original). This view finds support in the decisions of international tribunals. Georg Schwarzenberger, *International Law,* 3rd ed. (London: Stevens and Sons, 1957), vol. 1, pp. 352–53. States thus are free to derogate from high-seas freedoms as well as from the application of zones of offshore jurisdiction and from rights of innocent passage, but only *inter se.* Alexidze, p. 253. See nn. 406, 408 below and accompanying text; and also Gaja, p. 280 n. 10. The exclusive economic zone "derogates *massively* from the traditional principle of freedom of the high seas, one of whose most basic constitutional elements being the freedom to fish." Gotlieb, p. 199 (emphasis in original). Theoretically states may also derogate from the norm prohibiting piracy, although the difficulty of limiting victims to nationals of the derogating states has led to a view of this norm as peremptory. See Alexidze, p. 231; and *Third Report* (U.N. Doc. A/CN.4/115), in *Yearbook of the International Law Commission, 1958,* vol. 2, p. 40 (para. 76).

405. See n. 298 et seq. above and accompanying text. Because the *res communis* principle requires the establishment of a regulatory regime in cases of exhaustibility, it is inaccurate to argue that the creation of such a regime by the process of customary international law would be inconsistent with the *jus cogens* status of the freedom of the seas doctrine. See n. 407 below and accompanying text; and also Christenson, "Guarding Interests," p. 613.

406. *Third Report* (U.N. Doc. A/CN.4/115), in *Yearbook of the International Law Commission, 1958*, vol. 2, p. 40 n. 54. See Sztucki, *Jus Cogens*, p. 72 n. 282; and Verdross, "Forbidden Treaties," p. 572. Significantly, the Commission has concluded that "massive pollution . . . of the high seas" would be an international crime prohibited by the principle of *jus cogens*. See Sinclair, pp. 213, 215; and also n. 400 above. "If . . . a State entered into a convention with another State not to interfere in case the latter should appropriate a certain part of the open sea . . . , such treaty would be null and void." L. Oppenheim, *International Law*, 3rd ed. (London: Longmans, Green, 1920), vol. 1, p. 663. See also Kelsen, *Principles of International Law*, p. 344. In contrast to the principle of the freedom of the seas, nonappropriation is a true peremptory norm. Thus, Grotius maintained that appropriation of the sea would be invalid, whether claimed by virtue of custom, prescription, convention, or papal donation. Grotius, pp. 15–16, 46–47, 52. De Castro and Vasquez agreed. Fenn, *The Right of Fishery*, pp. 150–51. See also Justice, p. 7.

407. See nn. 50, 202, 284 above and accompanying text. Thus, where conditions of exhaustibility call for implementation of a regulatory regime, a state may not make use of the sea in a manner opposed by a large portion of the international community, any more than it may continue to rely on the freedom of the seas doctrine, notwithstanding implications to the contrary in the 1951 Anglo-Norwegian Fisheries case and the 1974 Fisheries Jurisdiction case. See Charney, "The United States and the Law of the Sea After UNCLOS III," pp. 45–46; n. 286 above and accompanying text; and also Jennings, p. 46. Significantly, the International Law Commission has indicated that the common heritage of mankind principle "expresses . . . a collective interest," infringement of which would cause injury to all states (assuming the principle became binding customary law; otherwise its *erga omnes* application would be limited to states parties). *Yearbook of the International Law Commission, 1985*, vol. 2, p. 27. See nn. 378, 445 above and below and accompanying text.

408. Because derogation may be given effect only *inter se*, such norms may be described as quasi-peremptory. See nn. 399, 406 above. Thus, "no state can avoid, vis à vis other states, the new prohibition on fishing within 200 miles of the coast without the permission of the coastal state." Meijers, p. 22 n. 50. Nor may two states agree to partition another state's continental shelf among themselves. Stefan A. Riesenfeld, "The Stability of Customary International Law," in *United States Law of the Sea Policy*, ed. Shutler, p. 95 n. 17.

409. Azuni, vol. 1, p. 11. See Roxburgh, pp. 32–33; C. Wolff, vol. 2, p. 70; and Pufendorf, vol. 2, p. 562. "The sovereign himself will bring war upon himself, if he refuses the sea to others; and those will be justified in making war who are refused a privilege of nature." Gentili, vol. 2, p. 92. Similarly, conflict may also be expected to result where exhaustible uses of the sea are not governed by a generally accepted regulatory regime. See nn. 43, 441–43 above and below and accompanying text.

410. Fulton, p. 3. See n. 343 above and accompanying text.

411. Reppy, p. 249 n. 23. See nn. 323–26 above and accompanying text; and also Azuni, vol. 1, p. 24 et seq. Before destroying Carthage, Rome had its consul inform the Carthaginians of the justification for the attack: "It is the sea, it is the power that you have there acquired, it is those treasures which you have drawn from it, that hasten your ruin." C. Marcius Figulus, quoted in Azuni, vol. 1, p. 172. Similarly, Nebuchadnezzar sacked Tyre after the Phoenicians had ruthlessly interfered with high-seas navigation throughout the ancient world. See Reppy, p. 247 n. 15. "Many other nations have, in their turn, disputed the empire of the sea, and its waves have often been tinged with blood, which the unbridled ambition of ruling over that element has produced." Azuni, vol. 1,, p. 24, and see pp. 12, 23.

412. Azuni, vol. 1, p. 20. "It is this which formerly dissolved the government of our ancestors, the happiest that was ever devised, which now involves us in much confusion and distress, and which, in one word, is the occasion of all the calamities we have either suffered or inflicted." Isocrates, quoted in ibid.

413. John Dos Passos, *The Portugal Story* (Garden City, N.Y.: Doubleday, 1969), p. 48. See nn. 340, 343 above; and Fulton, pp. 4, 105–11, 338–39, 527–29. "The Sound dues run like a thread through more than four centuries, and their history is intertwined with contests for the possession of the Sound and the dominion over the Baltic." Hill, p. vii. See also Orfield, pp. 4–7; and Newton, "Inexhaustibility," p. 383. The emergence of Sweden as a military power in the 16th century led to a renewed regional conflict, in which maritime dominance — and its

prevention — were a principal goal. Orfield, pp. 236–38. At the beginning of the 17th century, the Hanseatic League formed an alliance with the Dutch "for the preservation and maintenance of the freedom of Navigation," after Denmark sought to increase the Sound dues. Selden, pp. 121–22. The exclusion of Dutch fishermen from broad areas of the sea off the Scandinavian coast "called forth an energetic protest from the States-General, and affairs took a bellicose turn. . . . Hostilities were averted by the intercession of Sweden, and of the British and French Ministers at Copenhagen in favour of the Dutch Republic and the freedom of the seas." Fulton, p. 529. England successfully opposed subsequent attempts by Denmark and Sweden to exercise control over Baltic navigation, and Danish claims to exclusive fishing rights in the northern Atlantic were relinquished in the 18th century under threat of war from England and Holland. See Hall, pp. 185–86.

   414. See Gustav Droysen, "The Statesman of 'Realpolitik,'" in The Thirty Years' War, ed. Rabb, p. 90; and also Fulton, pp. 376–77. "The struggle for the 'dominium maris Baltici' set Gustavus in opposition to Protestant Denmark, Catholic Poland and Orthodox Russia. . . . Wallenstein's appointment as 'General of the Atlantic and Baltic Seas' (1628) threatened Sweden's maritime position: her vital interests demanded armed intervention against the Catholic Hapsburgs and alliance with Catholic France." S.H. Steinberg, "The Not So Destructive, Not So Religious, and Not Primarily German War," in The Thirty Years' War, ed. Rabb, p. 49. Gustavus Adolphus was greatly influenced by Grotius. See Lawrence, p. 42.

   415. See nn. 340, 343 above; and also Fulton, pp. 3–4, 339, 539–43. Genoa performed a similar police function. See Azuni, vol. 1, p. 84. Grotius, quoting Vasquez, nevertheless disapproved such claims. Grotius, p. 53. Gentili agreed. See Reppy, p. 278 n. 108.

   416. Fulton, pp. 10, 12, 112, 117, 150, 327–28, 337, 378, 518. L. Brown, pp. 35–40. See also Hall, pp. 183–84. "Had Charles [I] been able to give effect to his selfish and ambitious scheme, he would soon have been confronted with an overwhelming coalition of maritime Powers, to whom the free use of the sea was as necessary as it was to England." Fulton, p. 210. War with Holland erupted when James I claimed exclusive British rights to "enormous numbers" of whales off the island of Spitzbergen, and similar claims by Denmark provoked conflict later in the century. Fulton, pp. 181–85, 527–28. Although Holland signed a mutual protection treaty with France against Britain's exclusive fisheries claims in 1662, the Dutch "claimed the right of fishing in the open sea by the law of nations; that it was a right independent of any treaties, which merely illustrated and explained it." Fulton, pp. 451–53.

   417. L. Brown, pp. 88–92, 133, 164–65.

   418. Charles Maurice de Talleyrand, quoted in L. Brown, p. 135. "Talleyrand wrote French consuls how France was setting about to free the seas for the cause of all humanity." L. Brown, p. 148. England's broad assertion of belligerent rights provoked hostile reactions among other European maritime states. Only delicate diplomatic maneuvering averted the formation of a maritime league by Denmark, Sweden, Russia, and Holland. L. Brown, p. 80. See also Fulton, p. 21.

   419. President Franklin D. Roosevelt, "Freedom of the Seas," International Conciliation 373 (October 1941): 650–51, 655. "The Hitler government, in defiance of the laws of the sea, in defiance of the recognized rights of all other nations, has presumed to declare, on paper, that great areas of the seas, even including a vast expanse lying in the Western Hemisphere, are to be closed and that no ships may enter them for any purpose, except at risk of being sunk." Ibid., p. 651.

   420. See H.R. Trevor-Roper, The Crisis of the Seventeenth Century (New York: Harper and Row, 1968), pp. 48–49. Between them, the two Iberian powers sought to bar foreign vessels from the Pacific Ocean, the Indian Ocean, and most of the Atlantic Ocean. "Men of any nation who braved the terrors of excommunication and were found in forbidden waters were to be seized and treated as 'corsairs and violators of the peace.'" L. Brown, p. 8. Other maritime states continued to expand their expeditionary voyages, however, and international conflict intensified throughout the 16th century as competition for New World resources increased. See Hyma, p. 91.

   421. Diffie and Winius, pp. 173–74, 283. See Bull Romanus Pontifex (1455), reprinted in European Treaties, ed. F. Davenport, pp. 22–26; and also H. Vander Linden, "Alexander VI and the Demarcation of the Maritime and Colonial Domains of Spain and Portugal, 1493–1494," American Historical Review 22 (1916): 12–13. Alexander "owed his election in the

previous year largely to the support of Ferdinand and Isabella and was, on that account, willing to grant them whatever they asked." Bell, p. 183. Conflict between the Iberian kingdoms had occurred earlier in the century when Portugal sought to exclude Spanish ships from navigation in the Atlantic. See n. 341 above. After 1494, however, Spain and Portugal became allies in their efforts to enforce their exclusive claims. See Quinn, p. 127.

422. R.B. Mowat, *A History of European Diplomacy 1451-1789* (London: Edward Arnold, 1928), pp. 26–27. See A.J. Grant, *The French Monarchy (1483-1789)* Cambridge: Cambridge University Press, 1914, vol. 1, pp. 21–24; and also Hyma, p. 91. Although Charles VIII of France claimed Naples by hereditary right — as did the illegitimate relations of Ferdinand — and although Ludovico Sforza of Milan had requested French intervention in order to advance his own parochial interests, the attack was still unexpected and unexplained, Charles having consulted only two of his advisers on the matter. Philip Comines, *Historical Memoirs*, orig. pub. 1524 (London: J. Davis, 1817), p. 422. "Ever since the summer of 1494, when the first clumps of French lances trotted down into the Lombard plain, people have puzzled over why Charles accepted Ludovico's invitation.... The only explanation for the Neopolitan adventure that Commynes could arrive at was that the king was young and silly and had bad counsellors." Mattingly, pp. 135–36. See also Mowat, p. 11. Reconstruction of the exact motivation of the French king has, moreover, been rendered virtually impossible by the destruction of most of the diplomatic records of the time. See J. Williamson, p. 58.

Circumstantial evidence nevertheless points to Spain's discoveries in the New World — and the exclusive legal rights thereto which Spain was able to secure from "the Spanish Pope" — as a chief cause of French belligerence. In January 1493, two months before the return of Columbus prompted the Spanish monarchs to seek a swift ratification of exclusive rights from Pope Alexander, Spain had agreed not to contest Charles's claim to Naples. See Grant, pp. 20–21. In May, when the first papal bulls were issued, Charles informed Alexander's envoy that he would march to Naples; notwithstanding the agreement, Spain became the leading member of a league formed in early 1495 by the Pope to repel the French invasion. John S.C. Bridge, *A History of France from the Death of Louis XI* (Oxford: Clarendon, 1924), vol. 2, p. 66. Although incomplete historical documentation makes it difficult to determine the reason for Spain's reversal — commonly attributed to King Ferdinand's Machiavellian instincts — the timing strongly suggests that the discovery of the New World precipitated an immediate reevaluation of the importance of relations with the papacy — in Spain as well as in France, although from the opposite perspective — and that the invasion was perceived as an alarmed response to the newly changed geopolitical situation resulting from papal endorsement of the Spanish claims. Significantly, at a time when he was working to persuade England to enter the Italian Wars against France, Ferdinand regarded Henvy VII's support for John Cabot's 1496 western voyage as a product of Charles's diplomatic machinations. Michael van Cleave Alexander, *The First of the Tudors* (Totowa, N.J.: Rowman and Littlefield, 1980), p. 147. "From Henry's point of view, Charles VIII's involvement in Italy was something to be encouraged rather than opposed." M. Alexander, p. 111. Without an ocean-going navy, France was unable to challenge Spain's closed-sea policies directly, but in the early sixteenth century French corsairs began to sail into the western hemisphere in search of private gain. Herbert Ingram Priestley, *France Overseas Through the Old Regime* (New York: D. Appleton-Century, 1939), pp. 36–37. See n. 427 below and accompanying text.

423. See M. Alexander, pp. 144–55; and also Wood, pp. 4–5. "They were to find new lands anywhere to the west of England and Iceland, or to the north or the east (but not to the south, where Columbus was entrenched and Spanish and Portuguese authority effective), and to govern them in the king's name." Quinn, p. 93. The patents were renewed, despite the vehement protests of the Spanish ambassador. See Wood, p. 8. Similar expeditions to "the new found land" were authorized in the early years of the 16th century in an effort to discover a north-west passage to the Orient. See Kenneth R. Andrews, *Trade, Plunder and Settlement* (New York: Cambridge University Press, 1984), pp. 48–49. Henry "refused to agree that the two Iberian kingdoms were entitled to a monopoly of trade with the Far East because of a bilateral treaty between them or a papal bull of 1493 to that effect." M. Alexander, p. 150.

424. Quinn, pp. 130–31, 143–44, 160–61. See Andrews, *Trade, Plunder and Settlement*, pp. 55–56; and also Wood, pp. 10–11. Although there is no hard evidence of a direct causal relationship, it is noteworthy that the expeditions which Henry VIII did send westward

departed in 1527 and 1536, after "the diplomatic revolution of 1525," which marked the end of England's cultivated alliance with Emperor Charles V—ruler of Spain, patron of the Catholic Church and nephew of Henry's wife, Catherine of Aragon—and after the English king had decided to seek an end to his marriage from an uncooperative Pope. Alan G.R. Smith, *The Emergence of a Nation State* (New York: Longman, 1984), pp. 18, 47–48. "The problem of the succession was no doubt uppermost in Henry's mind when he brooded about his personal problems, but there were other factors which helped to convince him that both his own happiness and the welfare of his kingdom would be best served by repudiating Catherine." A. Smith, p. 18. "Henry's serious involvement with Charles V . . . is likely to have underlined for him the unwisdom of penetrating the Spanish monopoly of sea and land power in the west until it could be done with adequate force." Quinn, p. 182.

425. A. Smith, p. 49. "Oversea dominions are nothing without sufficient sea power, naval and mercantile, to win, to hold, and foster them." Wood, p. 20.

426. A.L. Rowse, *The Elizabethans and America* (New York: Harper and Brothers, 1959), p. 7. See Henry Folmer, *Franco-Spanish Rivalry in North America* (Glendale, Calif.: Arthur H. Clark, 1953), pp. 35–41, 45–61; and also Priestley, p. 41. Cartier's commission "leaves no doubt that the voyage was a royal challenge to Spanish rights in America." Folmer, p. 37. Although Francis, like other European rulers, was sometimes willing to compromise navigation rights in order to accommodate other geopolitical objectives, he became recognized as the first modern monarch to adopt the freedom of the seas as an official basis for foreign policy. L. Brown, p. 9. Francis "did all he could to break the Iberian monopoly." Priestley, p. 43.

427. L. Brown, pp. 9–11. See also John Lynch, *Spain Under the Hapsburgs*, 2nd ed. (New York: New York University Press, 1981), vol. 1, p. 165. "The French were the first vigorously to make their way into the distant regions, from which the Pope, Portugal, and Spain desired to exclude them." *European Treaties*, ed. Davenport, p. 2. See W.J. Eccles, *France in America* (New York: Harper and Row, 1972), pp. 2–3; and also McAlister, pp. 200, 259, 268.

With authorization from the admiral of Normandy, Jean Ango led dozens of raids on the Spanish Indies during Francis's reign. Priestley, pp. 39–40. See Folmer, pp. 33–34; and also J. Williamson, p. 28. "In the thought of Ango's men, when they took a Spanish or Portuguese prize, whether they had letters or not the act was just reprisal, inasmuch as their comrades, taken in forbidden waters, had been treated as pirates." L. Brown, pp. 11–12. See also David B. Quinn and A.N. Ryan, *England's Sea Empire, 1550–1642* (London: George Allen and Unwin, 1983), p. 79. In 1535 Charles V offered to concede the duchy of Milan to France in an unsuccessful effort to induce Francis to acknowledge the papal bulls of 1493 and to renounce French rights to sail to the Indies. Folmer, pp. 43–44. "The long campaign of pillage reached its climax in the 1550s in an offensive of unprecedented force and range, culminating in the sack of Havana by Jacques de Sores in 1555." Andrews, *Trade, Plunder and Settlement*, p. 118. See Folmer, pp. 65–66; and also McAlister, pp. 201, 235.

428. *European Treaties*, ed. Davenport, pp. 3, 219–21. See Folmer, pp. 67–71; J. Williamson, pp. 90, 165, 198–99; McAlister, pp. 200–1, 235; and also Lynch, vol. 1, pp. 152–53. "The policy thus inaugurated long remained a rule of the sea; Spaniards and Portuguese treating as pirates all foreigners found beyond the Line, and French authorities ruling that Spanish and Portuguese vessels might be seized beyond the Line until the rulers in question should admit the right of the French to trade freely in Indian and American seas." L. Brown, pp. 13–14.

Even when French monarchs refrained from challenging the division directly, "the absence of letters patent of discovery, of formal state sanction, or royal rewards did not deter the individual Frenchmen from asserting their rights to the ocean and to the liberty of the seas, in spite of papal excommunications and Spanish legal speculations." Folmer, p. 33. The Huguenot effort, directed by Admiral Gaspard de Coligny, was encouraged by Catherine de Medici. Edward Dwight Salmon, *Imperial Spain* (Westport, Conn.: Greenwood, 1971), p. 102. The massacre of Huguenot settlers in Florida by Spanish troops sent by Philip II "added fuel to the growing Protestant detestation of Spain," as did Spanish mistreatment of English merchantmen captured beyond the line. Rowse, pp. 9, 11. See Folmer, pp. 78–115; and also Salmon, pp. 127–28.

In 1530 merchants from Plymouth had begun commercial voyages to Guinea and Brazil, prompting Portuguese countermeasures. Andrews, *Trade, Plunder and Settlement*, pp.

59–60. Although this trade halted temporarily in the 1540s, by the late 1550s "it was English doctrine and established practice that West Africa was free to all." J. Williamson, pp. 33, 45. See also Kenneth R. Andrews, *Drake's Voyages* (New York: Charles Scribner's Sons, 1967), pp. 7–8, 13. Before his death in 1553, Edward VI approved a trading voyage by Thomas Wyndham to Guinea, but with the accession of Queen Mary and her husband, Philip, royal support for such ventures was replaced by active opposition. Andrews, *Trade, Plunder and Settlement*, pp. 106–8. "The moment Elizabeth came to the throne the government reversed Mary's prohibition upon trading to Guinea." Rowse, p. 5.

The first English capture of a Spanish treasure ship in 1545 began a long period of private warfare against the Spanish maritime monopoly. "The lawlessness of the privateersmen of the 1540s was a feature of the movement, as was the emergence among them of anti–Spanish sentiments, though England was officially at war with France." Quinn and Ryan, p. xii. See Wood, p. 32; and also Andrews, *Drake's Voyages*, p. 8. By the 1560s there had emerged a "regular pattern of Anglo-French, or at any rate Anglo-Huguenot, co-operation against the monolithic power of Spain." Rowse, p. 8. See also Quinn and Ryan, p. 77. In 1569, following new efforts by Portugal and Spain to enforce their exclusive maritime claims against foreigners, Dutch "sea-beggars" joined French and English privateers in an unofficial four-year war against all Catholic shipping. Andrews, *Trade, Plunder and Settlement*, p. 111. See Mitchell B. Garrett, *European History 1500–1815* (New York: American Book, 1940), pp. 205–6.

429. See Lynch, vol. 1, pp. 165–67. Although she was careful to avoid direct confrontation with Spain early in her reign, Elizabeth and her advisers frequently provided official and unofficial support for such voyages. Quinn and Ryan, pp. 24, 32. See also Andrews, *Drake's Voyages*, p. 13. Hawkins left Plymouth on his first slave-trading voyage in 1562, crossing the Atlantic without Spanish permission and returning with a profitable cargo, part of which was promptly seized when the English merchant sought to dispose of it in Seville. Elizabeth defended Hawkins against Portuguese charges of trespass and intervened unsuccessfully on his behalf with Philip II. J. Williamson, pp. 54–57. See also Andrews, *Trade, Plunder and Settlement*, pp. 109–10, 117. In 1564, when Hawkins departed again for Africa and the Caribbean, he sailed under the royal standard on one of the Queen's own ships, although without an official patent. J. Williamson, pp. 62–64, 202. Elizabeth "contributed a ship because this was her expedition, but her motive could not have been trading profit." J. Williamson, p. 65. See also Walter Oakeshott, *Founded Upon the Seas* (Cambridge: Cambridge University Press, 1942), pp. 35–36; and Quinn and Ryan, pp. 28–29. Hawkins's third such voyage, which was ended by a Spanish attack at San Juan de Ulua, included two ships belonging to the Queen, who issued specific sailing instructions. See J. Williamson, p. 110.

After sailing with Hawkins, Drake drew upon growing anti-Spanish sentiment in organizing his own expeditions in 1570, 1571, and 1577, the latter voyage occurring with the open support of the Queen, who contributed one of her own ships to the predatory circumnavigation venture. Rowse, pp. 26–27. See also Quinn and Ryan, pp. 30–31, 82–83. "Piracy against Spain was now normally legitimized by Drake's success and the Queen's acceptance of the major part of his spoils." Quinn and Ryan, p. 38. See also Andrews, *Drake's Voyages*, p. 85. Drake's 1585 foray was designed to damage Spain's financial and political security further by attacking the plate fleet and Spanish garrisons en route. See Andrews, *Trade, Plunder and Settlement*, pp. 280–81. "Drake was not only a pirate; he was also a national enemy striking at strategic positions within the imperial economy, and the subsequent loss of confidence among Spanish merchants was profound." Lynch, vol. 1, p. 172.

More than a dozen other private ventures into forbidden waters were organized during the 1570s, and Elizabeth supported a 1578 expedition to prey on Spanish shipping in the Caribbean. Lynch, vol. 1, pp. 129, 188–89. Several other raiders followed Drake into the Atlantic in the mid-1580s, and Thomas Cavendish further harassed Spain on his own successful voyage of circumnavigation. See n. 432 below; and also Quinn and Ryan, pp. 134–35, 139. "There were others, how many we do not know." J. Williamson, p. 199.

430. Rowse, pp. 5, 11, 24. See also Quinn and Ryan, p. 91. The revolt in the Netherlands was the other principal issue of contention between England and Spain. Soon after Elizabeth acceded to the throne, the Spanish ambassador was informed that England did not accept the validity of the papal bulls supporting the exclusive claims of Spain and Portugal. Rowse, pp. 5–6. In 1568 Elizabeth granted letters of reprisal against Portuguese shipping after Portugal

interfered with English merchants, and such letters were freely issued against Spain after 1585. Andrews, *Trade, Plunder and Settlement*, pp. 111, 245. See Bell, p. 140.

Drake's 1577 voyage of circumnavigation, intended by Elizabeth as a means of retaliating against Spain's maritime countermeasures, "immensely raised the prestige of England on the threshold of the conflict, and helped to produce it – as the Queen knew it might." Rowse, pp. 26–27. See also Wood, p. 146. Upon Drake's return in 1580, Elizabeth told Spain's protesting ambassador that the Spanish and Portuguese claims were contrary to natural law: "The use of the sea and air is common to all; neither can any title to the ocean belong to any people or private man, forasmuch as neither nature nor regard of the public use permitteth any possession thereof." H. Smith, p. 44. See Azuni, vol. 1, pp. 130–31; and also Rowse, pp. 31–32. By 1582 the Queen was refusing even to hear the ambassador's remonstrances. Quinn, p. 267. "As Elizabeth's position grew stronger we observe her coming out more openly on the side of her seamen, fighting their battle politically and diplomatically, as she well knew how." Rowse, p. 6. Drake's 1585 expedition proved so successful that Philip felt constrained to organize much of Spain's far-flung naval forces for a direct attack upon the source of the escalating raids. Andrews, *Drake's Voyages*, p. 85. See Salmon, pp. 132–33; J. Williamson, p. 295; and also Oakeshott, p. 103. "Although English intervention in the Low Countries was one item in the series of provocations which caused Philip II to accept the project of invading his enemy, the Armada was regarded primarily as part of the defence system of the Spanish empire, which would repulse the English attack on it by striking at its roots." Lynch, vol. 1, p. 153. See also Quinn, p. 268. The huge invasion force was turned away by the English and by a storm which wrecked dozens of Spanish ships and scattered the remainder. See Reppy, p. 254. The war continued in the Atlantic, where English privateers captured more than 1,000 ships, severely damaging Spain's maritime capability and fostering a new sense of English national identity. Andrews, *Trade, Plunder and Settlement*, pp. 247–49, 253. See also Quinn and Ryan, pp. 115, 123.

431. J. Williamson, p. 193. "In Drake, as among the crews of the ships that sailed under his leadership, the flame of Puritanism burned fiercely, as fiercely as there glowed in Philip of Spain the conviction that Protestantism was frightful heresy, and that to exterminate it was to do God's work." Oakeshott, p. 31.

432. See Quinn, pp. 207–8, 225–26; and also *European Treaties*, ed. Davenport, pp. 5, 246–48. "[A]lthough the presence of interlopers was almost permanent, it was only from 1620 and still more from 1630 that they really began to make nonsense of the monopoly." Lynch, vol. 1, p. 163. See also Andrews, *Trade, Plunder and Settlement*, p. 300. Despite his own assertions of sovereignty over the British Seas, Charles I authorized the capture of shipping in foreign waters where freedom of navigation was prohibited, and the seizure of the island of Jamaica in 1655 was intended by Cromwell as a direct challenge to the continuing Spanish claims. L. Brown, pp. 33–34, 69–70. "The prevailing mood was one of national assertiveness and resentment of the exclusive claims of Spain in the New World." Andrews, *Trade, Plunder and Settlement*, p. 314.

433. McAlister, pp. 306, 429. See also *European Treaties*, ed. Davenport, pp. 5, 235–36, 329. The most difficult issue in peace negotiations from 1607–9 was that of access to the Indies, as Spain sought unsuccessfully to exclude Dutch merchants in exchange for the independence of the Netherlands. *European Treaties*, ed. Davenport, pp. 5, 258, 260–62. "Both sides regarded the question as vital. The Dutch believed the trade necessary to their existence. . . . To withdraw from a traffic which was allowed them by the laws of nature and of nations would prejudice their status as a sovereign power, and the principle of the freedom of the seas." *European Treaties*, ed. Davenport, p. 260. Maritime conflict over the issue continued after expiration of the truce in 1621. See *European Treaties*, ed. F. Davenport, pp. 353–54.

The disputes engendered by the maritime conflicts of the 16th and 17th centuries gave rise to a great deal of nationalistic argumentation regarding the legal status of the sea; the ultimate acceptance of *Mare Liberum* was "materially facilitated by the exploits of Drake, Hawkins, and Cavendish on the part of the English, and of Jakob von Heemskerk on the part of the Dutch." Fulton, pp. 5, 340. Grotius specifically denied the validity of papal bulls – or of customary or prescriptive rights – as a basis for the Spanish and Portuguese maritime claims, declaring "that a *pronunciamento* of this sort has no force" vis-à-vis third states. Grotius, pp. 46–47. See n. 440 below and accompanying text.

434. Charles Gibson, *Spain in America* (New York: Harper and Row, 1966), pp. 164–65.
435. See L. Brown, pp. 66–68. As the 19th century began, "Spain was still holding to her theory of colonial monopoly, having, indeed, little but the theory to which to cling." Ibid., p. 173.
436. See nn. 345, 422, 426, 428, 431 above and accompanying text; and also Pardo, "The Law of the Sea: Its Past and Its Future," pp. 9–10. The division created "real and lasting antagonism" throughout Europe. D. Johnston, p. 164. There was "serious renewed conciliar agitation" the following year. John A.F. Thomson, *Popes and Princes, 1417–1517* (Boston: George Allen and Unwin, 1980), p. 19. "The whole structure of the Church seemed to totter as the fifteenth century drew to a close." Hyma, p. 55. In 1540, in response to the Spanish ambassador's protests against Cartier's third voyage, Francis I expressly rejected the authority of the papal donation and defiantly dispatched his ships into the regions claimed by Spain. Folmer, pp. 35, 50–51. Although Henry VII of England initially respected the papal decree, Henry VIII and Elizabeth I viewed their dispatch of expeditions westward as a direct challenge to papal authority. Ward, vol. 2, pp. 112, 471–72. Colombos, pp. 46–47.
The Reformation could not have succeeded in the absence of the popular disenchantment with the workings of the Catholic Church which spread throughout Europe in the wake of Alexander VI's corrupt practices. Garrett, p. 167. The Donation of 1493 was regarded as a reassertion of the medieval canon law doctrine that the Pope was supreme and infallible — "in some way like a god on earth." Heinrich Boehmer, *Road to Reformation*, trans. John W. Doberstein and Theodore G. Tappert (Philadelphia: Muhlenberg, 1946), p. 377. The papacy had claimed for itself peremptory authority to invalidate agreements adverse to its interests. Ward, vol. 2, p. 125. "Christianity, by being abused, gave rise to maxims, too long received, which it would be an outrage upon common sense to endeavour to justify." Ward, vol. 2, p. 77. The special relationship between Alexander and the Spanish monarchs and its consequences for non-Iberian states in the New World has been well documented. See Linden, pp. 13–15; and also Thomson, p. 123. "The sixteenth century presents us with an extensive and important change, not only in the affairs, but in the public opinions of Europe. . . . All that deference which had been paid, and all those privileges which had been allowed to the Pope, as the head and father of the Christian Republic, were completely annihilated among nations adhering to the *Protestant Faith*." Ward, vol. 2, pp. 466–67. See also Holdsworth, p. 588.
The three papal bulls issued in May 1493 only prohibited foreigners from visiting land territories beyond the line of demarcation. A fourth bull issued several months later, however, while confirming the Bull *Inter Caetera* of May 4, expressly prohibited "all persons" from going beyond the demarcation line "for the purpose of navigating or of fishing, or of searching for islands or mainlands." Bull *Dudum Siquidem* (September 26, 1493), reprinted in *European Treaties*, ed. Davenport, p. 82. Portugal claimed similar exclusive maritime rights pursuant to a papal bull issued in 1455. See n. 421 above and accompanying text. Papal authority to issue bulls relating to undiscovered territories was based upon the Pope's power to regulate relations with heathen peoples and his power to mediate disputes between Christian rulers. Gibson, p. 15. The latter of these bases provided no grounds for an *erga omnes* exclusion of non-Iberian states, however, and the former provided no authority for an allocation of exclusive navigation rights.
437. See Thomson, pp. 24, 27. "Luther nailed his 95 theses to the cathedral door of the Schlosskirche in Wittenberg, scarcely a generation after the Treaty of Tordesillas." Pardo, "The Law of the Sea: Its Past and Its Future," p. 9.
438. Thomson, p. 21.
439. Before becoming Pope Alexander VI through the support of his native Spain, Rodrigo Borgia had prevailed upon Pope Sixtus IV to authorize the Spanish Inquisition. Lynch, vol. 1, p. 23. "Theoretically the Spanish inquisition existed only to combat heresy, but whether the Inquisitors were concerned with domestic heretics or with Protestant English merchants might vary in relation to how well English Catholics were treated in England, to whether Dutch privateers were allowed to use English ports . . . and to how active English privateers were in the Indies." J. Leitch Wright, Jr., *Anglo-Spanish Rivalry in North America* (Athens: University of Georgia Press, 1971), p. 20. In exchange for papal support for exclusive Spanish control over access to the West Indies, Spain undertook the religious indoctrination of native American populations. Lynch, vol. 1, p. 158. See McAlister, p. 437. "Charles V

aimed at establishing the religious unification of Christendom under his political supremacy and his son accepted that aim as the great object of his life, the very reason for his existence." Salmon, pp. 99–100. See also Lynch, vol. 1, pp. 273, 275. Frequently, the political interests of Philip II and the papacy converged: "there is no doubt that the Anglo-Spanish War of 1585–1604 was a decisive event in the struggle of Counter-Reformation Catholicism to suppress the Reformation." A. Smith, p. 236. See n. 431 above. After acceding to the Spanish throne in 1621, Philip IV emulated Philip II's policies. See Pages, *The Thirty Years' War*, pp. 86–87.

440. Grotius, p. 66. See Potter, p. 59. In the spirit of his times, Grotius affirmed that "the Pope is neither civil nor temporal Lord of the whole world." Grotius, p. 16.

441. See nn. 262, 279–86, 407 above and accompanying text.

442. See Shinaver, pp. 172–73; and also Fulton, pp. 10, 14, 170–75, 286–308, 529–30, 581, 584–85, 604–49. During the 20th century, in addition to the seizure of U.S. fishermen by South American states, open conflict over fisheries emerged between Japan and a number of American states and between Britain and Iceland (the "Cod Wars"). See Sanger, p. 141.

443. See n. 368 above and accompanying text; *Law of the Sea* (U.N. Doc. A/41/742), p. 27; and also Fulton, pp. 604–49, 662. Although the IWC established a moratorium on commercial whaling in 1982, its enforcement powers are limited. See Elizabeth A. Wehrmeister, "Giving the Cat Claws: Proposed Amendments to the International Whaling Convention," *Loyola of Los Angeles International and Comparative Law Journal* 11 (1989): 417–38. The United States has sought to enforce IWC standards through bilateral sanctions, and in 1986 a private conservation organization attempted to enforce the ban by scuttling two Icelandic whaling ships in Reykjavik Harbor. See *Oceans Policy News* (October/November 1986), p. 2.

To avoid a moratorium on driftnet fishing, fishing states such as Japan were required to agree to take effective conservation measures based upon "the best available scientific data." U.N. General Assembly, Resolution 44/225. See n. 368 above. "The states most directly involved in [driftnet] fishing are agreed to the principle that conservation measures are to be imposed when sound statistical analysis of the available evidence shows they are needed, but they are not prematurely required to cease their operations nor necessarily cease them at all, if adequate conservation measures can be devised." Burke, "Driftnets and Nodules," pp. 239–40.

444. See n. 35 above and accompanying text. Significantly, states parties are permitted to derogate only from provisions "applicable solely to the relations between them," and no derogating agreements may be concluded by states parties with respect to "the basic principles" of the Convention or to "the basic principle relating to the common heritage of mankind set forth in article 136." U.N. Convention on the Law of the Sea, art. 311 (3), (6). Any treaty language which could be construed as implicitly authorizing derogation from a peremptory norm by nonparty states would, to that extent, itself be invalid.

445. See nn. 64, 204, 216 above; and also Oxman, "The Eighth Session," p. 38. Although the 1987 agreement contained a Soviet commitment to "respect" the claims of the private consortia for an indefinite period — an agreement which could constitute an invalid derogation from the *res communis* principle's requirement of multilateral regulation for use of exhaustible marine resources, should the consortia ultimately decide to proceed with unilateral nodule mining — the preamble indicated that the parties were "[d]esirous of removing impediments to the universal adherence to the United Nations Convention on the Law of the Sea of 1982." Agreement on the Resolution of Practical Problems with Respect to Deep Seabed Mining Areas, and Exchange of Notes Between the United States and the Parties to the Agreement, preamble, art. 2, reprinted in *International Legal Materials* 26 (1987): 1505, 1506.

446. See nn. 216, 392 above. A new state coming into existence is bound by the rules of international law in existence at the time. Kelsen, *Principles of International Law*, p. 433. See *Rest. 3rd, Restatement of the Foreign Relations Law of the United States*, §§102, comments (d), (e), 206, comment (a). In any event, the United States has not persistently objected to the emergence of a norm requiring — eventually, at least — a multilateral regulatory regime. See nn. 225–26, 238–45 above and accompanying text.

Chapter 6

1. See Chap. 3 nn. 415–16 and accompanying text, and Chap. 5 nn. 155–214 and accompanying text. But see Chap. 5 n. 139 and accompanying text. The revised *Restatement*

essentially endorses the U.S. position. See American Law Institute, *Restatement 3rd, Restatement of the Foreign Relations Law of the United States*, Part V. "In most of Part V, the *Restatement* makes no attempt to provide either the details of the alleged state practice which would support its assertions about customary law or to cite the evidence that might confirm that the designated practices occur as claimed." W.T. Burke, "Customary Law of the Sea: Advocacy or Disinterested Scholarship?" *Yale Journal of International Law* 14 (1989): 508-27. In 1972 the United States had endorsed the negotiation of a single, comprehensive UNCLOS III convention in order to prevent states from accepting and rejecting regimes on an individual basis. American Bar Association, "Natural Resources of the Sea: Report of Section of Natural Resources Law," in *Status Report on the Law of the Sea Conference*, U.S. 93rd Congress, 1st Session, Senate, Committee on Interior and Insular Affairs, Subcommittee on Minerals, Materials and Fuels (Washington: U.S. Government Printing Office, 1973), pp. 670–71.

2. See Chap. 5 nn. 108–10 and accompanying text. Cf. *Rest. 3rd, Restatement of the Foreign Relations Law of the United States*, Introductory Note to Part V.

3. Luke T. Lee, "The Law of the Sea Convention and Third States," *American Journal of International Law* 77 (1983): 552. See Chap. 5 n. 106. "[P]rospective parties to the Law of the Sea Convention which have not denied third states' rights or have remained silent may intend that rights be conferred on nonparties and may tacitly agree to that." L. Lee, p. 549.

4. See L. Lee, p. 549; and Chap. 5 nn. 134–35 and accompanying text.

5. See L. Lee, p. 548; Chap. 5 nn. 149–50 and accompanying text; and also James K. Sebenius, *Negotiating the Law of the Sea* (Cambridge, Mass.: Harvard University Press, 1984), pp. 92–93.

6. José Luis Jesus, "Statement on the Issue of the Universality of the Convention," University of Kiel, July 1990, reprinted in *Oceans Policy News*, Special Report (July 1990), p. 2.

7. See *Oceans Policy News* (December 1989), p. 3; and also Satya N. Nandan, "The 1982 UN Convention on the Law of the Sea: At a Crossroad," *Ocean Development and International Law* 20 (1989): 516.

8. Council on Ocean Law, "The United States and the 1982 Convention on the Law of the Sea: A Synopsis of the Status of the Treaty and Its Expanded Role in the World Today," September 1989, p. 5. See Chap. 1 n. 26 and accompanying text. Arvid Pardo was particularly concerned about the potential impact of such "creeping baselines" upon the size of the international Area.

The 1982 Convention provides that the low-water line is normally the baseline, and would permit the use of straight baselines only "where the coastline is deeply indented and cut into, or if there is a fringe of islands along the coast," although archipelagic states may employ straight baselines as long as 125 miles to connect the outermost points of their outermost islands. U.N. Convention on the Law of the Sea, arts. 5, 7, 47.

9. See *Oceans Policy News* (October/November 1987), pp. 3–4; and also nn. 52–53, 63, 66 below and accompanying text. Indonesia has also objected to efforts within IMO and ICAO to implement provisions of the 1982 Convention. Council on Ocean Law, "The United States and the 1982 Convention on the Law of the Sea," p. 6. See also *Oceans Policy News* (June 1989), pp. 2–3.

10. Elliot L. Richardson, "Power, Mobility and the Law of the Sea," *Foreign Affairs* 58 (1980): 909–10. "Foreign ocean law and practice increasingly depart from the Convention's non-seabeds provisions." Council on Ocean Law, "The United States and the 1982 Convention on the Law of the Sea," p. 8.

Ironically, the *opinio juris* necessary for the formation of new rules of customary international law may be lacking where state practice is a product of coercion or negotiated bilateral arrangements rather than of a sense of legal obligation. See Chap. 5 n. 157.

11. Letter from Charles N. Brower, Acting Legal Adviser, Department of State, to Senator Henry M. Jackson, July 26, 1973, reprinted in *Status Report*, U.S. Senate, pp. 395–96.

12. U.S. Commission on Marine Science, Engineering, and Resources, *Our Nation and the Sea: A Plan for National Action* (Washington: U.S. Government Printing Office, 1969), pp. 84–86.

13. See Chap. 3 nn. 381, 428, 474, 511–12 and accompanying text. "Countries apparently feel that only guaranteed access of their private or State companies to the sea-bed resources

Notes — Chapter 6

can provide the efficiency of development and security of supply necessary." UNCLOS III, *Report by Mr. P.B. Engo, Chairman of the First Committee on the Work of the First Committee* (U.N. Doc. A/CONF.62/L.16, September 6, 1976), reprinted in *Official Records, UNCLOS III,* vol. VI, p. 133.

    14. James L. Malone, "Who Needs the Sea Treaty?" *Foreign Policy* 54 (Spring 1984): 45. See Chap. 4 n. 603 and accompanying text. "Free market economists believe that American foreign policy has too long been dominated by geopolitical and strategic considerations, as well as by the same confidence in the efficacy of governmental regulation that has characterized domestic policy." Bernard H. Oxman, "Introduction: On Evaluating the Draft Convention on the Law of the Sea," *San Diego Law Review* 19 (1982): 455.

    15. Panel on the Law of the Ocean Uses, Council on Ocean Law, "Deep Seabed Mining and the 1982 Convention on the Law of the Sea," September 25, 1987, p. 5. The Panel nevertheless agreed on the importance of achieving universal agreement on a comprehensive convention.

    16. See Chap. 2 n. 208. "The 1986 oil price collapse was so severe that, after adjustment for inflation and tax changes, the price of crude oil received by domestic producers was about the same as the prices received in the late 1960s." G. Kevin Jones, "Outer Continental Shelf Petroleum Resources and the Nation's Future Energy Needs," *Syracuse Journal of International Law and Commerce* 15 (1989): 325. In June 1990 the inflation-adjusted price of oil was still lower than it had been in 1974. "The gloom-and-doomers paid little attention to the reality that soaring oil prices set by OPEC would trigger efficiency, conservation and discovery of new sources of oil." Hobart Rowen, "Oils Well That Ends Well," *Washington Post National Weekly Edition,* June 18–24, 1990, p. 5. The same, economically rational response by consumers is expected to increase the nonseabed supply of nodule minerals as demand grows. See James M. Broadus, "Seabed Materials," *Science* 235 (February 20, 1987): 859.

    17. Invoking Chapter VII of the U.N. Charter, the United States successfully persuaded the Security Council to implement a costly embargo — and subsequent military action — against Iraq following its invasion of Kuwait, demonstrating the willingness of members of the international community to cooperate and to sacrifice immediate national economic interests, in large part to assure long-term access to regional oil reserves.

    18. See Chap. 2 nn. 205–7 and accompanying text; and also Finn Laursen, *Superpower at Sea* (New York: Praeger, 1983), pp. 104–5.

    19. Raymond F. Mikesell, *NonFuel Minerals: Foreign Dependence and National Security* (Ann Arbor, Mich.: The University of Michigan Press, 1987), p. 205.

    20. Telephone conversation with LeRon Bielak, Chief of Staff, Program Analysis and Issues Management, Office of Strategic and International Minerals, Minerals Management Service, U.S. Department of the Interior, November 9, 1990. The United States has been building a strategic stockpile of cobalt — traditionally regarded as the most strategically important nodule mineral because of its use in making jet engines — and supplies are now regarded as adequate, particularly in light of improved U.S. relations with former Soviet-bloc states. Ibid. See Chap. 2 n. 164 and accompanying text, and Chap. 5 n. 172.

    The Reagan administration had classified cobalt and manganese — but not copper — as "first tier" strategic minerals, but the Office of Technology Assessment concluded that U.S. strategic needs could be more readily met from alternative sources such as substitution and conservation than from seabed mining. Broadus, p. 858. See U.S. General Accounting Office, *Impediments to U.S. Involvement in Deep Ocean Mining Can Be Overcome: Report to the Congress* (Washington: U.S. Government Printing Office, 1983), Summary, p. 3.

    21. See Chap. 2 n. 195 and accompanying text, and Chap. 3 nn. 397–402 and accompanying text. "Before investing, banks normally require borrowers to obtain exclusive rights to an ore body." Richard A. Frank, "Jumping Ship," *Foreign Policy* 43 (1981): 134.

    The strategic value of seabed minerals should normally be reflected in larger and more rapid investment in deep-sea mining. Broadus, p. 859. Because, under the U.S. economic system, the decisionmaking procedures of private banks — or, for that matter, of private mining firms — are not designed to reflect broader public policy objectives, a system of tax incentives would probably be necessary to promote national strategic goals effectively. See Chap. 4 nn. 124–25 and accompanying text. State-run mining operators, on the other hand, tend to consider a much broader range of socioeconomic factors in determining the "profitability" of seabed mining activities. See U.N. Office for Ocean Affairs and the Law of the Sea, "Seminar on the Current Status of Developments in Deep Sea-Bed Mining Technology (New York, 18 and 19 August 1988)," *Law of the Sea Bulletin* 12 (December 1988): 63; and also Chap. 3 n. 406 and accompanying text.

22. See Chap. 3 nn. 407–18 and accompanying text; and also Peter E. Halbach et al., "Cobalt-Rich and Platinum-Bearing Manganese Crust Deposits on Seamounts: Nature, Formation, and Metal Potential," *Marine Mining* 8 (1989): 37–39. Alternatively, U.S. seabed mining operators may conclude contractual agreements with other coastal states having such offshore deposits. See Clyde Sanger, *Ordering the Oceans: The Making of the Law of the Sea* (London: Zed Books, 1986), p. 195.

23. See Chap. 3 nn. 406, 412 and accompanying text. The issue had caused "not a minimum amount of debate" within the Interior Department's Office of Strategic and International Minerals. Telephone conversation with LeRon Bielak, November 9, 1990.

24. Giulio Pontecorvo, "Musing About Seabed Mining, or Why What We Don't Know Can Hurt Us," *Ocean Development and International Law* 21 (1990): 118.

25. American Bar Association, *Revised Report with Recommendation* (as adopted by the House of Delegates, February 13, 1990), p. 4. The Association "strongly" recommended U.S. ratification of the Convention following the negotiation of unspecified "changes and clarifications" to Part XI. Ibid., p. 5.

26. U.S. General Accounting Office, *Impediments to U.S. Involvement*, Summary, p. 6. See Chap. 4 n. 417 and accompanying text. "Some oil company representatives and others in the mining consortia privately acknowledge that the treaty really isn't all that bad compared with many business deals in the Third World." Tom Alexander, "The Reaganites' Misadventure at Sea," *Fortune* 106 (August 23, 1982): 143–44. The American Mining Congress has nevertheless declared "that ocean mining by U.S. companies could not and would not take place under the LOS regime." Statement of Theodore Kronmiller, Deputy Assistant Secretary of State for Oceans and Fisheries Affairs, in *Law of the Sea Negotiations*, U.S. 97th Congress, 2nd Session, Senate, Committee on Foreign Relations, Subcommittee on Arms Control, Oceans, International Operations, and Environment (Washington: U.S. Government Printing Office, 1983), p. 4.

27. Panel on the Law of Ocean Uses, p. 5. See also Mikesell, p. 142.

28. See Stephen Vasciannie, "Part XI of the Law of the Sea Convention and Third States: Some General Observations," *Cambridge Law Journal* 48 (1989): 95–96.

29. See, e.g., n. 67 below and accompanying text. "Even when a concept is accepted as customary law, . . . there often exist significant differences in the provisions adopted by states to implement it. Many instances of state practice have been protested. Cumulatively, these variations give cause for concern, in particular for the exercise of rights of navigation and overflight. Such discrepancies are unlikely to decrease with time and, as long as the Convention is not in force, none of the institutional mechanisms it provides to resolve disputes are available." Philippe Kirsch and Douglas Fraser, "The Law of the Sea Preparatory Commission After Six Years: Review and Prospects," *Canadian Yearbook of International Law* 26 (1988): 125.

30. Closing statement by President Koh, April 30, 1982, quoted in Burke, "Customary Law of the Sea," p. 523, n. 57. See Chap. 3 nn. 26, 154 and accompanying text. "In fact very few states have declared that the margin beyond 200 nautical miles is subject to coastal state sovereign rights." Burke, "Customary Law of the Sea," p. 523.

31. U.N. Convention on the Law of the Sea, art. 82 (2). See Burke, "Customary Law of the Sea," p. 523. The revenues would not be available to the Enterprise, but would be distributed among states parties to the Convention "on the basis of equitable sharing criteria, taking into account the interests and needs of developing States, particularly the least developed and the land-locked among them." U.N. Convention on the Law of the Sea, art. 82 (4).

32. Secretary of State Henry Kissinger, address before American Bar Association, Montreal, August 11, 1975, reprinted in *U.S. Department of State Bulletin* 73 (1975): 356. See Chap. 3 n. 64. U.S. oil companies would thereby be assured of internationally recognized rights and could be spared the financial burden of production fees by the use of tax credits or other adjustments to federal royalty payments. See Chap. 4 nn. 124–25 and accompanying

text. Revenue sharing within this zone was "important to the U.S. and to the success of the Conference." Statement of Ambassador T. Vincent Learson, May 20, 1976, in *Law of the Sea*, U.S. 94th Congress, 2nd Session, Senate, Committee on Foreign Relations, Subcommittee on Oceans and International Environment (Washington: U.S. Government Printing Office, 1976), p. 28.

33. Ann L. Hollick, *U.S. Foreign Policy and the Law of the Sea* (Princeton, N.J.: Princeton University Press, 1981), p. 345. Cf. U.N. Convention on the Law of the Sea, art. 82 (3). "The United States had sought to replace this exemption with a scheme that would have permitted developing countries to opt out of the revenue-sharing system for a fixed period." Jeffrey Lee Gertler and Paul Wayne, "Synopsis: Recent Developments in the Law of the Sea 1979–1980," *San Diego Law Review* 18 (1981): 545 n. 85.

34. See Chap. 4 n. 119 and accompanying text. The offer was made, with full knowledge that the U.S. government was in fact forgoing revenues which it would otherwise receive, "as part of an overall settlement despite our conclusion from previous exploitation patterns that a significant portion of the total international revenues w[ould] come from the continental margin off the United States in early years." Statement of Ambassador John R. Stevenson before Seabed Committee, August 10, 1972, in *Status Report*, U.S. Senate, p. 410. See H. Gary Knight, "The Draft United Nations Conventions on the International Seabed Area: Background, Description, and Some Preliminary Thoughts," *San Diego Law Review* 8 (1971): 533–34. Landlocked states, faced with expansive coastal-state jurisdictional claims which were diminishing the size of the international Area, had originally sought direct access to the outer-margin resources of neighboring states. Jonathan I. Charney, "United States Interests in a Convention on the Law of the Sea: The Case for Continued Efforts," *Vanderbilt Journal of Transnational Law* 11 (1978): 52, n. 29.

35. Letter from Charles N. Brower, July 26, 1973 (Attachment 5: V.E. McKelvey, Director, U.S. Geological Survey, "Summary of Procedures in Oil and Gas Leasing and Regulation on the U.S. Outer Continental Shelf"), in *Status Report*, U.S. Senate, p. 464. See Robert B. Krueger, "International and National Regulation of Pollution from Offshore Oil Pollution Production," *San Diego Law Review* 7 (1970): 561–62 n. 77; and also Jonathan I. Charney, "The Equitable Sharing of Revenues from Seabed Mining," in *Policy Issues in Ocean Law*, American Society of International Law (St. Paul, Minn,: West, 1975), p. 108 n. 109. In 1970 alone, oil companies paid $2.1 billion in bonus bids for tract leases in the Gulf of Mexico, and by 1976 the United States had collected total revenues of more than $15 billion. Charney, "The Equitable Sharing of Revenues," pp. 71–72 n. 64. U.S. oil companies nevertheless "were well satisfied with the Outer Continental Shelf Lands Act and its administration and did not see it as a deterrent to further offshore exploration and exploitation activities." Knight, "The Draft United Nations Conventions," p. 548.

36. Charney, "The Equitable Sharing of Revenues," p. 71 n. 64. See James L. Johnston, "Geneva Update," in *The Law of the Sea: U.S. Interests and Alternatives*, ed. Ryan C. Amacher and Richard James Sweeney (Washington: American Enterprise Institute for Public Policy Research, 1976), p. 161. It was once estimated that less than 5 percent of U.S. offshore petroleum reserves were located in this area. Charney, "United States Interests," p. 52 n. 29. By 1990 tracts had been leased for oil drilling operations more than 100 miles from shore in the Gulf of Mexico. Telephone conversation with LeRon Bielak, November 9, 1990. See Joseph R. Vadus, "Ocean Technology in the United States: Recent Advances, Future Needs and International Collaboration," *Marine Technology Society Journal* 24 (March 1990): 79. Recently discovered oil deposits in the "deep Gulf" could reverse the decline in U.S. offshore petroleum production. See Jean-Luc Gadon, "Offshore Oil Prospects for the Year 2000," *Natural Resources Forum* 11 (1987): 302.

37. Alexander F. Holser, "Offshore Lands of the USA: The US Exclusive Economic Zone, Continental Shelf and Outer Continental Shelf," *Marine Policy* 12 (1988): 7, table 1. The amount of outer continental margin proposed for such leasing has been identified elsewhere as 215 million acres. Michael Cruichshank, Joseph P. Flanagan, Buford Holt and John W. Padan, *Marine Mining on the Outer Continental Shelf* (Washington: U.S. Department of the Interior, 1987), p. 66.

38. Telephone conversation with LeRon Bielak, November 9, 1990. Because of the inter-

nationally disputed status of the outer continental margin, leasing beyond 200 miles was "out," so far as the Department was concerned. Ibid.

The jurisdictional authority exercised by the Interior Department has thus been based upon the doctrine of the exclusive economic zone rather than the traditional doctrine of the continental shelf, in accordance with a 1985 opinion issued by the Department's Solicitor. See Holser, pp. 2–3. The distinction is significant in light of official U.S. adherence to the 1958 definition of the continental shelf, apparently in an effort to avoid the revenue-sharing obligations imposed under article 82 of the 1982 Convention as a condition of exercising jurisdiction over the resources of the outer margin. See n. 40 below.

39. See Charney, "The Equitable Sharing of Revenues," pp. 91–92. High demand for petroleum products may actually favor imposition of royalty charges for purposes of conservation, as well as revenue maximization. See Charney, "The Equitable Sharing of Revenues," p. 91; and Chap. 4 nn. 189–90 and accompanying text.

40. Charney, "The Equitable Sharing of Revenues," p. 114. Cf. *Rest. 3rd, Restatement of the Foreign Relations Law of the United States*, §515, comments (a), (b), reporters' note 1.

Federal regulations adopted in 1988 pursuant to the Seabed Act still defined the continental shelf by the exploitability criterion, which was set forth in the 1958 Continental Shelf Convention but has since been rejected by the international community. Deep Seabed Mining Regulations for Commercial Recovery Permits, 15 CFR §971.101 (f) (1). See Chap. 1 nn. 43, 84 et seq. and accompanying text, Chap. 3 n. 242 and accompanying text, and Chap. 5 nn. 29–31, 190 and accompanying text.

41. Victor D. Comras, "U.S. Interests in an LOS Treaty," in *Current Issues in the Law of the Sea*, ed. Christopher C. Joyner (Allentown, Penn.: Muhlenberg College, 1980), p. 139. See U.N. Convention on the Law of the Sea, art. 62.

42. U.N. Convention on the Law of the Sea, arts. 73 (2), (3), (4), 292. Similar provisions apply with respect to vessels detained for violating environmental standards. Ibid., arts. 226 (1) (b), (c), 230, 231.

43. Hollick, *U.S. Foreign Policy*, pp. 236–37. See Chap. 2 n. 79 and accompanying text, and Chap. 5 nn. 145–48 and accompanying text. "[T]he ambiguous doctrine of innocent passage would otherwise apply, and states bordering straits would . . . decide which ships and planes should, and which should not, pass." President Richard M. Nixon, "U.S. Foreign Policy for the 1970s: The Emerging Structure of Peace: A Report to the Congress," reprinted in *U.S. Department of State Bulletin* 66 (1972): 410.

44. Comras, p. 138. See John A. Knauss, "Factors Influencing a U.S. Position in a Future Law of the Sea Conference," Occasional Paper No. 10 (Kingston, R.I.: Law of the Sea Institute, 1971), p. 17.

45. See James A. Hazlett, "Strait Shooting," *U.S. Naval Institute Proceedings* 108 (June 1982): 71; William L. Chaze, Col. H.G. Summers, Jr., Robert A. Kittle and Nicholas Daniloff, "Rust to Riches: The Navy Is Back," *U.S. News and World Report*, August 4, 1986, p. 28; and also T. Alexander, p. 144. For a table detailing missile acquisitions by states bordering the Straits of Gibraltar and Indonesian straits since 1966, see Hazlett, p. 72.

46. See Melissa Healy, "Army May Bear Brunt of Budget Cuts, Cheney Hints," *Los Angeles Times*, November 25, 1989, pp. A1, A30; and also statement of James Malone in *Nomination of James L. Malone*, U.S. 97th Congress, 1st Session, Senate, Committee on Foreign Relations (Washington: U.S. Government Printing Office, 1981), p. 31.

47. Hazlett, p. 73.

48. Sanger, p. 199.

49. Cf. U.N. Convention on the Law of the Sea, arts. 18–21, 34–39, 42–45. See Thomas A. Clingan, Jr., "Freedom of Navigation in a Post-UNCLOS III Environment," *Law and Contemporary Problems* 46 (Spring 1983): 117–18. The situation is similar with respect to archipelagic sea lanes passage. See Frank B. Swayze, "Negotiating a Law of the Sea," *U.S. Naval Institute Proceedings* 106 (July 1980): 36. Cf. U.N. Convention on the Law of the Sea, arts. 49, 52–54.

50. See statement of William T. Burke in *Consensus and Confrontation: The United States and the Law of the Sea Convention*, ed. Jon Van Dyke (Honolulu: Law of the Sea Institute, 1985), pp. 294–95. Cf. Presidential Proclamation No. 5928, December 27, 1988, reprinted in *U.S. Department of State Bulletin* 89 (March 1989): 72. "A mark of the weakness

of the U.S. position on the customary law of navigation rights is that the *Restatement* not only makes no attempt to cite evidence of state practice but is apparently untroubled with the lack of any real information on the issue." Burke, "Customary Law of the Sea," p. 513.

51. Swayze, p. 35.

52. Hazlett, p. 71. See also Sanger, p. 89.

53. See "US Security Arrangements in the Persian Gulf," *Gist* (July 1987), pp. 1–2; L. Lee, p. 548; and also Joseph L. Galloway et al., "The Gulf War – Among Friends," *U.S. News and World Report*, June 15, 1987, p. 26. Such expenditures on behalf of Kuwait, a prominent OPEC member and strong supporter of the Law of the Sea Convention, seem ironic when it is recalled that the hysteria engendered by the OPEC oil embargoes of the 1970s contributed substantially, albeit indirectly, to U.S. rejection of the Convention. It seems even more ironic that in 1990 the United States was prepared to sacrifice billions of additional dollars in order to ensure that embargoed Kuwaiti oil *not* be exported through the Straits of Hormuz during the Iraqi occupation which followed Iraq's long, U.S.-backed war against Iran. See n. 17 and accompanying text above. In 1974 it had been believed that U.S. maritime interests in the Persian Gulf could be secured by providing $2 billion in arms to Iran. Sebenius, p. 87.

In 1985 approximately 20 percent of non-Communist oil supplies traversed the Straits of Hormuz. G. Jones, p. 336.

54. See Sebenius, pp. 87–91; and also Peter Polomka, *Ocean Politics in Southeast Asia* (Singapore: Institute of Southeast Asian Studies, 1978), pp. 46–47. "The distinctive function of a widely accepted LOS treaty, in contrast with country-by-country understandings, is its tendency to lift transit rights out of the domain of the discretionary, tradable, and disputable." Sebenius, p. 88. Even U.S.-Canadian relations were threatened in 1985 by U.S. navigation through the Northwest Passage. See *Oceans Policy News* (January/February 1988), p. 8. "Whether the price is lack of support of the claimant nation on some esoteric issue before the United Nations or the failure to gain approval for a nuclear-powered warship visit, the assertion of U.S. rights in the face of these excessive claims does carry a price." Dennis R. Neutze, "Whose Law of the Sea?" *U.S. Naval Institute Proceedings* 109 (January 1983): 45.

U.S. efforts to preserve the customary three-mile limit to the territorial sea through state practice proved unsuccessful in the face of expansive 20th-century coastal-state claims. See Chap. 5 n. 187 and accompanying text; and also Henry M. Arruda, "The Extension of the United States Territorial Sea: Reasons and Effects," *Connecticut Journal of International Law* 4 (1989): 720. During the 1980s U.S. naval vessels and aircraft challenged some three dozen objectionable coastal-state claims each year "as a matter of routine." U.S. Department of State, "Rights and Freedoms in International Waters," *U.S. Department of State Bulletin* 86 (May 1986): 79. See Chap. 3 n. 416 and accompanying text; and also John D. Negroponte, Assistant Secretary of State for Oceans and International Environmental and Scientific Affairs, "Who Will Protect Freedom of the Seas," *U.S. Department of State Bulletin* 86 (October 1986): 41–43.

55. See Chap. 3 n. 23; and Burke, "Customary Law of the Sea," p. 514 n. 19. For example, the key requirement that coastal-state environmental regulations comport with "generally accepted international regulations, procedures and practices" might not find its way into customary law. See n. 61 below. Cf. U.N. Convention on the Law of the Sea, arts. 39 (2) (b), 42 (1) (b), 54. See Sanger, p. 93. Maritime states would also be unable to enforce the regime through the Convention's binding dispute settlement procedures. See nn. 111–13 below and accompanying text.

56. Cf. U.N. Convention on the Law of the Sea, arts. 55–58, 73, 223–33, 292. See Chap. 3 nn. 31, 102, and nn. 100–6 below and accompanying text; and also Sanger, pp. 111–12.

57. American Bar Association, "Natural Resources of the Sea," p. 643. See n. 42 above and accompanying text.

58. "U.S. Delegation Report," in *Law of the Sea*, U.S. Senate, 1976, p. 40.

59. See nn. 108–13 below and accompanying text.

60. John Temple Swing, "A Treaty for the Taking," *Oceans* 14 (January-February 1981): 2–3. "Although gunboat diplomacy may work in isolated incidents, the [United States] doubtless does not want to employ the confrontational approach as its official ocean policy." William Turbeville, "American Ocean Policy Adrift: An Exclusive Economic

Zone as an Alternative to the Law of the Sea Treaty," *University of Florida Law Review* 35 (1983): 515.

61. Eduardo Lachica, "U.S. Intends to Back Request by Indonesia for $2 Billion in Assistance During 1983," *Wall Street Journal*, June 6, 1983, p. 24. "Of the two Malay states, Indonesia tends to adopt the more flexible position in accommodating superpower interests." Polomka, p. 43. See also Sebenius, p. 87.

Ultimately, Indonesia and Malaysia were unable to maintain their 1971 claims, which were withdrawn two years later after objecting maritime states provided assurances that the UNCLOS III transit passage regime would include specific provisions to control vessel-source pollution. See Polomka, pp. 71–74; and Sanger, pp. 86–88.

62. Statement of Daniel O'Donohue, Deputy Assistant Secretary of State for East Asian and Pacific Affairs, before Subcommittee on Asian and Pacific Affairs, House Foreign Affairs Committee, March 18, 1983, reprinted in *U.S. Department of State Bulletin* 83 (May 1983): 42.

63. Council on Ocean Law, "The United States and the 1982 Convention on the Law of the Sea," pp. 5–6. See also Marian Nash Leich, "Contemporary Practice of the United States Relating to International Law," *American Journal of International Law* 83 (1989): 559–61.

64. See "Vital Sea-Lanes, Vital Bases," *U.S. News and World Report*, February 10, 1986, p. 30; and also statement of President Ronald W. Reagan, February 11, 1986, reprinted in *New York Times*, February 12, 1986, p. 11. By 1988, the United States was paying $481 million annually to lease air and naval bases from the Philippine government, and the Bush administration found it necessary to seek to negotiate an extension of lease rights for the bases despite a diminished Soviet military threat and despite agreement among several former Secretaries of State and national security advisers that the United States "should not ordinarily be dependent on" such bases to defend its Third World interests. See Jim Mann, "U.S. Move to Back Aquino Could Prove Costly Later," *Los Angeles Times*, December 2, 1989, p. A16. It was expected that much of the U.S. naval presence would be relocated to Singapore.

65. Statement of General George S. Brown, Chairman, Joint Chiefs of Staff, in *Extending Jurisdiction of the United States Over Certain Ocean Areas*, U.S. 93rd Congress, 2nd Session, Senate, Committee on Armed Services (Washington: U.S. Government Printing Office, 1974), pp. 35–37.

66. Burke, "Customary Law of the Sea," pp. 515–16. See Chap. 3 nn. 287–88, and nn. 239, 274 and accompanying text; and also Sanger, pp. 84–89. "Warships can and have been excluded." Knauss, p. 6.

Following the U.S. lead in "picking and choosing" the provisions of the 1982 Convention with which it would comply, the Soviet Union in 1987 sought to restrict the navigation of U.S. warships in the Black Sea. See *Oceans Policy News* (March 1988), pp. 1–2; and Lt. Commander Ronald D. Neubauer, "The Right of Innocent Passage for Warships in the Territorial Sea: A Response to the Soviet Union," *Naval War College Review* 41 (Spring 1988): 49–56. The issue was resolved bilaterally two years later — at an unknown cost to the United States — when the two states agreed upon a Uniform Interpretation of the Rules of International Law Governing Innocent Passage and issued a joint statement recognizing that the provisions of the Convention "generally" represent customary international law "with respect to traditional uses of the ocean." Marian Nash Leich, "Contemporary Practice of the United States Relating to International Law," *American Journal of International Law* 84 (1990): 239–40.

67. U.N. Convention on the Law of the Sea, arts. 56 (1), 58 (1), 95, 96. Although articles 95 and 96 refer only to ships navigating "on the high seas," they "apply to the exclusive economic zone in so far as they are not incompatible with" Part V. Ibid., art. 58 (2).

68. See ibid., arts. 19 (1), 30. Noninnocent passage would include the threat or use of force, use of weaponry "of any kind," intelligence-gathering activities, acts of propaganda "aimed at affecting the defence or security of the coastal State," the launching or landing of aircraft or military devices, interference with coastal-state facilities and communications, or "any other activity not having a direct bearing on passage." Ibid., art. 19 (2). In exercising rights of innocent passage in the territorial sea, warships and other state-run ships retain their immunity from coastal-state jurisdiction, but flag states "shall bear international responsibility for any loss or damage to the coastal State resulting from ... non-compliance ... with the laws and regulations of the coastal State." Ibid., arts. 30, 31.

69. Ibid., arts. 38 (1), 44, 53 (2), 54. Transiting ships and aircraft are required to "refrain

from any activities other than those incident to their normal modes of continuous and expeditious transit," including in particular any threat or use of force against bordering states. Ibid., art. 39 (1), 54.

70. See ibid., art. 20. The importance of submerged passage to the U.S. Navy has decreased with the commissioning of Trident-class submarines with longer-range missile capability. See Sanger, p. 98.

71. Cf. U.N. Convention on the Law of the Sea, art. 53 (5). See Sanger, p. 97.

72. U.N. Convention on the Law of the Sea, arts. 297 (1) (a), 298 (1) (b). See n. 110 below and accompanying text. Whether a particular dispute related to military activities would be a jurisdictional issue which the court or tribunal would have jurisdiction to resolve. Cf. U.N. Convention on the Law of the Sea, art. 286 (4).

73. Sanger, p. 127. C. Barry Raleigh, "The Internationalism of Ocean Science vs. International Politics," *Marine Technology Society Journal* 23 (March 1989): 45. The seizure of the USS *Pueblo* by North Korea in early 1968 was one of several instances where a purported research vessel has been revealed to be engaged in espionage. See Sanger, pp. 126–27; and Knauss, pp. 19–20. "The Government assigns a high priority to the military applications of marine science." U.S. Commission on Marine Science, Engineering, and Resources, *Our Nation and the Sea*, p. 29.

74. "U.S. Delegation Report," p. 38. The Truman Proclamation originated with President Roosevelt's concern that a European power might undertake oil drilling operations in the Gulf of Mexico. Hollick, *U.S. Foreign Policy*, p. 35.

75. Convention on the Continental Shelf, art. 5 (8), reprinted in *United Nations Treaty Series*, vol. 499, no. 7302, p. 316. See *Rest. 3rd, Restatement of the Foreign Relations Law of the United States*, §521, reporters' note 5.

76. John Temple Swing, "Address," in *Current Issues in the Law of the Sea*, ed. Joyner, p. 113.

77. See United States, "Reports of the United States Delegation to the Third United Nations Conference on the Law of the Sea," Occasional Paper No. 33, ed. Myron H. Nordquist and Choon-ho Park (Honolulu: Law of the Sea Institute, 1983), pp. 52–53; and also Sanger, p. 129.

78. U.N. Convention on the Law of the Sea, arts. 246 (2), (5), (6), 253. Consent is to be granted, "in normal circumstances," with respect to purely scientific research conducted "for the benefit of all mankind." Ibid., art. 246 (3). "Although these . . . provisions do not represent compelling improvements over the no-convention alternative, they do provide benefits to the marine scientists which may be most valuable in the future." Charney, "United States Interests," p. 47.

79. U.N. Convention on the Law of the Sea, arts. 248, 252. "[T]he rule of implied consent is itself a significant benefit of the Convention." Bernard H. Oxman, "The Third United Nations Conference on the Law of the Sea: The Ninth Session (1980)," *American Journal of International Law* 75 (1981): 237.

80. U.N. Convention on the Law of the Sea, art. 253.

81. Ibid., arts. 246, 253, 297 (2). Disputes relating to the exercise of discretionary coastal-state authority under articles 246 and 253 are subject to compulsory nonbinding conciliation. Ibid., art. 297 (2) (b).

82. Statement of President Ronald W. Reagan, March 10, 1983, reprinted in *U.S. Department of State Bulletin* 83 (June 1983): 70. "Access under regional and bilateral agreements has been available to marine scientists for many years. Such arrangements have not been sufficient in the past, and there is no tangible evidence supporting the view that they will be any more productive in the future." Jonathan I. Charney, "The United States and the Law of the Sea After UNCLOS III – The Impact of General International Law," *Law and Contemporary Problems* 46 (Spring 1983): 41.

83. See Sebenius, p. 89.

84. Comras, p. 140.

85. See Sanger, pp. 132–33.

86. Raleigh, p. 45. "Now we are besieged by research vessels waiting at the dock while bureaucratic formalities are observed, and scientists on both sides of the waters fume over the expense and delays while awaiting permission to conduct research." Ibid., p. 44. Although it

is unclear to what extent such failures to comply with the implied consent procedure may be a response to U.S. rejection of the Convention, it is noteworthy that Raleigh's private oceanographic research firm only began to encounter such difficulties in the late 1980s when it became necessary for authorization requests to be processed "through the State Department because of the law of the sea." Ibid., pp. 44-45. In 1988 the State Department reported that almost one third of the authorization requests submitted by U.S. researchers were denied. Council on Ocean Law, "The United States and the 1982 Convention on the Law of the Sea," p. 6. In fact, rather than working to facilitate scientific access within the exclusive economic zone generally, the Reagan administration promulgated burdensome new restrictions upon research activities within U.S. offshore jurisdiction. See *Oceans Policy News* (August 1988), pp. 1-2.

87. Council on Ocean Law, "The United States and the 1982 Convention on the Law of the Sea, p. 7. Cf. U.N. Convention on the Law of the Sea, art. 247; and Chap. 3 n. 245 and accompanying text.

88. "Reports of the United States Delegation," ed. Nordquist and Park, p. 357.

89. Elliot L. Richardson, "The Politics of the Law of the Sea," *Ocean Development and International Law* 11 (1982): 22. See Don Walsh, "Comment," *Law and Contemporary Problems* 46 (Spring 1983): 170.

90. See Chap. 1 n. 66 and accompanying text, and Chap. 2 nn. 73, 140, 178 and accompanying text; and Douglas D. Busch and Edward I. Mears, "Ocean Pollution: An Examination of the Problem and an Appeal for International Cooperation," *San Diego Law Review* 7 (1970): 582, 587-88, 591, 603.

91. S. Houston Lay, "Book Review," *California Western International Law Journal* 10 (1980): 153-55. See Norman Wulf, "Protecting the Marine Environment," in *Current Issues in the Law of the Sea*, ed. Joyner, pp. 91-92.

92. "The major shortcoming of the [1954 International Convention for the Prevention of Oil Pollution] was that all enforcement was left to the country in which the vessel was registered, not to the country aggrieved. This . . . encouraged vessels to sail under 'flags of convenience,' that is, to register either in those countries not bound by the [1954] Convention or in countries which place few restrictions on their vessels." John Warren Kindt, "Special Claims Impacting Upon Marine Pollution Issues at the Third U.N. Conference on the Law of the Sea," *California Western International Law Journal* 10 (1980): 433.

93. Letter from Charles N. Brower, July 26, 1973, in *Status Report*, U.S. Senate, p. 411. For a general discussion of the problem of marine pollution see Daniel Behrman, *Exploring the Ocean* (Paris: United Nations Educational, Scientific and Cultural Organization, 1970), pp. 47-54.

94. See Wulf, p. 95.

95. See Kindt, pp. 433-34; and Kreuger, pp. 544-45.

96. See Comras, p. 138.

97. Working Paper on Competence to Establish Standards for the Control of Vessel Source Pollution, April 2, 1973, reprinted in *Status Report*, U.S. Senate, pp. 541 43. "Since ocean currents carry some amounts of pollution from one ocean area to another and from far offshore to inshore areas, individual coastal state standards could not as effectively reduce such pollution." Ibid., p. 543.

98. Letter from Charles N. Brower, July 26, 1973, in *Status Report*, U.S. Senate, p. 413. "Uniform environmental standards for vessel operation can help to reduce the expenses inherent in varying regulations which might be applicable. . . . In this connection the stability necessary to ensure integrity of investment is promoted by private users knowing what to expect and having recourse to compulsory dispute mechanisms in the event of disagreement." Ibid.

99. Nixon, "U.S. Foreign Policy for the 1970s," p. 410. "The United States is vigorously seeking to bring ocean pollution under effective international regulation in a number of different forums." Statement of John R. Stevenson, Legal Adviser, Department of State, before Subcommittee on International Organizations and Movements, House Committee on Foreign Affairs, April 10, 1972, reprinted in *U.S. Department of State Bulletin* 66 (1972): 675.

100. U.N. Convention on the Law of the Sea, arts. 208 (3), (5), 209, 210 (4), (6), 211 (1), (2), 286, 297 (1) (c). See Wulf, p. 96. These provisions have been described as "a neat balance" between national sovereignty and international community. Sanger, p. 110.

101. U.N. Convention on the Law of the Sea, arts. 213–15, 216 (1) (b), 217 (1), 222, 235. See also Chap. 5, n. 400, and text accompanying n. 110 below. Apparently for political reasons related to national sovereignty, states have not been required to establish minimum international standards and practices with respect to land-based and atmospheric pollution, although they are required to "endeavour" to do so. Cf. U.N. Convention on the Law of the Sea, arts. 207 (4), 212 (3).

102. U.N. Convention on the Law of the Sea, arts. 216 (b), 217, 235 (1). Because of relatively low detection and policing costs, assignment of flag-state liability is an economically preferred method of allocating responsibility for damage caused by marine pollution, particularly with respect to oil discharges, which may be traced to the offending vessel. Statement of Robert D. Tollison in *The Law of the Sea,* ed. Amacher and Sweeney, p. 120. See Busch and Mears, p. 600 n. 87.

103. U.N. Convention on the Law of the Sea, arts. 216 (1) (a), 218, 220 (2), (3), (5), (6). Proceedings may be instituted with respect to violations occurring within the territorial sea where there are "clear grounds" to believe that a violation of the applicable standards and practices has occurred. Ibid., art. 220 (2). Institution of proceedings with respect to violations occurring within the exclusive economic zone requires "clear objective evidence" that the violation has resulted in "a discharge causing major damage or threat of major damage to the coastline or related interests of the coastal State." Ibid., art. 220 (6). Port-state jurisdiction may be exercised "where the evidence so warrants," and proceedings may be instituted at the request of other coastal states suffering environmental damage. Ibid., art. 218 (1), (2). Where the violation occurs outside the territorial waters of the coastal or port state, the flag state may institute preemptive enforcement proceedings within six months unless the violation involves "major damage to the coastal State or the flag State in question has repeatedly disregarded its obligations to enforce effectively the applicable international rules and standards in respect of violations committed by its vessels." Ibid., art. 228 (1). See Wulf, pp. 96–98.

104. U.N. Convention on the Law of the Sea, art. 210 (5). Under the 1972 Dumping Convention, states may only exercise enforcement powers within their territorial seas. See *Oceans Policy News* (January/February 1988), p. 6.

105. *Oceans Policy News* (August 1987), p. 6. See "Reports of the United States Delegation," ed. Nordquist and Park, p. 48.

106. *Rest. 3rd, Restatement of the Foreign Relations Law of the United States,* §603. The obligation applies only with respect to "significant" pollution, and only "to the extent practicable under the circumstances." Ibid., §603 (1), (2). Section 603 is purportedly "based on" article 220 of the 1982 Convention, but it makes no mention of coastal-state enforcement rights, much less port-state jurisdiction. Ibid., §603, source note. See n. 103 above and accompanying text.

107. Vasciannie, p. 97. The power of port states to bring enforcement actions in cases of discharges which have occurred on the high seas or within the jurisdictional zones of other states is "a major departure in international law." Wulf, p. 98. See also Australia, Department of Foreign Affairs, *Third United Nations Conference on the Law of the Sea, Seventh Session, Geneva: Report of the Australian Delegation* (Canberra: Australian Government Publishing Service, 1978), pp. 68–69.

108. "U.S. Delegation Report," p. 39. See Chap. 2 n. 185 and accompanying text; and John E. Noyes, "Compulsory Third-Party Adjudication and the 1982 United Nations Convention on the Law of the Sea," *Connecticut Journal of International Law* 4 (1989): 682.

109. Bernard H. Oxman, "Navigation, Pollution, and Compulsory Settlement of EEZ Disputes," *Oceanus* 27 (Winter 1984–85): 53. See Noyes, pp. 677–78. "Probably no more thorough system could have been agreed among sovereign states at this stage of the development of international law." Sanger, p. 202. Developing states have been particularly averse to international legal institutions which had been employed by European powers to perpetuate colonialism. Hollick, *U.S. Foreign Policy,* p. 136.

110. U.N. Convention on the Law of the Sea, arts. 286, 296 (1), 297 (2), (3), 298 (1). See Noyes, pp. 679–80, 683–84. "[T]he Convention's rules governing arbitration eliminate many of the techniques that parties to international arbitration treaties traditionally could use to opt out of an arbitration. . . . Since arbitration is likely to be the most commonly used form of third-party adjudication for disputes arising under the . . . Convention, such provisions are significant." Noyes, p. 680.

111. See *Rest. 3rd, Restatement of the Foreign Relations Law of the United States*, Introductory Note to Part V, n. 6. Although it is conceivable that some arrangements might be made for partial U.S. participation in the dispute settlement procedures if "incentives" were offered to states parties, important procedures established by the Convention would remain unavailable. Oxman, "Navigation, Pollution, and Compulsory Settlement," p. 56. "This fact would contribute to the pressures pushing against uniform observation and interpretation of Convention norms." Noyes, p. 695.

112. Oxman, "The Ninth Session," p. 247. See also Swayze, p. 37. "States likely perceive that the threat of adjudication will constrain unreasonable unilateral interpretations of substantive Convention provisions." Noyes, p. 683.

113. See nn. 6, 29 above and accompanying text; Noyes, p. 695; American Bar Association, *Revised Report with Recommendation*, p. 2; and also Knauss, p. 23.

114. Malone, "Who Needs the Sea Treaty?" p. 44.

115. Secretary of State Kissinger, address before American Bar Association, August 11, 1975, p. 353. See generally Daniel Patrick Moynihan, *On the Law of Nations* (Cambridge, Mass.: Harvard University Press, 1990).

116. Robert A. Taft, *A Foreign Policy for Americans*, quoted in John Foster Dulles, "The Challenge of Our Time: Peace with Justice," *ABA Journal* 39 (1953): 1066.

117. Statement of Under Secretary Elliot L. Richardson before Special Subcommittee on Outer Continental Shelf, Senate Committee on Interior and Insular Affairs, May 27, 1970, reprinted in *U.S. Department of State Bulletin* 62 (1970): 738. "The system . . . cannot be designed to reflect the interests of any one nation or group of nations." Ibid. "The law of the sea lies at the heart of modern international law as it emerged in the 17th century. Should it collapse under the weight of conflicting unilateral actions based almost exclusively on immediate national interests, the result will be a severe blow to the prospects for the rule of law not only in the oceans but in the international community generally." Statement of John R. Stevenson, April 10, 1972, in *U.S. Department of State Bulletin* 66 (1972): 672. See Secretary of State William P. Rogers, "The Rule of Law and the Settlement of International Disputes," *U.S. Department of State Bulletin* 62 (1970): 623.

118. See Council on Ocean Law, "The United States and the 1982 Convention on the Law of the Sea," p. 6. "Within a few years the United Nations was finding it increasingly difficult to muster the necessary funds for the ocean-related activities of FAO, UNEP, UNDP, and other organizations." Lewis M. Alexander, "The Cooperative Approach to Ocean Affairs: Twenty Years Later," *Ocean Development and International Law* 21 (1990): 107.

119. For an official explanation of the withdrawal, see Abraham D. Sofaer, Legal Adviser, Department of State, "The United States and the World Court," Current Policy No. 769 (Washington: U.S. Department of State, 1985). The administration displayed "a contempt for international organizations," reflecting a view of the world "as one filled with enemies . . . a worldview that says we too must be ruthless — unconstrained by the niceties of international law." Senator Claiborne Pell, address before U.S. Institute of Peace, July 1987, quoted in *Ocean Science News* 29 (July 14, 1987): 6.

The Reagan administration did demonstrate pragmatism with regard to international organizations in some areas of foreign policy, turning, for example, to the World Bank, which "it once suspected . . . of fostering socialism," for help in resolving the global debt crisis. Robert A. Manning, "Making Foreign Policy Fit a Changing World," *U.S. News and World Report*, March 10, 1983, p. 34. But see Chap. 4 nn. 620, 640 and accompanying text.

120. See Jeff Trimble and Maia Wechsler, "Look Who's Playing Peacekeeper Now," *U.S. News and World Report*, November 2, 1987, p. 47.

121. See John M. Goshko, "Suddenly, the U.N. Matters," *Washington Post National Weekly Edition*, October 3–9, 1988, p. 17; and also *Law of the Sea: Report of the Secretary-General* (U.N. Doc. A/44/650, November 1, 1989), p. 5 (para. 2). "The United States recognizes that — in order to meet the challenges facing us as we look toward the next century — we will have to have recourse to a variety of multilateral forums, of which the United Nations is going to be just one." Richard S. Williamson, "Toward the 21st Century: The Future of Multilateral Diplomacy," *U.S. Department of State Bulletin* 88 (December 1988): 53.

122. President George H.W. Bush, "The UN: World Parliament of Peace," Current Policy No. 1303 (Washington: U.S. Department of State, 1990), p. 2. The Bush administration's insistence that the United States and its Western allies retain control over international decisionmaking in such a partnership may have undermined international support for the "New World Order."

123. President Richard M. Nixon, "United States Policy for the Seabed," *U.S. Department of State Bulletin* 62 (1970): 737. See Chap. 2 n. 203 and accompanying text.

124. Nixon, "U.S. Foreign Policy for the 1970s," p. 409.

125. Secretary of State Kissinger, address before American Bar Association, August 11, 1975, pp. 354–55, 358. "On a planet marked by interdependence, unilateral action and unrestrained pursuit of the national advantage inevitably provoke counteraction and therefore spell futility and anarchy. . . . If the politics of ideological confrontation and strident nationalism become pervasive, broad and humane international agreement will grow ever more elusive and unilateral actions will dominate. In an environment of widening chaos the stronger will survive and may even prosper temporarily. But the weaker will despair, and the human spirit will suffer." Ibid., pp. 354, 362.

126. See American Bar Association, *Revised Report with Recommendation*, p. 2.

127. Secretary of State Henry Kissinger, address before University of Wisconsin Institute of World Affairs, Milwaukee, July 14, 1975, reprinted in *U.S. Department of State Bulletin* 73 (1975): 156. "Throughout almost the entire post–World War II era, a fundamental objective of U.S. foreign policy has been to assist the emerging developing countries with their economic development, and thus to improve standards of living for their peoples." American Bar Association, "Natural Resources of the Sea," p. 682.

128. American Bar Association, *Revised Report with Recommendation*, p. 4. See also Swayze, p. 37; and Hollick, *U.S. Foreign Policy*, pp. 383, 389. Conversely, commencement of unilateral nodule mining operations "could invite the majority of the coastal States – which would not be in a financial or technological position that would allow them to also follow the path of unilateral activities – to take steps to extend their national claims of a continental shelf beyond the limits agreed in the Convention, as a compensatory measure." Jesus, "Statement on the Issue," p. 2.

129. "There would be no guarantee that the United States and its allies would be able to achieve as satisfactory a balance between the rights of coastal states and distant water ocean users as is reflected in the 1982 treaty, nor to negotiate a seabed mining regime that differed substantially from what could be obtained at this time." Council on Ocean Law, "The United States and the 1982 Convention on the Law of the Sea," p. 7. The Chairman of the Preparatory Commission has indicated that "it would be next to impossible" to reopen renegotiations only with respect to the seabed mining provisions of the Convention. Jesus, "Statement on the Issue," p. 3.

130. "Without the Law of the Sea Convention in place as treaty law, . . . a significant risk exists that nations will not follow its provisions. . . . The great convern [sic] is that, unless the Convention enters into force and achieves widespread acceptance, the world will find itself back in the unstable pre–Convention situation where unilateral state assertions of extensive national jurisdictional [sic] were rampant." American Bar Association, *Revised Report with Recommendation*, p. 3. See Swing, "Address," p. 117; and also Sebenius, pp. 93–94, 103, 108–9.

Bibliography

Books

Alexander, Lewis M., ed. *The Law of the Sea: International Rules and Organization for the Sea.* Kingston: University of Rhode Island, 1967.
_____. *The Law of the Sea: Offshore Boundaries and Zones.* Columbus: Ohio State University Press, 1967.
Alexander, Michael van Cleave. *The First of the Tudors.* Totowa, N.J.: Rowman and Littlefield, 1980.
Amacher, Ryan C., and Richard James Sweeney, eds. *The Law of the Sea: U.S. Interests and Alternatives.* Washington: American Enterprise Institute for Public Policy Research, 1976.
American Bar Association. *Revised Report with Recommendation.* February 13, 1990.
American Law Institute. *Restatement 3rd, Restatement of the Foreign Relations Law of the United States.* (St. Paul, Minn.: American Law Institute, 1987).
American Society of International Law. *Policy Issues in Ocean Law.* St. Paul, Minn.: West, 1975.
_____. *Proceedings of the 62nd Annual Meeting.* 1968.
_____. *Proceedings of the 70th Annual Meeting.* 1976.
_____. *Proceedings of the 75th Anniversary Convocation.* 1981.
_____. *Proceedings of the 76th Annual Meeting.* 1982.
Anand, R.P. *Origin and Development of the Law of the Sea.* Boston: Martinus Nijhoff, 1983.
Andrews, Kenneth R. *Drake's Voyages.* New York: Charles Scribner's Sons, 1967.
_____. *Trade, Plunder and Settlement.* New York: Cambridge University Press, 1984.
Ayliffe, John. *A New Pandect of the Roman Civil Law.* London: Thomas Osborne, 1734.
Azuni, M.D.A. *The Maritime Law of Europe.* Translated by William Johnson. New York: I. Riley, 1806. Volume 1.
Barkenbus, Jack N. *Deep Seabed Resources: Politics and Technology.* New York: Free Press, 1979.
Bell, Christopher. *Portugal and the Quest for the Indies.* New York: Harper and Row, 1974.
Boehmer, Heinrich. *Road to Reformation.* Translated by John W. Doberstein and Theodore G. Tappert. Philadelphia: Muhlenberg Press, 1946.
Booth, Ken. *Law, Force and Diplomacy at Sea.* London: George Allen and Unwin, 1985.
Borgese, Elisabeth Mann, ed. *Pacem in Maribus.* New York: Dodd, Mead, 1972.
Bouchez, J., and L. Kaijen, eds. *The Future of the Law of the Sea.* The Hague: Martinus Nijhoff, 1973.
Bowett, D.W. *The Law of the Sea.* Dobbs Ferry, N.Y.: Oceana Publications, 1967.
Bridge, John S.C. *A History of France from the Death of Louis XI.* Oxford: Clarendon, 1924. Volumes 2–3.

Brown, Louise Fargo. *The Freedom of the Seas*. New York: E.P. Dutton, 1919.

Brown, Seyom, Nina M. Cornell, Larry L. Fabian, and Edith Brown Weiss. *Regimes for the Ocean, Outer Space, and Weather*. Washington: Brookings Institution, 1977.

Brownlie, Ian. *Principles of Public International Law*. 3rd edition. Oxford: Clarendon, 1979.

Brownstein, Ronald, and Nina Easton. *Reagan's Ruling Class*. New York: Pantheon Books, 1983.

Burke, William T. *Ocean Sciences, Technology, and the Future International Law of the Sea*. Columbus: Ohio State University Press, 1966.

Butler, William E. *The Soviet Union and the Law of the Sea*. Baltimore: Johns Hopkins University Press, 1971.

Buzan, Barry. *Seabed Politics*. New York: Praeger, 1976.

Christy, Francis T., Jr., Thomas A. Clingan, Jr., John King Gamble, Jr., H. Gary Knight, and Edward Miles, eds. *Law of the Sea: Caracas and Beyond*. Cambridge, Mass.: Ballinger, 1975.

Clingan, Thomas A., Jr., ed. *The Law of the Sea: What Lies Ahead?* Honolulu: Law of the Sea Institute, 1988.

Cohen, Ronald, and Elman R. Service, eds. *Origins of the State: The Anthropology of Political Evolution*. Philadelphia: Institute for the Study of Human Issues, 1978.

Colombos, C. John. *The International Law of the Sea*. 4th revised edition. London: Longmans, Green, 1959.

Comines, Philip. *Historical Memoirs*. Originally published 1524. London: J. Davis, 1817.

Cruichshank, Michael, Joseph P. Flanagan, Buford Holt, and John W. Padan. *Marine Mining on the Outer Continental Shelf*. Washington: U.S. Department of the Interior, 1987.

Davenport, Frances Gardiner, ed. *European Treaties Bearing on the History of the United States and Its Dependencies to 1648*. Washington, D.C.: Carnegie Institution of Washington, 1917.

Diffie, Bailey W., and George D. Winius. *Foundations of the Portuguese Empire, 1415–1580*. Minneapolis: University of Minnesota Press, 1977.

Dupuy, René-Jean. *The Law of the Sea*. Dobbs Ferry, N.Y.: Oceana Publications, 1974.

Eccles, W.J. *France in America*. New York: Harper and Row, 1972.

Edwards, Charles S. *Hugo Grotius: The Miracle of Holland*. Chicago: Nelson-Hall, 1981.

Fenn, Percy Thomas, Jr. *The Right of Fishery in Territorial Waters*. Cambridge, Mass.: Harvard University Press, 1926.

Folmer, Henry. *Franco-Spanish Rivalry in North America*. Glendale, Calif.: Arthur H. Clark, 1953.

Fulton, Thomas Wemyss. *The Sovereignty of the Sea*. Edinburgh: William Blackwood and Sons, 1911.

Gamble, John King, Jr., ed. *Law of the Sea: Neglected Issues*. Cambridge, Mass.: Ballinger, 1979.

————, and Edward Miles, eds. *Law of the Sea: Conference Outcomes and Problems of Implementation*. Cambridge, Mass.: Ballinger, 1977.

Garcia-Amador, F.V. *The Exploitation and Conservation of the Resources of the Sea*. 2nd edition. Leyden: A.W. Sijthoff, 1963.

Garner, J.F., ed. *Compensation for Compulsory Purchase*. London: United Kingdom National Committee of Comparative Law, 1975.

Garrett, Mitchell B. *European History 1500–1815*. New York: American Book, 1940.

Gentili, Alberico. *De Juri Belli Libri Tres*. Translated by John C. Rolfe. Oxford: Clarendon Press, 1933. Volume 3.

Gibson, Charles. *Spain in America*. New York: Harper and Row, 1966.

Gierke, Otto. *Political Theories of the Middle Age*. Translated by Frederic William Maitland. Cambridge: Cambridge University Press, 1938.

Gold, Edgar. *Maritime Transport*. Lexington, Mass.: D.C. Heath, 1981.

Grant, A.J. *The French Monarchy (1483–1789)*. Cambridge: Cambridge University Press, 1914. Volume 1.

Gregg, Robert W., and Michael Barkun, eds. *The United Nations System and Its Functions.* Princeton, N.J.: D. Van Nostrand, 1968.

Grotius, Hugo. *The Freedom of the Seas.* Translated by Ralph van Deman Magoffin. New York: Oxford University Press, 1916.

Hall, Edward. *A Treatise on International Law.* Oxford: Clarendon, 1924.

Hauser, Wolfgang. *The Legal Regime for Deep Seabed Mining Under the Law of the Sea Convention.* Translated by Frances Bunce Dielmann. Deventer, Netherlands: Kluwer, 1983.

Hay, Peter. *Federalism and Supranational Organizations.* Urbana: University of Illinois Press, 1966.

Hill, Charles E. *The Danish Sound Dues and the Command of the Baltic.* Durham, N.C.: Duke University Press, 1926.

Holdsworth, William. *A History of English Law.* London: Methisen, 1956, Volume 1.

Hollick, Ann L. *U.S. Foreign Policy and the Law of the Sea.* Princeton: Princeton University Press, 1981.

_____, and Robert E. Osgood. *New Era of Ocean Politics.* Baltimore: Johns Hopkins University Press, 1974.

Hyma, Albert. *A Short History of Europe, 1500-1815.* New York: F.S. Crofts, 1929.

International Law Association. American Branch. *Proceedings and Committee Reports, 1977-78.* New York: n.p., 1979.

_____. *Report of the Sixtieth Conference.* London: International Law Association, 1983.

Jacobsen, Harold K. *Networks of Interdependence.* New York: Alfred A. Knopf, 1979.

Janis, Mark W. *Sea Power and the Law of the Sea.* Lexington, Mass.: D.C. Heath, 1976.

Jennings, R.Y. *The Acquisition of Territory in International Law.* New York: Oceana Publications, 1963.

Johnston, Douglas M. *The International Law of Fisheries.* New Haven, Conn.: Yale University Press, 1965.

_____, and Norman G. Letalik, eds. *The Law of the Sea and Ocean Industry: New Opportunities and Restraints.* Honolulu: Law of the Sea Institute, 1983.

Joyner, Christopher C., ed. *Current Issues in the Law of the Sea.* Allentown, Penn.: Muhlenberg College, 1980.

Justice, Alexander. *A General Treatise of the Dominion of the Sea.* 2nd edition. London: D. Leach, 1709.

Kelsen, Hans. *Peace Through Law.* Chapel Hill: University of North Carolina Press, 1944.

_____. *Principles of International Law.* New York: Rinehart, 1952.

Keohane, Robert O., and Joseph S. Nye. *Power and Interdependence: World Politics in Transition.* Boston: Little, Brown, 1977.

Kirgis, Frederic L., Jr. *International Organizations in Their Legal Setting.* St. Paul, Minn.: West, 1977.

Knight, H. Gary, J.D. Nyhart, and Robert E. Stein, eds. *Ocean Thermal Energy Conversion.* Lexington, Mass.: D.C. Heath, 1977.

Koers, Albert W., and Bernard H. Oxman, eds. *The 1982 Convention on the Law of the Sea.* Honolulu: Law of the Sea Institute, 1984.

Kooijmans, P.H. *The Doctrine of the Legal Equality of States.* Leyden: A.W. Sijthoff, 1964.

Kronmiller, Theodore G. *The Lawfulness of Deep Seabed Mining.* New York: Oceana Publications, 1980. Volumes 1-3.

Lapidoth, Ruth. *Freedom of Navigation with Special Reference to International Waterways in the Middle East.* Jerusalem: Jerusalem Post, 1975.

Laursen, Finn. *Superpower at Sea.* New York: Praeger, 1983.

Lawrence, T.J. *The Principles of International Law.* 2nd edition. Boston: D.C. Heath, 1898.

Lipsky, George A., ed. *Law and Politics in the World Community.* Los Angeles: University of California Press, 1953.

Locke, John. *Second Treatise of Civil Government.* Edited by Lester de Koster. Originally published 1690. Grand Rapids, Mich.: William B. Eardmans, 1978.

Logue, John J., ed. *The Fate of the Oceans.* Villanova, Penn.: Villanova University Press, 1972.

Luard, Evan. *The Control of the Sea-Bed.* New York: Taplinger, 1977.

Lynch, John. *Spain Under the Hapsburgs.* 2nd edition. New York: New York University Press, 1981. Volume. 1.

McAlister, Lyle N. *Spain and Portugal in the New World 1492-1700.* Minneapolis: University of Minnesota Press, 1984.

McDougal, Myres S., and William T. Burke. *The Public Order of the Oceans.* New Haven, Conn.: Yale University Press, 1962.

McNair, Lord. *The Law of Treaties.* Oxford: Clarendon, 1961.

Mattingly, Garrett. *Renaissance Diplomacy.* Boston: Houghton Mifflin, 1955.

Mikesell, Raymond F. *NonFuel Minerals: Foreign Dependence and National Security.* Ann Arbor, Mich.: The University of Michigan Press, 1987.

Miles, Edward, and John King Gamble, Jr., eds. *Law of the Sea: Conference Outcomes and Problems of Implementation.* Cambridge, Mass.: Ballinger, 1977.

Morenoff, Jerome. *World Peace Through Space Law.* Charlottesville, Va.: Michie, 1967.

Morley, Felix. *The Charter of the United Nations: An Analysis.* New York: American Enterprise Association, 1946.

Mowat, R.B. *A History of European Diplomacy 1451-1789.* London: Edward Arnold, 1928.

Moynihan, Daniel Patrick. *On the Law of Nations.* Cambridge, Mass.: Harvard University Press, 1990.

Nys, Ernest. *Les Origines du Droit International.* Bruxelles: Alfred Castaigne, 1894.

Oakeshott, Walter. *Founded Upon the Seas.* Cambridge: Cambridge University Press, 1942.

Ocean Association of Japan, ed. *Marine Technology and Law: Development of Hydrocarbon Resources in Offshore Structures.* Tokyo: Ocean Association of Japan, 1978.

O'Connell, D.P. *The International Law of the Sea.* Oxford: Clarendon, 1982. Volumes 1-2.

Oda, Shigeru. *The Law of the Sea in Our Time – I: New Developments, 1966-1975.* Leyden: A.W. Sijthoff, 1977.

Oppenheim, L. *International Law.* 3rd edition. London: Longmans, Green, 1920. Volume 1.

Orfield, Lester Bernhardt. *The Growth of Scandinavian Law.* Philadelphia: University of Pennsylvania Press for Temple University Publications, 1953.

Oxman, Bernard H., with Charles L.O. Buderi and David D. Caron, eds. *Law of the Sea: U.S. Policy Dilemma.* San Francisco: ICS, 1983.

Padelford, Norman J., and Leland M. Goodrich, eds. *The United Nations in the Balance: Accomplishments and Prospects.* New York: Praeger, 1965.

Pages, Georges. *The Thirty Years War 1618-1648.* Translated by David Maland and John Hooper. New York: Harper and Row, 1970.

Passos, John Dos. *The Portugal Story.* Garden City, N.Y.: Doubleday & Co., 1969.

Plucknett, Theodore F.T. *A Concise History of the Common Law.* Rochester, N.Y.: The Lawyers Co-Operative Publishing Co., 1929.

Polomka, Peter. *Ocean Politics in Southeast Asia.* Singapore: Institute of Southeast Asian Studies, 1978.

Post, Alexandra M. *Deepsea Mining and the Law of the Sea.* Boston: Martinus Nijhoff, 1983.

Potter, Pitman B. *The Freedom of the Seas in History, Law, and Politics.* New York: Longmans, Green, 1924.

Priestley, Herbert Ingram. *France Overseas Through the Old Regime.* New York: D. Appleton-Century, 1939.

Pufendorf, Samuel. *De Jure Naturae et Gentium Libri Octo.* Translated by C.H. Oldfather and W.A. Oldfather. Originally published 1688. Oxford: Clarendon, 1934. Volume 2.

Quinn, David B. *England and the Discovery of America, 1481-1620.* New York: Alfred A. Knopf, 1974.

_____, and A.N. Ryan. *England's Sea Empire, 1550-1642.* London: George Allen and Unwin, 1983.

Rabb, Theodore K., ed. *The Thirty Years' War.* Lexington, Ky.: D.C. Heath, 1972.

Ramberg, Bennett. *The Seabed Arms Control Negotiations: A Study of Multilateral Arms Control Conference Diplomacy.* University of Denver Monograph Series in World Affairs, 1978. Volume 15, book 2.

Rembe, Nasila S. *Africa and the International Law of the Sea.* Germantown, Md.: Sijthoff and Noordhoff, 1980.

Rocky Mountain Mineral Law Foundation. *The American Law of Mining.* New York: Mathew Bender & Co., 1982. Volume 1.

Rowse, A.L. *The Elizabethans and America.* New York: Harper and Brothers, 1959.

Roxburgh, Ronald F. *International Conventions and Third States.* New York: Longmans, Green, 1917.

Rozakis, Christos L. *The Concept of Jus Cogens in the Law of Treaties.* New York: North-Holland, 1976.

_____, and Constantine A. Stephanou, eds. *The New Law of the Sea.* New York: Elsevier Science Publishers B.V., 1983.

Salmon, Edward Dwight. *Imperial Spain.* Westport, Conn.: Greenwood Press, 1971.

Sanger, Clyde. *Ordering the Oceans: The Making of the Law of the Sea.* London: Zed Books, 1986.

Schachter, Oscar. *Sharing the World's Resources.* New York: Columbia University Press, 1977.

Schwarzenberger, Georg. *International Law.* London: Stevens and Sons, 1957. Volume 1.

Scott, James Brown. *The Spanish Origin of International Law.* Washington: Georgetown University School of Foreign Service, 1928.

Sebenius, James K. *Negotiating the Law of the Sea.* Cambridge, Mass.: Harvard University Press, 1984.

Selden, John. *Of the Dominion, Or, Ownership of the Sea.* Translated by Marchamont Nedham. Originally published 1652. New York: Arno, 1972.

Shutler, Sharon K., ed. *United States Law of the Sea Policy: Options for the Future.* Oceans Policy Study Series, number 6. New York: Oceana Publications, 1985.

Sinclair, Ian. *The Vienna Convention on the Law of Treaties.* 2nd edition. Manchester: Manchester University Press, 1984.

Smith, Alan G.R. *The Emergence of a Nation State.* New York: Longman, 1984.

Smith, George P., III. *Restricting the Concept of Free Seas.* Huntington, N.Y.: Robert E. Krieger Publishing Co., 1980.

Smith, H.A. *The Law and Custom of the Sea.* London: Stevens and Sons, 1950.

Sohn, Louis B., and Kristen Gustafson. *The Law of the Sea in a Nutshell.* St. Paul, Minn.: West, 1984.

Stone, Christopher D. *Where the Law Ends.* New York: Harper & Row, 1975.

Sztucki, Jerzy. *Jus Cogens and the Vienna Convention on the Law of Treaties.* New York: Springer, 1974.

_____, ed. *Symposium on the International Regime of the Sea-Bed.* Rome: Accademi Nazionale dei Lincei, 1970.

Textor, Johann Wolfgang. *Synopsis of the Law of Nations.* Translated by John Pawley Bate. Originally published 1680. Washington: Carnegie Institution of Washington, 1916.

Third World Conference on World Peace Through Law, Geneva, 1967. *Proceedings.* Geneva: World Peace Through Law Center, 1969.

Thomson, John A.F. *Popes and Princes, 1417–1517.* Boston: George Allen and Unwin, 1980.

Trevor-Roper, H.R. *The Crisis of the Seventeenth Century.* New York: Harper and Row, 1968.

Tully, James. *A Discourse on Property: John Locke and His Adversaries.* Cambridge: Cambridge University Press, 1980.

Tunstall, Marion David. *The Third United Nations Conference on the Law of the Sea.* Charlottesville, Va.: Michie, 1980.

Van Dyke, Jon, ed. *Consensus and Confrontation: The United States and the Law of the Sea Convention.* Honolulu: Law of the Sea Institute, 1985.

Vattel, M.D. *The Law of Nations.* Translated by Joseph Chitty. Originally published 1758. Northampton, Mass.: Thomas M. Pomroy, 1805.

Vitzthum, Wolfgang Graf. *Aspekte der Seerechtsentwicklung.* Munich: Hochochule der Bundeswehr, 1980.

Waldheim, Kurt. *Building the Future Order: The Search for Peace in an Interdependent World.* Edited by Robert L. Schiffer. New York: Free Press, 1980.

Ward, Robert. *An Enquiry Into the Foundation and History of the Law of Nations in Europe from the Time of the Greeks and Romans to the Age of Grotius.* London: A. Strahan and W. Woodfall, 1795. Volumes 1–2.

Weiss, Thomas George. *International Bureaucracy.* Lexington, Mass.: Lexington, 1975.
Welwood, William. *An Abridgement of All Sea-Lawes.* Originally published 1613. New York: Da Capo, 1971.
Wenk, Edward, Jr. *The Politics of the Ocean.* Seattle: Washington University Press, 1972.
Westbury-Jones, J. *Roman and Christian Imperialism.* London: MacMillan, 1939.
Wilcox, Francis O., and Carl M. Marcy. *Proposals for Changes in the United Nations.* Washington: Brookings Institution, 1955.
Williamson, James A. *Hawkins of Plymouth.* 2nd edition. New York: Barnes and Noble, 1969.
Wolff, Christian. *Jus Gentium Methodo Scientifica Pertractatum.* Translated by Joseph H. Drake. Originally published 1764. Oxford: Clarendon Press, 1934. Volume 2.
Wood, William. *Elizabethan Sea-Dogs.* New Haven, Conn.: Yale University Press, 1920.
Wright, J. Leitch, Jr. *Anglo-Spanish Rivalry in North America.* Athens: University of Georgia Press, 1971.
Yemin, Edward. *Legislative Powers in the United Nations and Specialized Agencies.* Leyden: A.W. Sijthoff, 1969.

Articles

Agrawala, S.K. "The Role of General Assembly Resolutions as Trend-Setters of State Practice." *Indian Journal of International Law* 21 (1981): 513–33.
Akehurst, Michael. "Custom as a Source of International Law." *British Year Book of International Law* 47 (1974–75): 1–53.
Akesson, Rolf. "The Law of the Sea Conference." *Journal of World Trade Law* 8 (1974): 283–97.
Aldrich, George H. "A System of Exploitation." *Syracuse Journal of International Law and Commerce* 6 (1978–79): 245–56.
Alexander, Lewis M. "The Cooperative Approach to Ocean Affairs: Twenty Years Later." *Ocean Development and International Law* 21 (1990): 105–9.
_____. "National Jurisdiction and the Use of the Sea." *Natural Resources Journal* 8 (1968): 373–400.
_____. "The Ocean Enclosure Movement: Inventory and Prospect." *San Diego Law Review* 20 (1983): 561–94.
Alexander, Tom. "The Reaganites' Misadventure at Sea." *Fortune* 106 (August 23, 1982): 129–44.
Alexandrowicz, C.H. "Freitas *Versus* Grotius." *British Year Book of International Law* 35 (1959): 162–82.
Alexidze, Levan. "Legal Nature of *Jus Cogens* in Contemporary International Law." *Recueil des Cours* 172 (1981): 219–70.
Allott, Philip. "Power Sharing in the Law of the Sea." *American Journal of International Law* 77 (1983): 1–30.
American Mining Congress. "Statement." *Congressional Record* 126 (June 9, 1980): 13679–81.
Amin, S.H. "The Regime of the Sea-Bed and Ocean Floor: A Legal Analysis." *Juridical Review* (1983), pages 51–72.
Anand, R.P. "Maritime Practice in South-East Asia Until 1600 A.D. and the Modern Law of the Sea." *International and Comparative Law Quarterly* 30 (1981): 440–54.
_____. "UN Convention on the Law of the Sea and the United States." *Indian Journal of International Law* 24 (1984): 153–99.
Anderson, Chandler P. "Editorial Comment: Exploitation of the Products of the Sea." *American Journal of International Law* 20 (1926): 752–53.
Armstrong, Patrick. "Mare Raptum: Government Proposals for Ocean Mismanagement, Caracas, 1974." *Ocean Management* 2 (1974): 75–88.
Armstrong, S.W. "The Doctrine of the Equality of Nations in International Law and the Relation of the Doctrine to the Treaty of Versailles." *American Journal of International Law* 14 (1920): 540–62.
Arnold, Rudolph Preston. "The Common Heritage of Mankind as a Legal Concept." *International Lawyer* 9 (1975): 153–58.

Arrow, Dennis W. "The Customary Norm Process and the Deep Seabed." *Ocean Development and International Law Journal* 9 (1981): 1–59.

_____. "The Proposed Regime for the Unilateral Exploitation of Deep Seabed Mineral Resources by the United States." *Harvard International Law Journal* (1980): 337–417.

Arruda, Henry M. "The Extension of the United States Territorial Sea: Reasons and Effects." *Connecticut Journal of International Law* 4 (1989): 697–727.

Astrachan, Anthony. "Who Owns Seabed Riches?" *Washington Post*, December 2, 1973, page A22.

Auburn, F.M. "The Deep Seabed Hard Mineral Resources Bill." *San Diego Law Review* 9 (1972): 491–513.

Auerbach, Stuart. "Trade Talks Collapse Over Farm Issue." *Washington Post*, December 8, 1990, pages A1, A13.

Avery, Dennis T. "International Commodity Agreements." Special Report No. 83. Washington: U.S. Department of State, 1981.

Bailey, John S. "The Future of the Exploitation of the Resources of the Deep Seabed and Subsoil." *Law and Contemporary Problems* 46 (Spring 1983): 71–76.

Baker, Stewart A., and Mark D. Davis. "Arbitral Proceedings Under the UNCITRAL Rules — The Experience of the Iran-United States Claims Tribunal." *George Washington University Journal of International Law and Economics* 23 (1989): 267–347.

Ball, Milner S. "Law of the Sea: Expression of Solidarity." *San Diego Law Review* 19 (1982): 461–73.

Bandow, Doug. "UNCLOS III: A Flawed Treaty." *San Diego Law Review* 19 (1982): 475–92.

Banks, Susan M. "Protection of Investment in Deep Seabed Mining: Does the United States Have a Viable Alternative to Participating in UNCLOS?" *Boston University International Law Journal* 2 (1983): 267–97.

Barnes, William Sprague. "Technology Transfer Rules: A Study in Comparative Law." *Boston College International and Comparative Law Review* 3 (1979): 1–28.

Barrett, Carol, and Hanna Newcombe. "Weighted Voting in International Organizations." *Peace Research Reviews* 2 (April 1968): 1–110.

Basiuk, Victor. "Marine Resources Development, Foreign Policy, and the Spectrum of Choice." *Orbis* 12 (1968): 39–72.

Baxter, R.R. "Multilateral Treaties as Evidence of Customary International Law." *British Year Book of International Law* 41 (1965–66): 275–300.

Beesley, Alan. "The Negotiating Strategy of UNCLOS III: Developing and Developed Countries as Partners — A Pattern for Future Multilateral International Conferences?" *Law and Contemporary Problems* 46 (Spring 1983): 183–94.

Behrman, Daniel. *Exploring the Ocean*. Paris: United Nations Educational, Scientific and Cultural Organization, 1970.

Beier, Friedrich-Karl, and Joseph Straus. "The Patent System and Its Informational Function — Yesterday and Today." *IIC International Review of Industrial Property and Copyright Law* 8 (1977): 387–406.

Belsky, Martin A. "Management of Large Marine Ecosystems: Developing a New Rule of Customary International Law." *San Diego Law Review* 22 (1985): 733–63.

Berlin, Michael J. "Some Diverse Views at UN on the Sea Bed." *New York Post*, March 23, 1973, page 3.

Bevans, Charles I. "Contemporary Practice of the United States: U.S. 12-Mile Fishing Zone." *American Journal of International Law* 60 (1966): 824–37.

Biblowit, Charles E. "Deep Seabed Mining: The United States and the United Nations Convention on the Law of the Sea." *St. John's Law Review* 58 (1984): 267–305.

Biggs, Gonzalo. "Deep Seabed Mining and Unilateral Legislation." *Ocean Development and International Law Journal* 8 (1980): 223–57.

Blair, Homer O. "Technology Transfer as an Issue in North/South Negotiations." *Vanderbilt Journal of Transnational Law* 14 (1981): 301–26.

Bodenheimer, Edgar. "The Influence of Roman Law on Early Medieval Culture." *Hastings International and Comparative Law Review* 3 (1979): 9–27.

Borchard, Edwin. "Resources of the Continental Shelf." *American Journal of International Law* 40 (1946): 53–70.

Borgese, Elisabeth Mann. "The Law of the Sea." *Scientific American* 248 (March 1983): 42–49.
Boyle, Francis A. "The Irrelevance of International Law: The Schism Between International Law and International Politics." *California Western International Law Journal* 10 (1980): 193–219.
Breaux, Representative John. "The Diminishing Prospects for an Acceptable Law of the Sea Treaty." *Virginia Journal of International Law* 19 (1979): 257–97.
————. "What Course to a Law of the Sea?" *Sea Power* 26, number 5 (Special Edition, April 15, 1983): 34–35, 37.
Brewer, William C., Jr. "Deep Seabed Mining: Can an Acceptable Regime Ever Be Found?" *Ocean Development and International Law Journal* 11 (1982): 25–67.
————. "Transfer of Mining Technology to the International Enterprise." Oceans Policy Study 2:4. Charlottesville, Va.: The Michie Co., 1980.
Brierly, J.L. "The *Lotus* Case." *Law Quarterly Review* 44 (1928): 154–63.
Briggs, Herbert W. "Editorial Comment: Jurisdiction Over the Sea Bed and Subsoil Beyond Territorial Waters." *American Journal of International Law* 45 (1951): 338–42.
Broadus, J.M., and Porter Hoagland, III. "Conflict Resolution in the Assignment of Area Entitlements for Seabed Mining." *San Diego Law Review* 21 (1984): 541–76.
Broadus, James M. "Seabed Materials." *Science* 235 (February 20, 1987): 853–60.
Brodd, Mary Julie Ann. "A 'Common Heritage' Approach to Fisheries Through Regional Controls." *New York University Journal of International Law & Politics* 10 (1977): 171–202.
Brooks, David B. "Deep Sea Manganese Nodules: From Scientific Phenomenon to World Resource." *Natural Resources Journal* 8 (1968): 401–23.
Brown, E.D. "Freedom of the High Seas Versus the Common Heritage of Mankind: Fundamental Principles in Conflict." *San Diego Law Review* 20 (1983): 521–60.
————. "Seabed Mining: From UNCLOS to Prep Com." *Marine Policy* 8 (1984): 151–64.
————. "The United Nations Convention on the Law of the Sea 1982: The British Government's Dilemma." *Current Legal Problems* 37 (1984): 259–93.
Brown, Richard P., Jr. "Changing the Rules: International Law and the Developing Countries: The ABA Workshops of 1977." *International Lawyer* 12 (1978): 265–92.
Brownlie, Ian. "Legal Status of Natural Resources in International Law (Some Aspects)." *Recueil des Cours* 162 (1975): 245–317.
Burke, William T. "Comments on Issues Relating to the Law of the Sea." *National Resources Lawyer* 4 (1971): 660–67.
————. "Customary Law of the Sea: Advocacy or Disinterested Scholarship?" *Yale Journal of International Law* 14 (1989): 508–27.
————. "Driftnets and Nodules: Where Goes the United States?" *Ocean Development and International Law* 20 (1989): 237–40.
————. "The Law of the Sea Treaty, Customary Law, and the United States." *Water Log* 5 (April-June 1985): 2–6.
————. "A Negative View of a Proposal for United Nations Ownership of Ocean Mineral Resources." *Natural Resources Lawyer* 1 (1968): 42–62.
Burton, Steven J. "Freedom of the Seas: International Law Applicable to Deep Seabed Mining Claims." *Stanford Law Review* 29 (1977): 1135–80.
Busch, Douglas D., and Edward I. Mears. "Ocean Pollution: An Examination of the Problem and an Appeal for International Cooperation." *San Diego Law Review* 7 (1970): 574–604.
Buzan, Barry. "Negotiating by Consensus: Developments in Technique at the United Nations Conference on the Law of the Sea." *American Journal of International Law* 75 (1981): 324–48.
Cairns, Robert D. "A Reconsideration of Ontario Nickel Policy." *Canadian Public Policy* 7 (1981): 526–33.
Caminos, Hugo, and Michael R. Molitor. "Progressive Development of International Law and the Package Deal." *American Journal of International Law* 79 (1985): 871–90.
Carlson, Sevinc. "Soviet Policy on the Sea-Bed and the Ocean Floor." *Syracuse Journal of International Law and Commerce* 1 (1972): 104–9.
Caron, David D. "Municipal Legislation for Exploitation of the Deep Seabed." *Ocean Development and International Law Journal* 8 (1980): 259–97.

Champ, Michael A., William P. Dillon, and David G. Howell. "Non-Living EEZ Resources: Minerals, Oil and Gas." *Oceanus* 27 (Winter 1984-85): 28-34.

Charney, Jonathan I. "The Delimitation of Lateral Seaward Boundaries Between States in a Domestic Context." *American Journal of International Law* 75 (1981): 28-68.

_____. "International Agreements and the Development of Customary International Law." *Washington Law Review* 61 (1986): 971-96.

_____. "The Law of the Deep Seabed Post UNCLOS III." *Oregon Law Review* 63 (1984): 19-52.

_____. "The Persistent Objector Rule and the Development of Customary International Law." *British Year Book of International Law* 56 (1985): 1-24.

_____. "The United States and the Law of the Sea After UNCLOS III – The Impact of General International Law." *Law and Contemporary Problems* 46 (Spring 1983): 37-54.

_____. "United States Interests in a Convention on the Law of the Sea: The Case for Continued Efforts." *Vanderbilt Journal of Transnational Law* 11 (1978): 39-75.

Chaze, William L., Col. H.G. Summers, Jr., Robert A. Kittle, and Nicholas Daniloff. "Rust to Riches: The Navy Is Back." *U.S. News and World Report*, August 4, 1986, page 28.

Cheever, Daniel S. "The Politics of the UN Convention on the Law of the Sea." *Journal of International Affairs* 37 (1984): 247-52.

_____. "The Role of International Organization in Ocean Development." *International Organization* 22 (1968): 629-48.

Christenson, Gordon A. "*Jus Cogens:* Guarding Interests Fundamental to International Society." *Virginia Journal of International Law* 28 (1988): 585-648.

_____. "The World Court and *Jus Cogens.*" *American Journal of International Law* 81 (1987): 93-101.

Christol, Carl Q. "The American Bar Association and the 1979 Moon Treaty: The Search for a Position." *Journal of Space Law* 9 (1981): 77-91.

_____. "The Common Heritage of Mankind Provision in the 1979 Agreement Governing the Activities of States on the Moon and Other Celestial Bodies." *International Lawyer* 14 (1980): 429-83.

_____. "Fisheries and the New Conventions on the Law of the Sea." *San Diego Law Review* 7 (1970): 455-68.

_____. "Marine Resources and the Freedom of the Seas." *Natural Resources Journal* 8 (1968): 424-33.

_____. "Telecommunications, Outer Space, and the New International Information Order (NIIO)." *Syracuse Journal of International Law and Commerce* 8 (1981): 343-64.

Christy, Francis T., Jr. "Alternative Regimes for Marine Resources Underlying the High Seas." *Natural Resources Lawyer* 1 (1968): 63-77.

Ciotti, Thomas E. "A Comparative Overview of the Patenting Process." Paper presented at State Bar of California conference on The Internationalization of Technology and the Law, San Francisco, January 19, 1990.

Clarkson, Kenneth W. "International Law, U.S. Seabeds Policy and Ocean Resource Development." *Journal of Law and Economics* 17 (1974): 117-42.

Clary, Mike. "Navy Foils Greenpeace Ship to Fire Trident 2." *Los Angeles Times*, December 5, 1989, page A4.

Clingan, Thomas A., Jr. "Freedom of Navigation in a Post-UNCLOS III Environment." *Law and Contemporary Problems* 46 (Spring 1983): 107-23.

Cocca, Aldo Armando. "The Advances in International Law Through the Law of Outer Space." *Journal of Space Law* 9 (1981): 13-20.

Colino, Richard R. "A Chronicle of Policy and Procedure: The Formulation of the Reagan Administration Policy of International Satellite Telecommunications." *Journal of Space Law* 13 (1985): 103-56.

Collins, Harry M. "Deep Seabed Hard Mineral Resources Act – Matrix for United States Deep Seabed Mining." *Natural Resources Lawyer* 13 (1981): 571-80.

Colson, David A. "How Persistent Must the Persistent Objector Be?" *Washington Law Review* 61 (1986): 957-70.

_____. "The United States Position on Antarctica." *Cornell International Law Journal* 19 (1986): 291-301.

Conforti, Benedeto. "Notes on the Unilateral Exploitation of the Deep Seabed." *Italian Year-book of International Law* 4 (1978–79): 3–19.

Council on Ocean Law. Panel on the Law of Ocean Uses. "Deep Seabed Mining and the 1982 Convention on the Law of the Sea." September 25, 1987.

Council on Ocean Law. "The United States and the 1982 UN Convention on the Law of the Sea: A Synopsis of the Status of the Treaty and Its Expanded Role in the World Today." September 1989.

D'Amato, Anthony. "Editorial Comment: An Alternative to the Law of the Sea Convention." *American Journal of International Law* 77 (1983): 281–85.

Dangler, Edward. "An Ocean Miner's View of the Draft Convention." *New York Law School Journal of International and Comparative Law* 3 (1981): 27–37.

Danzig, Aaron L. "A Funny Thing Happened to the Common Heritage on the Way to the Sea." *San Diego Law Review* 12 (1975): 655–64.

Darman, Richard G. "Legislative Policy for Deepsea Mining in the Context of International Law of the Sea Negotiations." *Congressional Record* 125 (December 14, 1979): 36049–52.

de Muralt, R.W.G. "The Military Aspects of the UN Law of the Sea Convention." *Netherlands International Law Review* 32 (1985): 78–99.

Denman, D.R. "Minerals, Mining and Maritime Resource Management." *Ocean Management* 7 (1981): 25–40.

Deporov, Y. "Antarctica: A Zone of Peace and Cooperation." *International Affairs* (Moscow) (1983, number 11): 29–37.

DeSousa, Gregory. "Ocean Management and World Order." *Columbia Journal of World Business* 9 (Summer 1974): 123–28.

Detoit, Bernard. "Joint Ventures and Intellectual Property." *IIC International Review of Industrial Property and Copyright Law* 20 (1989): 439–66.

Djalal, Hasjim. "Law of the Sea Conference: Other Alternatives for Seabed Mining?" *New York Law School Journal of International and Comparative Law* 3 (1981): 39–49.

Doyle, Stephen E. "Permanent Arrangements for the Global Commercial Communication Satellite System of INTELSAT." *International Lawyer* 6 (1972): 258–91.

_____. "Regulating the Geostationary Orbit: ITU's WARC-ORB – '85–'88." *Journal of Space Law* 15 (1987): 1–23.

_____. "Significant Developments in Space Law: A Projection for the Next Decade." *Journal of Space Law* 9 (1981): 105–10.

Dubow, Marlene. "The Third United Nations Conference on the Law of the Sea: Questions of Equity for American Business." *Northwestern Journal of International Law and Business* 4 (1982): 172–202.

Dulles, John Foster. "The Challenge of Our Time: Peace with Justice." *ABA Journal* 39 (1953): 1063–66.

Eaton, S.K., Jr., and Janet Judy. "Seamounts and Guyots: A Unique Resource." *San Diego Law Review* 10 (1973): 599–637.

Eckert, Ross D. "Exploitation of Deep Ocean Minerals: Regulatory Mechanisms and United States Policy." *Journal of Law and Economics* 17 (1974): 143–77.

Eichelberger, Clark M. "A Case for the Administration of Marine Resources Underlying the High Seas by the United Nations." *Natural Resources Lawyer* 1, number 2 (1968): 85–94.

Ellis, Mark E. "The New International Economic Order and General Assembly Resolutions: The Debate Over the Legal Effects of General Assembly Resolutions Revisited." *California Western International Law Journal* 15 (1985): 647–704.

Ely, Northcutt. "A Case for the Administration of Mineral Resources Underlying the High Seas by National Interests." *Natural Resources Lawyer* 1 (1968): 78–84.

Emshwiller, John R. "Material Change: Commodity Price Rises May Augur a Recovery But Not for Everyone." *Wall Street Journal*, February 11, 1983, page 1.

Faber, Mike, and Roland Brown. "Changing the Rules of the Game: Political Risk, Instability and Fairplay in Mineral Concession Contracts." *Third World Quarterly* 2 (1980): 100–19.

Falk, Richard A. "The Quest for World Order: The Legacy of Optimism Re-Examined." *Dalhousie Law Journal* 9 (1984): 132–48.

Fatouros, A.A. "The International Law of the New International Economic Order: Problems and Challenges for the United States." *Willamette Law Review* 17 (1980): 93–110.

Fenn, Percy Thomas, Jr. "Justinian and the Freedom of the Sea." *American Journal of International Law* 19 (1925): 716–27.
_____. "Origins of the Theory of Territorial Waters." *American Journal of International Law* 20 (1926): 465–82.
Ferguson, C. Clyde. "The New International Economic Order." *University of Illinois Law Forum* (1980): 693–705.
Finch, George A. "The Nuremberg Trial and International Law." *American Journal of International Law* 41 (1947): 20–37.
Fincham, Charles. "Upheaval in the Law of the Sea." *Nuclear Active* (July 1972): 20–25.
Finlay, Luke W. "The Proposed New Convention on the Law of the Sea—A Candid Appraisal." *Syracuse Journal of International Law and Commerce* 7 (1979–80): 137–52.
Fitzmaurice, Gerald. "The General Principles of International Law." *Recueil des Cours* 92 (1957): 1–227.
_____. "The Law and Procedure of the International Court of Justice: General Principles and Substantive Law." *British Year Book of International Law* 27 (1950): 1–41.
Flanagan, Joseph P. "Manganese Nodules: A Huge Resource for the Future." *Sea Technology* 25 (August 1984): 18–21.
Foders, Federico. "International Organizations and Ocean Use: The Case of Deep-Sea Mining." *Ocean Development and International Law* 20 (1989): 519–30.
Forster, Malcolm J. "Law of the Sea Convention: Signatories Express Problems." *Environmental Policy and Law* 15 (1985): 2–3.
_____. "The Question of Antarctica." *Environmental Policy and Law* 14 (1985): 2–4.
Francioni, Francesco. "Legal Aspects of Mineral Exploitation in Antarctica." *Cornell International Law Journal* 19 (1986): 163–88.
Frank, Richard A. "Jumping Ship." *Foreign Policy* 43 (1981): 121–38.
Fraser, Malcolm. "The Third World and the West." *Atlantic Community Quarterly* 20 (1982): 99–108.
Freudenschuss, Helmut. "Legal and Political Aspects of the Recognition of National Liberation Movements." *Millennium* 11 (1982): 115–29.
Friedheim, Robert L. "Understanding the Debate on Ocean Resources." Occasional Paper Number 1. Kingston, R.I.: Law of the Sea Institute, 1969.
_____, and J.B. Kadane, with John King Gamble, Jr. "Quantitative Content Analysis of the United Nations Seabed Debate." *International Organization* 24 (1970): 479–502.
_____, and Judith T. Kildow. "Report of the Ocean Policy Research Workshop." Occasional Paper Number 26. Kingston, R.I.: Law of the Sea Institute, 1975.
Fuandez, Julio. "The Sea-Bed Negotiations: Third World Choices." *Third World Quarterly* 2 (1980): 487–99.
Gadon, Jean-Luc. "Offshore Oil Prospects for the Year 2000." *Natural Resources Forum* 11 (1987): 301–23.
Gaertner, Larianne P. "The Disputes Settlement Provisions of the Convention on the Law of the Sea: Critique and Alternatives to the International Tribunal for the Law of the Sea." *San Diego Law Review* 19 (1982): 577–97.
Gaja, Giorgio. "*Jus Cogens* Beyond the Vienna Convention." *Recueil des Cours* 172 (1981): 271–316.
Galligan, David M. "Wrapping Up the UNCLOS III 'Package': At Long Last the Final Clauses." *Virginia Journal of International Law* 20 (1980): 347–416.
Galloway, Eilene. "Perspectives of Space Law." *Journal of Space Law* 9 (1981): 21–29.
Galloway, Joseph L., et al. "The Gulf War—Among Friends." *U.S. News and World Report*, June 15, 1987, page 26.
Gamble, John King, Jr. "Assessing the Reality of the Deep Seabed Regime." *San Diego Law Review* 22 (1985): 779–92.
Geddes, Roger A. "The Future of United States Deep Seabed Mining: Still in the Hands of Congress." *San Diego Law Review* 19 (1982): 613–30.
Gerstle, Margaret Lynch. "The UN and the Law of the Sea: Prospects for the United States Seabeds Treaty." *San Diego Law Review* 8 (1971): 573–83.
Gertler, Jeffrey Lee, and Paul Wayne. "Synopsis: Recent Developments in the Law of the Sea 1979–1980." *San Diego Law Review* 18 (1981): 533–81.

Glasby, G.P. "The Three-Million-Tons-Per-Year Manganese Nodule 'Mine Site': An Optimistic Assumption?" *Marine Mining* 4 (1983): 73–77.

Glassner, Martin Ira. "Developing Land-Locked States and the Resources of the Seabed." *San Diego Law Review* 11 (1974): 633–35.

Glickman, Stephen K. "Enforcement Mechanisms of the Law of the Sea Treaty." *Suffolk Transnational Law Journal* 1 (1977): 1–22.

Goldie, L.F.E. "The Exploitability Test—Interpretation and Potentialities." *Natural Resources Journal* 8 (1968): 434–77.

————. "A General International Law Doctrine for Seabed Regimes." *International Lawyer* 7 (1973): 796–824.

————. "A Selection of Books Reflecting Perspectives in the Seabed Mining Debate: Part I." *International Lawyer* 15 (1981): 293–337.

————. "A Selection of Books Reflecting Perspectives in the Seabed Mining Debate: Part II." *International Lawyer* 15 (1981): 445–98.

————. "Title and Use (and Usufruct)—An Ancient Distinction Too Oft Forgot." *American Journal of International Law* 79 (1985): 689–714.

Goldschneider, Robert. "The Technology Transfer Process: A Vehicle for Continuity and Change." *Vanderbilt Journal of Transnational Law* 14 (1981): 255–68.

Goldwin, Robert A. "Locke and the Law of the Sea." *Commentary* 71 (June 1981): 46–50.

Gorove, Stephen. "The Concept of 'Common Heritage of Mankind': A Political, Moral or Legal Innovation." *San Diego Law Review* 9 (1972): 390–403.

Goshko, John M. "Suddenly, the U.N. Matters." *Washington Post National Weekly Edition,* October 3–9, 1988, page 17.

Gotlieb, A.E. "The Impact of Technology on th Development of Contemporary International Law." *Recueil des Cours* 170 (1981): 115–329.

Greenblatt, Gershon D., James Robert Miller, and Alfred J. Waldchen. "Recent Developments in the Law of the Sea IV: A Synopsis." *San Diego Law Review* 10 (1973): 559–98.

Gregg, Donna C. "Capitalizing on National Self-Interest: The Management of International Telecommunication Policy by the International Telecommunication Union." *Law and Contemporary Problems* 45 (1982): 37–52.

Grier, Peter. "Staking a Claim to the Ocean's Bed." *Christian Science Monitor,* April 9, 1981, pages 12–13.

Griffin, William L. "The Emerging Law of Ocean Space." *International Lawyer* 1 (1966): 548–87.

Grolin, Jesper. "The Future of the Law of the Sea: Consequences of a Non-Treaty of Non-Universal Treaty Situation." *Ocean Development and International Law Journal* 13 (1983): 1–31.

Gross, Leo. "The Peace of Westphalia, 1648–1948." *American Journal of International Law* 42 (1948): 20–41.

Haggin, Joseph. "Guest Editorial: An External View of Marine Mining." *Marine Mining* 5 (1986): 349–56.

————. "Marine Mining to Improve Its Organization, Direction, Financing." *Chemical and Engineering News,* November 18, 1985, pages 63–67.

Haight, G. Winthrop. "Comments on Judge Oda's Approach to the Common Heritage of Mankind." *New York Law School Journal of International and Comparative Law* 3 (1981): 15–20.

————. "United Nations Affairs: Ad Hoc Committee on Sea-Bed and Ocean Floor." *International Lawyer* 3 (1968–69): 22–30.

Halbach, Peter, et al. "Cobalt-Rich and Platinum-Bearing Manganese Crust Deposits on Seamounts: Nature, Formation, and Metal Potential." *Marine Mining* 8 (1989): 23–39.

————, and F.T. Manheim. "Potential of Cobalt and Other Metals in Ferromanganese Crusts in Seamounts of the Central Pacific Basin." *Marine Mining* 4 (1984): 319–36.

Hardin, Garrett. "The Tragedy of the Commons." *Science* 162 (December 13, 1968): 1243–48, 1268.

Hardy, Michael. "The Law of the Sea and the Prospects for Deep Seabed Mining: The Position of the European Community." *Ocean Development and International Law* 17 (1986): 309–23.

Hargrove, John Lawrence. "New Concepts in the Law of the Sea." *Ocean Development and International Law Journal* 1 (1973): 5–12.

Harlow, Lt. Commander Bruce A., USN. "The Fourth Dimension of Seapower—Ocean Technology and International Law: Introduction." *JAG Journal* 22 (1967): 27–30, 38.

Harry, John R. "Deep Seabed Mining in the Law of the Sea Negotiation (I): The Contours of a Compromise." Oceans Policy Study 1:2. Charlottesville, Va.: Michie, 1978.

Hassan, Najmul. "Staking Its Claim: India Seeks Mineral Riches on Sea Floor." *Los Angeles Times*, September 15, 1983, part I-B, page 4.

Hayashi, Moritaka. "Japan and Deep Seabed Mining." *Ocean Development and International Law* 17 (1986): 351–65.

—————. "Registration of the First Group of Pioneer Investors by the Preparatory Commission for the International Sea-Bed Authority and for the International Tribunal for the Law of the Sea." *Ocean Development and International Law* 20 (1989): 1–33.

Hazlett, James A. "Strait Shooting." *U.S. Naval Institute Proceedings* 108 (June 1982): 70–73.

Healy, Melissa. "Army May Bear Brunt of Budget Cuts, Cheney Hints." *Los Angeles Times*, November 25, 1989, pages A1, A30.

Hegwood, David. "Deep Seabed Mining: Alternative Schemes for Protecting Developing Countries from Adverse Impacts." *Georgia Journal of International and Comparative Law* 12 (1982): 173–92.

Heimsoeth, Harald. "Antarctic Mineral Resources." *Environmental Policy and Law* 11 (1983): 61–63.

Henkin, Louis. "Seabed Pact Would Help U.S." *Washington Post*, August 8, 1971, page B2.

Hicks, Guy M. "The Law of the Sea Treaty: A Review of the Issues." *Journal of Social, Political and Economic Studies* 6 (1981): 107–18.

Hollick, Ann L. "United States Oceans Politics." *San Diego Law Review* 10 (1973): 467–501.

Holser, Alexander F. "Offshore Lands of the USA: The US Exclusive Economic Zone, Continental Shelf and Outer Continental Shelf." *Marine Policy* 12 (1988): 2–8.

Honnold, Edward E. "Thaw in International Law? Rights in Antarctica Under the Law of Common Spaces." *Yale Law Journal* 87 (1978): 804–59.

Hoyt, Monty. "Sea Panel Races Time to Form International Control." *Christian Science Monitor*, October 13, 1971, page 6.

Hufbauer, Gary C., and George N. Carlson. "United States Policy Toward the Transfer of Proprietary Technology: Licenses, Taxes, and Finance." *Vanderbilt Journal of Transnational Law* 14 (1981): 337–61.

Hull, E.W. Seabrook. "The International Law of the Sea: A Case for a Customary Approach." Occasional Paper Number 30. Kingston, R.I.: Law of the Sea Institute, 1976.

Hurst, Cecil J.B. "Whose Is the Bed of the Sea?" *British Year Book of International Law* 4 (1923–24): 34–43.

Iguchi, Takeo. "Japan and the New Law of the Sea: Facing the Challenge of Deep Seabed Mining." *Virginia Journal of International Law* 27 (1987): 527–49.

Jackson, Jon Gregory. "Deepsea Ventures: Exclusive Mining Rights to the Deep Seabed as a Freedom of the Sea." *Baylor Law Review* 28 (1976): 170–86.

Jacobsen, Mark P. "Recent Developments: International Agreements." *Harvard International Law Journal* 30 (1989): 237–47.

Jacobson, Jon L. "Bridging the Gap to International Fisheries Agreement: A Guide for Unilateral Action." *San Diego Law Review* 9 (1972): 454–90.

Jagota, S.P. "Developments in the UN Conference on the Law of the Sea: A Third World Review." *Third World Quarterly* 3 (1981): 287–319.

Janis, Mark W. "The Nature of *Jus Cogens.*" *Connecticut Journal of International Law* 3 (1988): 359–63.

—————. "The Seas and International Law: Rules and Rulers." *St. John's Law Review* 58 (1984): 306–17.

Jenisch, Uwe. "Bridging the Gap for Seabed Mining: Preparatory Instruments for the New Law of the Sea Convention." *San Diego Law Review* 18 (1981): 409–13.

Jesus, José Luis. Comments before Law of the Sea Institute, Tokyo, July 1990. Reprinted in *Oceans Policy News* (September 1990), page 7.

_____. "Statement on the Issue of the Universality of the Convention." University of Kiel, July 1990. Reprinted in *Oceans Policy News*, Special Report (July 1990), pages 1–5.

Johnson, Charles J., and Allen L. Clark. "Potential of Pacific Ocean Nodule, Crust, and Sulfide Mineral Deposits." *Natural Resources Forum* 9 (1985): 179–86.

Johnson, D.H.N. "Acquisitive Prescription in International Law." *British Year Book of International Law* 27 (1950): 332–54.

_____. "The Conclusions of International Conferences." *British Year Book of International Law* 35 (1959): 1–33.

Jones, G. Kevin. "Outer Continental Shelf Petroleum Resources and the Nation's Future Energy Needs." *Syracuse Journal of International Law and Commerce* 15 (1989): 305–60.

Jones, Robert A. "Dividing the Pie: History in Making at South Pole." *Los Angeles Times*, January 24, 1985, part I, page 24.

Jones, William B. "The International Sea-Bed Authority Without U.S. Participation." *Ocean Development and International Law Journal* 12 (1983): 151–71.

_____. "Risk Assessment: Corporate Ventures in Deep Seabed Mining Outside the Framework of the UN Convention on the Law of the Sea." *Ocean Development and International Law* 16 (1986): 341–51.

Joyner, Christopher C. "Antarctica and the Law of the Sea: An Introductory Overview." *Ocean Development and International Law Journal* 13 (1983): 277–89.

_____. "Antarctica and the Law of the Sea: Rethinking the Current Legal Dilemmas." *San Diego Law Review* 18 (1981): 415–42.

_____. "The Exclusive Economic Zone and, Antarctica." *Virginia Journal of International Law* 21 (1981): 691–725.

_____. "Towards a Legal Regime for the International Seabed: The Soviet Union's Evolving Perspective." *Virginia Journal of International Law* 15 (1975): 871–901.

_____. "U.N. General Assembly Resolutions and International Law: Rethinking the Contemporary Dynamics of Norm-Creation." *California Western International Law Journal* 11 (1981): 445–78.

Jumers, Peter A. "Limits on Predicting and Detecting Benthic Community Responses to Manganese Nodule Mining." *Marine Mining* 2 (1981): 213–29.

Junker, John M. "The Structure of the Fourth Amendment: The Scope of the Protection." *Journal of Criminal Law and Criminology* 79 (1989): 1105–84.

Katz, Ronald D. "A Method for Evaluating the Deep Seabed Mining Provisions of the Law of the Sea Treaty." *Yale Journal of World Public Order* 7 (1980): 114–30.

Kay, David A. "Operational Aspects of Managing the Oceans." *Columbia Journal of World Business* 10 (1975): 29–33.

Keen, E.A. "Common Property in Fisheries: Is Sole Ownership an Option?" *Marine Policy* 7 (1983): 197–211.

Keith, Kent M. "Laws Affecting the Development of Ocean Resources in Hawaii." *University of Hawaii Law Review* 4 (1982): 227–329.

Kent, George. "Fisheries and the Law of the Sea: A Common Heritage Approach." *Ocean Management* 4 (1978): 1–20.

Khan, Rahmatullah. "Editorial Comment: Ocean Resources Development – India's Options." *Indian Journal of International Law* 22 (1982): 448–58.

Kimball, Lee. "Implications of the Arrangements Made for Deep Sea Mining for Other Joint Exploitations." *Columbia Journal of World Business* 15 (1980): 52–61.

_____. "Introductory Note." *International Legal Materials* 26 (1987): 1502–4.

_____. "Turning Points in the Future of Deep Seabed Mining." *Ocean Development and International Law* 17 (1986): 367–98.

_____, and A.R.H. Schneider. "A Viable Convention?" *Environmental Policy and Law* 9 (1982): 66–73.

Kindt, John Warren. "Special Claims Impacting Upon Marine Pollution Issues at the Third U.N. Conference on the Law of the Sea." *California Western International Law Journal* 10 (1980): 397–449.

Kingham, J.D., and D.M. McCrae. "Competent International Organizations and the Law of the Sea." *Marine Policy* (April 1979): 106–32.

Kirgis, Frederic L., Jr. "Editorial Comment: Standing to Challenge Human Endeavors That Could Change the Climate." *American Journal of International Law* 84 (1990): 523-30.

Kirsch, Philippe, and Douglas Fraser. "The Law of the Sea Preparatory Commission After Six Years: Review and Prospects." *Canadian Yearbook of International Law* 26 (1988): 119-53.

Kleid, Bernice R. "Synopsis: Recent Developments in the Law of the Sea 1980-1981." *San Diego Law Review* 19 (1982): 631-58.

Knauss, John A. "Factors Influencing a U.S. Position in a Future Law of the Sea Conference." Occasional Paper Number 10. Kingston, R.I.: Law of the Sea Institute, 1971.

Knight, H. Gary. "The Deep Seabed Hard Mineral Resources Act — A Negative View." *San Diego Law Review* 10 (1973): 446-60.

_____. "The Draft United Nations Conventions on the International Seabed Area: Background, Description, and Some Preliminary Thoughts." *San Diego Law Review* 8 (1971): 459-550.

_____. "Foreword: Law of the Sea Negotiations 1971-1972 — From Internationalism to Nationalism." *San Diego Law Review* 9 (1972): 383-89.

Koh, Tommy T.B. "Negotiating a New World Order for the Sea." *Virginia Journal of International Law* 24 (1984): 761-84.

Kooijmans, P.H. "How to Handle the Grotian Heritage." *Netherlands International Law Review* 30 (1983): 81-92.

_____. "Protestantism and the Development of International Law." *Recueil des Cours* 152 (1976): 79-117.

Krueger, Robert B. "International and National Regulation of Pollution from Offshore Oil Pollution Production." *San Diego Law Review* 7 (1970): 541-73.

Kryzla, Bill F. "Joint Ventures and Technology Transfers." *Case Western Reserve Journal of International Law* 12 (1980): 549-73.

Kunz-Hallstein, Hans Peter. "Patent Protection, Transfer of Technology and Developing Countries — A Survey of the Present Situation." *IIC International Review of Industrial Property and Copyright Law* 6 (1975): 427-55.

Lachica, Eduardo. "Copper Imports Hurt U.S. Firms, ITC Decides." *Wall Street Journal*, June 15, 1984, page 2.

Lachs, Manfred. "Some Reflections on the State of the Law of Outer Space." *Journal of Space Law* 9 (1981): 3-11.

Lampe, William H. "The 'New' International Maritime Organization and Its Place in Development of International Maritime Law." *Journal of Maritime Law and Commerce* 14 (1983): 305-29.

Lancaster, John. "The 'Great Terrain Robbery.'" *Washington Post National Weekly Edition*, September 10-16, 1990, page 6.

Langevad, E.J. "Exploitation of the Mineral Resources of the Oceans as Affected by the Provisions of the Convention on the Law of the Sea." *Natural Resources Forum* 7 (1983): 227-38.

Lansing, Robert. "A Unique International Problem." *American Journal of International Law* 11 (1917): 763-71.

Larson, David L. "Deep Seabed Mining: A Definition of the Problem." *Ocean Development and International Law* 17 (1986): 271-308.

_____. "The Reagan Administration and the Law of the Sea." *Ocean Development and International Law Journal* 11 (1982): 297-320.

_____. "The United States Position on the Deep Seabed." *Suffolk Transnational Law Journal* 3 (1979): 1-33.

_____. "When Will the UN Convention on the Law of the Sea Come Into Effect?" *Ocean Development and International Law* 20 (1989): 175-202.

Lauterpacht, H. "The Grotian Tradition in International Law." *British Year Book of International Law* 23 (1946): 1-53.

_____. "Sovereignty Over Submarine Areas." *British Year Book of International Law* 27 (1950): 376-433.

"Law of the Sea Treaty: Pro & Con." *Congressional Digest* 62 (January 1983): 2-32.

Lay, S. Houston. "Book Review." *California Western International Law Journal* 10 (1980): 153-56.

Laylin, John G. "The Law to Govern Deep Sea Mining Until Superseded by International Agreement." *San Diego Law Review* 10 (1973): 433-45.

Lee, Luke T. "The Law of the Sea Convention and Third States." *American Journal of International Law* 77 (1983): 541-68.

Lee, Roy S. "Deep Seabed Mining and Developing Countries." *Syracuse Journal of International Law and Commerce* 6 (1978): 213-19.

_____. "The Enterprise: Operational Aspects and Implications." *Columbia Journal of World Business* 15 (1980): 62-74.

Legault, L.H.J. "The Freedom of the Seas: A License to Pollute?" *University of Toronto Law Journal* 21 (1971): 211-21.

Leich, Marian Nash. "Contemporary Practice of the United States Relating to International Law." *American Journal of International Law* 83 (1989): 559-61.

_____. "Contemporary Practice of the United States Relating to International Law." *American Journal of International Law* 84 (1990): 239-42.

Leitzell, Terry L. "The Ocean Dumping Convention – A Hopeful Beginning." *San Diego Law Review* 10 (1973): 502-13.

Leive, David M. "Essential Features of INTELSAT: Applications for the Future." *Journal of Space Law* 9 (1981): 45-51.

Lenoble, J.-P. "Polymetallic Nodule Resources and Reserves in the North Pacific from the Data Collected by AFERNOD." *Ocean Management* 7 (1981): 9-24.

Levin, Harvey J. "Spectrum Negotiations and the Geostationary Satellites." *New York Law School Journal of International and Comparative Law* 4 (1982): 77-81.

Lewis, Paul. "After 6 Years Law of the Sea Parley Nears Mining Pact." *New York Times*, August 30, 1980, page A4.

Linden, H. Vander. "Alexander VI and the Demarcation of the Maritime and Colonial Domains of Spain and Portugal, 1493-1494." *American Historical Review* 22 (1916): 1-20.

Logue, John J. "The Receding Heritage: UN Law of the Sea Conference." *Transnational Perspectives*, November 2, 1976.

Løvald, John Ludvik. "In Search of an Ocean Regime: The Negotiations in the General Assembly's Seabed Committee 1968-70." *International Organization* 29 (1975): 681-709.

McCormick, James M. "Intergovernmental Organizations and Cooperation Among Nations." *International Studies Quarterly* 24 (1980): 75-98.

McDade, Paul V. "The Interim Obligation Between Signature and Ratification of a Treaty." *Netherlands International Law Review* 32 (1985): 5-47.

McDorman, Ted L. "The 1982 Law of the Sea Convention: The First Year." *Journal of Maritime Law and Commerce* 15 (1984): 211-32.

_____. "Thailand and the 1982 Law of the Sea Convention." *Marine Policy* 9 (1985): 292-309.

McDougal, Myres S., and Norbert A. Schlei. "The Hydrogen Bomb Tests in Perspective: Lawful Measures for Security." *Yale Law Journal* 64 (1955): 648-710.

McDowell, Eleanor C. "Contemporary Practice of the United States Relating to International Law." *American Journal of International Law* 69 (1975): 861-86.

MacRae, Leslie N. "Customary International Law and the United Nations Law of the Sea Treaty." *California Western International Law Journal* 13 (1983): 181-222.

Malahoff, Alexander. "A Comparison of the Massive Submarine Polymetallic Sulfides of the Galapagos Rift with Some Continental Deposits." *Marine Technology Society Journal* 16, number 3 (1982): 39-45.

Mally, Gerhard. "Technology Transfer Controls." *U.S. Department of State Bulletin* 82 (November 1982): 52-55.

Malone, James L. "The United States and the Law of the Sea." *Virginia Journal of International Law* 24 (1984): 785-807.

_____. "The United States and the Law of the Sea After UNCLOS III." *Law and Contemporary Problems* 46 (Spring 1983): 29-36.

_____. "Who Needs the Sea Treaty?" *Foreign Policy* 54 (Spring 1984): 44-63.

Mann, Jim. "U.S. Move to Back Aquino Could Prove Costly Later." *Los Angeles Times*, December 2, 1989, page A16.

Manning, Robert A. "Making Foreign Policy Fit a Changing World." *U.S. News and World Report*, March 10, 1983, page 34.

Margolis, Emanuel. "The Hydrogen Bomb Experiments and International Law." *Yale Law Journal* 64 (1955): 629–47.

Martin, Jochen. "Deep-Sea Mining Between Convention and National Legislation." *Ocean Development and International Law Journal* 10 (1981): 175–85.

Mashayekhi, Mina. "The Present Legal Status of Deep Sea-Bed Mining." *Journal of World Trade Law* 19 (1985): 229–50.

Meijers, H. "How Is International Law Made? — The Stages of Growth of International Law and the Use of Its Customary Rules." *Netherlands Yearbook of International Law* 9 (1978): 3–26.

Menter, Martin. "Commercial Participation in Space Activities." *Journal of Space Law* 9 (1981): 53–68.

Mero, John L. "A Legal Regime for Deep Sea Mining." *San Diego Law Review* 7 (1970): 488–503.

_____. "Manganese." *North Dakota Engineer* 27 (1952): 28.

Miles, Edward L. "Preparing for UNCLOS IV?" *Ocean Development and International Law* 19 (1988): 421–30.

Minola, Patricia. "The Moon Treaty and the Law of the Sea." *San Diego Law Review* 18 (1981): 455–72.

Molitor, Michael R. "The U.S. Deep Seabed Mining Regulations: The Legal Basis for an Alternative Regime." *San Diego Law Review* 19 (1982): 599–612.

Moore, John Norton. "The Regime of Straits and the Third United Nations Conference on the Law of the Sea." *American Journal of International Law* 74 (1980): 77–121.

Mosler, Hermann. "Supra-National Judicial Decisions and National Courts." *Hastings International and Comparative Law Review* 4 (1980): 425–72.

Nandan, Satya N. "The 1982 UN Convention on the Law of the Sea: At a Crossroad." *Ocean Development and International Law* 20 (1989): 515–18.

Nash, Marian L. "Contemporary Practice of the United States Relating to International Law." *American Journal of International Law* 74 (1980): 418–32.

Negroponte, John D., Assistant Secretary of State for Oceans and International Environmental and Scientific Affairs. "Who Will Protect Freedom of the Seas?" *U.S. Department of State Bulletin* 86 (October 1986): 41–43.

Neubauer, Ronald D. "The Right of Innocent Passage for Warships in the Territorial Sea: A Response to the Soviet Union." *Naval War College Review* 41 (Spring 1988): 49–56.

Neutze, Dennis R. "Whose Law of the Sea?" *U.S. Naval Institute Proceedings* 109 (January 1983): 43–48.

"A New Ocean Treasure?" *Environmental Policy and Law* 10 (1983): 57.

Newcombe, Hanna. "National Patterns in International Organizations." *Peace Research Reviews* 6 (November 1975): 1–367.

Newton, W. Frank. "Inexhaustibility as a Law of the Sea Determinant." *Texas International Law Journal* 16 (1981): 369–432.

_____. "Seabed Resources: The Problems of Adolescence." *San Diego Law Review* 8 (1971): 551–72.

Nielson, Fred K. "Editorial Comment: The Solution of the Spitzbergen Question." *American Journal of International Law* 14 (1930): 232–35.

Note. "Exclusion of Ships from Nonterritorial Weapons Testing Zones." *Harvard Law Review* 99 (1986): 1040–58.

_____. "Power of a State to Extend Its Boundary Beyond the Three-Mile Limit." *Columbia Law Review* 39 (1939): 317–26.

Noyes, John E. "Compulsory Third-Party Adjudication and the 1982 United Nations Convention on the Law of the Sea." *Connecticut Journal of International Law* 4 (1989): 675–96.

Nys, Ernest. "The Development and Formation of International Law." *American Journal of International Law* 6 (1912): 1–29, 279–315.

Oberdorfer, Don. "George Shultz's Roller-Coaster Ride with Yasser Arafat." *Washington Post National Weekly Edition*, December 26, 1988–January 1, 1989, pages 16–17.

Oceans Policy News. 1984–90.

Oda, Shigeru. "Fisheries Under the United Nations Convention on the Law of the Sea." *American Journal of International Law* 77 (1983): 739–55.

"Oddfellows." *Economist* 293 (December 22, 1984): 30.

Onuf, N.G., and Richard K. Birney. "Peremptory Norms of International Law: Their Source, Function and Future." *Denver Journal of International Law and Policy* 4 (1974): 187–98.

Orlove, Mark. "Spaced Out: The Third World Looks for a Way in to Outer Space." *Connecticut Journal of International Law* 4 (1989): 597–634.

Orr, Samuel C. "Soviet, Latin Opposition Blocks Agreement on Seabeds Treaty." *CPR National Journal* (1970), pages 1974–77.

Orrego Vicuña, Francisco. "The Deep Seabed Mining Regime: Terms and Conditions for Its Renegotiation." *Ocean Development and International Law* 20 (1989): 531–39.

_____. "National Laws on Seabed Exploitation: Problems of International Law." *Lawyer of the Americas* 13 (1981): 139–56.

Oxman, Bernard H. "Antarctica and the New Law of the Sea." *Cornell International Law Journal* 19 (1986): 211–47.

_____. "The High Seas and the International Seabed Area." *Michigan Journal of International Law* 10 (1989): 526–42.

_____. "Introduction: On Evaluating the Draft Convention on the Law of the Sea." *San Diego Law Review* 19 (1982): 453–60.

_____. "Navigation, Pollution, and Compulsory Settlement of EEZ Disputes." *Oceanus* 27 (Winter 1984–85).

_____. "The Third United Nations Conference on the Law of the Sea." *American Journal of International Law* 76 (1982): 1–23.

_____. "The Third United Nations Conference on the Law of the Sea: The Seventh Session (1978)." *American Journal of International Law* 73 (1979): 1–41.

_____. "The Third United Nations Conference on the Law of the Sea: The Eighth Session." *American Journal of International Law* 74 (1980): 1–47.

_____. "The Third United Nations Conference on the Law of the Sea: The Ninth Session (1980)." *American Journal of International Law* 75 (1981): 211–56.

_____. "The Third United Nations Conference on the Law of the Sea: The 1976 New York Sessions." *American Journal of International Law* 71 (1977): 247–69.

_____. "The World Outlook for the International Law of the Sea." Keynote address to Marine Technology Society, February 19, 1971.

Pagano, Penny. "Men's First Steps on Moon Commemorated by Reagan." *Los Angeles Times*, July 21, 1984, part 1, page 14.

Pal, Mati L. "Financial Arrangements." *Syracuse Journal of International Law and Commerce* 6 (1979): 295–99.

_____, and James K. Sebenius. "Evolving Financial Terms of Mineral Agreements: Risks, Rewards and Participation in Deep Seabed Mining." *Columbia Journal of World Business* 15 (1980): 75–83.

Pardo, Arvid. "Before and After." *Law and Contemporary Problems* 46 (Spring 1983): 95–105.

_____. "The Convention on the Law of the Sea: A Preliminary Appraisal." *San Diego Law Review* 20 (1983): 489–503.

_____. "Development of Ocean Space—An International Dilemma." *Louisiana Law Review* 31 (1970): 45–72.

_____. "Foreword." *San Diego Law Review* 14 (1977): 507–22.

_____. "The Law of the Sea: Its Past and Its Future." *Oregon Law Review* 63 (1984): 7–17.

_____. "A Statement on the Future Law of the Sea in Light of Current Trends in Negotiations." *Ocean Development and International Law Journal* 1 (1974): 315–35.

_____, and Elisabeth Mann Borgese. "The New International Economic Order and the Law of the Sea." Occasional Paper Number 5. Malta: International Ocean Institute, 1976.

Pasho, David W. "Canada and Ocean Mining." *Marine Technology Society Journal* 19, number 4 (1985): 26–30.

Patey, Jacques. "La Conférence des Nations Unies sur le Droit de la Mer." *Revue Générale de Droit International Public* 62 (1958): 446–67.

Paulsen, Gordon W. "An Historical Overview of the Development of Uniformity in International Maritime Law." *Tulane Law Review* 57 (1983): 1065–87.

Pell, Senator Claiborne. "Introduction." *San Diego Law Review* 18 (1981): 387–94.

Pendley, William Perry. "It Ain't Broke—Don't Fix It: Mining in America's Exclusive

Economic Zone Requires No New Legislation." *Marine Technology Society Journal* 23 (March 1989): 48–50.

Peterson, M.J. "Antarctica: The Last Great Land Rush on Earth." *International Organization* 34 (1980): 377–403.

Pietrowski, Robert F., Jr. "Hard Minerals on the Deep Ocean Floor: Implications for American Law and Policy." *William and Mary Law Review* 19 (1977): 43–75.

———. "International Law Applicable to Deep Sea Mining." *Congressional Record* 125 (December 14, 1979): 36065–68.

Piper, Don C. "On Changing or Rejecting the International Legal Order." *International Lawyer* 12 (1978): 293–307.

Pontecorvo, Giulio. "Contribution of the Ocean Sector to the United States Economy: Estimated Values for 1987 – A Technical Note." *Marine Technology Society Journal* 23 (June 1989): 7–14.

———. "Musing About Seabed Mining, or Why What We Don't Know Can Hurt Us." *Ocean Development and International Law Journal* 21 (1990): 117–18.

———, and Maurice Wilkinson. "From Cornucopia to Scarcity: The Current Status of Ocean Resource Use." *Ocean Development and International Law Journal* 5 (1978): 383–95.

Powers, Richard J. "United Nations Voting Alignments: A New Equilibrium." *Western Political Quarterly* 33 (1980): 167–84.

Prewo, Wilfried. "Ocean Fishing: Economic Efficiency and the Law of the Sea." *Texas International Law Journal* 15 (1980): 261–85.

Proudfoot, David W. "Guarding the Treasures of the Deep: The Deep Seabed Hard Mineral Resources Act." *Harvard Journal on Legislation* 10 (1973): 596–620.

Pruitt, Virginia A. "Unilateral Deep Seabed Mining and Environmental Standards: A Risky Venture." *Brooklyn Journal of International Law* 8 (1982): 345–63.

"Questions in the European Parliament." *Environmental Policy and Law* 15 (1985): 96.

Quigley, John. "Law for a World Community." *Syracuse Journal of International Law and Commerce* 16 (1989): 1–38.

Radway, Robert J. "Comparative Evolution of Technology Transfer Policies in Latin America: The Practical Realities." *Denver Journal of International Law and Policy* 9 (1980): 197–215.

Raleigh, C. Barry. "The Internationalism of Ocean Science vs. International Politics." *Marine Technology Society Journal* 23 (March 1989): 44–47.

Rao, K. Narayana. "Editorial Comment: Common Heritage of Mankind and the Moon Treaty." *Indian Journal of International Law* 21 (1981): 275–78.

Rao, P. Chandrasekhara. "Editorial Comment: The UN Convention on the Law of the Sea: Some Reflections." *Indian Journal of International Law* 22 (1982): 459–67.

Ratiner, Leigh S. "The Law of the Sea: A Crossroads for American Foreign Policy." *Foreign Affairs* 60 (1982): 1006–21.

———, and Rebecca L. Wright. "United States Ocean Mineral Resource Interests and the United Nations Conference on the Law of the Sea." *Natural Resources Lawyer* 6 (1973): 1–43.

Reiley, Eldon H. "Introduction to a Tempest: The Legal, Technological and Political Dimensions of the 1984 Law of the Sea Conference in San Francisco." *University of San Francisco Law Review* 18 (1984): 415–31.

Reppy, Alison. "The Grotian Doctrine of the Freedom of the Seas Reappraised." *Fordham Law Review* 19 (1950): 243–85.

Rich, Roland. "A Minerals Regime for Antarctica." *International and Comparative Law Quarterly* 31 (1982): 709–25.

Richardson, Elliot L. "Law in the Making: A Universal Regime for Deep Seabed Mining?" *New York State Bar Journal* 53 (1981): 408–11, 440–46.

———. "Law of the Sea: A Reassessment of U.S. Interests." *Mediterranean Quarterly* 1 (Spring 1990): 1–13.

———. "The Politics of the Law of the Sea." *Ocean Development and International Law Journal* 11 (1982): 9–23.

———. "Power, Mobility and the Law of the Sea." *Foreign Affairs* 58 (1980): 902–19.

_____. "Superpowers Need Law: A Response to the United States Rejection of the Law of the Sea Treaty." *George Washington Journal of International Law and Economics* 17 (1982): 1–15.

_____. "The United States Posture Toward the Law of the Sea Convention: Awkward But Not Irreparable." *San Diego Law Review* 20 (1983): 505–13.

Riesenfeld, Stefan A. "Comment." *Law and Contemporary Problems* 46 (Spring 1983): 11–15.

Ritchey, Joseph L. "Assessment of Cobalt-Rich Manganese Crust Resources on Horizon and S.P. Guyots, U.S. Exclusive Economic Zone." *Marine Mining* 6 (1987): 231–43.

Roat, Rachel. "Promulgation and Enforcement of Minimum Standards for Foreign Flag Ships." *Brooklyn Journal of International Law* 6 (1980): 54–87.

Robertson, G. David, and Gaylene Vasatura. "Recent Developments in the Law of the Sea 1981–1982." *San Diego Law Review* 20 (1983): 679–711.

Robinson, Glen O. "Regulating International Airwaves: The 1979 WARC." *Virginia Journal of International Law* 21 (1980): 1–54.

Rona, Peter A. "Mineral Deposits from Sea-Floor Hot Springs." *Scientific American* 254 (January 1986): 84–92.

_____. "Perpetual Seafloor Metal Factory." *Sea Frontiers* 30 (1984): 132–41.

Rotberg, Robert I. "Shultz: Time to 'Deal In' the ANC?" *Christian Science Monitor*, January 16, 1987, page 13.

Rothblatt, Martin A. "ITU Regulation of Satellite Communication." *Stanford Journal of International Law* 18 (1982): 1–25.

Rowen, Hobart. "Oils Well That Ends Well." *Washington Post National Weekly Edition*, June 18–24, 1990, page 5.

Rozakis, Christos L. "The Greek-Turkish Dispute Over the Aegean Continental Shelf." Occasional Paper Number 28. Kingston, R.I.: Law of the Sea Institute, 1975.

Saffo, Paul Lawrence. "The Common Heritage of Mankind: Has the General Assembly Created a Law to Govern Seabed Mining?" *Tulane Law Review* 53 (1979): 492–520.

Santa Cruz, Fernando Zegers. "Deep Sea-Bed Mining Beyond National Jurisdiction in the 1982 UN Convention on the Law of the Sea: Description and Prospects." *German Yearbook of International Law* 31 (1988): 107–19.

Scheltens, D.F. "Grotius' Doctrine of the Social Contract." *Netherlands International Law Review* 30 (1983): 43–60.

Schneider, A.R.H. "The Law of the Sea in the 1980s." *Environmental Policy and Law* 7 (1981): 7–12.

_____. "UNCLOS III Revisited—Recent Events in the Law of the Sea." *Environmental Policy and Law* 9 (1982): 108–9.

Schneider, Jan. "Codification and Progressive Development of International Environmental Law at the Third United Nations Conference on the Law of the Sea: The Environmental Aspects of the Treaty Review." *Columbia Journal of Transnational Law* 20 (1981): 243–75.

Schweisfurth, Theodore. "The Influence of the Third United Nations Conference on the Law of the Sea on International Customary Law." *Zeitschrift für ausländisches öffentliches Recht und Völkerrecht* 43 (1983): 566–84.

Shaw, Malcolm. "The International Status of National Liberation Movements." *Liverpool Law Review* 5 (1983): 19–34.

Shinaver, Gary M. "Fishery Conservation: Is the Categorical Exclusion of Foreign Fleets the Next Step?" *California Western International Law Journal* 12 (1982): 154–203.

Shingleton, Brad. "UNCLOS III and the Struggle for Law: The Elusive Customary Law of Seabed Mining." *Ocean Development and International Law Journal* 13 (1983): 33–63.

Shusterich, Kurt. "Mining the Deep Seabed: A Complex and Innovative Industry." *Marine Policy* 6 (1982): 175–92.

Shyam, Manjula R. "Deep Seabed Mining: An Indian Perspective." *Ocean Development and International Law* 17 (1986): 325–49.

Siapno, William D. "Manganese Nodules: Overcoming the Constraints." *Marine Mining* 5 (1986): 457–65.

Sielen, Alan B. "Sea Changes? Ocean Dumping and International Regulation." *Georgetown International Environmental Law Review* 1 (1988): 1–32.

Silverstein, David. "Proprietary Protection for Deepsea Mining Technology in Return for Technology Transfer: New Approach to the Seabeds Controversy." *Journal of the Patent Office Society* 60 (1978): 135–70.

Simma, Bruno. "The Antarctic Treaty as a Treaty Providing for an 'Objective Regime.'" *Cornell International Law Journal* 19 (1986): 189–209.

Slouka, Zdenek J. "United Nations and the Deep Ocean: From Data to Norms." *Syracuse Journal of International Law* 1 (1972): 61–90.

Smith, Brian. "Innocent Passage as a Rule of Decision: Navigation v. Environmental Protection." *Columbia Journal of Transnational Law* 21 (1982): 49–102.

Smith, Milton L. "The Space WARC Concludes." *American Journal of International Law* 83 (1989): 596–99.

Smith, Wayne R. "Law of the Sea Treaty: Report on the Enterprise." *New York Law School Journal of International and Comparative Law* 3 (1981): 51–71.

Sohn, Louis B. "'Generally Accepted' International Rules." *Washington Law Review* 61 (1986): 1073–80.

————. "The Law of the Sea Crisis." *St. John's Law Review* 58 (1984): 237–66.

————. "The Law of the Sea: Customary International Law Developments." *American University Law Review* 34 (1985): 271–80.

————. "Peaceful Settlement of Disputes in Ocean Conflicts: Does UNCLOS III Point the Way?" *Law and Contemporary Problems* 46 (Spring 1983): 195–200.

Soroos, Marvin S. "The Commons in the Sky: The Radio Spectrum and Geosynchronous Orbit as Issues in Global Policy." *International Organization* 36 (1982): 665–77.

Spagni, Daniel, Luke Giorghiou and Michael Gibbons. "French Marine Technology Policy." *Marine Policy* 9 (1985): 280–91.

Spak, David G. "The Need for a Ban on All Radioactive Waste Disposal in the Ocean." *Northwestern Journal of International Law and Business* 7 (1986): 803–32.

Stang, David P. "Political Cobwebs Beneath the Sea." *International Lawyer* 7 (1972): 1–15.

Stauder, Dieter. "Patent Protection in Extraterritorial Areas (Continental Shelf, High Seas, Air Space, and Outer Space)." *IIC International Review of Industrial Property and Copyright Law* 7 (1976): 470–79.

Stein, Ted L. "The Approach of the Different Drummer: The Principle of the Persistent Objector in International Law." *Harvard International Law Journal* 26 (1985): 457–82.

Stevenson, John R. "Department Discusses Progress Toward 1973 Conference on the Law of the Sea." *U.S. Department of State Bulletin* 66 (1972): 672–79.

————. "Lawmaking for the Seas." *American Bar Association Journal* 61 (1975): 185–90.

————. "Legal Regulation of Mineral Exploitation in the Deep Seabed." *U.S. Department of State Bulletin* 65 (1971): 48–55.

————. "The Search for Equity in the Seabeds." *U.S. Department of State Bulletin* 64 (April 19, 1971): 529–33.

Stringer, Jack. "Mineral Treasure from the Galapagos Ridge." *NOAA* 12 (Winter 1982): 14–15.

"Summary of Detailed Ocean Mining Industry Comments." *Congressional Record* 126 (June 9, 1980): 13682–83.

Surace-Smith, Kathryn. "United States Activity Outside of the Law of the Sea Convention: Deep Seabed Mining and Transit Passage." *Columbia Law Review* 84 (1984): 1032–58.

Swayze, Frank B. "Negotiating a Law of the Sea." *U.S. Naval Institute Proceedings* 106 (July 1980): 33–39.

Sweeney, Richard James, Robert D. Tollison, and Thomas D. Willett. "Market Failure, the Common-Pool Problem, and Ocean Resource Exploitation." *Journal of Law and Economics* 17 (1974): 179–92.

Swing, John Temple. "A Treaty for the Taking." *Oceans* 14 (January-February 1981): 2–3.

"Symposium Panel Discussion, February 24, 1979." *Syracuse Journal of International Law and Commerce* 6 (1979): 300–9.

Tagliabue, John. "Bonn Cabinet Said to Oppose Sea Treaty." *New York Times*, November 24, 1984, page 3.

"Taking a Dive." *Economist* 278 (March 21, 1981): 85.

Temko, Ned. "Shultz Meeting with ANC Chief Lends Validity to S. Africa Rebels." *Christian Science Monitor*, January 28, 1987, page 7.

Torreh-Bayouth, Lilliana. "UNCLOS III: The Remaining Obstacles to Consensus on the Deepsea Mining Regime." *Texas International Law Journal* 16 (1981): 79–115.

Treves, Tullio. "The Adoption of the Law of the Sea Convention: Prospects for Seabed Mining." *Marine Policy* 7 (1983): 3–13.

_____. "The EEC and the Law of the Sea: How Close to One Voice?" *Ocean Development and International Law* 12 (1983): 173–89.

_____. "Seabed Mining and the United Nations Law of the Sea Convention." *Italian Yearbook of International Law* 5 (1980–81): 22–51.

Trimble, Jeff, and Maia Wechsler. "Look Who's Playing Peacekeeper Now." *U.S. News and World Report*, November 2, 1987, page 47.

Tsamenyi, B. Martin. "The South Pacific States, the USA and Sovereignty Over Highly Migratory Species." *Marine Policy* 10 (1986): 29–41.

Tsur, Yoel. "Compulsory Licensing in Israel Patents Law." *IIC International Review of Industrial Property and Copyright Law* 16 (1985): 541–49.

Turbeville, William. "American Ocean Policy Adrift: An Exclusive Economic Zone as an Alternative to the Law of the Sea Treaty." *University of Florida Law Review* 35 (1983): 492–520.

Turbyville, Linda. "Plumbing the Depths." *Sea Power* (January 1981), pages 23–28.

Turpel, Mary Ellen, and Philippe Sands. "Peremptory International Law and Sovereignty: Some Questions." *Connecticut Journal of International Law* 3 (1988): 364–69.

"UNCLOS: Preparatory Commission Meets." *Environmental Policy and Law* 15 (1985): 80.

United Nations Department of Public Information. "Commission Preparing Sea-Bed Authority Recesses First Session." *UN Chronicle* 20 (June 1983): 11–13.

_____. "Commission Studies Sea-Bed Mining Rules." *UN Chronicle* 21 (April 1984): 44–50.

_____. "Lack of Space: Geostationary Orbit." *UN Chronicle* 19 (July 1982): 58.

_____. "Law of the Sea: Range of Organizational Matters Decided by Preparatory Body." *UN Chronicle* 20 (November 1983): 36–39.

_____. "No Consensus on Antarctica." *UN Chronicle* 25 (March 1988): 77.

_____. "Outer Space Conference to Focus on Potential Benefits to Mankind." *U.N. Monthly Chronicle* 19 (January 1982): 43–45.

_____. "Pioneer Investor Duties; Rules for Sea-Bed Institutions Considered." *UN Chronicle* 26 (June 1989): 38–39.

_____. "Preparatory Commission Adopts 'Understanding' on Obligations of Pioneer Investors." *UN Chronicle* 27 (December 1990): 51–52.

_____. "Preparatory Commission Agrees on Procedures for Registering Pioneer Investors in Deep Sea-Bed Mining." *UN Chronicle* 23 (November 1986): 89–91.

_____. "Preparatory Commission Welcomes China as Fifth Pioneer Investor." *U.N. Monthly Chronicle* 28 (June 1991): 33.

_____. "Preparing for Sea-Bed Regime: An Agreement on Claims Procedures." *UN Chronicle* 21 (July 1984): 28–34.

_____. "Sea-Bed Commission Condemns Issuing of Licenses for Exploration of International Area." *UN Chronicle* 23 (August 1986): 107–9.

_____. "Sea-Bed Commission Postpones Registration of Pioneer Investors." *UN Chronicle* 24 (August 1987): 40–41.

_____. "Sea Law Commission Focuses on Pioneer Investor Duties, Training Programme Approved." *UN Chronicle* 26 (December 1989): 34–35.

_____. "Sea Law Commission Registers India as First Pioneer Investor." *UN Chronicle* 24 (November 1987): 56–58.

_____. "Training Programme for Enterprise to Be in Place by 1991." *UN Chronicle* 27 (June 1990): 38–39.

_____. "Unique Ceremony Marks End to Long Sea Law Conference." *UN Chronicle* 20 (February 1983): 3–9.

_____. "UNISPACE '82: Getting a New Perspective on Earth (Vienna, 9–21 August)." *U.N. Chronicle* 19 (October 1982): 26–31.

United Nations Office for Ocean Affairs and the Law of the Sea. "Registration of Pioneer Investors in the International Sea-Bed Area in Accordance with Resolution II of the Third

United Nations Conference on the Law of the Sea." *Law of the Sea Bulletin*, Special Issue II (April 1988), pages 1–169.

_____. "Report on the Fifth Session of the Preparatory Commission for the International Sea-Bed Authority and for the International Tribunal for the Law of the Sea, Kingston, 30 March–16 April 1987." *Law of the Sea Bulletin* 10 (November 1987): 115–19.

_____. "Report on the Meeting of the Preparatory Commission for the International Sea-Bed Authority and for the International Tribunal for the Law of the Sea, New York, 27 July–21 August 1987." *Law of the Sea Bulletin* 10 (November 1987): 120–23.

_____. "Report on the Sixth Session of the Preparatory Commission for the International Sea-Bed Authority and for the International Tribunal for the Law of the Sea: Kingston, 14 March–8 April 1988; New York, 15 August–2 September 1988." *Law of the Sea Bulletin* 12 (December 1988): 47–53.

_____. "Seminar on the Current Status of Developments in Deep Sea-Bed Mining Technology (New York, 18 and 19 August 1988)." *Law of the Sea Bulletin* 12 (December 1988): 57–63.

"U.S. Opposes Antarctic Mining Ban Now." *New York Times International*, June 23, 1991, page 3.

Vadus, Joseph R. "Ocean Technology in the United States: Recent Advances, Future Needs and International Collaboration." *Marine Technology Society Journal* 24 (March 1990): 76–86.

Van Dyke, Jon, and Christopher Yuen. "'Common Heritage' v. 'Freedom of the High Seas': Which Governs the Seabed?" *San Diego Law Review* 19 (1982): 493–551.

Vasciannie, Stephen. "Part XI of the Law of the Sea Convention and Third States: Some General Observations." *Cambridge Law Journal* 48 (1989): 85–97.

Verdross, Alfred. "Forbidden Treaties in International Law." *American Journal of International Law* 31 (1937): 571–77.

_____. "*Jus Dispositivum* and *Jus Cogens* in International Law." *American Journal of International Law* 60 (1966): 55–63.

"Vital Sea-Lanes, Vital Bases." *U.S. News and World Report*, February 10, 1986, page 30.

Walkate, J.A. "Developments in Special Commission 3 of the Preparatory Commission for the International Sea-Bed Authority and for the International Tribunal for the Law of the Sea: Drafting the Future Deep Seabed Mining Code." *Netherlands International Law Review* 36 (1989): 152–78.

Walker, Craig W. "Jurisdictional Problems Created by Artificial Islands." *San Diego Law Review* 10 (1973): 638–63.

Walsh, Don. "Comment." *Law and Contemporary Problems* 46 (Spring 1983): 167–73.

Wasserstrom, Silas J., and Louis Michael Seidman. "The Fourth Amendment as Constitutional Theory." *Georgetown Law Journal* 77 (1988): 19–112.

Watt, Donald Cameron. "Background to the Law of the Sea." *Marine Policy* 11 (1987): 242–44.

_____. "The Law of the Sea Conference and the Deep Sea Mining Issue: The Need for an Agreement." *International Affairs* (London) 58 (1981–82): 78–94.

_____. "Towards a New Order for the World's Oceans?" *The Round Table* 285 (January 1983): 25–33.

Wehrmeister, Elizabeth A. "Giving the Cat Claws: Proposed Amendments to the International Whaling Convention." *Loyola of Los Angeles International and Comparative Law Journal* 11 (1989): 417–38.

Weidner, Helen E. "The United States and North-South Technology Transfer: Some Practical and Legal Obstacles." *Wisconsin International Law Journal* (1983): 205–28.

Weil, Prosper. "Towards Relative Normativity in International Law?" *American Journal of International Law* 77 (1983): 413–42.

Weissberg, Guenter. "International Law Meets the Short-Term National Interest: The Maltese Proposal on the Sea-Bed and Ocean Floor — Its Fate in Two Cities." *International and Comparative Law Quarterly* 18 (1969): 41–102.

Weisskopf, Michael. "Running Hot and Cold on the Issue of Global Warming." *Washington Post National Weekly Edition*, November 26–December 2, 1990, page 32.

Welling, Conrad G. "Ocean Minerals and World Politics." *Stockton's Port Soundings* (November 1980): 11–13.

_____. "Polymetallic Sulfides: An Industry Viewpoint." *Marine Technology Society Journal* 16, number 3 (1982): 5–7.

Wendt, Allan. "A Restructured COCOM." Current Policy No. 1290. Washington: U.S. Department of State, 1990.

White, Mary Victoria. "The Common Heritage of Mankind: An Assessment." *Case Western Reserve Journal of International Law* 14 (1982): 509–42.

Wijkman, Per Magnus. "Managing the Global Commons." *International Organization* 36 (1982): 511–36.

Williams, Sylvia Maureen. "International Law Before and After the Moon Agreement." *International Relations* 7 (1981): 1168–93.

Williamson, Richard S. "Toward the 21st Century: The Future of Multilateral Diplomacy." *U.S. Department of State Bulletin* 88 (December 1988): 53–56.

Willsey, F. Patterson. "The Deep Seabed Hard Mineral Resources Act and the Third United Nations Conference on the Law of the Sea: Can the Conference Meet the Mandate Embodied in the Act?" *San Diego Law Review* 18 (1981): 509–32.

Work, Clemens P. "Crumpets in Cleveland." *U.S. News & World Report*, April 6, 1987, page 41.

Wulf, Norman A. "Comment." *Law and Contemporary Problems* 46 (Spring 1983): 155–66.

Yakovlev, I. "The World Ocean and International Law." *International Affairs* (Moscow) (1983, no. 8), pages 76–83.

Yost, Kathryn E. "The International Sea-Bed Authority Decision-Making Process: Does It Give a Proportionate Voice to the Participant's Interests in Deep Sea Mining?" *San Diego Law Review* 20 (1983): 659–78.

Young, Richard. "Inducement for Exploration by Companies." *Syracuse Journal of International Law and Commerce* 6 (1978–79): 199–205.

_____. "The Legal Status of Submarine Areas Beneath the High Seas." *American Journal of International Law* 45 (1951): 225–39.

Zuleta, Bernardo. "The Law of the Sea After Montego Bay." *San Diego Law Review* 20 (1983): 475–88.

Public Documents

Australia. Department of Foreign Affairs. *Third United Nations Conference on the Law of the Sea, Seventh Session, Geneva: Report of the Australian Delegation*. Canberra: Australian Government Publishing Service, 1978.

Council of Europe. "UN Convention on the Law of the Sea" (Resolution adopted May 11, 1984). Reprinted in *Environmental Policy and Law* 13 (1984): 80.

European Parliament. "Law of the Sea Resolution." June 9, 1983. Reprinted in *Environmental Policy and Law* 11 (1983): 81.

_____. Resolution C101/65–68, June 1, 1981. Reprinted in *Environmental Policy and Law* 7 (1981): 184.

First United Nations Conference on the Law of the Sea. *Official Records*. Volume VI.

International Law Commission. *Yearbook of the International Law Commission, 1950*. Volumes 1–2.

_____. *Yearbook of the International Law Commission, 1951*. Volumes 1–2.

_____. *Yearbook of the International Law Commission, 1953*. Volume 2.

_____. *Yearbook of the International Law Commission, 1955*. Volumes 1–2.

_____. *Yearbook of the International Law Commission, 1956*. Volumes 1–2.

_____. *Yearbook of the International Law Commission, 1958*. Volume 2.

_____. *Yearbook of the International Law Commission, 1985*. Volume 2.

League of Nations. Committee of Experts for the Progressive Development of International Law. "Questionnaire No. 7 – Exploitation of the Products of the Sea: Annex: Report on the Exploitation of the Products of the Sea." Reprinted in *American Journal of International Law* 20 (Special Supplement, 1926): 230–41.

_____. *Declaration Adopted by the Preparatory Commission on 30 August 1985*. U.N. Doc. LOS/PCN/72, September 2, 1985.

_____. *Declaration Adopted by the Preparatory Commission on 11 April 1986*. U.N. Doc. LOS/PCN/78, April 21, 1986.

Preparatory Commission for the International Sea-Bed Authority and for the International Tribunal for the Law of the Sea. *Declaration of the Group of 77.* U.N. Doc. LOS/PCN/5, April 11, 1983.

_____. *Letter Dated 3 August 1984 from the Government of the Netherlands to the Chairman of the Preparatory Commission.* U.N. Doc. LOS/PCN/46, August 16, 1974.

_____. *Letter Dated 4 November 1985 from the Chairman of the Delegation of the United Kingdom of Great Britain and Northern Ireland Addressed to the Chairman of the Preparatory Commission.* U.N. Doc. LOS/PCN/74, January 9, 1986.

_____. *Statement by the Chairman of the Delegation of the Netherlands on Behalf of the Delegations of Belgium, France, Federal Republic of Germany, Italy, Japan and the United Kingdom of Great Britain and Northern Ireland Delivered on 14 August 1984.* U.N. Doc. LOS/PCN/52, August 24, 1984.

_____. *Statement by the Chairman of the Group of East European Socialist Countries Delivered on 13 August 1984.* U.N. Doc. LOS/PCN/49, August 17, 1984.

_____. *Statement by the Chairman of the Group of 77 Delivered on 13 August 1984.* U.N. Doc. LOS/PCN/48, August 16, 1984.

_____. *Statement Made by the Acting Chairman of the Preparatory Commission.* U.N. Doc. LOS/PCN/L.41/Rev.1, September 5, 1986.

_____. *Statement of Understanding on the Implementation of Resolution II Made by the Chairman of the Preparatory Commission at the 34th Plenary Meeting, Held on 10 April 1987.* U.N. Doc. LOS/PCN/L.43/Rev.1, April 10, 1987.

Third United Nations Conference on the Law of the Sea. Final Act, Annex I, Resolution I: "Establishment of the Preparatory Commission for the International Sea-Bed Authority and for the International Tribunal for the Law of the Sea." Reprinted in *International Legal Materials* 21 (1982): 1253–54.

_____. Final Act, Annex I, Resolution II: "Governing Preparatory Investment in Pioneer Activities Relating to Polymetallic Nodules." Reprinted in *International Legal Materials* 21 (1982): 1254–57.

_____. Final Act, Annex I, Resolution IV. Reprinted in *International Legal Materials* 21 (1982): 1258.

_____. *Informal Composite Negotiating Text/Revision 1.* U.N. Doc. A/CONF.62/WP.10/Rev.1, April 28, 1979.

_____. *Informal Composite Negotiating Text/Revision 2.* U.N. Doc. A/CONF.62/WP.10/Rev.2, May 11, 1980.

_____. *Informal Composite Negotiating Text/Revision 3.* U.N. Doc. A/CONF.62/WP.10/Rev.3, September 22, 1980.

_____. Third United Nations Conference on the Law of the Sea. *Official Records.* Volumes I–XVI.

_____. *Second Committee: Main Trends.* U.N. Doc. A/CONF.62/C.2/WP.1, October 15, 1974.

_____. *The U.S. Proposal for Amendments to the Draft Convention on the Law of the Sea.* U.N. Doc. A/CONF.62/WG.21/Informal Paper 18, 1982.

United Nations. *Draft Ocean Space Treaty: Working Paper Submitted by Malta.* U.N. Doc. A/AC.138/53, August 5, 1971.

_____. *Draft United Nations Convention on the International Seabed Area.* U.N. Area. U.N. Doc. A/AC.138/25, August 3, 1970.

_____. Economic and Social Council, 40th Session. *Official Records.*

_____. Economic and Social Council, 40th Session. Resolution 1112, March 7, 1966.

_____. Economic and Social Council, 45th Session. Resolution 1380. August 2, 1968.

_____. *First Committee of Conference on Sea Law Ends Work for Year.* U.N. Doc. SEA/145, August 28, 1974.

_____. General Assembly. Ad Hoc Committee to Study the Peaceful Uses of the Sea-Bed and the Ocean Floor Beyond the Limits of National Jurisdiction. Economic and Technical Working Group. *Summary Records.*

_____. General Assembly. Ad Hoc Committee to Study the Peaceful Uses of the Sea-Bed and the Ocean Floor Beyond the Limits of National Jurisdiction. Legal Working Group. *Summary Records.*

_____. General Assembly. Ad Hoc Committee to Study the Peaceful Uses of the Sea-Bed and the Ocean Floor Beyond the Limits of National Jurisdiction. *Summary Records.*

_____. General Assembly, 13th Session. Resolution 1307. December 10, 1958.

_____. General Assembly, 16th Session. Resolution 1721A. December 20, 1961.

_____. General Assembly, 17th Session. Resolution 1803. December 14, 1962.

_____. General Assembly, 18th Session. Resolution 1962. December 13, 1963.

_____. General Assembly, 21st Session. *Provisional Verbatim Records.*

_____. General Assembly, 21st Session. Resolution 2158. November 15, 1966.

_____. General Assembly, 21st Session. Resolution 2172. December 6, 1966.

_____. General Assembly, 22nd Session. First Committee. *Official Records.*

_____. General Assembly, 22nd Session. Resolution 2340. December 18, 1967.

_____. General Assembly, 22nd Session. First Committee. *Provisional Verbatim Records.*

_____. General Assembly, 23rd Session. First Committee. *Provisional Verbatim Records.*

_____. General Assembly, 23rd Session. Resolutions 2467A-C. December 21, 1968.

_____. General Assembly, 24th Session. Resolution 2560. December 13, 1969.

_____. General Assembly, 24th Session. Resolution 2566. December 13, 1969.

_____. General Assembly, 24th Session. Resolutions 2574A-D. December 15, 1969.

_____. General Assembly, 25th Session. First Committee. *Official Records.*

_____. General Assembly, 25th Session. Resolution 2749. December 17, 1970.

_____. General Assembly, 25th Session. Resolutions 2750A-B. December 17, 1970.

_____. General Assembly, 27th Session. Resolution 3029A. December 18, 1972.

_____. General Assembly, 28th Session. Resolution 3067. December 16, 1973.

_____. General Assembly, 29th Session. Resolution 3237. November 22, 1974.

_____. General Assembly, 29th Session. Resolution 3821. December 12, 1974.

_____. General Assembly, 35th Session. Resolution 35/167. December 15, 1980.

_____. General Assembly, 39th Session. Resolution 39/73. December 13, 1984.

_____. General Assembly, 40th Session. Resolution 40/63. December 10, 1985.

_____. General Assembly, 40th Session. Resolutions 40/156A-B. December 16, 1985.

_____. General Assembly, 41st Session. Resolution 41/34. November 5, 1986.

_____. General Assembly, 41st Session. Resolution 41/68A. December 3, 1986.

_____. General Assembly, 42nd Session. Resolution 42/20. November 18, 1987.

_____. General Assembly, 43rd Session. Resolution 43/18. November 1, 1988.

_____. General Assembly, 43rd Session. Resolution 43/53. December 6, 1988.

_____. General Assembly, 44th Session. Resolution 44/26. November 20, 1989.

_____. General Assembly, 44th Session. Resolution 44/124A. December 15, 1989.

_____. General Assembly, 44th Session. Resolution 44/225. December 22, 1989.

_____. General Assembly, 45th Session. Resolution 45/78A. December 12, 1990.

_____. General Assembly, 45th Session. Resolution 45/145. December 14, 1990.

_____. *International Regime: Working Paper Submitted by the United Kingdom.* U.N. Doc. A/AC.138/26, August 5, 1970.

_____. *Law of the Sea: Report of the Secretary-General.* U.N. Doc. A/41/742, October 28, 1986.

_____. *Law of the Sea: Report of the Secretary-General.* U.N. Doc. A/43/718, October 20, 1988.

_____. *Law of the Sea: Report of the Secretary-General.* U.N. Doc. A/44/650, November 1, 1989.

_____. *Laws and Regulations on the Regime of the Territorial Sea.* U.N. Doc. ST/LEG/SER.B/6, November 1957.

_____. *Legal Aspects of the Question of the Reservation Exclusively for Peaceful Purposes of the Sea-Bed and the Ocean Floor, and the Sub-Soil Thereof, Underlying the High Seas Beyond the Limits of Present National Jurisdiction, and the Use of Their Resources in the Interests of Mankind: Study Prepared by the Secretariat.* U.N. Doc. A/AC.135/19/Add.2 June 21, 1968.

_____. *Marine Science and Technology: Survey and Proposals: Report of the Secretary-General.* U.N. Doc. E/4487, April 24, 1968.

_____. *A New United Nations Structure for Global Economic Cooperation: The Report* of the Group of Experts on the Structure of the United Nations System. U.N. Doc. E/AC.62/9, May 28, 1975.

_____. The Nickel Industry and the Developing Countries. U.N. Doc. ST/ESA/100, 1980.

_____. Note by the Secretary-General. U.N. Doc. A/C.1/952, October 31, 1967.

_____. Organization of African Unity Declaration on the Issues of the Law of the Sea — CM/Res. 289 (XIX). U.N. Doc. A/AC.138/89, July 2, 1973.

_____. Preparatory Commission for Sea-Bed Authority and Law of the Sea Tribunal Concludes Session at Kingston, 29 March–16 April. U.N. Doc. SEA/819, April 20, 1987.

_____. Press Release. U.N. Doc. SEA/494, April 30, 1982.

_____. Proposals Concerning the Establishment of a Regime for the Exploration and Exploitation of the Seabed. U.N. Doc. A/AC.138/27, August 5, 1970.

_____. Report of the Ad Hoc Committee to Study the Peaceful Uses of the Sea-Bed and the Ocean Floor Beyond the Limits of National Jurisdiction. U.N. Doc. A/7230, 1968.

_____. Report of the Committee on the Peaceful Uses of the Sea-Bed and the Ocean Floor Beyond the Limits of National Jurisdiction. U.N. Doc. A/7622, 1969.

_____. Report of the Committee on the Peaceful Uses of the Sea-Bed and the Ocean Floor Beyond the Limits of National Jurisdiction. U.N. Doc. A/8021, 1970.

_____. Report of the Committee on the Peaceful Uses of the Sea-Bed and the Ocean Floor Beyond the Limits of National Jurisdiction. U.N. Doc. A/8421, 1971.

_____. Report of the International Law Commission to the General Assembly. U.N. Doc. A/1316, July 1950.

_____. Report of the International Law Commission to the General Assembly. U.N. Doc. A/3159, 1956.

_____. Request for the Inclusion of a Supplementary Item in the Agenda of the Twenty-Second Session: Note Verbale Dated 17 August 1967 from the Permanent Mission of Malta to the United Nations Addressed to the Secretary-General. U.N. Doc. A/6695, August 18, 1967.

_____. Resources of the Sea: Report of the Secretary-General. U.N. Doc. E/4449/Add.1, February 19, 1968.

_____. Resources of the Sea: Report of the Secretary-General. U.N. Doc. E/4449, February 21, 1968. Summary.

_____. Study on International Machinery: Report of the Secretary-General. U.N. Doc. A/AC.138/23, May 26, 1970.

_____. Union of Soviet Socialist Republics: Provisional Draft Articles of a Treaty on the Use of the Sea-Bed for Peaceful Purposes. U/M/Doc. A/AC.138/43, July 22, 1971.

_____. United Nations Ad Hoc Committee to Study the Peaceful Uses of the Sea-Bed and the Ocean Floor Beyond the Limits of National Jurisdiction, Replies from Governments. U.N. Doc. A/AC.135/1, March 11, 1968.

_____. Uses of the Sea: Study Prepared by the Secretary-General. U.N. Doc. E/5120, April 28, 1972.

_____. Working Paper on the Regime for the Sea Bed and Ocean Floor and Its Subsoil Beyond the Limits of National Jurisdiction, Submitted by Chile, Colombia, Ecuador, El Salvador, Guatemala, Guyana, Jamaica, Mexico, Panama, Peru, Trinidad and Tobago, Uruguay, and Venezuela. U.N. Doc. A/AC.138/49, August 4, 1971.

United States. Aldrich, George H. "Law of the Sea." U.S. Department of State Bulletin 81 (February 1981): 56–59.

_____. Bush, George. "The UN: World Parliament of Peace." Current Policy Number 1303. Washington: U.S. Department of State, 1990.

_____. Church, Frank. Letter to Secretary of State Cyrus Vance, October 30, 1979. Reprinted in Congressional Record 125 (December 14, 1979): 36070.

_____. Commission on Marine Science, Engineering and Resources. Our Nation and the Sea: A Plan for National Action. Washington: U.S. Government Printing Office, 1969.

_____. Congressional Research Service (George A. Doumani). Exploiting the Resources of the Seabed. Washington: U.S. Government Printing Office, 1971.

_____. Constitution.

_____. Department of Commerce. National Oceanic and Atmospheric Administration.

Deep Seabed Mining: Report to Congress. Washington: U.S. Government Printing Office, 1983.
_____. _____. *U.S. Ocean Policy in the 1970s: Status and Issues.* Washington: U.S. Government Printing Office, 1978.
_____. Department of State. "Rights and Freedoms in International Waters." *U.S. Department of State Bulletin* 86 (May 1986): 79.
_____. _____. Statement of November 26, 1980: "Access to Seabed Minerals Under the Draft Convention on the Law of the Sea." *Reprinted in Environmental Policy and Law* 7 (1981): 39–40.
_____. _____. Statement of December 2, 1988. Reprinted in *U.S. Department of State Bulletin* 89 (1989): 23.
_____. General Accounting Office. *Impediments to U.S. Involvement in Deep Ocean Mining Can Be Overcome: Report to the Congress.* Washington: U.S. Government Printing Office, 1983. Summary.
_____. Kissinger, Henry. Address before American Bar Association, Montreal, August 11, 1975. Reprinted in *U.S. Department of State Bulletin* 73 (1975): 353–62.
_____. _____. Address before University of Wisconsin Institute of World Affairs, Milwaukee, July 14, 1975. Reprinted in *U.S. Department of State Bulletin* 73 (1975): 149–57.
_____. _____. "The Law of the Sea: A Test of International Cooperation." *U.S. Department of State Bulletin* 74 (1976): 533–42.
_____. _____. Remarks of September 1, 1976. Reprinted in *U.S. Department of State Bulletin* 75 (1976): 396–99.
_____. _____. "Secretary Kissinger Discusses U.S. Position on the Law of the Sea Conference." *U.S. Department of State Bulletin* 75 (1976): 397–99.
_____. Malone, James L. "Law of the Sea and Oceans Policy." *U.S. Department of State Bulletin* 82 (October 1982): 48–50.
_____. _____. Speech at Montego Bay, January 6, 1983. Reprinted in *Environmental Policy and Law* 10 (1983): 97–98.
_____. _____. "U.S. Participation in the Law of the Sea Conference." *U.S. Department of State Bulletin* 82 (May 1982): 61–63.
_____. _____. "U.S. Policy and the Law of the Sea." *U.S. Department of State Bulletin* 81 (July 1981): 48–51.
_____. 90th Congress, 2nd Session. U.S. House of Representatives. *The 22nd Session of the United Nations General Assembly: Report by Hon. William S. Broomfield and Hon. L.H. Fountain.* Washington: U.S. Government Printing Office, 1968.
_____. 92nd Congress, 2nd Session. House of Representatives. Committee on Foreign Affairs. Subcommittee on International Organizations and Movements. *Law of the Sea and Peaceful Uses of the Seabed.* Washington: U.S. Government Printing Office, 1972.
_____. 92nd Congress, 2nd Session. Senate. Committee on Commerce. Subcommittee on Oceans and Atmosphere. *Law of the Sea.* Washington: U.S. Government Printing Office, 1972.
_____. 93rd Congress, 1st Session. House of Representatives. Committee on Foreign Affairs. Subcommittee on International Organizations and Movements. *Law of the Sea Resolution.* Washington: U.S. Government Printing Office, 1973.
_____. 93rd Congress, 1st Session. House of Representatives. Committee on Merchant Marine and Fisheries. Subcommittee on Oceanography. *Ocean Affairs in the 93rd Congress.* Washington: U.S. Government Printing Office, 1975.
_____. 93rd Congress, 1st Session. Senate. Committee on Interior and Insular Affairs. Subcommittee on Minerals, Materials, and Fuels. *Status Report on the Law of the Sea Conference.* Washington: U.S. Government Printing Office, 1973.
_____. 93rd Congress, 1st Session. Senate. Resolution 82, July 16, 1973.
_____. 93rd Congress, 1st and 2nd Sessions. House of Representatives. Committee on Merchant Marine and Fisheries. Subcommittee on Oceanography. *Deep Seabed Hard Minerals.* Washington: U.S. Government Printing Office, 1974.
_____. 93rd Congress, 2nd Session. Senate. Committee on Armed Services. *Extending*

Jurisdiction of the United States Over Certain Ocean Areas. Washington: U.S. Government Printing Office, 1974.

_____. 94th Congress, 1st Session. Senate. Committee on Foreign Relations. Subcommittee on Oceans and International Environment. *Law of the Sea.* Washington: U.S. Government Printing Office, 1975.

_____. 94th Congress, 2nd Session. House of Representatives. Committee on Merchant Marine and Fisheries. Subcommittee on Oceanography. *National Ocean Policy.* Washington: U.S. Government Printing Office, 1976.

_____. 94th Congress, 2nd Session. Senate. Committee on Foreign Relations. Subcommittee on Oceans and International Environment. *Law of the Sea.* Washington: U.S. Government Printing Office, 1976.

_____. 95th Congress, 1st Session. House of Representatives. Committee on Merchant Marine and Fisheries. Subcommittee on Oceanography. *Deep Seabed Mining.* Washington: U.S. Government Printing Office, 1977.

_____. 95th Congress, 2nd Session. Senate. Committee on Commerce, Science and Transportation. *Report: The Third United Nations Law of the Sea Conference.* Washington: U.S. Government Printing Office, 1978.

_____. 97th Congress, 1st Session. Senate. Committee on Foreign Relations. *Nomination of James L. Malone.* Washington: U.S. Government Printing Office, 1981.

_____. 97th Congress, 2nd Session. Senate. Committee on Foreign Relations. Subcommittee on Arms Control, Oceans, International Operations, and Environment. *Law of the Sea Negotiations.* Washington: U.S. Government Printing Office, 1983.

_____. Nixon, Richard. "U.S. Foreign Policy for the 1970s: The Emerging Structure of Peace: A Report to the Congress." Reprinted in *U.S. Department of State Bulletin* 66 (1972): 311–418.

_____. _____. "United States Policy for the Seabed." *U.S. Department of State Bulletin* 62 (1970): 737–38.

_____. *Note Dated 13 January 1986 from the United States Mission to the United Nations Addressed to the Secretary-General of the United Nations.* Reprinted in *Law of the Sea Bulletin* 7 (1986): 74–86.

_____. *Note Presented to U.N. Secretary-General from the United States, June 12, 1970.* Reprinted in *International Legal Materials* 9 (1970): 833–37.

_____. Phillips, Christopher H. Statements of December 2, 1969, and December 15, 1969. Reprinted in *U.S. Department of State Bulletin* 62 (1970): 89–93.

_____. Presidential Proclamation Number 2667, September 28, 1945.

_____. Presidential Proclamation Number 2668, September 28, 1945.

_____. Presidential Proclamation Number 5030, March 10, 1983.

_____. Presidential Proclamation Number 5928, December 27, 1988.

_____. Reagan, Ronald W. Address before U.N. General Assembly, September 24, 1984. Reprinted in *New York Times,* September 25, 1984, page A10.

_____. _____. Address before World Affairs Council, Philadelphia, October 15, 1981. Reprinted in *U.S. Department of State Bulletin* 81 (December 1981): 14–17.

_____. _____. Statement of January 29, 1982. Reprinted in *U.S. Department of State Bulletin* 82 (March 1982): 54–55.

_____. _____. Statement of July 9, 1982. Reprinted in *U.S. Department of State Bulletin* 82 (August 1982): 71.

_____. _____. Statement of March 10, 1983. Reprinted in *U.S. Department of State Bulletin* 83 (June 1983): 70–71.

_____. _____. Statement of February 11, 1986. Reprinted in *New York Times,* February 12, 1986, page 11.

_____. "Reports of the United States Delegation to the Third United Nations Conference on the Law of the Sea." Occasional Paper Number 33. Edited by Myron H. Nordquist and Choon-ho Park. Honolulu: Law of the Sea Institute, 1983.

_____. Richardson, Elliot R. Statement before Special Subcommittee on Outer Continental Shelf, Senate Committee on Interior and Insular Affairs, May 27, 1970. Reprinted in *U.S. Department of State Bulletin* 62 (1970): 738–41.

_____. Rogers, William P. "The Rule of Law and the Settlement of International Disputes." *U.S. Department of State Bulletin* 62 (1970): 623–27.
_____. Roosevelt, Franklin D. "Freedom of the Seas." *International Conciliation* 373 (October 1941): 649–56.
_____. Schneider, William, Jr. Statement before Subcommittee on International Finance and Monetary Policy, Senate Committee on Banking, Housing, and Urban Affairs, March 2, 1983. Reprinted in *U.S. Department of State Bulletin* 83 (June 1983): 71–74.
_____. Sofaer, Abraham D., Legal Adviser, Department of State. "The United States and the World Court." Current Policy No. 769. Washington: U.S. Department of State, 1985.
_____. Stevenson, John R. Letter to Senator Lee Metcalf, January 16, 1970. Reprinted in *International Legal Materials* 9 (1970): 831–32.
_____. _____. Statement before Subcommittee on International Organizations and Movements, House Committee on Foreign Affairs, April 10, 1972. Reprinted in *U.S. Department of State Bulletin* 66 (1972): 672–79.
_____. Vance, Cyrus. Letter to Senator Jacob Javits, November 28, 1979. Reprinted in *Congressional Record* 125 (December 14, 1979): 36070.
_____. White House fact sheet. *U.S. Department of State Bulletin* 82 (March 1982): 55.
_____. White House news conference, May 23, 1970. Reprinted in *International Legal Materials* 9 (1970): 810–20.

International Agreements

Agreement Concerning Interim Arrangements Relating to Polymetallic Nodules of the Deep Seabed. Reprinted in *Environmental Policy and Law* 9 (1982): 134.
Agreement Governing the Activities of States on the Moon and Other Celestial Bodies. Reprinted in *International Legal Materials* 18 (1979): 1434–41.
Agreement on the Resolution of Practical Problems with Respect to Deep Seabed Mining Areas, and Exchange of Notes Between the United States and the Parties to the Agreement. Reprinted in *International Legal Materials* 26 (1987): 1504–15.
Antarctic Treaty. Reprinted in *Cornell International Law Journal* 19 (1986): 302–9.
Constitution of the World Health Organization. *United Nations Treaty Series*, volume 14, number I-221, pages 185–222.
Convention on Fishing and Conservation of the Living Resources of the High Seas. *United Nations Treaty Series*, volume 559, number 8164, pages 285–300.
Convention on International Civil Aviation. *United Nations Treaty Series*, volume 15, no. II-102, pages 295–360.
Convention on the Continental Shelf. *United Nations Treaty Series*, volume 499, number 7302, pages 311–20.
Convention on the High Seas. *United Nations Treaty Series*, volume 450, number 6465, pages 82–102.
Convention on the Regulation of Antarctic Mineral Resource Activities. Reprinted in *International Legal Materials* 27 (1988): 868–900.
Convention on the Territorial Sea and the Contiguous Zone. *United Nations Treaty Series*, volume 516, number 7477, pages 205–24.
Provisional Understanding Regarding Deep Seabed Matters. Reprinted in *International Legal Materials* 23 (1984): 1354–65.
Statute of the International Atomic Energy Agency. *United Nations Treaty Series*, volume 276, number 3988, pages 3–44.
Statute of the International Court of Justice.
Treaty Banning Nuclear Weapons Tests in the Atmosphere, in Outer Space and Under Water. Reprinted in *United Nations Treaty Series*, volume 480, number 6964, pages 45–49.
Treaty on Principles Governing the Activities of States in the Exploration and Use of Outer Space, Including the Moon and Other Celestial Bodies. Reprinted in *International Legal Materials* 6 (1967): 386–90.
Understanding of the Preparatory Commission for the International Sea-Bed Authority and for the International Tribunal for the Law of the Sea for Proceeding with Deep Sea-Bed

Mining Applications and Resolving Disputes of Overlapping Claims of Mine Sites. Reprinted in *International Legal Materials* 25 (1986): 1326–30.

United Nations Charter.

United Nations Convention on the Law of the Sea. Reprinted in *International Legal Materials* 21 (1982): 1261–1354.

Vienna Convention on the Law of the Treaties. Reprinted in *International Legal Materials* 8 (1969): 686–713.

Cases

International Court of Justice. Barcelona Traction, Light and Power Company, Limited, Judgment. *I.C.J. Reports 1970*, pages 3–357.

————. Colombian-Peruvian Asylum Case, Judgment of November 20th, 1950. *I.C.J. Reports 1950*, pages 266–389.

————. Continental Shelf (Malta v. Libyan Arab Jamahiriya), Judgment. *I.C.J. Reports 1985*, pages 13–187.

————. Continental Shelf (Tunisia v. Libyan Arab Jamahiriya), Judgment. *I.C.J. Reports 1982*, pages 18–323.

————. Delimitation of the Maritime Boundary in the Gulf of Maine Area, Judgment. *I.C.J. Reports 1984*, pages 3–160.

————. Fisheries Case, Judgment of December 18th, 1951. *I.C.J. Reports 1951*, pages 116–206.

————. Fisheries Jurisdiction (United Kingdom v. Iceland), Merits, Judgment. *I.C.J. Reports 1974*, pages 3–173.

————. Military and Paramilitary Activities in and Against Nicaragua (Nicaragua v. United States of America), Merits, Judgment. *I.C.J. Reports 1986*, pages 14–546.

————. North Sea Continental Shelf, Judgment. *I.C.J. Reports 1969*, pages 3–257.

————. Nuclear Tests (Australia v. France), Judgment of 20 December 1974. *I.C.J. Reports 1974*, pages 253–528.

————. Reparation for Injuries Suffered in the Service of the United Nations, Advisory Opinion. *I.C.J. Reports 1949*, pages 174–220.

————. South West Africa, Second Phase, Judgment. *I.C.J. Reports 1966*, pages 6–58.

New York Court of Appeals. People ex rel. Burhans et al. v. City of New York. *North Eastern Reporter* 92 (1910): 19–20.

Permanent Court of International Justice. The Case of the S.S. *Lotus. P.C.I.J. Reports 1927*, ser. A, no. 10.

United States District Court. Jayman-Ruby, Inc. v. Federal Trade Commission. *Federal Supplement* 496 (1980): 838–47.

UUnited States Supreme Court. Dable Grain Shovel Company v. Flint. *United States Reports* 137 (1980): 41–43.

————. James v. Campbell. *United States Reports* 104 (1881): 356–85.

Statutes and Regulations

Germany, Federal Republic of. Act on Interim Regulation of Deep Seabed Mining. Reprinted in *International Legal Materials* 19 (1980): 1330–39.

Union of Soviet Socialist Republics. "USSR Mining Decree." *Environmental Policy and Law* 9 (1982): 96.

United Kingdom of Great Britain and Northern Ireland. "U.K.: Deep Sea Mining (Temporary Provisions) Bill." *Environmental Policy and Law* 7 (February 1981): 41–42.

United States. Antarctic Protection Act of 1990. Public Law Number 101–594.

————. Communications Satellite Act of 1962. *United States Code*, title 47, §§701–757.

————. Deep Seabed Hard Minerals Resources Act. *United States Code*, title 30, §§1401–73.

_____. Deep Seabed Mining Regulations for Commercial Recovery Permits. *Code of Federal Regulations*, title 15, §§971.100–971.1007.

_____. Deep Seabed Mining Regulations for Exploration Licenses. *Code of Federal Regulations*, title 15, §§970.100–970.2601.

_____. Marine Resources and Engineering Development Act. *United States Code*, title 33, §§1101–8.

_____. National Aeronautics and Space Act of 1958. *United States Code*, title 42, §200 et seq.

_____. Outer Continental Shelf Lands Act. *United States Code*, title 43, §§1331–56.

_____. Outer Continental Shelf Minerals and Rights-of-Way Management. *Code of Federal Regulations*, title 30, §§256.0–256.82.

_____. Patent Law, 35 U.S.C. §100 et seq.

_____. Trade Act of 1974. *United States Code*, title 19, §§2461–66.

Index

Indexing includes the main text and the Notes.

www.ingramcontent.com/pod-product-compliance
Lightning Source LLC
Chambersburg PA
CBHW030855270326
41929CB00008B/425